AN INTRODUCTION TO
FLUID DYNAMICS

AN INTRODUCTION TO
FLUID DYNAMICS

BY

G . K . BATCHELOR, F.R.S.

Professor of Applied Mathematics in the University of Cambridge

CAMBRIDGE
AT THE UNIVERSITY PRESS
1967

Published by the Syndics of the Cambridge University Press
Bentley House, 200 Euston Road, London, N.W. 1
American Branch: 32 East 57th Street, New York, N.Y. 10022

Library of Congress Catalogue Card Number: 67-21953

Printed in Great Britain
at the University Printing House, Cambridge
(Brooke Crutchley, University Printer)

CONTENTS

Chapter 4. Flow of a Uniform Incompressible Viscous Fluid

Chapter 5. Flow at Large Reynolds Number: Effects of Viscosity

Appendices

Plates 1 to 24 are between pages 352 and 353

PREFACE

While teaching fluid dynamics to students preparing for the various Parts of the Mathematical Tripos at Cambridge I have found difficulty over the choice of textbooks to accompany the lectures. There appear to be many books intended for use by a student approaching fluid dynamics with a view to its application in various fields of engineering, but relatively few which cater for a student coming to the subject as an applied mathematician and none which in my view does so satisfactorily. The trouble is that the great strides made in our understanding of many aspects of fluid dynamics during the last 50 years or so have not yet been absorbed into the educational texts for students of applied mathematics. A teacher is therefore obliged to do without textbooks for large parts of his course, or to tailor his lectures to the existing books. This latter alternative tends to emphasize unduly the classical analytical aspects of the subject, and the mathematical theory of irrotational flow in particular, with the probable consequence that the students remain unaware of the vitally important physical aspects of fluid dynamics. Students, and teachers too, are apt to derive their ideas of the content of a subject from the topics treated in the textbooks they can lay hands on, and it is undesirable that so many of the books on fluid dynamics for applied mathematicians should be about problems which are mathematically solvable but not necessarily related to what happens in real fluids.

I have tried therefore to write a textbook which can be used by students of applied mathematics and which incorporates the physical understanding and information provided by past research. Despite its bulk this book is genuinely an introduction to fluid dynamics; that is to say, it assumes no previous knowledge of the subject and the material in it has been selected to introduce a reader to the important ideas and applications. The book has grown out of a number of courses of lectures, and very little of the material has not been tested in the lecture room. Some of the material is old and well known, some of it is relatively new; and for all of it I have tried to devise the presentation which appears to be best from a consistent point of view. The book has been prepared as a connected account, intended to be read and worked on as a whole, or at least in large portions, rather than to be referred to for particular problems or methods.

I have had the needs of second-, third- and fourth-year students of applied mathematics in British universities particularly in mind, these being the needs with which I am most familiar, although I hope that engineering students will also find the book useful. The true needs of applied mathematicians and engineers are nowadays not far apart. Both require above all an understanding of the fundamentals of fluid dynamics; and this can be achieved without the use of advanced mathematical techniques. Anyone who is familiar with vector analysis and the notation of tensors should have little difficulty with the purely mathematical parts of this work. The book is fairly heavily weighted with theory, but not with mathematics.

Attention is paid throughout the book to the correspondence between observation and the various conceptual and analytical models of flow systems. The photographs of flow systems that are included are an essential part of the book, and will help the reader, I hope, to develop a sense of the reality that lies behind the theoretical arguments and analysis. This is particularly important for students who do not have an opportunity of seeing flow phenomena in a laboratory. The various books and lectures by L. Prandtl seem to me to show admirably the way to keep both theory and observation continually in mind, and I have been greatly influenced by them. Prandtl knew in particular the value of a clear photograph of a well-designed experimental flow system, and many of the photographs taken by him are still the best available illustrations of boundary-layer phenomena.

A word is necessary about the selection of topics in this book and the order in which they have been placed. My original intention was to provide between two covers an introduction to all the main branches of fluid dynamics, but I soon found that this comprehensiveness was incompatible with the degree of thoroughness that I also had in mind. I decided therefore to attempt only a partial coverage, at any rate so far as this volume is concerned. The first three chapters prepare the ground for a discussion of any branch of fluid dynamics, and are concerned with the physical properties of fluids, the kinematics of a flow field, and the dynamical equations in general form. The purpose of these three introductory chapters is to show how the various branches of fluid dynamics fit into the subject as a whole and rest on certain idealizations or assumptions about the nature of the fluid or the motion. A teacher is unlikely to wish to include all this preliminary material in a course of lectures, but it can be adapted to suit a specialized course and will I hope be useful as background. In the remaining four chapters the fluid is assumed to be incompressible and to have uniform density and viscosity. I regard flow of an incompressible viscous fluid as being at the centre of fluid dynamics by virtue of its fundamental nature and its practical importance. Fluids with unusual properties are fashionable in research, but most of the basic dynamical ideas are revealed clearly in a study of rotational flow of a fluid with internal friction; and for applications in geophysics, chemical engineering, hydraulics, mechanical and aeronautical engineering, this is still the key branch of fluid dynamics. I regret that many important topics such as gas dynamics, surface waves, motion due to buoyancy forces, turbulence, heat and mass transfer, and magneto-fluid dynamics, are apparently ignored, but the subject is simply too large for proper treatment in one volume. If the reception given to the present book suggests that a second volume would be welcome, I may try later to make the coverage more nearly complete.

As to the order of material in chapters 4 to 7, the description of motion of a viscous fluid and of flow at large Reynolds number precedes the discussion of irrotational flow (although the many purely kinematical properties of an irrotational velocity distribution have a natural place in chapter 2) and of motion of an inviscid fluid with vorticity. My reason for adopting this unconventional arrangement is not that I believe the 'classical' theory of irrotational flow is less important than is commonly supposed. It is simply that results concerning the flow of inviscid fluid can be applied realistically only if the circumstances in which the approximation of zero viscosity is valid are first made clear. The mathematical theory of irrotational flow is a powerful weapon for the solution of problems, but in itself it gives no information about whether the whole or a part of a given flow field at large Reynolds number will be approximately irrotational. For that vital information some understanding of the effects of viscosity of a real fluid and of boundary-layer theory is essential; and, whereas the understanding was lacking when Lamb wrote his classic treatise *Hydrodynamics*, it is available today. I believe that the first book, at least in English, to show how so many common flow systems could be understood in terms of boundary layers and separation and vorticity movement was *Modern Developments in Fluid Dynamics*, edited by Sydney Goldstein. That pioneering book published in 1938 was aimed primarily at research workers, and I have tried to take the further step of making the understanding of the flow of real fluids accessible to students at an early stage of their study of fluid dynamics.

Desirable though it is for study of the flow of viscous fluids to precede consideration of an inviscid fluid and irrotational flow, I appreciate that a lecturer may have his hand forced by the available lecturing time. In the case of mathematics students who are to attend only one course on fluid dynamics, of length under about 30 lectures, it would be foolish to embark on a study of viscous fluid flow and boundary layers in preparation for a description of inviscid-fluid flow and its applications, since too little time would be left for this topic; the lecturer would need to compromise with scientific logic, and could perhaps take his audience from chapters 2 and 3 to chapter 6, with some of the early sections of chapters 5 and 7 included. It is a difficulty inherent in the teaching of fluid dynamics to mathematics undergraduates that a partial introduction to the subject is unsatisfactory, tending to leave them with analytical procedures and results but no information about when they are applicable. Furthermore, students do take some time to grasp the principles of fluid dynamics, and I suggest that 40 to 50 lectures are needed for an adequate introduction of the subject to non-specialist students. However, a book is not subject to the same limitations as a course of lectures. I hope lecturers will agree that it is desirable for students to be able to see all the material set out in logical order, and to be able to improve their own understanding of the subject by reading, even if in a course of lectures many important topics such as boundary-layer separation must be ignored.

Exercises are an important part of the process of understanding and mastering so analytical a subject as fluid dynamics, and the reading of this text should be accompanied by the working of illustrative exercises. I should have liked to be able to provide many suitable questions and exercises, but a search among those already published in various places did not produce many in keeping with the approach adopted in this book. Moreover, the published exercises are concentrated on a small number of topics. The lengthy task of devising and compiling suitable exercises over the whole field of 'modern' fluid dynamics has yet to be undertaken. Consequently only a few exercises will be found at the end of sections. To some extent exercises ought to be chosen to suit the particular background and level of the class for which they are intended, and it may be that a lecturer can turn into exercises for his class many portions of the text not included explicitly in his course of lectures, as I have done in my own teaching.

It is equally important that a course of lectures on the subject matter of this book should be accompanied by demonstrations of fluid flow. Here the assistance of colleagues in a department of engineering may be needed. The many films on fluid dynamics that are now available are particularly valuable for classes of applied mathematicians who do not undertake any laboratory work. By one means or another, a teacher should show the relation between his analysis and the behaviour of real fluids; fluid dynamics is much less interesting if it is treated largely as an exercise in mathematics.

I am indebted to a large number of people for their assistance in the preparation of this book. Many colleagues kindly provided valuable comments on portions of the manuscript, and enabled me to see things more clearly. I am especially grateful to Philip Chatwin, John Elder, Emin

Erdogan, Ken Freeman, Michael McIntyre, Keith Moffatt, John Thomas and Ian Wood who helped with the heavy task of checking everything in the proof. My thanks go also to those who supplied me with diagrams or photographs or who permitted reproduction from an earlier publication; to Miss Pamela Baker and Miss Anne Powell, who did the endless typing with patience and skill; and to the officers of Cambridge University Press, with whom it is a pleasure to work.

G. K. B.

Cambridge
April 1967

CONVENTIONS AND NOTATION

Bold type signifies vector character.

$\mathbf{x}, \mathbf{x}'$ position vectors; $|\mathbf{x}| = r$

$\mathbf{s} = \mathbf{x} - \mathbf{x}'$ relative position vector

$\mathbf{u}$ velocity at a specified time and position in space; $|\mathbf{u}| = q$

$\dfrac{D}{Dt} = \dfrac{\partial}{\partial t} + \mathbf{u}.\nabla$ operator giving the material derivative, or rate of change at a point moving with the fluid locally; applies only to functions of $\mathbf{x}$ and t

System	Co-ordinates	Velocity components
Rectilinear	x, y, z or x_1, x_2, x_3	u, v, w or u_1, u_2, u_3
Polar, two dimensions	r, θ	u, v or u_r, u_θ
Spherical polar	r, θ, ϕ	u, v, w or u_r, u_θ, u_ϕ
Cylindrical	x, σ, ϕ $(\sigma^2 = y^2 + z^2)$	u, v, w or u_x, u_σ, u_ϕ

$\Delta = \nabla.\mathbf{u}$ rate of expansion (fractional rate of change of volume of a material element)

$\boldsymbol{\omega} = \nabla \times \mathbf{u}$ vorticity (twice the local angular velocity of the fluid)

$e_{ij} = \dfrac{1}{2}\left(\dfrac{\partial u_j}{\partial x_i} + \dfrac{\partial u_i}{\partial x_j}\right)$ rate-of-strain tensor

ϕ scalar potential of an irrotational velocity distribution $(\mathbf{u} = \nabla\phi)$

$\mathbf{B}$ vector potential of a solenoidal velocity distribution $(\mathbf{u} = \nabla \times \mathbf{B})$

ψ stream function for a solenoidal velocity distribution;

(a) two-dimensional flow: $\mathbf{B} = (0, 0, \psi)$

$$u = \frac{\partial \psi}{\partial y}, \quad v = -\frac{\partial \psi}{\partial x} \quad \text{or} \quad u_r = \frac{1}{r}\frac{\partial \psi}{\partial \theta}, \quad u_\theta = -\frac{\partial \psi}{\partial r}$$

(b) axisymmetric flow:

cylindrical co-ordinates $B_\phi = \dfrac{\psi}{\sigma}, \quad u_x = \dfrac{1}{\sigma}\dfrac{\partial \psi}{\partial \sigma}, \quad u_\sigma = -\dfrac{1}{\sigma}\dfrac{\partial \psi}{\partial x}$

polar co-ordinates $B_\phi = \dfrac{\psi}{r \sin \theta}, \quad u_r = \dfrac{1}{r^2 \sin \theta}\dfrac{\partial \psi}{\partial \theta}, \quad u_\theta = -\dfrac{1}{r \sin \theta}\dfrac{\partial \psi}{\partial r}$

$\mathbf{n}$ unit normal to a surface, usually outward if the surface is closed

$\delta V, \mathbf{n}\delta A, \delta\mathbf{x}$ volume, surface and line elements with a specified position in space

$\delta\tau, \mathbf{n}\delta S, \delta\mathbf{l}$ material volume, surface and line elements

σ_{ij} stress tensor; $\sigma_{ij} n_j \delta A$ is the i-component of the force exerted across the surface element $\mathbf{n}\delta A$ by the fluid on the side to which $\mathbf{n}$ points

$\mathbf{F} = -\nabla\Psi$ conservative body force per unit mass

Inertia force (per unit mass)   minus the local acceleration

Vortex-line   line whose tangent is parallel to $\boldsymbol{\omega}$ locally

Line vortex   singular line in vorticity distribution round which the circulation is non-zero

Books which may provide collateral reading are cited in detail in the text, usually in footnotes. A comparatively small number of original papers are also referred to, sometimes for historical interest, sometimes because a precise acknowledgement is appropriate, and sometimes, although only rarely, as a guide to further reading on a particular topic. These papers are cited in the text as 'Smith (1950)', and the full references for both papers and books are listed at the end of the book.

1

THE PHYSICAL PROPERTIES
OF FLUIDS

1.1. Solids, liquids and gases

The defining property of fluids, embracing both liquids and gases, lies in the ease with which they may be deformed. A piece of solid material has a definite shape, and that shape changes only when there is a change in the external conditions. A portion of fluid, on the other hand, does not have a preferred shape, and different elements of a homogeneous fluid may be rearranged freely without affecting the macroscopic properties of the portion of fluid. The fact that relative motion of different elements of a portion of fluid can, and in general does, occur when forces act on the fluid gives rise to the science of fluid dynamics.

The distinction between solids and fluids is not a sharp one, since there are many materials which in some respects behave like a solid and in other respects like a fluid. A 'simple' solid might be regarded as a material of which the shape, and the relative positions of the constituent elements, change by a small amount only, when there is a small change in the forces acting on it. Correspondingly, a 'simple' fluid (there is no one term in general use) might be defined as a material such that the relative positions of the elements of the material change by an amount which is not small when suitably chosen forces, however small in magnitude, are applied to the material. But, even supposing that these two definitions could be made quite precise, it is known that some materials do genuinely have a dual character. A thixotropic substance such as jelly or paint behaves as an elastic solid after it has been allowed to stand for a time, but if it is subjected to severe distortion by shaking or brushing it loses its elasticity and behaves as a liquid. Pitch behaves as a solid normally, but if a force is imposed on it for a very long time the deformation increases indefinitely, as it would for a liquid. Even more troublesome to the analyst are those materials like concentrated polymer solutions which may simultaneously exhibit solid-like and fluid-like behaviour.

Fortunately, most common fluids, and air and water in particular, are quite accurately simple in the above sense, and this justifies a concentration of attention on simple fluids in an introductory text. In this book we shall suppose that the fluid under discussion cannot withstand any tendency by applied forces to deform it in a way which leaves the volume unchanged. The implications of this definition will emerge later, after we have examined the nature of forces that tend to deform an element of fluid. In the meantime it should be noted that a simple fluid may offer resistance to attempts to deform it; what the definition implies is that the resistance cannot prevent the deformation from occurring, or, equivalently, that the resisting force vanishes with the rate of deformation.

Since we shall be concerned exclusively with the kind of idealized material described here as a simple fluid, there is no need to use the term further. We shall therefore refer only to fluids in subsequent pages.

The distinction between liquids and gases is much less fundamental, so far as dynamical studies are concerned. For reasons related to the nature of intermolecular forces, most substances can exist in either of two stable phases which exhibit the property of fluidity, or easy deformability. The density of a substance in the liquid phase is normally much larger than that in the gaseous phase, but this is not in itself a significant basis for distinction since it leads mainly to a difference in the magnitudes of forces required to produce given magnitudes of acceleration rather than to a difference in the types of motion. The most important difference between the mechanical properties of liquids and gases lies in their bulk elasticity, that is, in their compressibility. Gases can be compressed much more readily than liquids, and as a consequence any motion involving appreciable variations in pressure will be accompanied by much larger changes in specific volume in the case of a gas than in the case of a liquid. Appreciable variations in pressure in a fluid must be reckoned with in meteorology, as a result of the action of gravity on the whole atmosphere, and in very rapid motions, of the kind which occur in ballistics and aeronautics, resulting from the motion of solid bodies at high speed through the fluid. It will be seen later that there are common circumstances in which motions of a fluid are accompanied by only slight variations in pressure, and here gases and liquids behave similarly since in both cases the changes in specific volume are slight.

The gross properties of solids, liquids and gases are directly related to their molecular structure and to the nature of the forces between the molecules. We may see this superficially from a consideration of the general form of the force between two typical molecules in isolation as a function of their separation. At small values of the distance d between the centres of the molecules, of order 10^{-8} cm for molecules of simple type, the mutual reaction is a strong force of quantum origin, being either attractive or repulsive according to the possibility of 'exchange' of electron shells. When exchange is possible, the force is attractive and constitutes a chemical bond; when exchange is not possible, the force is repulsive, and falls off very rapidly as the separation increases. At larger distances between the centres, say of order 10^{-7} or 10^{-6} cm, the mutual reaction between the two molecules (assumed to be un-ionized, as is normally the case at ordinary temperatures) is a weakly attractive force. This cohesive force is believed to fall off first as d^{-7} and ultimately as d^{-8} when d is large, and may be regarded, crudely speaking, as being due to the electrical polarization of each molecule under the influence of the other.† The mutual reaction as a function of d for two molecules not forming a chemical bond thus has the form shown in figure 1.1.1. At separation d_0, at which the reaction changes sign, one molecule is clearly in a position of stable equilibrium relative to that of the other. d_0 is of order 3–4×10^{-8} cm for most simple molecules.

From a knowledge of the mass of a molecule and the density of the corresponding substance, it is possible to calculate the average distance between the centres of adjoining molecules. For substances composed of simple molecules, the calculation shows that the average spacing of the molecules in a gaseous phase at normal temperature and pressure is of the order of $10 d_0$, whereas the average spacing in liquid and solid phases is of

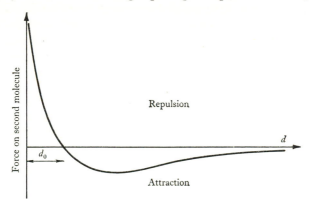

Figure 1.1.1. Sketch of the force exerted by one (un-ionized) simple molecule on another as a function of the distance d between their centres.

order d_0. In gases under ordinary conditions the molecules are thus so far apart from each other that only exceedingly weak cohesive forces act between them, except on the rare occasions when two molecules happen to come close together; and in the kinetic theory of gases it is customary to postulate a 'perfect gas', for which the potential energy of a molecule in the force fields of its neighbours is negligible by comparison with its kinetic energy; that is, a gas in which each molecule moves independently of its neighbours except when making an occasional 'collision'. In liquid and solid phases, on the other hand, a molecule is evidently well within the strong force fields of several neighbours at all times. The molecules are here packed together almost as closely as the repulsive forces will allow. In the case of a solid the arrangement of the molecules is virtually permanent, and may have a simple periodic structure, as in a crystal; the molecules oscillate about their stable positions (the kinetic energy of this oscillation being part of the thermal energy of the solid), but the molecular lattice remains intact until the temperature of the solid is raised to the melting point.

† See, for instance, *States of Matter*, by E. A. Moelwyn-Hughes (Oliver and Boyd, 1961).

The density of most substances falls by several per cent on melting (the increase in density in the transition from ice to water being exceptional), and it is paradoxical that such a small change in the molecular spacing is accompanied by such a dramatic change in the mobility of the material. Knowledge of the liquid state is still incomplete, but it appears that the arrangement of the molecules is partially ordered, with groups of molecules as a whole having mobility, sometimes falling into regular array with other groups and sometimes being split up into smaller groups. The arrangement of the molecules is continually changing, and, as a consequence, any force applied to the liquid (other than a bulk compression) produces a deformation which increases in magnitude for so long as the force is maintained. The manner in which some of the molecular properties of a liquid stand between those of a solid and a gas is shown in the following table. In the matter of the simplest macroscopic quantity, viz. density, liquids stand much closer to solids; and in the matter of fluidity, liquids stand wholly with gases.

	Intermolecular forces	Ratio of amplitude of random thermal movement of molecules to d_0	Molecular arrangement	Type of statistics needed
solid	strong	$\ll 1$	ordered	quantum
liquid	medium	of order unity	partially ordered	quantum + classical
gas	weak	$\gg 1$	disordered	classical

The molecular mechanism by which a liquid resists an attempt to deform it is not the same as that in a gas, although, as we shall see, the differential equation determining the rate of change of deformation has the same form in the two cases.

1.2. The continuum hypothesis

The molecules of a gas are separated by vacuous regions with linear dimensions much larger than those of the molecules themselves. Even in a liquid, in which the molecules are nearly as closely packed as the strong short-range repulsive forces will allow, the mass of the material is concentrated in the nuclei of the atoms composing a molecule and is very far from being smeared uniformly over the volume occupied by the liquid. Other properties of a fluid, such as composition or velocity, likewise have a violently non-uniform distribution when the fluid is viewed on such a small scale as to reveal the individual molecules. However, fluid mechanics is normally concerned with the behaviour of matter in the large, on a macroscopic scale large compared with the distance between molecules, and it will not often happen that the molecular structure of a fluid need be taken into account explicitly. We shall suppose, throughout this book, that the macroscopic behaviour of fluids is the same as if they were perfectly continuous in

structure; and physical quantities such as the mass and momentum associated with the matter contained within a given small volume will be
regarded as being spread uniformly over that volume instead of, as in strict
reality, being concentrated in a small fraction of it.

The validity of the simpler aspects of this *continuum hypothesis* under the
conditions of everyday experience is evident. Indeed the structure and
properties of air and water are so obviously continuous and smoothly-
varying, when observed with any of the usual measuring devices, that no
different hypothesis would seem natural.

When a measuring instrument is inserted in a fluid, it responds in some
way to a property of the fluid within some small neighbouring volume,
and provides a measure which is effectively an average of that property over

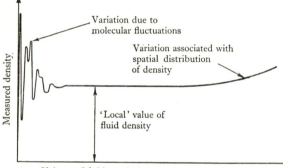

Figure 1.2.1. Effect of size of sensitive volume on the density
measured by an instrument.

the 'sensitive' volume (and sometimes also over a similar small sensitive
time). The instrument is normally chosen so that the sensitive volume is
small enough for the measurement to be a 'local' one; that is, so that further
reduction of the sensitive volume (within limits) does not change the reading
of the instrument. The reason why the particle structure of the fluid is
usually irrelevant to such a measurement is that the sensitive volume that
is small enough for the measurement to be 'local' relative to the macroscopic scale is nevertheless quite large enough to contain an enormous
number of molecules, and amply large enough for the fluctuations arising
from the different properties of molecules to have no effect on the observed
average. Of course, if the sensitive volume is made so small as to contain only
a few molecules, the number and kind of molecules in the sensitive volume
at the instant of observation will fluctuate from one observation to another
and the measurement will vary in an irregular way with the size of the sensitive volume. Figure 1.2.1 illustrates the way in which a measurement of
density of the fluid would depend on the sensitive volume of the instrument.

We are able to regard the fluid as a continuum when, as in the figure, the measured fluid property is constant for sensitive volumes small on the macroscopic scale but large on the microscopic scale.

One or two numbers will indicate the great difference between the length scale representative of the fluid as a whole and that representative of the particle structure. For most laboratory experiments with fluids, the linear dimensions of the region occupied by the fluid is at least as large as 1 cm and very little variation of the physical and dynamical properties of the fluid occurs over a distance of 10^{-3} cm (except perhaps in special places such as in a shock wave); thus an instrument with a sensitive volume of 10^{-9} cm³ would give a measurement of a local property. Small though this volume is, it contains about 3×10^{10} molecules of air at normal temperature and pressure (and an even larger number of molecules of water) which is large enough, by a very wide margin, for an average over the molecules to be independent of their number. Only under extreme conditions of low gas density, as in the case of flight of a missile or satellite at great heights above the earth's surface, or of very rapid variation of density with position, as in a shock wave, is there difficulty in choosing a sensitive volume which gives a local measurement and which contains a large number of molecules.

Our hypothesis implies that it is possible to attach a definite meaning to the notion of value 'at a point' of the various fluid properties such as density, velocity and temperature, and that in general the values of these quantities are continuous functions of position in the fluid and of time. On this basis we shall be able to establish equations governing the motion of the fluid which are independent, so far as their form is concerned, of the nature of the particle structure—so that gases and liquids are treated together—and indeed, independent of whether any particle structure exists. A similar hypothesis is made in the mechanics of solids, and the two subjects together are often designated as continuum mechanics.

Natural though the continuum hypothesis may be, it proves to be difficult to deduce the properties of the hypothetical continuous medium that moves in the same way as a real fluid with a given particle structure. The methods of the kinetic theory of gases have been used to establish the equations determining the 'local' velocity (defined as above) of a gas, and, with the help of simplifying assumptions about the collisions between molecules, it may be shown that the equations have the same form as for a certain continuous fluid although the values of the molecular transport coefficients (see §1.6) are not obtained accurately. The mathematical basis for the continuum treatment of gases in motion is beyond our scope, and it is incomplete for liquids, so that we must be content to make a hypothesis. There is ample observational evidence that the common real fluids, both gases and liquids, move as if they were continuous, under normal conditions and indeed for considerable departures from normal conditions, but some of the properties of the equivalent continuous media need to be determined empirically.

1.3. Volume forces and surface forces acting on a fluid

It is possible to distinguish two kinds of forces which act on matter in bulk. In the first group are long-range forces like gravity which decrease slowly with increase of distance between interacting elements and which are still appreciable for distances characteristic of natural fluid flows. Such forces are capable of penetrating into the interior of the fluid, and act on all elements of the fluid. Gravity is the obvious and most important example, but two other kinds of long-range force of interest in fluid mechanics are electromagnetic forces, which may act when the fluid carries an electric charge or when an electric current passes through it, and the fictitious forces, such as centrifugal force, which appear to act on mass elements when their motion is referred to an accelerating set of axes. A consequence of the slow variation of one of these long-range forces with position of the element of fluid on which it is acting is that the force acts equally on all the matter within a small element of volume and the total force is proportional to the size of the volume element. Long-range forces may thus also be called volume or body forces.

When writing equations of motion in general form, we shall designate the total of all body forces acting at time t on the fluid within an element of volume δV surrounding the point whose position vector is $\mathbf{x}$ by

$$\mathbf{F}(\mathbf{x}, t)\rho\,\delta V;\qquad(1.3.1)$$

the factor ρ has been inserted because the two common types of body force per unit volume—gravity and the fictitious forces arising from the use of accelerating axes—are in fact proportional to the mass of the element on which they act. In the case of the earth's gravitational field the force per unit mass is

$$\mathbf{F} = \mathbf{g},$$

the vector $\mathbf{g}$ being constant in time and directed vertically downwards.

In the second group are short-range forces, which have a direct molecular origin, decrease extremely rapidly with increase of distance between interacting elements, and are appreciable only when that distance is of the order of the separation of molecules of the fluid. They are negligible unless there is direct mechanical contact between the interacting elements, as in the case of the reaction between two rigid bodies, because without that contact none of the molecules of one of the elements is sufficiently close to a molecule of the other element. The short-range forces exerted between two masses of gas in direct contact at a common boundary are due predominantly to transport of momentum across the common boundary by migrating molecules. In the case of a liquid the situation is more complex because there are contributions to the short-range or contact forces from transport of momentum across the common boundary by molecules in oscillatory motion about some quasi-stationary position and from the forces between molecules on the two sides of the common boundary; both these contributions have large magnitude, but they act approximately in opposite directions and their resultant normally has a much smaller magnitude than either. However, as already remarked, the laws of continuum mechanics do not depend on the nature of the molecular origin of these contact forces and we need not enquire into the details of the origin in liquids, at this stage.

If an element of mass of fluid is acted on by short-range forces arising from reactions with matter (either solid or fluid) outside this element, these short-range forces can act only on a thin layer† adjacent to the boundary of the fluid element, of thickness equal to the 'penetration' depth of the forces. The total of the short-range forces acting on the element is thus determined by the surface area of the element, and the volume of the element is not directly relevant. The different parts of a closed surface bounding an element of fluid have different orientations, so that it is not useful to specify the short-range forces by their total effect on a finite volume element of fluid; instead we consider a *plane* surface element in the fluid and specify the local short-range force as the total force exerted on the fluid on one side of the element by the fluid on the other side. Provided the penetration depth of the short-range forces is small compared with the linear dimensions of the plane surface element, this total force exerted across the element will be proportional to its area δA and its value at time t for an element at position $\mathbf{x}$ can be written as the vector

$$\mathbf{\Sigma}(\mathbf{n}, \mathbf{x}, t)\,\delta A. \qquad (1.3.2)$$

The force per unit area, $\mathbf{\Sigma}$, is called the local *stress*. The way in which it depends on the unit normal to the surface element ($\mathbf{n}$) is determined below. The force exerted across the surface element on the fluid on its other side is of course $-\mathbf{\Sigma}(\mathbf{n}, \mathbf{x}, t)\,\delta A$, and since this is also the force represented by $\mathbf{\Sigma}(-\mathbf{n}, \mathbf{x}, t)\,\delta A$ we see that $\mathbf{\Sigma}$ must be an odd function of $\mathbf{n}$. The convention about the direction of $\mathbf{n}$ to be adopted here is that $\mathbf{\Sigma}$ is the stress exerted *by* the fluid on the side of the surface element to which $\mathbf{n}$ points, *on* the fluid on the side which $\mathbf{n}$ points away from; that is, a normal component of $\mathbf{\Sigma}$ having the same sense as $\mathbf{n}$ represents a tension.

In chapter 3 we shall formulate equations describing the motion of a fluid which is subject to long-range or body forces represented by (1.3.1) and short-range or surface forces represented by (1.3.2). Forces of these two kinds act also on solids, and their existence is perhaps more directly evident to the senses for a solid than for a fluid medium. In the case of a solid body which is rigid, only the short-range forces acting at the surface of the body (say, as a result of mechanical contact with another rigid body) are relevant, and it is a simple matter to determine the body's motion when the total body force and the total surface force acting on it are known. When the solid body

† Unless the element is chosen to have such small linear dimensions that the short-range forces exerted by external matter are still significant at the centre of the element; but the element would then contain only a few molecules at most, and representation of the fluid as a continuum would not be possible.

is deformable, and likewise in the case of a fluid, the different material elements are capable of different movements, and the distribution of the body and surface forces throughout the matter must be considered; moreover, both the body and surface forces may be affected by the relative motion of material elements. The way in which body forces depend on the local properties of the fluid is evident, at any rate in the cases of gravity and the fictitious forces due to accelerating axes, but the dependence of surface forces on the local properties and motion of the fluid will require examination.

Representation of surface forces by the stress tensor

Some information about the stress Σ may be deduced from its definition as a force per unit area and the law of motion for an element of mass of the fluid. First we determine the dependence of Σ on the direction of the normal to the surface element across which it acts.

Consider all the forces acting instantaneously on the fluid within an element of volume δV in the shape of a tetrahedron as shown in figure 1.3.1. The three orthogonal faces have areas δA_1, δA_2, δA_3, and unit (outward) normals $-\mathbf{a}$, $-\mathbf{b}$, $-\mathbf{c}$, and the fourth inclined face has area δA and unit normal $\mathbf{n}$. Surface forces will act on the fluid in the tetrahedron across each of the four faces, and their sum is

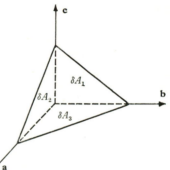

$$\Sigma(\mathbf{n})\,\delta A + \Sigma(-\mathbf{a})\,\delta A_1$$
$$+ \Sigma(-\mathbf{b})\,\delta A_2 + \Sigma(-\mathbf{c})\,\delta A_3;$$

Figure 1.3.1. A volume element in the shape of a tetrahedron with three orthogonal faces.

the dependence of Σ on $\mathbf{x}$ and t is not displayed here, because these variables have the same values (approximately, in the case of $\mathbf{x}$) for all four contributions. In view of the orthogonality of three of the faces, three relations like

$$\delta A_1 = \mathbf{a} \cdot \mathbf{n}\, \delta A$$

are available, and the i-component of the sum of the surface forces can therefore be written as

$$[\Sigma_i(\mathbf{n}) - \{a_j \Sigma_i(\mathbf{a}) + b_j \Sigma_i(\mathbf{b}) + c_j \Sigma_i(\mathbf{c})\}\, n_j]\, \delta A. \tag{1.3.3}†$$

Now the total body force on the fluid within the tetrahedron is proportional to the volume δV, which is of smaller order than δA in the linear

---

† The suffix notation for vector components has been used here, with the usual convention of vector and tensor analysis that terms containing a repeated suffix are to be regarded as summed over all three possible values of the suffix. Both the suffix form and the representation in bold-face type without suffixes will be used for vectors in this book, the choice being made usually with an eye to neatness of the formulae.

dimensions of the tetrahedron. The mass of the fluid in the tetrahedron is also of order δV, and so too is the product of the mass and the acceleration of the fluid in the tetrahedron, provided that both the local density and acceleration are finite. Thus if the linear dimensions of the tetrahedron are made to approach zero without change of its shape, the first two terms of the equation

mass × acceleration = resultant of body forces + resultant of surface forces

approach zero as δV, whereas the third term apparently approaches zero only as δA. In these circumstances the equation can be satisfied only if the coefficient of δA in (1.3.3) vanishes identically (with the implication that information about the resultant surface force on the element requires a higher degree of approximation which takes account of the difference between the values of Σ at different positions on the surface of the element), giving

$$\Sigma_i(\mathbf{n}) = \{a_j \Sigma_i(\mathbf{a}) + b_j \Sigma_i(\mathbf{b}) + c_j \Sigma_i(\mathbf{c})\} n_j. \qquad (1.3.4)$$

Thus the component of stress in a given direction represented by the suffix i across a plane surface element with an arbitrary orientation specified by the unit normal $\mathbf{n}$ is related to the same component of stress across any three orthogonal plane surface elements at the same position in the fluid in the same way as if it were a vector with orthogonal components $\Sigma_i(\mathbf{a})$, $\Sigma_i(\mathbf{b})$, $\Sigma_i(\mathbf{c})$.

The vectors $\mathbf{n}$ and Σ do not depend in any way on the choice of the axes of reference, and the expression within curly brackets in (1.3.4) must represent the (i,j)-component of a quantity which is similarly independent of the axes. In other words the expression within curly brackets is one component of a second-order tensor,[†] σ_{ij} say, and

$$\Sigma_i(\mathbf{n}) = \sigma_{ij} n_j. \qquad (1.3.5)$$

σ_{ij} is the i-component of the force per unit area exerted across a plane surface element normal to the j-direction, at position $\mathbf{x}$ in the fluid and at time t, and the tensor of which it is the general component is called the *stress tensor*. Specification of the local stress in the fluid is now provided by σ_{ij}, which is independent of $\mathbf{n}$, in place of $\Sigma(\mathbf{n})$.

A similar argument can be used to demonstrate that the nine components of the stress tensor are not all independent. This time we consider the moments of the various forces acting on the fluid within a volume V of arbitrary shape. The i-component of the total moment, about a point O

[†] A general familiarity with the elementary properties of tensors will be assumed in this book. Only Cartesian tensors (that is, tensors for which the suffixes denote components with respect to rectangular co-ordinate axes) will be used. Two special tensors which will appear often are the Kronecker delta tensor δ_{ij}, such that $\delta_{ij} = 1$ when $i = j$ and $\delta_{ij} = 0$ when $i \neq j$, and the alternating tensor ϵ_{ijk}, with value zero unless i, j, k are all different, in which case the value is $+1$ or -1 according as i, j, k are or are not in cyclic order.

within this volume, exerted by the surface forces at the boundary of the volume is

$$\int \epsilon_{ijk} r_j \sigma_{kl} n_l dA,$$

where $\mathbf{r}$ is the position vector of the surface element $\mathbf{n}\delta A$ relative to O. This integral over a closed surface can be transformed by the divergence theorem to the volume integral

$$\int \epsilon_{ijk} \frac{\partial (r_j \sigma_{kl})}{\partial r_l} dV, \quad = \int \epsilon_{ijk} \left(\sigma_{kj} + r_j \frac{\partial \sigma_{kl}}{\partial r_l} \right) dV. \qquad (1.3.6)$$

If now the volume V is reduced to zero in such a way that the configuration made up of the boundary of the volume and the fixed point O retains the same shape, the first term on the right-hand side of (1.3.6) becomes small as V whereas the second term approaches zero more quickly as $V^{\frac{4}{3}}$. The total moment about O exerted on the fluid element by the body forces is clearly of order $V^{\frac{4}{3}}$ when V is small,† and so too is the rate of change of the angular momentum of the fluid instantaneously in V. Thus $\int \epsilon_{ijk} \sigma_{kj} dV$ is apparently of larger order in V than all the other terms in the moment equation, and as a consequence it must be identically zero. This is possible for all choices of the position of O and the shape of V, when σ_{ij} is continuous in $\mathbf{x}$, only if

$$\epsilon_{ijk} \sigma_{kj} = 0 \qquad (1.3.7)$$

everywhere in the fluid; for if $\epsilon_{ijk} \sigma_{kj}$ were non-zero in some region of the fluid, we should be able to choose a small volume V for which the integral is non-zero, giving a contradiction.‡ The relation (1.3.7) shows that the stress tensor is symmetrical, that is, $\sigma_{ij} = \sigma_{ji}$, and has only six independent components.

The three diagonal components of σ_{ij} are *normal stresses* in the sense that each of them gives the normal component of surface force acting across a plane surface element parallel to one of the co-ordinate planes. The six non-diagonal components of σ_{ij} are *tangential stresses*, sometimes also called shearing stresses, since in both fluids and solids they are set up by a shearing motion or displacement in which parallel layers of matter slide relative to each other. Figure 1.3.2 shows the first approximation to the various surface forces acting in the (x_1, x_2)-plane on a small rectangular element with sides δx_1 and δx_2 and unit depth in the x_3-direction; the components of the stress do not have exactly the same values on opposite sides of the rectangle, and the differences, of order δx_1 or δx_2, will need to be taken into account when the equation of motion of an element of fluid is formulated.

It is always possible to choose the directions of the orthogonal axes of reference so that the non-diagonal elements of a symmetrical second-order

† In the absence of any 'body couple' of order V, like the couple exerted on a polarized dielectric medium by an imposed electric field.
‡ This deduction about the integrand of an integral which is zero for all choices of the range of integration will be needed often, for volume, surface and line integrals.

tensor are all zero. Referred to such *principal axes* of the stress tensor σ_{ij} at a given point $\mathbf{x}$, the diagonal elements of the stress tensor become *principal stresses*, σ'_{11}, σ'_{22}, σ'_{33} say; and it is a well-known property of second-order tensors that changes of directions of orthogonal axes of reference leave the sum of the diagonal elements unchanged, so that

$$\sigma'_{11} + \sigma'_{22} + \sigma'_{33} = \sigma_{ii}. \tag{1.3.8}$$

Relative to these new axes the components of the force per unit area acting across an element of area with normal (n'_1, n'_2, n'_3) are

$$\sigma'_{11} n'_1, \quad \sigma'_{22} n'_2, \quad \sigma'_{33} n'_3.$$

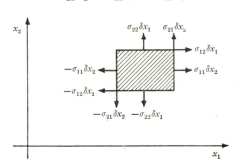

Figure 1.3.2. The surface forces acting on a rectangular element of fluid of unit depth.

A force per unit area with components $(\sigma'_{11} n'_1, 0, 0)$ corresponds to a state of tension (or compression if σ'_{11} is negative) in the direction of the first of the new axes, and similarly for forces $(0, \sigma'_{22} n'_2, 0)$ and $(0, 0, \sigma'_{33} n'_3)$. Thus the general state of the fluid near any given point may be regarded as a superposition of tensions in three orthogonal directions.

The stress tensor in a fluid at rest

We have defined a fluid as being unable to withstand any tendency by applied forces to deform it without change of volume. This definition has consequences for the form of the stress tensor in a fluid at rest. To see this, consider the surface forces exerted on the fluid within a sphere by the surrounding fluid, the radius of the sphere being small so that σ_{ij} is approximately uniform over the surface. We choose axes coinciding (locally) with principal axes of σ_{ij}, and take the further step of writing the stress tensor, which now has zero non-diagonal elements, as the sum of the two tensors

$$\begin{pmatrix} \tfrac{1}{3}\sigma_{ii} & 0 & 0 \\ 0 & \tfrac{1}{3}\sigma_{ii} & 0 \\ 0 & 0 & \tfrac{1}{3}\sigma_{ii} \end{pmatrix} \quad \text{and} \quad \begin{pmatrix} \sigma'_{11} - \tfrac{1}{3}\sigma_{ii} & 0 & 0 \\ 0 & \sigma'_{22} - \tfrac{1}{3}\sigma_{ii} & 0 \\ 0 & 0 & \sigma'_{33} - \tfrac{1}{3}\sigma_{ii} \end{pmatrix}. \tag{1.3.9}$$

The first of these tensors has spherical symmetry, or isotropy, and the corresponding contribution to the force per unit area exerted on the surface of the sphere at a point where the normal is $\mathbf{n}$ is $\frac{1}{3}\sigma_{ii}\mathbf{n}$. This uniform compression (for the sign of $\frac{1}{3}\sigma_{ii}$ is usually negative) of the fluid in the sphere tends to change its volume and can certainly be withstood by the fluid in the sphere while at rest.

The second of the tensors in (1.3.9) is the departure of the stress tensor from an isotropic form. The diagonal elements of this tensor have zero sum, in view of (1.3.8), and thus represent normal stresses of which at least one is a tension and at least one a compression. The corresponding contribution to the force per unit area exerted on the surface of the sphere at a point where the normal vector is (n'_1, n'_2, n'_3) has components (relative to the new axes)

$$(\sigma'_{11} - \tfrac{1}{3}\sigma_{ii})n'_1, \quad (\sigma'_{22} - \tfrac{1}{3}\sigma_{ii})n'_2, \quad (\sigma'_{33} - \tfrac{1}{3}\sigma_{ii})n'_3. \tag{1.3.10}$$

In other words, the sphere is embedded in fluid which is in a state of uniform tension in the direction of one axis, together with uniform compression in the (orthogonal) direction of another axis, and uniform tension or compression in the third orthogonal direction (the algebraic sum of the three

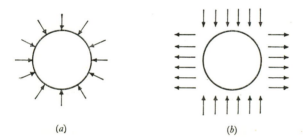

(a)   (b)

Figure 1.3.3. Two contributions to the stress at the surface of a spherical element of fluid; (a) an isotropic compression, and (b) uniform tension in the direction of one principal axis of the stress tensor together with uniform compression in the direction of another principal axis.

tensions and compressions being zero), as indicated in figure 1.3.3. This second contribution thus tends to deform the spherical element of fluid into an ellipsoid, without any necessary change of volume; nor can this deforming surface force be balanced by any volume force on the fluid, because the latter is of a different order of magnitude in the small volume of the spherical element. The spherical element of fluid cannot withstand such a tendency to deform it by applied forces (that is, by forces due to agencies external to the element), so that a state of rest is not compatible with the existence of non-zero values of any of the force components (1.3.10). Hence, in a fluid at rest, the principal stresses σ'_{11}, σ'_{22}, σ'_{33} are all the same and equal to $\frac{1}{3}\sigma_{ii}$, at all points in the fluid; that is, the stress tensor in a fluid at rest is everywhere isotropic, all orthogonal axes of reference are principal axes for the stress tensor, and only *normal stresses* act.

Fluids at rest are normally in a state of compression, and it is therefore convenient to write the stress tensor in a fluid at rest as

$$\sigma_{ij} = -p\,\delta_{ij}, \tag{1.3.11}$$

where $p\,(=-\frac{1}{3}\sigma_{ii})$ may be termed the *static-fluid pressure*† and is in general a function of $\mathbf{x}$.

It follows that in a fluid at rest the contact force per unit area exerted across a plane surface element in the fluid with unit normal $\mathbf{n}$ is $-p\mathbf{n}$, and is a normal force of the same magnitude for all directions of the normal $\mathbf{n}$ at a given point. This well-known property of the static-fluid pressure, of 'acting equally in all directions', is often established as a consequence of an assumption that in a fluid at rest the tangential stresses are zero; the argument is simply a consideration of the balance of forces on an element of fluid of simple geometrical shape, such as the tetrahedron with three orthogonal faces‡ or a portion of a cylinder with one plane section normal to the generators and one inclined to them. An assumption that tangential stresses are zero in a fluid at rest is reasonable, for in the absence of any bulk motion it seems unlikely that the random molecular configuration and motion could have any statistical directional preferences, in which event the reaction due to molecular forces and flux of momentum across a surface element would be purely normal. However, it seems preferable to derive the properties of the stress tensor in fluid at rest from the more primitive assertion that fluids cannot withstand any attempt to change their shape.

1.4. Mechanical equilibrium of a fluid

A rigid body is in equilibrium when the resultant force and the resultant couple exerted on it by external agencies are both zero. The conditions for equilibrium of a fluid are less simple, because the different elements of fluid can move relative to each other and must separately be in equilibrium.

The forces acting on any given portion of fluid are, as stated in the previous section, volume forces due to external agencies and surface forces exerted across the boundary by the surrounding matter. These volume and surface forces must balance if the fluid is to remain at rest. In the notation of the previous section, the total body force acting on the fluid lying within a volume V is

$$\int \rho\mathbf{F}\,dV,$$

in which both ρ and $\mathbf{F}$ may be functions of position in the fluid. The total

† The term hydrostatic pressure is often used, but the implied association with water has only historical justification and may be misleading. The terms 'hydrodynamics' and 'aerodynamics' are likewise unnecessarily restrictive, and are being superseded by the more general term 'fluid dynamics'.
‡ Put $\Sigma_i(\mathbf{n}) = n_i\Sigma(\mathbf{n})$, $\Sigma_i(\mathbf{a}) = a_i\Sigma(\mathbf{a})$, etc., in (1.3.4) and then take the scalar product of both sides of the equation with $\mathbf{a}$, $\mathbf{b}$, and $\mathbf{c}$ in turn.

contact force exerted by the surrounding matter at the surface A bounding the volume V (when the fluid is at rest) is

$$-\int p\mathbf{n}\,dA,$$

in which p is also in general a function of the position vector $\mathbf{x}$ and $\mathbf{n}$ is the unit outward normal to the surface A. This latter integral may be transformed to an integral over the volume V by the analogue of the divergence theorem for a scalar quantity, giving $-\int \nabla p\,dV$. Hence a necessary condition for equilibrium of the fluid is that

$$\int (\rho \mathbf{F} - \nabla p)\,dV = 0, \tag{1.4.1}$$

for all choices of the volume V lying entirely in the fluid, which is possible only if the integrand itself (assumed to be continuous in $\mathbf{x}$) is zero everywhere in the fluid. The necessary condition for equilibrium is then that

$$\rho \mathbf{F} = \nabla p \tag{1.4.2}$$

everywhere in the fluid.

If (1.4.1) holds for all choices of V, the resultant force on each element of the fluid is zero. Moreover, our use of a symmetrical stress tensor ensures that the couple on each volume element of fluid is zero, so that when (1.4.2) is satisfied the resultant couple on the fluid within a volume V of arbitrary shape and size is zero (in the absence of any body couple acting on the fluid), as may be verified directly. Equation (1.4.2) is therefore the necessary and sufficient condition for the fluid to be in equilibrium. In the case of a solid, for which the tangential stresses are not necessarily zero, the corresponding condition is an equation like (1.4.2) in which the (i-component of the) right-hand side has the more general form $-\partial \sigma_{ij}/\partial x_j$.

The restriction imposed by equation (1.4.2) lies in the fact that only for certain distributions of ρ and $\mathbf{F}$, viz. those for which $\rho \mathbf{F}$ (the body force *per unit volume*) can be expressed as the gradient of a scalar quantity, does there exist a pressure distribution satisfying (1.4.2). When the distribution of $\rho \mathbf{F}$ does have the form required for equilibrium, p is constant over any surface which is everywhere normal to the body force.

The nature of the restriction on ρ and $\mathbf{F}$ takes a more specific form in the common case in which the body force per unit mass ($\mathbf{F}$) represents a conservative field and can be written as $-\nabla \Psi$, where Ψ is the potential energy per unit mass associated with this field. In this case the condition for equilibrium is

$$-\rho \nabla \Psi = \nabla p, \tag{1.4.3}$$

or, on taking the curl of both sides,

$$(\nabla \rho) \times (\nabla \Psi) = 0.$$

Thus the level-surfaces of ρ and Ψ must coincide, and, when this condition is satisfied, these are also the level surfaces of p and we may write

$$dp/d\Psi = -\rho(\Psi). \tag{1.4.4}$$

The particular case in which $\nabla \Psi$ has the same direction everywhere, so that Ψ, ρ and p are constant on each one of a family of parallel planes, occurs in discussions of the earth's atmosphere.

The density of an element of fluid may be affected by the pressure to which it is subjected, and also by other factors, so that further discussion of the implications of (1.4.3) requires information about ρ. However, in a case in which the fluid has *uniform* density ρ, the solution of (1.4.3) is simply

$$p = p_0 - \rho \Psi, \tag{1.4.5}$$

where p_0 is a constant.

A body 'floating' in fluid at rest

The common notion of floating relates to a rigid body partially immersed at a horizontal free surface of liquid at rest under gravity, but the term may be used more generally. A body may be said to float when it is wholly immersed in fluid (some of the fluid may be liquid and some gaseous, giving partial immersion in everyday terminology) and both it and the fluid are at rest under the action of volume forces.

The primary result for a floating body is Archimedes' theorem, which is usually stated and proved for the case of a body supported by the buoyancy force due to the action of gravity on a uniform liquid. This is the most important field of application of the theorem, but the additional generality of the form of the theorem to be established here has some value. Suppose that a body of volume V and bounding surface A is immersed in fluid and that the body and the fluid are at rest. The resultant force on the body due entirely to the presence of the fluid is

$$-\int p\mathbf{n}\,dA,$$

where $\mathbf{n}$ is the outward normal to the body surface. The pressure p in the fluid is determined by the equilibrium relation (1.4.2), and, taking our cue from the conventional form of Archimedes' theorem, we wish to use (1.4.2) to express this resultant surface force in terms of the total volume force on fluid which in some sense is able to take the place of the body. We need to know how fluid can replace the body without disturbing the equilibrium and without changing conditions in the surrounding fluid.

A definite answer may be given in a case in which $\mathbf{F} = -\nabla \Psi$ and Ψ is a prescribed function of position in space. The level surfaces of Ψ may be continued through the region occupied by the body, and the uniform value that the density ρ must have on each level surface of Ψ for fluid in this region to be equilibrium is the same as the value of ρ on the same level surface outside the region. In other words, we have a specification for the distribution of density of fluid which can take the place of the body. The total volume force on this replacement fluid is

$$-\int \rho \nabla \Psi\,dV,$$

where the integral is taken over the region which was occupied by the body, and this force is balanced by the contact force at the boundary A, which is unchanged by the replacement of the body by fluid. Thus the 'buoyancy' force on an immersed body due to the action of a volume force on the surrounding fluid (at rest) is

$$\int \rho \nabla \Psi \, dV, \quad = - \int \rho \mathbf{F} \, dV,$$

where the density ρ at a point within the region occupied by the body is determined by continuation of the distribution in the surrounding fluid in the manner described above. A body immersed in fluid loses 'weight' equal to the 'weight' of the fluid 'displaced', where 'weight' and 'displaced' can both be given rather more general meanings than those intended by Archimedes.

The practical implications of these principles are examined in textbooks on hydrostatics† and need not be recounted here. However, the reader may be interested to consider briefly the application of the principles to one problem different from those involving only gravity and uniform liquids. Suppose, for instance, that a vessel containing fluid of non-uniform density is rotating steadily about the vertical z-axis and that the fluid has taken up the same steady rotation. Relative to axes rotating with the vessel, with angular velocity Ω say, the fluid is at rest and is acted on by a body force per unit mass with vertical component $-g$ due to gravity and with radial component $\Omega^2(x^2+y^2)^{\frac{1}{2}}$ in a horizontal plane due to the effective centrifugal force. Thus we have

$$\mathbf{F} = -\nabla \Psi, \quad \Psi = gz - \tfrac{1}{2}\Omega^2(x^2+y^2),$$

and the level surfaces of Ψ are equal paraboloids of revolution, with vertical axes, translated vertically from each other (figure 1.4.1). For equilibrium it is necessary that ρ be constant on each of these paraboloids; and then p is also constant on each paraboloid.

If now a solid body, say a sphere of uniform density, is immersed in the fluid in this vessel and is at rest relative to it, the fluid exerts a certain buoyancy force on the body. There arises the question: can this buoyancy be balanced by the same volume forces (gravity and centrifugal force) acting on the body itself? In other words, if the body is placed at a certain position in the fluid, will it remain there? We need to find a position for the centre of the sphere such that the sphere displaces its own mass of fluid, which selects (approximately) a certain value of Ψ (figure 1.4.1), and such that the same centrifugal force acts on the displaced fluid as on the solid sphere. It is evident that such a position cannot be found off the axis of rotation, because the tilting of the surfaces of equal density implies a greater centrifugal force on the displaced fluid, given that it has the same mass as the sphere, than on the sphere. Hence a uniform sphere would 'fall' down

---

† See, for instance, *Statics*, by H. Lamb (Cambridge University Press, 1933).

the paraboloid of revolution on which it must lie to displace its own mass of fluid and would come to rest at the axis. The same is true of a sphere at a free surface of rotating liquid, since this is simply a particular distribution of density with respect to Ψ.

On the other hand, if the sphere is sufficiently non-uniform in density, say by being weighted on one side, it is clearly possible for the total centrifugal force on the sphere to be greater than that on displaced fluid of the same total mass, in which case the sphere moves outward on a paraboloid of revolution until it meets the wall of the vessel.

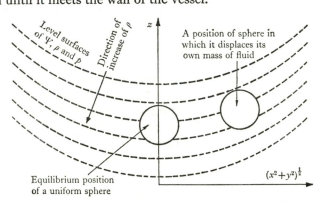

Figure 1.4.1. Non-uniform fluid at rest under the action of gravity and centrifugal force.

Fluid at rest under gravity

The case in which gravity is the only volume force acting on the fluid is both important and simple. Two extreme situations may be distinguished. In the first one, the mass of fluid concerned is large and isolated so that the gravitational attraction of other parts of the fluid provides the volume force on any element of the fluid, as in the case of a gaseous star. At the other extreme, the mass of fluid concerned is much smaller than that of neighbouring matter and the gravitational field is approximately uniform over the region occupied by the fluid.

In the case of a self-gravitating fluid, we have $\mathbf{F} = -\nabla\Psi$, where the gravitational potential Ψ is related to the distribution of density by the equation
$$\nabla^2\Psi = 4\pi G\rho, \qquad (1.4.6)$$

G being the constant of gravitation. On combining (1.4.6) with equation (1.4.3) for the pressure in a fluid at rest, we obtain
$$\nabla \cdot \left(\frac{\nabla p}{\rho}\right) = -4\pi G\rho. \qquad (1.4.7)$$

It is also necessary, as found earlier, that the level-surfaces of Ψ, ρ and p coincide. On expressing the differential operator in (1.4.7) in terms of curvilinear co-ordinates (not necessarily orthogonal) such that the level-surfaces of ρ coincide with one set of parametric surfaces, we see that the kinds of solution are severely restricted. Rigorous enumeration of the solutions is difficult, but the only possibilities seem to be solutions in which ρ and p are functions only of (i) one co-ordinate of a rectilinear system, or (ii) the radial co-ordinate of a cylindrical polar system, or (iii) the radial co-ordinate r of a spherical polar system, corresponding to symmetrical 'stars' in one, two or three dimensions.

In the last case, describing a spherically symmetrical distribution of density and pressure, (1.4.7) becomes

$$\frac{d}{dr}\left(\frac{r^2}{\rho}\frac{dp}{dr}\right) = -4\pi G r^2 \rho, \qquad (1.4.8)$$

and further progress cannot be made without information about the distribution of density. In real stars the density is in general not a function of p alone, but solutions of (1.4.8) corresponding to an assumed simple relationship between ρ and p are sometimes useful for comparison with more complicated models. If we assume for instance that

$$p \propto \rho^{1+1/n} \quad (n \geqslant 0),$$

it is possible to integrate (1.4.8) numerically for any value of n. Two analytical and representative solutions are also available. When $n = 0$, corresponding to a fluid of uniform density, ρ_0 say, we have

$$p = \tfrac{2}{3}\pi G \rho_0^2 (a^2 - r^2),$$

where $r = a$ may be interpreted as the outer boundary of the star. When $n = 5$, it may be verified that

$$p = C\rho^{\frac{6}{5}} = \frac{27 a^3 C^{\frac{5}{2}}}{(2\pi G)^{\frac{3}{2}}(a^2 + r^2)^3};$$

the pressure and density here are non-zero for all r and there is no definite outer boundary, but the total mass of the star is finite.

In the case of a uniform body force due to gravity, we have

$$\mathbf{F} = \mathbf{g}\,(= \text{const.}), \quad \Psi = -\mathbf{g}.\mathbf{x}, \qquad (1.4.9)$$

and the equation for the pressure in a fluid at rest is

$$\nabla p = \rho\mathbf{g}. \qquad (1.4.10)$$

The three functions Ψ, ρ and p are constant on each horizontal plane normal to $\mathbf{g}$, and hence depend only on $\mathbf{g}.\mathbf{x}$. If we choose the z-axis of a rectilinear co-ordinate system to be vertical (positive upwards) so that $\mathbf{g}.\mathbf{x} = -gz$, (1.4.10) becomes

$$dp/dz = -g\rho(z). \qquad (1.4.11)$$

Again this is as much as we can deduce from the condition of mechanical equilibrium alone.

When the fluid is of uniform density, we obtain from (1.4.11) the linear relation between pressure and height well-known in the study of hydrostatics:

$$p = p_0 - \rho g z. \qquad (1.4.12)$$

In the case of the earth's atmosphere, ρ decreases with decrease of the pressure owing to the compressibility of the air, although thermal effects are usually present and no single functional relation between ρ and p is adequate. As a crude approximation one may put

$$p/\rho = \text{const.}, \quad = gH \text{ say,}$$

corresponding to Boyle's law for a perfect gas of uniform temperature and constitution (§ 1.7). The pressure in an atmosphere for which this relation holds is found from (1.4.11) to be

$$p = p_0 e^{-z/H},$$

where p_0 is the pressure at ground level, $z = 0$. Thus both p and ρ diminish by a factor e^{-1} over a height interval H, and the constant H may be termed the 'scale-height' of the atmosphere. For air at 0 °C, $H = 8 \cdot 0$ km. When the temperature is not uniform, $p/\rho g$ may still be regarded as a local scale-height. Observed average values of the pressure, density and temperature at different heights in the atmosphere will be found in appendix 1 (b).

Exercises

1. A closed vessel full of water is rotating with constant angular velocity Ω about a horizontal axis. Show that the surfaces of equal pressure are circular cylinders whose common axis is at a height g/Ω^2 above the axis of rotation.

2. Obtain an expression for the pressure at the centre of a self-gravitating spherical star of which the density at distance r from the centre is

$$\rho = \rho_c(1 - \beta r^2).$$

Show that if the mean density be twice the surface density, the pressure at the centre is greater, by a factor $\frac{13}{8}$, than if the star had uniform density with the same total mass.

1.5. Classical thermodynamics

In our subsequent discussion of the dynamics of fluids we shall need to make use of some of the concepts of classical thermodynamics and of the relations between various thermodynamic quantities, such as temperature and internal energy. Classical thermodynamics is concerned, at any rate as the bulk of the subject stands, with equilibrium states of uniform matter, that is, with states in which all local mechanical, physical and thermal quantities are virtually independent of both position and time. Thermodynamical results may be applied directly to fluids at rest when their properties are uniform. Comparatively little is known of the thermodynamics of non-equilibrium states. However, observation shows that results for equilibrium states are approximately valid for the non-equilibrium non-uniform states common in practical fluid dynamics; large though the departures from equilibrium in a moving fluid may appear to be, they are apparently small in their effect on thermodynamical relationships.

The purpose of this section is to recapitulate briefly the laws and results of equilibrium thermodynamics and to set down for future reference the relations that will be needed later. For a proper account of the fundamentals of the subject the reader should refer to one of the many text-books available.[†]

The concepts of thermodynamics are helpful to the student of fluid mechanics for the additional reason that in both subjects the objective is a set of results which apply to matter as generally as possible, without regard for the different molecular properties and mechanisms at work. Additional results may of course be obtained by taking into account any known molecular properties of a fluid, as proves to be possible for certain gases with the aid of kinetic theory (see §1.7).

It is taken as a fact of experience that the state of a given mass of fluid in equilibrium (the word being used here and later to imply spatial as well as temporal uniformity) under the simplest possible conditions is specified uniquely by two parameters, which for convenience may be chosen as the specific volume v ($= 1/\rho$, where ρ is the density) and the pressure p as defined above. All other quantities describing the state of the fluid are thus functions of these two *parameters of state*. One of the most important of these quantities is the temperature. A mass of fluid in equilibrium has the same temperature as a test mass of fluid also in equilibrium if the two masses remain in equilibrium when placed in thermal contact (that is, when separated only by a wall allowing transmission of heat); and the second law of thermodynamics provides an absolute measure of the temperature of a fluid, as we shall note later. The relation between the temperature T and the two parameters of state, which we may write as

$$f(p, v, T) = 0, \qquad (1.5.1)$$

thereby exhibiting formally the arbitrariness of the choice of the two parameters of state, is called an *equation of state*. For every quantity like temperature which describes the fluid, but excluding the two parameters of state of course, there is an equation of state.

Another important quantity describing the state of the fluid is the *internal energy* per unit mass, E say.[‡] Work and heat are regarded as equivalent forms

---

† See, for instance, *Classical Thermodynamics*, by A. B. Pippard (Cambridge University Press, 1957).

‡ The usual practice in the literature of thermodynamics is to use a capital letter for the total amount of some extensive quantity like internal energy in the system under consideration, and a small letter for the amount per unit mass. Introduction of the latter quantity alone is sufficient in fluid dynamics, and the use of a capital letter for it is conventional.

of energy, and the change in the internal energy of a mass of fluid at rest consequent on a change of state is defined, by the first law of thermo-dynamics, as being such as to satisfy conservation of energy when account is taken of both heat given to the fluid and work done on the fluid. Thus if the state of a given uniform mass of fluid is changed by a gain of heat of amount Q per unit mass and by the performance of work on the fluid of amount W per unit mass, the consequential increase in the internal energy per unit mass is

$$\Delta E = Q + W. \tag{1.5.2}$$

The internal energy E is a function of the parameters of state, and the change ΔE, which may be either infinitesimal or finite, depends only on the initial and final states; but Q and W are measures of external effects and may separately (but not in sum) depend also on the particular way in which the transition between the two states is made. If the mass of fluid is thermally isolated from its surroundings so that no exchange of heat can occur, $Q = 0$ and the change of state of the fluid is said to be *adiabatic*.

There are many ways of performing work on the system, although com-pression of the fluid by inward movement of the bounding walls is of special relevance in fluid mechanics. An analytical expression for the work done by compression is available in the important case in which the change occurs *reversibly*. This word implies that the change is carried out so slowly that the fluid passes through a succession of equilibrium states, the direction of the change being without effect. At each stage of a reversible change the pressure in the fluid is uniform,† and equal to p say, so that the work done on unit mass of the fluid as a consequence of compression leading to a small decrease in volume‡ is $-p\,\delta v$. Thus for a reversible transition from one state to another, neighbouring, state we have

$$\delta E = \delta Q - p\,\delta v. \tag{1.5.3}$$

A finite reversible change of this kind can be described by summing (1.5.3) over the succession of infinitesimal steps making up the finite change; the particular path by which the initial and final equilibrium states are joined is relevant here, because p is not in general a function of v alone.

A practical quantity of some importance is the *specific heat* of the fluid, that is, the amount of heat given to unit mass of the fluid per unit rise in tem-perature in a small reversible change. A complete discussion of specific heat is best preceded by the second law of thermodynamics, but we may first see a direct consequence of the first law. The specific heat may be written as

$$c = \delta Q/\delta T, \tag{1.5.4}$$

† If a fluid at rest is acted on by a body force, the pressure varies throughout the fluid, as we have seen, but the pressure variation may be made negligibly small by considera-tion of a portion of fluid of small volume. The thermodynamical arguments refer to local properties of the fluid when a body force acts.

‡ Note that our definition of a simple fluid in §1.1 implies that no work is done on a fluid during a reversible change if only the shape and not the volume changes.

and is not determined uniquely until we specify further the conditions under which the reversible change occurs. An equilibrium state of the fluid may be represented as a point on a (p, v)-plane, or indicator diagram, and a small reversible change $(\delta p, \delta v)$ starting from a point A (see figure 1.5.1) may proceed in any direction. If the only work done on the fluid is that done by compression, the heat δQ which must be supplied to unit mass is determined by (1.5.3) as

$$\delta Q = \left(\frac{\partial E}{\partial p}\right)_v \delta p + \left(\frac{\partial E}{\partial v}\right)_p \delta v + p\, \delta v,$$

and the change in temperature is

$$\delta T = \left(\frac{\partial T}{\partial p}\right)_v \delta p + \left(\frac{\partial T}{\partial v}\right)_p \delta v.$$

The specific heat thus depends on the ratio $\delta p/\delta v$, that is, on the choice of direction of the change from A.

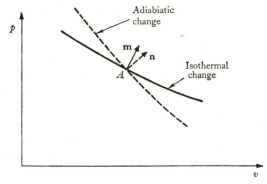

Figure 1.5.1. Indicator diagram for the equilibrium states of a fluid.

Two well-defined particular choices are changes parallel to the axes of the indicator diagram, giving the *principal* specific heats

$$c_p = \left(\frac{\delta Q}{\delta T}\right)_{\delta p = 0} = \left(\frac{\partial E}{\partial T}\right)_p + p\left(\frac{\partial v}{\partial T}\right)_p, \quad c_v = \left(\frac{\delta Q}{\delta T}\right)_{\delta v = 0} = \left(\frac{\partial E}{\partial T}\right)_v.$$

$$(1.5.5)$$

Now δT varies sinusoidally as the point representing the final state moves round a circle of small radius centred on A, being zero on the isotherm through A and a maximum in a direction $\mathbf{m}$ normal to the isotherm. Likewise δQ varies sinusoidally, being zero on the adiabatic line through A and a maximum in a direction $\mathbf{n}$ normal to it. Thus if (m_v, m_p) and (n_v, n_p) be the components of the two unit vectors,

$$c_p = \frac{n_v(\delta Q)_{\text{max.}}}{m_v(\delta T)_{\text{max.}}}, \quad c_v = \frac{n_p(\delta Q)_{\text{max.}}}{m_p(\delta T)_{\text{max.}}},$$

and, more usefully, for the ratio of the principal specific heats, commonly denoted by γ, we obtain

$$\gamma = \frac{c_p}{c_v} = \frac{m_p}{m_v} \bigg/ \frac{n_p}{n_v}$$

$$= \left(\frac{\partial p}{\partial v}\right)_{\text{adiab.}} \bigg/ \left(\frac{\partial p}{\partial v}\right)_{\text{isoth.}} \quad \text{or} \quad \left(\frac{1}{v}\frac{\partial v}{\partial p}\right)_{\text{isoth.}} \bigg/ \left(\frac{1}{v}\frac{\partial v}{\partial p}\right)_{\text{adiab.}}. \quad (1.5.6)$$

The ratio $-\delta p/\delta v$ of the increments in p and v in a small reversible change is the *bulk modulus of elasticity* of the fluid; a related, and more useful quantity for fluid dynamical purposes, is the *coefficient of compressibility* $-\delta v/(v\,\delta p)$, or $\delta\rho/(\rho\,\delta p)$. Like the specific heat, the bulk modulus takes a different value for each direction of the change in the indicator diagram. Adiabatic and isothermal changes correspond to two particular directions, with special physical significance, and, somewhat surprisingly, the first law requires the ratio of the two corresponding bulk moduli to equal the ratio of the principal specific heats.

It is clearly possible to draw an adiabatic line (defining the direction of a small reversible change involving no gain of heat) through each point of the indicator diagram, and it is natural to enquire if these adiabatic lines are the lines of equal value of some function of state. This question is answered in the affirmative by the second law of thermodynamics. The second law can be stated in a number of apparently different but equivalent ways, none of which is easy to grasp. Our use of the law will be indirect, and will not require any of the usual statements. It will in fact be sufficient for our purpose (although not very enlightening physically) to know that the second law of thermodynamics implies the existence of another extensive (or size-dependent) property of the fluid in an equilibrium state, termed the entropy, such that, in a reversible transition from an equilibrium state to another, neighbouring, equilibrium state, the increase in entropy is proportional to the heat given to the fluid; moreover, the constant of proportionality is itself a function of state and can be identified with the reciprocal of the temperature. Thus, with entropy per unit mass of a fluid denoted by S, we have

$$T\,\delta S = \delta Q, \quad (1.5.7)$$

where δQ is the infinitesimal amount of heat given to unit mass of the fluid. This is the means by which the thermodynamic or absolute scale (unrelated to the properties of any particular material) of temperature is defined. The entropy is constant in an adiabatic reversible transition, which is therefore said to be *isentropic*. It is shown in textbooks on thermodynamics that the entropy *cannot diminish* in an adiabatic irreversible change (when T is chosen to be positive); any change in the entropy must be an increase.

Since both (1.5.3) and (1.5.7) apply to reversible changes, it follows that, for a small reversible change in which work is done on the fluid by compression,

$$T\,\delta S = \delta E + p\,\delta v. \quad (1.5.8)$$

Now the initial and final values of S and E, as of all other functions of state, are fully determined by the initial and final states, and consequently the relation (1.5.8), which contains only functions of state, must be valid for any infinitesimal transition in which work is done by compression, whether reversible or not. If the transition is irreversible, the equality (1.5.7) is not valid, and neither is that between δW and $-p\,\delta v$.

Another function of state which, like internal energy and entropy, proves to be convenient for use in fluid mechanics, particularly when effects of compressibility of the fluid are important, is the *enthalpy*, or heat function. The enthalpy of unit mass of fluid, I say, is defined as

$$I = E + pv, \qquad (1.5.9)$$

and has the dimensions of energy per unit mass. A small change in the parameters of state corresponds to small changes in the functions I, E and S which are related by

$$\delta I = \delta E + p\,\delta v + v\,\delta p,$$
$$= T\delta S + v\,\delta p \qquad (1.5.10)$$

in view of (1.5.8). The relation (1.5.10), like (1.5.8), involves only functions of state, and is consequently independent of the manner in which the fluid might be brought from one to the other of the two neighbouring states. For a reversible small change at constant pressure, it appears from (1.5.7) that $\delta I = \delta Q$.

Yet another important function of state with the dimensions of energy is the *Helmholtz free energy*, of which the amount per unit mass is defined as

$$F = E - TS.$$

The small change in F consequent on small changes in the parameters of state is given by

$$\delta F = -p\,\delta v - S\,\delta T,$$

showing that the gain in free energy per unit mass in a small isothermal change, whether reversible or not, is equal to $-p\,\delta v$. Alternatively, it follows from (1.5.2) and (1.5.7) that the gain in free energy in a small reversible isothermal change is equal to the work done on the system.

Four useful identities, known as *Maxwell's thermodynamic relations*, follow from the above definitions of the various functions of state. To obtain the first of these relations, we note from (1.5.8) that, if v and S are now regarded as the two independent parameters of state on which all functions of state depend, the two partial derivatives of E are

$$\left(\frac{\partial E}{\partial v}\right)_S = -p, \quad \left(\frac{\partial E}{\partial S}\right)_v = T, \qquad (1.5.11)$$

where the subscript serves as a reminder of the variable held constant. The double derivative $\partial^2 E/\partial v\,\partial S$ may now be obtained in two different ways, yielding the relation

$$\left(\frac{\partial p}{\partial S}\right)_v = -\left(\frac{\partial T}{\partial v}\right)_S. \qquad (1.5.12)$$

The other three identities are

$$\left(\frac{\partial v}{\partial S}\right)_p = \left(\frac{\partial T}{\partial p}\right)_S, \tag{1.5.13}$$

$$\left(\frac{\partial v}{\partial T}\right)_p = -\left(\frac{\partial S}{\partial p}\right)_T, \tag{1.5.14}$$

$$\left(\frac{\partial p}{\partial T}\right)_v = \left(\frac{\partial S}{\partial v}\right)_T, \tag{1.5.15}$$

and may be obtained similarly by forming the double derivative, in two different ways, of the functions $E+pv$, $E-TS$ and $E+pv-TS$ respectively. Alternatively they may be shown to follow from (1.5.12) and the rules for partial differentiation of implicit functions. For example, since T may be regarded as a function of p and S, the derivative on the right-hand side of (1.5.12) may be written as

$$\left(\frac{\partial T}{\partial v}\right)_S = \left(\frac{\partial T}{\partial p}\right)_S \left(\frac{\partial p}{\partial v}\right)_S;$$

and for the left-hand side of (1.5.12) we make use of the well-known identity

$$\left(\frac{\partial p}{\partial S}\right)_v = -\left(\frac{\partial p}{\partial v}\right)_S \left(\frac{\partial v}{\partial S}\right)_p$$

for three quantities p, v, S subject to a single functional relationship, whence (1.5.13) is obtained.

One of the derivatives in Maxwell's thermodynamic relations specifies the *coefficient of thermal expansion* of the fluid, defined as

$$\beta = \frac{1}{v}\left(\frac{\partial v}{\partial T}\right)_p, \tag{1.5.16}$$

which plays an important role in considerations of the action of gravity on a fluid of non-uniform temperature.

Introduction of the entropy provides alternative expressions for the specific heat. For the general specific heat we have

$$c = \frac{\delta Q}{\delta T} = T\frac{\delta S}{\delta T},$$

and for the two principal specific heats (compare (1.5.5))

$$c_p = T\left(\frac{\partial S}{\partial T}\right)_p, \quad c_v = T\left(\frac{\partial S}{\partial T}\right)_v. \tag{1.5.17}$$

Moreover, on regarding S as a function of T and v, we find

$$\delta S = \left(\frac{\partial S}{\partial T}\right)_v \delta T + \left(\frac{\partial S}{\partial v}\right)_T \delta v,$$

so that
$$\left(\frac{\partial S}{\partial T}\right)_p = \left(\frac{\partial S}{\partial T}\right)_v + \left(\frac{\partial S}{\partial v}\right)_T \left(\frac{\partial v}{\partial T}\right)_p;$$

and it then follows from (1.5.17) and the Maxwell relation (1.5.15) that

$$c_p - c_v = T\left(\frac{\partial p}{\partial T}\right)_v \left(\frac{\partial v}{\partial T}\right)_p, \qquad (1.5.18)$$

the right-hand side of which can be calculated when the equation of state connecting p, v and T is known. An alternative expression for $c_p - c_v$ involving measurable quantities may be obtained by using the identity

$$\left(\frac{\partial p}{\partial T}\right)_v = -\left(\frac{\partial p}{\partial v}\right)_T \left(\frac{\partial v}{\partial T}\right)_p$$

for three quantities p, v, T subject to a single functional relationship; it is

$$c_p - c_v = -T\left(\frac{\partial p}{\partial v}\right)_T \left(\frac{\partial v}{\partial T}\right)_p^2. \qquad (1.5.19)$$

Finally, we obtain an expression for the increments in S and E consequent on small changes in two parameters of state, which will be needed in later considerations of flow of a fluid with non-uniform temperature. We may regard S as a function of T and p, whence

$$\delta S = \left(\frac{\partial S}{\partial T}\right)_p \delta T + \left(\frac{\partial S}{\partial p}\right)_T \delta p,$$

or, from (1.5.17) and (1.5.14),

$$= \frac{c_p}{T} \delta T - \left(\frac{\partial v}{\partial T}\right)_p \delta p.$$

Hence, on making use of the notation of (1.5.16), we have

$$T \delta S = \delta E + p\, \delta v = c_p\, \delta T - \beta v T\, \delta p. \qquad (1.5.20)$$

The usefulness of this relation lies in the fact that all terms except $T\delta S$ and δE contain only directly observable quantities. The increments δT and δp on the right-hand side of (1.5.20) are independent, and the relative importance of the two terms containing them will of course depend on the circumstances. We see from (1.5.19) that the ratio of the two terms on the far right-side of (1.5.20) can be written as

$$-\frac{c_p\,\delta T}{\beta v T\,\delta p} = \frac{c_p}{c_p - c_v}\left(\frac{\partial p}{\partial v}\right)_T \left(\frac{\partial v}{\partial T}\right)_p \frac{\delta T}{\delta p}$$

$$= \frac{\gamma}{\gamma - 1}\frac{\left(\frac{\partial v}{\partial T}\right)_p \delta T}{\left(\frac{\partial v}{\partial p}\right)_T \delta p}, \qquad (1.5.21)$$

from which it will often be possible to see at a glance whether one term is dominant. When the factor $\gamma/(\gamma-1)$ is of order unity, as it is for gases and most liquids, a comparison of the two terms reduces essentially to a comparison of the increments in v that would be caused by the given increments in T and p acting separately.

1.6. Transport phenomena

Equilibrium states of matter are characterized by a uniform spatial distribution of each of the various properties of the material, each element of the material then being in mechanical and thermal balance with neighbouring elements. If certain properties of the material are not uniform initially, it is observed that exchanges of mechanical or thermal properties occur between adjoining elements and that the exchanges always tend to bring the material towards an equilibrium state, that is, to smooth out the non-uniformity. The existence of this tendency to equilibrium in material with non-uniform properties, which is taken for granted in classical thermodynamics, appears to require only that contiguous portions of matter interact in some way. The nature of this interaction may depend on the molecular structure of the contiguous portions of matter, and the physical consequences of the exchanges depend on the particular property which is distributed non-uniformly, but the tendency to equilibrium between interacting portions of matter exists quite generally and is independent, like the results of classical thermodynamics, of the existence of a particular structure of matter.

An important and common outcome of the exchange between two elements of matter with different properties is that the amount of some quantity satisfying a conservation law associated with one element decreases and the amount associated with the other increases. The whole group of such exchanges constitute what are called *transport phenomena*. Three basic kinds of transport phenomena, to which familiar names are attached, correspond to transfer of matter, energy and momentum. Our main concern in this section will be the general features of these three kinds of transport. No appeal will be made in this section to particular molecular properties of the material, although for convenience and clarity the nature of the molecular mechanism of the transport in fluids will be referred to briefly.

Transfer of matter of a specific kind occurs in a fluid mixture† of which the composition varies with position. We may suppose that the molecules belonging to one constituent of the mixture are marked in some manner. All molecules are in continual motion of a random kind, and as a consequence have a tendency to migrate away from any initial position. Then if at any instant the proportion of marked molecules immediately on one side of an

---

† And also in a solid such as an alloy composed of different kinds of molecule, since a molecule in a solid is not held in the same lattice position absolutely permanently; but the rates of transfer are very much smaller in a solid than in a fluid.

element of surface drawn in the fluid is larger than that on the other side, random migration of marked molecules in both directions across the surface element will lead in general to a non-zero flux of marked molecules across the element, of such a sign as to tend to make the proportion of marked molecules more nearly equal on the two sides.† This non-zero flux of a constituent of the fluid due to migration of molecules constitutes *diffusion of matter*. Our discussion of this rather complex phenomenon will be limited to cases of small concentration of the diffusing constituent.

Transfer of kinetic energy of molecular motion occurs as a consequence of the interaction of neighbouring molecules, either through the molecules being so close as to lie within each other's force fields, as in the case of solids and liquids, or through the occurrence of occasional collisions as in the case of a gas. The circumstances under which a net transfer of molecular energy— that is, of heat—occurs are known empirically. Two masses separated by a thin rigid wall permeable to heat are in thermal equilibrium when the function of state called the temperature has the same value for the two masses; and if the two temperatures are not equal, there is a net flux of heat across the boundary in the direction of decreasing temperature. Removal of the wall separating two masses at the same pressure clearly does not affect the condition for thermal equilibrium or the direction of net flux of heat when the two temperatures are different, although the consequences of the flux of heat are altered inasmuch as the pressures must remain equal. This net flux of molecular energy when the temperature is non-uniform constitutes *conduction of heat*.

Transport of momentum of molecules across an element of surface moving with the local 'continuum' velocity of a fluid occurs when molecules cross the surface, and it occurs, in effect, if a force is exerted between the two groups of molecules instantaneously on the two sides of the surface element. The combined effect of momentum flux by passage of molecules across the element and forces exerted between molecules on the two sides is represented by the local stress in the fluid. The stress at any point is a consequence of molecular motions and interactions in the neighbourhood of the point, so that, if the fluid velocity is uniform in this neighbourhood, the stress has the form appropriate to a fluid at rest and is normal to the surface element for all orientations of the element. On the other hand, if the continuum velocity is not uniform in this neighbourhood the tangential components of stress may not be zero. The manner in which a vector function of position

† It might be thought that the condition for a non-zero flux of marked molecules across the element to occur is that the number density of marked molecules be different on the two sides. When the fluid density is uniform, a choice between these two conditions for a non-zero flux is immaterial. But when the fluid density is non-uniform (which normally will require non-uniformity of temperature also), the tendency for the marked molecules to flow—by random migration—relative to the unmarked molecules arises more from non-uniformity of the proportion of marked molecules than from non-uniformity of their absolute number density.

like the fluid velocity may vary in the neighbourhood of any point is not obvious, and will be considered in the kinematical analysis of chapter 2; and the form of the stress associated with this velocity variation will be described fully in chapter 3. However, in the meantime we may bring momentum transport partially within the scope of the present discussion by confining attention to the case (which appears to be rather special, but which we shall find later to be fundamental) in which the fluid velocity relative to the surface element, which is itself moving with the fluid, has a direction in the plane of the surface element and a magnitude which varies only with respect to the position co-ordinate normal to the surface element; this is a *simple shearing motion* in which planes of fluid parallel to the surface element slide rigidly over one another. In these circumstances, it is evident that, if the fluid velocities on the two sides of the surface element are different, any random molecular interaction across the element will result in the establishment of a tangential component of stress, and that the sign of the stress will be such as to tend to eliminate the difference between the velocities on the two sides. Transport of momentum thus constitutes *internal friction*, and a fluid exhibiting internal friction is said to be *viscous*.

The main common features of these three kinds of transport phenomena are, first, that zero net transfer of some quantity (number of marked molecules, heat, momentum) occurs when an associated quantity representing local intensity (proportion of marked molecules, temperature, fluid velocity) is spatially uniform, and, second, that the direction of a non-zero net transfer across an element of surface in the material is such as to tend to equalize the values of the intensity on the two sides.

We proceed now to consider a quantitative relation between the net transfer and the non-uniformity of the associated intensity. As a preliminary, we note that, although the existence or absence of equilibrium in classical thermodynamics is represented in terms of the consequences of bringing into contact two masses, each of which is uniform within itself, continuum mechanics presents us with situations in which the intensity is normally a continuous function of position. It is evident that molecular transport will still lead to a net transfer across an element of surface in the material when the distribution of intensity in the neighbourhood of the surface element is non-uniform, but, instead of representing this local non-uniformity as a difference between the values of the intensity on the two sides of the element, we must take a more general point of view and represent it as a vector gradient of intensity at the position of the surface element.

The linear relation between flux and the gradient of a scalar intensity

Consider first the cases in which the relevant intensity is a scalar quantity (viz. proportion of marked molecules, or temperature), which we shall denote by C (standing for concentration). C will be assumed to be a continuous function of position $\mathbf{x}$ in the material, and possibly also of time t

although that will not affect the instantaneous transfer. Now the net transfer of the quantity associated with C across a surface element in the material, per unit area of that element, is a local quantity which varies with the direction $\mathbf{n}$ of the normal to the surface element in the same manner as the component, in the direction $\mathbf{n}$, of a vector. This follows formally from an argument like that leading to (1.3.4) for the stress; the sum of the inward transfers across the three orthogonal faces of a small tetrahedron differs from the outward transfer across the inclined face only by a quantity of the order of the volume. Thus the net transfer per second across a surface element of area δA and normal $\mathbf{n}$ may be written as

$$\mathbf{f.n}\,\delta A,$$

where the *flux vector* $\mathbf{f}$ is a function of $\mathbf{x}$ (and perhaps also of t) but not of $\mathbf{n}$.

Our objective is a relation between the two functions of position in the material, C and $\mathbf{f}$. Direct attempts to calculate the flux vector from considerations of the molecular processes involved are almost out of the question for liquids and solids, and meet only limited success (which will be described in the next section) in the case of a gas. Some hypothesis is needed, and it should preferably be independent of the exact nature of the underlying molecular mechanism, so as to be applicable to a wide range of materials. The hypothesis that will now be described was initially based on measurements of the flux vector in particular physical contexts and used only in those contexts, but has since been recognized as having more general significance.

The first part of our hypothesis is that, for a sufficiently smooth or gradual variation of the intensity C with respect to position in the material, the flux vector depends only on the local properties of the medium and the local values of C and ∇C. The idea here is simply that the transport across a surface element is determined by molecular motions and interactions in the neighbourhood of the surface element, and that over this region C can be approximated by a linear function of position provided some condition of the type

$$\left|\frac{\partial C}{\partial x}\right| \bigg/ \left|\frac{\partial^2 C}{\partial x^2}\right| \gg \text{length representative of molecular motion or interaction}$$

is satisfied, as is normally so in practice. The second part of the hypothesis is that, for sufficiently small values of the magnitude $|\nabla C|$ the flux vector varies linearly with the components of ∇C. The flux vector is known to vanish with $|\nabla C|$, so that the hypothesis may be expressed as

$$f_i = k_{ij}\frac{\partial C}{\partial x_j}. \tag{1.6.1}$$

Both f_i and $\partial C/\partial x_j$ are vectors, and the requirement that (1.6.1) be valid for all choices of the co-ordinate system shows that the *transport coefficient* k_{ij}

is a second-order tensor. k_{ij} depends on the local properties of the material (that is, on the local state of the material in the thermodynamic sense) and possibly also on the local value of C, but not on ∇C. Mathematically (1.6.1) can be regarded as an assumption that, when the flux vector is written as a Taylor series in the components of ∇C, terms of second and higher degree are negligible.

This general hypothesis may be supplemented by further assumptions based on the known properties of certain materials. For a *homogeneous* material, k_{ij} can depend on position only through any dependence on the local value of C; and reversal of the direction of ∇C must here lead to reversal of the direction of $\mathbf{f}$, so that in this case the terms of second and other even degree in the Taylor series for $\mathbf{f}$ are identically zero. In many materials† the molecular structure is statistically *isotropic*, in which case k_{ij} must have a form from which all directional distinction is absent. All sets of orthogonal axes must then be principal axes of the coefficient k_{ij}, which is possible only if

$$k_{ij} = -k\,\delta_{ij}. \tag{1.6.2}$$

(Alternatively, we may argue that in an isotropic medium $\mathbf{f}$ must be parallel to ∇C, since no basis exists for selection of a different direction of $\mathbf{f}$, and (1.6.2) is again the necessary form for k_{ij}.) The scalar coefficient k is positive, as defined by (1.6.1) and (1.6.2), if we reckon flux as positive when it is in the direction $\mathbf{n}$, since the quantity associated with C is transported *down* the gradient of intensity.

The testing of the hypothesis represented by (1.6.1), together with (1.6.2) where appropriate, and the determination of the range of values of $|\nabla C|$ over which (1.6.1) is accurate are primarily matters for experiment. The nature of the test varies according to the physical quantity being transported, but it appears that for all such quantities the linear relation (1.6.1) is remarkably accurate for most normal or practical values of $|\nabla C|$. An examination of the reasons why departures from uniformity of the intensity C are usually so small as to ensure the accuracy of (1.6.1) requires consideration of the particular molecular mechanisms involved, and we are content to leave it on an empirical basis for the moment. The various relations corresponding to (1.6.1) for different choices of the meaning of the intensity C are known as *constitutive relations*, since they express physical properties of the material concerned.

The equations for diffusion and heat conduction in isotropic media at rest

The expression for the flux vector is here

$$\mathbf{f} = -k\nabla C \tag{1.6.3}$$

---

† A solid with a regular and anisotropic crystal structure is a notable exception, and heat is observed in some crystals to spread more rapidly in certain directions than in others. But exceptions among fluids, which do not have a permanent molecular arrangement, are rare.

at all points of the medium. It follows that the total transfer per second of the quantity concerned out of the material enclosed by a closed surface A with unit (outward) normal $\mathbf{n}$ is

$$-\int k\mathbf{n}.\nabla C\,dA, \quad = -\int \nabla.(k\nabla C)\,dV, \qquad (1.6.4)$$

where V is the volume of the enclosed region. If the quantity being transferred is known to satisfy a conservation law, it may now be possible to obtain an equation governing the dependence of the intensity C on position and time. This we shall do for the separate cases in which C represents proportion of marked molecules or temperature. The medium will be assumed to be at rest, and later (§3.1) we shall see what modification is needed when the molecular transport takes place in a moving fluid.

When C represents the proportion of marked molecules in a fluid mixture, a simple conservation law holds. The number of marked molecules in the volume V of the fluid is $\int CN\,dV$ (where N is the total number of molecules per unit volume) and can change only as a consequence of molecular transport across the surface, so that

$$\frac{\partial}{\partial t}\int CN\,dV = \int \nabla.(k_D\nabla C)\,dV, \qquad (1.6.5)$$

and

$$\int \left\{\frac{\partial(CN)}{\partial t} - \nabla.(k_D\nabla C)\right\}dV = 0,$$

where k_D is the value of k appropriate to diffusion of marked molecules. The total number density of molecules does not itself change as a consequence of the exchange of marked and unmarked molecules and may be regarded as constant. This relation is valid for all choices of the volume V lying entirely in the fluid, and the integrand must therefore be zero everywhere, that is,

$$N\frac{\partial C}{\partial t} = \nabla.(k_D\nabla C). \qquad (1.6.6)$$

The parameter k_D depends on the local state of the material and perhaps on the concentration C (inasmuch as the magnitude of C may affect the molecular environment of any one marked molecule); k_D is thus in general a function of position in the fluid. However, it happens often in practice that the gradient of k_D is sufficiently small for (1.6.6) to take the approximate form

$$\frac{\partial C}{\partial t} = \kappa_D\nabla^2 C, \qquad (1.6.7)$$

which is known as the *diffusion equation*. The new parameter

$$\kappa_D = k_D/N \qquad (1.6.8)$$

is the *coefficient of diffusion* of the marked constituent in the ambient fluid composed of the unmarked molecules, and has dimensions

$$(\text{length})^2 \times (\text{time})^{-1}.$$

When N is independent of position, κ_D is equal to the flux of marked molecules per unit gradient of number density of marked molecules. In the particular case in which marked and unmarked molecules are dynamically similar and thus have the same migratory behaviour, k_D and κ_D are independent of C and κ_D is then the *coefficient of self-diffusion*.

Equation (1.6.7) is one of the archetypes of linear partial differential equations of the second order, and a good deal is known about solutions for various types of boundary and initial conditions.[†]

When C represents temperature of the material, we may use the law of conservation of energy, taking account if necessary of both heat and work. The quantity transported here is heat, and according to (1.6.4) the rate of gain of heat by the material lying within the small volume δV due to transport of heat across the bounding surface is (on reverting to the use of T for temperature)

$$\nabla.(k_H \nabla T)\delta V.$$

k_H is the value of k appropriate to conduction of heat, and is termed the *thermal conductivity*. The thermodynamic state of the material is changing continually due to this heat flux, but provided the rate of change is slow (a condition which has already been assumed to be satisfied in the argument underlying (1.6.1)) we may regard the gain of heat in a small time δt, per unit mass of material, as the heat addition δQ postulated in the discussion in §1.5 of reversible changes from one equilibrium state of the material to another; that is,

$$\delta Q = \frac{\delta t}{\rho} \nabla.(k_H \nabla T). \tag{1.6.9}$$

Some of this heat addition may be manifested as an increase in the internal energy per unit mass, and some as work done by unit mass of the material, as represented by (1.5.3) for the case of work done by expansion against external pressure (which is by far the most important case of interest in fluid dynamics). In any event, there is an increase in the entropy per unit mass of amount $\delta Q/T$ (see (1.5.7)), and the outcome of the heat addition may be expressed in terms of increments in both T and p by means of (1.5.20). Thus on combining (1.6.9) and (1.5.20) (with ρ now written in place of $1/v$), and on passing to rates of change instead of increments, we obtain

$$T\frac{\partial S}{\partial t} = c_p \frac{\partial T}{\partial t} - \frac{\beta T}{\rho}\frac{\partial p}{\partial t} = \frac{1}{\rho}\nabla.(k_H \nabla T). \tag{1.6.10}$$

This is the general equation representing the effect of conduction of heat in a medium at rest (apart from the small movements due to thermal expansion). The medium may be solid, liquid or gaseous, provided that the stress at interior points is purely normal. The derivatives of T and p with respect to t are independent, like the increments δT and δp in (1.5.20), and the

† See *The Conduction of Heat in Solids*, by H. S. Carslaw and J. C. Jaeger (Oxford University Press, 1947).

relative importance of the two terms containing them will depend on the circumstances. We saw, in (1.5.21), that the ratio of the two terms is of the same order of magnitude as the ratio of the changes in v (or ρ) that would result from the given increments in T and p acting separately. In a gas in motion, it is quite possible for changes in T and p associated with the motion to be of such magnitudes as to correspond to changes in ρ (at constant p and T respectively) of the same order of magnitude. Also, for a solid, liquid or gas whose volume is fixed by rigid enclosing walls and in which the temperature is changing with time more-or-less uniformly over the whole of the material, it is evident that the pressure and temperature changes separately would lead to comparable changes in ρ. However, for media at rest and free to expand, in which case p is constant, and for confined media at rest in which the average temperature, and hence also the pressure, remain approximately constant, (1.6.10) reduces to

$$T\frac{\partial S}{\partial t} = c_p \frac{\partial T}{\partial t} = \frac{1}{\rho}\nabla.(k_H \nabla T). \tag{1.6.11}$$

We see from (1.5.10) that the term on the far left can also be written as the rate of change of enthalpy I, in these circumstances of constant pressure.

When the thermal conductivity k_H is approximately uniform throughout the material, the equation for T becomes

$$\frac{\partial T}{\partial t} = \kappa_H \nabla^2 T, \tag{1.6.12}$$

where

$$\kappa_H = k_H/\rho c_p; \tag{1.6.13}$$

this 'heat conduction equation' is thus identical in form with the diffusion equation for media at rest. The parameter κ_H here may be termed the *thermal diffusivity*, being sometimes known also as the thermometric conductivity.

Since places in the material where the temperature is low tend to gain heat by conduction and vice versa, the effect of the factor T in the term containing entropy in (1.6.11) is to add weight to the gains of entropy. The outcome is that the total entropy of a thermally isolated mass of material within which the temperature is non-uniform increases. We may see this formally by rewriting (1.6.11) as

$$\rho\frac{\partial S}{\partial t} = k_H \left(\frac{1}{T}\nabla T\right)^2 + \nabla.\left(\frac{k_H}{T}\nabla T\right); \tag{1.6.14}$$

integration over the various elements of mass $\rho\,\delta V$ then gives

$$\frac{\partial}{\partial t}\int S\rho\,dV = \int k_H \left(\frac{1}{T}\nabla T\right)^2 dV, \quad > 0,$$

since $\mathbf{n}.\nabla T = 0$ everywhere on the bounding surface. This is an irreversible change for the system formed by the whole isolated mass of material, inasmuch as no variation of the external conditions can cause the reverse change, and the increase of entropy accompanying the internal conduction of heat is an illustration of the general proposition stated in §1.5 that the entropy cannot diminish in an adiabatic irreversible change. However, it is possible to regard the gain of heat, due to conduction, by a small element of the material as a reversible change for the system formed by the element alone, as was done in the argument leading to (1.6.10).

Molecular transport of momentum in a fluid

The case of transport of momentum requires a different analytical description owing to the vector character of the transported quantity. However, as stated earlier in this section, we may exhibit the common features of transport of marked molecules, heat and momentum by imposing a restriction on the local velocity distribution, viz. by considering only a simple shearing motion in a definite direction. The fluid velocity in a simple shearing motion has components $U(y)$, o, o relative to orthogonal rectilinear axes at a point whose position co-ordinates are x, y, z, and we enquire into the stress exerted across an element of surface lying in the (z, x)-plane. The tangential component of this stress is non-zero wholly as a consequence of the existence, first, of non-uniformity of the fluid velocity, and, second, of interaction of the molecules on the two sides of the surface element, either through movement of molecules across the element or through intermolecular forces exerted across the element.

The arguments leading to the hypothesis of a linear relation between the flux vector and the local gradient of a scalar intensity may now be applied, with changes only in notation. The molecular interactions extend over a small distance only, and the molecular transport of momentum across the element will normally depend on the distribution of fluid velocity only through dependence on the local gradient dU/dy (dependence on U being impossible, since U is affected by the use of moving axes). Furthermore, it is to be expected that, for sufficiently small values of $|dU/dy|$, the tangential component of stress across the surface element (i.e. the net transfer of x-component of momentum across the element, per unit time and per unit area of the element) varies linearly with dU/dy. With the notation of §1.3 for the stress, this implies that

$$\sigma_{12} = \mu \frac{dU}{dy}, \qquad (1.6.15)$$

where the parameter μ, the *viscosity* of the fluid, depends on the local properties of the fluid. This momentum flux results from disordered or random interaction of the molecules and is inevitably in such a direction as to tend to eliminate the non-uniformity of fluid velocity. μ is therefore positive, as defined by (1.6.15) (and with the convention of §1.3 that $\sigma_{ij} n_j$ is the force per unit area exerted by the fluid on the side of the surface element to which the normal **n** points), and is a measure of the internal friction opposing deformation of the fluid.

The linear relation (1.6.15) is well known as an empirical expression for the tangential stress set up in common fluids by a simple shearing motion, and is found to be accurate over a surprisingly large range of values of $|dU/dy|$ which includes the values commonly met in practice.

Considerations of the effect of this stress due to viscosity on the distribution of fluid velocity must await the more complete analysis of chapter 3. However, it is evident in advance that the quantity which, like the diffusivities κ_D and κ_H, measures the ability of the molecular transport to eliminate the non-uniformity of intensity (fluid velocity, here) which gives rise to the transport is

$$\nu = \mu/\rho. \tag{1.6.16}$$

ν is termed the *kinematic viscosity* since its dimensions (length)$^2 \times$ (time)$^{-1}$ do not include mass. κ_D, κ_H and ν are the *diffusivities* for matter, heat and momentum respectively.

1.7. The distinctive properties of gases

The feature of a gas to which most of its distinctive properties are attributable is the wide separation of the molecules and the dynamical isolation of each molecule during most of its life. At temperature 0 °C and a pressure of one atmosphere, the number of molecules in one cubic centimetre of gas is $2 \cdot 69 \times 10^{19}$ (known as Loschmidt's number, and the same for all gases, as stated by Avogadro's law), so that if the molecules were placed at the corners of a cubical lattice the distance between neighbours would be $3 \cdot 3 \times 10^{-7}$ cm. The diameter of a molecule is not a well-defined quantity, but one reasonably definite measure is provided by the distance between centres of two molecules in isolation at which the intermolecular force changes sign (§ 1.1). For many simple molecules this effective diameter d_0 lies in the range 3–4×10^{-8} cm, so that the average separation of molecules in the above sense is something like $10 d_0$. At this distance apart the cohesive force between molecules is completely negligible, so that for most of their life molecules move freely, in straight lines with constant speed (provided they are electrically neutral, as we shall assume). A collision between two molecules is likewise not a precise concept, but, if we define a collision as occurring whenever one molecule comes so close to another that the mutual force between them is repulsive, the average distance travelled by a molecule between collisions may be calculated to be $8 \cdot 3 \times 10^{-21} d_0^{-2}$ cm, that is, about 7×10^{-6} cm, or $200 d_0$, with the above estimate of d_0.

The notion of a gas as an assemblage of molecules moving almost freely except at occasional collisions is the basis of the *kinetic theory* of gases. It is found convenient in that theory to consider the properties of a *perfect gas* whose molecules exert no force on each other except at collisions and have negligible volume. (The frequency of collision of a molecule diminishes to zero with the volume of the molecules, but the frequency of collisions plays little part in the theory and it is sufficient to know that *some* collisions occur.)

It seems likely, from the figures quoted above, that under normal conditions real gases have properties which approximate closely to those of the hypothetical perfect gas, and observation shows this to be so; indeed some of the empirical laws found early in the investigation of properties of gases, such as Boyle's and Charles's laws, can be deduced as properties of a perfect gas. An elementary account of the properties of gases therefore begins appropriately with derivation of the properties of a perfect gas.

We shall naturally make full use of the powerful ideas and results of classical thermodynamics (§ 1.5) and of the less rigorous results of transport theory (§ 1.6). The arguments of the two preceding sections are independent of the molecular structure of the material, and part of the price paid for this generality is that very few detailed and particular results can be obtained. If we now relax the generality, and confine ourselves to a particular material with a comparatively simple molecular structure, such as a perfect gas, it becomes possible to take the results considerably further.

A perfect gas in equilibrium

We consider first a perfect gas at rest in a state of thermodynamic equilibrium in the sense of § 1.5; all properties of the gas are independent of position and time, so that transport phenomena are absent. We shall also assume, to begin with, that the molecules are identical, with mass m, $= \rho/N$, where N is the number density of molecules. Although the molecules obey dynamical laws, there are so many of them as to make a statistical description of their motion appropriate. We therefore introduce probability density functions, that for the molecular velocity $\mathbf{u}$ being denoted by $f(\mathbf{u})$. The product $f(\mathbf{u})\,\delta u\,\delta v\,\delta w$ is the probability that the velocity of a given molecule at any instant has components whose values lie between u and $u+\delta u$, v and $v+\delta v$, w and $w+\delta w$; alternatively $f(\mathbf{u})\,\delta u\,\delta v\,\delta w$ can be regarded as the fraction of the molecules in a given volume which at any instant have velocity components in these ranges. The function f satisfies the relation

$$\int\int\int_{-\infty}^{\infty} f(\mathbf{u})\,du\,dv\,dw = 1$$

identically. We shall assume that collisions destroy any initial directional features of the velocity distribution, so that in the equilibrium state f is a function of $|\mathbf{u}|$ alone, irrespective of the shape of the molecules.

The simple relation between pressure and the average properties of a molecule provides a good example of the developments which are possible for a perfect gas. For consider the flux of momentum by molecules moving in both directions across a stationary element of surface δA with normal $\mathbf{n}$ in the gas. The number of molecules crossing with velocity components in the ranges δu, δv, δw about $\mathbf{u}$ in unit time (a crossing to the side to which $\mathbf{n}$ points being reckoned as positive) is $\mathbf{n}.\mathbf{u}\,\delta A\,Nf(\mathbf{u})\,\delta u\,\delta v\,\delta w$, and each of these molecules carries momentum $m\mathbf{u}$ across the surface element. The total flux of momentum across the surface element due to molecular movement is thus

$$\rho\,\delta A \int\int\limits_{-\infty}^{\infty}\!\!\int \mathbf{u}\,\mathbf{n}.\mathbf{u}f(\mathbf{u})\,du\,dv\,dw.$$

By symmetry of the velocity distribution, the component of this momentum flux tangential to the surface element is zero, and only the normal component need be considered. Morever, the force exerted directly between molecules on either side of the surface element is zero for a perfect gas, so that the stress in the gas is due wholly to flux of momentum. Hence the stress in a perfect gas is a normal pressure of magnitude

$$p = \rho\!\int\!\!\int\!\!\int (\mathbf{n}.\mathbf{u})^2 f(\mathbf{u})\,du\,dv\,dw, \quad = \rho\overline{(\mathbf{n}.\mathbf{u})^2}, \qquad (1.7.1)$$

where the overbar denotes an average over all the molecules in unit volume. The mean value in $(1.7.1)$ is independent of the direction of $\mathbf{n}$, so that we may write

$$p = \tfrac{1}{3}\rho\overline{\mathbf{u}^2}. \qquad (1.7.2)$$

If several different kinds of molecule are present in the gas, the same argument holds for each constituent separately, so that, with an obvious notation,

$$p = \tfrac{1}{3}\sum_r \rho_r \overline{\mathbf{u}_r^2}$$

$$= \tfrac{2}{3}\times \text{total kinetic energy of translation}$$
$$\text{of the molecules in unit volume.} \qquad (1.7.3)$$

This argument does not give the pressure exerted by the gas on a rigid boundary, but $(1.7.3)$ is valid there since the conditions for mechanical equilibrium of the gas require p to be continuous throughout the gas (see $(1.4.2)$) and uniform in the absence of a body force on the gas.

The velocity distribution function $f(\mathbf{u})$ for each constituent of the gas is one of the fundamental quantities in kinetic theory, and many attempts have been made to derive its form for a perfect gas in equilibrium. None of the available derivations is entirely free from assumptions, although they all yield the same result, which is also in good agreement with observation, and there is little doubt about its validity. Perhaps the most satisfactory derivation is that employing the concepts of statistical mechanics. This 'method of most probable state' gives useful additional information, and we shall quote here the general result obtained.†

The result to be stated concerns the probability distribution of all the parameters needed to specify the state of a molecule, including the three components of momentum of the molecule as a whole. If we suppose the

† For an account of the method, see, for instance, chapter 9 of *Kinetic Theory of Gases*, by E. H. Kennard (McGraw-Hill, 1938).

rotational and internal modes of motion of a molecule to be describable by classical laws (in fact, quantum theory is needed under some conditions, as we shall note later), the whole state of a molecule with s degrees of freedom at any instant can be represented by s generalized co-ordinates $q_1, q_2, ..., q_s$ and the corresponding s generalized momenta $p_1, p_2, ... p_s$; q_1, q_2, q_3 can be taken as rectilinear position co-ordinates of the centre of mass of the molecule, in which case $p_1 = mu$, $p_2 = mv$, $p_3 = mw$, and the other $s - 3$ degrees of freedom concern rotational and internal modes of motion. In a perfect gas the molecules are dynamically independent of each other, so that, corresponding to given values of the generalized co-ordinates and momenta, there is a total energy ϵ of the molecule, of which energy of translational motion forms a part. Then the result is that the probability of a given molecule at any instant having values of its generalized co-ordinates and momenta which lie within the ranges q_1 to $q_1 + \delta q_1, ... p_s$ to $p_s + \delta p_s$, is of the form

$$C e^{-\alpha\epsilon} \delta q_1 ... \delta q_s \, \delta p_1 ... \delta p_s, \tag{1.7.4}$$

where the constants C and α are independent of $q_1, ..., p_s$. C may be determined from the identity

$$C \int ... \int e^{-\alpha\epsilon} dq_1 ... dp_s = 1 \tag{1.7.5}$$

and a knowledge of ϵ as a function of $q_1, ..., p_s$, the integral in (1.7.5) being taken over all possible values of $q_1, ..., p_s$; C thus depends only on the type of molecule and on α. The expression (1.7.4) is known as the classical (i.e. non-quantum) *Boltzmann distribution*, and is widely used.

In the absence of any body force acting on the gas, ϵ and hence also the expression (1.7.4) are independent of the position co-ordinates q_1, q_2, q_3, so that the density of the gas is uniform, as we had already supposed.

Whatever the internal structure of the molecule may be, we may write

$\epsilon = \frac{1}{2}m\mathbf{u}^2 +$ energy associated with the $s - 3$ non-translational degrees
of freedom

$$= \frac{1}{2}(p_1^2 + p_2^2 + p_3^2)/m + F(q_4, ..., q_s, p_4, ..., p_s), \tag{1.7.6}$$

and the range of possible values of p_1, p_2 and p_3 is $-\infty$ to ∞. The integration with respect to p_1, p_2, p_3 in (1.7.5) can be carried out separately, giving

$$C(2\pi m/\alpha)^{\frac{3}{2}} \int ... \int e^{-\alpha F} dq_1 ... dq_s \, dp_4 ... dp_s = 1 \tag{1.7.7}$$

as the relation determining C. The probability of a molecule having velocity components in the ranges $\delta u, \delta v, \delta w$ about $\mathbf{u}$, divided by $\delta u \, \delta v \, \delta w$, is then

$$f(\mathbf{u}) = m^3 \int ... \int C e^{-\alpha\epsilon} dq_1 ... dq_s \, dp_4 ... dp_s$$

$$= m^3 C e^{-\frac{1}{2}\alpha m \mathbf{u}^2} \int ... \int e^{-\alpha F} dq_1 ... dq_s \, dp_4 ... dp_s$$

$$= \left(\frac{\alpha m}{2\pi}\right)^{\frac{3}{2}} e^{-\frac{1}{2}\alpha m \mathbf{u}^2}. \tag{1.7.8}$$

This is the well-known *Maxwell distribution of molecular velocities*, first obtained by Maxwell from an assumption (which is apparently correct but is difficult to justify strictly) that the three components u, v, w are statistically independent. The single parameter α that is needed for the complete specification of the distribution of molecular velocities is related to the average energy of translational motion of the molecule, for

$$\tfrac{1}{2}m\overline{\mathbf{u}^2} = \tfrac{1}{2}m \int\limits_{-\infty}^{\infty}\!\!\int\!\!\int \mathbf{u}^2 f(\mathbf{u})\, du\, dv\, dw$$

$$= \frac{3}{2\alpha}. \tag{1.7.9}$$

If the gas is a mixture of molecules of different type, the Boltzmann distribution (1.7.4) and the Maxwell distribution (1.7.8) apply to each constituent separately. Moreover, it is a consequence of the argument leading to (1.7.4) that the parameter α has the same value for all the molecules of which the gas is composed—as we may see to be needed for consistency, by supposing that $q_1, \ldots, q_t, p_1, \ldots, p_t$ are the generalized co-ordinates and momenta of a pair of molecules of different type, in which case ϵ is the sum of the energies of the two molecules separately and there are six translational degrees of freedom. Thus if $\mathbf{u}_1$ and $\mathbf{u}_2$ are the velocities of two molecules of mass m_1 and m_2 respectively, we have, from (1.7.9),

$$\tfrac{1}{2}m_1\overline{\mathbf{u}_1^2} = \tfrac{1}{2}m_2\overline{\mathbf{u}_2^2} = \frac{3}{2\alpha}, \tag{1.7.10}$$

showing that all the molecules in a mixed gas have equal amounts of energy of translational motion on the average. We saw earlier that molecules make a contribution to the pressure in proportion to their translational energy; and (1.7.3) may now be written

$$p = \sum_r N_r/\alpha = N/\alpha, \tag{1.7.11}$$

where N_r is the number density of the molecules of one type, showing that the contribution to the pressure from each constituent is proportional to the number of molecules of that constituent in unit volume.

The relation (1.7.10) is one manifestation of the important *principle of equipartition of energy*. This principle is applicable to any one of the generalized co-ordinates and momenta of a molecule which appears in the expression for the molecular energy ϵ as an additive square term and of which the range of possible values is $-\infty$ to ∞. For suppose that in (1.7.6) F is of such a form that

$$\epsilon = \tfrac{1}{2}m\mathbf{u}^2 + aq_4^2 + G(q_5, \ldots, q_s, p_4, \ldots, p_s),$$

where a is independent of q_4. Then, according to (1.7.4) and (1.7.5), the mean value of aq_4^2 with fixed values of all other generalized co-ordinates and momenta is

$$\frac{\int_0^\infty aq_4^2 \, e^{-\alpha\varepsilon} \, dq_4}{\int_0^\infty e^{-\alpha\varepsilon} \, dq_4}, \quad = \frac{1}{2\alpha}. \qquad (1.7.12)$$

This mean value is independent of the chosen values of the co-ordinates and momenta other than q_4, and so is valid generally. Hence the average energy associated with any generalized co-ordinate or momentum which occurs in as an additive square term is $\frac{1}{2}\alpha^{-1}$. This additive square term may represent kinetic energy of translation in one of three orthogonal directions, or kinetic energy of rotation of the molecule about one of its principal axes, or kinetic energy of a vibrational mode, or potential energy associated with a small deformation of the molecule from its equilibrium form. If classical laws of dynamics apply to molecules, it follows that the average total energy of a monatomic molecule with mass but no extension is $\frac{3}{2}\alpha^{-1}$, of a rigid diatomic molecule with non-zero moment of inertia about two principal axes is $\frac{5}{2}\alpha^{-1}$, of a diatomic molecule capable of vibration of the atoms along the line joining them is $\frac{7}{2}\alpha^{-1}$, and so on.

In a mixed gas, the distribution of molecular co-ordinates and momenta for the molecules of one type is determined by the value of the parameter α in (1.7.4); and α has the same value for each constituent. This can be regarded as a statement that when two different gases are in thermal equilibrium with each other, the corresponding values of α are equal. Temperature is a quantity defined as having this same property, and it is therefore natural to seek a connection between the parameter α and the temperature of the gas. We may do this by comparing (1.7.11) with an expression for p derived from thermodynamics. We saw in § 1.5 that for any material

$$T \, \delta S = \delta E + p \, \delta(1/\rho)$$

and

$$T \left(\frac{\partial S}{\partial \rho} \right)_T = \left(\frac{\partial E}{\partial \rho} \right)_T - \frac{p}{\rho^2}.$$

The Maxwell relation (1.5.15) allows this to be written as

$$T \left(\frac{\partial p}{\partial T} \right)_\rho = p - \rho^2 \left(\frac{\partial E}{\partial \rho} \right)_T. \qquad (1.7.13)$$

Now a perfect gas by definition is a material for which the internal energy is the sum of the separate energies of the molecules in unit mass and is independent of the distances between the molecules, that is, independent of ρ. Hence for a perfect gas

$$E \equiv E(T) \quad \text{and} \quad \left(\frac{\partial p}{\partial T} \right)_\rho = \frac{p}{T}.$$

It appears that at constant density p is proportional to T (Charles's law) and that, from (1.7.11),

$$1/\alpha = kT, \tag{1.7.14}$$

where k is an absolute constant known as *Boltzmann's constant*. When the unit of temperature is defined so that $T = 273 \cdot 15$ deg at the temperature of melting ice (corresponding to o degC) k is found to have the value

$$1 \cdot 381 \times 10^{-16} \text{ cm dyn/degC.}$$

The expression (1.7.11) for the pressure now becomes the *equation of state for a perfect gas*:

$$p = NkT = \frac{k}{\overline{m}} \rho T$$

$$= R\rho T, \quad \text{say,} \tag{1.7.15}$$

where $\overline{m}$ is the average mass of the molecules of the gas and $R, = k/\overline{m}$, is known as the gas constant ($R = 2 \cdot 870 \times 10^3 \text{ cm}^2/\text{sec}^2 \text{ degC}$ for dry air). One of the consequences of this equation of state is that the coefficient of thermal expansion for a perfect gas is

$$\beta = -\frac{1}{\rho}\left(\frac{\partial\rho}{\partial T}\right)_p = \frac{1}{T}. \tag{1.7.16}$$

For the isothermal coefficient of compressibility we have

$$\frac{1}{\rho}\left(\frac{\partial p}{\partial\rho}\right)_T = \frac{1}{p}, \tag{1.7.17}$$

showing that fractional variations of p and ρ at constant T are equal; the adiabatic coefficient of compressibility is γ^{-1} times the isothermal value (see (1.5.6)).

The fact that for a perfect gas E is a function of T alone simplifies some of the general expressions for specific heat. The two principal specific heats defined in (1.5.5) become

$$c_p = \frac{dE}{dT} + R, \quad c_v = \frac{dE}{dT}, \tag{1.7.18}$$

and we have *Carnot's law* $\quad c_p - c_v = R, \tag{1.7.19}$

which is found to be satisfied to better than 1% by air at normal temperatures and pressures. c_p and c_v, like E, are now functions of T alone, and the functions of state E, I and S may be written as

$$E = \int c_v\, dT, \quad I = \int c_p\, dT, \tag{1.7.20}$$

$$S = \int\frac{dE}{T} + \int\frac{p}{T}d\left(\frac{1}{\rho}\right) = \int\frac{c_v}{T}dT - (c_p - c_v)\log\rho. \tag{1.7.21}$$

When the molecules of a perfect gas have a simple structure we can go further with the evaluation of the internal energy and the specific heats. For point-mass molecules with only translational energy, the average energy of a molecule is $\frac{3}{2}\alpha^{-1}$, $= \frac{3}{2}kT$, so that

$$E = \tfrac{3}{2}kT, \quad c_p = \tfrac{5}{2}R, \quad c_v = \tfrac{3}{2}R.$$

More generally, if for each molecule of the gas the expression for the molecular energy ϵ is the sum of n terms, each of which is proportional to the square of a generalized co-ordinate or momentum, the average value of ϵ is $\frac{1}{2}nkT$ and

$$E = \frac{\bar{\epsilon}}{m} = \tfrac{1}{2}nRT, \quad c_v = \tfrac{1}{2}nR, \quad \gamma = \frac{c_p}{c_v} = \frac{n+2}{n}. \qquad (1.7.22)$$

The measured values of c_p and c_v for the rare gases, which are known to be monatomic, agree closely with these classical formulae with $n = 3$. There are also several gases with diatomic molecules, including oxygen and nitrogen, for which the measured values at normal temperatures and

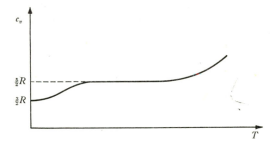

Fig. 1.7.1. Variation of specific heat with temperature
for a polyatomic molecule.

pressures are described accurately by the formulae with $n = 5$ (for instance, for dry air at 15 °C and one atmosphere pressure, the measured value of γ is 1·40 to within 1 %). However, in many other circumstances no choice of n gives good agreement with the observed specific heats, and the contributions to E from the non-translational modes of motion of a molecule are evidently not always of the form to be expected from classical laws.

It is known that the variation of c_v with T for some common polyatomic gases is as sketched in figure 1.7.1 and that it can be accounted for with the aid of quantum considerations. The amount of energy associated with a non-translational mode is quantized, and takes one of a set of discrete values. Only when $\frac{1}{2}kT$ is appreciably larger than the smallest of these energy levels will the continuous distribution (1.7.4) provide an approximate representation of the equilibrium; and only then will the average energy associated with the mode be $\frac{1}{2}kT$ approximately. At very low temperatures of the gas, none of the non-translational modes of a molecule is 'excited' and the internal energy is made up almost wholly of translational energy, so that $c_v = \frac{3}{2}R$. As the temperature is raised, the smallest energy level of some non-translational mode is reached, usually a rotational mode first, and E increases with T more rapidly than linearly. The flat portion of the curve in figure 1.7.1 corresponds to an intermediate range of temperatures such that $\frac{1}{2}kT$ exceeds the smallest energy level of a rotational mode by a sufficient margin for rotational energy to contribute its full amount to E (viz. kT for diatomic molecules) but nevertheless is still much less than the smallest energy level of a vibrational mode. For air, energy of vibrational modes does not begin to be significant until the temperature reaches about 600 °K, and c_p and c_v are quite accurately constant and equal to $\frac{7}{2}R$ and $\frac{5}{2}R$ respectively over the range 250–400 °K which conveniently includes 'normal' temperatures. At even higher temperatures, above about 20,000 °K, a significant contribution to E is made by the energy of the electronic system of a molecule (which can be regarded as including energy of rotation of monatomic molecules).

For the important special case of a *perfect gas with constant specific heats* over some range of T, to which air approximates closely at normal temperatures and pressures, we may obtain explicit expressions for E, I and S. The relations (1.7.20) become

$$E - E_0 = c_v\,T, \quad I - I_0 = c_p\,T, \qquad (1.7.23)$$

and, for (1.7.21), on using the equation of state we find

$$S - S_0 = c_v \log(p\rho^{-\gamma}), \qquad (1.7.24)$$

where the constants E_0, I_0, S_0 do not have any absolute significance unless c_v is constant for all temperatures less than T. It follows that the relation between p and ρ for an isentropic change of state is

$$p \propto \rho^\gamma; \qquad (1.7.25)$$

this is often referred to as the relation between p and ρ for adiabatic change of a perfect gas, although it is also necessary that the change be a reversible one (since otherwise S may not be constant) and that c_v be constant.

Departures from the perfect-gas laws

We shall not be concerned in this book with conditions in which the relations found above do not apply with fair accuracy to the common real gases, but it is of interest to note briefly here the kind of departure from the perfect-gas laws which may occur under extreme conditions. These departures are of two main types. The first type occurs at large densities and is due to the proximity of the molecules. The second occurs at high temperatures, and is due to changes in the structure of the molecules.

At large densities the dynamical behaviour of a molecule is influenced by the presence of the other molecules, and the basic formula (1.7.2) for the pressure exerted by the gas needs modification. This formula takes account only of the flux of normal momentum across an element of surface in the gas, and we must now add a contribution from the forces acting between pairs of molecules instantaneously on the two sides of the surface element. The force exerted on one molecule at any instant by all the molecules on the other side of the surface element is proportional to the number density of molecules N, so that the total force exerted across the surface element (which will be in the direction of the normal, for reasons of symmetry) will be proportional to N^2. The expression for p can thus be written as

$$p = \begin{pmatrix} \text{flux of normal momentum} \\ \text{per unit area} \end{pmatrix} - a\rho^2,$$

where a is a constant for a given gas which depends on the intermolecular force. We expect cohesive forces to be dominant, in which case $a > 0$. However, observation suggests that the effective value of a diminishes with increase of T, essentially because repulsive forces play an increasingly important part as the molecular speeds rise and molecules penetrate more deeply into each other's force fields.

The expression for the momentum flux also needs correction, for it assumes that the chance of a molecule crossing a surface element is independent of the presence of other molecules. If the volume occupied by the molecules themselves is no longer a negligible fraction of the total volume, the rate at which a given molecule crosses the surface element is greater than was assumed, because the space accessible to the molecule is less. To the first order in the molecular volume, we can obtain the increased rate at which molecules cross the surface element by dividing by a factor $1 - b\rho$ (corresponding to the same number of molecules moving independently in a volume which is diminished by b for each unit mass of gas present), whence the contribution to p from momentum flux becomes

$$\frac{NkT}{1-b\rho}, \quad = \frac{R\rho T}{1-b\rho}.$$

Again b is not an absolute constant, but diminishes as T increases because molecules penetrate closer to each other at higher speeds.

The amended form of the equation of state is thus

$$p = \frac{R\rho T}{1-b\rho} - a\rho^2. \tag{1.7.26}$$

This is *van der Waals' equation* and is the best known of the various attempts to take account of the 'imperfections' of real gases. The arguments on which it is based are not rigorous, but it is found to be useful in describing small departures from the equation of state for a perfect gas. For air the empirical values of a and b are about $3 \times 10^{-3} p_0/\rho_0^2$ and $3 \times 10^{-3}/\rho_0$ respectively, where p_0 and ρ_0 refer to standard conditions. The equation is not adequate for gases near the point of condensation.

Departures from the perfect-gas relations of a rather different kind occur at very high temperatures, when some collisions are so violent that polyatomic molecules may be dissociated into their constituent atoms. For instance, at normal pressures an appreciable fraction of the diatomic molecules of oxygen are dissociated at 3,000 °K, and of nitrogen at 6,000 °K. At temperatures of this order, air is thus a mixture of the constituents O, O_2, N, N_2. It may also happen at even higher temperatures that ionization occurs and that free electrons are added to the mixture. It is possible for the particles of this kind of mixed gas to be approximately dynamically independent (at any rate, provided electrostatic forces, which fall off only as the inverse square of the separation, are unimportant), so that in a sense the gas may still be 'perfect'. The expression for pressure, and the relation between translational energy and temperature, still stand, whence, as before,

$$p = \frac{k}{\bar{m}} RT.$$

However, $\bar{m}$ is the average mass of the particles of which the gas is composed and is now a function of temperature and density (since both these quantities affect the equilibrium between molecules and atoms, or between atoms and electrons), so that the equation of state has the perfect-gas form only superficially. The energy relations also need modification, since dissociation of a molecule or ionization of an atom absorbs energy. The internal energy of the gas thus depends on the composition of the mixture, and does not have the dependence on temperature alone that characterizes a perfect gas.

Transport coefficients in a perfect gas

If certain properties of the gas are spatially non-uniform, and if these properties are such that the amount associated with single molecules is conserved in some sense, random migration of the molecules and persistence of their properties tends to smooth out the spatial variations. Molecular transport effects of this kind exist in all fluids, as remarked in §1.6. In the case of a perfect gas it is possible to estimate the magnitude of the transport coefficients (represented by the parameter k_{ij} in (1.6.1)) from an actual calculation of the transport due to molecules acting separately. Accurate calculation of the transport coefficients for a gas is difficult, both conceptually and mathematically, and a proper account of the methods and results is beyond our scope here.†

In a perfect gas in equilibrium, the distribution of molecular velocities (see (1.7.8)) has an isotropic form. Thus the response of the gas to an imposed departure from equilibrium, of the kind represented by spatial non-uniformity of gas properties, is without directional distinctions, and the

---

† Reference may be made to *The Mathematical Theory of Non-uniform Gases*, by S. Chapman and T. G. Cowling (Cambridge University Press, 1952). More elementary accounts will be found in books on the kinetic theory of gases.

tensor transport coefficient k_{ij} is determined by the single scalar parameter k as in (1.6.2).

Let us suppose that the non-uniformity of the gas is associated with some quantity which is conserved during collisions and of which the amount contributed by a given molecule is q; the various possible interpretations of q will be given in a moment. Then the amount of the quantity concerned transferred across an element of surface in the gas, per unit area and per unit time, due to free flights across the element of those molecules with property q and velocity $\mathbf{u}.\mathbf{n}$ in the direction of the normal to the element is $\mathbf{u}.\mathbf{n}q$ multiplied by the number of such molecules in unit volume; and the total flux per unit area is

$$\overline{N\mathbf{u}.\mathbf{n}q},$$

the average being taken over all molecules in the neighbourhood. If the local average (in this same sense) of q is uniform throughout the gas, there can be no statistical tendency for a certain value of q to be associated with either sign of the velocity component $\mathbf{u}.\mathbf{n}$, and the flux is zero. On the other hand, if $\bar{q}$ is non-uniform, molecules moving in a direction of increasing values of $\bar{q}$ will tend to be associated with values of q smaller than the local average. The direction of motion of a molecule becomes completely random within several collisions after having a prescribed direction of motion, so that the flux can be influenced only by the variation of $\bar{q}$ within a few molecular path lengths of the point concerned; that is, it can depend only on the local gradient of $\bar{q}$, as in the more general circumstances discussed in §1.6.

If we make the crude assumption that during a free flight a molecule always has a value of q equal to $\bar{q}$ at the position of its last collision, the value of q at the surface element can be written as

$$\bar{q} - t\mathbf{u}.\nabla\bar{q},$$

where $\bar{q}$ and $\nabla\bar{q}$ are evaluated at the position of the element and t is the time interval since the element last made a collision. A further rough approximation to be made is that t can be replaced by τ, the average time between collisions. The flux per unit area then becomes

$$-N\tau\overline{\mathbf{u}.\mathbf{n}\mathbf{u}.\nabla\bar{q}}.$$

In this case of an isotropic medium the flux vector $\mathbf{f}$ is in the direction of the local gradient of $\bar{q}$, and the magnitude of $\mathbf{f}$ is therefore equal to the flux per unit area across a surface element with normal $\mathbf{n}$ in the direction of $\nabla\bar{q}$. Hence the expression for the flux vector is

$$\mathbf{f} = -\tfrac{1}{3}N\tau\overline{\mathbf{u}^2}\nabla\bar{q}, \tag{1.7.27}$$

in which we have ignored any small difference between the values of the mean squares of the components of $\mathbf{u}$. This expression cannot be expected to be correct numerically, but it reveals the molecular parameters relevant to transport coefficients for a perfect gas. The distinctive part of this expression is the product $\overline{\tau \mathbf{u}^2}$, which enters into all diffusivities for a perfect gas. We may write $\overline{\tau \mathbf{u}^2}$ alternatively as $l(\overline{\mathbf{u}^2})^{\frac{1}{2}}$, where l is a kind of average path length between collisions.

We now give q various specific meanings. If q takes either the value unity when the molecule is one of a certain type or the value zero otherwise, in a gaseous mixture of molecules, then $\bar{q}$ is the local proportion of marked molecules (by number) and is identical with the concentration C employed in § 1.6. The flux represented by (1.7.27) is a flux of number of marked molecules, and (1.7.27) is therefore equivalent in this case to an estimate of the coefficient of diffusion κ_D (see (1.6.3) and (1.6.8)) as being of order $\overline{\tau \mathbf{u}^2}$, where τ and $\overline{\mathbf{u}^2}$ are average properties of the marked molecules. The value of $\overline{\tau \mathbf{u}^2}$ for the marked molecules may be different from that for the unmarked molecules, and decreases with increasing mass of the marked molecules in a gas at a given temperature.

If we take q to represent the total energy of the molecule, in which case $\bar{q}/\bar{m}$ is the internal energy per unit mass of the gas, $E(T)$, the flux represented by (1.7.27) is effectively a flux of heat. The thermal conductivity k_H, defined in § 1.6 as the heat flux per unit area divided by (minus) the local temperature gradient, is therefore estimated by (1.7.27) to be of order

$$\overline{\tau \mathbf{u}^2} \rho \frac{dE}{dT}, \quad = \overline{\tau \mathbf{u}^2} \rho c_v,$$

and the thermal diffusivity κ_H (see (1.6.13)) to be of order $\overline{\tau \mathbf{u}^2} c_v / c_p$.

If we take q to represent the component of momentum of the molecule in a given direction in the plane of the surface element, in a gas undergoing a simple shearing motion such that the gas velocity varies only in the direction of the normal to the surface element, $\bar{q}/\bar{m}$ is the gas velocity U and the flux represented by (1.7.27) (which in this case is to be regarded as applying to only one choice of the direction of the normal to the surface element) is a tangential component of stress across the surface element.† The viscosity of the fluid, μ, defined in (1.6.15) as the tangential stress divided by the gradient of fluid velocity, is therefore estimated by (1.7.27) to be of order $\rho \overline{\tau \mathbf{u}^2}$, and the kinematic viscosity ν (see (1.6.16)), or diffusivity for momentum, to be of order $\overline{\tau \mathbf{u}^2}$.

It seems that for a perfect gas all three diffusivities are of the form

$$\text{number of order unity} \times \overline{\tau \mathbf{u}^2} \ (\text{or } l(\overline{\mathbf{u}^2})^{\frac{1}{2}}) \tag{1.7.28}$$

(although in the case of diffusion of one constituent the quantities τ and $\overline{\mathbf{u}^2}$

---

† It will be noted that the ratio of the tangential to the normal component of stress is estimated to be of order $\tau dU/dy$, which, in view of the fact that τ is about 10^{-10} sec in air at normal temperature and pressure, is very much smaller than unity for all common practical values of the velocity gradient dU/dy. But the effect of stresses on motion of the gas is determined by their spatial gradients (as implied by (1.4.2), for example), and not by their absolute values, and the normal and tangential stresses may in fact have comparable influences on the motion of a gas, as we shall see.

refer to the marked molecules and not to the gas as a whole). Unfortunately τ and l are not very definite quantities for real molecules, since they demand some arbitrary decision about what constitutes a collision.†As a consequence, the above simple theory is not able to make accurate *a priori* predictions of the absolute magnitudes of the diffusivities; indeed it is more usual to work in the opposite direction and to infer the value of τ or l from the theory and observed values of the diffusivities. However, the prediction that the diffusivities for marked molecules (in a case of self-diffusion), heat and momentum are of the same order of magnitude is borne out by observation. A simple formula which describes the measurements of ν and κ_H for many gases quite well is

$$\frac{\nu}{\kappa_H} = \frac{4\gamma}{9\gamma - 5};\qquad (1.7.29)$$

for air this formula gives 0·74 for ν/κ_H, and the observed ratio is 0·72. The ratio ν/κ_D, with κ_D referring to self-diffusion, is observed to lie between 0·6 and 0·8 for most simple gases.

The dependence of all the diffusivities on absolute temperature and density indicated by the estimate (1.7.28) is also of interest. $\overline{\mathbf{u}^2}$ depends only on T and varies linearly with it. At a given temperature, and hence with a given character of collisions, the number of molecules which at any instant lie in unit length of the cylinder swept out by a molecule in free flight is proportional to ρ, so that $l\rho$ is constant; and as T (and the average molecular speed) increases the effective number of collisions made by a molecule per unit path length may be expected to decrease a little because the more distant encounters then cease to qualify as collisions. Hence

$$l(\overline{\mathbf{u}^2})^{\frac{1}{2}} \propto T^{\frac{1}{2}+a}/\rho,\qquad (1.7.30)$$

where a takes account of the effect of temperature on collisions and which for air is observed to be about 0·25 for temperatures in the range 200–400 °K.

Observed values of the diffusivities and other relevant parameters for air at various values of T and ρ are set out in appendix 1, at the end of this volume.

Other manifestations of departure from equilibrium of a perfect gas

If uniform and steady conditions are maintained at the boundary of a mass of perfect gas, the gas comes into equilibrium with its surroundings through the effect of molecules colliding with each other and with the boundary. Collisions provide the only means by which molecules of a perfect gas can be influenced by conditions at the boundary. The fact that the average time between collisions by a given molecule (τ) is non-zero, although exceedingly small (normally about 10^{-10} sec in air), implies that equilibrium is not

---

† In a more refined version of the theory, one uses the notion of a *collision cross-section*, which is the effective area of a molecule presented to an incident molecule and which depends on the intermolecular force and the speed of the incident molecule as well as on the geometry of the collision.

attained instantaneously, and that if conditions at the boundary are changing continually there is a permanent small departure from equilibrium in the gas. (The way in which the related kind of spatial persistence of molecular properties, arising from movement of molecules between collisions and retention of some influence of conditions at their earlier positions, leads to transport phenomena has already been described.) Here we consider briefly some possible consequences of the departures from equilibrium in the illustrative case of a perfect gas which is being compressed by a piston in a cylinder. Like the simple shearing motion introduced in § 1.6, this is a special kind of motion of the gas whose place in the general analysis of fluid motions will be seen later; the purpose of considering it at this stage is to reveal the physical nature of the response of the gas.

We suppose that a mass of gas in a cylinder is being compressed uniformly and adiabatically by a sliding piston. The internal energy of the gas is increasing, as a consequence of the work being done by external forces at the boundary, and the effect of collisions is to tend to distribute the instantaneous total internal energy over the available modes of molecular motion in the manner described by the equilibrium Boltzmann distribution (1.7.4). Clearly collisions cause energy changes more rapidly in some modes than in others. The first and direct effect of the displacement of the piston is to cause an increase in the energy of translation of molecules in the direction of movement of the piston. Collisions then spread some of this excess energy into the other two translational modes, and into rotational and vibrational modes. Detailed calculations with particular laws of force between colliding molecules show that equipartition of energy between the three translational modes is achieved very quickly after the piston comes to rest, in fact within a few collision 'intervals', as one might expect. The translational modes thus have a *time of relaxation* to equilibrium which is normally of the order of 10^{-10} or 10^{-9} sec, and it will not often happen in practice that conditions are changing so rapidly as to lead to appreciably different† values of $\overline{u^2}$, $\overline{v^2}$ and $\overline{w^2}$.

If the length of the column of gas in the cylinder is decreasing with a steady (negative) rate of extension e, the maintained differences between $\overline{u^2}$, $\overline{v^2}$ and $\overline{w^2}$ will be of the order of magnitude of the differences which movement of the piston would produce, in the absence of restoration of the equilibrium distribution by collisions, in the time of relaxation for the translational modes, which is of order τ. In a time τ the length of the column of gas changes by the small fraction τe and the work done against a gas pressure of order $\rho\overline{u^2}$ during this movement provides energy per unit volume of order

$$\rho\overline{u^2}\tau e, \tag{1.7.31}$$

all of which goes, in the absence of collisions, to the translational mode in the direction of piston movement (u, say). This gives the magnitude of the

† But it would be wrong to infer that the differences are without significance, for reasons of the kind given in the footnote to p. 49.

maintained difference between $\overline{\rho u^2}$ and $\overline{\rho v^2}$ or $\overline{\rho w^2}$, which are the normal stresses in three orthogonal directions. These small differences, which are of such a sign as to give a greater resistance to continued movement of the piston than an equilibrium pressure at each stage, and which are proportional to the velocity gradient in the gas (e), represent contributions to the normal components of the stress due to internal friction. Although the connection is not obvious, they are related to the tangential stress component which is set up by internal friction in a gas undergoing a simple shearing motion. As we shall see later, the departures of the stress tensor from the isotropic form (appropriate to fluid at rest) in any fluid undergoing deformation of general type can be written, with the kind of hypothesis employed in §1.6, as a linear function of the local velocity gradients which involves a single scalar parameter, viz. the viscosity μ. Thus the estimate (1.7.31) of the difference between two normal components of stress is equivalent to an estimate of the viscosity as being of order $\overline{\rho u^2}\tau$, in agreement with the estimate from molecular transport theory.

Adjustment of the energy of the rotational and vibrational modes also lags behind the supply of energy by the piston, although with slightly different consequences. Rotational modes of polyatomic molecules are not quite as directly involved in encounters between molecules as are translational modes, and a few more collisions are needed for the achievement, when the piston is stationary, of equipartition of energy between the translational and rotational modes. Vibrational modes appear to be much less affected by collisions, and experimental evidence shows that they need a very much longer time for the achievement of equilibrium; on the other hand, as remarked earlier, the average energy of a vibrational mode for air molecules in equilibrium at ordinary temperatures is much less than the classical equipartition value $\frac{1}{2}kT$ because of the large energy level of the ground state of these modes. Thus, at temperatures at which rotational but not vibrational modes make a significant contribution to the internal energy, when the piston movement is maintained the energy in translational modes is a greater fraction of the total internal energy than in equilibrium, the excess energy (per unit volume) at any stage being also of the order of magnitude given by (1.7.31). In these non-equilibrium situations, our definition of temperature does not have a precise meaning, but that of internal energy does (for see (1.5.2)) and it is possible to specify the amount of energy associated with a mode as a fraction of the internal energy. Whereas for a perfect gas with constant specific heats we have the equilibrium relations

$$p = \tfrac{1}{3}\rho\overline{\mathbf{u}^2} = NkT = (\gamma - 1)\rho E,$$

the disturbed state may be represented by

$$\frac{\tfrac{1}{3}\overline{\mathbf{u}^2} - (\gamma - 1)E}{\tfrac{1}{3}\overline{\mathbf{u}^2}} \propto \tau e, \qquad (1.7.32)$$

with a constant of proportionality of order unity. Note that again the departure of the normal stress, against which the piston is working, from its equilibrium value at any stage has such a sign as to provide a resistance to the attempt to deform the gas.

Effects of this kind associated with relaxation of the rotational and vibrational modes of a molecule are important in situations in which a mass of gas is subjected to rapid changes of pressure, as when it is traversed by a sound wave of high frequency or by a shock wave.

1.8. The distinctive properties of liquids

Much less is known about the structure of liquids than about that of gases. No simple model, like that of a perfect gas with dynamically independent molecules, is available for the derivation of approximate results for liquids. As a consequence, it is not possible to set many of the observed values of properties of a liquid within a logical framework or to account for them in terms of properties of the individual molecules. A further handicap to exposition is that water, which is the particular liquid of greatest practical importance, is anomalous in many of its properties. In this section many of the known properties of the common liquids that are relevant in fluid dynamics will be described without much comment. Numerical data for pure water in particular will be found in appendix 1 at the end of this volume.

The primary property of the liquid and solid phases of matter is that they are *condensed* phases in which a molecule is continually within the strong cohesive force fields of several of its neighbours. But liquids have in common with gases the property of fluidity and ability to change shape freely. In a uniform liquid at rest, tangential components of stress are zero, the work done on unit mass of liquid in changing its density slowly by the small amount $\delta\rho$ is $(p/\rho^2)\,\delta\rho$, and an equation of state relates the three variables ρ, p, T, all as in the case of a gas.

A given material may exist in the liquid phase for some values of the two parameters of state (p and T say) and in the gaseous phase for other values. The reason for this occurrence in two distinct fluid phases with widely different values of the density is of course to be found in the variation of intermolecular force with molecular spacing, and warrants consideration for a moment. If a mass of gas is compressed isothermally, the average translational energy of a molecule remains constant and the average distance between neighbouring molecules decreases. When the specific volume of the gas is so small that the average spacing of molecules is only a few times their diameter, attractive forces between molecules are significant and the pressure exerted by the gas is less than the normal momentum flux represented by (1.7.1). Provided the temperature is less than a critical value T_c, further decrease of the specific volume leads to an unstable situation in which molecules may be unable to escape from the attractive forces of their neighbours and tend to form clusters. The formation of some clusters of closely packed molecules reduces the number density of the molecules still moving freely and separately, so that a new equilibrium is established with, at a given overall density, a definite (average) proportion of the mass in the condensed or liquid phase and the remainder in the dispersed or vapour phase. This polyphase equilibrium is highly sensitive to changes in pressure; a small increase leads to a completely condensed or homogeneous liquid phase with large density, or a small decrease to the homogeneous vapour phase with rather smaller density.

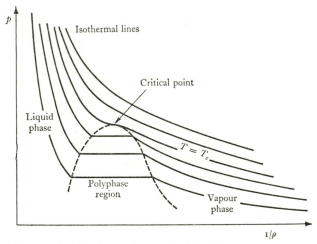

Figure 1.8.1. Isotherms for a typical liquid–vapour system.

The isotherms on an indicator diagram for a typical liquid–vapour system are sketched in figure 1.8.1. The approximately constant pressure on an isotherm through the polyphase region is the 'saturated vapour pressure' p_v, that is, the pressure existing in pure vapour which is in contact with liquid at the given temperature. At temperatures above the critical value T_c, the translational energy of molecules is large enough to prevent the formation of clusters, and there is a continuous transition along the isotherm from a material with gas-like properties at low densities to one with liquid-like properties at high densities. The Van der Waals equation of state (1.7.26) represents qualitatively the main features of observed isotherms, but as already remarked it does not have quantitative validity when the material is in or near the condensed phase.

For water $T_c = 374\,°C$, and the pressure and density at the point where the critical isotherm touches the polyphase region (the 'critical point') are 218 atmospheres and about 0·4 gm/cm³ respectively. The isotherm corresponding to the normal temperature 15 °C runs approximately horizontally across the polyphase region at a pressure of $1·7 \times 10^4$ dyn/cm² or 0·017 of an atmosphere and meets the boundaries of the polyphase region at densities of 1 gm/cm³ (the homogeneous liquid) and 1.28×10^{-5} gm/cm³ (the homogeneous vapour); the liquid and gaseous phases of water are thus quite different at normal temperature.

The effect of reducing the pressure of a liquid below the saturated vapour pressure is of special relevance to fluid dynamics, since the pressure variations in flowing water (which are likely to be adiabatic rather than isothermal) may readily exceed one atmosphere. When the pressure of a liquid is reduced to a value slightly less than the saturated vapour pressure at the temperature of the liquid, the liquid is in an unstable state and normally tends to form vapour pockets distributed throughout the liquid.† The appearance of such pockets, termed *cavitation*, has important mechanical consequences in a flowing liquid, as will be seen in §§ 6.12, 6.13.

Equilibrium properties

The pressure exerted on an element of surface in the interior of a liquid may be regarded as the sum of the normal momentum flux per unit area and the resultant force between molecules on the two sides of the element.‡ The normal momentum flux per unit area is found, by the same calculation as for a gas, to be $\frac{1}{3}\rho \overline{u^2}$. Moreover, the classical Boltzmann distribution (1.7.4) may be applied, with certain extensions of the argument used in § 1.7, to the molecules of a liquid or a liquid–vapour system in thermal equilibrium to show that the mean translational energy of a molecule is $\frac{3}{2}kT$, where k is Boltzmann's constant and T is the absolute temperature as before. Thus the contribution to pressure in a liquid from momentum flux is NkT, and is greater than that in a gas at the same temperature and total pressure by the ratio of the number densities of their molecules. This ratio is normally large; for instance, the contribution to the pressure in water at $15\,°C$ from momentum flux is found on this basis to be $1,312$ atmospheres. This large contribution is evidently nearly balanced, under normal conditions, by a large tension resulting from intermolecular forces. The available data about intermolecular forces suggests that, in water at $15\,°C$ and one atmosphere total pressure, the resultant of the cohesive forces operating between molecules with average spacing is a tension of the order of $10,000$ atmospheres, and it seems that there is also a large contribution, of opposite sign, from the repulsive forces acting between those molecules that happen to be very close together.

---

† It has been found that a homogeneous liquid can be maintained if great care is taken to rid the liquid initially of minute pockets of undissolved gas (perhaps trapped in crevices of small dust particles in the liquid); water at $15\,°C$ can be brought in this way to a *negative* pressure of many atmospheres and is then in a highly unstable state of tension.

‡ This and other statements in this paragraph have mainly qualitative validity; the wave mechanical treatment that is appropriate when the motions of neighbouring molecules are strongly coupled shows that a precise separation of the two contributions cannot be made.

This picture of the pressure in a liquid as the sum of a large positive contribution from momentum flux and an almost equally large (at standard total pressure) negative contribution which is itself the difference between two even larger terms resulting from cohesive and repulsive force fields provides a rough guide to the molecular effects determining the observed properties of liquids. It accounts in particular for the extreme sensitivity of the pressure in the liquid to molecular spacing. Quite small changes of density correspond, at either constant temperature or at constant entropy, to enormous changes in pressure; that is, the coefficient of compressibility for liquids is exceedingly small, and both the isothermal and adiabatic lines through any point on the indicator diagram (figure 1.5.1) are nearly vertical. For instance, the density of water increases by only $\frac{1}{2}$ % when the pressure is increased from one to 100 atmospheres at constant (normal) temperature. This great resistance to compression is the important characteristic of liquids, so far as fluid dynamics is concerned, and it enables us to regard them for most purposes as being incompressible with high accuracy.

Pressures in the deep oceans may be as large as several hundred atmospheres, and in these and other circumstances it may be necessary to take account of the small variation of density with change of pressure. An equation which represents the observed isentropic (p, ρ)-relation for water over a wide range of pressures (Cole 1948) is

$$\frac{p+B}{1+B} = \left(\frac{\rho}{\rho_0}\right)^n, \tag{1.8.1}$$

where ρ_0 is the density at atmospheric pressure. If the constants n (dimensionless) and B (measured in atmospheres) are chosen as 7 and 3,000 respectively, this relation agrees with the data for water to within a few per cent for pressures less than 10^5 atmospheres. The parameter n appears not to depend on the entropy, but B and ρ_0 are both slowly varying functions of S.

When the temperature of a liquid is increased, with the pressure held constant, the liquid (usually) expands. If the momentum flux alone contributed to the pressure, the consequent fall in density would be such as to keep ρT constant, as in the case of a gas. But the contribution to the pressure from intermolecular forces is more important, and has a less predictable dependence on temperature. The example of water at temperatures near 4 °C also shows that a positive expansion with increase of temperature is not inevitable, as it is for gases. In general, measurements show rather smaller values of the coefficient of thermal expansion β (defined as in (1.5.16)) for liquids than the value T^{-1} appropriate to a perfect gas. For water at 15 °C, β is $1 \cdot 5 \times 10^{-4}$/degC. Values of β for other common liquids tend to be larger, and range up to about 16×10^{-4}/degC.

Direct measurement of only one of the two principal specific heats, viz. c_p, is feasible, since enormous pressures develop in a liquid which is heated at constant density. Observation shows that for most liquids at normal temperatures c_p does not vary much with either temperature or pressure, and is of the same order of magnitude as for gases. When c_p has been measured, c_v may be calculated by means of the thermodynamical relation (1.5.19), viz.

$$c_p - c_v = \frac{T\beta^2}{\rho} \bigg/ \frac{1}{\rho}\left(\frac{\partial\rho}{\partial T}\right)_T. \tag{1.8.2}$$

For several common liquids $c_p - c_v$ is of the order of $0.1c_p$. Water is untypical, and has small values of β and consequently also of $c_p - c_v$ at normal temperature; for instance, for water at 15 °C it is found that the value of $c_p - c_v$ is $0.003c_p$. The value of γ ($= c_p/c_v$) may be taken as unity for water at temperatures and pressures near the normal values.

When a small amount of heat is added reversibly to unit mass of fluid at constant pressure, the fraction of this energy gain that is used in expansion against the external pressure is

$$\frac{-(p/\rho^2)\,\delta\rho}{c_p\,\delta T}, \quad = \frac{p\beta}{\rho c_p},$$

which is of order unity for gases but is much smaller for liquids chiefly owing to the much larger values of ρ in the latter case. Thus heat added reversibly to liquids is manifested almost wholly as a gain in internal energy, irrespective of the associated changes in p and ρ, and we may write

$$\delta Q = T\,\delta S \approx \delta E. \tag{1.8.3}$$

We may also show that the small changes in the functions of state S and E consequent on a small reversible change involving addition of heat are normally determined mainly by the change in T. For if we take T and p as the two independent parameters of state, the ratio of the contributions to $T\,\delta S$ from the changes in T and p is of the same order of magnitude, according to (1.5.21), as the ratio of the changes in ρ that would be produced by the given increments in T and p taken separately; and this ratio is very large for a liquid, except for a change which happens to be a direction nearly parallel to the isothermal line in the indicator diagram. Thus for liquids the relation (1.5.20) takes the approximate form

$$T\,\delta S \approx \delta E \approx c_p\,\delta T \tag{1.8.4}$$

for changes of state which are not nearly isothermal.

Transport coefficients

The phenomenon of transport of matter, heat and momentum in a liquid whose properties are spatially non-uniform is well-documented experimentally, although theoretical analysis is even more difficult than it is for equilibrium properties of a liquid.

Whereas in a gas transport of any molecular property takes place primarily by the random movement of the molecules themselves to a different position, in a liquid exchange of energy and momentum between molecules through the action of intermolecular forces plays an important role. The random motion of a molecule in a liquid may be regarded roughly as a combination of a rapid translational oscillation with an amplitude of the same order as the molecular diameter, and a slower migratory motion in co-ordination with a number of other molecules held together (temporarily) by strong cohesive forces. Transfer of marked molecules, which takes place solely by bodily migration of molecules, is thus comparatively weak in a liquid. The coefficient of diffusion, defined in § 1.6 as the flux of marked molecules per unit gradient of number density of marked molecules, can be measured directly, and for several different types of marked molecules belonging to solutes such as NaCl in water is found to be of order 10^{-5} cm^2/sec (compared with 0·2 for the coefficient of self-diffusion of nitrogen) at 15 °C. The coefficient of diffusion of solutes is found to vary appreciably with the concentration for solutes like $KMnO_4$ with molecules considerably larger than those of water, the usual variation being a decrease, at a diminishing rate, as the concentration increases from zero.

Transport of heat in a liquid, on the other hand, is achieved primarily by the direct exchange of translational energy between molecules lying within each other's force fields, and is consequently not as weak a process as diffusion of marked molecules in liquids. The thermal diffusivity for water is $1·4 \times 10^{-3}$ cm^2/sec at 15 °C, which is much smaller than for air (by a factor of 145), as would be expected from a rough assessment of the effectiveness, per molecule, of the two different transporting mechanisms concerned; however, the thermal conductivity k_H weights the diffusivity with the number density of molecules, and as a consequence the flux of heat per unit temperature gradient is greater in water than in air. For most other liquids the thermal diffusivity at normal temperature is likewise of order 10^{-3} cm^2/sec, except that in the case of liquid metals κ_H is considerably larger (0·042 cm^2/ sec for mercury at 15 °C) owing to the existence of an additional and important contribution to heat transport from free electrons which are not restrained by intermolecular forces and move throughout the liquid roughly like molecules in a gas.

The thermal diffusivity of liquids is practically independent of pressure, as was to be expected, but there appears to be some variation with temperature. Such data as are available show a slow decrease with rise of temperature for most liquids, although for water there is a slow change in the reverse direction.

The mechanism of momentum transport in liquids is a complex one about which little is known. Momentum is evidently not transferred primarily by migration of molecules across an element of surface within the liquid, for that would yield a momentum diffusivity of the same order of magnitude as the coefficient of self-diffusion whereas the measured values of ν are larger by a factor of order 10^3. It seems likely that coherent groups of molecules in a liquid resist deformation in some manner involving the direct action of intermolecular forces, and that the primary effect of a simple shearing motion (for instance) of the liquid is to tear apart some of the existing groups against this resistance. Coherent groups constantly reform in the liquid, with consequent release of energy of molecular motion, and in this way some of the energy of the ordered bulk motion of the liquid is converted (or 'dissipated') to disordered molecular motion, or heat. It is not easy to see with certainty why the tangential stress component should be proportional to the velocity gradient in a simple shearing motion, but we are relieved of the necessity of doing so by the formal argument of § 1.6 and the abundant experimental verification of the linear relation for nearly all homogeneous liquids not having very long chain-like molecules. However, the formal argument says only that a linear relation between the tangential stress component and velocity gradient is to be expected for sufficiently small magnitudes of the velocity gradient, and an explanation of the observation that the linear relation holds for virtually all 'practical' magnitudes of the velocity gradient in liquids not of complicated molecular structure must be sought in the mechanism of transport of momentum in liquids. All that can be said with confidence is that the times characteristic of the formation of a coherent group of molecules in a liquid and of other changes in a group are no doubt very small, and that only when the reciprocal of the velocity gradient is comparably small is the hypothesis of a linear relation between tangential stress and velocity gradient likely to break down.

There is a wide variation in the values of the kinematic viscosity (i.e. the momentum diffusivity) for different liquids; for example, $0.0012 \text{ cm}^2/\text{sec}$ for mercury and $1.0 \text{ cm}^2/\text{sec}$ for olive oil at normal temperature. This variation cannot be accounted for in terms of molecular structure in any simple way. Most liquids show a marked variation of viscosity with temperature. Increase of temperature leads to smaller coherent groups of molecules, owing to the increased agitation of individual molecules, and as a consequence there is *less* resistance to deformation of the liquid. This is in contrast to the behaviour of a gas, which resists deformation more as the temperature rises owing to the more rapid migratory movement of molecules at higher temperatures. The data given in appendix 1 (c) show that the kinematic viscosity of water decreases by about 50 % as the temperature rises from 10 to 40 °C; the assumption of uniform viscosity may therefore be unacceptable when there are moderate variations of temperature in a mass of water, which is unfortunate because the mathematical difficulties of determining the velocity distribution in a moving fluid increase greatly if variation of viscosity must be taken into account.

As remarked in § 1.7, transport phenomena are not the only consequences of a departure from equilibrium which are relevant to fluid dynamics. In a gas there may be relaxation effects arising from a lag in the adjustment of the distribution of molecular energy between the translational modes on the one hand and the rotational and vibrational modes on the other. Molecular relaxation effects occur also in liquids, although no doubt with a different mechanism, and are known to lead to additional attenuation of sound waves of high frequency. However, the data are scanty and uncertain, and we shall simply acknowledge the existence of relaxation effects in liquids under certain conditions.

1.9. Conditions at a boundary between two media

The conditions occurring at a boundary between a fluid and some other medium warrant special consideration, since they play a part in the dynamical problems described later and since they give rise directly to several important phenomena. The boundary may separate two different phases, solid, liquid or gaseous, or it may separate two media of the same phase but different constitution.

As noted in § 1.5, two masses in contact and in thermodynamic equilibrium with each other have the same temperature, and any departure from equilibrium involving a difference between the temperatures of the two media is accompanied by the existence of a flux of heat across the boundary, the direction of the flux being such as to tend to bring the two media into equilibrium. Temperature is thus uniform across the boundary in an equilibrium state, just as it is uniform within each medium separately. The same is true of velocity, which also is the intensity of a conserved quantity (viz. momentum) which is transferred across the boundary as a result of interaction of the matter on the two sides. However, molecular constitution is in a special category inasmuch as there exist some types of boundary at which interaction of the matter on the two sides does not yield a tendency to uniformity of constitution. An obvious example is a liquid–solid interface; the molecules of the solid are bound in a lattice, and although some of the molecules of the liquid occasionally come within the force fields of the molecules of the solid, thereby transferring heat and momentum, they return to the liquid and produce no change in constitution. It is thus useful to postulate a boundary between two media in equilibrium with each other at which there may be a discontinuity in molecular structure and constitution. This is the type of boundary under consideration in this section.

Surface tension

The fact that small liquid drops in air and small gas bubbles in water take up a spherical form, and a host of other phenomena, may be accounted for by the hypothesis that a boundary between two media in equilibrium is the seat of a special form of energy of amount proportional to the area of the interface. Thermodynamical relations were presented in § 1.5 in terms of amounts of energy and work per unit mass of fluid, on the implicit assumption that the total amounts of energy and work are proportional to the volume of the fluid in all cases. We need now to make a correction in cases involving a mass of fluid of small volume to surface ratio, and to include a surface-dependent contribution.

The hypothesis that is found to fit all the facts is that an interface of area A in a system in equilibrium makes a contribution $A\gamma$ to the total Helmholtz free energy (§ 1.5) of the system, where the constant of proportionality γ is a function of state of the system; the total free energy of a system consisting of two uniform media of densities ρ_1 and ρ_2 and volumes V_1 and V_2, with interfacial area A, is then of the form

$$\rho_1 V_1 F_1 + \rho_2 V_2 F_2 + A\gamma, \qquad (1.9.1)$$

where F_1 and F_2 are, as in §1.5, the free energies per unit mass of the two media. It is a consequence of the definition of free energy that, in any small reversible isothermal change in the system, the total work done on the system is equal to the gain in total free energy. Thus, if the change in the system is such as to leave the densities, as well as the common temperature, of the two media unchanged, the total work done on the system is equal to $\gamma \, \delta A$. This expression for the work which must be done on the system to change only the area of the interface is exactly the same as if we had supposed that a film at the interface is in a state of uniform tension like a uniformly stretched membrane; indeed this is an equivalent form of the hypothesis made above. We see moreover that γ may be interpreted both as free energy per unit area of the interface and as the *surface tension*, the sense of the latter phrase being that across any line drawn on the interface there is exerted a force of magnitude γ per unit length in a direction normal to the line and tangential to the interface.

The molecular origin of the phenomenon of surface tension evidently lies in the intermolecular cohesive forces described in §1.1. The average free energy associated with a molecule of a medium is independent of its position, provided it is in the interior of the matter, but at distances from the bounding surface less than the range of action of the cohesive forces (of order 10^{-7} cm for simple molecules) the free energy is affected by proximity to the surface; and since this depth of action of the surface is so small, all parts of the surface contribute equally to the term in $(1.9.1)$ correcting the total free energy for the presence of the interface. When only one of the two media concerned is a condensed phase, it is easily seen that the parameter γ is likely to be positive. For the molecules of a liquid are subject to predominantly attractive forces of neighbouring molecules and the molecules that are near a boundary with a gas lack neighbours on one side and experience an unbalanced cohesive force directed away from the interface; this tendency for all liquid molecules near the interface to move inwards, consistent with a given total volume of the liquid, is equivalent to a tendency for the interface to *contract*. When the interface separates a liquid and either a solid or another liquid, the sign of γ is not predictable by this argument, and in reality both signs do occur.

Measured values of the surface tension γ for various pairs of fluids are given in appendix 1. For an interface separating air and pure water at $15\ ^\circ$C, $\gamma = 73\cdot5$ dyn/cm or erg/cm^2. The circumstances in which a surface tension of this magnitude is significant in its mechanical effects will depend in part on the nature of other forces acting on the system, but we can obtain some idea of its thermodynamic importance by noting that the surface energy of a spherical drop of pure water in air at $15\ ^\circ$C is equal to the latent heat of vaporization of the drop at that temperature for a drop radius of about 10^{-8} cm. For an interface between air and a liquid metal, γ is much larger, as might be expected in view of the larger density of this liquid. For an oil–water interface γ is typically positive and less than that for an air–water interface. For some other pairs of liquids, such as alcohol and water, an interface cannot be observed without special precautions because it is in compression (corresponding to a negative value of γ) and tends to become as large as possible, leading rapidly to complete mixing of the two liquids; only pairs of liquids for which $\gamma > 0$ are immiscible.

The surface tension for a given pair of media normally diminishes with increase of temperature. For a liquid in contact with its vapour, an empirical rule which is accurate over a wide range of temperature is that γ is proportional to $T - T_c$, where T_c is the critical temperature.

The value of γ at a liquid–fluid interface in equilibrium may be affected significantly by the presence of adsorbed (or 'surface-active') material at the surface of the liquid.[†] For mechanical reasons to be described below, a drop of lubricating oil placed on the free surface of water spreads out into a very thin layer covering the whole surface. The minute amounts of oil and grease and some other contaminating substances which inevitably are present in water under normal conditions likewise spread out over any free surface, and, as may be expected from the molecular origin of the phenomenon, have an appreciable effect on the surface tension. The normal effect of adsorbed contaminant molecules at a free water surface is to diminish[‡] the surface tension (essentially because the large contaminant molecules take up a preferred orientation in the surface layer, and exert on each other a repulsive force which partially balances the tension of the pure water surface) by an amount which increases with the surface concentration of the adsorbed material; at a free surface of ordinary tap water, the surface tension may be close to the value for pure water immediately after formation of the surface but usually falls quickly to something like half this value. The effect of contamination of a free surface of mercury is similar.

[†] For an extensive account of this important practical matter, see *Interfacial Phenomena*, by J. T. Davies and E. K. Rideal (*Academic Press*, 1961).

[‡] The usefulness of household detergents depends on this reduction in γ and the consequent increased ability of the water to 'wet' solid surfaces with which it comes into contact.

Inasmuch as the concentration of adsorbed material may vary over a liquid surface in some (non-equilibrium) circumstances, the surface tension may not be uniform and may give rise to unbalanced forces on an element of the surface. This may have dynamical consequences, as in the case of a small boat with a piece of camphor stuck at the back which propels itself on a dish of water.†

The mechanical properties of an interface between two fluids at which material has been adsorbed and which is not in equilibrium are not well understood. It is sometimes supposed that the surface is both elastic, exerting a tension which varies linearly with the strain (as is suggested by the fact that when a contaminated surface is stretched the concentration of adsorbed material falls and the surface tension rises, at any rate until more contaminant is adsorbed from the neighbouring liquid), and viscous, exerting frictional stresses which vary linearly with the rate of strain. In this book an interface between two media will normally be assumed to have only the equilibrium property of a uniform surface tension.

Equilibrium shape of a boundary between two stationary fluids

We turn now to a brief consideration of the consequences of the tension which exists at the boundary of a liquid. Only the case in which the boundary separates the liquid from another fluid need concern us, since only then is the boundary free to move. We suppose the two fluids to be stationary and in thermodynamic equilibrium, so that the tension γ is uniform over the interface. The problem is then to determine the geometrical form of the interface compatible with mechanical equilibrium. This proves to be quite difficult, except in a small number of special cases.

As a preliminary we note that a curved surface in a state of tension exerts a normal stress across the surface, as is apparent from experience with a stretched rubber sheet. For consider the effect of the tension near a point O of the surface, the tangent plane at which will be taken as the (x, y)-plane of a rectilinear co-ordinate system (x, y, z). The equation to the surface may then be written as

$$z - \zeta(x, y) = 0,$$

where the function ζ and its first derivatives are zero at O. At points near O the unit normal $\mathbf{n}$ to the surface has components

$$-\frac{\partial \zeta}{\partial x}, \quad -\frac{\partial \zeta}{\partial y}, \quad 1,$$

correct to the first order in the small quantities $\partial \zeta / \partial x$, $\partial \zeta / \partial y$. Now the resultant of the tensile forces exerted on a portion of the surface containing O is

$$-\gamma \oint \mathbf{n} \times d\mathbf{x},$$

---

† Many striking illustrations of the dynamical effect of non-uniform surface tension are shown in the film entitled *Surface Tension*, made by L. Trefethen under the auspices of the U.S. National Committee for Fluid Mechanics Films.

where $\delta\mathbf{x}$ is a line element of the closed curve bounding the portion of surface. For a portion of a plane surface (for which $\mathbf{n}$ is uniform) this is zero, the tension then being self-balancing; and for a portion of a curved surface of small area δA the resultant is clearly of smaller order than a linear dimension of the surface element. Correct to the second order in this linear dimension the resultant is a force parallel to the z-axis, i.e. parallel to the normal at O, of magnitude

$$-\gamma\oint\left(-\frac{\partial\zeta}{\partial x}\,dy+\frac{\partial\zeta}{\partial y}\,dx\right),\quad =\gamma\left(\frac{\partial^2\zeta}{\partial x^2}+\frac{\partial^2\zeta}{\partial y^2}\right)_0\delta A.$$

In other words, the tension exerted across the curve bounding the surface element is equivalent, in its effect on the surface element, to a *pressure* on the surface of magnitude

$$\gamma\left(\frac{\partial^2\zeta}{\partial x^2}+\frac{\partial^2\zeta}{\partial y^2}\right)_0,\quad =\gamma\left(\frac{1}{R_1}+\frac{1}{R_2}\right),$$

where R_1 and R_2 are the radii of curvature of the intercepts of the surface by two orthogonal planes containing the axis Oz. The sum $R_1^{-1}+R_2^{-1}$ is known to be independent of the particular choice of the orthogonal intercepts, and it is often convenient to take R_1 and R_2 as the principal radii of curvature. R_1 and R_2 must of course be regarded as quantities with an appropriate sign, the contribution to the equivalent pressure on the surface being directed towards the centre of curvature in each case.

Since the interface has zero mass (ideally), a curved interface can be in equilibrium only if the effective pressure due to surface tension is balanced by an equal and opposite difference between the pressures in the fluids on the two sides of the interface. Thus at any point of the interface there must be a jump in the fluid pressure of magnitude

$$\Delta p=\gamma\left(\frac{1}{R_1}+\frac{1}{R_2}\right) \tag{1.9.2}$$

when passing towards the side of the surface on which the centre of curvature lies.

A case in which the equilibrium shape of the interface is obvious is that of a mass of one fluid immersed in a second fluid, e.g. a mist droplet in air or a gas bubble in water. Provided that either the volume of the drop or bubble or the difference between the densities on the two sides of the interface is sufficiently small, we may ignore the effect of gravity. The pressure is then uniform in each fluid and the pressure jump (1.9.2) is constant over the interface. An unbounded surface with a constant sum of the principal curvatures is spherical, and this must be the equilibrium shape of the surface. This result also follows from the fact that in a state of (stable) equilibrium the energy of the surface must be a minimum consistent with a given value of the volume of the drop or bubble, and the sphere is the shape which has least surface area for given volume.

Suppose now that the interface separates a gas, in which the pressure may be regarded as constant, and a liquid of uniform density ρ, in which the pressure variation with height z due to gravity is given by the formula (1.4.12) for an incompressible fluid. The condition for equilibrium at any point of the interface is then

$$\rho g z - \gamma \left(\frac{1}{R_1} + \frac{1}{R_2} \right) = \text{const.,} \qquad (1.9.3)$$

R_1 and R_2 here being taken as positive when the respective centres of curvature are on the gas side of the interface. Equation (1.9.3) is difficult to solve for the surface shape, but it has the merit of showing that the only relevant parameter is $(\gamma/\rho g)^{\frac{1}{2}}$, with the dimensions of length. For pure water this parameter is about 0·27 cm at normal temperatures, which indicates the length scale on which effects of surface tension on the shape of an air–water interface are likely to be comparable with effects of gravity.

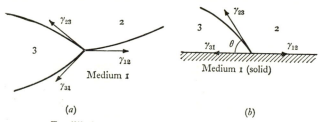

Figure 1.9.1. Equilibrium at the line of contact of three different media.

Liquid–gas interfaces to which (1.9.3) applies are necessarily open surfaces, and in practice will usually be bounded by a line along which three media are in contact—as when a drop of mercury rests on a table. The known properties of such a line of triple contact serve as boundary conditions in the integration of (1.9.3) for the surface shape. The line of contact is subject to the tensions of three different surfaces and, since it is without mass, the vector resultant of the three tensions must have zero component in any direction in which it is free to move (figure 1.9.1 *a*); if the direction of the normal of one of the three surfaces meeting at the line of contact be given, the other two directions are then determinate. In cases in which

$$|\gamma_{12}| > |\gamma_{23}| + |\gamma_{31}|,$$

it is evident that the conditions for equilibrium at the line of contact cannot be satisfied. Thus it is possible for a lens-shaped globule of fat to float on the surface of soup, but, in the case of a drop of mineral oil at the free surface of water, the air–water interfacial tension is too strong for the tensions of the two oil surfaces and the drop is pulled out indefinitely until either it covers the whole surface or the thickness of the oil layer reaches molecular dimensions. Likewise petrol or water containing a 'wetting' agent cannot form an isolated drop on some solid surfaces and spreads out in a very thin layer.

5

When one of the three media is a solid,† say medium 1, the local surface of which will normally be a plane (figure 1.9.1 *b*), the line of contact is free to move only in a direction parallel to the solid surface. The single scalar condition for equilibrium is then

$$\gamma_{12} = \gamma_{31} + \gamma_{23}\cos\theta,$$

which determines the angle of contact θ. When medium 2 is air and medium 3 is a liquid, the liquid is sometimes said to 'wet' the solid if $\theta < \frac{1}{2}\pi$ (as it is for pure water on most solids, such as glass, unlike mercury which has a contact angle of about 150° on many solids), although there is no special significance in the value $\frac{1}{2}\pi$ and it is more appropriate to regard the degree of wetting as increasing as θ decreases to zero.

The complete problem of determining an interface shape may now be illustrated by the case of a free liquid meeting a plane vertical rigid wall, sketched in figure 1.9.2. In this two-dimensional field, the equation to the

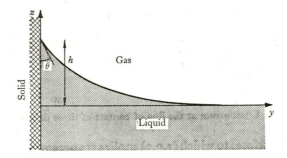

Figure 1.9.2. A free liquid surface meeting a vertical plane wall.

liquid–gas interface is $z = \zeta(y)$ and the principal curvatures of the interface are

$$\frac{1}{R_1} = 0, \quad \frac{1}{R_2} = \frac{\zeta''}{(1+\zeta'^2)^{\frac{3}{2}}},$$

where the dashes denote differentiation with respect to y. Equation (1.9.3) is then

$$\frac{\rho g}{\gamma}\zeta - \frac{\zeta''}{(1+\zeta'^2)^{\frac{3}{2}}} = 0,$$

the constant on the right-hand side being zero because the interface becomes plane far from the wall, where $\zeta = 0$. One integration gives

$$\frac{1}{2}\frac{\rho g}{\gamma}\zeta^2 + \frac{1}{(1+\zeta'^2)^{\frac{1}{2}}} = C,$$

† The concept of a tension acting in the surface of a solid meets some difficulties, but an equivalent and rigorous argument in terms of surface energy can be given; see *Interfacial Phenomena*, by Davies & Rideal.

and the same boundary condition shows that $C = 1$. It follows that the height to which the liquid climbs at the rigid wall is given by

$$h^2 = 2\frac{\gamma}{\rho g}(1 - \sin\theta), \tag{1.9.4}$$

the contact angle θ for the liquid being known from the properties of the media. The boundary condition $y = 0$, $\zeta = h$ may now be used to determine the constant in the second integration, the result being

$$\frac{y}{d} = \cosh^{-1}\frac{2d}{\zeta} - \cosh^{-1}\frac{2d}{h} + \left(4 - \frac{h^2}{d^2}\right)^{\frac{1}{2}} - \left(4 - \frac{\zeta^2}{d^2}\right)^{\frac{1}{2}}, \tag{1.9.5}$$

where $d^2 = \gamma/\rho g$.

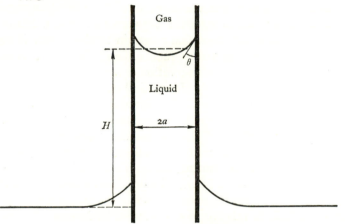

Figure 1.9.3. Capillary rise of liquid in a small tube.

The fact that the free surface of a liquid rises or falls to meet a rigid wall (depending on the inclination of the wall to the vertical and on the contact angle) is the basis of the phenomena, known generally as *capillarity*, which manifest themselves in small tubes and crevices. Consider, for instance, a circular tube of small radius a containing liquid with a free surface (figure 1.9.3). The liquid surface meets the wall at the contact angle θ, and it is evident that, when $a \ll d$, the radius of curvature of an axial section of the free surface is approximately uniform and equal to $a/\cos\theta$ (the departure of the surface from a spherical form being due only to the relatively small variation of liquid pressure over the surface due to gravity). The tension in this highly curved surface causes a large jump in pressure across the interface, and, if the tube is open and dips vertically into a larger free surface of the liquid, a considerable column of liquid will be supported in the tube against gravity. The condition for equilibrium of a column of height H is approximately

$$\rho g H = \Delta p = \frac{2\gamma\cos\theta}{a},$$

i.e.
$$H = \frac{2d^2 \cos\theta}{a}.$$
(1.9.6)

Thus H may be very large in the case of the very small tubes in porous materials such as blotting paper, bricks or soil, which are known to exert a strong 'suction' on a 'wetting' fluid like water. In the case of a liquid which does not 'wet' the wall of the tube, $\theta > \frac{1}{2}\pi$ and $H < 0$, corresponding to a depression of the free surface in a tube. Note that when the tube is not vertical, (1.9.6) gives the vertical displacement of the free surface.

Transition relations at a material boundary

We note here for future use a number of relations between conditions on the two sides of a material interface between two media. Many of these relations amount to a statement that a certain local quantity is continuous across the interface, either as a consequence of equilibrium (exact or approximate) or as a consequence of conservation of some quantity.

There is first of all the purely kinematical condition associated with the fact that, unless rupture occurs at the interface, the boundary remains a material surface for both media. The component of velocity locally normal to the boundary must be continuous across the boundary.

For two media in contact with each other at a boundary which allows transport of heat and momentum through molecular interaction at the boundary (as virtually all real boundaries do), both temperature and velocity must be continuous across the boundary when the media are in equilibrium.

However, a fluid in relative motion cannot be in exact thermodynamic and mechanical equilibrium, and it is necessary to ask if the departure from equilibrium may be accompanied by a discontinuity in temperature or velocity at a boundary between two media. As noted in § 1.6, the spatial gradient of a quantity like temperature or velocity provides one measure of the local departure from equilibrium; and an internal discontinuity in such a quantity would constitute a violent departure from equilibrium. The effect of the transport of heat or momentum accompanying a departure from equilibrium is to tend to make the temperature or velocity uniform, and to do so at a rate which increases with the magnitude of the departure from equilibrium. We may therefore expect that quantities to which the transport relations apply are continuous everywhere in a fluid in most real non-equilibrium situations. Molecular migration and interaction are likely to be as effective in equalizing the local temperatures or velocities of two different media at an interface as in equalizing the temperatures or velocities at two neighbouring points in a fluid, and approximate equilibrium will be established everywhere. All the available evidence does show that, under common conditions of moving fluids, temperature and velocity (both tangential and normal components) are continuous across a material boundary between a fluid and another medium.

In the particular case of a liquid–gas interface, there is also the possibility of mass transport across the boundary, by evaporation of the liquid, the result of which is to produce, in approximate equilibrium, not continuity of constitution at the interface but a jump from the stable liquid phase to a gas 'saturated' with vapour.

The conservation properties that are associated with transport relations also yield boundary conditions, when the departures from equilibrium are small in the sense of § 1.6. For consider the balance of heat for a small right cylinder whose generators are parallel to the local normal **n** to a boundary between two media and whose end faces lie one in each medium. If the length of the cylinder is made much smaller than any lateral dimension, conservation of heat requires equality of the fluxes across the two end faces, that is,

$$(k_H \mathbf{n}.\nabla T)_{\text{medium 1}} = (k_H \mathbf{n}.\nabla T)_{\text{medium 2}} \qquad (1.9.7)$$

at each point of the boundary. The values of the thermal conductivity k_H may be different on the two sides of the boundary, and (1.9.7) in general implies a discontinuity in the temperature gradient $\mathbf{n}.\nabla T$ across the boundary.

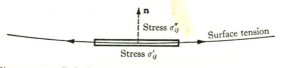

Figure 1.9.4. Relation between the stresses on the two sides of a boundary between two fluids.

Similar considerations apply to the flux of momentum across a boundary between two fluids, although it is necessary here to allow for the effect of surface tension. We have not yet established a general expression for the momentum transport in a moving fluid, but we can write down our boundary condition in terms of the stress tensor σ_{ij} described in § 1.3. When the length of the same right cylinder is reduced to zero, the sum of the forces exerted on the two end faces of the cylinder must be balanced by the resultant tensile force exerted on the cylinder by the part of the interface outside the cylinder (figure 1.9.4). As already seen, this resultant tensile force is equivalent (when the surface tension is uniform) to a pressure on the interface towards the centre of curvature, so that the boundary condition is

$$\sigma''_{ij} n_j - \sigma'_{ij} n_j = -\gamma \left(\frac{1}{R_1} + \frac{1}{R_2} \right) n_i, \qquad (1.9.8)$$

where R_1 and R_2 are the radii of curvature of the interface in any two orthogonal planes containing **n**, being reckoned here as positive when the corresponding centre of curvature lies on the side of the interface to which **n** points. When the two fluids are stationary, the stress tensor has the purely normal form $-p\,\delta_{ij}$, and (1.9.8) reduces to the simpler relation (1.9.2).

In more general circumstances in which the surface tension varies over an interface between two fluids, owing to non-uniformity of either the temperature or (more commonly) the concentration of adsorbed material at the interface, there is a resultant tangential force on an element of the interface due to surface tension. It is not difficult to show that a term $(\nabla\gamma)_i$ must then be added to the right-hand side of (1.9.8), where $\nabla\gamma$ is the vector gradient of γ in the interfacial surface. Such a tangential force on the interface cannot be balanced by stresses when the two fluids are stationary.

Exercises

1. Two spherical soap bubbles of radii a_1 and a_2 are made to coalesce. Show that the radius r of the resulting soap bubble is given by

$$p_0 r^3 + 4\gamma r^2 = p_0(a_1^3 + a_2^3) + 4\gamma(r_1^2 + r_2^2),$$

where p_0 is the ambient pressure and γ is the tension of the air–liquid surfaces.

2. A rigid sphere of radius a rests on a flat rigid surface, and a small amount of liquid surrounds the point of contact making a concave-planar lens whose diameter is small compared with a. The angle of contact of the liquid with each of the solid surfaces is zero, and the tension in the air–liquid surface is γ. Show that there is an adhesive force of magnitude $4\pi a\gamma$ acting on the sphere. (The fact that this adhesive force is independent of the volume of liquid is noteworthy.)

3. A number of small solid bodies are floating on the surface of a liquid. Show that the effect of surface tension is to make two neighbouring bodies approach each other, either when both bodies are wet by the liquid or when neither body is wet, and to make them move away from each other when one is wet by the liquid and the other is not.

Further reading relevant to chapter 1

The Mechanical Properties of Matter, by A. H. Cottrell (John Wiley & Sons, 1964).

2

KINEMATICS OF THE FLOW FIELD

2.1. Specification of the flow field

The continuum hypothesis enables us to use the simple concept of local velocity of the fluid, and we must now consider how the whole field of flow may be specified as an aggregate of such local velocities. Two distinct alternative kinds of specification are possible. The first, usually called the Eulerian type, is like the specification of an electromagnetic field in that the flow quantities are defined as functions of position in space ($\mathbf{x}$) and time (t). The primary flow quantity is the (vector) velocity of the fluid, which is thus written as $\mathbf{u}(\mathbf{x}, t)$. This Eulerian specification can be thought of as providing a picture of the spatial distribution of fluid velocity (and of other flow quantities such as density and pressure) at each instant during the motion.

The second, or Lagrangian type of specification, makes use of the fact that, as in particle mechanics, some of the dynamical or physical quantities refer not only to certain positions in space but also (and more fundamentally) to identifiable pieces of matter. The flow quantities are here defined as functions of time and of the choice of a material element of fluid, and describe the dynamical history of this selected fluid element. Since material elements of fluid change their shape as they move, we need to identify the selected element in such a way that its linear extension is not involved; one suitable method is to specify the element by the position ($\mathbf{a}$) of its centre of mass at some initial instant (t_0), on the understanding that the initial linear dimensions of the element are so small as to guarantee smallness at all relevant subsequent instants in spite of distortions and extensions of the element. Thus the primary flow quantity according to the Lagrangian specification is the velocity $\mathbf{v}(\mathbf{a}, t)$.

The Lagrangian type of specification is useful in certain special contexts, but it leads to rather cumbersome analysis and in general is at a disadvantage in not giving directly the spatial gradients of velocity in the fluid. We shall not need to use it in any systematic way, and it will be taken for granted in the following pages that an Eulerian specification is being employed. Nevertheless, the notion of *material* volumes, surfaces and lines which consist always of the same fluid particles and move with them is indispensable, and will often be employed within the framework of an Eulerian specification of the flow field.

The function $\mathbf{u}(\mathbf{x}, t)$ will thus be the primary dependent variable in our analysis, and other flow quantities such as pressure will likewise be regarded as being functions of $\mathbf{x}$ and t.

When **u** is independent of t, the flow is said to be *steady*.

A line in the fluid whose tangent is everywhere parallel to **u** instantaneously is a line of flow, or a *streamline*; the family of streamlines at time t are solutions of

$$\frac{dx}{u(\mathbf{x}, t)} = \frac{dy}{v(\mathbf{x}, t)} = \frac{dz}{w(\mathbf{x}, t)}, \qquad (2.1.1)$$

where u, v, w are the components of **u** parallel to rectilinear axes and x, y, z are the components of **x**. When the flow is steady, the streamlines have the same form at all times. A related concept is a *stream-tube*, which is the surface formed instantaneously by all the streamlines that pass through a given closed curve in the fluid.

The *path* of a material element of fluid does not in general coincide with a streamline, although it does so when the motion is steady. In addition to a streamline and a path line, it is useful for observational purposes to define a *streak line*, on which lie all those fluid elements that at some earlier instant passed through a certain point of space; thus when dye or some other marking material is discharged slowly at some fixed point in a moving fluid, the visible line produced in the fluid is a streak line. When the flow is steady, streak lines, streamlines and path lines all coincide.

A flow field is said to be *two-dimensional* when the velocity $\mathbf{u}(\mathbf{x}, t)$ is everywhere at right angles to a certain direction and independent of displacements parallel to that direction. Thus it is always possible to choose rectilinear co-ordinates (x, y, z) so that the components of **u** in a two-dimensional flow are $(u, v, 0)$, where u and v are independent of z. A flow field is said to be *axisymmetric* when the velocity components (u, v, w) with respect to cylindrical co-ordinates (x, σ, ϕ) (with a suitable choice of the direction of the line $\sigma = 0$) are all independent of the azimuthal angle ϕ. In some axisymmetric flow fields, the azimuthal or 'swirl' component of velocity w is zero everywhere and the velocity vector lies in a plane through the axis of symmetry. In some others, w is the only non-zero component and all the streamlines are circles about the axis of symmetry.

Differentiation following the motion of the fluid

It will be evident that in a steady flow field a material element of fluid may nevertheless experience acceleration through moving to a position where **u** has a different value. The derivative $\partial \mathbf{u}/\partial t$ is not the acceleration of an element at position **x** at time t, because the element is at that position only instantaneously. The correct expression for the acceleration of a material element may be found by noting that an element at position **x** at time t is at position $\mathbf{x} + \mathbf{u}\,\delta t$ at time $t + \delta t$, and that the change in its velocity in the small interval δt is

$$\mathbf{u}(\mathbf{x} + \mathbf{u}\,\delta t, t + \delta t) - \mathbf{u}(\mathbf{x}, t), \quad = \delta t \left(\frac{\partial \mathbf{u}}{\partial t} + \mathbf{u}.\nabla \mathbf{u} \right) + O(\delta t^2).$$

Thus the acceleration of an element of fluid at $(\mathbf{x}, t)$ is

$$\frac{\partial \mathbf{u}}{\partial t} + \mathbf{u} . \nabla \mathbf{u}. \qquad (2.1.2)$$

(The acceleration is of course a quantity which can be expressed very simply in the Lagrangian type of specification of the flow field; if $\mathbf{v}$ is the velocity of a certain fluid element, the acceleration of the element is $\partial \mathbf{v}/\partial t$.)

Similar considerations may be applied to any other dynamical or physical point quantity, θ say, which is specified as a function of $\mathbf{x}$ and t in the above way and which represents a property of the fluid that is located at position $\mathbf{x}$ at time t; θ might be a scalar quantity, such as the local density or temperature of the fluid, or a vector quantity, such as the local effective angular velocity of the fluid. $\partial \theta/\partial t$ is the *local* rate of change due to temporal changes at position $\mathbf{x}$, and to find the rate of change of θ for a material element we must add the *convective* rate of change $\mathbf{u} . \nabla \theta$ due to transport of the element to a different position.

It is convenient to introduce the notation

$$\frac{D}{Dt} = \frac{\partial}{\partial t} + \mathbf{u} . \nabla, \qquad (2.1.3)$$

so that in particular the acceleration of a fluid element may be written as $D\mathbf{u}/Dt$. The operator D/Dt has meaning only when applied to a field variable (that is, a function of $\mathbf{x}$ and t), and is said to give a time derivative following the motion of the fluid, or a material derivative. It makes a frequent appearance in differential equations expressing conservation laws, the first example of which is conservation of fluid mass (§2.2).

If a material surface in the fluid is specified geometrically by the equation

$$F(\mathbf{x}, t) = \text{constant},$$

F is a quantity which is invariant for a fluid particle on the surface, so that

$$\frac{DF}{Dt} = 0. \qquad (2.1.4)$$

In particular, the equation to any surface bounding the fluid must satisfy (2.1.4).

2.2. Conservation of mass

The requirement of conservation of mass of the fluid imposes certain restrictions on the velocity field, and although these restrictions are not strictly 'kinematical' it is convenient to consider them at this stage. Sometimes the character of the flow is such that the consequences of conservation of mass can be recognized directly, as for instance when the flow has spherical symmetry or is effectively one-dimensional, but in many cases the following general condition in the form of a differential equation will be needed.

Consider a closed surface A whose position is fixed relative to the co-ordinate axes and which encloses a volume V entirely occupied by fluid. If ρ is the density of the fluid at position $\mathbf{x}$ and time t, the mass of fluid enclosed by the surface at any instant is $\int \rho \, dV$ and the net rate at which mass is flowing outwards across the surface is $\int \rho \mathbf{u} . \mathbf{n} \, dA$, where δV and δA are elements of the enclosed volume and of the area of the surrounding surface, the latter having the unit outward normal $\mathbf{n}$. Conservation of mass of the fluid requires

$$\frac{d}{dt} \int \rho \, dV = - \int \rho \mathbf{u} . \mathbf{n} \, dA,$$

which, on differentiation under the integral sign (remembering that the volume V is fixed in space) and transformation of the surface integral, may be written as

$$\int \left\{ \frac{\partial \rho}{\partial t} + \nabla . (\rho \mathbf{u}) \right\} dV = 0. \qquad (2.2.1)$$

This relation (2.2.1) is valid for all choices of the volume V, provided only that it lies entirely in the fluid; and so the integrand, if continuous in $\mathbf{x}$, must be identically zero everywhere in the fluid. Hence

$$\frac{\partial \rho}{\partial t} + \nabla . (\rho \mathbf{u}) = 0 \qquad (2.2.2)$$

at all points in the fluid when the quantity on the left-hand side is a continuous function of position. This latter restriction is of little importance, since discontinuities in ρ or $\mathbf{u}$ will appear in our analysis only at isolated points, lines or surfaces, in which case (2.2.2) holds everywhere except possibly at these points, lines or surfaces.

The mass-conservation equation (2.2.2) is one of the fundamental equations of fluid mechanics. It has been called the 'equation of continuity' for many years, although not for any evident good reason.

A different form of the equation is obtained by expanding the divergence term and noting from (2.1.3) that two of the terms together make up a material derivative:

$$\frac{1}{\rho} \frac{D\rho}{Dt} + \nabla . \mathbf{u} = 0. \qquad (2.2.3)$$

In this form the equation may be interpreted in terms of the changes in the volume of a given mass of fluid. The volume τ of a material body of fluid changes as a result of movement of each element $\mathbf{n} \, \delta S$ of the bounding material surface (where $\mathbf{n}$ is the outward normal vector),† and

$$\frac{d\tau}{dt} = \int \mathbf{u} . \mathbf{n} \, dS,$$

$$= \int \nabla . \mathbf{u} \, d\tau$$

† In order to distinguish between geometrical and material elements, δV, $\mathbf{n} \delta A$, $\delta \mathbf{x}$ will be used as consistently as possible throughout this book to denote elements of volume, surface and line defined by their (fixed) positions in space and $\delta \tau$, $\mathbf{n} \delta S$, $\delta \mathbf{l}$ to denote material elements of volume, surface and line which move with the fluid.

on use of the divergence theorem. Hence the rate at which the volume of a material element† instantaneously enclosing the point $\mathbf{x}$ is changing, divided by that volume, is

$$\lim_{\tau \to 0} \frac{1}{\tau} \frac{d\tau}{dt} = \lim_{\tau \to 0} \frac{1}{\tau} \int \nabla . \mathbf{u} \, d\tau = \nabla . \mathbf{u}.$$

This fractional rate of change of the volume of a material element is called the (local) *rate of expansion* or rate of dilatation, and will sometimes be denoted by the single symbol Δ. The mass-conservation equation in the form (2.2.3) is then seen to be equivalent to the statement that the fractional rates of change of density and volume of a material element of fluid are equal in magnitude and opposite in sign, which could well have been made a starting-point for a derivation of the equation.

A fluid is said to be *incompressible* when the density of an element of fluid is not affected by changes in the pressure. We shall see that the pressure variations in some common flow fields are such a small fraction of the absolute pressure that even gases may behave as if they were almost completely incompressible. The density of the fluid in a mass element may also change as a consequence of molecular conduction of heat (or, rarely, of a solute) into the element; however, circumstances in which the effect of conduction of heat in the fluid is negligible are common, and a statement that a fluid is effectively incompressible is usually taken to imply, in the absence of any explicit qualification about heat conduction, that the density of each mass element of the fluid remains constant (see § 3.6). Thus, for an incompressible fluid, the rate of change of ρ following the motion is zero, that is,

$$\frac{D\rho}{Dt} = 0. \tag{2.2.4}$$

The mass-conservation equation then takes the simple form

$$\nabla . \mathbf{u} = 0. \tag{2.2.5}$$

In this case the rate of expansion is everywhere zero, and, as explained in books on vector analysis, a stream-tube cannot end in the interior of the fluid; it must either be closed, or end on the boundary of the fluid, or extend 'to infinity'. A vector $\mathbf{u}$ having zero divergence is said to be *solenoidal*.

Use of a stream function to satisfy the mass-conservation equation

In the cases of flow of an incompressible fluid, and of steady flow of a compressible fluid, the mass-conservation equation (2.2.2) reduces to the statement that a vector divergence is zero, the divergences being of $\mathbf{u}$ and $\rho\mathbf{u}$ respectively. If we impose the further restriction that the flow field either is two-dimensional or has axial symmetry, this vector divergence is the sum of

† The word 'element' is used here and elsewhere to imply infinitesimal size and (usually) a passage to an appropriate limit.

only two derivatives, and the mass-conservation equation can then be regarded as defining a scalar function from which the components of **u** or $\rho\mathbf{u}$ are obtained by differentiation. The procedure will be described here for the case of an incompressible fluid.

Assume first that the motion is two-dimensional, so that $\mathbf{u} = (u, v, 0)$ and u and v are independent of z. The mass-conservation equation for an incompressible fluid then has the form

$$\frac{\partial u}{\partial x} + \frac{\partial v}{\partial y} = 0, \tag{2.2.6}$$

from which it follows that $u\,\delta y - v\,\delta x$ is an exact differential, equal to $\delta\psi$ say. Then

$$u = \frac{\partial \psi}{\partial y}, \quad v = -\frac{\partial \psi}{\partial x}, \tag{2.2.7}$$

and the unknown scalar function $\psi(x, y, t)$ is defined by

$$\psi - \psi_0 = \int (u\,dy - v\,dx), \tag{2.2.8}$$

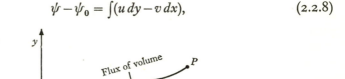

Figure 2.2.1. Calculation of the flux of fluid volume across a curve joining the reference point O to the point P with co-ordinates (x, y).

where ψ_0 is a constant and the line integral is taken along an arbitrary curve joining some reference point O to the point P with co-ordinates x, y. In this way we have used the mass-conservation equation to replace the two dependent variables u, v by the single dependent variable ψ, which is a very valuable simplification in many cases of two-dimensional flow.

The physical content of the above argument also proves to be of interest. The flux of fluid volume across a curve joining the points O and P in the (x, y)-plane (by which is meant the flux across the open surface swept out by translating this curve through unit distance in the z-direction), the flux being reckoned positive when it is in the anti-clockwise sense about P, is given exactly by the right-hand side of (2.2.8) (see figure 2.2.1). Now the flux of volume across the closed curve formed from any two different paths joining O to P is necessarily zero when the region between the two paths is wholly occupied by incompressible flow. The flux represented by the integral in (2.2.8) is therefore independent of the choice of the path joining

O to P, provided it is one of a set of paths of which any two enclose only incompressible fluid, and therefore defines a function of the position of P, which we have written as $\psi - \psi_0$.

Since the flux of volume across any curve joining two points is equal to the difference between the values of ψ at these two points, it follows that ψ is constant along a streamline, as is also apparent from (2.2.7) and the equation (2.1.1) that defines the streamlines. ψ is termed the *stream function*, and is associated (in this case of two-dimensional flow) with the name of Lagrange. The function ψ can also be regarded as the only non-zero component of a 'vector potential' for $\mathbf{u}$ (analogous to the vector potential of the magnetic induction, which also is a solenoidal vector, in electromagnetic field theory), since (2.2.7) can be written as

$$\mathbf{u} = \nabla \times \mathbf{B}, \quad \mathbf{B} = (0, 0, \psi). \tag{2.2.9}$$

It is common practice in fluid mechanics to provide a picture of a flow field by drawing various streamlines, and if these lines are chosen so that the two values of ψ on every pair of neighbouring streamlines differ by the same amount, ϵ say, the eye is able to perceive the way in which the velocity magnitude q, as well as its direction, varies over the field, since

$$q \approx \epsilon/(\text{distance between neighbouring streamlines}).$$

Examples of families of streamlines describing two-dimensional flow fields, with equal intervals in ψ between all pairs of neighbouring streamlines, will be found in figures 2.6.2 and 2.7.2.

Expressions for the velocity components parallel to any orthogonal co-ordinate lines in terms of ψ may be obtained readily, either by the use of (2.2.9) or with the aid of the relation between ψ and the volume flux 'between two points'. For flow referred to polar co-ordinates (r, θ), we find, by evaluating the flux between pairs of neighbouring points on the r- and θ-co-ordinate lines and equating it to the corresponding increments in ψ (allowance being made for the signs in the manner required by (2.2.8)), that

$$u_r = \frac{1}{r} \frac{\partial \psi}{\partial \theta}, \quad u_\theta = -\frac{\partial \psi}{\partial r}. \tag{2.2.10}$$

The reader may find useful the general rule for two-dimensional flow, that differentiation of ψ in a certain direction gives the velocity component $90°$ in the clockwise sense from that direction.

Finally, for this case of two-dimensional flow of incompressible fluid, we should note the possibility that ψ is a many-valued function of position. For suppose that across some closed inner geometrical boundary there is a net volume flux m; this flux might be due to an effective creation of fluid within the inner boundary (as when a tube discharges fluid into this region) or to change of volume of the part of the enclosed region not occupied by the fluid (as when a gaseous cavity surrounded by water expands or contracts).

If now we choose two different paths joining the two points O and P which together make up a closed curve enclosing the inner boundary, the fluxes of volume across the two joining curves differ by an amount m (or, more generally, by pm, where p is the number of times the combined closed curve passes round the inner boundary). The value of $\psi - \psi_0$ at the point P thus depends on the choice of path joining it to the reference point O, and may take any one of a number of values differing by multiples of m. This kind of many-valuedness of a scalar function related to the velocity distribution in a region which is not singly-connected will be described more fully in §2.8. It is not confined to two-dimensional flow, although that is the context in which it occurs most often.

If now the flow has symmetry about an axis, the mass-conservation equation for an incompressible fluid takes the form

$$\nabla \cdot \mathbf{u} = \frac{\partial u}{\partial x} + \frac{1}{\sigma} \frac{\partial(\sigma v)}{\partial \sigma} = 0$$

in terms of cylindrical co-ordinates† (x, σ, ϕ) with corresponding velocity components (u, v, w), the axis of symmetry being the line $\sigma = 0$. This relation ensures that $\sigma u\, \delta\sigma - \sigma v\, \delta x$ is an exact differential, equal to $\delta\psi$ say. Then

$$u = \frac{1}{\sigma} \frac{\partial\psi}{\partial\sigma}, \quad v = -\frac{1}{\sigma} \frac{\partial\psi}{\partial x}, \tag{2.2.11}$$

and the function $\psi(x, \sigma, t)$ is defined by

$$\psi - \psi_0 = \int \sigma(u\, d\sigma - v\, dx), \tag{2.2.12}$$

where the line integral is taken along an arbitrary curve in an axial plane joining some reference point O to the point P with co-ordinates (x, σ). It will be noticed that the azimuthal component of velocity w does not enter into the mass-conservation equation in a flow field with axial symmetry and cannot be obtained from ψ.

Again it is possible to interpret ψ both as a measure of volume flux and as one component of a vector potential. The flux of fluid volume across the surface formed by rotating an arbitrary curve joining O to P in an axial plane, about the axis of symmetry, the flux again being reckoned as positive when it is in the anti-clockwise sense about P, is 2π times the right-hand side of (2.2.12). Lines in an axial plane on which ψ is constant are everywhere parallel to the vector $(u, v, 0)$, and can be described as 'streamlines of the flow in an axial plane'. ψ is here termed the *Stokes stream function*. A sketch of lines on which ψ is constant, with the same increment in ψ between all pairs of neighbouring lines (see figure 2.5.2 for an example), does not give quite as direct an impression of the distribution of velocity magnitude here

† Expressions for the divergence and other vector operators in terms of general orthogonal curvilinear co-ordinates, and cylindrical co-ordinates in particular, will be found in appendix 2.

as in two-dimensional flow, owing to the occurrence of the factor $1/\sigma$ in the expressions for u and v in (2.2.11). The relations (2.2.11) are readily seen to be equivalent to

$$\mathbf{u} = \nabla \times \mathbf{B}, \quad B_\phi = \psi/\sigma, \qquad (2.2.13)$$

the components of the vector potential $\mathbf{B}$ referred to cylindrical co-ordinate lines here being independent of the azemuthal angle ϕ.

The relation between ψ and the volume flux 'between two points' may be used to obtain expressions for the velocity components referred to other orthogonal systems of co-ordinates in terms of ψ. For instance, for flow with axial symmetry referred to spherical polar co-ordinates (r, θ, ϕ), we find, by evaluating the flux between pairs of neighbouring points on the r- and θ-co-ordinate lines and equating it to 2π times the corresponding increments in ψ (allowance being made for the signs in the manner required by (2.2.12)), that

$$u_r = \frac{1}{r^2 \sin\theta} \frac{\partial \psi}{\partial \theta}, \quad u_\theta = -\frac{1}{r \sin\theta} \frac{\partial \psi}{\partial r}. \qquad (2.2.14)$$

With this co-ordinate system, the vector potential for the velocity has the azimuthal component

$$\mathbf{B}_\phi = \frac{\psi}{r \sin\theta}. \qquad (2.2.15)$$

Exercise

At time t_0 the position of a material element of fluid has Cartesian co-ordinates (a, b, c) and the density of the fluid is ρ_0. At a subsequent time t the position co-ordinates and density of the element are (X, Y, Z) and ρ. Show that with this Lagrangian specification of the flow field the equation of mass conservation is

$$\frac{\partial(X, Y, Z)}{\partial(a, b, c)} = \frac{\rho_0}{\rho}.$$

2.3. Analysis of the relative motion near a point

The force exerted by one portion of fluid on an adjacent portion depends on the way in which the fluid is being deformed by the motion, and it is necessary, as a preliminary to dynamical considerations, to make an analysis of the character of the motion in the neighbourhood of any point. This analysis is similar to that used in the theory of local deformation of an elastic solid, rate of strain and rate of rotation of the fluid taking the place of total strain and total rotation of the solid.

The velocity of the fluid at position $\mathbf{x}$ and time t is $\mathbf{u}(\mathbf{x}, t)$, and the simultaneous velocity at a neighbouring position $\mathbf{x}+\mathbf{r}$ is $\mathbf{u}+\delta\mathbf{u}$, where, for rectangular co-ordinates,

$$\delta u_i = r_j \frac{\partial u_i}{\partial x_j} \qquad (2.3.1)$$

correct to the first order in the small distance r between the two points. The geometrical character of the relative velocity $\delta\mathbf{u}$, regarded as a (linear) function of $\mathbf{r}$, can be recognized by decomposing $\partial u_i/\partial x_j$, which is a second-order tensor, into parts which are symmetrical and anti-symmetrical in the suffices i and j. Thus we write

$$\delta u_i = \delta u_i^{(s)} + \delta u_i^{(a)},$$

where $\qquad \delta u_i^{(s)} = r_j e_{ij}, \quad \delta u_i^{(a)} = r_j \xi_{ij},$ (2.3.2)

and $\qquad e_{ij} = \frac{1}{2}\left(\frac{\partial u_i}{\partial x_j} + \frac{\partial u_j}{\partial x_i}\right), \quad \xi_{ij} = \frac{1}{2}\left(\frac{\partial u_i}{\partial x_j} - \frac{\partial u_j}{\partial x_i}\right).$ (2.3.3)

The two terms $\delta u_i^{(s)}$ and $\delta u_i^{(a)}$ make distinct and basically different contributions to the relative velocity which we proceed to interpret in turn.

The first contribution may evidently be written as

$$\delta u_i^{(s)} = r_j e_{ij} = \frac{\partial \Phi}{\partial r_i},$$ (2.3.4)

where $\qquad \Phi = \frac{1}{2} r_k r_l e_{kl},$ (2.3.5)

since e_{ij} is a symmetrical second-order tensor. The surfaces on which Φ, regarded as a function of $\mathbf{r}$, is constant form a family of similar quadrics, and $\delta \mathbf{u}^{(s)}$ is parallel to the local normal to the quadric passing through position $\mathbf{r}$. The nature of this contribution to $\delta \mathbf{u}$ becomes clearer if we choose the directions of the orthogonal axes of reference so that the non-diagonal elements of e_{ij} are zero, as is always possible. The axes of reference then coincide with the principal axes of the tensor e_{ij} and of the family of quadrics, and

$$\Phi = \frac{1}{2}(ar_1'^2 + br_2'^2 + cr_3'^2),$$ (2.3.6)

where r_1', r_2', r_3' are the components of $\mathbf{r}$ referred to the new axes. a, b and c are the diagonal components of the tensor e_{ij}' obtained from the general transformation formula

$$e_{ij}' = \frac{\partial r_k}{\partial r_i'}\frac{\partial r_l}{\partial r_j'} e_{kl},$$ (2.3.7)

and satisfy the invariant relation

$$a+b+c = e_{ii}' = e_{ii} = \frac{\partial u_i}{\partial x_i}.$$ (2.3.8)

The contribution $\delta \mathbf{u}^{(s)}$ to the relative velocity has the three components (ar_1', br_2', cr_3') with reference to the new axes. Thus any material line element near position $\mathbf{x}$ which is parallel to the r_1'-axis (so that for all points on the line element the values of r_2' and r_3' are the same) continues to have that orientation and is being stretched at a rate $e_{11}', = a$. Likewise all material line elements parallel to the r_2'- and r_3'-axes are being stretched at rates b and c without rotation (so far as the contribution $\delta \mathbf{u}^{(s)}$ alone is concerned). Material line elements not parallel to any one of the r_1'-, r_2'-, r_3'-axes in general experience both stretching and rotation, but only to the extent that is needed for consistency with the pure stretching motion for line elements parallel to any one of the orthogonal axes.

The contribution $\delta\mathbf{u}^{(s)}$ is said to represent a *pure straining motion*. e_{ij} is called the rate-of-strain tensor, and is determined completely by the directions of its principal axes and by the three principal rates of strain a, b, c. Another description of the relative velocity field $\delta\mathbf{u}^{(s)}$ is that it converts a material element near $\mathbf{x}$ which initially is spherical into an ellipsoid with principal diameters which do not rotate and whose rates of extension are a, b, c. For an incompressible fluid, this ellipsoid has constant volume and e_{ii} is zero (for see (2.3.8)). For a compressible fluid, the pure straining motion may be regarded as a superposition of an isotropic expansion in which the rate of extension of all line elements is $\frac{1}{3}e_{ii}$, for which the contribution to Φ is $\frac{1}{6}r_k r_k e_{ii}$, and a straining motion without change of volume, for which the contribution to Φ is $\frac{1}{2}r_k r_l(e_{kl} - \frac{1}{3}e_{ii}\,\delta_{kl})$.

Turning now to the contribution $\delta\mathbf{u}^{(a)}$, we see that ξ_{ij} is an anti-symmetrical tensor with only three independent components and may quite generally be written in the form

$$\xi_{ij} = -\tfrac{1}{2}\epsilon_{ijk}\,\omega_k, \tag{2.3.9}$$

where ω_1, ω_2 and ω_3 evidently are the components of a vector $\boldsymbol{\omega}$; the factor $(-\frac{1}{2})$ is put in to simplify the subsequent relation (2.3.10). The contribution to δu_i is then

$$\delta u_i^{(a)} = r_j\,\xi_{ij} = -\tfrac{1}{2}\epsilon_{ijk}\,r_j\,\omega_k,$$

which is the i-component of the vector $\frac{1}{2}\boldsymbol{\omega}\times\mathbf{r}$. Therefore $\delta\mathbf{u}^{(a)}$ is the velocity produced at position $\mathbf{r}$ relative to a point about which there is a *rigid-body rotation* with angular velocity $\frac{1}{2}\boldsymbol{\omega}$.

The explicit expression for the components of $\boldsymbol{\omega}$ follow from (2.3.3) and (2.3.9):

$$\omega_1 = \frac{\partial u_3}{\partial x_2} - \frac{\partial u_2}{\partial x_3}, \quad \omega_2 = \frac{\partial u_1}{\partial x_3} - \frac{\partial u_3}{\partial x_1}, \quad \omega_3 = \frac{\partial u_2}{\partial x_1} - \frac{\partial u_1}{\partial x_2},$$

or, in vector notation,

$$\boldsymbol{\omega} = \nabla\times\mathbf{u}. \tag{2.3.10}$$

This vector $\boldsymbol{\omega}$ plays an important part in fluid mechanics, and is termed the local *vorticity* of the fluid. It is common practice in general vector analysis to describe a vector function of position having zero curl as *irrotational* in view of the above connection between $\nabla\times\mathbf{u}$ and the local rotation of the fluid.

We can see why $\nabla\times\mathbf{u}$ should appear as twice the effective local angular velocity of the fluid. By Stokes's theorem in vector analysis, we have

$$\int(\nabla\times\mathbf{u}).\mathbf{n}\,dA = \oint\mathbf{u}.\,d\mathbf{r}$$

for any open surface A bounded by a closed curve of which $\delta\mathbf{r}$ is an element. The choice of a plane surface bounded by a circle of small radius a centred at $\mathbf{x}$ and having the unit normal $\mathbf{n}$ then shows that

$$\left(\begin{matrix}\text{tangential component of velocity}\\ \text{averaged over the circumference}\end{matrix}\right)\bigg/a = \frac{1}{2\pi a^2}\oint\mathbf{u}.\,d\mathbf{r}$$

$$\approx \tfrac{1}{2}(\nabla\times\mathbf{u}).\mathbf{n}. \tag{2.3.11}$$

The fluid is not rotating as a rigid body about the point **x**, so that it cannot be said to have a local angular velocity in the simple sense; for a fluid which is being deformed, some extended definition of angular velocity is needed and the expression on the left-hand side of (2.3.11) seems a natural choice for the component of local angular velocity parallel to **n**. The pure straining motion represented by $\delta\mathbf{u}^{(s)}$ makes no contribution to this effective angular velocity of the fluid.

It is useful to look also at the expression for the angular momentum of a spherical element of fluid with centre at **x**, viz.

$$\int \epsilon_{ijk} r_j \left(u_k + r_l \frac{\partial u_k}{\partial x_l} \right) \rho \, dV(\mathbf{r}).$$

u_k and $\partial u_k/\partial x_l$ are constants so far as the integration over the volume of the element is concerned, so that the first term makes no contribution and the angular momentum is

$$\epsilon_{ijk} \frac{\partial u_k}{\partial x_l} \int r_j r_l \rho \, dV,$$

$$= \tfrac{1}{2} \epsilon_{ijk} \frac{\partial u_k}{\partial x_l} I \delta_{jl} = \tfrac{1}{2} \omega_i I, \qquad (2.3.12)$$

where I is the moment of inertia of the element about any axis through the centre. This is exactly the angular momentum that the spherical element would have if it were rotating as a rigid body with angular velocity $\tfrac{1}{2}\boldsymbol{\omega}$. It will be noted that this result does not hold for an element of arbitrary shape, because the angular momentum then in general depends on the straining motion represented by $\delta\mathbf{u}^{(s)}$ (as is clear from a consideration of the angular momentum of an element in the form of a long thin ellipsoid with its maximum diameter not parallel to one of the principal axes of strain), as well as on the rotational motion represented by $\delta\mathbf{u}^{(a)}$.

In summary, we have seen that, to the first order in the linear dimensions of a small region surrounding the position **x**, the velocity field in this region consists, in effect, of the superposition of

(*a*) a uniform translation with velocity **u(x)**,

(*b*) a pure straining motion characterized by the rate-of-strain tensor e_{ij}, which itself may be decomposed into an isotropic expansion and a straining motion without change of volume, and

(*c*) a rigid-body rotation with angular velocity $\tfrac{1}{2}\boldsymbol{\omega}$.

In analytical terms, the conclusion is that the velocity at the position **x**+**r** may be written approximately as

$$u_i(\mathbf{x}) + \frac{\partial}{\partial r_i} \left(\tfrac{1}{2} r_j r_k e_{jk} \right) + \tfrac{1}{2} \epsilon_{ijk} \omega_j r_k, \qquad (2.3.13)$$

where e_{ij} and ω_j are evaluated at the point **x**.

Simple shearing motion

A type of relative velocity field which occurs often in practice is a simple shearing motion, in which plane layers of fluid slide over one another. The relative velocity $\delta\mathbf{u}$ here has the same direction everywhere and varies with respect to distance in one direction perpendicular to $\delta\mathbf{u}$. With an appropriate choice of axes $\partial u_i/\partial x_j$ is non-zero only when $i = 1$, $j = 2$, and then

$$\delta\mathbf{u} = \left(r_2\frac{\partial u_1}{\partial x_2}, 0, 0 \right),$$

and
$$\Phi = \tfrac{1}{2}r_1 r_2\frac{\partial u_1}{\partial x_2}, \quad \boldsymbol{\omega} = \left(0, 0, -\frac{\partial u_1}{\partial x_2} \right). \tag{2.3.14}$$

The principal rates of strain are $\dfrac{1}{2}\dfrac{\partial u_1}{\partial x_2}$, $-\dfrac{1}{2}\dfrac{\partial u_1}{\partial x_2}$, and 0, and act along principal axes whose directions with respect to the original axes are given by the unit vectors $(1/\sqrt{2}, 1/\sqrt{2}, 0)$, $(-1/\sqrt{2}, 1/\sqrt{2}, 0)$ and $(0, 0, 1)$ respectively. Figure 2.3.1 shows how the contributions from the straining motion and rotation combine to produce the simple shearing motion at points on a circle in the (r_1, r_2)-plane.

This representation of a simple shearing motion as a superposition of a pure straining motion (with zero rate of expansion) and a rigid rotation allows us to choose a simple shearing motion as a basic element in the representation of a general relative velocity field, and it is sometimes useful to do this. Before seeing how this may be done in general, we show that any two-dimensional local relative velocity field may be represented as the superposition of a symmetrical expansion, a simple shearing motion, and a rigid rotation. First resolve the relative velocity field into a pure straining motion and a rigid rotation as above and rotate the axes of reference so that they coincide with the principal axes of the rate-of-strain tensor. Then we have

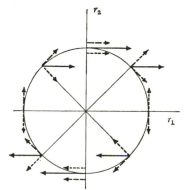

Figure 2.3.1. Simple shearing motion near a point decomposed into a straining motion and a rotation. The principal axes of the straining motion are at 45° to the r_1- and r_2-axes.

---->, straining motion;
— —→, rotation;
———→, resultant.

$$\Phi = \tfrac{1}{2}(ar_1'^2 + br_2'^2)$$
$$= \tfrac{1}{4}(a+b)\,r^2 + \tfrac{1}{4}(a-b)\,(r_1'^2 - r_2'^2),$$

where $r^2 = r_1'^2 + r_2'^2$. Now rotate the axes through a further 45°, so that $\mathbf{r}$ has components (r_1'', r_2'') and

$$\Phi = \tfrac{1}{4}(a+b)\,r^2 - \tfrac{1}{2}(a-b)\,r_1''r_2''. \tag{2.3.15}$$

The first term represents a symmetrical expansion, in which all material lines are being extended at the rate $\frac{1}{2}(a+b) = \frac{1}{2}\nabla.\mathbf{u}$, and it is evident from (2.3.14) that the second term, when combined with a rigid rotation with angular velocity $\mp \frac{1}{2}(a-b)$ about the normal to the plane of motion, represents a simple shearing motion; whence the result stated follows.

The corresponding result in the general three-dimensional case is that any local relative velocity field may be represented as the superposition of a symmetrical expansion, two simple shearing motions, and a rigid rotation. Again we proceed by resolving the motion into a rigid rotation and a pure straining motion, for the latter of which, relative to principal axes, we have

$$\Phi = \tfrac{1}{2}(ar_1'^2 + br_2'^2 + cr_3'^2)$$
$$= \tfrac{1}{6}r^2 e_{ii} + \tfrac{1}{2}(a - \tfrac{1}{3}e_{ii})(r_1'^2 - r_3'^2) + \tfrac{1}{2}(b - \tfrac{1}{3}e_{ii})(r_2'^2 - r_3'^2).$$

The first term represents a spherically symmetrical expansion in which the volume of a material element is increasing at the rate $\nabla.\mathbf{u}$ per unit volume, and the second and third terms each represent a two-dimensional pure straining motion with zero rate of expansion which, as already seen, can be represented as the superposition of a simple shearing motion and a suitably chosen rigid rotation; whence the result follows. It will be noticed that the two simple shearing motions that go to represent a given relative velocity field can be chosen in a number of different ways (corresponding to the fact that a non-diagonal element of the rate-of-strain tensor yields a simple shearing motion when combined with either of two rigid rotations of equal and opposite angular velocity).

2.4. Expression for the velocity distribution with specified rate of expansion and vorticity

The divergence and the curl of a vector function of position are fundamental differential operators in vector analysis which yield quantities independent of the choice of co-ordinate system. Applied to the velocity field they give the local rate of expansion Δ and the local vorticity $\boldsymbol{\omega}$:

$$\nabla.\mathbf{u} = \Delta, \quad \nabla \times \mathbf{u} = \boldsymbol{\omega}. \tag{2.4.1}$$

It was seen in §2.3 that the instantaneous relative motion of the fluid near any point is a combination of (i) an isotropic expansion such that the rate of increase of volume of a material element, per unit volume, is Δ, (ii) a pure straining motion without change of volume, and (iii) a rigid-body rotation with an angular velocity $\frac{1}{2}\boldsymbol{\omega}$. Evidently a good deal of information about the whole velocity distribution is conveyed by the distributions of Δ and $\boldsymbol{\omega}$ throughout the fluid. It sometimes happens that the distributions of Δ and $\boldsymbol{\omega}$ are prescribed, or can be inferred from the circumstances of the fluid motion, and it is useful to examine analytically the extent to which the velocity distribution is then determined. So far as the relative motion near any point is concerned, a pure straining motion without change of volume is left undetermined; but we shall see that there are quite strong restrictions on the distribution of these pure straining motions over the fluid as a whole.

Our plan is to construct a velocity distribution whose divergence and curl have specified values Δ and $\boldsymbol{\omega}$ at all points of the fluid, and then to consider (in §2.7 *et seq.*) the properties of velocity fields with zero rate of expansion and vorticity. We begin with the specified distribution of the rate of expansion Δ and seek any hypothetical velocity, $\mathbf{u}_e$ say, such that

$$\nabla \cdot \mathbf{u}_e = \Delta, \quad \nabla \times \mathbf{u}_e = 0 \qquad (2.4.2)$$

everywhere, without regard for other properties of $\mathbf{u}_e$. One way of choosing $\mathbf{u}_e$ to satisfy (2.4.2) is to put

$$\mathbf{u}_e = \nabla \phi_e, \quad \nabla^2 \phi_e = \Delta. \qquad (2.4.3)$$

(This choice is of course not arbitrary; it is known in vector analysis that under fairly general conditions any vector function of position may be written as the sum of two vectors of the form $\nabla \phi$ and $\nabla \times \mathbf{B}$, of which only the first can have non-zero divergence and zero curl.) A solution of this Poisson-type equation for ϕ_e is known to be†

$$\phi_e(\mathbf{x}) = -\frac{1}{4\pi} \int \frac{\Delta'}{s} \, dV(\mathbf{x}'), \qquad (2.4.4)$$

where s is the magnitude of the vector $\mathbf{s} = \mathbf{x} - \mathbf{x}'$, the prime denotes evaluation at the point $\mathbf{x}'$, and the volume integral is taken over the region occupied by fluid; the specified distribution of Δ must of course be such that the integral in (2.4.4) exists.

The corresponding expression for $\mathbf{u}_e$ is

$$\mathbf{u}_e(\mathbf{x}) = -\frac{1}{4\pi} \int \Delta' \nabla_\mathbf{x}\left(\frac{1}{s}\right) dV(\mathbf{x}') = \frac{1}{4\pi} \int \frac{\mathbf{s}}{s^3} \Delta' \, dV(\mathbf{x}'). \qquad (2.4.5)\ddagger$$

The velocity $\mathbf{u}_e$ at the point $\mathbf{x}$ may be regarded formally as the sum of contributions from the different elements of volume of the fluid, that from a volume element δV at point $\mathbf{x}'$ being

$$\delta \mathbf{u}_e(\mathbf{x}) = \frac{\mathbf{s}}{s} \frac{\Delta' \, \delta V(\mathbf{x}')}{4\pi s^2}. \qquad (2.4.6)$$

This is simply the irrotational velocity distribution in an infinite fluid that is consistent with a volume flux $\Delta' \delta V(\mathbf{x}')$ across all closed surfaces enclosing the point $\mathbf{x}'$. The velocity field (2.4.6) has zero rate of expansion everywhere except within the volume element $\delta V(\mathbf{x}')$, containing the point

† See chap. 6 of *Methods of Mathematical Physics*, by H. Jeffreys and B. S. Jeffreys, 3rd ed. (Cambridge University Press, 1956).

‡ Where there is ambiguity about the position vector with respect to which the differential operator ∇ applies, as in the case of the quantity $\nabla(1/s)$, the relevant position vector is indicated by a suffix. $\nabla_\mathbf{x} = -\nabla_{\mathbf{x}'}$ for operation on a function of $\mathbf{s}$ alone.

x', where the rate of expansion is Δ'; and so the velocity field (2.4.5) has the specified rate of expansion everywhere. We may say that each volume element $\delta V(\mathbf{x}')$ acts like a source of volume in an otherwise expansion-free fluid, the rate of emission of volume (or 'strength' of the source) being $\Delta(\mathbf{x}') \, \delta V(\mathbf{x}')$.

Now let us suppose that the distribution of vorticity $\boldsymbol{\omega}$ is specified (with $\nabla \cdot \boldsymbol{\omega} = 0$ everywhere), and seek a hypothetical velocity, $\mathbf{u}_v$, say, such that

$$\nabla \times \mathbf{u}_v = \boldsymbol{\omega}, \quad \nabla \cdot \mathbf{u}_v = 0, \tag{2.4.7}$$

without regard for other properties for $\mathbf{u}_v$. The natural substitution here is

$$\mathbf{u}_v = \nabla \times \mathbf{B}_v, \tag{2.4.8}$$

and the equation for the 'vector potential' $\mathbf{B}_v$ is then

$$\nabla \times (\nabla \times \mathbf{B}_v) = \nabla(\nabla \cdot \mathbf{B}_v) - \nabla^2 \mathbf{B}_v = \boldsymbol{\omega}. \tag{2.4.9}$$

If it happens that $\nabla \cdot \mathbf{B}_v = 0$ everywhere, the equation for $\mathbf{B}_v$ is

$$\nabla^2 \mathbf{B}_v = -\boldsymbol{\omega},$$

of which a solution is $\quad \mathbf{B}_v(\mathbf{x}) = \dfrac{1}{4\pi} \displaystyle\int \dfrac{\boldsymbol{\omega}'}{s} \, dV(\mathbf{x}'), \tag{2.4.10}$

where the volume integral is taken over the region occupied by the fluid as before. We now test this solution to see if it is such as to make $\nabla \cdot \mathbf{B}_v = 0$. We find

$$\nabla \cdot \left\{ \frac{1}{4\pi} \int \frac{\boldsymbol{\omega}'}{s} \, dV(\mathbf{x}') \right\} = \frac{1}{4\pi} \int \boldsymbol{\omega}' \cdot \nabla_\mathbf{x} \left(\frac{1}{s} \right) dV(\mathbf{x}')$$

$$= -\frac{1}{4\pi} \int \nabla_{\mathbf{x}'} \cdot \left(\frac{\boldsymbol{\omega}'}{s} \right) dV(\mathbf{x}')$$

$$= -\frac{1}{4\pi} \int \frac{\boldsymbol{\omega}' \cdot \mathbf{n}}{s} \, dA(\mathbf{x}'),$$

the surface integral being taken over the entire boundary of the fluid. This surface integral vanishes when the prescribed vorticity has zero normal component at each point of the boundary, as would be so, in particular, at the hypothetical exterior boundary of a fluid which extends to infinity in all directions and is at rest there. In cases in which $\boldsymbol{\omega} \cdot \mathbf{n} \neq 0$ at some points of the boundary of the fluid we may adopt the artifice of imagining that the fluid and the vorticity distribution extend beyond the actual boundary and that the vorticity is so distributed there as to make a new region of fluid with a new boundary at which $\boldsymbol{\omega} \cdot \mathbf{n} = 0$. All the lines whose tangents are everywhere parallel to $\boldsymbol{\omega}$ are then closed, and none end at the new boundary. There are many ways in which the vorticity distribution may be extended, since the problem of finding a solenoidal vector $\boldsymbol{\omega}$ such that $\boldsymbol{\omega} \cdot \mathbf{n}$ takes given values at the boundary of the region concerned is under-determined, but the choice is immaterial to our present purpose since for all choices the velocity $\mathbf{u}_v$ given by (2.4.8) and (2.4.10) has the specified vorticity at all points of the actual fluid. (For one possible choice, $\nabla \times \boldsymbol{\omega} = 0$ in the region of extension, in which case determination of $\boldsymbol{\omega}$ there is the problem taken up in §2.7.)

On the understanding, then, that the volume integral in (2.4.10) (and in (2.4.11) below) is taken over an extended region in cases in which $\boldsymbol{\omega}.\mathbf{n} \neq 0$ at the real boundary of the fluid, we have from (2.4.8) and (2.4.10)

$$\mathbf{u}_v(\mathbf{x}) = \frac{1}{4\pi}\int \nabla_\mathbf{x} \times \left(\frac{\boldsymbol{\omega}'}{s}\right) dV(\mathbf{x}')$$

$$= -\frac{1}{4\pi}\int \frac{\mathbf{s} \times \boldsymbol{\omega}'}{s^3} dV(\mathbf{x}'). \qquad (2.4.11)$$

$\mathbf{u}_v$ may be regarded formally as the sum of contributions from the different volume elements of the fluid, that from $\delta V(\mathbf{x}')$ being

$$\delta\mathbf{u}_v = -\frac{\mathbf{s} \times \boldsymbol{\omega}' \, \delta V(\mathbf{x}')}{4\pi s^3}. \qquad (2.4.12)$$

The vorticity cannot be uniform and non-zero inside a volume element and zero in surrounding fluid, since such a distribution of $\boldsymbol{\omega}$ would not have zero divergence, so that (2.4.12), unlike (2.4.6), is a velocity distribution which cannot exist by itself.

There is an analogy between (2.4.11) and the formula in electromagnetic theory that relates a steady volume distribution of electric current (the current density replacing $\boldsymbol{\omega}$) and the accompanying magnetic field. Just as the electric current can be said to produce a distribution of magnetic field (given by a formula like (2.4.11)) in the space under consideration, so vorticity can be said to produce the velocity distribution (2.4.11) in the surrounding fluid. The word 'produce' (which is sometimes replaced by 'induce') does not imply here a mechanical cause and effect; what is meant, strictly speaking, is that (2.4.11) is the solenoidal velocity whose curl has the specified value everywhere and which is therefore associated with the given distribution of vorticity.

The conclusion is that, if $\mathbf{u}$ is a velocity field which is consistent with specified values of the rate of expansion Δ and the vorticity $\boldsymbol{\omega}$ at each point of the fluid, $\mathbf{u} - \mathbf{u}_e - \mathbf{u}_v$ is both solenoidal and irrotational, where $\mathbf{u}_e$ and $\mathbf{u}_v$ are given by (2.4.5) and (2.4.11). Thus we may write

$$\mathbf{u} = \mathbf{u}_e + \mathbf{u}_v + \mathbf{v}, \qquad (2.4.13)$$

where $\mathbf{v}$ is a vector satisfying the equations

$$\nabla.\mathbf{v} = 0, \quad \nabla \times \mathbf{v} = 0 \qquad (2.4.14)$$

at all points of the fluid. We shall see later (§2.7) that $\mathbf{v}$ is determined by the conditions to be satisfied at the boundary of the fluid.

2.5. Singularities in the rate of expansion. Sources and sinks

We consider in this section the irrotational velocity field $\mathbf{u}_e(\mathbf{x})$, as defined by (2.4.3) and (2.4.4), that is associated with a distribution of rate of expansion containing singularities of certain kinds. The basic type of singularity is simply an isolated peak in the value of Δ at a given point in the fluid. We suppose that Δ has a large magnitude in a region of small volume ϵ containing the point $\mathbf{x}'$ and is zero elsewhere (and if it should not be zero elsewhere, the additional contribution to $\mathbf{u}_e$ is linearly additive, as (2.4.5) shows). Since the non-zero values of Δ are now concentrated near $\mathbf{x}'$, the expression (2.4.5) becomes

$$\mathbf{u}_e(\mathbf{x}) \approx \frac{1}{4\pi} \frac{\mathbf{s}}{s^3} \int_\epsilon \Delta'' \, dV(\mathbf{x}''), \qquad (2.5.1)$$

where $\mathbf{s} = \mathbf{x} - \mathbf{x}'$ as before. Only the volume integral of Δ is relevant in this approximation, other details of the distribution of Δ near $\mathbf{x}'$ having no effect on $\mathbf{u}_e$. By supposing ϵ now to contract on to the point $\mathbf{x}'$ while $\int_\epsilon \Delta'' \, dV(\mathbf{x}'')$ remains constant (implying $|\Delta'| \to \infty$ as $\epsilon \to 0$), and equal to m say, we arrive at the mathematical concept of a *point source* of fluid volume, for which the associated irrotational velocity field is given exactly by

$$\phi_e = -\frac{m}{4\pi s}, \quad \mathbf{u}_e(\mathbf{x}) = \frac{m}{4\pi} \frac{\mathbf{s}}{s^3}. \qquad (2.5.2)$$

m is termed the 'strength' of the source (which becomes a sink when m is negative), and is equal to the total outward flux of fluid volume across any closed surface enclosing the point $\mathbf{x}'$.

The concept of a point source has some value as a direct representation of one aspect of real flow fields, although this value is limited since peaks in the distribution of Δ are not readily produced in the interior of the fluid by dynamical effects. When something like a point source does occur, it will normally be a direct consequence of some external action. For instance, a pipe of small diameter sucking in fluid at one end produces a flow like that due to a point sink located at the end of the pipe (figure 2.5.1), and the flow on one side of a rigid plane containing a small hole through which fluid is sucked† at a rate M units of volume per second is approximately the same as the flow produced in an unbounded fluid by a point source of strength $-2M$. However, the more important role of the point source in theoretical fluid mechanics is as one of a set of mathematical units from which more complex and interesting flow fields can be constructed. The remainder of this section

---

† The reason why the fluid must flow into, and not out of, the pipe and the hole in the plane in order to be capable of representation by flow due to a point source or sink is connected with the effect of fluid viscosity at a rigid boundary. Fluid which is forced out of the pipe or the hole in the plane usually emerges as a concentrated jet. A match can be extinguished by blowing, but not by sucking!

illustrates the possibility of 'synthesizing' flow fields with the point source as the basic unit.

An aspect of the concept of a point source which makes it useful mathematically is its localization to a point. We may obtain a different singularity, similarly localized, by imagining a source and a sink, with strengths of equal magnitude m, to be placed at the points $\mathbf{x}' + \frac{1}{2}\delta\mathbf{x}'$ and $\mathbf{x}' - \frac{1}{2}\delta\mathbf{x}'$ respectively. The separation $\delta\mathbf{x}'$ is now allowed to approach zero, and the source strength m to approach infinity, in such a way that the product tends to a finite limit:

$$\boldsymbol{\mu} = \lim_{\delta\mathbf{x}' \to 0} m\,\delta\mathbf{x}'.$$

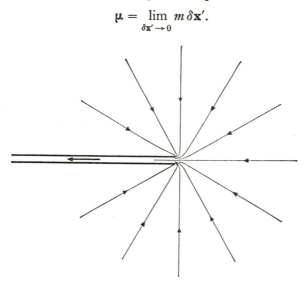

Figure 2.5.1. Flow due to suction at the open end of a pipe—approximately the same as flow due to a point sink.

This gives a singularity termed a *source doublet* of strength $\boldsymbol{\mu}$ at the point $\mathbf{x}'$. The irrotational velocity field associated with a source doublet (or dipole) is a linear superposition of the fields associated with the source and sink separately, and is therefore given by

$$\phi_e(\mathbf{x}) = \lim_{\delta\mathbf{x}' \to 0} \frac{m}{4\pi}\left\{ -\frac{1}{|\mathbf{x} - \mathbf{x}' - \frac{1}{2}\delta\mathbf{x}'|} + \frac{1}{|\mathbf{x} - \mathbf{x}' + \frac{1}{2}\delta\mathbf{x}'|} \right\}$$

$$= -\frac{1}{4\pi}\boldsymbol{\mu}.\nabla_{\mathbf{x}'}\frac{1}{s} = \frac{1}{4\pi}\boldsymbol{\mu}.\nabla_{\mathbf{x}}\frac{1}{s}, \tag{2.5.3}$$

and

$$\mathbf{u}_e(\mathbf{x}) = \nabla\phi_e(\mathbf{x}) = \frac{1}{4\pi}\boldsymbol{\mu}.\nabla_{\mathbf{x}}\left(\nabla_{\mathbf{x}}\frac{1}{s}\right)$$

$$= \frac{1}{4\pi}\boldsymbol{\mu}.\nabla_{\mathbf{x}}\left(-\frac{\mathbf{s}}{s^3}\right)$$

$$= \frac{1}{4\pi}\left\{ -\frac{\boldsymbol{\mu}}{s^3} + 3\frac{\boldsymbol{\mu}.\mathbf{s}}{s^5}\mathbf{s} \right\}. \tag{2.5.4}$$

The velocity field (2.5.4) has axial symmetry (with zero azimuthal component) about the direction of $\mathbf{\mu}$, and the components of $\mathbf{u}_e$, which is a solenoidal vector except at $\mathbf{x} = \mathbf{x}'$, can therefore be written in terms of a stream function ψ (§2.2). In terms of spherical polar co-ordinates (s, θ, ϕ) with origin at the point $\mathbf{x}'$ and with $\theta = 0$ in the direction of $\mathbf{\mu}$, the radial component of $\mathbf{u}_e$ is, according to (2.2.14),

$$\frac{1}{s^2 \sin\theta} \frac{\partial\psi}{\partial\theta} = \frac{\mathbf{s} . \mathbf{u}_e}{s}$$

$$= \frac{1}{2\pi} \frac{\mathbf{\mu} . \mathbf{s}}{s^4} = \frac{\mu}{2\pi} \frac{\cos\theta}{s^3},$$

where $\mu = |\mathbf{\mu}|$; hence

$$\psi = \frac{\mu}{4\pi} \frac{\sin^2\theta}{s},$$

(2.5.5)

the arbitrary function of integration being determined (as zero) by the need for the correct transverse component of $\mathbf{u}_e$ also to be derivable from (2.5.5). Figure 2.5.2 shows the pattern of streamlines in an axial plane for the flow associated with a source doublet.

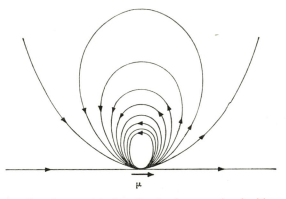

Figure 2.5.2. Streamlines in an axial plane for the flow associated with a source doublet. The stream function increases by the same amount between each pair of neighbouring streamlines.

An important point to notice for later applications is that as $s \to \infty$ the velocity associated with a single source at $\mathbf{x}'$ decreases to zero as s^{-2}, whereas that for a source doublet does so as s^{-3}. Another important property is that, since the strengths of the source and sink making up the doublet are equal, there is no net flux of fluid volume across a surface enclosing the source doublet; this property renders the source doublet more useful than the single source as a direct representation of actual flow fields.

It is possible to obtain other point singularities, of increasing degree of complexity, by the same procedure as was used to construct a source doublet from a single source. If a source doublet of strength $\mathbf{\mu}$ is placed at the point

$\mathbf{x}' + \frac{1}{2}\delta\mathbf{x}'$ and another of strength $-\mu$ at $\mathbf{x}' - \frac{1}{2}\delta\mathbf{x}'$, and if $|\delta\mathbf{x}'|$ is allowed to approach zero and $|\mu|$ to increase, in such a way as to make $\mu_i \delta x_j'$ tend to a finite limit, ν_{ij} say, we obtain a point singularity for which the associated velocity distribution is derivable from

$$\phi_e(\mathbf{x}) = \nu_{ij}\frac{\partial^2}{\partial x_i' \partial x_j'}\left(-\frac{1}{4\pi s}\right) = -\frac{\nu_{ij}}{4\pi}\frac{\partial^2}{\partial x_i \partial x_j}\left(\frac{1}{s}\right). \qquad (2.5.6)$$

This singularity can also be regarded as (the limiting case of) the super-position of two equal sources at opposite corners of a small parallelogram of certain shape near the point $\mathbf{x}'$ together with equal sinks of the same strength as the sources at the other two corners. The form of the velocity distribution associated with singularities of even higher order will be evident from (2.5.6).

It is also possible to imagine peaks in the distribution of the rate of expansion on certain lines or surfaces in the fluid, and to define line and surface singularities. Whereas the total flux of fluid volume from a point source has a given non-zero value, the flux *per unit length* of a line source is non-zero and the amount measures the line 'density' of source strength (which may not be the same at all points on the line). Likewise the flux per unit area of a surface source is non-zero and measures the surface density of source strength. Source doublets and higher order singularities may also be distributed over lines and surfaces with non-zero and finite density.

If the line density of source strength has the value m at all points of a line parallel to the z-axis at (x', y'), each element $\delta z'$ of the line may be regarded as acting as a point source of strength $m\,\delta z'$, and the irrotational velocity field $(u_e, v_e, 0)$ associated with the whole line is given by

$$\left.\begin{aligned} u_e(x, y) &= \frac{m}{4\pi}\int_{-\infty}^{\infty}\frac{x-x'}{s^3}\,dz' = \frac{m}{2\pi}\frac{x-x'}{\sigma^2}, \\ v_e(x, y) &= \frac{m}{4\pi}\int_{-\infty}^{\infty}\frac{y-y'}{s^3}\,dz' = \frac{m}{2\pi}\frac{y-y'}{\sigma^2}, \end{aligned}\right\} \qquad (2.5.7)$$

where $\sigma^2 = (x-x')^2 + (y-y')^2$. The scalar function whose gradient has the above components is

$$\phi_e(x, y) = \frac{m}{2\pi}\log\sigma. \qquad (2.5.8)$$

It will be noticed that an attempt to obtain (2.5.8) directly by integrating the expression for ϕ_e for a point source over all values of z' fails because the integral diverges; however, the divergence is independent of (x, y) (and gives rise to an infinite constant) and the expression (2.5.8) represents the finite part of the integral which does depend on (x, y) and which is therefore relevant to the velocity field.

This uniform and straight line source in a three-dimensional field is of course equivalent to a point source in a two-dimensional field. The velocity components (2.5.7) could have been derived by beginning with the concept of a point source of strength m at a point (x', y') in a two-dimensional field, the net flux of fluid area (or volume, per unit depth of the flow field) across all curves in the (x, y)-plane enclosing the point (x', y') then being m.

2.6. The vorticity distribution

There are many purposes for which it is more convenient to think about the fluid motion in terms of vorticity rather than in terms of velocity, despite the simpler physical character of the latter quantity. It also proves to be possible and useful, in many important cases of fluid flow, to divide the flow field into two regions with different properties, one of them being characterized by the vorticity being approximately zero everywhere. Considerations of the way in which changes in the distribution of vorticity take place will therefore be given often in later chapters. We are not yet in a position to describe the effect that the various forces acting on the fluid have on the vorticity, but we can note the purely kinematical consequences of the definition of $\boldsymbol{\omega}$ as $\nabla \times \mathbf{u}$—or, equivalently, as twice the effective local angular velocity of the fluid. One consequence is of course the identity

$$\nabla . \boldsymbol{\omega} = 0. \tag{2.6.1}$$

A line in the fluid whose tangent is everywhere parallel to the local vorticity vector is termed a *vortex-line*, and the family of such lines at any instant is defined by an equation analogous to (2.1.1). The surface in the fluid formed by all the vortex-lines passing through a given reducible† closed curve drawn in the fluid is said to be a *vortex-tube*. The flux of vorticity across an open surface bounded by this same closed curve and lying entirely in the fluid is

$$\int \boldsymbol{\omega} . \mathbf{n} \, dA,$$

where $\mathbf{n} \, \delta A$ is an element of area of the surface, and we can use the relation (2.6.1) to show that this integral has the same value for *any* such open surface lying in the fluid and bounded by *any* closed curve which lies on the vortex-tube and passes round it once. For if $\mathbf{n}' \, \delta A'$ and $\mathbf{n}'' \, \delta A''$ are elements of area of two such open surfaces (the directions of $\mathbf{n}'$ and $\mathbf{n}''$ having the same sense relative to the vortex-tube), the divergence theorem applied to the volume

† This useful term, which will be used again later, implies that the closed curve can be reduced to a point by a process of continuous deformation, *without passing outside the fluid*. Thus for a reducible closed curve drawn in the fluid it is always possible to find an open surface which is bounded by the curve and which lies entirely in the fluid (this surface being that traced out by the closed curve during its continuous reduction to a point). When the region occupied by the fluid is singly-connected, all closed curves in the fluid are reducible; and when this region is not singly-connected, some closed curves in the fluid are irreducible. A flow field which we shall wish to discuss is that due to a cylinder of infinite length moving in an infinite body of fluid; the region of space occupied by the fluid is here doubly-connected, closed curves in the fluid which pass round the cylinder are irreducible, and those which do not link the cylinder are reducible.

of fluid enclosed by these two surfaces and the connecting portion of the vortex-tube shows that

$$\int \boldsymbol{\omega}' . \mathbf{n}' \, dA' - \int \boldsymbol{\omega}'' . \mathbf{n}'' \, dA'' = \int \nabla . \boldsymbol{\omega} \, dV = 0,$$

there being no contribution to the surface integral from the portion of the vortex-tube. The flux of vorticity along a vortex-tube is thus independent of the choice of open surface used to measure it, and is termed the *strength* of the vortex-tube. In the case of a vortex-tube of infinitesimal cross-section, the strength is equal to the product of the cross-sectional area and the magnitude of the local vorticity, being the same at all stations along the tube. We note that a vortex-tube cannot end in the interior of the fluid.

An application of Stokes's theorem to a closed curve lying entirely on the vortex-tube and passing round it once gives

$$\int \boldsymbol{\omega} . \mathbf{n} \, dA = \oint \mathbf{u} . d\mathbf{x}, \qquad (2.6.2)$$

where $\mathbf{n} \, \delta A$ is an element of an open surface bounded by that closed curve. The line integral of fluid velocity round a closed curve is termed the *circulation*; thus the circulation round any reducible closed curve is equal to the flux of vorticity across an open surface bounded by the curve and, equivalently, is equal to the strength of the vortex-tube formed by all the vortex-lines passing through the curve.

Line vortices

A number of flow fields are characterized by values of the magnitude of the vorticity in the neighbourhood of a certain line in the fluid which are much larger than those elsewhere† (this line of necessity being parallel to $\boldsymbol{\omega}$ everywhere, since it would not otherwise be possible to satisfy $\nabla . \boldsymbol{\omega} = 0$). A useful mathematical idealization is derived from such cases by supposing a vortex-tube in which $\boldsymbol{\omega} \neq 0$ to contract on to a curve with the strength of the vortex-tube remaining constant, and equal to κ say. We then have a line singularity of the vorticity distribution which is specified entirely, so far as the contribution to the flux of vorticity across any surface is concerned, by the value of κ and the position of the line; it may be called a *line vortex* of strength κ (and should not be confused with a vortex-line, or line of vorticity). The solenoidal velocity distribution that is associated with the existence of a single line vortex and with zero vorticity elsewhere in the fluid is readily found from (2.4.11). For if $\delta \mathbf{l}$ be a vector element of length of the line vortex which lies in the volume element δV, we have

$$\int_{\delta V} \boldsymbol{\omega} \, dV = \kappa \, \delta \mathbf{l},$$

so that (2.4.11) becomes

$$\mathbf{u}_v = -\frac{\kappa}{4\pi} \oint \frac{\mathbf{s} \times d\mathbf{l}(\mathbf{x}')}{s^3}, \qquad (2.6.3)$$

† Tornadoes, whirl-pools, and vapour trails from the tips of the wing of an aircraft making a sharp turn, are all phenomena associated with such a concentration of vorticity.

where $\mathbf{s} = \mathbf{x} - \mathbf{x}'$, and the line integral is taken over a closed path extended beyond the fluid if necessary, as explained in §2.4. The corresponding expression in electromagnetic theory for the magnetic field 'due to' a steady current round a closed line conductor is called the Biot–Savart law.

In the very simple case of a straight line vortex of infinite length (and with zero vorticity elsewhere), the velocity $\mathbf{u}_v$ is everywhere in the azimuthal direction about the line vortex, with direction corresponding to positive circulation about the line vortex, and has magnitude

$$|\mathbf{u}_v| = \frac{\kappa\sigma}{4\pi}\int_{-\infty}^{\infty}\frac{dl}{(\sigma^2+l^2)^{\frac{3}{2}}} = \frac{\kappa}{2\pi\sigma} \qquad (2.6.4)$$

at distance σ from the line vortex (see figure 2.6.1); the two 'ends' of the straight line vortex at infinity can be regarded as being joined by a line

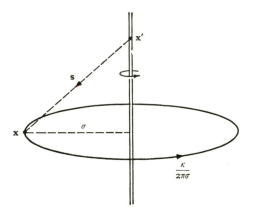

Figure 2.6.1. The solenoidal velocity distribution associated with a straight line vortex of strength κ.

vortex in the form of a semi-circle of radius R, say, the contribution to $\mathbf{u}_v$ from this curved path being of order R^{-1} and thus negligible. The velocity distribution (2.6.4) can also be obtained directly from the axial symmetry of the vorticity distribution and the application of (2.6.2) to a circular path centred on the line vortex. Even when the line vortex is curved, the value of $\mathbf{u}_v$ at points near the line vortex will be given approximately by (2.6.4) because the integral in (2.6.3) is then dominated by the neighbouring, approximately straight, portion of the line vortex (see §7.1).

We may note also that this two-dimensional solenoidal flow associated with a straight line vortex may be described in terms of a stream function; on comparing (2.2.10) and (2.6.4) we see that

$$\psi = -\frac{\kappa}{2\pi}\log\sigma. \qquad (2.6.5)$$

In a wholly two-dimensional flow field the appropriate term for the singularity is 'point vortex'.

Another formula for the solenoidal velocity field associated with a single curved line vortex of strength κ (with zero vorticity elsewhere) can be obtained by returning to the expression (2.4.10) for the vector potential $\mathbf{B}_v$. We have

$$\mathbf{u}_v(\mathbf{x}) = \nabla \times \mathbf{B}_v = \nabla \times \oint \frac{\kappa \, d\mathbf{l}(\mathbf{x}')}{4\pi s},$$

$$= -\frac{\kappa}{4\pi} \nabla \times \int \left(\nabla_{\mathbf{x}'} \frac{1}{s}\right) \times \mathbf{n} \, dA(\mathbf{x}')$$

by the analogue of Stokes's theorem for a scalar quantity integrated round a closed curve, where $\mathbf{n} \, \delta A$ is an element of area of any open surface bounded by the line vortex. On making use of the fact that

$$\nabla_{\mathbf{x}} \cdot \nabla_{\mathbf{x}'} \frac{1}{s} = -\nabla_{\mathbf{x}}^2 \frac{1}{s} = 0,$$

we find

$$\mathbf{u}_v(\mathbf{x}) = -\frac{\kappa}{4\pi} \int \mathbf{n} \cdot \nabla_{\mathbf{x}} \left(\nabla_{\mathbf{x}'} \frac{1}{s}\right) dA(\mathbf{x}').$$

This can be written as

$$\mathbf{u}_v(\mathbf{x}) = -\frac{\kappa}{4\pi} \nabla\Omega, \tag{2.6.6}$$

where

$$\Omega(\mathbf{x}) = \int \frac{\mathbf{s} \cdot \mathbf{n}}{s^3} \, dA(\mathbf{x}')$$

is the solid angle subtended by the line vortex at the point $\mathbf{x}$; the positive sense of $\mathbf{n}$ here is the same as the positive sense of the circulation round the line vortex. The corresponding formula in electromagnetic theory is also well known.

Just as a point source doublet and other more complicated singularities in the expansion distribution can be constructed by an appropriate superposition of single point sources, so other line singularities can be constructed from line vortices. We obtain a line *vortex doublet* by placing a straight line vortex of strength κ at position $\mathbf{x}' + \frac{1}{2}\delta\mathbf{x}'$ and another of strength $-\kappa$ at $\mathbf{x}' - \frac{1}{2}\delta\mathbf{x}'$ (where $\mathbf{x}'$ and $\delta\mathbf{x}'$ now represent—temporarily—vectors in the plane normal to the line vortices), and by allowing κ to increase and $|\delta\mathbf{x}'|$ to approach zero in such a way that $\kappa\,\delta\mathbf{x}'$ tends to the finite limit $\boldsymbol{\lambda}$. The associated two-dimensional solenoidal velocity distribution may be represented by the stream function

$$\psi(\mathbf{x}) = -\frac{1}{2\pi} \boldsymbol{\lambda} \cdot \nabla_{\mathbf{x}}(\log \sigma) = \frac{1}{2\pi} \frac{\boldsymbol{\lambda} \cdot (\mathbf{x} - \mathbf{x}')}{\sigma^2}, \tag{2.6.7}$$

where $\sigma = |\mathbf{x} - \mathbf{x}'|$. The streamlines in the plane normal to the line vortices are thus all circles† passing through the point $\mathbf{x}'$ with their centres on the line through $\mathbf{x}'$ parallel to $\boldsymbol{\lambda}$ (figure 2.6.2). It may be shown readily that the . solenoidal velocity distribution associated with a vortex doublet in two dimensions is identical with the irrotational distribution due to a source doublet (in two dimensions) located at the same point and perpendicular to the vortex doublet.

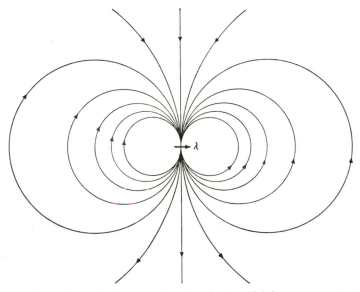

Figure 2.6.2. Streamlines for the two-dimensional solenoidal flow associated with a line vortex doublet. The stream function increases by the same amount between each pair of neighbouring streamlines.

Sheet vortices

Cases in which the magnitude of the vorticity is large everywhere in the neighbourhood of a surface in the fluid (which likewise must be a surface on which lines of $\boldsymbol{\omega}$ lie) also occur in practice, for example in flow fields involving aeroplane wings and other lifting bodies (§ 7.8) and in some involving movement of bluff bodies (§ 5.11). The local properties of such a surface concentration of vorticity are evidently specified by the vector

$$\boldsymbol{\Gamma} = \int \boldsymbol{\omega} \, dx_n,$$

where x_n denotes distance normal to the surface and the integral is taken

† This is true of the streamlines associated with two parallel line vortices of equal and opposite strength at arbitrary positions $\mathbf{x}_1'$ and $\mathbf{x}_2'$ in the plane orthogonal to the line vortices, as may be seen from the fact that the stream function is then

$$\frac{\kappa}{2\pi} \log \frac{\sigma_2}{\sigma_1},$$

where $\sigma_1 = |\mathbf{x} - \mathbf{x}_1'|$ and $\sigma_2 = |\mathbf{x} - \mathbf{x}_2'|$.

over a small range ϵ containing the surface. If now we suppose that $\epsilon \to 0$ and that $\int \omega \, dx_n$ remains constant and equal to Γ, we arrive at the concept of a *sheet vortex* characterized (locally) by the parameter Γ. The strength of a vortex-tube which encloses a narrow strip of the sheet parallel to Γ is $\Gamma \, (= |\Gamma|)$ per unit width of the strip, and Γ may be termed the strength density of the sheet vortex.

When the vorticity is zero everywhere except on a given sheet vortex, the expression (2.4.11) for the velocity distribution associated with the vorticity becomes

$$\mathbf{u}_v(\mathbf{x}) = -\frac{1}{4\pi} \int \frac{\mathbf{s} \times \Gamma'}{s^3} \, dA(\mathbf{x}'), \tag{2.6.8}$$

where $\mathbf{s} = \mathbf{x} - \mathbf{x}'$ as before and the integral is taken over the area of the sheet. In the particularly simple case of a single plane sheet vortex over which Γ is uniform, we have

$$\mathbf{u}_v(\mathbf{x}) = \frac{1}{4\pi} \Gamma \times \int \frac{\mathbf{s}}{s^3} \, dA(\mathbf{x}')$$

$$= \frac{1}{4\pi} \Gamma \times \int \frac{\mathbf{n} \cdot \mathbf{s}}{s^3} \mathbf{n} \, dA(\mathbf{x}')$$

$$= \tfrac{1}{2} \Gamma \times \mathbf{n}, \tag{2.6.9}$$

where $\mathbf{n}$ is the unit normal to the sheet directed towards the side on which the point $\mathbf{x}$ lies. The fluid velocity associated with the sheet vortex is thus uniform on each side of the sheet, with magnitude $\tfrac{1}{2}\Gamma$ and direction parallel to the sheet and perpendicular to Γ, but with opposite senses on the two sides. This result could also have been obtained, apart from a numerical factor, from the facts that no length scale is provided by this given distribution of vorticity and that the parameter Γ specifying the (uniform) strength density of the sheet has the dimensions of a velocity.

A related result holds for a sheet vortex in the form of a cylinder of arbitrary cross-section, over which Γ is uniform and Γ is everywhere at right angles to the generators of the cylinder (so that the vortex-lines are plane curves, all of the same shape, passing round the cylinder). The integral (2.6.8) becomes

$$\mathbf{u}_v(\mathbf{x}) = -\frac{\Gamma}{4\pi} \int_{-\infty}^{\infty} \oint \frac{\mathbf{s} \times d\mathbf{l}(\mathbf{x}')}{s^3} \, dm(\mathbf{x}'),$$

where $\delta m(\mathbf{x}')$ is an element of length of a generator and $\delta \mathbf{l}(\mathbf{x}')$ a vector element of length of a vortex-line, both at position $\mathbf{x}'$. The component of $\mathbf{s}$ parallel to the generators makes no net contribution to the integral with respect to m, in view of the anti-symmetry of the integrand, so that

$$u_v(\mathbf{x}) = -\frac{\Gamma}{2\pi} \oint \frac{\mathbf{p} \times d\mathbf{l}(\mathbf{x}')}{p^2}, \tag{2.6.10}$$

where $\mathbf{p}$ is the projection of $\mathbf{s}$ on a cross-sectional plane (figure 2.6.3). $|\mathbf{p} \times \delta \mathbf{l}|/p^2$ is the angle subtended at $\mathbf{x}$ by the length element $\delta \mathbf{l}$ in a cross-

7

sectional plane, and we see therefore that at any point $\mathbf{x}$ within the cylinder $\mathbf{u}_v$ is parallel to the generators and has uniform magnitude Γ, while at any point $\mathbf{x}$ outside the cylinder $\mathbf{u}_v$ is zero.† Thus a sheet vortex of uniform strength density again separates two regions in each of which the associated velocity is uniform.

In these two cases of sheet vortices over which the strength density is constant, there has been seen to be a discontinuity at the sheet, of magnitude Γ, in the component of $\mathbf{u}_v$ parallel to the sheet and perpendicular to Γ. This

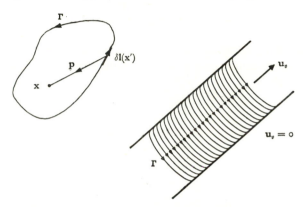

Figure 2.6.3. Calculation of the solenoidal velocity distribution associated with a cylindrical sheet vortex.

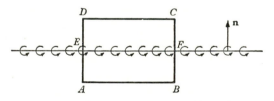

Figure 2.6.4. A small portion of a non-uniform sheet vortex.

may be shown to be true of any sheet vortex, with Γ not uniform, the relation between Γ and the velocity jump then being a local one. We consider the circulation round a circuit in the form of a small rectangle with two opposite sides AB and CD which lie on either side of the sheet vortex and which are parallel to the sheet and perpendicular to Γ (figure 2.6.4). The sheet may be supposed to be plane and Γ approximately uniform over the intercept of the rectangle at the sheet, and $\mathbf{u}_v$ is likewise uniform over the rectangle on each side of the sheet where the distribution of vorticity has a singularity. Then the contribution to $\oint \mathbf{u}_v.d\mathbf{x}$ from the path element EA cancels with that from BF (with an error of the second order in the linear

† There is a well-known corresponding result in electromagnetic theory, that the magnetic field due to a steady current in a solenoid (a long wire in the form of a closely-wound helix) is uniform and parallel to the axis within the solenoid but is zero outside it.

dimensions of the rectangle), and the contribution from *FC* cancels with that from *DE*, so that the general relation (2.6.2) gives

$$\int_A^B \mathbf{u}_v . d\mathbf{x} + \int_C^D \mathbf{u}_v . d\mathbf{x} = \Gamma \times EF.$$

Thus the component of $\mathbf{u}_v$ parallel to the sheet and perpendicular to Γ has a discontinuity of amount Γ across the sheet. The same argument for a rectangle with sides *AB* and *CD* parallel to Γ shows that there is no jump in the component of $\mathbf{u}_v$ parallel to Γ; nor can there be any jump in the component of $\mathbf{u}_v$ normal to the sheet in view of the requirement $\nabla . \mathbf{u}_v = 0$. The jump in $\mathbf{u}_v$ accompanying passage across the sheet in the direction of the normal $\mathbf{n}$ can therefore be written as

$$[\mathbf{u}_v] = \Gamma \times \mathbf{n}. \tag{2.6.11}$$

The local jump in $\mathbf{u}_v$ is thus the same as if the whole sheet were plane with uniform strength density equal to the local value of Γ. When the sheet is plane and the strength density uniform, $\mathbf{u}_v$ simply reverses direction across the sheet, but this property does not hold in general.

2.7. Velocity distributions with zero rate of expansion and zero vorticity

It has been shown that a velocity distribution of the form (2.4.13) is consistent with specified values of the rate of expansion Δ and the vorticity $\boldsymbol{\omega}$ at all points of the fluid. The terms $\mathbf{u}_e$ and $\mathbf{u}_v$ in (2.4.13) are obtainable from the distributions of Δ and $\boldsymbol{\omega}$ respectively, but the remaining term $\mathbf{v}$ was left undetermined. It is the purpose of this section to consider the properties of a velocity field $\mathbf{v}$ satisfying the equations (2.4.14), viz.

$$\nabla . \mathbf{v} = 0, \quad \nabla \times \mathbf{v} = 0. \tag{2.7.1}$$

The velocity $\mathbf{u}$ of a fluid which is effectively incompressible satisfies the equation $\nabla . \mathbf{u} = 0$, so that the equations (2.7.1) are satisfied, not only by a function $\mathbf{v}$ which is one of three contributions to the velocity of a fluid in which the rate of expansion and vorticity take specified values, but also by the actual velocity of an incompressible fluid in which for some reason the vorticity is zero. We shall see that most fluids behave, under a wide range of flow conditions, as if they were nearly incompressible (§3.6), and also that flow fields with the seemingly restrictive property of zero vorticity over large parts of the field are, for dynamical reasons, remarkably common (chapter 5). Study of *irrotational solenoidal vector fields* therefore has great practical value in fluid mechanics. The simplicity of the equations (2.7.1) has also made possible extensive mathematical developments and the employment of powerful analytical techniques. Actual flow fields in which the fluid velocity is irrotational and solenoidal will be considered in chapter 6, but it is desirable to establish here some of the more general results concerning the vector function $\mathbf{v}$ (which will be spoken of as a velocity for convenience, even though it may be only one of three contributions to the real velocity of the fluid) satisfying (2.7.1).

In a fluid in which the instantaneous distribution of velocity is $\mathbf{v}(\mathbf{x})$, material elements are being subjected to translation and a pure straining motion without change of volume, and without superposed rotation.

Since $\nabla \times \mathbf{v}$ is zero at all points of the fluid, Stokes's theorem shows that

$$\oint \mathbf{v} \cdot d\mathbf{x} = 0 \qquad (2.7.2)$$

for all reducible closed curves lying within the fluid, because it is always possible to find an open surface bounded by any such reducible curve and lying entirely in the fluid. If O and P are two points in a connected region of fluid, and C_1 and C_2 are two different curves joining O to P in such a way that the two together form a reducible closed curve lying entirely in the fluid, we see from (2.7.2) that

$$\int_{C_1} \mathbf{v} \cdot d\mathbf{x} = \int_{C_2} \mathbf{v} \cdot d\mathbf{x}.$$

The line integral of $\mathbf{v}$ over a curve joining O to P and lying within the fluid thus has the same value for all members of a set of paths of which any two make a reducible closed curve, and depends only on the position vectors $\mathbf{x}_0$ and $\mathbf{x}$ of O and P respectively. It is therefore possible to define a function $\phi(\mathbf{x})$ such that

$$\phi(\mathbf{x}) = \phi(\mathbf{x}_0) + \int_0^P \mathbf{v} \cdot d\mathbf{x}, \qquad (2.7.3)$$

in which the integral is taken over one of the paths in the set mentioned. The vector gradient of $\phi(\mathbf{x})$ is found by varying the position of P, giving

$$\nabla\phi(\mathbf{x}) = \mathbf{v}(\mathbf{x}). \qquad (2.7.4)$$

$\phi(\mathbf{x})$ is termed the *velocity potential* for the field $\mathbf{v}$ (although there is no question here of an interpretation of ϕ as a potential energy function). It is customary to leave the position $\mathbf{x}_0$ unspecified, since the difference between the values of ϕ corresponding to two different choices of $\mathbf{x}_0$ is independent of $\mathbf{x}$ and so without effect on $\nabla\phi(\mathbf{x})$.

We notice in passing the converse of the result represented by (2.7.2) since it will be needed in later discussions of the dynamical equations for a fluid of small viscosity: if the circulation associated with a velocity field $\mathbf{v}$ round all closed reducible curves lying in a region of fluid is zero, $\nabla \times \mathbf{v} = 0$ everywhere within that region. This result follows from the fact that for points P within the region the function ϕ given by (2.7.3) can be defined and $\mathbf{v}$ then has the irrotational form (2.7.4). Alternatively, we may argue that by Stokes's theorem

$$\int (\nabla \times \mathbf{v}) \cdot \mathbf{n} \, dA = 0$$

for all open surfaces A lying in the region and bounded by reducible curves, which is possible, when the integrand is continuous in $\mathbf{x}$, only if $\nabla \times \mathbf{v} = 0$ at all points of the region.

The introduction of the function ϕ by means of the relation (2.7.4) ensures that the equation $\nabla \times \mathbf{v} = 0$ is satisfied identically, and the three unknown scalar components of $\mathbf{v}$ are thereby determined by a single unknown scalar function ϕ. The first of the equations (2.7.1) then requires

$$\nabla^2 \phi = 0 \tag{2.7.5}$$

at all points of the fluid. This equation for ϕ, known as *Laplace's equation*, appears in many branches of mathematical physics, and many general results about functions satisfying the equation (often referred to as *harmonic functions*) are known. The linearity of the equation is noteworthy, and accounts for the relative simplicity of analysis of irrotational solenoidal flow; the dynamical equations governing the change of the velocity distribution in a fluid from one instant to the next are in general non-linear (chapter 3), but in the particular case of irrotational solenoidal flow the constraints on the velocity distribution are so strong as to require the spatial distribution of $\mathbf{v}$ to satisfy the simple linear equations (2.7.4) and (2.7.5) independently of temporal changes.†

Equation (2.7.5) is a second-order linear partial differential equation with constant coefficients, and is of the type designated in the theory of such equations‡ as *elliptic*. It is known that the solutions of equations of this type, and all their derivatives with respect to components of $\mathbf{x}$, are finite and continuous at all points, except possibly at some points on the boundary of the field. (This is in contrast to solutions of corresponding equations of hyperbolic type, such as the wave equation, which may have discontinuities at interior points.) Thus smoothness of the velocity distribution is ensured at all points of the fluid, except at those points of the boundary where a singularity of some kind—for example, an abrupt change of the tangent plane to the boundary, as at a corner or edge—is prescribed as a part of the boundary conditions.

The properties of solutions of (2.7.5) are strongly dependent on the topology of the region of space in which the equation holds. When the region occupied by the fluid is *singly-connected*, any pair of paths joining two points O and P and lying in the fluid together make a reducible closed curve, round which the circulation is zero, so that the function ϕ defined by (2.7.3) is a *single-valued* function of $\mathbf{x}$. When the region occupied by the fluid is

---

† Whether the dynamical equations allow the velocity distribution to remain solenoidal and irrotational is of course a matter for investigation. In fact, they do, under certain conditions (see §5.3). In this section, concerned with kinematics, we are examining the properties of a function $\mathbf{v}(\mathbf{x})$ which by definition satisfies the equations (2.7.1) at a given instant at which the expansion Δ and vorticity $\boldsymbol{\omega}$ are prescribed.

‡ For general accounts of second-order partial differential equations, see *Partial Differential Equations in Physics*, by A. Sommerfeld (Academic Press Inc., 1949), and *Methods of Mathematical Physics*, Volume 2, by R. Courant (Interscience, 1962).

multiply-connected, $\phi(\mathbf{x}) - \phi(\mathbf{x_0})$ has the same value for all paths belonging to a set of which any two together made a reducible closed curve, but it may have different values for paths belonging to different sets and may thus be many-valued. For the moment we assume that the fluid occupies a singly-connected region; the less important case of flow in a multiply-connected region will be investigated in §2.8.

Conditions for $\nabla\phi$ to be determined uniquely

An important result concerning the conditions under which the function ϕ is determined uniquely, apart from an arbitrary additive constant, may be established in the following way. We note first the identity

$$\nabla.(\phi\mathbf{v}) = \mathbf{v}.\nabla\phi + \phi\nabla.\mathbf{v} = \mathbf{v}.\mathbf{v},$$

and use it to rewrite the following integral over the volume occupied by the fluid:
$$\int \mathbf{v}.\mathbf{v}\,dV = \int \nabla.(\phi\mathbf{v})\,dV.$$

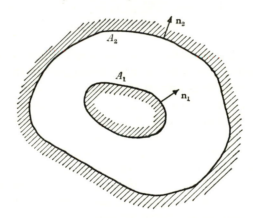

Figure 2.7.1. Definition sketch for fluid bounded internally (A_1) and externally (A_2).

When $\phi\mathbf{v}$ is a single-valued function of position, as it certainly is when the fluid occupies a singly-connected region of space, this volume integral may be transformed, by the divergence theorem, to an integral over the surface A bounding the fluid. Hence for a region of fluid bounded externally by a surface A_2, and perhaps also internally by a surface A_1, we have

$$\int \mathbf{v}.\mathbf{v}\,dV = \int \phi\mathbf{v}.\mathbf{n_2}\,dA - \int \phi\mathbf{v}.\mathbf{n_1}\,dA_1, \qquad (2.7.6)$$

where $\mathbf{n_1}$ and $\mathbf{n_2}$ are unit vectors normal to the surface elements δA_1 and δA_2 and are both drawn in the outward direction relative to the closed surfaces A_1 and A_2 (figure 2.7.1).

The relation (2.7.6) yields the remarkable result that, in any case in which the normal component of **v** is zero at all points of the inner and outer boundaries,

$$\int \mathbf{v} \cdot \mathbf{v} \, dV = 0$$

and so **v** must be zero everywhere in the fluid. This means that no irrotational motion of an incompressible fluid contained in a singly-connected region within rigid boundaries (across which the flux of fluid mass must be zero) can occur unless at least part of the boundary is moving with a non-zero component of velocity in the direction of the local normal.

The fact that only one solution of the equations (2.7.1) (viz. **v** = 0) is compatible with a zero normal component of velocity everywhere on the boundaries suggests that prescribed values of the normal component of **v** at the boundaries might determine uniquely the value of **v** everywhere. This is in fact so, and may be established quite simply by noting that, if $\mathbf{v} \, (= \nabla \phi)$ and $\mathbf{v}^* \, (= \nabla \phi^*)$ are two solutions of the equations (2.7.1), their difference $\mathbf{v} - \mathbf{v}^*$ is likewise a solution and the relation (2.7.6) may be rewritten with $\mathbf{v} - \mathbf{v}^*$ in place of **v** and $\phi - \phi^*$ in place of ϕ. The conditions under which not more than one solution exists, that is, under which $\mathbf{v} - \mathbf{v}^* = 0$ everywhere, are then the same as those that make the quantity

$$\int (\phi - \phi^*)(\mathbf{v} - \mathbf{v}^*) \cdot \mathbf{n}_2 \, dA_2 - \int (\phi - \phi^*)(\mathbf{v} - \mathbf{v}^*) \cdot \mathbf{n}_1 \, dA_1 \qquad (2.7.7)$$

equal to zero. If the normal components of **v** and $\mathbf{v}^*$ have the same prescribed value at each point of the boundaries A_1 and A_2, we have

$$(\mathbf{v} - \mathbf{v}^*) \cdot \mathbf{n} = 0$$

on A_1 and A_2, in which event the quantity (2.7.7) is zero, and $\mathbf{v} = \mathbf{v}^*$ at all points of the fluid. Similarly the quantity (2.7.7) vanishes if ϕ and ϕ^* have the same prescribed value at each point of the boundaries, although this condition for uniqueness is less relevant to practical problems. Equality of **v** and $\mathbf{v}^*$ everywhere is also ensured if we require that $\phi = \phi^*$ at some points of the boundaries and that $\mathbf{n} \cdot \mathbf{v} = \mathbf{n} \cdot \mathbf{v}^*$ at the remaining points.

Many of the flow fields considered in fluid mechanics are of large extent, by comparison with representative linear dimensions of the region of interest, and a useful mathematical idealization in such cases is that the fluid 'extends to infinity'. A particularly common type of flow is that produced by a rigid body moving through a large expanse of fluid which would otherwise be at rest,† and it is desirable to establish for this kind of flow a uniqueness theorem like that given above. The proof makes use of (2.7.6) in the same way, with the surface A_2 chosen to be a sphere of sufficiently large radius to enclose all the interior boundaries. However, the evaluation of the integral of $\phi \mathbf{v} \cdot \mathbf{n}$ over the surface A_2 requires a careful consideration of the behaviour of ϕ 'at infinity', to be given in §§ 2.9, 2.10, and we therefore postpone a demonstration of the uniqueness theorem for fluid extending to infinity and

---

† When the speed of the body is steady, this flow is of course mechanically identical with that produced by the same body held fixed in a stream of fluid whose velocity would otherwise be uniform and equal, but oppositely directed, to that of the body in the original flow.

at rest there. The result is that the solution of the equations (2.7.1) for **v** is unique when certain alternative conditions are imposed at each point of the inner boundary alone, one such condition—the most important one—being that the normal component of **v** at the boundary takes a prescribed value.

These uniqueness theorems have very important consequences for irrotational flow of an incompressible fluid. The whole velocity distribution in such a flow (in a singly-connected region of space) is determined uniquely by prescribed values of the normal component of velocity at whatever inner and outer boundaries are present, and hence, in cases in which these boundaries are the surfaces of rigid bodies, by a prescribed motion of the rigid bodies. Thus when a rigid body moves through fluid which would otherwise be stationary, the flow field is determined uniquely by the instantaneous velocity of the body (together with its geometry); neither the acceleration nor the past history of the motion of the body is relevant.† In particular, when the fluid is bounded by stationary rigid boundaries, the fluid is necessarily stationary everywhere. The instantaneous motions of the body and of the fluid are evidently completely 'locked' together (which suggests that equations (2.7.1) are likely to govern fluid flow only in the absence of elastic and dissipative properties of the fluid).

We may now close our general account of the way in which the complete distribution of velocity of fluid in a singly-connected region is determined when the distributions of the expansion and the vorticity are specified. There are three contributions to the velocity distribution, as stated in (2.4.13), one of which ($\mathbf{u}_e$) is associated with the specified distribution of expansion and is given explicitly by (2.4.5), and another ($\mathbf{u}_v$) which is associated with the specified distribution of vorticity and is given explicitly by (2.4.11). The remaining contribution ($\mathbf{v}, = \nabla\phi$) is such that ϕ satisfies the equation (2.7.5) and **v** is determined uniquely by specified values of the normal component **v** at all points of the boundary of the fluid (or by specified values of ϕ at the boundary). It will usually happen that the expressions (2.4.5) for $\mathbf{u}_e$ and (2.4.11) for $\mathbf{u}_v$ have non-zero normal components at the boundary of the fluid. Hence the value of the normal component of **v** that is to be prescribed at the boundary is not simply equal to the normal component of the actual fluid velocity at the boundary, but is equal to the difference between that actual velocity component and the sum of the contributions from $\mathbf{u}_e$ and $\mathbf{u}_v$. In a case in which the boundary of the fluid is a rigid body moving with purely translational velocity **U**, the prescribed value of the normal component of **v** at the boundary is

$$\mathbf{n}.\mathbf{U} - \mathbf{n}.(\mathbf{u}_e + \mathbf{u}_v), \tag{2.7.8}$$

where **n** is the local unit normal to the surface of the body.

---

† This striking result has its mathematical origin in the fact that the equations (2.7.1) and (2.7.5) for **v** and ϕ are differential equations with respect to **x** only and do not contain the time explicitly; any boundary conditions which determine the solutions uniquely will necessarily involve only instantaneous quantities.

Irrotational solenoidal flow near a stagnation point

As a simple example of a velocity distribution satisfying the equations (2.7.1), we consider conditions in the neighbourhood of a point O where $\mathbf{v} = 0$. Such a point is commonly referred to as a *stagnation* point, and may occur either in the interior of the fluid or at the boundary. The velocity potential ϕ has finite and continuous derivatives near O, unless O is a point on the boundary where there is a geometrical singularity, and so ϕ can be expanded near O as a Taylor series in Cartesian co-ordinates x_i with origin at O:

$$\phi = \phi_0 + a_i x_i + \tfrac{1}{2} a_{ij} x_i x_j + O(r^3),$$

where $r^2 = x_i x_i$ and the tensor a_{ij} is symmetric. Since $\nabla\phi = 0$ at O, all the coefficients a_i vanish; and since $\nabla^2\phi = 0$ everywhere we have $a_{ii} = 0$. Thus the motion near O is a pure straining motion without change of volume characterized by the rate-of-strain tensor a_{ij}, and a linear velocity distribution

$$v_i = a_{ij} x_j,$$

apart from the small error term of order r^2.

It follows that there are three orthogonal lines through O, parallel to principal axes of the tensor a_{ij}, on each of which the velocity is parallel to that line, being towards O on at least one line and away from O on at least one line. The streamline through O evidently has three orthogonal branches, in general. If we use axes parallel to the principal axes of a_{ij}, with position co-ordinates (x, y, z), we have for the corresponding velocity components

$$u = ax, \quad v = by, \quad w = -(a+b)z, \tag{2.7.9}$$

where a and b are unknown constants relating to the flow field of which the region near O forms a part.

When the flow near O is either two-dimensional or axisymmetric, it is possible also to describe the motion in terms of a stream function (§2.2). In the case of two-dimensional flow, and with axes parallel to the principal axes of the rate-of-strain tensor at O, we evidently have

$$\phi = \tfrac{1}{2} k(x^2 - y^2), \quad \psi = kxy \tag{2.7.10}$$

near O, where k is a constant. The streamlines near O are rectangular hyperbolae, all of which asymptote to the two orthogonal branches of the streamline through O, as sketched in figure 2.7.2; and the equipotential lines form an identical and orthogonal family with asymptotes at $45°$ to the axes. Likewise, in the case of flow near O which is symmetrical about the x-axis of a cylindrical co-ordinate system (x, σ, θ), we have

$$\phi = k(x^2 - \tfrac{1}{2}\sigma^2), \quad \psi = kx\sigma^2. \tag{2.7.11}$$

Each streamline here lies in a plane through the axis of symmetry, and the whole family of streamlines in one such axial plane has the qualitative appearance of the family shown in figure 2.7.2.

These results apply to a stagnation point at a boundary, provided there is no geometrical singularity of the boundary there, and the tangent plane to the boundary at O will then contain two of the principal axes of the rate-of-strain tensor a_{ij} at O. For instance, in figure 2.7.2 either the x-axis or the y-axis could be a boundary. On the other hand, the results do not apply to a stagnation point at a point on the boundary where the tangent to the boundary is discontinuous, as at the apex of either a conical or wedge-shaped boundary. In such a case, some branches of the streamline through the stagnation point must coincide with the boundary and must therefore intersect at angles determined by the geometry of the boundary.

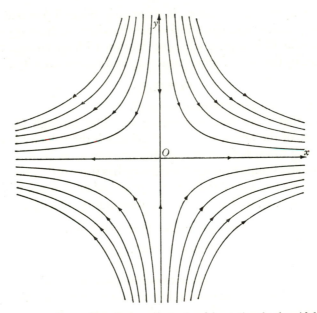

Figure 2.7.2. Streamlines in two-dimensional irrotational solenoidal flow near a stagnation point; $\psi = kxy$.

The complex potential for irrotational solenoidal flow in two dimensions

In the particular case of a two-dimensional field, and in that case alone, **v** satisfies relations of such a form that it is possible to make use of the theory of functions of a complex variable in a way which is both elegant and effective. Applications of complex variable theory to particular two-dimensional flow fields will be made in chapter 6; here we simply set down the basic mathematical relations.

The components v_x, v_y of a vector **v** in two dimensions which is irrotational can be written as

$$v_x = \frac{\partial \phi}{\partial x}, \quad v_y = \frac{\partial \phi}{\partial y}.$$

On the other hand, we have seen that the components of a solenoidal vector **v** in two dimensions can be expressed in terms of a stream function ψ (see §2.2) thus:

$$v_x = \frac{\partial \psi}{\partial y}, \quad v_y = -\frac{\partial \psi}{\partial x}.$$

The two scalar functions $\phi(x, y)$ and $\psi(x, y)$ provide alternative specifications of a vector **v** which is both irrotational and solenoidal, and are evidently related by

$$\frac{\partial \phi}{\partial x} = \frac{\partial \psi}{\partial y}, \quad \frac{\partial \phi}{\partial y} = -\frac{\partial \psi}{\partial x}. \tag{2.7.12}$$

Two relations of precisely the same form as (2.7.12) are well known in the theory of functions of a complex variable as the Cauchy–Riemann conditions that the complex quantity $\phi + i\psi$ should be a function of x and y of such special form as to depend only on the combination $x + iy$ in the sense that $\phi + i\psi$ has a unique derivative with respect to $x + iy$.[†] In the usual terminology, the relations (2.7.12) are conditions, which are both necessary and sufficient when the four partial derivatives in (2.7.12) are finite and continuous throughout a region, for $\phi + i\psi$ to be an *analytic* (or 'regular') function of the complex argument $z = x + iy$ in that region, the real functions ϕ and ψ then being *conjugate* functions.[‡]

We shall write

$$w(z) = \phi + i\psi$$

and term $w(z)$ the *complex potential* for the flow described by ϕ and ψ. It is an immediate consequence of this link with complex variable theory that any analytic function of z, irrespective of its form, may be interpreted as a complex potential and as a description of a possible irrotational solenoidal flow field in two dimensions. Moreover, if f is an analytic function of z, so too is if, so that in effect two flow fields can be obtained from f; for one of them ϕ and ψ are equated to $\mathscr{R}(f)$ and $\mathscr{I}(f)$ respectively (where $\mathscr{R}(f)$ and $\mathscr{I}(f)$ denote real and imaginary parts of f) and for the other to $-\mathscr{I}(f)$ and $\mathscr{R}(f)$ respectively.

Several other conjugate properties of ϕ and ψ are implied by the relations (2.7.12). Both ϕ and ψ satisfy Laplace's equation:

$$\frac{\partial^2 \phi}{\partial x^2} + \frac{\partial^2 \phi}{\partial y^2} = 0, \quad \frac{\partial^2 \psi}{\partial x^2} + \frac{\partial^2 \psi}{\partial y^2} = 0.$$

Since

$$(\nabla \phi).(\nabla \psi) = \frac{\partial \phi}{\partial x}\frac{\partial \psi}{\partial x} + \frac{\partial \phi}{\partial y}\frac{\partial \psi}{\partial y} = 0,$$

the equipotential lines on which ϕ is constant are orthogonal in general to the

---

[†] It may easily be verified that, when the relations (2.7.12) hold, the ratio of the differential of $\phi + i\psi$ to the differential $\delta x + i\delta y$ tends to a limit, as $(\delta x^2 + \delta y^2)^{\frac{1}{2}} \to 0$, which is independent of $\delta y/\delta x$.

[‡] For an account of complex variable theory generally, see, for instance, *Theory of Functions of a Complex Variable*, by E. T. Copson (Oxford, 1935).

streamlines on which ψ is constant; the deduction fails at a point where $|\mathbf{v}|$ is zero, and the result is not valid there (as is indeed evident from the example in figure 2.7.2).

Since the derivative

$$\frac{dw}{dz} = \lim_{|\delta z| \to 0} \frac{\delta w}{\delta z}$$

is independent of the direction of the differential δz in the (x, y)-plane, for convenience we may imagine the limit to be taken with δz remaining parallel to the x-axis, giving

$$\frac{dw}{dz} = \frac{\partial \phi}{\partial x} + i \frac{\partial \psi}{\partial x} = v_x - iv_y.$$

Choosing δz to be parallel to the y-axis (so that $\delta z = i\,\delta y$), with equal convenience, gives

$$\frac{dw}{dz} = \frac{1}{i} \frac{\partial \phi}{\partial y} + \frac{\partial \psi}{\partial y} = v_x - iv_y.$$

If v is written for the magnitude of $\mathbf{v}$ and θ for the angle between the direction of $\mathbf{v}$ and the x-axis, the expression for dw/dz becomes

$$\frac{dw}{dz} = v_x - iv_y = v\,e^{-i\theta}. \tag{2.7.13}$$

All these relations will be found useful later in various particular contexts.

2.8. Irrotational solenoidal flow in doubly-connected regions of space

When the region occupied by the fluid is not singly-connected, not all pairs of paths joining two points O and P in the fluid together make a closed reducible curve; putting it crudely, one path might go round one side of a boundary and the other member of the pair might go round the other. In these circumstances, the line integral of $\mathbf{v}$ (the irrotational solenoidal part of a general velocity field, as before) over a path joining O to P cannot be shown to be independent of the path chosen,† and the line integral may not be single-valued. The uniqueness result established in the preceding section is valid only when the function $\phi(\mathbf{x})$ defined by the line integral is single-valued, and it is desirable now to consider the changes required when ϕ may not be single-valued.

First we recall the way in which regions of space are classified topologically. A singly-connected region of space is distinguished by the facts that any two points in the region can be joined by paths lying entirely in the region

† We take it as self-evident that, when a closed curve is not reducible, it is not possible to find an open surface bounded by the curve and lying entirely in the fluid, so that Stokes's theorem cannot be used to show that the line integral of $\mathbf{v}$ round the curve is zero. Despite the 'obviousness' of this statement, it is in fact correct only for regions of space of fairly simple topological character (including those normally enountered in fluid mechanics). It is possible to construct peculiar regions of space, with a high degree of connectivity, containing some curves which are irreducible *and* which bound open surfaces lying entirely in the fluid.

and that any two such paths together made a reducible closed curve. In a multiply-connected region, it is still possible to join any two points in the region by paths lying entirely in the region, but some pairs of such paths together make irreducible closed curves. The degree of connectivity of a multiply-connected region is determined by the number of different barriers, in the form of open surfaces whose bounding curves lie entirely on the boundary of the region, which it is possible to insert in the region without dividing the region into unconnected parts; if $n-1$ such barriers can be inserted, the region is said to be *n-ply connected*. For example, the region external to a torus is doubly-connected, because only one barrier (say extending across the central opening of the torus) can be inserted without the region losing connectivity altogether. The insertion of each barrier creates a new region (for which both sides of the barrier are now part of the boundary) whose degree of connectivity is one less than that of the region without the barrier.

The degree of connectivity may also be stated in terms of the number of irreconcilable closed curves which may be drawn in the region. Two circuits in the region are said to be *reconcilable* if they can be made to coincide by continuous deformation without passing out of the region; sometimes the reconciliation will be such that there is a one-to-one correspondence between points on the two circuits (that is, such that each point of one circuit coincides with only one point of the other circuit), and sometimes one of the circuits will have become double, or multiple, during the reconciliation. In a singly-connected region, all circuits are reconcilable (and reducible). In the doubly-connected region external to a torus, all the reducible circuits are reconcilable, one with another, and all the irreducible circuits which thread the torus are likewise reconcilable with another; however, no circuit of the former group is reconcilable with any circuit of the latter group. There are thus just two irreconcilable circuits which can be drawn in a doubly-connected region. In an n-ply connected region, n irreconcilable circuits can be drawn, one of which will be reducible and $n-1$ of which will be irreducible. Each of the $n-1$ barriers which can be inserted in an n-ply connected region, without dividing it into unconnected parts, excludes one of the $n-1$ irreconcilable irreducible circuits which can be drawn in the region.

The case of a doubly-connected region is important in fluid mechanics. The flow generated by a long solid cylinder moving normal to its length takes place in such a region and the fact that some closed curves are then not reducible is at the basis of the theory of lift (§§ 6.6, 6.7). The region outside a torus is doubly-connected, and this is relevant to analysis of the kind of flow typified by motion of a smoke-ring (§ 7.2). Flow in regions with a higher degree of connectivity than two does not occur often, and in any event it is not difficult to infer the results for regions with connectivity of degree three and four when those for a doubly-connected region are known. The discussion in the remainder of this section will therefore refer to flow in doubly-connected regions of space.

It is convenient, for purposes of exposition, to use phrases relating to the concrete case of flow in the doubly-connected region outside a solid cylinder of infinite length. Consider the various closed curves which can be drawn in the fluid. A number of these circuits are reducible curves, round which the line integral of $\mathbf{v}$ is zero, by Stokes's theorem. Some of the circuits are irreducible curves which pass completely round (or 'loop') the cylinder once. Now any two of the circuits which loop the cylinder once are reconcilable, with a one-to-one correspondence of points on the two curves, and the surface traced out by the two curves during the reconciliation is a strip lying in the fluid and bounded by the two closed curves. Stokes's theorem for this open surface in the form of a strip† shows that the line integrals of $\mathbf{v}$ round the two closed curves, taken in the same sense relative to the cylinder, are equal; hence

$$\oint \mathbf{v} . d\mathbf{x} = \kappa \qquad (2.8.1)$$

for all circuits looping the cylinder once, the unknown quantity κ being called the *cyclic constant*‡ of the velocity field $\mathbf{v}$.

Other irreducible curves loop the cylinder more than once, p times say. Any two of the circuits which loop the cylinder p times are reconcilable, with a one-to-one correspondence, and Stokes's theorem applied as before to the strip swept out by the deformation leading to reconciliation again shows that the line integrals of $\mathbf{v}$ round the two closed curves have the same value. But included among the circuits linking the cylinder p times is one which repeats p times a closed curve which loops the cylinder once. Hence, for all circuits looping the cylinder p times, we have

$$\oint \mathbf{v} . d\mathbf{x} = p\kappa. \qquad (2.8.2)$$

This relation gives the circulation (associated with $\mathbf{v}$) round *any* closed curve drawn in the fluid, provided p is taken as zero for a curve which does not loop the cylinder.

If now we define a function $\phi(\mathbf{x})$ such that

$$\phi(\mathbf{x}) = \phi(\mathbf{x}_0) + \int_0^P \mathbf{v} . d\mathbf{x}, \qquad (2.8.3)$$

where the integral is taken over some path lying in the fluid and joining the point O with position vector $\mathbf{x}_0$ to the point P with position vector $\mathbf{x}$, the value of $\phi(\mathbf{x})$ is seen to depend on the choice of path. The difference between the two values of ϕ corresponding to two choices of path from O to P is equal to the line integral of $\mathbf{v}$ round the closed curve formed by the two paths together, and this, as (2.8.2) shows, must be an integral multiple of the cyclic

† When Stokes's theorem is applied to an open surface whose boundary consists of two or more unconnected closed curves, the sense of the line integral round the boundary is determined by the rule that it must everywhere be anticlockwise about the normal to the adjacent element of surface.

‡ In a n-ply connected region of space, $n-1$ cyclic constants are associated with the velocity field $\mathbf{v}$.

constant κ. Thus, in a doubly-connected region, ϕ is in general a many-valued function of position, the difference between possible values of ϕ being an integral multiple of κ. Irrotational solenoidal flow in a doubly-connected region is said to be *cyclic* when κ is non-zero; when $\kappa = 0$, the flow is *acyclic* and ϕ is a single-valued function of position, as in the case of flow in a singly-connected region.

It should be noticed that ϕ, as defined by (2.8.3), is still a continuous function of $\mathbf{x}$ (when $|\mathbf{v}|$ is finite). As the point P moves continuously round the cylinder in an anti-clockwise sense, ϕ changes continuously and is greater by an amount κ when P returns to its starting point after completing one loop of the cylinder. At all points in the fluid, an infinitesimal change $\delta\mathbf{x}$ in the position of P gives rise to an infinitesimal change $\mathbf{v}.\delta\mathbf{x}$ in the value of ϕ, and the relation

$$\mathbf{v}(\mathbf{x}) = \nabla\phi(\mathbf{x})$$

holds as before. $\mathbf{v}$ is of course a single-valued function of $\mathbf{x}$ in all circumstances.

An example of cyclic irrotational solenoidal flow is provided, in effect, by a line vortex of the kind described in § 2.6. The velocity field $\mathbf{u}_v$ associated with a single line vortex (which is necessarily closed or extends to infinity at both ends) in an infinite region of fluid, is, by definition, solenoidal everywhere and irrotational everywhere except on the line vortex itself; hence $\mathbf{u}_v = \nabla\phi$ everywhere in the doubly-connected region outside the line vortex and the cyclic constant for ϕ is equal to the strength of the line vortex. We did in fact determine explicitly in (2.6.6) the velocity potential for the flow associated with a closed line vortex of strength κ, viz.

$$\phi(\mathbf{x}) = -\frac{\kappa}{4\pi}\Omega, \tag{2.8.4}$$

where Ω is the solid angle subtended at the point $\mathbf{x}$ by the closed line vortex. As expected, this expression increases by an amount κ as the point $\mathbf{x}$ is taken once round any closed path linking the line vortex once in the positive sense relative to the vorticity of the line vortex. In the limiting case of a straight line vortex of infinite length, the line vortex can be imagined as being closed by a semi-circle of infinite radius, so that

$$\phi(\mathbf{x}) = \frac{\kappa}{2\pi}\theta, \tag{2.8.5}$$

where θ is the anti-clockwise polar angle, in the plane normal to the line vortex, of the point $\mathbf{x}$ relative to the line vortex, the direction $\theta = 0$ being arbitrary. It will be noted that this velocity potential gives the same flow field as the stream function (2.6.5), and that the expression for the complex potential of the flow in the z-plane normal to the line vortex is

$$w(z) = -(i\kappa/2\pi)\log z.$$

Conditions for $\nabla\phi$ to be determined uniquely

The argument in §2.7 that led to a statement of the boundary conditions under which a solution of Laplace's equation for ϕ is unique (apart from an additive constant) involved a use of the divergence theorem, in the step leading to (2.7.6), which is valid only when ϕ is a single-valued function of position. The argument therefore fails in the case of flow in a doubly-connected region, unless the cyclic constant κ happens to be zero.† However, there is a simple way of using the earlier results to obtain sufficient conditions for uniqueness of $\nabla\phi$ when ϕ is a many-valued velocity potential. For if ϕ and ϕ^* are two solutions of Laplace's equation which are known to have the same cyclic constant, $\phi - \phi^*$ is the velocity potential of an acyclic motion and is a single-valued function of position to which the earlier deductions do apply. Hence irrotational solenoidal flow in a doubly-connected region is determined uniquely when the boundary conditions needed for uniqueness of flow in a singly-connected region are imposed *and* the cyclic constant is specified.

Despite the fact that this simple argument gives immediately a useful uniqueness theorem, it is illuminating to consider in detail the way in which relations like (2.7.6) must be modified when ϕ is a many-valued function of position. As before, we begin with the identity

$$\int \mathbf{v} \cdot \mathbf{v} \, dV = \int \nabla \cdot (\phi \mathbf{v}) \, dV,$$

the integrals being taken over the doubly-connected region occupied by the fluid. In order to be able to transform to a surface integral, we imagine a barrier (without thickness) of the kind described earlier in this section‡ to be inserted in the fluid. If the two sides of this barrier are regarded as part of the boundary of the fluid, the flow now takes place in a singly-connected region within which ϕ is a single-valued function of position; the path used to join the reference point O to the current point $P(\mathbf{x})$ must not cross the barrier (since it must not pass outside the fluid) and consequently all pairs of paths together make reducible closed curves.

Use of the divergence theorem, which is permissible now that the volume V is singly-connected, gives

$$\int \mathbf{v} \cdot \mathbf{v} \, dV = \int \phi \mathbf{v} \cdot \mathbf{n} \, dA + \int \phi_- \mathbf{v} \cdot \mathbf{n} \, dS - \int \phi_+ \mathbf{v} \cdot \mathbf{n} \, dS, \qquad (2.8.6)$$

where A is the real boundary of the fluid, including interior and exterior boundaries A_1 (with $\mathbf{n} = -\mathbf{n}_1$) and A_2 (with $\mathbf{n} = \mathbf{n}_2$) where they exist, and the normal $\mathbf{n}$ to the barrier S has the same sense, relative to the real boundaries, as

---

† The way in which κ is determined by dynamical processes in flow generated by a certain type of moving cylinder is considered in § 6.7, and it will be seen there that the case $\kappa \neq 0$ is common and important.

‡ The word 'barrier' has topological, but not mechanical, significance. Insertion of the barrier has no effect on the flow and should be thought of as the drawing of a surface in the fluid.

that used to define a positive value of κ. ϕ_+ and ϕ_- are the values of ϕ on the two sides of the barrier, ϕ_+ referring to the side towards which the normal **n** points. Figure 2.8.1 specifies the notation for the case of a doubly-connected region between two infinitely long cylinders. Now when a point $P(\mathbf{x})$ moves, in a positive sense, from a position on one side of the barrier to a neighbouring position on the other side of the barrier without crossing it, the change in ϕ is

$$\phi_- - \phi_+ = \oint \mathbf{v} . d\mathbf{x} = \kappa. \tag{2.8.7}$$

Hence

$$\int \mathbf{v} . \mathbf{v} \, dV = \int \phi \mathbf{v} . \mathbf{n} \, dA + \kappa \int \mathbf{v} . \mathbf{n} \, dS. \tag{2.8.8}$$

The last integral is equal to the flux of fluid volume across the barrier.

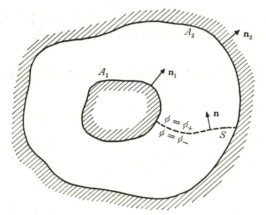

Figure 2.8.1. Insertion of a barrier S in the space between two cylinders.

It now follows that the whole of the right-hand side of (2.8.8) is zero for the 'difference' motion represented by $\phi - \phi^*$, provided the cyclic constants for the motions represented by ϕ and ϕ^* separately are equal and provided conditions of the kind already described are imposed on both ϕ and ϕ^* at the bounding surface A. We also see that an alternative way of making the second term on the right-hand side of (2.8.8) vanish for the difference motion is to specify that the motions represented by ϕ and ϕ^* separately produce the same flux of volume across the barrier. However, this prescription for uniqueness is not as useful in practice as specification of the cyclic constant.

In cases of cyclic flow in which the normal component of **v** is prescribed over the entire boundary A of the fluid, it is possible, and useful for later work, to divide the velocity **v** into two uniquely determined parts which make separate contributions to the right-hand side of (2.8.8). One part, $\mathbf{v}_1$ say, is derived from a single-valued potential ϕ_1 such that $\mathbf{n} . \nabla \phi_1$ has the specified value of $\mathbf{v} . \mathbf{n}$ at all points of the boundary A, and the other, $\mathbf{v}_2$, is derived from a many-valued potential ϕ_2 which has the prescribed cyclic constant κ and satisfies

$$\mathbf{n} . \nabla \phi_2 = 0$$

8

at all points of the boundary A. Then

$$\int \mathbf{v}_1 \cdot \mathbf{v}_2 \, dV = \int \nabla \cdot (\phi_1 \nabla \phi_2) \, dV$$

$$= \int \phi_1 \mathbf{n} \cdot \nabla \phi_2 \, dA = 0, \qquad (2.8.9)$$

showing that the two contributions $\mathbf{v}_1$ and $\mathbf{v}_2$ are orthogonal in an integral sense, and the relation (2.8.8) becomes

$$\int \mathbf{v} \cdot \mathbf{v} \, dV = \int \mathbf{v}_1 \cdot \mathbf{v}_1 \, dV + \int \mathbf{v}_2 \cdot \mathbf{v}_2 \, dV$$

$$= \int \phi_1 \mathbf{v}_1 \cdot \mathbf{n} \, dA + \kappa \int \mathbf{v}_2 \cdot \mathbf{n} \, dS. \qquad (2.8.10)$$

Exercise

Show that the integral over the barrier S in (2.8.10) is independent of the choice of barrier, whereas in general that in (2.8.8) is not.

2.9. Three-dimensional flow fields extending to infinity

Asymptotic expressions for $\mathbf{u}_e$ and $\mathbf{u}_v$

When the fluid extends to infinity in all directions and is at rest there, as we shall suppose to be the case, the rate of expansion Δ and vorticity $\boldsymbol{\omega}$ normally also vanish at infinity. The integral expressions (2.4.5) and (2.4.11) for the contributions to the velocity $\mathbf{u}(\mathbf{x})$ due to specified distributions of Δ and $\boldsymbol{\omega}$ are still solutions of the governing equations (2.4.2) and (2.4.7), provided only that the integrals over the infinite region of fluid are convergent. In many cases of practical interest $|\Delta|$ and $|\boldsymbol{\omega}|$ diminish quite rapidly with increasing distance from the interior boundary of the fluid, and we may reasonably make strong assumptions about their order of magnitude in order to obtain useful results about the asymptotic expressions for $\mathbf{u}_e$ and $\mathbf{u}_v$ when $|\mathbf{x}|$ is large.

Consider first the contribution $\mathbf{u}_e(\mathbf{x})$ representing the irrotational velocity field associated with the specified distribution of Δ and given by (2.4.5). When $|\Delta(\mathbf{x}')|$ decreases rapidly as $r' \to \infty$, the value of the integral in (2.4.5) is likely to be dominated by contributions from the central region surrounding the origin; and since for all these contributions

$$\frac{1}{s} \approx \frac{1}{r}$$

when r is large (where $s = |\mathbf{x} - \mathbf{x}'|$, $r = |\mathbf{x}|$), with an error of order r^{-2}, it is a plausible speculation that

$$\mathbf{u}_e(\mathbf{x}) \sim -\frac{1}{4\pi} \{\int \Delta' \, dV(\mathbf{x}')\} \nabla \frac{1}{r} \qquad (2.9.1)$$

as $r \to \infty$. This can be proved by considering separately the contributions to the integral in (2.4.5) from the regions $r' \leqslant \alpha r$ (yielding an integral I_1 say) and $r' \geqslant \alpha r$ (yielding I_2), where $\alpha < 1$. Provided $\Delta(\mathbf{x}')$ varies as r'^{-n} when

r' is large, I_2 is seen to be proportional to r^{1-n} when r is large. In the integrand of I_1, $r' < r$ and so it is possible to write s^{-1} as a Taylor series in $\mathbf{x}'$ with remainder, the series in this case consisting only of the first term r^{-1} and a remainder of order r^{-2}. With a suitable restriction on n, viz. $n > 3$, the integral I_2 is negligible and (2.9.1) follows.

The asymptotic form (2.9.1) represents the irrotational velocity field associated with a single source at the origin emitting volume at a rate $\int \Delta(\mathbf{x}') \, dV(\mathbf{x}')$. If this effective source strength is zero, the second term of the Taylor series for s^{-1} must be retained, s^{-1} then being replaced in the integrand of I_1 by

$$\frac{1}{r} - \mathbf{x}' \cdot \nabla \frac{1}{r}$$

with an error of order r^{-3}, whence it is found that

$$\mathbf{u}_e(\mathbf{x}) \sim \frac{1}{4\pi} \{\textstyle\int \mathbf{x}' \Delta' \, dV(\mathbf{x}')\} . \nabla \left(\nabla \frac{1}{r} \right) \tag{2.9.2}$$

as $r \to \infty$ with the stronger restriction $n > 4$. The asymptotic form here represents the irrotational velocity field associated with a source doublet (§ 2.5) of strength $\int \mathbf{x}' \Delta' \, dV(\mathbf{x}')$ at the origin. If this latter integral is zero, an approximation of even higher order is sought in the same way.

Similar remarks may be made about the contribution $\mathbf{u}_v(\mathbf{x})$ representing the solenoidal velocity field associated with the specified distribution of $\boldsymbol{\omega}$ and given by (2.4.11). It may be shown in the same way that, provided $|\boldsymbol{\omega}(\mathbf{x})|$ is of order r^{-n} ($n > 3$) when r is large,

$$\mathbf{u}_v(\mathbf{x}) \sim -\frac{1}{4\pi} \{\textstyle\int \boldsymbol{\omega}' \, dV(\mathbf{x}')\} \times \nabla \frac{1}{r} \tag{2.9.3}$$

as $r \to \infty$. This asymptotic form represents the solenoidal velocity distribution associated with uniform vorticity in a volume element at the origin (compare (2.4.12)), the product of the vorticity and the volume of the element being equal to $\int \boldsymbol{\omega}' \, dV(\mathbf{x}')$, or, equivalently, associated with an element of a line vortex at the origin, the product of the (vector) element of length and the strength of the line vortex being equal to $\int \boldsymbol{\omega}' \, dV(\mathbf{x}')$. However, the vortex-lines are all closed curves lying in the fluid (or in some extended region, going beyond the inner boundary, over which the volume integral in (2.4.11) and (2.9.3) must be taken, as explained in § 2.4), which suggests that the integral in (2.9.3) vanishes; we see formally that this is so from the identity

$$\textstyle\int \omega_i(\mathbf{x}) \, dV(\mathbf{x}) = \int \nabla . (x_i \boldsymbol{\omega}) \, dV(\mathbf{x}),$$

and use of the divergence theorem, and the supposed smallness of $|\boldsymbol{\omega}|$ when r is large.

It is therefore necessary to obtain a higher-order approximation to $\mathbf{u}_v$, by developing the Taylor series for s^{-1} by one more term in the manner that led to (2.9.2). With the stronger restriction that $|\boldsymbol{\omega}|$ is of order r^{-n} $(n > 4)$ when r is large, we find that, as $r \to \infty$,

$$\mathbf{u}_v(\mathbf{x}) \sim \frac{1}{4\pi} \int \boldsymbol{\omega}' \times \left\{ \mathbf{x}' . \nabla \left(\nabla \frac{1}{r} \right) \right\} dV(\mathbf{x}')$$

$$= -\frac{1}{4\pi} \nabla \times \int \boldsymbol{\omega}' \mathbf{x}' . \nabla \frac{1}{r} \, dV(\mathbf{x}'). \tag{2.9.4}$$

This expression can be interpreted more readily by noting that

$$\int (x_i \omega_j + x_j \omega_i) \, dV(\mathbf{x}) = \int \nabla . (x_i x_j \boldsymbol{\omega}) \, dV(\mathbf{x}),$$

$$= 0$$

from the divergence theorem and the supposed smallness of $|\boldsymbol{\omega}|$ when r is large. As a consequence,

$$\int \boldsymbol{\omega}' \mathbf{x}' . \nabla \frac{1}{r} \, dV(\mathbf{x}') = \frac{1}{2} \int \left(\boldsymbol{\omega}' \mathbf{x}' . \nabla \frac{1}{r} - \mathbf{x}' \boldsymbol{\omega}' . \nabla \frac{1}{r} \right) dV(\mathbf{x}')$$

$$= -\frac{1}{2} \left(\nabla \frac{1}{r} \right) \times \int \mathbf{x}' \times \boldsymbol{\omega}' \, dV(\mathbf{x}'). \tag{2.9.5}$$

The asymptotic expression for $\mathbf{u}_v(\mathbf{x})$ is thus

$$\mathbf{u}_v(\mathbf{x}) \sim \frac{1}{8\pi} \nabla \left\{ \left(\nabla \frac{1}{r} \right) . \int \mathbf{x}' \times \boldsymbol{\omega}' \, dV(\mathbf{x}') \right\}, \tag{2.9.6}$$

which may be seen to be of the same form as (2.9.2). Now for a single closed line vortex of strength κ of which a line element is $\delta\mathbf{l}$ we have

$$\frac{1}{2} \int \mathbf{x} \times \boldsymbol{\omega} \, dV(\mathbf{x}) = \frac{1}{2}\kappa \oint \mathbf{x} \times d\mathbf{l}(\mathbf{x})$$

$$= \kappa \int \mathbf{n} \, dA = \kappa \mathbf{A}, \tag{2.9.7}$$

where $\mathbf{n} \, \delta A$ is a vector element of any open surface bounded by the line vortex (with the direction of $\mathbf{n}$ defined relative to the sense of $\boldsymbol{\omega}$ round the line vortex), and $\mathbf{A}$, the total vectorial area of this surface, depends only on the shape of the closed line vortex. Hence the asymptotic form of $\mathbf{u}_v$ represents the solenoidal velocity distribution associated with a single closed line vortex of infinitesimal linear dimensions located at the origin such that the product of the strength and the vectorial area bounded by the vortex is equal to $\frac{1}{2} \int \mathbf{x} \times \boldsymbol{\omega} \, dV(\mathbf{x})$.

In summary, we have found that, in a case in which the total rate of expansion $\int \Delta \, dV(\mathbf{x})$ is zero, $\mathbf{u}_e$ and $\mathbf{u}_v$ have a common asymptotic form as $r \to \infty$, which is of order r^{-3}, and which represents the velocity field associated with either a source doublet or a single closed line vortex located at the origin.

The behaviour of ϕ at large distances

When the velocity of the fluid vanishes at infinity, and when the distributions of rate of expansion and vorticity are such that $\mathbf{u}_e$ and $\mathbf{u}_v$ vanish at infinity, the remaining contribution $\mathbf{v}(\mathbf{x})$ $(=\nabla\phi)$ must also vanish there. We shall now use the supposition that $\mathbf{v} \to$ o as $r \to \infty$ to determine the functional forms of $\mathbf{v}$ and ϕ at large values of r, the information obtained being useful in later considerations of solenoidal flow which is known to be irrotational in the outer parts of fluid of infinite extent. It will be shown first that ϕ tends to a constant value, in a particular way, as $r \to \infty$, as a direct consequence of the fact that ϕ satisfies the equation $\nabla^2\phi =$ o. For the moment we assume that ϕ is a single-valued function of $\mathbf{x}$, which is ensured when the

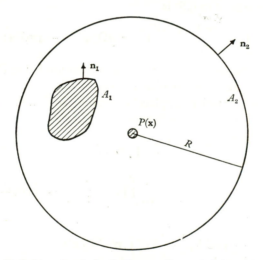

Figure 2.9.1. Definition sketch for fluid extending to infinity and at rest there.

region occupied by the fluid is singly-connected; the necessary modifications when ϕ is not single-valued are considered in the next section.

The inner bounding surface of the fluid will be denoted as before by A_1, with $\mathbf{n}_1$ the unit (outward) normal to an element of this surface. A_2 will denote the surface of a sphere with centre at a point $P(\mathbf{x})$ in the fluid and sufficiently large radius R to enclose all the interior boundaries, with $\mathbf{n}_2$ the unit (outward) normal to the sphere; the region outside and including A_2 is wholly occupied by fluid (figure 2.9.1). We make use of Green's theorem,† one form of which states that, if F and G are scalar functions of position which, together with their spatial derivatives, are single-valued, finite and

---

† Well known in vector analysis and potential theory. The relation (2.9.8) can be obtained by applying the divergence theorem for the volume V to the vector $F\nabla G - G\nabla F$.

continuous throughout the volume V bounded by A_1 and A_2,

$$\int (F\nabla G - G\nabla F).\mathbf{n}_2\, dA_2 - \int (F\nabla G - F\nabla G).\mathbf{n}_1\, dA_1 = \int (F\nabla^2 G - G\nabla^2 F)\, dV.$$
(2.9.8)

The particular choice of functions F and G to be made here is

$$F(\mathbf{x}') = \phi(\mathbf{x}'), \quad G(\mathbf{x}') = s^{-1},$$

where $s = |\mathbf{x} - \mathbf{x}'|$ is the distance between the point $P(\mathbf{x})$ and the point $\mathbf{x}'$ at which the element of integration lies. ϕ has the requisite properties of being single-valued, finite and continuous throughout V, but s^{-1} is not finite at P; P must therefore be surrounded by a sphere of small radius ϵ which is excluded from V and the surface of which must be regarded as included in the inner boundary. This additional contribution to the surface integration on the left-hand side of (2.9.8) is

$$-\int \left(\phi' \frac{\partial s^{-1}}{\partial s} - \frac{1}{s}\frac{\partial \phi'}{\partial s} \right)_{s=\epsilon} \epsilon^2\, d\Omega(\mathbf{x}'), \quad \rightarrow 4\pi\phi(\mathbf{x})$$
(2.9.9)

as $\epsilon \rightarrow 0$, where $\delta\Omega$ is an element of solid angle subtended at P and the prime denotes evaluation at the point $\mathbf{x}'$ as before.

Now both ϕ and s^{-1} satisfy Laplace's equation, so that the right-hand side of (2.9.8) vanishes and we are left with

$$\phi(\mathbf{x}) = \frac{1}{4\pi}\int \left(\phi'\nabla_{\mathbf{x}'}\frac{1}{s} - \frac{1}{s}\nabla\phi' \right).\mathbf{n}_1\, dA_1(\mathbf{x}')$$

$$- \frac{1}{4\pi}\int \left(\phi'\nabla_{\mathbf{x}'}\frac{1}{s} - \frac{1}{s}\nabla\phi' \right).\mathbf{n}_2\, dA_2(\mathbf{x}'),$$

and, since $s = R$ on A_2,

$$= -\frac{1}{4\pi}\int \left(\phi'\nabla_{\mathbf{x}}\frac{1}{s} + \frac{1}{s}\nabla\phi' \right).\mathbf{n}_1\, dA_1(\mathbf{x}')$$

$$+ \frac{1}{4\pi R^2}\int \phi'\, dA_2(\mathbf{x}') + \frac{1}{4\pi R}\int \mathbf{n}_2.\nabla\phi'\, dA_2(\mathbf{x}').$$

Since $\nabla.\mathbf{v} = 0$ everywhere within V, we have

$$\int \mathbf{n}_2.\nabla\phi'\, dA_2(\mathbf{x}') = \int \mathbf{n}_1.\nabla\phi'\, dA_1(\mathbf{x}'), \quad = m \quad \text{say}, \quad (2.9.10)$$

m being the flux of fluid volume across the internal boundary A_1 (in the outward direction) associated with the velocity field $\mathbf{v}$. Also we may write

$$\frac{1}{4\pi R^2}\int \phi'\, dA_2(\mathbf{x}') = \bar{\phi}(\mathbf{x}, R),$$
(2.9.11)

representing the mean value of ϕ over the spherical surface A_2, of radius R and centred at $\mathbf{x}$. Then

$$\phi(\mathbf{x}) = \bar{\phi} + \frac{m}{4\pi R} - \frac{1}{4\pi}\int \left(\phi'\nabla_{\mathbf{x}}\frac{1}{s} + \frac{1}{s}\nabla\phi' \right).\mathbf{n}_1\, dA_1(\mathbf{x}').$$
(2.9.12)

This relation is in a form suitable for the determination of the behaviour of ϕ at large values of r, since all terms on the right-hand side except the first tend to zero as $r\,(=|\mathbf{x}|)$, and hence also s, becomes large (with R also being made large in such a way that the sphere A_2 centred at $\mathbf{x}$ always encloses the internal boundaries). However, we need now to know more about the surviving term $\overline{\phi}$. This information is supplied by a theorem first established by Gauss for a gravitational potential, which also satisfies Laplace's equation in free space. Gauss's result is the relation (2.9.14) below, and the proof is as follows.

The flux relation (2.9.10) may be written as

$$R^2 \int \left(\frac{\partial \phi'}{\partial s}\right)_{s=R} d\Omega(\mathbf{x}') = R^2 \frac{\partial}{\partial R} \int (\phi')_{s=R} \, d\Omega(\mathbf{x}') = m, \qquad (2.9.13)$$

where $\delta\Omega$ is again an element of solid angle subtended at P. Hence, integration of (2.9.13) with respect to R gives

$$\overline{\phi}(\mathbf{x}, R) = \frac{1}{4\pi} \int (\phi')_{s=R} \, d\Omega(\mathbf{x}') = C - \frac{m}{4\pi R}, \qquad (2.9.14)$$

where C is independent of R. To see whether C depends on the position $\mathbf{x}$ of the centre of the sphere A_2, we calculate the derivative of C with respect to any component of $\mathbf{x}$, say x_1, with R kept constant:

$$\frac{\partial C}{\partial x_1} = \frac{\partial \overline{\phi}}{\partial x_1} = \frac{1}{4\pi R^2} \frac{\partial}{\partial x_1} \int \phi' \, dA_2(\mathbf{x}')$$

$$= \frac{1}{4\pi R^2} \int \frac{\partial \phi'}{\partial x_1'} \, dA_2(\mathbf{x}'). \qquad (2.9.15)$$

This last expression is the mean value of the velocity component v_1 over the surface of the sphere A_2, which we know to be zero for large values of R since $\mathbf{v}$ is zero everywhere at infinity. Hence C is independent of both R and $\mathbf{x}$.

On substituting (2.9.14) in (2.9.12) we find

$$\phi(\mathbf{x}) = C - \frac{1}{4\pi} \int \left(\phi' \nabla_{\mathbf{x}} \frac{1}{s} + \frac{1}{s} \nabla\phi'\right) . \mathbf{n}_1 \, dA_1(\mathbf{x}'), \qquad (2.9.16)$$

which is an expression dependent only on the position $\mathbf{x}$ and the conditions at the internal boundary. As $r \to \infty$, s also becomes large and the integrand in (2.9.16) becomes small everywhere on the finite surface A_1; hence

$$\phi(\mathbf{x}) \to C \quad \text{as} \quad r \to \infty.$$

Conditions for $\nabla\phi$ to be determined uniquely

The fact that ϕ tends to a constant value at infinity may now be used, together with (2.7.6), to establish the conditions for uniqueness of $\nabla\phi$. For, on choosing as external boundary A_2 a sphere of large radius R enclosing all the interior boundaries, the integral of $\mathbf{v}.\mathbf{v}$ over the whole volume of the fluid becomes

$$\int \mathbf{v}.\mathbf{v}\,dV = \lim_{R\to\infty} \int \phi \mathbf{v}.\mathbf{n}_2\,dA_2 - \int \phi \mathbf{v}.\mathbf{n}_1\,dA_1.$$

The value of $\int \mathbf{v}.\mathbf{n}_2\,dA_2$ is finite (and equal to the flux m across the inner boundary), so that

$$\int \mathbf{v}.\mathbf{v}\,dV = \lim_{R\to\infty} \int (\phi - C)\mathbf{v}.\mathbf{n}_2\,dA - \int (\phi - C)\mathbf{v}.\mathbf{n}_1\,dA_1$$

$$= -\int (\phi - C)\mathbf{v}.\mathbf{n}_1\,dA_1. \tag{2.9.17}$$

This relation takes the place of (2.7.6), for a fluid extending to infinity, and we see from it that the conditions under which two solutions $\nabla\phi$ and $\nabla\phi^*$ are necessarily identical are those such that

$$-\int (\phi - \phi^*)(\mathbf{v} - \mathbf{v}^*).\mathbf{n}_1\,dA_1 + (C - C^*)(m - m^*) = 0,$$

where C and C^* are the constant values of ϕ and ϕ^* at infinity and m and m^* are the fluxes of volume across the inner boundary corresponding to these two solutions. Again we see that, as stated in § 2.7, $\nabla\phi$ is determined uniquely when the value of the normal component of $\nabla\phi$ at each point of the boundaries of the fluid (the boundaries being wholly internal here) is prescribed, since this requires $\mathbf{v}.\mathbf{n}_1 = \mathbf{v}^*.\mathbf{n}_1$ at each point of A_1 and $m = m^*$. Again there is another, although less important, way of ensuring uniqueness of $\nabla\phi$, viz. to prescribe the value of ϕ at each point of A_1 and either the value of the flux m or the constant C to which ϕ tends at infinity.

The expression of φ as a power series

The exact relation (2.9.16) has been used as a means of showing that ϕ tends to a constant value at infinity. The relation is also of interest in itself, in that it shows explicitly how ϕ is determined throughout the fluid by conditions at the inner boundary. (Note, however, that (2.9.16) does not give $\phi(\mathbf{x})$ explicitly in terms of the normal component of $\nabla\phi$ alone at the inner boundary; the distribution of ϕ over the inner boundary is also involved. At first sight this does not seem to be consistent with the uniqueness theorem, which shows that $\phi(\mathbf{x})$ is determined uniquely, apart from an additive constant, by a prescribed distribution of $\mathbf{n}.\nabla\phi$ over the inner boundary. The explanation lies in the fact that the distributions of $\mathbf{n}.\nabla\phi$ and ϕ over the inner boundary are not independent; and in principle one of them may be eliminated.)

We shall use (2.9.16) to obtain a representation of $\phi(\mathbf{x})$ as a power series in r^{-1}, of which the constant C is the first term. The first step is to write s^{-1} as a Taylor series in $\mathbf{x}'$,

$$\frac{1}{s} = \frac{1}{r} - x_i' \frac{\partial}{\partial x_i}\left(\frac{1}{r}\right) + \tfrac{1}{2} x_i' x_j' \frac{\partial^2}{\partial x_i\,\partial x_j}\left(\frac{1}{r}\right) + \dots, \tag{2.9.18}$$

which is clearly convergent, when $r' < r$, for the cases $\mathbf{x}.\mathbf{x}' = \mp rr'$ and thence for all values of the angle between $\mathbf{x}$ and $\mathbf{x}'$. The series (2.9.18) may be substituted in (2.9.16) and the integration carried out term-by-term, provided r is greater than the largest value of r' involved in the integration, giving

$$\phi(\mathbf{x}) = C + \frac{c}{r} + c_i \frac{\partial}{\partial x_i}\left(\frac{1}{r}\right) + c_{ij}\frac{\partial^2}{\partial x_i\,\partial x_j}\left(\frac{1}{r}\right) + ..., \qquad (2.9.19)$$

where

$$c = -\frac{1}{4\pi}\int \mathbf{n}.\nabla\phi\,dA = -\frac{m}{4\pi}, \quad c_i = \frac{1}{4\pi}\int (x_i\,\mathbf{n}.\nabla\phi - n_i\phi)\,dA, \\[2mm] c_{ij} = \frac{1}{4\pi}\int (-\tfrac{1}{2}x_i x_j\,\mathbf{n}.\nabla\phi + x_i n_j\phi)\,dA, \quad .... \qquad\Big\} \quad (2.9.20)$$

These integrals are taken over the whole of the inner boundary of the fluid, an element of which is now denoted by $\mathbf{n}\,\delta A$, the suffix 1 being superfluous.

This interesting series shows that, in the region external to a sphere centred at the origin and enclosing the inner boundary, the potential ϕ may be written as the sum of a number of contributions of different integral degree in r^{-1}, each of which satisfies $\nabla^2\phi = 0$ (for $\phi = r^{-1}$ satisfies this equation and hence so do all spatial derivatives of r^{-1}), and each of which represents the potential due to a point singularity, located at the origin and constructed from point sources in the manner described in §2.5. The set of independent solutions of $\nabla^2\phi = 0$

$$\frac{1}{r}, \quad \frac{\partial}{\partial x_i}\left(\frac{1}{r}\right), \quad \frac{\partial^2}{\partial x_i\,\partial x_j}\left(\frac{1}{r}\right), \quad .... \qquad (2.9.21)$$

play a fundamental part in the theory of harmonic functions,[†] and are known as *spherical solid harmonics* of degree $-1, -2, -3, ...$ The corresponding coefficients of $r^{-1}, r^{-2}, ...$, of the general form

$$S_n = r^{n+1}\frac{\partial^n}{\partial x_i\,\partial x_j...}\left(\frac{1}{r}\right) \quad (n = 0, 1, 2, ...), \qquad (2.9.22)$$

depend only on the direction of the vector $\mathbf{x}$—or, equivalently, on position on a sphere centred at the origin—and are known as spherical *surface* harmonics of integral order. It follows immediately from the form of Laplace's equation in spherical polar co-ordinates (see appendix 2) that if $r^{-n-1}S_n$ is a solution, so too is

$$\phi(\mathbf{x}) = r^n S_n;$$

that is, to every spherical solid harmonic of degree $-n-1$ there corresponds one of degree n (n being a positive integer). The spherical solid harmonics of negative degree are needed, and are sufficient, for the representation of ϕ as a power series in a region exterior to a sphere and extending to infinity

† See, for instance, chapter 24 of *Methods of Mathematical Physics*, by H. Jeffreys and
   B. S. Jeffreys, 3rd ed. (Cambridge University Press, 1956).

where the fluid is at rest, whereas those of positive degree suffice in a region of fluid interior to a sphere; both are needed in a region bounded both internally and externally.

It will be noticed that the second term on the right-hand side of (2.9.19)—which is the first in the corresponding series for $\mathbf{v} = \nabla\phi$—represents the velocity field associated with a point source of strength $\int \mathbf{n} . \nabla\phi \, dA \ (=m)$ at the origin. In other words, the effect of the net flux of $\mathbf{v}$ across the inner boundary dominates the expression for $\mathbf{v}$ at large distances from the boundary and $\mathbf{v}$ is there the same as if the flux emanated from a single point (the exact location of this point being arbitrary, so far as the leading term in the series for $\mathbf{v}$ is concerned). As already remarked, the commonest case is that in which the fluid is bounded internally by a rigid boundary, and for irrotational flow of an incompressible fluid (for which $\mathbf{v}$ represents the actual fluid velocity) outside such a boundary the net flux m is necessarily zero; thus $\mathbf{v}$ is here of order r^{-3} at large distances from the boundary. The same is true for the more general situation in which Δ and $\boldsymbol{\omega}$ are non-zero and $\mathbf{v}$ is one of three contributions to the actual fluid velocity, because the net volume flux across a rigid inner boundary corresponding to the contributions $\mathbf{u}_e$ and $\mathbf{u}_v$ may be shown to be zero. This latter flux is

$$\int (\mathbf{u}_e + \mathbf{u}_v) . \mathbf{n} \, dA = \int \nabla . (\mathbf{u}_e + \mathbf{u}_v) \, dV,$$

where the volume integral is taken over the region within the closed rigid boundary; $\mathbf{u}_e$ and $\mathbf{u}_v$ do not have direct physical meaning in this region, but they are defined mathematically at points $\mathbf{x}$ in this region by the expressions (2.4.5) and (2.4.11) and are solenoidal there. Thus m, as defined by (2.9.10), is the actual net flux of fluid volume across the interior boundary, and must be zero when the interior boundary is rigid.

Irrotational solenoidal flow due to a rigid body in translational motion

$\mathbf{v}$ takes a particular form when the condition to be satisfied at the interior boundary is

$$\mathbf{n} . \mathbf{v} = \mathbf{n} . \mathbf{U}$$

at all points of a given closed surface, where $\mathbf{U}$ is a given vector constant, as would be required if $\mathbf{v}$ represented the actual velocity in irrotational solenoidal flow due to translational motion of a rigid body with velocity $\mathbf{U}$ through fluid which is at rest at infinity.† The determination of the velocity $\mathbf{v}$ here reduces to the problem of finding a solution of $\nabla^2\phi = 0$ which satisfies the conditions

$$\phi(\mathbf{x}) \rightarrow \text{constant} \quad \text{as} \quad r \rightarrow \infty,$$

$$\mathbf{n} . \nabla\phi = \mathbf{n} . \mathbf{U} \quad \text{at the surface of the body.}$$

---

† The reader is reminded again that the conditions under which actual flow due to a moving rigid body is solenoidal and irrotational have yet to be determined from the dynamical equations.

We know from the uniqueness theorem that there is only one solution for $\nabla\phi$ which can satisfy these conditions. An arbitrary constant may be added to ϕ without affecting $\mathbf{v}$ or the equation for ϕ or the inner boundary condition; consequently the value of the constant to which ϕ tends at infinity can here be chosen arbitrarily, and will be taken as zero for convenience. The differential equation and the relations to be satisfied at the boundaries are now linear and homogeneous in ϕ and $\mathbf{U}$, and since the solution is to be valid for all choices of $\mathbf{U}$ it must be of the form

$$\phi(\mathbf{x}) = \mathbf{U}\cdot\boldsymbol{\Phi}(\mathbf{x}). \qquad (2.9.23)$$

Here $\boldsymbol{\Phi}(\mathbf{x})$ is an unknown vector function *independent* of both the magnitude and direction of $\mathbf{U}$.† Since $\boldsymbol{\Phi}$ is determined by the inner boundary condition, it follows that $\boldsymbol{\Phi}$ depends only on position in the fluid *relative to the body*, that is, only on $\mathbf{x} - \mathbf{x}_0$, where $\mathbf{x}_0$ is the instantaneous position vector of some material point of the body. The form (2.9.23) for ϕ is found (by the same argument) also to hold when a rigid body moves through fluid which does not extend to infinity but is bounded externally by a rigid stationary boundary, although here $\boldsymbol{\Phi}$ does not depend solely on position relative to the rigid body.

The relation (2.9.23) is useful in a number of contexts, and may even be an aid to the direct determination of ϕ. For instance, we see immediately that in the case of a spherical rigid body with centre instantaneously at the origin, no vector or direction occurs in the specification of the shape of the boundary and $\mathbf{x}$ is the only vector which can occur in the expression for $\boldsymbol{\Phi}$. It follows that the only one of the set of independent solutions (2.9.21) which can be combined with $\mathbf{U}$ to give a solution of the form (2.9.23) is the second, and that

$$\phi(\mathbf{x}) = \alpha\mathbf{U}\cdot\nabla\frac{1}{r} = -\alpha\frac{\mathbf{U}\cdot\mathbf{x}}{r^3}, \qquad (2.9.24)$$

where α is a constant, is a solution of the required form. In terms of spherical polar co-ordinates with $\theta = 0$ in the direction of $\mathbf{U}$, the corresponding fluid velocity has components

$$\frac{\partial\phi}{\partial r} = \frac{\partial}{\partial r}\left(-\alpha\frac{U\cos\theta}{r^2}\right) = \alpha\frac{2U\cos\theta}{r^3}, \quad \frac{1}{r}\frac{\partial\phi}{\partial\theta} = \alpha\frac{U\sin\theta}{r^3}. \qquad (2.9.25)$$

The inner boundary condition is satisfied for a sphere of radius a if

$$\frac{\partial\phi}{\partial r} = U\cos\theta \quad\text{at}\quad r = a,$$

---

† Some readers may find it evident that ϕ must be a linear and homogeneous function of the three components of $\mathbf{U}$, but be unaccustomed to this kind of argument for vectors. The equation and boundary conditions determining ϕ have been expressed in a form independent of the co-ordinate system, and the expression for ϕ in terms of the components of $\mathbf{U}$ must likewise be independent of the co-ordinate system; that is, the three components of $\mathbf{U}$ can occur only in the combination required to make up the vector $\mathbf{U}$, giving (2.9.23).

which requires
$$\alpha = \tfrac{1}{2}a^3;$$

thus
$$\phi(\mathbf{x}) = \tfrac{1}{2}a^3\mathbf{U}.\nabla\frac{1}{r} = -\tfrac{1}{2}a^3\frac{\mathbf{U}.\mathbf{x}}{r^3} = -\tfrac{1}{2}a^3 U\frac{\cos\theta}{r^2}. \qquad (2.9.26)$$

These formulae refer to the velocity at a position defined by axes fixed relative to the fluid at infinity and with the origin of the co-ordinate system at the instantaneous position of the centre of the sphere. With a different origin such that the centre of the sphere is at position $\mathbf{x}_0$, the velocity distribution is obviously identical provided the position vector is measured relative to $\mathbf{x}_0$, so that we have

$$\phi(\mathbf{x}) = \tfrac{1}{2}a^3\mathbf{U}.\nabla\frac{1}{|\mathbf{x}-\mathbf{x}_0|} = -\tfrac{1}{2}a^3\frac{\mathbf{U}.(\mathbf{x}-\mathbf{x}_0)}{|\mathbf{x}-\mathbf{x}_0|^3}. \qquad (2.9.27)$$

We note also for future use that the velocity potential for the flow relative to axes moving with the sphere and with origin at the centre is

$$\phi(\mathbf{x}) = -\mathbf{U}.\mathbf{x}\left(1+\frac{1}{2}\frac{a^3}{r^3}\right). \qquad (2.9.28)$$

2.10. Two-dimensional flow fields extending to infinity

Provided the fluid does not extend to infinity in the plane of the motion, the formulae of earlier sections (and in particular of §2.8 when the fluid is bounded internally and hence occupies a multiply-connected region) are easily adapted to apply to a case of two-dimensional motion. The fluid necessarily extends to an infinite distance (in the mathematical version of the problem) in the direction normal to the plane of motion, but the behaviour of the velocity 'at infinity' is here known and no difficulties arise; where a surface integral has to be taken over the boundary of the fluid, it will often be useful to imagine the flow field to be bounded by two planes parallel to the plane of motion on which the normal component of the fluid velocity is zero.

However, two-dimensional flow in a fluid bounded internally and extending to infinity in all directions in the plane of motion does have some special features requiring separate consideration. We shall suppose that the fluid is at rest at large distances from the origin in the plane of motion, the origin being located near the interior boundaries of the fluid. The proofs of the formulae of §2.9, and in particular of those giving the behaviour of ϕ at large distances from the inner boundary, need modification, since they are based on the assumption that the fluid velocity is small everywhere on a *sphere* of large radius centred near the inner boundaries. The modifications required do not present much difficulty and can therefore be given in outline only.

Arguments like those leading to the relations (2.9.1) and (2.9.2) lead again to the conclusion that, provided $|\Delta|$ is suitably small at large distances from the inner boundary, the velocity field associated with a prescribed distribution of the rate of expansion Δ behaves asymptotically as if all the expansion were located at the origin; and if it happens that $\int \Delta' \, dV(\mathbf{x}') = 0$ (where the element of volume is now a cylinder of unit depth normal to the plane of motion and with cross-section of area δV), the velocity field far from the inner boundary is the same as if a source doublet were located at the origin. Results corresponding to those represented by (2.9.3) and (2.9.6) may also be obtained for the velocity $\mathbf{u}_v$ associated with a given vorticity distribution in two dimensions.

The important problem of determining the behaviour of $\phi(\mathbf{x})$ at large values of $r \, (= |\mathbf{x}|)$, given that $\nabla\phi \to 0$ as $r \to \infty$, may be investigated by the same general method. The relation (2.9.8) from Green's theorem holds in two dimensions, under the same conditions on F and G, provided δA_1 and δA_2 are now elements of length and δV is an element of area in the plane of motion (exactly as if they referred to a layer of fluid of unit depth normal to the plane of motion); the outer boundary A_2 is now a circle with centre at $P(\mathbf{x})$ and of sufficiently large radius R to enclose all the interior boundaries. We choose

$$F(\mathbf{x}') = \phi(\mathbf{x}'), \quad G(\mathbf{x}') = \log s,$$

both of which satisfy Laplace's equation in two dimensions, where $s = |\mathbf{x} - \mathbf{x}'|$ as before and ϕ is a *single-valued* velocity potential. (Flows with many-valued potentials do occur since the region concerned is multiply-connected, but we exclude such cases for the moment in order to be able to use Green's theorem.) The point $P(\mathbf{x})$ is surrounded by a small circle, the interior of which must be excluded from the integration with respect to V, and in place of (2.9.9) we find an additional contribution $-2\pi\phi(\mathbf{x})$ to the integration with respect to A_1. The flux relation (2.9.10) stands without change, and we find, in place of (2.9.12),

$$\phi(\mathbf{x}) = \overline{\phi} - \frac{m}{2\pi}\log R + \frac{1}{2\pi}\int (\phi'\nabla_{\mathbf{x}}\log s + \log s \, \nabla\phi') . \mathbf{n}_1 \, dA_1(\mathbf{x}'),$$

in which

$$\overline{\phi}(\mathbf{x}, R) = \frac{1}{2\pi R}\int \phi' \, dA_2(\mathbf{x}'). \tag{2.10.1}$$

In place of (2.9.14) we find, by integration of the flux relation corresponding to (2.9.13),

$$\overline{\phi} = C + \frac{m}{2\pi}\log R; \tag{2.10.2}$$

C is a constant of integration independent of R, and an investigation as before shows that when $\nabla\phi$ vanishes at infinity C is also independent of the position $\mathbf{x}$ of the centre of the circle A_2. We then have, in place of (2.9.16),

$$\phi(\mathbf{x}) = C + \frac{1}{2\pi}\int (\phi'\nabla_{\mathbf{x}}\log s + \log s \, \nabla\phi') . \mathbf{n}_1 \, dA_1(\mathbf{x}'). \tag{2.10.3}$$

The asymptotic form of ϕ can now be found. For we have from (2.10.3)

$$\phi(\mathbf{x}) - C - \frac{m}{2\pi}\log r = \frac{1}{2\pi}\int\left(\frac{\mathbf{s}}{s^2}\phi' + \log\frac{s}{r}\nabla\phi'\right).\mathbf{n}_1\,dA_1(\mathbf{x}'),$$

$$\to 0$$

as $r \to \infty$. The result that ϕ does not tend to a constant at infinity here is associated with the fact that the 'source term' in the series for ϕ corresponding to (2.9.20), which is the term with the slowest rate of decrease as $r \to \infty$, does not decrease at all in a two-dimensional space but increases as $\log r$.

Despite this difference in the behaviour of ϕ as $r \to \infty$, the conditions for uniqueness of $\nabla\phi$ have the same form as in a three-dimensional flow field. The quantity $\phi - (m/2\pi)\log r$ can be regarded as the (single-valued) velocity potential of a flow field and is known to approach a constant value at infinity in the plane of motion; consequently the divergence theorem, applied in the manner that leads to (2.9.17), for the volume of fluid bounded by two planes parallel to the plane of motion, shows that the gradient of $\phi - (m/2\pi)\log r$ is determined uniquely everywhere when the value of the normal derivative of $\phi - (m/2\pi)\log r$ at each point of the internal boundary is prescribed. But if the value of the normal derivative of ϕ at each point of the inner boundary is prescribed, the value of m (the net flux of volume of fluid across the internal boundary) is known and the normal derivative of $\phi - (m/2\pi)\log r$ at each point of the inner boundary is known. Consequently, specification of the normal derivative of ϕ at each point of the internal boundary determines uniquely the value of $\nabla\phi$ everywhere. (Likewise there can be at most one solution for $\nabla\phi$ when the value of m and the value of ϕ at each point of the inner boundary are prescribed.)

The above remarks apply to single-valued velocity potentials, and thus apply to the difference between two many-valued velocity potentials which are known to have the same cyclic constants. Hence we may assert quite generally, in the manner of §2.8, that two-dimensional irrotational solenoidal flow in a region bounded internally and extending to infinity (where the fluid is at rest) is determined uniquely everywhere when the cyclic constant (or constants, if the degree of connectivity is greater than two) of the motion is given and the normal component of $\nabla\phi$ is given at each point of the internal boundary. We may go further in the case of flow in the doubly-connected region outside a single cylinder, with the cyclic constant κ; a simple solution of Laplace's equation having the same cyclic character (and making no contribution to m) is $\kappa\theta/2\pi$, where θ is the polar angle in the plane of motion relative to an origin within the internal boundary, so that in this case

$$\phi(\mathbf{x}) - \frac{\kappa}{2\pi}\theta$$

is a single-valued velocity function to which the above deductions, and in particular the exact relation (2.10.3), apply.

We may again see in more detail the variation of ϕ at large distances from the internal boundary by expanding ϕ as a power series in r^{-1}. When $r'/r < 1$, $\log s$ may be written as a Taylor series in $\mathbf{x}'$, like (2.9.18), and substitution of the series in (2.10.3) gives, for a single-valued ϕ,

$$\phi(\mathbf{x}) = C + c \log r + c_i \frac{\partial}{\partial x_i}(\log r) + c_{ij} \frac{\partial^2}{\partial x_i \, \partial x_j}(\log r) + ..., \quad (2.10.4)$$

where

$$c = \frac{1}{2\pi}\int \mathbf{n} . \nabla\phi \, dA = \frac{m}{2\pi}, \quad c_i = \frac{1}{2\pi}\int (-x_i \mathbf{n} . \nabla\phi + n_i \phi) \, dA,$$

$$c_{ij} = \frac{1}{2\pi}\int (\tfrac{1}{2} x_i x_j \mathbf{n} . \nabla\phi - x_i n_j \phi) \, dA, \quad ....$$

These integrals are taken over the inner boundary of the fluid, an element of which is now denoted by $\mathbf{n}\,\delta A$. The set of fundamental solutions of Laplace's equation in two dimensions generated by the terms of this series, viz.

$$\log r, \quad \frac{\partial}{\partial x_i}(\log r), \quad \frac{\partial^2}{\partial x_i \, \partial x_j}(\log r), \quad ... \quad (2.10.5)$$

are termed *circular harmonics* of integral degree, and play a part analogous in every way to that of the spherical harmonics of §2.9. The quantity

$$S_n = r^n \frac{\partial^n}{\partial x_i \, \partial x_j \, ...}(\log r) \quad (n = 0, 1, 2, ...)$$

depends only on the direction of $\mathbf{x}$, and it follows from the form of Laplace's equation in terms of (two-dimensional) polar co-ordinates (see appendix 2) that, if $r^{-n}S_n$ is a solution, so too is $r^n S_n$, giving a corresponding set of fundamental solutions of positive degree in r.

For a many-valued ϕ appropriate to flow in the doubly-connected region outside a cylinder, with cyclic constant κ, the series (2.10.4) is replaced by

$$\phi(\mathbf{x}) = C + \frac{\kappa}{2\pi}\theta + \frac{m}{2\pi}\log r + c_i \frac{\partial}{\partial x_i}(\log r)$$

$$+ c_{ij}\frac{\partial^2}{\partial x_i \, \partial x_j}(\log r) + ..., \quad (2.10.6)$$

in which the coefficients $mc_i, c_{ij}, ...$ are equal to corresponding integrals of $\phi - (\kappa/2\pi)\theta$ and its normal derivative over the internal boundary. The first variable term on the right-hand side of (2.10.6) represents the potential due to a 'point vortex' (that is, a straight line vortex in three-dimensional space— see (2.8.5)) of strength κ at the origin and accounts for the many-valued character of ϕ; the second represents the potential due to a point source of strength m at the origin, and accounts for the net flux across the internal boundary, as already remarked (and will vanish when the internal boundary is rigid, as in three-dimensional space); the third represents the potential due to a source doublet of (vector) strength $-2\pi c_i$ at the origin; and so on.

Many of these results may be expressed in a natural way in terms of the complex potential introduced in §2.7. The analytic function of $z\ (=x+iy)$ that has as its real part the sum of the 'point vortex' and 'point source' terms in (2.10.6) is

$$\frac{1}{2\pi}(m-i\kappa)\log z.$$

The stream function ψ corresponding to this complex potential is a many-valued function of position, owing to the existence of the non-zero volume flux across the inner boundary, as was to be expected from the definition of ψ in §2.2. The many-valuedness of ψ is similar in type to that of ϕ, with m taking the place of the cyclic constant κ, and is another manifestation of the conjugacy of the two functions ϕ and ψ in two-dimensional irrotational solenoidal flow. The complex potential corresponding to other terms in (2.10.6) may be recognized with the help of the relation

$$\frac{\partial^n \log r}{\partial x^m\, \partial y^{n-m}} = \mathscr{R}\left(\frac{\partial^n \log z}{\partial x^m\, \partial y^{n-m}}\right) = \mathscr{R}\left(i^{n-m}\frac{d^n \log z}{dz^n}\right),$$

which shows, incidentally, that there are only two independent circular harmonics of degree $-n$, viz. the real and imaginary parts of $d^n \log z/dz^n$, or $r^{-n}\cos n\theta$ and $r^{-n}\sin n\theta$. Thus the complex potential corresponding to the whole of (2.10.6) can be written as

$$w(z) = \frac{1}{2\pi}(m-i\kappa)\log z + C + \sum_{n=1}^{\infty} D_n \frac{d^n \log z}{dz^n},$$

$$= \frac{1}{2\pi}(m-i\kappa)\log z + \sum_{n=0}^{\infty} A_n z^{-n}, \tag{2.10.7}$$

in which the constants D_n and A_n are complex. The real and imaginary parts of A_n are related to the real coefficients $C, c_i, c_{ij}, \ldots$ in (2.10.6), e.g.

$$A_0 = C, \quad A_1 = c_1 + ic_2, \quad A_2 = c_{22} - c_{11} - ic_{12},$$

where the suffixes 1 and 2 denote components in the directions of the x- and y-axes respectively. The series in (2.10.7) is recognizable as the Laurent series for a function which is known to be analytic and single-valued in the region of the z-plane outside a circle centred at the origin and to tend to a constant at infinity.

Irrotational solenoidal flow due to a rigid body in translational motion

As in §2.9, we may obtain more specific results about ϕ when the normal derivative of ϕ at the internal boundary satisfies the simple condition

$$\mathbf{n}.\nabla\phi = \mathbf{n}.\mathbf{U},$$

where $\mathbf{U}$ is the velocity of the rigid body bounding the fluid internally. A single-valued ϕ which tends to zero as $r \to \infty$ then satisfies a differential equation and boundary conditions which are linear and homogeneous in ϕ and $\mathbf{U}$ and which determine it uniquely, and so it must be of the form

$$\phi(\mathbf{x}) = \mathbf{U} \cdot \mathbf{\Phi}(\mathbf{x}). \tag{2.10.8}$$

The unknown function $\mathbf{\Phi}(\mathbf{x})$ is independent of $\mathbf{U}$ and depends only on position in the fluid relative to the body.

In the particular case of a circular body of radius a and centre instantaneously at the origin, no vector or direction occurs in the specification of the shape of the boundary. Hence the only solution among the set (2.10.5) which can be combined with $\mathbf{U}$ to give a solution of the form (2.10.8) is the second. The solution is therefore of the form

$$\phi(\mathbf{x}) = \alpha \mathbf{U} \cdot \nabla(\log r) = \alpha \frac{\mathbf{U} \cdot \mathbf{x}}{r^2} = \alpha \frac{U \cos \theta}{r}, \tag{2.10.9}$$

where α is a constant and r, θ are polar co-ordinates with $\theta = 0$ in the direction of $\mathbf{U}$. The corresponding fluid velocity has components

$$\frac{\partial \phi}{\partial r} = -\alpha \frac{U \cos \theta}{r^2}, \quad \frac{1}{r} \frac{\partial \phi}{\partial \theta} = -\alpha \frac{U \sin \theta}{r^2}, \tag{2.10.10}$$

and satisfies both the outer and inner boundary conditions if

$$\frac{\partial \phi}{\partial r} = U \cos \theta \quad \text{at} \quad r = a,$$

i.e., if
$$\alpha = -a^2. \tag{2.10.11}$$

This is the only possible solution when ϕ is single-valued. We note for future use that the velocity potential for the flow relative to axes moving with the cylinder and with origin at the centre is

$$\phi(\mathbf{x}) = -\mathbf{U} \cdot \mathbf{x} \left(1 + \frac{a^2}{r^2} \right) = -U \cos \theta \left(r + \frac{a^2}{r} \right). \tag{2.10.12}$$

If now there is a circulation κ round the rigid body moving with velocity $\mathbf{U}$, we can write the velocity potential as the sum of the term ϕ_1 representing the flow due to the same body moving with velocity $\mathbf{U}$ and with zero circulation round it, and a term ϕ_2 representing the flow due to circulation κ round the same body at rest. For ϕ_1 we have the form (2.10.8). ϕ_2 does not depend on $\mathbf{U}$ in any way and is necessarily linear in κ, so that we may put

$$\phi_2 = \kappa \left\{ \frac{\theta}{2\pi} + \Psi(\mathbf{x}) \right\}, \tag{2.10.13}$$

where Ψ' is a single-valued function of $\mathbf{x}$ independent of κ. Ψ' satisfies Laplace's equation, has zero gradient at infinity, satisfies the condition

$$\mathbf{n}.\nabla\left(\frac{\theta}{2\pi}+\Psi'\right) = 0 \qquad (2.10.14)$$

at the surface of the body, and is therefore determined uniquely (apart from an additive constant). In the particular case of a circular body with centre instantaneously coincident with the origin, the one possible functional form for Ψ' is $\Psi' = \text{const.}$ ($= 0$, say), so that the total velocity potential is here

$$\phi = \frac{\kappa}{2\pi}\theta - a^2\frac{\mathbf{U}.\mathbf{x}}{r^2}. \qquad (2.10.15)$$

The streamlines and other properties of this and related flow fields will be described in chapter 6.

Exercises for chapter 2

1. Show that the rate of extension of a material line element at a point P in a fluid varies with direction in the same way as PQ^{-2}, where PQ is parallel to the line element and the point Q lies on a rate-of-strain quadric centred on P.

2. Show that the vector potential

$$\mathbf{B}_v(\mathbf{x}) = \frac{1}{4\pi}\int_S \frac{\boldsymbol{\omega}'}{s}\,dV(\mathbf{x}') - \frac{1}{4\pi}\int_S \frac{\mathbf{n}\times\mathbf{u}'}{s}\,dA(\mathbf{x}')$$

corresponds to the vorticity $\boldsymbol{\omega}$ everywhere in the volume V bounded by the surface A, where the notation is that of §2.4.

3. Use Green's theorem to show that any acyclic irrotational solenoidal motion with velocity potential ϕ in a given region may be regarded as being due (i) to a distribution of sources over the boundary of the region, with strength $\mathbf{n}.\nabla\phi(\mathbf{x})$ per unit area at position $\mathbf{x}$ at the boundary, together with source doublets of strength $-\mathbf{n}\phi(\mathbf{x})$ per unit area, where $\mathbf{n}$ is the unit normal to the boundary and is directed into the fluid; or (ii) to a distribution of sources over the boundary with strength density $\mathbf{n}.\nabla(\phi+\phi^*)$, where ϕ^* is the potential of that acyclic irrotational solenoidal motion in the remainder of infinite space for which $\phi^* = \phi$ at the common boundary of the two motions and $\nabla\phi^* = 0$ (or $\nabla\phi = 0$, as appropriate) at infinity; or (iii) to a distribution of source doublets with strength density $-\mathbf{n}(\phi-\phi^*)$, where ϕ^* is the potential of that acyclic motion in the remainder of space for which $\mathbf{n}.\nabla\phi^* = \mathbf{n}.\nabla\phi$ at the boundary and $\nabla\phi^*$ (or $\nabla\phi$) = 0 at infinity.

4. Show that the irrotational solenoidal motion due to a line vortex of strength κ is the same as that due to a distribution of source doublets over an open surface whose bounding curve coincides with the line vortex, the strength per unit area being $\kappa\mathbf{n}$, where $\mathbf{n}$ is the unit normal to the surface. A sheet vortex is thus equivalent to a distribution of source doublets over, and normal to, the surface coinciding with the sheet, provided the closed vortex-lines are reducible on the sheet; and conversely. Hence show that any irrotational solenoidal motion, whether acyclic or not, can be regarded as being due to a certain sheet vortex coinciding with the boundary of the region of motion.

3

EQUATIONS GOVERNING THE MOTION OF A FLUID

3.1. Material integrals in a moving fluid

Dynamical relations describing the motion of a fluid are concerned essentially with the response of a specified piece or mass of the fluid to external influences. It is useful therefore to develop ways of describing the physical history of a material portion of fluid which may be undergoing distortion as well as change of position.

As a preliminary piece of kinematics, we consider the changes in size and orientation of material volume, surface and line *elements* due to the movement of the fluid. The elements will be assumed to be so small in linear dimensions that at any instant they are being subjected to a pure straining motion and a rigid rotation (as well as translational motion), as indicated by (2.3.13). However, in a consideration of the change in the volume, vector area or vector length of the material element it proves to be more convenient not to make an explicit division of the change in the element into a pure strain and a rigid rotation.

Consider first a material element of fluid whose volume is $\delta\tau$. The rate of change of this volume is, as remarked in §2.2,

$$\frac{d\,\delta\tau}{dt} = \int_{\delta\tau} \nabla.\mathbf{u}\,d\tau = \nabla.\mathbf{u}\,\delta\tau + o(\delta\tau), \qquad (3.1.1)$$

in which the rate of expansion $\nabla.\mathbf{u}$ is evaluated at the instantaneous position of the material volume element and the symbol $o(\delta\tau)$ denotes a quantity of smaller order than $\delta\tau$. A convenient way of obtaining an exact relation from (3.1.1) is to consider the ratio of $\delta\tau$ to its value at some initial instant, t_0 say, $\delta\tau(t_0)$ then being made indefinitely small. Thus

$$\frac{d\tau^*}{dt} = \nabla.\mathbf{u}\tau^*, \qquad (3.1.2)$$

where
$$\tau^* = \lim_{\delta\tau(t_0)\to 0} \frac{\delta\tau(t)}{\delta\tau(t_0)}.$$

τ^* is a dimensionless form of the instantaneous specific volume of the fluid in a material element, and is evidently equal to $\rho(t_0)/\rho(t)$, where ρ is the density of the same portion of fluid.

The rate of change of the vector $\delta\mathbf{l}$ representing a material line element which remains approximately straight is simply the difference between the velocities at the two ends of the element, that is,

$$\frac{d\,\delta \mathbf{l}}{dt} = \delta \mathbf{l}.\nabla \mathbf{u} + o(|\delta \mathbf{l}|). \tag{3.1.3}$$

Again we can make this an exact relation by dividing by $|\delta \mathbf{l}(t_0)|$ and taking the limit as $|\delta \mathbf{l}(t_0)| \to 0$.

The rate of change of volume of a material element depends on the magnitude of that volume, but not on the shape of the surface bounding it. We may therefore choose a material volume element in the form of a cylinder whose two end faces are identical material surface elements with vector area represented by $\delta \mathbf{S}$ and of which a generator is the material line element $\delta \mathbf{l}$; such a material volume element remains cylindrical under the action of pure straining and rigid-body rotation, although $\delta \mathbf{l}$, $\delta \mathbf{S}$ and the angle between these vectors all change, and

$$\delta \tau = \delta \mathbf{l}.\delta \mathbf{S} + o(\delta \tau) \tag{3.1.4}$$

at all times. On substituting (3.1.4) in (3.1.1) we find with the help of (3.1.3) that

$$\delta l_i \left(\frac{d\,\delta S_i}{dt} + \delta S_j \frac{\partial u_j}{\partial x_i} - \delta S_i \frac{\partial u_j}{\partial x_j} \right) = o(\delta \tau)$$

(vector notation being less convenient here), and since this relation must hold for all choices of $\delta \mathbf{l}$ we have

$$\frac{d\,\delta S_i}{dt} = \delta S_i \frac{\partial u_j}{\partial x_j} - \delta S_j \frac{\partial u_j}{\partial x_i} + o(|\delta \mathbf{S}|). \tag{3.1.5}$$

Again an exact relation may be obtained by dividing by $|\delta \mathbf{S}(t_0)|$ and taking the limit as $|\delta \mathbf{S}(t_0)| \to 0$. An alternative way of writing this expression for the rate of a change of a material surface element which follows from the mass-conservation equation (2.2.3) is

$$\frac{d(\rho\,\delta S_i)}{dt} = -\rho\,\delta S_j \frac{\partial u_j}{\partial x_i} + o(|\delta \mathbf{S}|), \tag{3.1.6}$$

in which ρ is to be evaluated at the position of the moving element.

The interesting duality of the behaviours of $\delta \mathbf{l}$ and $\rho\,\delta \mathbf{S}$ is further exemplified by expressions for the rates of change of their magnitudes δl and $\rho\,\delta S$. We find from (3.1.3) and (3.1.6) that

$$\frac{1}{\delta l}\frac{d\,\delta l}{dt} \to m_i m_j \frac{\partial u_i}{\partial x_j} \quad \text{as} \quad \delta l \to 0, \tag{3.1.7}$$

and

$$\frac{1}{\rho\,\delta S}\frac{d(\rho\,\delta S)}{dt} \to -n_i n_j \frac{\partial u_i}{\partial x_j} \quad \text{as} \quad \delta S \to 0, \tag{3.1.8}$$

where $\mathbf{m}$ and $\mathbf{n}$ are unit vectors parallel to $\delta \mathbf{l}$ and $\delta \mathbf{S}$ respectively. The scalar quantity $m_i m_j\,\partial u_i/\partial x_j$ is the rate of extension of the fluid in the direction of $\delta \mathbf{l}$, whereas $-n_i n_j\,\partial u_i/\partial x_j$ is the rate of contraction of the fluid in the direction of $\delta \mathbf{S}$.

In the particular case of an incompressible fluid, ρ and $\delta \tau$ are each invariant for a material element, and the factor ρ drops out of the relations (3.1.6) and (3.1.8).

Rates of change of material integrals

A line integral of some quantity along a path which moves with the fluid and consists always of the same fluid particles may be termed a material integral. Surface and volume integrals may also be material integrals in the same sense. Material integrals occur often in fluid mechanics, sometimes through a need to represent the total amount of some quantity associated with a given body of fluid, and their rates of change with respect to time are also relevant. There will now be explained a simple and direct procedure for calculation of the rates of change of material integrals which will be used subsequently.

Consider first the line integral

$$\int_P^Q \theta \, d\mathbf{l}$$

taken over a material curve joining two material points P and Q, where θ represents some intensive property of the fluid and is specified in the usual way as a function of $\mathbf{x}$ and t. The integral is a function of t alone, once the material curve is specified, and there will be contributions to its time derivative due to both changes in the value of θ at a material point on the path of integration and changes in the shape and orientation of the material curve of integration. In order to calculate the latter contribution we may imagine the line integral to be defined at some instant in the usual elementary way as the limit, as $\epsilon \to 0$, of the sum of contributions from a large number of infinitesimal sub-ranges, each of length ϵ. If these sub-ranges, or line elements of integration, now be regarded as material elements, the line elements will change as they move with the fluid but can nevertheless continue to be used to form a sum whose limit as $\epsilon \to 0$ defines the integral at any subsequent instant. The lengths of the line elements are not equal at a later instant, but they are all proportional to ϵ and are all infinitesimal provided no sub-range experiences infinite extension during the relevant interval of time. Thus we write

$$\frac{d}{dt}\int_P^Q \theta \, d\mathbf{l} = \frac{d}{dt}\left\{ \lim_{\epsilon \to 0} \sum_n \theta_n \, \delta\mathbf{l}_n \right\},$$

in which θ_n is evaluated at the position of the material line element $\delta\mathbf{l}_n$ and so has a time derivative represented by $D\theta_n/Dt$. It follows then from (3.1.3) that

$$\frac{d}{dt}\int_P^Q \theta \, d\mathbf{l} = \lim_{\epsilon \to 0} \sum_n \left\{ \frac{D\theta_n}{Dt}\delta\mathbf{l}_n + \theta_n \delta\mathbf{l}_n . \nabla\mathbf{u} \right\}$$

$$= \int_P^Q \frac{D\theta}{Dt}d\mathbf{l} + \int_P^Q \theta \, d\mathbf{l} . \nabla\mathbf{u}. \tag{3.1.9}$$

The representation of the integral as the limit of the sum of contributions from many material elements of the curve joining Q to P is purely an intermediate step in the argument, and for working purposes we may think of the two terms in (3.1.9) as obtained directly by differentiation of the integrand θ (evaluated at a moving point) and the material element of integration $\delta \mathbf{l}$.

An equivalent procedure leading to (3.1.9) which is less directly related to the underlying physical processes makes use of a parametric specification of the material curve of integration. Let $\mathbf{y}(s, t)$ be the instantaneous position vector of a material point on the curve specified by the parameter s, which could represent, for example, distance from P along the curve of integration at some initial instant. Then we may write

$$\int_P^Q \theta \, d\mathbf{l} = \int_P^Q \theta(\mathbf{y}, t) \frac{\partial \mathbf{y}}{\partial s} \, ds.$$

Differentiation with respect to t may now be carried out in the conventional way, giving

$$\frac{d}{dt} \int_P^Q \theta \, d\mathbf{l} = \int_P^Q \left(\frac{\partial \theta}{\partial t} + \frac{\partial \mathbf{y}}{\partial t} \cdot \nabla \theta \right) \frac{\partial \mathbf{y}}{\partial s} \, ds + \int_P^Q \theta \frac{\partial^2 \mathbf{y}}{\partial t \, \partial s} \, ds,$$

and, since $\partial \mathbf{y} / \partial t$ is the fluid velocity $\mathbf{u}$ at the position $\mathbf{y}$,

$$= \int_P^Q \frac{D\theta}{Dt} \frac{\partial \mathbf{y}}{\partial s} \, ds + \int_P^Q \theta \frac{\partial \mathbf{u}}{\partial s} \, ds.$$

These two integrals are simply parametric versions of those on the right-hand side of (3.1.9).

The same direct procedure of differentiation of the material element of integration may be employed in evaluation of the rates of change of surface and volume integrals over a material range, using (3.1.5) and (3.1.1). Thus

$$\frac{d}{dt} \int \theta \, dS_i = \int \frac{D\theta}{Dt} \, dS_i + \int \theta \frac{\partial u_j}{\partial x_j} \, dS_i - \int \theta \frac{\partial u_j}{\partial x_i} \, dS_j \qquad (3.1.10)$$

and

$$\frac{d}{dt} \int \theta \, d\tau = \int \frac{D\theta}{Dt} \, d\tau + \int \theta \nabla . \mathbf{u} \, d\tau. \qquad (3.1.11)$$

An alternative useful form of this latter relation is obtained by replacing the arbitrary scalar quantity θ by $\theta \rho$ and simplifying the right-hand side by use of the mass-conservation equation (2.2.3):

$$\frac{d}{dt} \int \theta \rho \, d\tau = \int \frac{D\theta}{Dt} \rho \, d\tau. \qquad (3.1.12)$$

Equation (3.1.12) may of course be regarded as a direct consequence of the constancy of the mass $\rho \, \delta \tau$ of the material element of integration. Again these results may be recovered by the equivalent and now rather longer process of changing the variables of integration to parametric co-ordinates specifying position in the domain of integration at an initial instant.

Conservation laws for a fluid in motion

Many of the laws of continuum mechanics state that the total amount of some quantity associated with a material body of fluid either is invariant or changes in a certain way under the action of known external influences such as molecular transport across the bounding surface. Provided the net effect of these external influences can be expressed as a volume integral over the body of fluid, the differential equation governing the distribution of such quantities can be derived with the help of the above expressions for rates of change of material integrals.

The total mass of the fluid in the specified material volume is the most obvious conserved quantity. $\int \rho \, d\tau$ is invariant, and the necessity for the right-hand side of (3.1.11) to be zero when $\theta = \rho$, for all choices of the volume τ, leads at once to the mass-conservation equation (2.2.3).

Consider now an arbitrary extensive property of the fluid (e.g. kinetic energy, or momentum), the amount of which per unit mass of fluid is a local or intensive quantity to be denoted by $\theta(\mathbf{x}, t)$. The total amount of this extensive quantity associated with a material volume τ is $\int \theta \rho \, d\tau$, and will be supposed to change, under the action of external influences, at a rate given by $\int Q \, d\tau$, where Q is a function of $\mathbf{x}$ and t. Q is an effective density of source strength, and may depend on the (instantaneous) fluid motion in some way. The 'conservation' law for the extensive quantity corresponding to θ is then

$$\frac{d}{dt} \int \theta \rho \, d\tau = \int Q \, d\tau,$$

that is, in view of (3.1.12),

$$\int \frac{D\theta}{Dt} \rho \, d\tau = \int Q \, d\tau. \tag{3.1.13}$$

If this is valid for all choices of τ, the differential equation satisfied by θ follows as

$$\rho \frac{D\theta}{Dt} = Q. \tag{3.1.14}$$

It is also possible to derive (3.1.14) by considering changes in the total amount of the quantity associated with the fluid instantaneously enclosed by a surface A fixed in space. The two lines of argument have differences which are slight but which are worth notice. The total amount of the quantity is here $\int \theta \rho \, dV$, where V is the volume bounded by A, and this total amount changes as a consequence of both external influences and passage of fluid across the surface A. The flux of the quantity outward across A due to the fluid motion is $\int \theta \rho \mathbf{u} . \mathbf{n} \, dA$, so that the conservation law may be expressed as

$$\frac{d}{dt} \int \theta \rho \, dV = - \int \theta \rho \mathbf{u} . \mathbf{n} \, dA + \int Q \, dV,$$

i.e.

$$\int \frac{\partial (\theta \rho)}{\partial t} \, dV = - \int \nabla . (\theta \rho \mathbf{u}) \, dV + \int Q \, dV. \tag{3.1.15}$$

The requirement that this be valid for all choices of the volume V then gives the differential equation

$$\frac{\partial(\theta\rho)}{\partial t} + \nabla\cdot(\theta\rho\mathbf{u}) = Q. \tag{3.1.16}$$

We see that this is identical with (3.1.14) after making use of the mass-conservation equation (2.2.2), which of course had already been employed in the derivation of (3.1.14) through the use of (3.1.12).

The exact form of the function Q in (3.1.14) depends on the nature of the extensive quantity corresponding to θ, and need not be considered here. However, if the total amount of the quantity associated with a given body of fluid changes only as a consequence of molecular transport across the bounding surface, we can see immediately the change in the form of the differential equation for θ resulting from the motion of the fluid. It was established in § 1.6 that for a fluid at rest the rate of molecular transport of the physical quantity of which θ is the intensity is proportional to the local gradient of θ, the corresponding value of Q being given by (1.6.4). The argument used in § 1.6 is equally applicable to a fluid in motion,† and so the form taken by Q, when it represents the effect of molecular transport, is the same as in a fluid at rest. Consequently the effect of the fluid motion on the form of the differential equation (3.1.14) is confined to the term on the left-hand side, this term being $\rho\,\partial\theta/\partial t$ for a fluid at rest and $\rho\,D\theta/Dt$ for a fluid in motion.

The various special differential equations found in § 1.6 for a stationary medium for different physical quantities subject to molecular transport may now be adapted to the case of a moving fluid. Thus the differential equation (1.6.7) for the numerical fraction of marked molecules C becomes

$$\frac{DC}{Dt} = \kappa_D\nabla^2 C. \tag{3.1.17}$$

When conduction of heat is in question, the quantity of which the total amount remains unchanged in the absence of molecular conduction is entropy, and, provided that in the moving fluid conduction of heat is the only entropy-changing process,‡ (1.6.10) is evidently to be replaced, in the case of a moving fluid, by

$$T\frac{DS}{Dt} = c_p\frac{DT}{Dt} - \frac{\beta T}{\rho}\frac{Dp}{Dt} = \frac{1}{\rho}\nabla\cdot(k_H\nabla T). \tag{3.1.18}$$

The special forms (1.6.11) and (1.6.12) applicable in the circumstances stated are modified similarly.

---

† The existence of relative motion implies the absence of equilibrium of the fluid, but so too does the existence of non-uniformity of θ, and the only restriction required by the argument is that the departures from equilibrium of all kinds should be small.

‡ Internal friction due to molecular transport of momentum is another possibility, but in common circumstances it makes only a negligible contribution to the rate of change of entropy.

Differential equations for vector quantities satisfying conservation relations may also be deduced from (3.1.9) and (3.1.10); one example will arise in §§ 5.2 and 5.3.

3.2. The equation of motion

The 'equation of motion' for a fluid is, in its most fundamental form, a relation equating the rate of change of momentum of a selected portion of fluid and the sum of all forces acting on that portion of fluid. For the body of fluid of volume τ enclosed by the material surface S, the momentum is $\int \mathbf{u} \rho \, d\tau$ and its rate of change, according to (3.1.12), is

$$\int \frac{D\mathbf{u}}{Dt} \rho \, d\tau,$$

which is simply the sum of the products of mass and acceleration for all the elements of the material volume τ.

As explained in § 1.3, a portion of fluid is acted on, in general, by both volume and surface forces. We denote the vector resultant of the volume forces, per unit mass of fluid, by $\mathbf{F}$, so that the total volume force on the selected portion of fluid is

$$\int \mathbf{F} \rho \, d\tau.$$

The i-component of the surface or contact force exerted across a surface element of area δS and normal $\mathbf{n}$ may be represented as $\sigma_{ij} n_j \delta S$, where σ_{ij} is the stress tensor introduced in § 1.3, and the total surface force exerted on the selected portion of fluid by the surrounding matter is thus

$$\int \sigma_{ij} n_j \, dS, \quad = \int \frac{\partial \sigma_{ij}}{\partial x_j} \, d\tau.$$

Thus the momentum balance for the selected portion of fluid is expressed by

$$\int \frac{Du_i}{Dt} \rho \, d\tau = \int F_i \rho \, d\tau + \int \frac{\partial \sigma_{ij}}{\partial x_j} \, d\tau, \tag{3.2.1}$$

in which all three integrals are taken over the volume τ.

The relation (3.2.1) holds for all choices of the material volume τ, which is possible, when the integrands are continuous functions of $\mathbf{x}$, only if

$$\rho \frac{Du_i}{Dt} = \rho F_i + \frac{\partial \sigma_{ij}}{\partial x_j} \tag{3.2.2}$$

at all points of the fluid. This differential equation giving the acceleration of the fluid in terms of the local volume force and stress tensor is the relation usually understood by the term 'equation of motion'. It is a member of the class of conservation relations represented by (3.1.14), in which volume and surface forces lead to an effective generation of momentum per unit volume at a rate given by the right-hand side of (3.2.2). Surface forces contribute to the acceleration of the fluid only if the stress tensor varies with position in the fluid, or, more precisely, only if σ_{ij} has non-zero divergence with respect to the second suffix determining direction of the surface element; when

$\partial\sigma_{ij}/\partial x_j = 0$ the effect of surface forces on a material element of fluid is to tend to deform it without change of its momentum.

Equation (3.2.2) cannot be used for the determination of the distribution of fluid velocity until more is known about F_i and σ_{ij}. The volume force acting on a fluid in many cases is due simply to the earth's gravitational field, for which $\mathbf{F} = \mathbf{g}$; and in other cases the appropriate expression for $\mathbf{F}$ will usually be evident from the given circumstances. The stress tensor presents more of a problem, since it is a manifestation of internal reactions in the fluid and is itself affected by the motion of the fluid, in a manner to be discussed in the next section.

Use of the momentum equation in integral form

Although the majority of problems in fluid dynamics require use of the equation in the differential form (3.2.2), or some particular version of it, there are a few important cases in which an integral relation specifying the momentum balance for a certain region of fluid leads directly to the required information. If use of an integral relation for the momentum balance succeeds at all, it usually does so very simply and quickly, and is then preferable to use of the differential equation of motion. Consideration of the balance of momentum for the fluid contained within a surface A fixed in space is more convenient in practice than that for a material body of fluid, so that we begin with the integral relation which differs from (3.2.1) in the way in which (3.1.15) differs from (3.1.13), viz.

$$\int \frac{\partial(u_i\rho)}{\partial t}\, dV = -\int \rho u_i u_j n_j\, dA + \int F_i \rho\, dV + \int \sigma_{ij} n_j\, dA, \qquad (3.2.3)$$

the two volume integrals being taken over the volume V bounded by A.

The usual circumstances in which this momentum balance in integral form is useful are those in which all terms of (3.2.3) can be written as integrals over the bounding surface A, for then the details of the motion within the region enclosed by A are irrelevant. The contribution from the volume force can be put in the form of a surface integral when $\rho\mathbf{F}$ can be written as the gradient of a scalar quantity, as is possible when ρ is uniform and the body force per unit mass is conservative; in this latter case

$$\rho\mathbf{F} = -\nabla(\rho\Psi),$$

Ψ being the associated potential energy per unit mass. The remaining volume integral, on the left-hand side of (3.2.3), which normally prevents use of the integral relation, is zero in the important particular case of steady motion. In these special circumstances, (3.2.3) can be written as

$$\int \rho u_i u_j n_j\, dA = \int(-\rho\Psi n_i + \sigma_{ij} n_j)\, dA, \qquad (3.2.4)$$

which is an analytical statement of the fact that the convective flux of momentum out of the region bounded by A is equal to the sum of the resultant contact force exerted at the boundary by the surrounding matter and the resultant force at the boundary arising from the stress system equivalent to the body force.

The relation (3.2.4) for steady motion is often termed the *momentum theorem*, and the bounding surface A, which may be chosen freely, is referred to as the *control surface*. Examples of the use of the momentum theorem will be given in subsequent chapters to illustrate the point that, although the principle of the theorem is evident enough, thoughtful choice of the control surface can lead to surprisingly strong results which would otherwise be difficult to obtain. The particular flow fields to which the theorem is applied in §5.15 involve viscous forces in a significant way whereas those considered in §6.3 are cases of approximately irrotational flow of an incompressible fluid in which viscous forces are negligible.

Equation of motion relative to moving axes

If the external boundary to a fluid is in motion, it may be convenient to choose a frame of reference relative to which the boundary is at rest. The acceleration of an element of fluid relative to the moving frame of reference may then be different from the absolute acceleration in the Newtonian frame of reference, and the equation of motion must be modified accordingly. The common cases are axes in translational motion and axes in uniform rotational motion, but there is no difficulty in obtaining an expression for the acceleration of an element relative to axes in general motion. Any substantial book on mechanics of particles gives the required expression, but we include the derivation here for completeness.

We suppose that instantaneously the moving frame of reference is rotating with angular velocity $\mathbf{\Omega}$ about a point O which itself is moving relative to the Newtonian frame with acceleration $\mathbf{f_0}$. The absolute acceleration of an element is then

$$\mathbf{f_0} + \mathbf{f_1},$$

where $\mathbf{f_1}$ is the acceleration of the element relative to the point O. The relation between $\mathbf{f_1}$ and the acceleration of the element relative to the rotating frame is determined in the following way.

If $(\mathbf{i}, \mathbf{j}, \mathbf{k})$ is a triad of orthogonal unit vectors fixed in the moving frame, any vector $\mathbf{P}$ can be written as

$$\mathbf{P} = P_1 \mathbf{i} + P_2 \mathbf{j} + P_3 \mathbf{k}.$$

Change of $\mathbf{P}$ with respect to t then occurs as a result of both change of the components P_1, P_2, P_3 in the moving frame and change of the unit vectors $\mathbf{i}$, $\mathbf{j}$, $\mathbf{k}$, as the frame rotates about O; that is, the rate of change of $\mathbf{P}$ as it appears to an observer at O is

$$\sum_i \left(\frac{dP_1}{dt} \mathbf{i} + P_1 \frac{d\mathbf{i}}{dt} \right), \quad = \sum_i \left(\frac{dP_1}{dt} \mathbf{i} + P_1 \mathbf{\Omega} \times \mathbf{i} \right)$$

$$= \left(\frac{d\mathbf{P}}{dt} \right)_r + \mathbf{\Omega} \times \mathbf{P}, \qquad (3.2.5)$$

where $(d\mathbf{P}/dt)_r$ denotes the rate of change of $\mathbf{P}$ as it appears to an observer in
the rotating frame. This relation may be applied first with $\mathbf{P}$ taken as the
vector $\mathbf{y}$ representing the position of a material element of fluid relative to O,
and then with $\mathbf{P}$ as the vector $\mathbf{v}_1$ representing its velocity relative to a non-
rotating frame moving with O, giving

$$\mathbf{v}_1 = \left(\frac{d\mathbf{y}}{dt}\right)_r + \mathbf{\Omega} \times \mathbf{y}, \tag{3.2.6}$$

and $\qquad \mathbf{f}_1 = \left(\dfrac{d\mathbf{v}_1}{dt}\right)_r + \mathbf{\Omega} \times \mathbf{v}_1$

$$= \left(\frac{d^2\mathbf{y}}{dt^2}\right)_r + 2\mathbf{\Omega} \times \left(\frac{d\mathbf{y}}{dt}\right)_r + \left(\frac{d\mathbf{\Omega}}{dt}\right)_r \times \mathbf{y} + \mathbf{\Omega} \times (\mathbf{\Omega} \times \mathbf{y}). \tag{3.2.7}$$

Now $(d^2\mathbf{y}/dt^2)_r$, $= \mathbf{f}$ say, is the acceleration of the element relative to the
translating and rotating frame of reference, and $(d\mathbf{y}/dt)_r$, $= \mathbf{v}$ say, is the
velocity of the element in this frame; also the rate of change of $\mathbf{\Omega}$ is the same
in an absolute frame as in the rotating frame. The absolute acceleration of an
element is thus

$$\mathbf{f} + \mathbf{f}_0 + 2\mathbf{\Omega} \times \mathbf{v} + \frac{d\mathbf{\Omega}}{dt} \times \mathbf{y} + \mathbf{\Omega} \times (\mathbf{\Omega} \times \mathbf{y}). \tag{3.2.8}$$

This expression may be equated to the local force acting per unit mass of
fluid to give the equation of motion in the moving frame.

In terms of the velocity $\mathbf{u}(\mathbf{x}, t)$ in the Eulerian specification of the flow
field, relative to the moving frame, we have

$$\mathbf{f} = \frac{\partial \mathbf{u}}{\partial t} + \mathbf{u}.\nabla\mathbf{u} = \frac{D\mathbf{u}}{Dt},$$

and the element position $\mathbf{y}$ and velocity $\mathbf{v}$ in (3.2.8) may be replaced by $\mathbf{x}$
and $\mathbf{u}$. The equation of motion of a fluid in the moving frame is therefore
identical in form with that in an absolute frame provided we suppose that
the fictitious body force

$$-\mathbf{f}_0 - 2\mathbf{\Omega} \times \mathbf{u} - \frac{d\mathbf{\Omega}}{dt} \times \mathbf{x} - \mathbf{\Omega} \times (\mathbf{\Omega} \times \mathbf{x}) \tag{3.2.9}$$

per unit mass acts upon the fluid in addition to the real body and surface
forces. $-\mathbf{f}_0$ is simply the apparent body-force that compensates for the
translational acceleration of the frame; $-2\mathbf{\Omega} \times \mathbf{u}$ is the deflecting or Coriolis
force, which is perpendicular to both $\mathbf{u}$ and $\mathbf{\Omega}$; and $-\mathbf{\Omega} \times (\mathbf{\Omega} \times \mathbf{x})$ is the
centrifugal force. No name is in general use for the remaining term
$-d\mathbf{\Omega}/dt \times \mathbf{x}$.

The case of axes of reference which are rotating steadily relative to the
absolute frame and for which $\mathbf{f}_0 = 0$ is of particular interest, and will be
referred to in later sections. The fictitious body force (3.2.9) is then

$$-2\mathbf{\Omega} \times \mathbf{u} - \mathbf{\Omega} \times (\mathbf{\Omega} \times \mathbf{x}). \tag{3.2.10}$$

3.3. The expression for the stress tensor

Mechanical definition of pressure in a moving fluid

It was shown in §1.3 that, in a fluid at rest, only normal stresses are exerted, the normal stress is independent of the direction of the normal to the surface element across which it acts, and the stress tensor has the form

$$\sigma_{ij} = -p\,\delta_{ij}, \qquad (3.3.1)$$

where the parameter p is the static-fluid pressure and may be a function of position in the fluid. There is no reason to expect these results to be valid for a fluid in motion, and it is clear from observation that they are not; the tangential stresses are then non-zero, in general, and the normal component of the stress acting across a surface element depends on the direction of the normal to the element. The simple notion of a pressure acting equally in all directions is lost in most cases of a fluid in motion.

It is useful nevertheless to have available a scalar quantity characterizing a moving fluid which is analogous to the static-fluid pressure in the sense that it is a measure of the local intensity of the 'squeezing' of the fluid. Such a quantity is provided by (minus) the average of the three normal stresses for any orthogonal set of axes. It is known from tensor theory that $\frac{1}{3}\sigma_{ii}$ is an invariant under rotation of the axes of reference, and a physical interpretation of $\frac{1}{3}\sigma_{ii}$ which does not involve any particular axes can also be given. The average value of the normal component of the stress on a surface element at position $\mathbf{x}$ over all directions of the normal $\mathbf{n}$ to the element is

$$\frac{1}{4\pi}\sigma_{ij}\int n_i n_j \, d\Omega(\mathbf{n}), \quad = \tfrac{1}{3}\sigma_{ij}\delta_{ij} = \tfrac{1}{3}\sigma_{ii},$$

where $\delta\Omega(\mathbf{n})$ is an element of solid angle about $\mathbf{n}$; an equivalent interpretation is that $\frac{1}{3}\sigma_{ii}$ is the average value of the normal component of stress over the surface of a small sphere centred on $\mathbf{x}$. Thus the quantity $-\frac{1}{3}\sigma_{ii}$, which reduces to the static-fluid pressure when the fluid is at rest, has a mechanical significance which makes it an appropriate generalization of the elementary notion of 'pressure' for a situation in which the normal component of stress is not independent of direction of the normal to the surface element; we therefore define the *pressure at a point in a moving fluid* to be the mean normal stress with sign reversed, and denote it by p for convenience:

$$p = -\tfrac{1}{3}\sigma_{ii}. \qquad (3.3.2)$$

It will be noted that this is a purely mechanical definition of 'pressure', and that nothing is implied, for the moment, about the connection between this mechanical quantity and the term pressure used in thermodynamics. The precise connection is not a simple one, since thermodynamical relations such as the equation of state for a fluid refer to equilibrium conditions whereas the elements of a fluid in relative motion are not in exact thermodynamic equilibrium. The quantity to which we have chosen to give the name pressure in a moving fluid is a real parameter of the fluid system and is accessible to direct observation, whereas any quantity calculated from equilibrium relations is at best an approximation to a property of a moving fluid. We shall return to this question in § 3.4, when the use of relations referring to thermodynamic equilibrium is under discussion.

It is convenient now to regard the stress tensor σ_{ij} as the sum of an isotropic part $-p\,\delta_{ij}$, having the same form as the stress tensor in a fluid at rest (although the value of p for a moving fluid is not necessarily the same as in the same fluid at rest), and a remaining non-isotropic part, d_{ij} say, contributing the tangential stresses and also diagonal elements whose sum is zero:

$$\sigma_{ij} = -p\,\delta_{ij} + d_{ij}. \tag{3.3.3}$$

The non-isotropic part d_{ij} may be termed the *deviatoric stress tensor*, and has the distinctive property of being due entirely to the existence of the motion of the fluid.

The relation between deviatoric stress and rate-of-strain for a Newtonian fluid

Since the stress at any point in the fluid is an expression of the mutual reactions of adjacent parts of fluid near that point, it is natural to consider the connection between the stress and the local properties of the fluid. In the case of a fluid at rest, this is a simple matter since the stress is determined wholly by the one scalar quantity p, the static-fluid pressure, and p in turn is specified locally by the equilibrium equation of state when the values of the two parameters of state (e.g. density and temperature) are known; and if the distribution of the body force per unit volume acting on the fluid is known, there is no need to consider local variables of state at more than one point because the relative pressure is determined everywhere by the equation for mechanical equilibrium (1.4.2). In the case of a fluid in relative motion, the connection between the stress and the local properties of the fluid is more complicated, in two respects: first, the stress tensor contains a non-isotropic part as well as an isotropic part, and, second, the scalar quantity p specifying the isotropic part is not itself one of the variables of state used in equilibrium thermodynamics. The first of these two manifestations of the departure from equilibrium represents a transport of momentum, or internal friction, and is by far the more important, in the great majority of flow fields, as we shall see.

The argument to be used in establishing a relation between the deviatoric stress tensor d_{ij} and the local properties of the fluid is of the kind explained in § 1.6, and differs only in analytical details associated with the vectorial character of the quantity being transported (viz. fluid momentum). The reader is advised to look again at § 1.6, so as to be able to keep in mind the fact that internal friction in a moving fluid is only one of several similar kinds of transport phenomena arising from a departure from equilibrium, and to re-read in particular the discussion, at the end of § 1.6, of the molecular transport of momentum in the case of a simple shearing motion. The parts of §§ 1.7 and 1.8 concerned with the values of transport coefficients, such as viscosity, characteristic of gases and liquids are also relevant, although in this section, as in § 1.6, we shall adopt the *phenomenological* approach and seek relations whose forms are independent of the nature of the molecular mechanism of the internal friction.

The part of the flux of momentum across a material surface element which results from frictional interaction of the matter in relative motion on the two sides of the element and which is represented by the deviatoric stress is assumed, as in the general hypothesis of § 1.6, to depend only on the instantaneous distribution of fluid velocity in the neighbourhood of the element, or, more precisely, on the departure from uniformity of that distribution. The local velocity gradient, of which a typical component is $\partial u_i/\partial x_j$, is thus the parameter of the flow field with most relevance to the deviatoric stress, and since $\partial u_i/\partial x_j$ is normally uniform over distances large compared with distances characteristic of the mechanism of molecular transport of momentum we assume it is the *only* relevant parameter. Furthermore, d_{ij} is zero in a stationary fluid and so vanishes with $\partial u_i/\partial x_j$.

We have no way of deducing the dependence of d_{ij} on $\partial u_i/\partial x_j$ for fluids in general, and we therefore fall back on the hypothesis, introduced in § 1.6, that d_{ij} (which is the counterpart of the flux vector of § 1.6) is approximately a linear function of the various components of the velocity gradient for sufficiently small magnitudes of those components. Analytically the hypothesis is expressed as

$$d_{ij} = A_{ijkl}\frac{\partial u_k}{\partial x_l}, \qquad (3.3.4)$$

where the fourth-order tensor coefficient A_{ijkl} depends on the local state of the fluid, but not directly on the velocity distribution, and is necessarily symmetrical in the indices i and j like d_{ij}. This is the counterpart of the linear relation (1.6.1) for a scalar transportable quantity. It is convenient at this stage to write $\partial u_k/\partial x_l$, as in § 2.3, as the sum of its symmetrical part e_{kl} (the rate-of-strain tensor) and its antisymmetrical part $-\frac{1}{2}\epsilon_{klm}\omega_m$ (where $\boldsymbol{\omega}$ is the vorticity), so that (3.3.4) becomes

$$d_{ij} = A_{ijkl}e_{kl} - \tfrac{1}{2}A_{ijkl}\epsilon_{klm}\omega_m. \qquad (3.3.5)$$

The tensor coefficient A_{ijkl} takes a simple form when the molecular structure of the fluid is statistically isotropic, that is, when the deviatoric stress generated in an element of the fluid by a given velocity gradient is independent of the orientation of the element. All gases have isotropic structure, and so do simple liquids, although suspensions and solutions containing very long chain-like molecules may exhibit some directional preferences owing to alignment of these molecules in a manner which depends on the past history of the motion. We shall restrict attention to fluids of isotropic structure, in which case A_{ijkl} is an isotropic tensor, having a form from which all directional distinction is absent.

It is shown in books on Cartesian tensor analysis† that the basic isotropic tensor is the Kronecker delta tensor, and that all isotropic tensors of even order can be written as the sum of products of delta tensors. Thus

$$A_{ijkl} = \mu\, \delta_{ik}\, \delta_{jl} + \mu'\, \delta_{il}\, \delta_{jk} + \mu''\, \delta_{ij}\, \delta_{kl}, \qquad (3.3.6)$$

where μ, μ' and μ'' are scalar coefficients, and since A_{ijkl} is symmetrical in i and j we require

$$\mu' = \mu.$$

It will be observed that A_{ijkl} is now symmetrical in the indices k and l also, and that as a consequence the term containing $\boldsymbol{\omega}$ drops out of (3.3.5), giving

$$d_{ij} = 2\mu\, e_{ij} + \mu''\, \Delta\, \delta_{ij}, \qquad (3.3.7)$$

where Δ denotes the rate of expansion e_{kk}, $= \nabla \cdot \mathbf{u}$, as in chapter 2.

This expression for d_{ij} for a fluid of isotropic structure may be deduced from (3.3.5) in another way which does not make explicit use of the identity (3.3.6). Consider first a case of a fluid in pure rotation. It follows from (3.3.5) that reversal of the direction of $\boldsymbol{\omega}$ leads to change of sign of all components of the deviatoric stress, which is impossible in an isotropic fluid because this operation is equivalent to keeping $\boldsymbol{\omega}$ fixed and choosing a different orientation of the fluid; hence A_{ijkl} must have such a form that the term in $\boldsymbol{\omega}$ in (3.3.5) vanishes identically.‡ Then, for a pure straining motion, we can argue that, since the structure of the fluid does not distinguish any directions, the principal axes of d_{ij} must be determined by e_{ij} and must coincide with those of e_{ij}; and (3.3.7) is the only possible linear relation between the tensor d_{ij} and e_{ij} satisfying this condition.

Finally we recall that by definition d_{ij} makes zero contribution to the mean normal stress, whence

$$d_{ii} = (2\mu + 3\mu'')\, \Delta = 0$$

for all values of Δ, implying that

$$2\mu + 3\mu'' = 0. \qquad (3.3.8)$$

On choosing μ as the one independent scalar constant, we obtain for the deviatoric stress tensor the expression

$$d_{ij} = 2\mu(e_{ij} - \tfrac{1}{3}\Delta\, \delta_{ij}); \qquad (3.3.9)$$

the quantity within brackets is simply the non-isotropic part of the rate-of-strain tensor. This expression for d_{ij} was obtained by Saint-Venant (1843)

---

† See *Cartesian Tensors*, by H. Jeffreys (Cambridge University Press, 1931).

‡ It is taken for granted, in most expositions of fluid dynamics, that a deviatoric stress cannot be generated by pure rotation, irrespective of the structure of the fluid, simply on the grounds that there is then no deformation of the fluid; however, rigorous justification for this belief is elusive.

and Stokes (1845) in essentially the above way, after having been derived by Navier (1822) and Poisson (1829) from specific assumptions concerning the molecular mechanism of internal friction. There is an analogous linear relation between stress and amount of strain for isotropic elastic solids.

It will be noticed that a spherically symmetrical straining motion, for which $e_{ij} = \frac{1}{3}\Delta\,\delta_{ij}$, is associated with zero deviatoric stress. This is a simple consequence of the symmetry of the motion and of our definition of d_{ij} as the departure of the stress tensor from an isotropic form. This raises the question: are there any non-equilibrium effects in an isotropic expansion? The answer is that there may be, although they are only rarely of any importance, and that they are incorporated, in our analysis, in the quantity p defined as the mean normal stress from all causes. The manner in which the departure from equilibrium represented by an isotropic expansion may affect the mean normal stress will be examined in the next section.

The significance of the parameter μ, which depends on the local state of the fluid, can be seen from the form taken by the relation (3.3.9) in the special case of a simple shearing motion. With $\partial u_1/\partial x_2$ as the one non-zero velocity derivative, all components of d_{ij} are zero except the tangential stresses

$$d_{12} = d_{21} = \mu\frac{\partial u_1}{\partial x_2}. \qquad (3.3.10)$$

Thus μ is the constant of proportionality between rate of shear and the tangential force per unit area when plane layers of fluid slide over each other, already introduced in (1.6.15) and termed the *viscosity* of the fluid. The fact that μ is the only scalar constant needed in the above general expression for d_{ij} is associated with the result of §2.3 that a general relative motion near any point may be represented as the superposition of two simple shearing motions, each of which gives rise to a tangential stress determined by μ and the corresponding velocity gradient, together with a rigid rotation and an isotropic expansion, neither of which has any effect (in a fluid of isotropic structure) on the non-isotropic part of the stress tensor; and (3.3.9) may of course be regarded as the only possible linear tensorial relation, involving one scalar parameter, between e_{ij} and a symmetrical tensor d_{ij} whose diagonal elements have zero sum.

It is a matter of common experience that the force between layers of fluid in relative sliding motion is always a frictional force resisting the relative motion, corresponding to $\mu > 0$, as expected from the fact that molecular transfer of momentum resulting from random movement or arrangement of the molecules of the fluid tends to smooth out spatial variations of mean velocity irrespective of the mechanism of the transfer. The relation (3.3.9) shows that a positive value of μ also corresponds to principal stresses arising from d_{ij} of such signs as to resist the principal rates of strain (compare the discussion in §1.7 of the response of a gas to compression by a sliding piston); that is to say, a small material sphere being deformed into an ellipsoid exerts

10

on the surrounding fluid a frictional force whose normal component is outward (inward) at places on the surface where the surface is moving inward (outward) relative to a sphere of the same volume as the ellipsoid.

Experiments on a variety of fluids and flow fields have shown that the linearity of the relation between the rate of strain and the non-isotropic part of the stress holds over a remarkably wide range of values of the rate of strain. Observations of the flux of fluid volume along a circular tube of small radius with a maintained difference between the pressures at the two ends (see § 4.2) are particularly sensitive for this purpose. Although the exclusion of all but a linear term in the velocity gradient on the right-hand side of (3.3.4) has been proposed purely as a hypothesis likely to be accurate only for small magnitudes of the velocity gradient, it seems from observation that 'small' magnitudes of the velocity gradient may include those values normally encountered in practice. For water and most gases, the linear law appears to be accurate under all except possibly the most extreme conditions, such as within a shock wave. Fluids for which the linear relation (3.3.9) between the non-isotropic parts of the stress and rate-of-strain tensors does hold accurately are usually said to be *Newtonian* (in recognition of the fact that the simple relation (3.3.10) for a simple shearing motion was proposed by Newton). For liquids of elaborate molecular structure, and in particular for those consisting of long molecular chains, and for some emulsions and mixtures, the expression (3.3.9) for the deviatoric stress may cease to be accurate at only moderate rates of strain; and for some rubber-like liquids the stress evidently depends on the strain history as well as on its instantaneous rate of change. Little is known about how the expression should be modified for such liquids. Chemical engineers frequently encounter liquids which behave in a non-Newtonian manner under common operating conditions, but despite their industrial importance they lie outside the scope of this introductory text.

The observation that a linear relation between deviatoric stress and rate of strain holds over a large range of values of the rate of strain for many fluids becomes understandable when the molecular mechanism of internal friction is considered. The bulk relative motion of the fluid can cause only a small change in the statistical properties of the molecular motion when the characteristic time of the bulk motion, i.e. the reciprocal of the rate of strain, is long compared with the characteristic time of the molecular motion (which in the case of a gas would be given by the average time between collisions). These are the circumstances in which a perturbation assumption of the kind used in obtaining (3.3.9) might be expected to be valid. For air at normal temperature and pressure the average time between collisions is about 10^{-10} sec; and for gases at least, it is evident that common practical values of the bulk rate of strain are indeed 'small' in the sense used above. For liquids one cannot so readily estimate the relevant characteristic time of the molecular motion, but any time associated with the molecular movement is likely to be exceedingly small when measured against the reciprocal of a common value of the bulk rate of strain.

The typical values of the viscosity of gases and liquids under different conditions have been discussed in §§ 1.7 and 1.8, and observed values of μ for air, water and some other common fluids are listed in appendix 1. For air at normal temperature and pressure μ is 0·00018 gm/cm sec and for water 0·010 gm/cm sec. In neither of these cases does μ vary much with pressure, but for air μ increases with temperature at the rate of about 0·3 per cent per degree (Centigrade) rise in temperature and for water μ decreases at the rate of about 3 per cent per degree rise in the neighbourhood of normal temperature. The viscosities of air and water under all ordinary conditions are thus exceedingly small when expressed in units which are 'practical' for most other mechanical quantities, and it is natural to enquire if these common fluids may be regarded, for some purposes at least, as having zero viscosity, that is, as being *inviscid*. This is an important question which will be considered in chapter 5. For the moment, we need notice only that for an inviscid fluid the tangential stresses are zero everywhere and the stress tensor has the same isotropic form as for an arbitrary fluid at rest.

The Navier–Stokes equation

With the expression (3.3.9) for the deviatoric stress tensor, the total stress (3.3.3) becomes

$$\sigma_{ij} = -p\,\delta_{ij} + 2\mu(e_{ij} - \tfrac{1}{3}\Delta\,\delta_{ij}), \qquad (3.3.11)$$

where

$$e_{ij} = \frac{1}{2}\left(\frac{\partial u_i}{\partial x_j} + \frac{\partial u_j}{\partial x_i}\right) \quad \text{and} \quad \Delta = e_{ii}.$$

Substitution in the equation of motion (3.2.2) then gives

$$\rho\frac{Du_i}{Dt} = \rho F_i - \frac{\partial p}{\partial x_i} + \frac{\partial}{\partial x_j}\{2\mu(e_{ij} - \tfrac{1}{3}\Delta\,\delta_{ij})\}. \qquad (3.3.12)$$

This is usually called the *Navier–Stokes equation of motion*.

For many fluids, the viscosity μ depends significantly on the temperature (see §§ 1.7, 1.8), and when appreciable temperature differences exist in the flow field it is necessary to regard μ as a function of position. However, it happens often that the differences in temperature are small enough for μ to be taken as uniform over the fluid, in which case (3.3.12) becomes

$$\rho\frac{Du_i}{Dt} = \rho F_i - \frac{\partial p}{\partial x_i} + \mu\left(\frac{\partial^2 u_i}{\partial x_j\,\partial x_j} + \frac{1}{3}\frac{\partial\Delta}{\partial x_i}\right). \qquad (3.3.13)$$

A further special case of great importance is that of an incompressible fluid. The mass-conservation equation reduces here to $\nabla\cdot\mathbf{u} = 0$ and (3.3.13) becomes, in vector notation,

$$\rho\frac{D\mathbf{u}}{Dt} = \rho\mathbf{F} - \nabla p + \mu\nabla^2\mathbf{u}. \qquad (3.3.14)$$

Provided we may take the form of $\mathbf{F}$ and the value of μ as given, the momentum and mass-conservation equations provide four scalar equations for the determination of $\mathbf{u}$, ρ and p as functions of $\mathbf{x}$ and t. In general one further scalar equation is needed, and is usually sought in the equation of state of the fluid and, since one further variable (usually temperature) is introduced thereby, in considerations of the internal energy of the fluid (see § 3.4). However, in the event of the fluid behaving as if it were incompressible, as real fluids do in circumstances to be described in § 3.6, the density of each material element is unaffected by changes of pressure and is thus invariant when no other density-changing processes (such as molecular transport of heat or solute) are present. We then have the additional equation

$$D\rho/Dt = 0, \qquad (3.3.15)$$

which is of course simply a particular form of the equation of state for the fluid; explicit use of (3.3.15) is often rendered unnecessary by a statement that the density is initially uniform and consequently remains uniform. Thus for an incompressible fluid the set of equations is now sufficient for the determination of $\mathbf{u}$ and p, provided that adequate boundary conditions are known.

There is an apparent paradox in the form of the above expression for the net force on unit volume of fluid due to internal friction, which is revealed most clearly in the context of incompressible fluid of uniform viscosity. The net viscous force is then

$$2\mu \frac{\partial e_{ij}}{\partial x_j} = \mu \nabla^2 u_i, \quad = -\mu(\nabla \times \boldsymbol{\omega})_i. \qquad (3.3.16)$$

We have seen that the viscous stress is generated solely by deformation of the fluid and is independent of the local vorticity. It is therefore surprising, at first sight, to find that the net viscous force on unit volume is proportional to a spatial derivative of the vorticity. The explanation is wholly a matter of kinematics, and lies in the vector identity used in (3.3.16); e_{ij} and $\boldsymbol{\omega}$ play independent roles in the generation of stress, but certain spatial derivatives of e_{ij} are identically related to certain derivatives of $\boldsymbol{\omega}$.

It will be noted that the viscous force per unit volume of an incompressible uniform fluid vanishes when $\boldsymbol{\omega}$ has the same value everywhere, and in particular when $\boldsymbol{\omega} = 0$, that is, when the motion is irrotational; but the viscous stress is not then zero.

Conditions on the velocity and stress at a material boundary

As was remarked in § 1.9, there are in general two transition relations at a surface for each transportable quantity, one representing continuity of the appropriate intensity across the surface, based on the assumption that the local departure from equilibrium is not too violent, and one representing continuity of the normal component of the flux vector (with allowance for the effect of surface tension). Fluid momentum is one such transportable quantity, the associated intensity and flux vector being velocity and stress respectively. Now that we have an expression for the stress tensor, we can set out explicitly the boundary conditions to be used in subsequent mathematical determinations of the velocity distribution in a fluid.

The first of the above two transition relations is simply that the tangential component of velocity† is continuous across a material boundary separating a fluid and another medium. We recall from § 1.9 that the justification for this condition can be regarded as lying in the fact that any discontinuity in velocity across a material surface would lead almost immediately (through molecular transport) to a very large stress at the surface of such a direction as to tend to eliminate the relative velocity of the two masses; the condition of continuity of the velocity is thus not an exact law, but a statement of what may be expected to happen, approximately, in normal circumstances. The effectiveness of the viscous stress in smoothing out a discontinuity in fluid velocity depends on the magnitude of the viscosity and on other factors to be examined later. Clearly there will be some special circumstances in which the viscous stress is relatively weak, and in which a steep velocity gradient promoted by some other cause is able to persist; and in such cases it may be convenient to speak of a 'discontinuity' in velocity without the phrase being taken literally.

The case of a boundary separating a fluid and a solid is particularly important in practice. Continuity of the tangential component of velocity across the boundary is here referred to as the *no-slip condition*. The validity of the no-slip condition at a fluid–solid interface was debated for some years during the last century, there being some doubt about whether molecular interaction at such an interface leads to momentum transfer of the same nature as that at a surface in the interior of a fluid; but the absence of slip at a rigid wall is now amply confirmed by direct observation and by the correctness of its many consequences under normal conditions. The one important exception is flow of a gas at such low density that the mean velocity of the molecules varies appreciably over a distance of one mean free path. It seems that here there can be a non-zero jump in velocity and also in temperature at a rigid wall, which is understandable, since the number of collisions made by the molecules in an element of volume before they disperse to other places in the flow field is not large enough for even an approximate equilibrium to be established.

The second of the two transition relations is that the difference between the values of the stress on two surface elements parallel to the boundary and immediately on either side of it is a normal force due wholly to surface tension, as represented by (1.9.8). When making this relation more explicit by use of the expression (3.3.11) for the stress tensor σ_{ij} it is convenient to

---

† The normal component is of course continuous, as noted in §1.9, for kinematical reasons not involving molecular interaction.

take separately the components of the surface force normal to the boundary (direction **n**) and tangential to it (direction **t**, say). For the tangential component we have, in the notation of figure 1.9.4,

$$\mu'' e''_{ij} t_i n_j = \mu' e'_{ij} t_i n_j,\qquad(3.3.17)$$

and, for the normal component,

$$p'' - 2\mu''(e''_{ij} n_i n_j - \tfrac{1}{3}\Delta'') = p' - 2\mu'(e'_{ij} n_i n_j - \tfrac{1}{3}\Delta') + \gamma(R_1^{-1} + R_2^{-1})\quad(3.3.18)$$

at each point of a boundary between two fluids.

It is worthwhile to contrast the form taken by the two transition relations at a material boundary—continuity of velocity and continuity of stress with allowance for surface tension—in the two extreme cases in which the medium on one side of the boundary is either wholly rigid or of negligible density and viscosity. At a fluid–solid interface, both normal and tangential components of velocity are continuous across the boundary, so that if the velocity of the rigid boundary is given we have here a usable boundary condition on the distribution of velocity in the fluid. However, the stress in a rigid body is unknown and no usable boundary condition on the stress distribution in the fluid is available.

The other extreme case may be typified as a liquid–gas interface, the density and viscosity of a gas being much smaller than those of a liquid under normal conditions. It is evident from the form of the Navier–Stokes equation (3.3.12) that the magnitude of pressure variations in a fluid diminishes with ρ and μ, so that, provided the velocities and velocity derivatives in the gas and liquid are of comparable magnitude, the pressure variations in the gas are much smaller than those in the liquid; and the frictional stresses are likewise smaller in the gas. As an approximation the stress everywhere in the gas may be taken as $-p_0\,\delta_{ij}$, where p_0 is the uniform gas pressure. Equating the jump in stress across the interface to the normal force due to surface tension therefore yields the following approximate boundary conditions for the flow in the liquid (assumed to lie on the side of the interface which **n** points away from):

$$e_{ij} t_i n_j = 0,\qquad(3.3.19)$$

$$p - 2\mu(e_{ij} n_i n_j - \tfrac{1}{3}\Delta) = p_0 - \gamma(R_1^{-1} + R_2^{-1})\qquad(3.3.20)$$

at each point of the interface; and it will normally be possible to put $\Delta = 0$ in view of the effective incompressibility of liquids. The relations (3.3.19) and (3.3.20) are appropriate to what is called a *free surface* of the liquid. The condition of continuity of velocity across the interface is not normally useful here because, as is already implied by our approximate representation of the stress in the gas, the velocity distribution in the gas is not of interest and may be allowed to remain unknown.†

† There is an analogy between the extreme cases of perfectly conducting and perfectly insulating boundaries in problems of heat conduction and the above cases of rigid and free boundaries respectively in problems of momentum transport; surface tension of course lies outside the analogy.

Only when the velocity distributions in the fluids on both sides of a material boundary are to be determined will it be necessary to make use of both transition relations at the interface.

Exercise

Show that a material line element which initially is normal to the free surface of a liquid remains normal to it.

3.4. Changes in the internal energy of a fluid in motion

Further insight into the effect of surface forces on the motion of a fluid may be obtained by considering the energy balance for the fluid of volume τ contained within a material surface S. Work is being done on this mass of fluid by both volume and surface forces, and it may also be gaining heat by transfer across the boundary. Some of this total gain of energy is manifested as an increase in the kinetic energy of the fluid, and the remainder, according to the first law of thermodynamics (see § 1.5), appears as an increase in the internal energy of the fluid. We shall represent this balance analytically, obtaining a differential equation valid at each point of the fluid from the energy balance for a given mass of fluid in the usual way.

First a word is necessary about the definition of some of the thermodynamic quantities relating to a material element of fluid under conditions of non-equilibrium. As explained in §§ 1.6 and 3.3, under common conditions a material element of a fluid in which the velocity or temperature is non-uniform may be regarded as passing through a succession of states in each of which the departure from equilibrium is small. For some purposes the departure from equilibrium at any instant may be neglected; for others (as in a calculation of the deviatoric stress) it is significant. This points to a need for care in framing definitions of thermodynamic quantities, in order to ensure that they are not dependent on the existence of exact equilibrium. There is no difficulty in defining the density ρ as the ratio of mass to instantaneous volume of the element, but the definitions of some other quantities, such as temperature, are not so straightforward. The definition of internal energy per unit mass is of central importance, and will be considered first. The first law of thermodynamics, as represented by (1.5.2), is effectively a definition of the difference between the values of the internal energy per unit mass of a material element in two different equilibrium states. Now the amount of work done on a material element and the amount of heat added to the element, between two instants of time, are concrete 'observable' quantities, whose definitions are not dependent on the existence of equilibrium. We may therefore continue to define an internal energy E (per unit mass) for a material element at any instant by means of (1.5.2), on the understanding that the equilibrium state to which E refers instantaneously is achieved by suddenly isolating the material element from the surrounding fluid and allowing it to come to equilibrium without work being done on it and without gain of heat.

Now that we have defined two properties of state, ρ and E, in ways which are independent of the existence of equilibrium, it is possible to define other quantities by regarding ρ and E as the two parameters of state and using equilibrium equations of state (provided the fluid is homogeneous). Thus, the temperature T of a moving element of fluid may be defined as satisfying the equilibrium relation between ρ, E and T with the appropriate instantaneous values of ρ and E for the element, and similarly for the entropy per unit mass S. This is in fact what we mean by the symbols T and S in expressions like (1.6.10) which refer to a non-equilibrium situation.

We proceed now to the calculation of the internal energy balance for a mass of homogeneous fluid. The rate at which work is being done on the fluid in the material volume τ is the sum of a contribution

$$\int u_i F_i \rho \, d\tau$$

from the resultant body force, and a contribution

$$\int u_i \sigma_{ij} n_j \, dS, \quad = \int \frac{\partial(u_i \sigma_{ij})}{\partial x_j} d\tau,$$

from the surface forces exerted at the boundary by the surrounding matter. Thus the total rate of working on a material element, per unit mass of fluid, is

$$u_i F_i + \frac{u_i}{\rho} \frac{\partial \sigma_{ij}}{\partial x_j} + \frac{\sigma_{ij}}{\rho} \frac{\partial u_i}{\partial x_j},$$

$$= u_i \frac{Du_i}{Dt} + \frac{\sigma_{ij}}{\rho} \frac{\partial u_i}{\partial x_j} \tag{3.4.1}$$

on use of the equation of motion (3.2.2). It will be seen that the first of the two terms arising from the rate of working by surface forces on an element of fluid, viz. $\rho^{-1} u_i \, \partial \sigma_{ij} / \partial x_j$, is associated with the small difference between the stresses on opposite sides of the element and contributes (together with the rate of working by body forces) to the gain in kinetic energy of bulk motion of the element, and the second, viz. $\rho^{-1} \sigma_{ij} \, \partial u_i / \partial x_j$, is associated with the small difference between the velocities on opposite sides of the element and represents the work done in deforming the element without change of its velocity. This work done in deforming the element is manifested wholly as an increase in the internal energy of the fluid.

We shall assume that heat is being transferred in the fluid by molecular conduction, the rate of gain of heat by a mass of fluid by conduction across the material bounding surface S being

$$\int k \frac{\partial T}{\partial x_i} n_i \, dS, \quad = \int \frac{\partial}{\partial x_i} \left(k \frac{\partial T}{\partial x_i} \right) d\tau,$$

where T is the local temperature and k the thermal conductivity of the fluid (§ 1.6). Thus the rate of gain of heat by a material element of fluid, per unit mass of fluid, is

$$\frac{1}{\rho}\frac{\partial}{\partial x_i}\left(k\frac{\partial T}{\partial x_i}\right).\tag{3.4.2}$$

We may regard all terms of (1.5.2) as referring to the change per unit time for a material element of fluid. The quantity W is given here by the second term of (3.4.1) and the quantity Q by (3.4.2). Hence the rate of change of internal energy per unit mass of a material element of fluid is

$$\frac{DE}{Dt} = \frac{\sigma_{ij}}{\rho}\frac{\partial u_i}{\partial x_j} + \frac{1}{\rho}\frac{\partial}{\partial x_i}\left(k\frac{\partial T}{\partial x_i}\right),$$

$$= \frac{\sigma_{ij}e_{ij}}{\rho} + \frac{1}{\rho}\frac{\partial}{\partial x_i}\left(k\frac{\partial T}{\partial x_i}\right).\tag{3.4.3}$$

Substitution of the expression (3.3.11) for the stress tensor in (3.4.3) gives

$$\frac{DE}{Dt} = -\frac{p\Delta}{\rho} + \frac{2\mu}{\rho}(e_{ij}e_{ij} - \tfrac{1}{3}\Delta^2) + \frac{1}{\rho}\frac{\partial}{\partial x_i}\left(k\frac{\partial T}{\partial x_i}\right).\tag{3.4.4}$$

An alternative form useful for interpretation is

$$\frac{DE}{Dt} = \frac{1}{\rho}(-p\,\delta_{ij})(\tfrac{1}{3}\Delta\,\delta_{ij}) + \frac{2\mu}{\rho}(e_{ij} - \tfrac{1}{3}\Delta\,\delta_{ij})(e_{ij} - \tfrac{1}{3}\Delta\,\delta_{ij}) + \frac{1}{\rho}\frac{\partial}{\partial x_i}\left(k\frac{\partial T}{\partial x_i}\right),$$

which shows the separate contributions to the work done in deforming the element made by the isotropic or pressure part of the stress in association with the isotropic or expansion part of the rate of strain and by the deviatoric part of the stress in association with the non-isotropic or shearing part of the rate of strain. This latter contribution is non-negative, showing that any shearing motion in the fluid is inevitably accompanied by a unidirectional transfer of energy from the mechanical agencies causing the motion to internal energy of the fluid, as is to be expected from the frictional character of the associated stress. We write

$$\Phi = \frac{2\mu}{\rho}(e_{ij}e_{ij} - \tfrac{1}{3}\Delta^2)\tag{3.4.5}$$

for this rate of *dissipation* of mechanical energy, per unit mass of fluid, due to viscosity, and note that it is equivalent, in its effect on the fluid, to an irreversible addition of heat.

It is natural to suppose that the first term on the right-hand side of (3.4.4) represents (rate of change of) energy of compression, capable of being returned without loss to the mechanical system when the element expands. This is true, although only approximately owing to the existence (in general) of a first-order effect of the departure from equilibrium on the mechanical pressure p in (3.4.4), which we now examine. p is defined as (minus) the mean normal stress and is an observable quantity. Now, as explained earlier in this section, ρ and E are two functions of state of an element of fluid whose definitions need no modification, and which have definite values, when the element is not in equilibrium; and to given values of ρ and E there corresponds a certain value of the pressure obtained from the equilibrium equation of state for the fluid. We may call this latter quantity the 'equilibrium pressure', and denote it by p_e. In the absence of any relative motion of the fluid, the values of p and p_e for an element of the fluid are identical, but when relative motion occurs they may differ.

The approximate value of $p - p_e$ for an element of a moving fluid may be determined by exactly the same kind of argument as was used to determine the deviatoric stress tensor. We assume that $p - p_e$ depends only on the instantaneous local velocity gradient and that, for sufficiently small magnitudes of the velocity gradient, $p - p_e$ is a linear function of the various components of the tensor $\partial u_i / \partial x_j$; that is,

$$p - p_e = B_{ij} \frac{\partial u_i}{\partial x_j}, \quad = B_{ij} e_{ij} - \tfrac{1}{2} B_{ij} \epsilon_{ijk} \omega_k, \qquad (3.4.6)$$

where the tensor coefficient B_{ij} depends on the local state of the fluid but not directly on the velocity distribution. We also assume, as before, that the response of the fluid to an imposed velocity gradient is without directional preferences, so that B_{ij} is an isotropic tensor. An isotropic tensor of the second order must have all axes as principal axes, which is possible only if

$$B_{ij} = -\kappa \, \delta_{ij}, \qquad (3.4.7)$$

where κ is a scalar coefficient (with the same dimensions as the viscosity μ) dependent on the local state of the fluid. The relation (3.4.6) then reduces to

$$p - p_e = -\kappa \Delta, \qquad (3.4.8)$$

showing that again rigid rotation of the fluid has no effect.

The rate at which the isotropic part of the stress tensor does work which contributes to the internal energy of the fluid, per unit mass of fluid, can now be written as

$$-\frac{p\Delta}{\rho} = -\frac{p_e \Delta}{\rho} + \frac{\kappa \Delta^2}{\rho}. \qquad (3.4.9)$$

The first term on the right-hand side of (3.4.9) represents a reversible transformation of energy, involving only the equilibrium pressure corresponding to the instantaneous values of ρ and E, whereas the second is one-signed and represents (provided we anticipate a positive value of κ) a dissipation of mechanical energy. The rate of expansion is the only part of the local velocity gradient that is relevant to the value of $p - p_e$, and κ is thus an expansion damping coefficient. κ might also be termed the *expansion viscosity* of the fluid,† with μ for distinction being the *shear viscosity*.

---

† Other terms in use are bulk viscosity and second coefficient of viscosity.

Under the assumed conditions the second of the two terms on the right-hand side of (3.4.9) is of small magnitude compared with the first, but since the second term is one-signed it may give rise to a significant total amount of dissipation when the rate of expansion is periodic and goes through many cycles.

The manner in which molecular transport of momentum leads to the establishment of tangential stresses in a simple shearing motion and to dissipation of mechanical energy is evident enough; this is 'friction' in the ordinary sense. The kind of molecular action which could account for the existence of expansion damping is less obvious, and, although the nature of the molecular mechanism plays no part in our phenomenological approach to (3.4.8), it is worthwhile to consider this question for a moment. Explicit consideration of molecular mechanism is necessary in any event if the magnitude of κ for different fluids is to be estimated.

Equation (3.4.8) may be interpreted as giving the lag in the adjustment of the mechanical pressure to the continually changing values of ρ and E in a motion involving expansion of the fluid; and presumably κ is non-zero for any fluid in which the mechanical pressure depends on aspects of the molecular motion and configuration different from those aspects determining ρ and E. We have in fact already seen, at the end of §1.7, how a lag in the adjustment of the mechanical pressure may occur in a perfect gas with polyatomic molecules. The mean normal stress is here proportional to the translational energy of the molecules, whereas the internal energy involves also the energy of rotational and (if the temperature is high enough) vibrational modes of motion of a molecule; and the delay in establishment (by collision between molecules) of equipartition of energy between the various modes leads to a larger-than-equilibrium value of the mean normal stress for given E and ρ when the gas is being compressed, that is, to a positive value of κ. Furthermore, since $\frac{1}{3}\rho\overline{\mathbf{u}^2}$ (where $\mathbf{u}$ is, as in §1.7, the velocity of a molecule) is the mean normal stress in a perfect gas, in or out of equilibrium, and $(\gamma - 1)\rho E$ is the equilibrium value of the mean normal stress in terms of ρ and E, we see that the relation (1.7.32) is simply the version of (3.4.8) appropriate to a perfect gas; it follows that, for a perfect gas with polyatomic molecules whose rotational modes have a relaxation time of the order of several collision intervals and make a significant contribution to the internal energy, μ/κ is a constant of order unity.

Observations of the attenuation of sound waves of fairly high frequency in some gases with diatomic molecules have shown the accuracy of the linear relation (3.4.8) and have yielded values of μ/κ of order unity. But for higher frequencies (for example, above 10^7 cycles/sec in nitrogen under standard conditions), the linear dependence on Δ breaks down. Under conditions in which vibrational modes of a molecule make an appreciable contribution to the internal energy of a gas, (3.4.8) is not normally accurate, owing to the very long relaxation time of these modes. In these cases, a different type of theory† which takes some account of the history of the motion is required. Little is known about the adequacy of the linear relation (3.4.8) and the magnitude of κ in liquids.

The rates of expansion typical of the great majority of flow systems are very much smaller than the rates of shear, for the obvious reason that the variation of mean normal stress accompanying a change of volume is enormously greater than the typical tangential stress. Consequently, circumstances in which the expansion viscosity plays a significant part are rare, being mostly confined to considerations of the damping of high frequency sound waves and the structure of shock waves. We shall not have occasion in this volume to refer again to the phenomenon of expansion damping, except in a discussion of the properties of a liquid containing small gas bubbles in suspension (§4.11), and in all subsequent sections p and p_e will be taken as identical without comment.

Finally, we note the expression for the rate of change of entropy per unit mass of fluid in a material element. The relations (1.5.20) between the various increments describing a change of state yield, in the present context,

$$T\frac{DS}{Dt} = \frac{DE}{Dt} + p_e \frac{D(1/\rho)}{Dt} = c_p \frac{DT}{Dt} - \frac{\beta T}{\rho}\frac{Dp_e}{Dt}. \qquad (3.4.10)$$

Equation (3.4.4) and the mass-conservation equation may now be used to obtain

$$T\frac{DS}{Dt} = c_p \frac{DT}{Dt} - \frac{\beta T}{\rho}\frac{Dp_e}{Dt} = \frac{\kappa\Delta^2}{\rho} + \Phi + \frac{1}{\rho}\frac{\partial}{\partial x_i}\left(k\frac{\partial T}{\partial x_i}\right), \qquad (3.4.11)$$

which is the counterpart of (1.6.10) for a moving fluid and is more general than (3.1.18).

All terms on the far right-hand side of (3.4.11) represent molecular transport effects. There are many flow fields in which all molecular transport effects may be neglected, as we shall see, and in those cases

$$T\frac{DS}{Dt} = c_p \frac{DT}{Dt} - \frac{\beta T}{\rho}\frac{Dp}{Dt}, \quad \approx 0. \qquad (3.4.12)$$

Flow fields of this kind, in which the entropy of a material element of fluid is constant, are said to be *isentropic*. Another useful term, not yet generally adopted, is *homentropic*, meaning that the entropy per unit mass S is uniform over the fluid.

3.5. Bernoulli's theorem for steady flow of a frictionless non-conducting fluid

The equation of motion for an isolated solid particle of mass m moving under the action of a force which is a function of position in space alone of

---

† Of the kind first put forward by Maxwell, and described by M. J. Lighthill in 'Viscosity effects in sound waves of finite amplitude', an article in *Surveys in Mechanics*, edited by G. K. Batchelor and R. M. Davies (Cambridge University Press, 1956).

the form $-m\nabla\Psi$ yields the relation

$$\dot{\mathbf{s}}.\frac{d\dot{\mathbf{s}}}{dt} = -\dot{\mathbf{s}}.\nabla\Psi = -\frac{d\Psi(\mathbf{s})}{dt},$$

where $\mathbf{s}(t)$ is the particle position and $\dot{\mathbf{s}}(t)$ its velocity. This equation may be integrated to give the 'energy integral'

$$\tfrac{1}{2}\dot{\mathbf{s}}^2 + \Psi(\mathbf{s}) = \text{constant},$$

and the force potential Ψ is designated as the 'potential energy' per unit mass of the particle. The requirements for a relation of this type to exist are that the force under which a particle of unit mass moves is equal to the spatial gradient of a scalar function $-\Psi$ *and* that Ψ depends only on position in space. The second requirement is so often met in particle mechanics that it is usually taken for granted.

Under certain conditions there exists a similar energy integral for the individual mass elements of a fluid. The total real energy (as distinct from fictitious 'potential energy') per unit mass of a material element of fluid moving with velocity $\mathbf{u}$ is the sum of the kinetic energy $\tfrac{1}{2}\mathbf{u}^2$ of the bulk motion and the internal energy E. This total energy may change as a result of work done by volume and surface forces acting on the element and transfer of heat (assumed to be by conduction only) across the surface bounding the element, and, as shown in §3.4,

$$\frac{D}{Dt}(\tfrac{1}{2}\mathbf{u}^2 + E) = u_i F_i + \frac{1}{\rho}\frac{\partial(u_i\sigma_{ij})}{\partial x_j} + \frac{1}{\rho}\frac{\partial}{\partial x_i}\left(k\frac{\partial T}{\partial x_i}\right). \tag{3.5.1}$$

When the body force per unit mass ($\mathbf{F}$) can be written in the form $-\nabla\Psi$, and provided Ψ is a function of position alone and not of time, we can write

$$u_i F_i = -u_i\frac{\partial\Psi}{\partial x_i} = -\frac{D\Psi}{Dt}$$

and regard Ψ as a 'potential energy' for the body-force field.

Now although the pressure acts as a normal stress at the surface bounding an element of fluid, it produces a resultant force on the element which is the same as a body force per unit volume equal to $-\nabla p$. This suggests that under certain conditions the pressure might play the part of a potential energy so far as integration of (3.5.1) is concerned. The pressure appears on the right-hand side of (3.5.1) in the term

$$-\frac{1}{\rho}\frac{\partial(u_i p\,\delta_{ij})}{\partial x_j}, \quad = \frac{p}{\rho^2}\frac{D\rho}{Dt} - \frac{u_i}{\rho}\frac{\partial p}{\partial x_i}$$

$$= -\frac{D(p/\rho)}{Dt} + \frac{1}{\rho}\frac{\partial p}{\partial t}, \tag{3.5.2}$$

and so, when the pressure field is steady, the direct effect of the pressure on the energy of the material element is the same as if the element moved in a body-force field of potential energy p/ρ per unit mass. Note that this representation of the effect of the pressure includes the work done by pressure both in compressing the element and in accelerating it as a whole.

The energy equation (3.5.1) can thus be written, when $\mathbf{F} = -\nabla\Psi$ and both Ψ and p are independent of t, in the form

$$\frac{D}{Dt}\left(\tfrac{1}{2}\mathbf{u}^2 + E + \frac{p}{\rho} + \Psi\right) = \frac{1}{\rho}\frac{\partial}{\partial x_j}\{2u_i\,\mu(e_{ij} - \tfrac{1}{3}\Delta\,\delta_{ij})\} + \frac{1}{\rho}\frac{\partial}{\partial x_i}\left(k\frac{\partial T}{\partial x_i}\right),$$

$$(3.5.3)$$

the expression for the stress tensor σ_{ij} being taken from (3.3.11). If it happens also that the two remaining terms on the right-hand side of (3.5.3) are zero, the energy equation for a material element may be integrated, as was done for an isolated solid particle. We then have the very important result that, for a frictionless non-conducting fluid in motion with a steady pressure distribution, the quantity H defined by

$$H = \tfrac{1}{2}q^2 + E + \frac{p}{\rho} + \Psi \qquad (3.5.4)$$

has the same value at all points of the path of a material element, where q stands for the speed $|\mathbf{u}|$. When the pressure field is steady the velocity field will usually be steady also, in which case the path of an element is a streamline. In terms of energy we may say that, for steady motion of a frictionless non-conducting fluid, the total energy per unit mass H is constant for a material element, provided this total includes, not only the kinetic and internal energies, but also the fictitious potential energies associated with the external body-force field and with the pressure field. In more general circumstances this total energy of a material element is not constant, usually because either (*a*) viscous stresses act on the boundary of the element and do work in accelerating the element (in which case the kinetic energy is changed), and in deforming it (in which case the internal energy is changed), or (*b*) heat is conducted into or out of the element, or (*c*) the pressure field is not steady and the associated 'potential energy' changes independently of changes in the other forms of energy of the material element.

The fact that H is constant along a streamline in steady motion of a frictionless non-conducting fluid is known as *Bernoulli's theorem*, having been established first by Daniel Bernoulli in 1738 for the particular case of an incompressible fluid.

An alternative derivation of the theorem proceeds by direct calculation of the energy balance for frictionless non-conducting fluid flowing along a stream-tube of small cross-section. If q, ρ, E, p and Ψ are values of the speed, etc., at a place in the stream-tube where the cross-section is δA, the rate at which energy of the fluid (including the conventional potential energy associated with the external body-force field) is convected past that cross-section is

$$(\tfrac{1}{2}q^2 + E + \Psi)\,q\rho\,\delta A,$$

and the rate at which the normal surface force is doing work at that cross-section is

$$pq\,\delta A.$$

But in a steady flow field the energy of the fluid contained between two fixed cross-sections of the stream-tube is constant, and the gain resulting from energy convected in and work done by pressure at one cross-section must be balanced exactly by the loss at the other. Thus

$$\left(\tfrac{1}{2}q^2 + E + \frac{p}{\rho} + \Psi\right)q\rho\,\delta A$$

is constant along the stream-tube, and, since the mass flux $q\rho\,\delta A$ also is constant, Bernoulli's theorem follows.

The particular fluid properties found to be sufficient for validity of Bernoulli's theorem—zero values of the viscosity μ and heat conductivity k—have implications for the rate of change of entropy of a material element. The complete absence of entropy-changing processes is often said to be a necessary condition for validity of Bernoulli's theorem. This is not strictly true, since lag in the adjustment of the mechanical pressure p to the changing internal energy E is an entropy-increasing process (at a rate given by the term in (3.4.11) containing the expansion viscosity κ) which nevertheless does not cause H to vary along a streamline in steady flow. However it seldom happens that the shear viscosity and heat conductivity of a fluid may be neglected but not the expansion viscosity, and we may for practical purposes assert that Bernoulli's theorem is valid when, and only when, the flow is isentropic (meaning that $DS/Dt = 0$) and steady.

Nothing has yet been said about the way in which the Bernoulli constant H varies from one streamline to another in steady isentropic flow; nor could we expect to be able to make general statements since the value of H for each streamline must depend on how the flow was established. The most we can hope to do is to find consistency relations between the variations of H and of S across streamlines. When the fluid has the same physical composition everywhere, each material element of the fluid has an equilibrium state† determined uniquely by two independent variables which we may take here to be E and ρ. The difference between the values of the entropy S for two different elements at any instant is then related to the difference between the corresponding values of E and ρ by (1.5.8), so that at any point in the fluid

$$T\nabla S = \nabla E + p\nabla(1/\rho). \qquad (3.5.5)$$

The quantity H given by (3.5.4) is also a function of position, and (3.5.5) can be written alternatively as

$$\nabla H = T\nabla S + \nabla(\tfrac{1}{2}q^2 + \Psi) + \frac{1}{\rho}\nabla p. \qquad (3.5.6)$$

---

† That is, the state obtained by isolating the element instantaneously and allowing it to come to equilibrium adiabatically and without work being done on it.

This relation is quite general, except for the assumption that $\mathbf{F} = -\nabla\Psi$. Now when the flow is steady and isentropic, the equation of motion (3.3.12) reduces to

$$\rho\mathbf{u}.\nabla\mathbf{u} = -\rho\nabla\Psi - \nabla p,$$

which in view of the vector identity

$$\mathbf{u} \times (\nabla \times \mathbf{u}) = \tfrac{1}{2}\nabla q^2 - \mathbf{u}.\nabla\mathbf{u}$$

may be rewritten as $\mathbf{u} \times \boldsymbol{\omega} = \nabla(\tfrac{1}{2}q^2 + \Psi) + \dfrac{1}{\rho}\nabla p.$ (3.5.7)

Substitution in (3.5.6) then gives

$$\nabla H = T\nabla S + \mathbf{u} \times \boldsymbol{\omega},\qquad(3.5.8)$$

a relation first found by Crocco (1937). We see that uniformity of both H and S across streamlines in steady isentropic flow, as well as along them, is possible only when either $\boldsymbol{\omega} = 0$ everywhere (that is, in irrotational flow) or—rather less probably—$\mathbf{u}$ and $\boldsymbol{\omega}$ are parallel vectors everywhere.

Just as a flow field in which S is uniform over the fluid has been termed homentropic (§ 3.4), so a flow field in which H is uniform over the fluid may be called *homenergic*. Steady flow of a frictionless non-conducting fluid may be either homentropic or homenergic, or both, or neither, depending on the circumstances.

The relation (3.5.8) yields the further result that in the case of steady homentropic flow

$$\nabla H = \mathbf{u} \times \boldsymbol{\omega}\qquad(3.5.9)$$

everywhere. Thus H is here constant along vortex-lines also, and the level-surfaces of H coincide with intersecting families of streamlines and vortex-lines. If in addition the velocity distribution is irrotational, H has the same value throughout the fluid.

We note now one or two alternative forms of Bernoulli's theorem. In § 1.5, the thermodynamic function I defined by

$$I = E + \frac{p}{\rho}$$

was introduced and termed the enthalpy or heat function of unit mass of the fluid. The quantity that is uniform along a streamline in steady isentropic flow may thus be written as

$$H = \tfrac{1}{2}q^2 + I + \Psi.\qquad(3.5.10)$$

In the case of flow of a gas it will often happen that variations of the potential energy Ψ are much smaller than those of both $\tfrac{1}{2}q^2$ and I, and then an approximate version of Bernoulli's theorem is that

$$H = \tfrac{1}{2}q^2 + I\qquad(3.5.11)$$

is constant along a streamline. In these circumstances, we may interpret the constant H for any one streamline as the *stagnation enthalpy*, that is, as the value of the enthalpy at a point on the streamline where $q = 0$, or, if no such point exists in a particular case, as the value that the enthalpy of any material element on the streamline would have if it were brought isentropically to rest.

An alternative form of Bernoulli's theorem which avoids the explicit appearance of the thermodynamic function E and which contains only mechanical quantities may be obtained by using the second law of thermodynamics (see (1.5.8)) in the form

$$\frac{D}{Dt}\left(E+\frac{p}{\rho}\right) = T\frac{DS}{Dt} + \frac{1}{\rho}\frac{Dp}{Dt}. \tag{3.5.12}$$

The pressure may be regarded as a function of the two parameters of state ρ and S, and since in an isentropic flow $DS/Dt = 0$ it follows that changes in p for a material element are here determined entirely by the changes in ρ. In these circumstances (3.5.12) may be written as

$$\frac{D}{Dt}\left(E+\frac{p}{\rho}\right) = \frac{D}{Dt}\int\frac{c^2}{\rho}d\rho, \tag{3.5.13}$$

where the integration is carried out at constant entropy and

$$c^2 = (\partial p/\partial \rho)_S \tag{3.5.14}$$

is a function of ρ alone for given S. Combination of (3.5.4) and (3.5.13) then shows that in steady isentropic flow the quantity

$$H' = \tfrac{1}{2}q^2 + \int\frac{c^2}{\rho}d\rho + \Psi \tag{3.5.15}$$

has the same value at all points of a streamline. This form of Bernoulli's theorem is commonly used when the isentropic relation between p and ρ for the fluid is known.

Special forms of Bernoulli's theorem

We record here for future reference the particular forms taken by Bernoulli's theorem in the important cases in which the fluid either (i) is incompressible, (ii) is a perfect gas, or (iii) is in steady motion relative to steadily rotating axes.

The density of a material element of an incompressible fluid is not affected by variations of pressure alone (§2.2). The internal energy of the element can then be changed only by the conduction of heat into or out of the element or by the doing of work against internal friction, and, in the absence of such changes, as in isentropic flow, we have

$$DE/Dt = 0.$$

Bernoulli's theorem then states that the quantity

$$H = \tfrac{1}{2}q^2 + \frac{p}{\rho} + \Psi \tag{3.5.16}$$

has the same value at all points of a streamline in steady isentropic flow. Since ρ does not vary along the streamline, and Ψ is a known function of position (being $-\mathbf{g} \cdot \mathbf{x}$ in the case of a uniform gravitational body force), Bernoulli's theorem here provides a simple relation between the two important flow variables q and p on a streamline. This simple relation is very useful indeed in circumstances in which the compressibility of real fluids may be ignored, as we shall see in subsequent chapters.

For a perfect gas we have the thermodynamic equation of state (see (1.7.15) and (1.7.19))

$$p = (c_p - c_v)\rho T, \tag{3.5.17}$$

and integral expressions for E, I and S are available in (1.7.20) and (1.7.21). With the use of the one for I, (3.5.10) becomes

$$H = \tfrac{1}{2}q^2 + \int c_p \, dT + \Psi. \tag{3.5.18}$$

In all these relations the specific heats c_p and c_v are functions of the temperature alone. In common circumstances, described in §1.7, c_p and c_v are approximately constant, in which case the relation between p and ρ for an isentropic change of state becomes $p \propto \rho^\gamma$, and

$$H = H' = \tfrac{1}{2}q^2 + c_p T + \Psi. \tag{3.5.19}$$

For a gas moving at speeds high compared with those obtained by free fall over the range of heights concerned, the effect of the gravitational body force may be neglected and we are left with a simple relation between q and T on any one streamline in steady isentropic flow. The gas is hotter at places on a streamline where the speed is smaller, and has its maximum temperature T_0 at the stagnation point (if it exists); $c_p T_0$ is the stagnation enthalpy for a perfect gas with constant specific heats.

When the motion is referred to axes rotating with steady angular velocity $\boldsymbol{\Omega}$, the additional body force given by (3.2.10) must be supposed to act on unit mass of the fluid. The Coriolis force has zero component in the direction of $\mathbf{u}$; and for the centrifugal force we may write

$$-\boldsymbol{\Omega} \times (\boldsymbol{\Omega} \times \mathbf{x}) = \tfrac{1}{2}\nabla(\boldsymbol{\Omega} \times \mathbf{x})^2.$$

Consequently the argument leading to (3.5.3) and (3.5.4) is applicable in a case of steady motion relative to steadily rotating axes, provided that the body force potential Ψ includes a term $-\tfrac{1}{2}(\boldsymbol{\Omega} \times \mathbf{x})^2$ arising from the centrifugal force. Cases of flow which is unsteady relative to absolute axes but steady relative to rotating axes, and to which Bernoulli's theorem may thus be applied, occur often in connexion with rotating machinery, such as a turbine. We note also that for isentropic steady flow relative to steadily rotating axes, the equation of motion (3.5.7) is modified by the addition of the Coriolis force term $-2\boldsymbol{\Omega} \times \mathbf{u}$ to the left-hand side, in addition to the above change in Ψ. Thus in the important case of homentropic flow, we have, in place of (3.5.9),

$$\nabla H = \mathbf{u} \times (\boldsymbol{\omega} + 2\boldsymbol{\Omega}), \tag{3.5.20}$$

where **u** and **ω** are now the velocity and vorticity relative to the rotating axes and the body force potential Ψ occurring in the expression for H includes a contribution from centrifugal force. The quantity **ω** + 2**Ω** in (3.5.20) is equal to the local vorticity of the fluid relative to an absolute frame of reference, and if this is zero everywhere H is constant throughout the fluid, just as it is in flow which is steady relative to non-rotating axes.

Constancy of H across a transition region in one-dimensional steady flow

We now note an important result which lies outside the scope of Bernoulli's theorem but which is used in conjunction with it in discussions of shock waves and other regions of rapid change of the flow variables. When the flow is steady and one-dimensional (with all flow variables a function of only one scalar position co-ordinate x and a velocity of magnitude u everywhere parallel to the x-axis), the complete energy equation (3.5.3) becomes

$$\rho \frac{DH}{Dt} = \rho u \frac{\partial H}{\partial x} = \frac{\partial}{\partial x}\left(\tfrac{4}{3}\mu u \frac{\partial u}{\partial x}\right) + \frac{\partial}{\partial x}\left(k \frac{\partial T}{\partial x}\right), \qquad (3.5.21)$$

while the mass-conservation equation reduces to

$$\frac{\partial}{\partial x}(\rho u) = 0.$$

Integration of (3.5.21) between any two positions x_1 and x_2 then gives

$$\rho u [H]_{x_1}^{x_2} = \left[\tfrac{4}{3}\mu u \frac{\partial u}{\partial x} + k \frac{\partial T}{\partial x}\right]_{x_1}^{x_2}, \qquad (3.5.22)$$

which shows that, even when the fluid is viscous and conducting, H has the same value at any two points in the flow at which the gradients of u and T vanish; although H here varies along a streamline, the increase and decrease of H at different parts of the streamline due to viscous forces and heat conduction exactly cancel over the particular range bounded by these two points.

The point of this result lies in its application to cases of two- and three-dimensional steady flow of a fluid for which μ and k are small and in which there exists a thin layer within which the flow variables change rapidly. There is no need for us to go into the matter in detail, and a brief statement will be sufficient. For some thin transition layers, the flow may be regarded as locally one-dimensional near the layer and the flow variables as locally uniform on each of the two sides of the layer. Outside the layer itself, where the gradients of velocity and temperature are not large, Bernoulli's theorem is approximately valid, and across the layer there is no net change in the value of H, as seen above, despite the large gradients and appreciable effects of viscosity and heat conduction within the layer. Thus H has the same value at all points on a streamline, apart from those points actually lying inside the transition layer. Note however that the entropy S does not have exactly the same value on the two sides of the transition layer, since it follows from (3.4.11) that in a steady one-dimensional flow

$$\rho u[S]_{x_1}^{x_2} = \left[\frac{k}{T}\frac{\partial T}{\partial x}\right]_{x_1}^{x_2} + \int_{x_1}^{x_2}\left\{\frac{\kappa\Delta^2}{T} + \frac{\rho\Phi}{T} + \frac{k}{T^2}\left(\frac{\partial T}{\partial x}\right)^2\right\}dx, \qquad (3.5.23)$$

the integral term being necessarily non-zero and representing an increase in the entropy per unit mass across the layer in the direction of flow.

Unless a shock wave is weak (meaning that the ratio of values of the pressure or density or fluid speed on the two sides of the shock is little different from unity), the width of the transition layer composing the shock may be so small as to invalidate the 'Newtonian' expressions for the viscous stress and heat transfer used above. However, the terms on the right-hand side of (3.5.21) are both divergences, and, irrespective of the form of the molecular momentum flux and molecular heat flux within the transition layer, we still have $H(x_1) = H(x_2)$ provided the stress and heat flux vanish at x_1 and x_2, as they do when x_1 and x_2 lie in the approximately uniform regions on the two sides of the shock wave.

3.6. The complete set of equations governing fluid flow

It may be useful to bring together the various equations which have been shown to represent the motion of a Newtonian fluid of uniform constitution. Conservation of mass of the fluid was found (see (2.2.3)) to require

$$\frac{1}{\rho}\frac{D\rho}{Dt} + \nabla\cdot\mathbf{u} = 0. \qquad (3.6.1)$$

The acceleration of the fluid produced by the various forces acting on it is given by (3.3.12):

$$\rho\frac{Du_i}{Dt} = \rho F_i - \frac{\partial p}{\partial x_i} + \frac{\partial}{\partial x_j}\{2\mu(e_{ij} - \tfrac{1}{3}\Delta\,\delta_{ij})\}, \qquad (3.6.2)$$

where e_{ij} is the rate-of-strain tensor defined in (2.3.3) and $\Delta = e_{ii} = \nabla\cdot\mathbf{u}$.

Consideration of the exchange between the internal energy and other forms of energy of the fluid led (see (3.4.11)—and note we are now ignoring effects of expansion damping) to the relation

$$T\frac{DS}{Dt} = c_p\frac{DT}{Dt} - \frac{\beta T}{\rho}\frac{Dp}{Dt} = \Phi + \frac{1}{\rho}\frac{\partial}{\partial x_i}\left(k\frac{\partial T}{\partial x_i}\right), \qquad (3.6.3)$$

where Φ, defined by (3.4.5), is the rate of dissipation of mechanical energy per unit mass of fluid due to shear viscosity, and

$$\beta = -\frac{1}{\rho}\left(\frac{\partial\rho}{\partial T}\right)_p$$

is the coefficient of thermal expansion of the fluid.

The molecular transport coefficients μ and k in (3.6.2) and (3.6.3) are functions of the local state of the fluid, the form of which may be regarded as known from previous observation of the fluid concerned (see §§ 1.7 and 1.8). With ρ and T as convenient choices of the two parameters of state, we have

$$\mu \equiv \mu(\rho, T), \quad k \equiv k(\rho, T). \tag{3.6.4}$$

The two quantities c_p and β in (3.6.3) are also functions of the local state, the form of which may be regarded as known from previous observation.

Equations (3.6.1), (3.6.2) and (3.6.3) contain $\mathbf{u}$, ρ, p and T as unknown dependent variables, and one further scalar equation is needed to make possible the determination of a flow field. This additional relation is provided by the equation of state for the fluid (§ 1.5), which may be written generally as

$$f(p, \rho, T) = 0. \tag{3.6.5}$$

The functional form of the equation of state depends on the nature of the fluid.

Isentropic flow

Flow fields from which all molecular transport effects are absent form an important special case to which constant reference is made in analytical fluid dynamics. We therefore temporarily set μ and k equal to zero in the above equations, without at this stage being aware of the conditions under which this may be a valid approximation.

Equation (3.6.3) shows that in these circumstances $DS/Dt = 0$; and as remarked in § 3.4 the flow is said then to be isentropic. The remainder of (3.6.3), viz.

$$c_p \frac{DT}{Dt} = \frac{\beta T}{\rho} \frac{Dp}{Dt}, \tag{3.6.6}$$

may be regarded as being combined with (3.6.5) to give an equation of state between ρ and p for isentropic changes of a material element:

$$\rho \equiv \rho(p, S), \tag{3.6.7}$$

in which the appearance of the entropy S is a reminder that when the flow field is not homentropic ρ is a different function of p for different material elements. Equation (1.7.24) shows the special form of (3.6.7) appropriate to a perfect gas with constant specific heats. Equations (3.6.1) and (3.6.2), when supplemented by this relation (3.6.7) between ρ and p, are now sufficient to determine the flow field, and (3.6.6) serves to determine the associated temperature distribution. The simplifying feature of isentropic flow is that exchanges between the internal energy and other forms of energy are reversible, and internal energy and temperature play passive roles, merely changing in response to the compression of a material element.

The equations governing isentropic flow may thus be written as

$$\frac{1}{\rho c^2}\frac{Dp}{Dt} + \nabla \cdot \mathbf{u} = 0,$$ (3.6.8)

$$\rho\frac{D\mathbf{u}}{Dt} = \rho\mathbf{F} - \nabla p,$$ (3.6.9)

together with (3.6.7), where $c^2 = (\partial p/\partial \rho)_s$ is a known function of ρ (or, alternatively, of p) of a form which may be different for different material elements.

The physical significance of the parameter c, which has the dimensions of velocity, may be seen in the following way. Suppose that a mass of fluid of uniform density ρ_0 is initially at rest, in equilibrium, so that the pressure p_0 is given by

$$\rho_0 \mathbf{F} = \nabla p_0.$$

The fluid is then disturbed slightly (all changes being isentropic), by some or all material elements being compressed and their density changed by small amounts, and is subsequently allowed to return freely to equilibrium and to oscillate about it.[†] The perturbation quantities $\rho_1 (=\rho - \rho_0)$ and $p_1 (=p - p_0)$ and $\mathbf{u}$ are all small in magnitude, and a consistent approximation to equations (3.6.8) and (3.6.9) is

$$\frac{1}{\rho_0 c_0^2}\frac{\partial p_1}{\partial t} + \nabla \cdot \mathbf{u} = 0,$$

$$\rho_0\frac{\partial \mathbf{u}}{\partial t} = \rho_1 \mathbf{F} - \nabla p_1,$$

where c_0 is the value of c at $\rho = \rho_0$. On eliminating $\mathbf{u}$ we have

$$\frac{1}{c_0^2}\frac{\partial^2 p_1}{\partial t^2} = \nabla^2 p_1 - \rho_1 \nabla \cdot \mathbf{F} - \frac{\mathbf{F} \cdot \nabla p_1}{c_0^2}.$$ (3.6.10)

The body force commonly arises from the earth's gravitational field, in which case $\mathbf{F} = \mathbf{g}$, $\nabla \cdot \mathbf{F} = 0$, and the last term in (3.6.10) is negligible except in the unlikely event of the length scale of the pressure variations not being small compared with c_0^2/g (which is about $1 \cdot 2 \times 10^4$ m for air under normal conditions and is even larger for water). Thus in these common circumstances (3.6.10) reduces to the wave equation[‡] for p_1, and ρ_1 satisfies the same equation. There exist solutions of this equation representing plane compression waves, which propagate with phase velocity c_0 and in which the fluid velocity $\mathbf{u}$ is parallel to the direction of propagation. In other words, c_0 is the speed of propagation of sound waves in a fluid whose undisturbed

† The fluid is elastic, and no energy is dissipated, so oscillations about the equilibrium are to be expected.
‡ About which many mathematical results are known; see, for example, *Partial Differential Equations in Physics*, by A. Sommerfeld (Academic Press, 1949) and *Methods of Mathematical Physics*, Volume 2, by R. Courant (Interscience, 1962).

density is ρ_0. Not all solutions of (3.6.8) and (3.6.9) represent compression waves of small amplitude, but it is useful nevertheless to keep in mind the interpretation of c as the *local* value of the speed with which sound waves would propagate through the fluid.

Conditions for the velocity distribution to be approximately solenoidal

It was remarked in §2.2 that in practice the rate at which the density of a material element changes is often negligibly small and that in these circumstances the mass-conservation equation (3.6.1) reduces to the statement that the velocity distribution is solenoidal. This is an important and valuable simplification, the conditions for which must now be examined carefully.

We shall suppose that the spatial distributions of $\mathbf{u}$ and other flow quantities are characterized by a length scale L (meaning that in general $\mathbf{u}$ varies only slightly over distances small compared with L), and that the variations of $|\mathbf{u}|$ with respect to both position and time have the magnitude U. Then in general the order of magnitude of spatial derivatives of components of $\mathbf{u}$ is U/L, and the velocity distribution can be said to be approximately solenoidal if

$$|\nabla . \mathbf{u}| \ll U/L,$$

i.e. if

$$\left|\frac{1}{\rho}\frac{D\rho}{Dt}\right| \ll \frac{U}{L}. \tag{3.6.11}$$

For a homogeneous fluid we may choose ρ and the entropy per unit mass S as the two independent parameters of state, in which case the rate of change of pressure experienced by a material element can be expressed as

$$\frac{Dp}{Dt} = c^2\frac{D\rho}{Dt} + \left(\frac{\partial p}{\partial S}\right)_\rho \frac{DS}{Dt}. \tag{3.6.12}$$

Thus the condition that $\mathbf{u}$ should be approximately solenoidal is

$$\left|\frac{1}{\rho c^2}\frac{Dp}{Dt} - \frac{1}{\rho c^2}\left(\frac{\partial p}{\partial S}\right)_\rho \frac{DS}{Dt}\right| \ll \frac{U}{L}. \tag{3.6.13}$$

The condition (3.6.13) will normally be satisfied only if each of the two terms on the left-hand side has a magnitude small compared with U/L, and we proceed to examine these subsidiary conditions in turn.

I. When the condition

$$\left|\frac{1}{\rho c^2}\frac{Dp}{Dt}\right| \ll \frac{U}{L} \tag{3.6.14}$$

is satisfied, the changes in density of a material element due to pressure variations are negligible, that is, *the fluid is behaving as if it were incompressible*. This is by far the more practically important of the two requirements for $\mathbf{u}$ to be a solenoidal vector. In estimating $|Dp/Dt|$ we shall lose little generality by assuming the flow to be isentropic, because the effects of viscosity and heat conductivity are normally to modify the pressure distribution rather than to control the magnitude of the pressure variation. We may then rewrite (3.6.14), with the aid of (3.6.9), as

$$\left| \frac{1}{\rho c^2} \frac{\partial p}{\partial t} - \frac{1}{2c^2} \frac{Dq^2}{Dt} + \frac{\mathbf{u}.\mathbf{F}}{c^2} \right| \ll \frac{U}{L}, \tag{3.6.15}$$

showing that in general (that is, in the absence of cancellation of terms on the left-hand side of (3.6.15) at all points of the flow field) three separate conditions, viz. that each term on the left-hand side should have a magnitude small compared with U/L, must be satisfied if the fluid is to be effectively incompressible.

I (i). Consider first the second term on the left-hand side of (3.6.15). The order of magnitude of Dq^2/Dt will be the same as that of $\partial q^2/\partial t$ or of $\mathbf{u}.\nabla q^2$ (i.e. U^3/L), whichever is the greater. There can exist cases of oscillatory flow in which the frequency of variation at a fixed point is considerably larger than U/L, but they need not be considered at the moment because, as we shall see, the requirement that the first term of (3.6.15) be small compared with U/L is then even more stringent. Thus the condition arising from the second term in (3.6.15) is

$$U^2/c^2 \ll 1. \tag{3.6.16}$$

The parameter c is a function of position in the fluid, and if the variation of c is appreciable some representative value must be chosen for use in (3.6.16).

In a steady flow, or one in which temporal rates of change do not play a dominant part, a change in the speed of a fluid element from zero to U requires a pressure variation of order ρU^2 (as Bernoulli's theorem also shows), and so $\delta\rho/\rho$, or $\delta p/\rho c^2$ ($\delta\rho$ and δp being changes experienced by a material element), is small compared with unity when $U^2/c^2 \ll 1$; this is the informal argument underlying (3.6.16). The ratio U/c is termed the *Mach number* of the flow field for which U and c are representative parameters, and plays an important role in the dynamics of gases. For air at 15 °C and one atmosphere pressure, $c = 340{\cdot}6$ m/sec, and, for water at 15 °C, $c = 1470$ m/sec. It is to be expected that the flow due to bodies moving steadily through the atmosphere at speeds below about 100 m/sec will show little, if any, effect of compressibility of the air, and that normal steady flows in water are most unlikely ever to be influenced by compressibility of the medium.

I (ii). The magnitude of the first term in (3.6.15) depends directly on the unsteadiness of the flow. Let us suppose that the flow field is oscillatory, in some rough sense if not exactly periodic, and that n is a measure of the dominant frequency. The magnitude of the pressure fluctuations can be estimated by noting that the fluid in a region with linear dimensions L has a roughly uniform velocity which changes sign in a time of order n^{-1}, and that the pressure differences over the boundary of the region causing this change of momentum must be of magnitude $\rho L U n$. The magnitude of $\partial p/\partial t$ is then

$\rho L U n^2$, and the condition that the first term of (3.6.15) be small compared with U/L is

$$\frac{n^2 L^2}{c^2} \ll 1. \qquad (3.6.17)$$

If the representative frequency of the temporal variations is U/L, this condition reduces to (3.6.16), showing that (3.6.17) is a more exacting condition than (3.6.16) in cases in which $n \gg U/L$, as anticipated earlier. Note also that nL/c is unity when L is equal to the wavelength of sound waves of frequency n, corresponding to the fact that compressibility cannot be irrelevant to the flow due to passage of a sound wave.

I (iii). If we regard the body force as arising from gravity, the third term on the left-hand side of (3.6.15), viz. $\mathbf{u} \cdot \mathbf{F}/c^2$ (which originates in the pressure variation required to balance the body force), has a magnitude of order gU/c^2, so that the condition that it be small compared with U/L is

$$\frac{gL}{c^2} \ll 1. \qquad (3.6.18)$$

In the case of air, which is the only working fluid for which there is any chance of this requirement not being met, we can use the isentropic equation of state (1.7.25) to find

$$\frac{gL}{c^2} = \frac{\rho g L}{\gamma p}.$$

This shows that the condition (3.6.18) is satisfied provided the difference between the static-fluid pressures at two points at vertical distance L apart is a small fraction of the absolute pressure, that is, provided the length scale L characteristic of the velocity distribution is small compared with $p/\rho g$, the 'scale-height' of the atmosphere (see §1.4), which is about 8·4 km for air under normal conditions. It is evident that this condition will be satisfied for all motions occurring in a laboratory or in layers of the atmosphere not exceeding a few hundred metres in depth.

The fluid will thus behave as if it were incompressible when the three conditions (3.6.16), (3.6.17) and (3.6.18) are all satisfied. The field of gas dynamics is concerned, for the most part, with circumstances in which (3.6.16) is not satisfied; (3.6.17) is not satisfied in the situations studied in acoustics; and circumstances in which (3.6.18) is not satisfied occur in dynamical meteorology. We shall not study any of these fields in this volume, interesting and important though they be. Subsequent chapters are concerned with circumstances in which all three conditions are satisfied, and in which the fluid may therefore be regarded as being incompressible.

II. We return now to (3.6.13) and consider the second subsidiary condition for $\mathbf{u}$ to be solenoidal, viz. that the magnitude of the second term on the left-hand side be small compared with U/L. Since there is a single functional relationship between p, S and ρ when the fluid is homogeneous, as we are assuming to be the case, we may write

$$\frac{1}{\rho c^2}\left(\frac{\partial p}{\partial S}\right)_\rho = -\frac{1}{\rho c^2}\left(\frac{\partial p}{\partial \rho}\right)_S\left(\frac{\partial \rho}{\partial S}\right)_p,$$

$$= -\frac{1}{\rho}\frac{(\partial \rho/\partial T)_p}{(\partial S/\partial T)_p}$$

$$= \frac{\beta T}{c_p}, \tag{3.6.19}$$

in view of the relations (1.5.16) and (1.5.17). Substitution for DS/Dt from equation (3.6.3) then gives the condition in the form

$$\left|\frac{\beta}{c_p}\left\{\Phi + \frac{1}{\rho}\frac{\partial}{\partial x_i}\left(k\frac{\partial T}{\partial x_i}\right)\right\}\right| \ll \frac{U}{L}, \tag{3.6.20}$$

the essential meaning of which is that variations of density of a material element due to internal dissipative heating or due to molecular conduction of heat into the element must be small.†

We suppose again that the effect of spatial differentiation of a quantity is to change its general magnitude by a factor L^{-1}, and that the two terms on the left-hand side of (3.6.20) do not cancel at all points of the flow field. Then it follows from the expression (3.4.5) for Φ that (3.6.20) is equivalent to the two subsidiary conditions

$$\frac{\beta U^2}{c_p}\frac{\mu}{\rho L U} \ll 1, \quad \beta\theta\frac{\kappa}{LU} \ll 1, \tag{3.6.21}$$

where $\kappa = k/\rho c_p$ is the thermal diffusivity, θ is a measure of the magnitude of temperature differences in the fluid, and β has been assumed to be positive. An indication of the circumstances in which the conditions (3.6.21) are not satisfied is provided by the following numbers for air and water at 15 °C and one atmosphere pressure in a case in which

$$L = 1\text{ cm}, \quad U = 10\text{ cm/sec}, \quad \theta = 10\text{ °C}:$$

	$\dfrac{\beta U^2}{c_p}\dfrac{\mu}{\rho L U}$	$\beta\theta\dfrac{\kappa}{LU}$
air	5×10^{-10}	7×10^{-4}
water	4×10^{-13}	3×10^{-7}

It is evidently most unlikely that heat of dissipation is ever large enough to prevent the first of the conditions (3.6.21) from being satisfied; and only

† If we allow the fluid to be inhomogeneous, changes in the density of a material element may also occur as a consequence of molecular diffusion, as, for instance, in the case of water containing salt in solution with non-uniform concentration. The condition that **u** be solenoidal in such circumstances can be determined in a similar way.

in rather improbable circumstances (for example, θ of order 100 °C and LU of order 10^{-1} cm²/sec, in a gas) will the conduction of heat into an element be sufficiently rapid to prevent the second condition from being satisfied.

For practical purposes we may therefore ignore the second term on the left-hand side of (3.6.13). In the absence of an explicit qualification, a statement that a fluid is effectively incompressible may be regarded as implying that the velocity distribution is solenoidal; and the conditions under which a fluid does behave as if it were incompressible are (3.6.16), (3.6.17) and (3.6.18), of which the first is the most important in practice.

3.7. Concluding remarks to chapters 1, 2 and 3

The end of this chapter marks a turning-point in our exposition of fluid dynamics. The first three chapters have described the general properties of fluids—physical, kinematical and dynamical respectively—and have concluded with the differential equations governing the flow of common fluids. Leaving aside some small areas of ignorance, for example about the precise conditions under which the deviatoric stress is a linear function of the rate of strain, it may be said that we now have a fairly reliable and understandable set of laws on which a study of the motion of fluids may be based. From the point of view of a 'pure' scientist concerned only with basic laws, there seems to be little need to go further. But such a point of view would be quite inappropriate for fluid dynamics. It is an intrinsic and characteristic feature of the subject that many different mechanical or physical processes are involved, and that although each of them separately may be regarded as well understood in the sense of fundamental physics, in combination they can produce many unexpected effects. It is one thing to know that the Navier–Stokes equation describes the motion of a fluid, and quite another to know, for instance, that thin boundary layers form on the upstream side, but not on the downstream side, of a large rigid sphere falling through fluid. Ability to predict what will happen in a given situation, in broad outline if not in numerical detail, is an essential part of knowledge, and, as we shall see, accurate predicition of flow properties demands much more than a mere knowledge of the governing equations. The motion of a given body of fluid is only weakly constrained by the shape and nature of the boundaries, and too many possible forms of the motion are consistent with the more obvious requirements, for example of conservation of mass, for the flow field that will actually occur in a given situation to be apparent. The pattern of fluid flow is sensitive to the circumstances giving rise to it, and exhibits a wide variety of properties of which only a hint may be seen in the governing equations.

The preceding three chapters thus do no more than set the stage. We are now prepared for an analysis of the motion of fluids, and a word is needed about the plan to be adopted. As remarked, many different physical or mechanical processes are represented in the governing equations set out in §3.6, and it is difficult to comprehend the way in which they affect the flow in given circumstances. Likewise the set of governing equations is much too complicated for a direct mathematical approach to be feasible. Progress in theoretical fluid dynamics has been made hitherto mainly by examination of the various physical or mechanical processes acting in as isolated a form as possible and by analysis of a large number of particular flow fields illustrating the effect of those separate processes. When assembled and interpreted appropriately, these special cases provide an understanding of the range and nature of the processes represented in the equations and of the effects to which they give rise. Many of the recognized branches of theoretical fluid dynamics correspond to the study of the separate mechanical processes represented in the general governing equations. 'Gas dynamics' is the study of flow fields in which appreciable variations of the absolute pressure, and hence also of the density, occur and it proceeds, for the most part, by the analytical examination of a number of representative or illustrative special cases; 'free convection' is the study of motion due wholly to buoyancy forces acting on a fluid; 'lubrication theory' refers to flow fields of a certain kind in which viscous stresses are dominant; 'magnetohydrodynamics' is the study of flow systems in which the electromagnetic field and the velocity distribution affect each other; and so on. There are also branches of fluid dynamics, such as 'hydrodynamic stability', which are not concerned with some new mechanical process but which have acquired a special character through the development of suitable analytical techniques.

A complete exposition of fluid dynamics would include an account of all such branches, many of which have yielded many recognizably distinct sub-branches. The preceding three chapters have been written with no particular selection of branches of fluid dynamics in view, and it is hoped that they provide a foundation for the study of any specialized branch, either in subsequent chapters of this book or elsewhere. Even if the generality of the first three chapters is not used directly in later study of a particular branch of fluid dynamics, it is nevertheless desirable to be able to see how the ideas and approximations of that particular branch fit into the framework of the subject as a whole.

In an introductory text intended for use by students, space is limited, and some selection is inevitable; to spread a reasonable total number of pages over the many important branches of modern fluid dynamics would lead to superficiality of the treatment of each one. The selection adopted in subsequent chapters of this book is based on the following views. First, it is essential to gain a proper understanding of the properties of flow of a fluid endowed with inertia but with no other physical properties.† A fluid with inertia as its sole physical property might seem to present a very simple

† That is to say, the fluid has zero compressibility, zero viscosity, zero heat conductivity, etc.

special case, but in fact a wide variety of corresponding flow properties will be seen to exist. As a special case it is absolutely fundamental, since only in rather exceptional circumstances is the effect of inertia of the fluid not important. In addition, study of the flow of a fluid with inertia alone has direct practical value, since real fluids often do move in a way which is not affected significantly by their other physical properties. Second, of all physical properties of a fluid other than inertia, the one which has the greatest effect on the flow in common circumstances is the internal friction, measured by the shear viscosity. Moreover, prediction of the effect of viscosity on the flow pattern proves to be quite difficult, for reasons which, mathematically speaking, derive from the fact that the viscosity μ multiplies the highest order derivative of the velocity in equation (3.6.2). The addition of viscosity to the physical properties possessed by the fluid may have a singular or discontinuous effect on the flow pattern, and considerable study of the motion of a viscous fluid is needed before the flow properties appear to be natural and explicable.

In the remainder of this book we therefore concentrate on the flow of a fluid which possesses inertia and viscosity but which is effectively incompressible. This programme might appear to be modest, but it lies at the centre of fluid dynamics and both deserves and requires serious study. We begin with a consideration of the effect of viscosity of a fluid, and later, after having established the conditions under which viscosity has little effect on the flow, we shall describe in detail the motion of a fluid with inertia alone.

4

FLOW OF A UNIFORM INCOMPRESSIBLE VISCOUS FLUID

4.1. Introduction

In this and the next chapter the effect of the stresses due to viscosity will be examined. In order to reveal this effect clearly and to allow the development of physical insight, we shall assume that the fluid is behaving as if it were incompressible; as already seen, this is a valid approximation for a wide range of conditions, the principal restriction being that the speed of the fluid should be small compared with the speed of sound everywhere. It will also be assumed that no other effects which could cause the density of a material element to vary significantly (see §3.6) are present. Under these conditions the equation describing the balance of internal energy and the thermodynamic equation of state are irrelevant, being replaced by the property of invariance of the density of an element of fluid:

$$D\rho/Dt = 0. \tag{4.1.1}$$

The mass-conservation equation reduces to

$$\nabla \cdot \mathbf{u} = 0 \tag{4.1.2}$$

in view of (4.1.1), and the expression for the stress in the fluid becomes

$$\sigma_{ij} = -p\,\delta_{ij} + \mu\left(\frac{\partial u_i}{\partial x_j} + \frac{\partial u_j}{\partial x_i}\right), \tag{4.1.3}$$

giving for the equation of motion

$$\rho\frac{Du_i}{Dt} = \rho F_i - \frac{\partial p}{\partial x_i} + \frac{\partial}{\partial x_j}\left\{\mu\left(\frac{\partial u_i}{\partial x_j} + \frac{\partial u_j}{\partial x_i}\right)\right\}. \tag{4.1.4}$$

These are the equations governing the flow of a Newtonian fluid with no approximation other than that the fluid is incompressible. The most common type of boundary condition to be imposed on the solution is that all components of the velocity are continuous across a fluid–solid boundary.

We note from (3.4.5) that the expression for the rate of dissipation of mechanical energy by viscosity, per unit mass of fluid, becomes

$$\Phi = \frac{2\mu}{\rho}e_{ij}e_{ij}, \quad \text{where} \quad e_{ij} = \frac{1}{2}\left(\frac{\partial u_i}{\partial x_j} + \frac{\partial u_j}{\partial x_i}\right). \tag{4.1.5}$$

This energy is lost from the mechanical system and appears as heat.

The viscosity of a fluid varies primarily with its temperature, so that when appreciable variations of temperature exist, either as a result of heat conducted from the boundary or heat supplied internally by viscous dissipation of mechanical energy, it is necessary to regard μ as a function of position obtained from the distribution of temperature. However, the viscosity μ will here be assumed to be uniform throughout the fluid.

The body force $\mathbf{F}$ will represent the action of the earth's gravitational field in the great majority of cases, and we shall put

$$\mathbf{F} = \mathbf{g}$$

hereafter in this chapter, with $\mathbf{g}$ assumed to be uniform over the fluid.

Under these conditions, the equation of motion for a viscous fluid becomes (on using vector notation, which is now neater)

$$\rho \frac{D\mathbf{u}}{Dt} = \rho\mathbf{g} - \nabla p + \mu \nabla^2 \mathbf{u}, \tag{4.1.6}$$

where ρ and μ are now given constants.

A further simplification to be adopted is that the density of the fluid is uniform. For cases in which 'buoyancy' forces due to the action of gravity on fluid with small variations of density are entirely responsible for the motion of the fluid, the approximation of uniform density is clearly inadmissible; but such cases of 'free convection' will not be considered in this volume.

The contribution to the acceleration of a material element due to viscous stresses arising from a given rate of strain is evidently determined by the ratio μ/ρ, and not by the viscosity μ alone. As remarked in §1.6, μ/ρ is given the special name *kinematic viscosity*, and is denoted by ν. Inasmuch as the equation of motion has the form

$$\frac{\partial \mathbf{u}}{\partial t} = \ldots + \nu \nabla^2 \mathbf{u},$$

ν is effectively a *diffusivity* for the velocity $\mathbf{u}$, having the dimensions (length) × (velocity) like all diffusivities, and playing the same role relative to the 'dynamic' viscosity μ as the thermal diffusivity $\kappa = k/\rho c_p$ plays relative to the thermal conductivity k. Values of μ and ν for air and water under different conditions are given in tables in appendix 1, and their values for some common fluids at 15 °C and one atmosphere pressure are reproduced below. It is noteworthy that the values of μ for air, water and mercury are in an increasing order, and that the values of ν are in the reverse order; when this latter quantity is relevant, mercury is effectively a much less viscous fluid than air.

	μ (gm/cm sec)	ν (cm²/sec)
air	0·00018	0·15
water	0·011	0·011
mercury	0·016	0·0012
olive oil	0·99	1·08
glycerine	23·3	18·5

The numerical values of both μ and ν for the two common fluids air and water are quite small. We do not yet know what these values should be compared with, in a consideration of the importance of the effects of viscosity, but there is a clear hint that in many contexts the effects of viscosity are negligible. It proves to be difficult to specify in a general way the conditions under which effects of viscosity may safely be neglected, and difficult to predict those effects in cases in which they are not negligible. For this reason we must study the effects of viscosity before taking advantage of the smallness of the numerical values of μ and ν for air and water.

Modification of the pressure to allow for the effect of the body force

It may be seen from (4.1.6) that when ρ is uniform the force per unit volume due to gravity is balanced exactly by a pressure equal to $\rho\mathbf{g}.\mathbf{x}$. This suggests that the pressure p should be written as

$$p = p_0 + \rho\mathbf{g}.\mathbf{x} + P, \qquad (4.1.7)$$

where p_0 is a constant and $p_0 + \rho\mathbf{g}.\mathbf{x}$ is the pressure which would exist in the same body of fluid at rest and which, in a moving fluid, gives rise to a pressure gradient in balance with the force of gravity. P is the remaining part of the pressure and evidently arises wholly from the effect of the motion of the fluid, being determined by the equation

$$\rho\frac{D\mathbf{u}}{Dt} = -\nabla P + \mu\nabla^2\mathbf{u}. \qquad (4.1.8)$$

No term for P is in general use, and it will be referred to here as the *modified pressure*. It should be noted that the modified pressure may be introduced only if the density is uniform, the gravitational body force *per unit volume* then being representable as the gradient of a scalar quantity.

Gravity no longer appears in the equation of motion when the modified pressure is introduced, and provided it is also absent from the boundary conditions we may infer that gravity has no effect on the velocity distribution in the fluid. However, if the absolute pressure occurs in the boundary conditions, as happens if part of the boundary is an interface with another fluid or in particular if it is a free surface, the full expression (4.1.7) must be used in the boundary conditions and in this way the effect of gravity re-enters the problem. The introduction of the modified pressure P for a fluid of uniform density is useful only when the boundary conditions involve the velocity alone. This will almost always be so when the fluid under discussion is air; for water, it may not be so.

In subsequent discussions of motions of a uniform fluid for which the boundary conditions involve only the velocity, and for which as a consequence gravity has no effect on the motion and does nothing more than make a contribution $\rho\mathbf{g}.\mathbf{x}$ to the pressure as in (4.1.7), we shall introduce the modified pressure and write the equation of motion in the form (4.1.8).

However, following the general practice, this modified pressure will be denoted by the same symbol p that in more general circumstances represents the absolute pressure.

The conditions under which the introduction of the modified pressure is a useful device are likewise those under which Archimedes' principle is valid for a body in a moving fluid. The net force exerted by fluid on an immersed body with surface A and volume V is

$$-\int p\mathbf{n}\,dA$$

together with a contribution from viscous stresses which depends only on the velocity distribution in the fluid. After substitution from (4.1.7), and use of the analogue of the divergence theorem for a scalar quantity, this integral becomes

$$-\int \nabla(p_0 + \rho\mathbf{g}.\mathbf{x})\,dV - \int P\mathbf{n}\,dA,$$
$$= -\rho V\mathbf{g} - \int P\mathbf{n}\,dA.$$

Thus if P and the velocity distribution in the fluid are independent of the action of gravity on the fluid, as they will be when the boundary conditions do not involve the absolute pressure, the whole consequence of the action of gravity on the fluid, so far as an immersed body is concerned, is that the body experiences a loss of weight equal to the weight of fluid displaced by the body. It may also be shown that the couple on the immersed body arising from the action of gravity on the surrounding fluid is equal and opposite to the couple which gravity would exert on the displaced fluid. Hence an immersed body which is free to move will do so in a way which is unaffected by the existence of gravity, provided it has the same total mass and centre of mass as the displaced fluid.

There are analogous although less important results concerning the effect of steady rotation on a two-dimensional flow field. When the motion of a fluid is referred to accelerating axes the form of the equation of motion is changed only by the addition of an apparent body force, given by (3.2.9). In the case of flow relative to rectilinear axes in steady rotation with angular velocity Ω about the z-axis, this apparent body force per unit mass lies in the (x, y)-plane and has components

$$2\Omega v + \Omega^2 x, \quad -2\Omega u + \Omega^2 y,$$

where (x, y) and (u, v) are corresponding components of $\mathbf{x}$ and $\mathbf{u}$. Now for two-dimensional flow of incompressible fluid in the (x, y)-plane we may satisfy the mass-conservation equation by writing

$$u = \frac{\partial \psi}{\partial y}, \quad v = -\frac{\partial \psi}{\partial x}$$

(see §2.2). It follows that the equation of motion in the rotating frame differs in form from that in an absolute frame only by the addition of a body force per unit volume

$$\rho \nabla\{-2\Omega\psi + \tfrac{1}{2}\Omega^2(x^2 + y^2)\}, \tag{4.1.9}$$

and when ρ is uniform the apparent body force may be balanced by a contribution to the pressure, like the effect of gravity; part of this addition to the pressure arises from the Coriolis force and part from the centrifugal force. Thus again the equation of motion can be written in the form (4.1.8) and, again provided the absolute pressure does not occur in the boundary conditions, the apparent body force does not enter into the determination of the velocity distribution. In other words, steady rotation of a two-dimensional flow system, without change of the boundary conditions relative to rotating axes, does not affect the velocity distribution.

The analogue of Archimedes' principle for the body force (4.1.9) may be established without difficulty. In the case of the contribution to (4.1.9) from centrifugal force the argument is identical with that for gravity, since both these forces are functions only of position in space. In order to be able to use the same kind of argument for the contribution to (4.1.9) from Coriolis force, we suppose that the function ψ is continued into the region occupied by an immersed rigid body in a manner corresponding to movement of fluid in that region as a rigid body. The additional force on (unit depth of) the body due to the action of the body force (4.1.9) on the fluid is then

$$- \int \nabla \{ -2\rho\Omega\psi + \tfrac{1}{2}\rho\Omega^2(x^2 + y^2) \}\, dA,$$

where the integral is taken over the area of the body in the (x, y)-plane. This force is equal and opposite to the resultant centrifugal and Coriolis force which would act on the displaced fluid if it were moving rigidly like the body. There is a corresponding result for the couple exerted on the body by the fluid. It follows that a rigid body which is free to move will do so in a way which is unaffected by steady rotation of the whole system, provided it has the same total mass and centre of mass as the displaced fluid.†

Exercises

1. Show that in an incompressible fluid the deviatoric stress acting across an element of a rigid boundary is wholly tangential. (Use a curvilinear co-ordinate system of which one co-ordinate surface coincides with the rigid boundary.)

2. Obtain the following alternative expression for the force per unit area exerted across a surface element with normal $\mathbf{n}$ in an incompressible fluid:

$$-p\mathbf{n} + \mu(2\mathbf{n}.\nabla\mathbf{u} + \mathbf{n} \times \boldsymbol{\omega}).$$

Show that at a rigid boundary this reduces to $-p\mathbf{n} - \mu\mathbf{n} \times \boldsymbol{\omega}$.

† The difference between two- and three-dimensional motions in this respect can be demonstrated by towing a body of the same density as water in a constant horizontal direction, relative to the fluid, through a dish of water in steady rotation (Taylor 1921). A circular cylinder with vertical generators is found to move in the direction of the applied force, as in a non-rotating system, whereas a sphere is found to veer off the straight path, in a sense opposite to that of the rotation of the dish (showing that the contribution to fluid pressure from the Coriolis force on the fluid is here not large enough to balance the Coriolis force on the body itself). A vortex ring propagated in a horizontal direction is also found to follow a curve path, relative to the fluid, with a sense opposite to that of the rotation of the system.

4.2. Steady unidirectional flow

The feature of the equation of motion (4.1.6) (or (4.1.8)) which causes most analytical difficulty is the non-linearity in the velocity $\mathbf{u}$ arising from the expression (2.1.2) for the acceleration of a fluid element in the Eulerian specification of the flow field. The mathematical difficulties presented by the complete equation of motion are so severe that most of the existing solutions are applicable only to circumstances in which for some reason the equation reduces to a linear form. Among the simplest of such cases are those in which the velocity vector has the same direction everywhere, and is independent of distance in the flow direction. The convective rate of change of velocity then vanishes identically and the acceleration of a fluid element is equal to $\partial \mathbf{u}/\partial t$, with only one component non-zero.

As a preliminary, we note the precise conditions under which the fluid velocity in a long cylindrical region is exactly unidirectional. Our interest is in situations in which a general streaming motion in the x-direction parallel to the generators has 'settled down' and is independent of end effects, so that all three velocity components u, v, w relative to a Cartesian co-ordinate system are independent of x. The equation of motion shows that ∇p also must be independent of x. Thus v and w are velocity components for a two-dimensional motion which is in no way affected by the motion in the x-direction. Such a motion in the cross-sectional plane can survive the viscous dissipation of energy only if there is a continual supply of energy to the fluid by tangential stresses exerted at a portion of the boundary in tangential motion, that is, as may readily be shown, only if

$$\mu \oint \{n_2(ve_{22} + we_{23}) + n_3(ve_{23} + we_{33})\}\, dl > 0,$$

where $(0, n_2, n_3)$ is the unit normal to the boundary and the integration is over the closed bounding curve in the cross-sectional plane. Hence if in particular the boundary be rigid and either stationary or moving in the x-direction, or be a 'free' surface across which the tangential stress is zero, the flow is ultimately everywhere in the x-direction.

We now take $v = w = 0$, whence the y and z-components of the equation of motion (4.1.8) reduce to

$$\frac{\partial p}{\partial y} = 0, \quad \frac{\partial p}{\partial z} = 0, \tag{4.2.1}$$

where p stands for the modified pressure as explained in § 4.1. The x-component of the equation of motion is

$$\rho \frac{\partial u}{\partial t} = -\frac{\partial p}{\partial x} + \mu \left(\frac{\partial^2 u}{\partial y^2} + \frac{\partial^2 u}{\partial z^2} \right), \tag{4.2.2}$$

and since neither the first nor the last term depends on x, we can write

$$\frac{\partial p}{\partial x} = -G(t). \tag{4.2.3}$$

When G is positive, the pressure gradient represents a uniform body force in the direction of the *positive x-axis*.

In the cases of steady flow to be considered in this section, $\partial u/\partial t = 0$, $-G$ is a constant pressure gradient, and (4.2.2) becomes

$$\frac{\partial^2 u}{\partial y^2} + \frac{\partial^2 u}{\partial z^2} = -\frac{G}{\mu}. \tag{4.2.4}$$

The fluid density does not appear in (4.2.4), since the total acceleration of every fluid element is zero. Each element of fluid is in equilibrium, so far as the x-components of the forces are concerned, under the action of normal stresses which vary with x (the pressure gradient) and tangential stresses due to viscosity which vary with y and z. (In addition, there is a normal stress—concealed by the use of the modified pressure—whose variation with position is such as to make it balance the force of gravity on the element.)

It is now necessary to solve (4.2.4), subject to the boundary conditions, which in general will prescribe the value of the pressure gradient $-G$ and the value of u at certain values of y and z.

Poiseuille flow

The case of flow in a long tube of circular section under the action of a difference between the pressures imposed at the two ends of the tube was studied by Hagen in 1839 and by Poiseuille in 1840. The flow may be assumed here to have the same axial symmetry as the boundary conditions, so that u is a function only of the distance r from the axis of the tube. The corresponding solution of (4.2.4) is

$$u = \frac{G}{4\mu}(-r^2 + A\log r + B).$$

There is a singularity at $r = 0$ (associated with the exertion of a finite force on the fluid over unit length of the axis) unless $A = 0$, and we therefore choose this value for the constant A. On choosing B to make $u = 0$ at the tube boundary $r = a$ we find

$$u = \frac{G}{4\mu}(a^2 - r^2). \tag{4.2.5}$$

A quantity of practical importance is the flux of volume past any section of the tube, the value of which is

$$Q = \int_0^a u 2\pi r\, dr$$

$$= \frac{\pi G a^4}{8\mu} = \frac{\pi a^4 (p_0 - p_1)}{8\mu l}, \tag{4.2.6}$$

where p_0 and p_1 are the (modified) pressures at the beginning and end of a portion of the tube of length l. Hagen and Poiseuille were able to establish by experiments with water that the flux varies as the first power of the pressure drop down the tube and as the fourth power of the tube radius (of which a contribution of two to the power comes from the dependence of cross-sectional area on a, with a further contribution of two from the higher value of u allowed by a given net viscous force in a larger tube). The accuracy with which the observations showed Q/a^4 to be constant is strong evidence for the assumption that no 'slipping' occurs at the wall of the tube and, less directly, for the hypothesis that the viscous stress varies linearly with rate of strain, under the conditions of the experiments.

The tangential stress at the wall of the tube is

$$\mu \left(\frac{du}{dr} \right)_{r=a} = -\tfrac{1}{2}Ga,$$

so that the total frictional force, in the direction of flow, on a length l of the tube is

$$2\pi a l (\tfrac{1}{2}Ga), \quad = \pi a^2 (p_0 - p_1).$$

Such an expression for the total frictional force on the tube was to be expected, because all elements of the fluid instantaneously within this portion of the tube are in steady motion under the action of normal forces $\pi a^2 p_0$ and $\pi a^2 p_1$ on the two end faces and the frictional force exerted by the tube wall. We see also from (4.1.5) that the rate at which mechanical energy is dissipated by viscosity, per unit mass of fluid, is here

$$\Phi = \frac{\mu}{\rho} \left(\frac{du}{dr} \right)^2 = \frac{G^2 r^2}{4\rho\mu}.$$

Thus the total rate of dissipation in the fluid instantaneously filling a length l of the circular tube is

$$\frac{\pi l G^2 a^4}{8\mu}, \quad = lQG.$$

In a case in which the fluid in the tube is a liquid and the tube is open to atmospheric pressure at the two end points (as it would be if liquid entered the tube from a shallow open tank and was discharged from the end of the tube), the pressure gradient down the tube is provided in effect by the force of gravity. The absolute pressure is here the same at the two ends and therefore constant throughout the fluid, so that the modified pressure is $-\rho g x \cos \alpha$ and

$$G = -\frac{dP}{dx} = \rho g \cos \alpha, \tag{4.2.7}$$

where α is the inclination of the tube to the downward vertical. The above expression for the rate of dissipation in the fluid occupying a length l of the pipe is here equal to the rate at which the same body of fluid is losing gravitational potential energy, as it should be.

Tubes of non-circular cross-section

Similar relations apply to steady unidirectional flow in tubes with a different cross-section, although only in a few particular cases can the velocity distribution be found analytically.

The velocity distribution

$$u = \frac{G}{2\mu(b^{-2}+c^{-2})}\left(1 - \frac{y^2}{b^2} - \frac{z^2}{c^2}\right) \tag{4.2.8}$$

is a solution of the equation (4.2.4), and satisfies the boundary condition required for flow in a tube of elliptical cross-section with semi-axes b and c.

To find the velocity distribution in a tube of rectangular cross-section whose sides are at $y = \mp b$, $z = \mp c$ ($c > b$, say), we note that the quantity

$$u - \tfrac{1}{2}G(b^2 - y^2)/\mu$$

is an even function of both y and z which satisfies Laplace's equation and is zero at $y = \mp b$. It may therefore be written as a Fourier series in y of the form

$$\sum_{n \text{ odd}} A_n \cosh\frac{n\pi z}{2b} \cos\frac{n\pi y}{2b},$$

and the coefficients A_n can be found from the boundary condition at $z = \mp c$. This kind of procedure is explained in the paint-brush problem below.

In both of these cases the relation between the volume flux and the pressure gradient can be found readily.

Two-dimensional flow

The governing equation is here

$$\frac{d^2u}{dy^2} = -\frac{G}{\mu}, \tag{4.2.9}$$

and we may suppose without loss of generality that the flow takes place between the planes $y = 0$ and $y = d$. If the two bounding planes are rigid, and move in the x-direction with speeds zero and U respectively, the appropriate solution of (4.2.9) is

$$u = \frac{G}{2\mu}y(d-y) + \frac{Uy}{d}. \tag{4.2.10}$$

When the two rigid planes are not in relative motion, the velocity profile is parabolic; and when the applied pressure gradient G is zero, we obtain a simple shearing motion with a linear profile, each thin layer of fluid being in steady motion under the action of equal and opposite frictional forces on its two faces (figure 4.2.1). It is clearly possible to superpose these parabolic and linear profiles also in a case in which the directions of the pressure gradient and of the relative motion of the two planes are not parallel, although the resulting flow is not then two-dimensional.

Steady unidirectional and two-dimensional flow may also occur when liquid runs down the upper face of an inclined plane in a layer of uniform thickness h. The boundary condition to be satisfied at the (plane) free surface is that the tangential stress vanishes (see § 3.3), so that $du/dy = 0$ there. Thus the velocity profile is

$$u = \frac{G}{2\mu} y(2h - y) \qquad (4.2.11)$$

(being of the same form as for two-dimensional flow between two parallel rigid planes without relative motion and distance $2h$ apart), and G has the value (4.2.7). In practice it may happen that the slope of the boundary and the volume flux Q across a plane normal to the flow and per unit width of this plane are the only given quantities; we find from (4.2.11) that

$$Q = \int_0^h u\, dy = \frac{Gh^3}{3\mu}, \qquad (4.2.12)$$

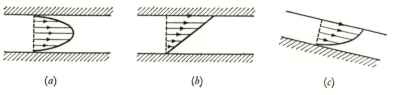

(a)   (b)   (c)

Figure 4.2.1. Steady two-dimensional flow, (a) under pressure gradient between two fixed rigid planes, (b) due to relative motion of two rigid planes, (c) in a layer on an inclined plane due to gravity.

showing that the gravitational and viscous forces on the layer will be in balance when its thickness is

$$h = \left(\frac{3\mu Q}{G}\right)^{\frac{1}{3}} = \left(\frac{3\mu Q}{\rho g \cos\alpha}\right)^{\frac{1}{3}}. \qquad (4.2.13)$$

If the rate at which liquid is poured on to an inclined plane is changed suddenly by a small amount δQ, a layer of thickness greater by an amount $(dh/dQ)\,\delta Q$ will be set up. The region of transition from one thickness to another advances down the plane, at a speed V say, and considerations of the steady volume flux, on each side of the transition, relative to axes moving with speed V show that

$$V\,\delta h = \delta Q,$$

that is,

$$V \approx \frac{dQ}{dh} = \left(\frac{9Q^2 \rho g \cos\alpha}{\mu}\right)^{\frac{1}{3}}. \qquad (4.2.14)$$

A model of a paint-brush

When a brush loaded with paint is drawn along in contact with a rigid wall, some of the paint is pulled from the brush by the frictional force at the rigid boundary and remains behind in a layer whose thickness is soon made uniform by the action of surface tension. The amount of paint left on the wall is a matter of practical importance, and it is useful to know how this quantity depends on the properties of the paint and of the brush. The following simple model of a brush (which has no pretensions to realism) provides a rough guide to the relevant factors, and serves here as an example of steady unidirectional flow.

We suppose that the brush is composed of a large number of parallel and equi-distant thin rigid plates which slide together over a plane wall in the direction of their line of contact with the wall (figure 4.2.2). The space between the plates is filled with liquid, and as the plates are drawn across

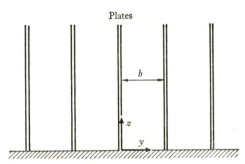

Figure 4.2.2. Sketch of model of a paint-brush.

the wall the liquid is set into motion relative to the plates by the tangential stress at the wall. To begin with, we assume the plates to be of infinite extent in the x- and z-directions, so that the resulting motion is unidirectional and steady. No pressure gradient is imposed here, and the governing equation is

$$\frac{\partial^2 u}{\partial y^2} + \frac{\partial^2 u}{\partial z^2} = 0. \tag{4.2.15}$$

It is convenient to take axes fixed in the plates, in which case the boundary conditions for flow in the channel between two adjoining plates are

$$u = 0 \quad \text{at} \quad y = 0 \quad \text{and} \quad y = b, \quad 0 < z < \infty,$$

$$u = U \quad \text{at} \quad z = 0, \quad 0 < y < b,$$

where U is the relative speed of brush and wall.

This mathematical problem is of the familiar boundary-value form,[†] and we can proceed by examining solutions of the equation in which the variables are separated. One such solution, which satisfies the homogeneous boundary conditions at $y = 0$ and $y = b$, is

$$u = \sin\frac{n\pi y}{b}(A_n e^{-n\pi z/b} + B_n e^{n\pi z/b}),$$

† See, for instance, *Fourier Series and Boundary Value Problems*, by R. V. Churchill (McGraw-Hill, 1941).

where n is an integer. The occurrence of large velocities as $z \to \infty$ is excluded by the physical conditions, so that the required solution is

$$u = \sum_{n=1}^{\infty} A_n e^{-n\pi z/b} \sin \frac{n\pi y}{b}, \qquad (4.2.16)$$

provided we can choose the constants A_n so as to satisfy the remaining condition at $z = 0$. This latter condition requires

$$A_n = \frac{2U}{b} \int_0^b \sin \frac{n\pi y}{b} \, dy = \begin{cases} 0 \ (n \text{ even}) \\ \dfrac{4U}{n\pi} \ (n \text{ odd}) \end{cases},$$

whence we obtain

$$u(y, z) = \frac{4U}{\pi} \sum_{n \text{ odd}} \frac{1}{n} e^{-n\pi z/b} \sin \frac{n\pi y}{b}. \qquad (4.2.17)$$

This series, and those obtained from it by term-by-term differentiation with respect to y or z, are convergent when $z > 0$; when $z = 0$, the series (4.2.17) is convergent, but there is of course a discontinuity in u at $y = 0$ and at $y = b$ and term-by-term differentiation does not give a convergent series.

This velocity distribution may now be used to obtain an estimate of the thickness of the layer of liquid which would be left behind on the wall if all the plates were imagined to have a rear edge at the same value of x. We suppose here that the liquid occupies only the space between the plates, except near the wall where it is dragged from this space by friction at the wall, and the above distribution is assumed to hold right to the rear edge of the plates. The volume flux out of the rear end of one channel is

$$Q = \int_0^b \int_0^\infty u \, dy \, dz$$

$$= \frac{8Ub^2}{\pi^3} \sum_{n \text{ odd}} \frac{1}{n^3} \approx 0.27 Ub^2,$$

and so the mean thickness of the layer left on the wall is $0.27b$. The part played by the spacing of the plates is evident. The viscosity of the fluid does not enter the expression for the layer thickness, as was to be expected in a model in which the net viscous force on each element of fluid is zero; in practice the properties of the paint may be relevant, no doubt because the way in which the paint leaves the brush is more complex than as assumed here, and perhaps also because the relation between stress and rate of strain for paint does not always have the Newtonian linear form assumed here.

A remark on stability

The conclusions of this (and of many subsequent) sections need to be qualified by the statement that most steady unidirectional flows are found in practice to be unstable under certain conditions. The simple case of

Poiseuille flow was in fact the first flow field to be used for the purpose of investigating systematically the phenomenon of hydrodynamic instability. Reynolds established experimentally in 1883 that, provided the flux of fluid along the tube was sufficiently small, the flow field described above was realized, and accidental disturbances were obliterated; at higher speeds, however, the flow showed intermittent oscillations and ultimately became permanently unsteady and wildly irregular (this being the phenomenon of turbulence). Although the conditions under which the various steady unidirectional flows are stable, and consequently capable of being set up in practice, are not known with precision in all cases, it is generally true that these flows are stable for small enough values of the dimensionless number UL/ν, where U and L are a velocity and a length representative of the motion and (lateral) extent of the flow in question. Reynolds's estimate of the critical value of this parameter for the case of Poiseuille flow (with L equal to the tube diameter and U equal to the mean velocity over the cross-section) was about 6,400, implying that he was able to observe steady unidirectional flow of water at 20 °C for values of UL less than about 64 cm²/sec. Thus here, as in other cases, the conditions under which stable steady flow occurs are within the range of conditions encountered in nature and in the laboratory, so that the above solutions have practical value; a knowledge of the properties of the turbulent flow that occurs under conditions for which a steady flow is unstable is also needed, but this lies beyond the scope of the present volume.

4.3. Unsteady unidirectional flow

As shown in §4.2, the pressure gradient in unidirectional flow is in the direction of the streamlines and is a function of t only. In nearly all the cases of unidirectional flow to be considered in this section the fluid motion is due entirely to some kind of unsteady motion of the boundaries, the pressures far upstream and far downstream being kept equal throughout the motion. Thus $G = 0$ for such cases, and (4.2.2) reduces to

$$\frac{\partial u}{\partial t} = \nu \left(\frac{\partial^2 u}{\partial y^2} + \frac{\partial^2 u}{\partial z^2} \right). \qquad (4.3.1)$$

This 'diffusion equation' is exactly the same as that governing the two-dimensional distribution of temperature in a stationary medium of thermal diffusivity ν, and we can make use of a number of mathematical results already established in that connection.† One of the most useful of these results is the solution describing the temperature distribution due to the release of a finite amount of heat at time $t = 0$ (an 'initial source') at the point $y = y'$, $z = z'$, in a medium of uniform temperature over the whole

† See *Conduction of Heat in Solids*, by H. S. Carslaw and J. C. Jaeger (Oxford, 1947).

(y, z)-plane; with u representing the excess temperature at position (y, z) at a subsequent time t, this solution is

$$u(y, z, t) = \frac{A}{4\pi\nu t} \exp\left\{ -\frac{(y-y')^2}{4\nu t} - \frac{(z-z')^2}{4\nu t} \right\}, \qquad (4.3.2)$$

where the constant A is a measure of the amount of heat released and is determined from the initial conditions by means of the relation

$$A = \int\!\!\!\int_{-\infty}^{\infty} u(y, z, t)\, dy\, dz = \int\!\!\!\int_{-\infty}^{\infty} u(y, z, 0)\, dy\, dz.$$

From this elementary solution we can construct an integral representing the temperature distribution at any time t in terms of the distribution over the whole (y, z)-plane at the initial instant, viz.

$$u(y, z, t) = \frac{1}{4\pi\nu t} \int\!\!\!\int_{-\infty}^{\infty} u(y', z', 0) \exp\left\{ -\frac{(y-y')^2}{4\nu t} - \frac{(z-z')^2}{4\nu t} \right\} dy'\, dz'. \quad (4.3.3)$$

The essential point underlying this solution is that since equation (4.3.1) is linear, and since there are no boundary conditions to be satisfied, the heat released initially in each element of area of the (y, z)-plane spreads out as if from an isolated initial point source with the factor A in (4.3.2) replaced by $u(y', z', 0)\, \delta y'\, \delta z'$. Analogous solutions are available for the diffusion equation in spaces of one and three dimensions, the former being obtained from (4.3.3) by choosing $u(y', z', 0)$ to be independent of z' and integrating with respect to z'.

The smoothing-out of a discontinuity in velocity at a plane

A simple and basic problem to which (4.3.3) can be applied is the effect of viscosity on the transition from the steady uniform velocity in one stream to the steady uniform velocity in another, adjacent, stream. We shall suppose that the transition layer is initially of zero thickness (which is not in itself a realistic initial condition, but there are ways of making use of the solution), and so is a sheet vortex, coinciding with the plane $y = 0$. Then with axes moving with the mean of the velocities of the two streams, and a jump in velocity of magnitude $2U$ across the layer, we should substitute

$$u(y', z', 0) = \mp U \quad \text{for} \quad y' \lessgtr 0$$

in (4.3.3), giving

$$u(y, t) = \frac{U}{(\pi\nu t)^{\frac{1}{2}}} \int_0^y \exp\left(-\frac{y'^2}{4\nu t} \right) dy' = U \operatorname{erf}\left\{ \frac{y}{(4\nu t)^{\frac{1}{2}}} \right\}. \qquad (4.3.4)$$

This velocity distribution is a function of $y/(\nu t)^{\frac{1}{2}}$ alone, and has the form shown in figure 4.3.1. Only the width of the transition layer changes with time, the variation being as $t^{\frac{1}{2}}$; if we choose to define the layer as extending from the place where $u = 0.99U$ to that where $u = -0.99U$, the width of the layer is $8.0(\nu t)^{\frac{1}{2}}$.

It was to be expected that the velocity distribution would be a function of $y/(\nu t)^{\frac{1}{2}}$ alone, because there are insufficient dimensional parameters in the problem to allow a dependence on y and t separately. By changing the dependent variable to $u/U = f(y, t)$, the problem to be solved becomes

$$\frac{\partial f}{\partial t} = \nu \frac{\partial^2 f}{\partial y^2}, \quad \text{with } f(y, 0) = \mp 1 \quad \text{for} \quad y \lessgtr 0, \qquad (4.3.5)$$

and the only dimensional quantity on which f can depend, other than y and t, is ν. From these three quantities only one dimensionless combination can be formed, viz. $y/(\nu t)^{\frac{1}{2}}$, so that the need for the velocity distribution to be independent of the units employed leads inevitably to a dependence on

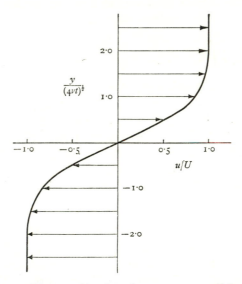

Figure 4.3.1. The transition layer between two parallel streams.

$y/(\nu t)^{\frac{1}{2}}$ alone. Acting on this conclusion, we could have converted $(4.3.5)$ into an ordinary differential equation with $\eta = y/(\nu t)^{\frac{1}{2}}$ as the sole independent variable, the expression $(4.3.4)$ again being the solution. There are many problems of fluid mechanics in which it pays to recognize on dimensional grounds that the various space and time variables will appear in the solution in certain combinations only and that the governing differential equation may be reduced from one of partial to one of ordinary type. Such a solution, involving the time only in combination with the space variable, is often termed a *similarity* solution, since the shape of the velocity distribution with respect to the space variable is similar at all times.

The expression $(4.3.4)$ describes the velocity in a transition layer which has developed from an initial discontinuity across the common boundary of the two streams, and it is not difficult to see that it also gives the asymptotic distribution of velocity (as $t \to \infty$) for an arbitrary initial form of the transition layer. For suppose that at $t = 0$ the velocity distribution in the layer is written as

$$u = \frac{y}{|y|} U + F(y), \tag{4.3.6}$$

where $F(y) \to 0$ as $y \to \mp \infty$, and $\int_{-\infty}^{\infty} F(y)\, dy = 0$ (the latter condition being achieved by appropriate choice of the position of the origin $y = 0$). Provided the function F is also such that $\int_{-\infty}^{\infty} |F|\, dy$ is bounded, we can now put

$$u(y, t) = U \operatorname{erf}\{y/(4\nu t)^{\frac{1}{2}}\} + \int_{-\infty}^{\infty} \{A(k, t) \sin ky + B(k, t) \cos ky\}\, dk,$$
$$\tag{4.3.7}$$

which satisfies the differential equation (4.3.1) provided

$$A(k, t) = A(k, 0) \exp(-k^2 \nu t), \quad B(k, t) = B(k, 0) \exp(-k^2 \nu t) \tag{4.3.8}$$

(differentiation within the integral sign in (4.3.7) being justified *a posteriori* by the evident uniform convergence of the differentiated integral for $t > 0$) and satisfies the initial condition (4.3.6) provided the functions $A(k, 0)$, $B(k, 0)$ are chosen as the coefficients in a representation of $F(y)$ as a Fourier integral. We see then that the integral in (4.3.7) tends to zero at least as rapidly as $t^{-\frac{3}{2}}$, as $t \to \infty$, since $A(0, 0) = 0$ by definition and $B(0, 0) = 0$ in view of the relation $\int_{-\infty}^{\infty} F(y)\, dy = 0$. Thus the velocity distribution in the layer is ultimately the same as if it had developed from a simple discontinuity.

This example of unsteady unidirectional flow illustrates several features which are characteristic of viscous flows of this and some other related types. First, there is the gradual spreading or diffusion of velocity variations across streamlines, which comes about through the tangential force exerted across planes normal to the y-axis. The distance of penetration of these velocity variations into regions of uniform velocity after a time t is of order $(\nu t)^{\frac{1}{2}}$. The rate of penetration decreases as t increases, because the velocity gradients and their spatial rates of change become progressively smaller, and the rate of penetration is smaller for fluids of smaller kinematic viscosity. Second, the velocity distribution in the layer tends asymptotically to a similarity form depending on $y/(\nu t)^{\frac{1}{2}}$ alone, irrespective of the initial form of the transition. This asymptotic form is set up because the departure of the initial velocity distribution from a simple discontinuity is equivalent to a set of initial 'sources' and 'sinks' with zero total source strength, the effects of which gradually cancel as they spread out and overlap.

Plane boundary moved suddenly in a fluid at rest

Suppose that a semi-infinite region of stationary fluid is bounded by a rigid plane (at $y = 0$, say) which is suddenly given a velocity U in its own plane and thereafter maintained at that speed. The fluid is brought into motion through the action of viscous stress at the plate, the velocity distribution being governed by the equation

$$\frac{\partial u}{\partial t} = \nu \frac{\partial^2 u}{\partial y^2}, \tag{4.3.9}$$

as in the preceding example, and by the boundary conditions

$$\left.\begin{array}{l} u(y, o) = o \quad \text{for} \quad y > o, \\ u(o, t) = U \text{ for} \quad t > o. \end{array}\right\} \tag{4.3.10}$$

The parameter U can be effectively removed from the problem by using u/U as the dependent variable, and we then recognize on dimensional grounds that u/U must be a function of $y/(\nu t)^{\frac{1}{2}}$ alone, as explained above. However, it is not necessary to go through the analytical details, because, by choosing axes fixed relative to the plate, we can transform the problem so that it becomes identical with the preceding problem. The boundary conditions are now

$$u(y, o) = -U \quad \text{for} \quad y > o,$$

$$u(o, t) = o \quad \quad \text{for} \quad t > o,$$

which are equivalent to the boundary conditions for one half of a transition layer which develops from a simple discontinuity at the common boundary of two streams of speeds $-U$ and $+U$, the velocity at $y = o$ being permanently zero on account of the anti-symmetry about $y = o$. Thus the solution to the problem represented by (4.3.9) and (4.3.10) is

$$u(y, t) = U - U \operatorname{erf}\{y/(4\nu t)^{\frac{1}{2}}\}, \tag{4.3.11}$$

the qualitative features of which can be interpreted as before. The frictional force per unit area exerted by the fluid on the plate is

$$\mu \left(\frac{\partial u}{\partial y}\right)_{y=0} = -\pi^{-\frac{1}{2}} \rho U^2 \left(\frac{\nu}{U^2 t}\right)^{\frac{1}{2}},$$

the decrease as $t^{-\frac{1}{2}}$ being a consequence of the thickening of the region of variable velocity.

One rigid boundary moved suddenly and one held stationary

Suppose now that the fluid is bounded by two rigid boundaries at $y = o$ and $y = d$ and is initially at rest, and that the fluid motion is due, as before, to the lower plate being brought suddenly to the steady velocity U in its own plane, the upper plate being held stationary. The governing differential equation is (4.3.9) again, with the boundary conditions

$$u(o, t) = U, \quad u(d, t) = o \quad \text{for} \quad t > o,$$

$$u(y, o) = o \quad \text{for} \quad o < y \leqslant d.$$

Since the dimensional parameter d enters the problem, $y/(\nu t)^{\frac{1}{2}}$ is no longer the only dimensionless combination of the available parameters, and we have no grounds for anticipating a similarity solution.

The appropriate solution of (4.3.9) may be found conveniently by first transforming to the new dependent variable

$$w(y, t) = U(1 - y/d) - u,$$

which satisfies the same differential equation and has homogeneous boundary conditions at $y = 0$ and $y = d$. A particular solution for w which satisfies these two boundary conditions is

$$\exp\left(-n^2\pi^2\frac{\nu t}{d^2}\right)\sin\frac{n\pi y}{d}, \qquad (4.3.12)$$

where n is an integer. We now try to satisfy the condition on w at $t = 0$ by using the whole set of such solutions, that is, we seek values of the constants A_n such that

$$\sum_{n=1}^{\infty} A_n \sin\frac{n\pi y}{d} = w(y, 0) = U\left(1 - \frac{y}{d}\right).$$

This requires
$$A_n = \frac{2}{d}\int_0^d U\left(1 - \frac{y}{d}\right)\sin\frac{n\pi y}{d}\,dy = \frac{2U}{\pi n}. \qquad (4.3.13)$$

Thus the velocity distribution is given by

$$u(y, t) = U\left(1 - \frac{y}{d}\right) - \frac{2U}{\pi}\sum_{n=1}^{\infty}\frac{1}{n}\exp\left(-n^2\pi^2\frac{\nu t}{d^2}\right)\sin\frac{n\pi y}{d}; \qquad (4.3.14)$$

the form of this Fourier series reflects the discontinuity in u with respect to y at $y = 0$ when $t = 0$. This series solution is not well suited to computation for $\nu t \ll d^2$, since the series is then only slowly convergent, and a solution in a form which is more convenient for that case has been found by the Laplace transform method in the context of heat conduction in a stationary medium.

The velocity profiles for different values of $\nu t/d^2$ in figure 4.3.2 show how the effect of the stationary upper boundary, although negligible at first, gradually influences the diffusion of the velocity variations. As was to be expected, the velocity tends asymptotically to that appropriate to steady flow between two rigid planes in relative motion (see §4.2); by contrast, in the previous case in which the upper boundary was absent the velocity variations continued to diffuse into the undisturbed fluid indefinitely. The rapidity with which the terms of the series in (4.3.14) tend to zero increases with n, and the first term ($n = 1$) survives longest. As soon as this first term dominates the series, the departure from the asymptotic steady state decays approximately exponentially, with a 'half-life' equal to $d^2/(\pi^2\nu)$.

Flow due to an oscillating plane boundary

A case which exhibits clearly the intrinsic damping or smoothing action of viscous diffusion is the two-dimensional flow produced by a rigid plane boundary moving in its own plane with a sinusoidal variation of velocity. We may suppose the upper half of the (x, y)-plane to be occupied by fluid, the rigid boundary being at $y = 0$ and having a velocity $U \cos nt$. In practice the fluid motion would be set up from rest, and, for some time after the initiation of the motion, the velocity field contains 'transients' determined by these initial conditions. It may be shown that the fluid velocity gradually becomes a harmonic function of t, with the same frequency as the velocity of the boundary, and only this steady periodic state will be considered here.

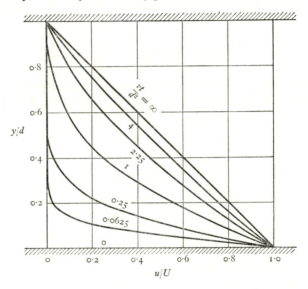

Figure 4.3.2. The development from rest of steady flow between parallel planes in relative motion.

We therefore put
$$u(y, t) = \mathscr{R}\{e^{int}F(y)\}, \tag{4.3.15}$$

where $\mathscr{R}$ denotes the real part of the expression that follows, the complex form being used for convenience. From the differential equation (4.3.9) we find
$$inF = \nu \frac{d^2F}{dy^2},$$

of which the only solution that remains finite as $y \to \infty$ is
$$F(y) = A \exp\{-(1+i)(n/2\nu)^{\frac{1}{2}}y\}.$$

In order to make the velocity of the fluid at $y = 0$ equal to that of the oscillating boundary, we must have $A = U$, whence the solution is
$$u(y, t) = U \exp\{-(n/2\nu)^{\frac{1}{2}}y\} \cos\{nt - (n/2\nu)^{\frac{1}{2}}y\}. \tag{4.3.16}$$

This velocity profile may be described as a damped transverse 'wave' of wavelength $2\pi(2\nu/n)^{\frac{1}{2}}$ 'propagating' in the y-direction with phase velocity

$(2\nu n)^{\frac{1}{2}}$, the damping being such as to make the amplitude of the oscillations fall off as $\exp\{-(n/2\nu)^{\frac{1}{2}}y\}$. The way in which n and ν enter into the formulae for the wavelength and phase velocity can be accounted for by recalling that the distance through which a velocity variation can be diffused in a time t is of order $(\nu t)^{\frac{1}{2}}$, but the damping formula is less easy to explain, except on general dimensional grounds. The ratio of the amplitudes of the oscillation at two points distance one wavelength apart, that is, at two neighbouring points at which the oscillations are in phase, is $e^{-2\pi}$ ($\approx$ 0·002), and in view of the smallness of this ratio the motion is effectively confined to a 'penetration depth' of order equal to part of a wavelength, that is, of order $(\nu/n)^{\frac{1}{2}}$.

It will be noticed that, since the differential equation and the boundary conditions are linear, the above solution for an arbitrary harmonic component of the velocity of the rigid boundary can be used to build up the solution for a general periodic motion of the rigid boundary.

The solution (4.3.16) has other direct applications, including the diurnal variation of temperature in the surface layers of the ground under the action of solar radiation to the surface. The value of the thermal diffusivity for soil may be taken very roughly to be 0·01 cm²/sec, showing that the wavelength of the diurnal temperature wave (the temperature at the surface being taken as simple harmonic in t) is about a metre; diurnal variations of temperature would be very small at depths of this order. We shall also see later (§ 5.14) that under certain conditions the flow in the neighbourhood of the surface of a rigid body in translational oscillation is described approximately by the solution (4.3.16), and that the work done against friction in one cycle of the body motion can be deduced from it.

Starting flow in a pipe

Finally, we consider a case in which the flow is due, not to a moving boundary, but to the application of a pressure gradient. The fluid contained in a long pipe of circular cross-section is initially at rest, and is set in motion by a difference between the pressures at the two ends of the pipe suddenly imposed and maintained by external means. This pressure difference produces immediately a uniform axial pressure gradient, $-G$ say, throughout the fluid, and so the equation to be satisfied by the axial velocity u is

$$\frac{\partial u}{\partial t} = \frac{G}{\rho} + \nu\left(\frac{\partial^2 u}{\partial r^2} + \frac{1}{r}\frac{\partial u}{\partial r}\right), \qquad (4.3.17)$$

in which G is a constant. The boundary and initial conditions are

$$u = 0 \quad \text{at} \quad r = a \quad \text{for all } t,$$
$$u = 0 \quad \text{at} \quad t = 0 \quad \text{for } 0 \leqslant r \leqslant a.$$

Equation (4.3.17) can be made homogeneous by using as dependent variable the departure of the velocity from its steady asymptotic value, which is of course that given in equation (4.2.5). With the new variable w given by

$$w(r, t) = \frac{G}{4\mu}(a^2 - r^2) - u,$$

the equation to be solved becomes

$$\frac{\partial w}{\partial t} = \nu \left(\frac{\partial^2 w}{\partial r^2} + \frac{1}{r} \frac{\partial w}{\partial r} \right),$$

with $w(a, t) = 0, \quad w(r, 0) = \frac{G}{4\mu}(a^2 - r^2).$

A particular solution of this equation which satisfies the boundary condition at $r = a$ is

$$J_0 \left(\lambda_n \frac{r}{a} \right) \exp \left(-\lambda_n^2 \frac{\nu t}{a^2} \right),$$

where J_0 is the Bessel function of the first kind of order zero and λ_n is one of the positive roots of $J_0(\lambda) = 0$. By using the whole set of these particular solutions, we can also satisfy the condition at $t = 0$. Thus w is given by the Fourier–Bessel series†

$$w(r, t) = \frac{G}{4\mu} \sum_{n=1}^{\infty} A_n J_0 \left(\lambda_n \frac{r}{a} \right) \exp \left(-\lambda_n^2 \frac{\nu t}{a^2} \right), \qquad (4.3.18)$$

where the coefficients A_n are such as to satisfy

$$a^2 - r^2 = \sum_{n=1}^{\infty} A_n J_0 \left(\lambda_n \frac{r}{a} \right),$$

that is, $A_n = \frac{2a^2}{J_1^2(\lambda_n)} \int_0^1 x(1 - x^2) J_0(\lambda_n x) \, dx$

$$= \frac{8a^2}{\lambda_n^3 J_1(\lambda_n)}.$$

The velocity distribution is then given by

$$u(r, t) = \frac{G}{4\mu}(a^2 - r^2) - \frac{2Ga^2}{\mu} \sum_{n=1}^{\infty} \frac{J_0 \left(\lambda_n \frac{r}{a} \right)}{\lambda_n^3 J_1(\lambda_n)} \exp \left(-\lambda_n^2 \frac{\nu t}{a^2} \right). \qquad (4.3.19)$$

The variation of u across the pipe is shown in figure 4.3.3 for several different values of $\nu t / a^2$. Initially the whole fluid has acceleration G/ρ, but as the velocity increases the restraining influence of the wall spreads further into the fluid. The central portion of fluid whose velocity is increasing as Gt/ρ becomes narrower as t increases, until when t is of order $a^2/(\nu \lambda_1^2)$ (where $\lambda_1 = 2 \cdot 41$) all parts of the fluid are subject to the effect of the wall and the velocity at $r = 0$ ceases to increase. As in previous cases, the

† See chap. 18 of *Theory of Bessel Functions*, by G. N. Watson (Cambridge University Press, 1958).

approach to the steady state is soon dominated by the first term of the series in (4.3.19).

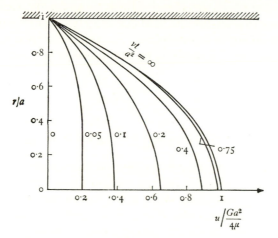

Figure 4.3.3. Starting flow in a circular pipe. Velocity profiles at different instants (from Szymanski 1932).

Exercises

1. A circular cylinder of radius a and infinite length is immersed in fluid at rest everywhere, and is suddenly made to move with steady velocity U parallel to its length. Show that the frictional force on unit length of the cylinder at time t after the motion has begun is

$$\frac{8\mu U}{\pi} \int_0^\infty \frac{\exp\left(-k^2 \nu t\right) dk}{k\{J_0^2(ka) + Y_0^2(ka)\}},$$

of which the asymptotic form, as $t \to \infty$, is $4\pi\mu U/\log\left(\nu t/a^2\right)$.

2. Obtain an expression in terms of Bessel functions for the velocity distribution in flow in a long straight circular pipe due to an oscillating axial pressure gradient $-G + C \cos nt$, where G, C and n are constants, and examine the limiting cases $n \to 0$ and $n \to \infty$.

4.4. The Ekman layer at a boundary in a rotating fluid

Consider a situation in which a large body of water, initially at rest under gravity, is set into motion by the action of a steady uniform tangential stress applied at the horizontal free surface. This is a simple problem of unsteady unidirectional flow, in which the velocity $u(y, t)$ satisfies the differential equation (4.3.9) and the boundary conditions

$$u(y, 0) = 0, \quad \text{and} \quad \left.\begin{aligned} \frac{\partial u}{\partial y} &= S \quad \text{at} \quad y = 0 \\[4pt] u &\to 0 \quad \text{as} \quad y \to -\infty \end{aligned}\right\} \quad \text{for all } t,$$

where μS is the constant stress at the free surface. On differentiating (4.3.9) with respect to y, we obtain an equation and set of boundary conditions for the variable $\partial u/\partial y$ which are identical with those for the variable u in the problem of flow in a semi-infinite fluid due to a plane rigid boundary moving with constant velocity in its own plane. Thus the velocity u in the problem under discussion here follows from an adaptation of (4.3.11):

$$\frac{\partial u}{\partial y} = S + S \operatorname{erf}\left\{\frac{y}{(4\nu t)^{\frac{1}{2}}}\right\}, \tag{4.4.1}$$

and

$$u = Sy + Sy \operatorname{erf}\left\{\frac{y}{(4\nu t)^{\frac{1}{2}}}\right\} + 2S\left(\frac{\nu t}{\pi}\right)^{\frac{1}{2}} \exp\left(-\frac{y^2}{4\nu t}\right). \tag{4.4.2}$$

This solution applies, under appropriate conditions, to the velocity distribution in water which is set into motion by air blowing steadily over its free surface; the speed of the water at the free surface increases with time and there may be a consequential change in the stress exerted by the wind on the water, but the air speeds are usually a good deal larger than the water speeds and the change in the stress is small. It is natural then to enquire if (4.4.2) applies to large-scale systems and in particular to the drift at the surface of the sea due to wind blowing over it. Flow systems in the atmosphere and ocean are seldom undisturbed or uniform over horizontal planes in the manner assumed above, and transport of momentum due to random fluctuations in the fluid velocity ('turbulence') is usually much more important than the viscous stress, but quite apart from these complicating features it can be seen that the rotation of the earth has a significant effect on surface drift motions. The equation of motion relative to axes fixed in the earth's surface contains fictitious forces which we see from (3.2.9) to be (*a*) the Coriolis force $-2\mathbf{\Omega} \times \mathbf{u}$ (per unit mass of fluid), where $\mathbf{\Omega}$ is the angular velocity of the earth, and (*b*) the centrifugal force, which is approximately uniform over quite large regions of the earth's surface and consequently equivalent in its effect to a (small) change in the gravitational acceleration. In a motion like that described by (4.4.2), the velocity changes appreciably over distances of order $(\nu t)^{\frac{1}{2}}$, and the viscous force per unit mass of fluid is of order $|\mathbf{u}|/t$; thus the Coriolis force becomes comparable in magnitude with the viscous force when t is of the order of one day, and is clearly relevant in geophysical drift motions.

It happens that inclusion of the effect of the Coriolis force on motions like that represented by (4.4.2) leads to a new and remarkably simple velocity distribution which plays an important part in many rotating systems, and which warrants description here. The key features of this new velocity distribution are that it is steady and that variation of the velocity is confined to a layer of finite thickness adjoining the boundary. Steadiness is made possible by the fact that if the fluid velocity is a function of the vertical co-ordinate alone, and varies in direction as well as in magnitude, the viscous force may be everywhere perpendicular to the local velocity, like the Coriolis force, which is the only other force acting on the fluid. A steady flow could be discovered by analysing the development of the motion from rest, but we shall take its existence for granted and simply find its properties.

The layer at a free surface

We suppose first that, as before, the fluid is bounded by a horizontal free surface at which a uniform and constant stress μS is applied. We use a rectilinear co-ordinate system rotating steadily with angular velocity Ω, with z the vertical co-ordinate now (positive direction upwards), and with the x-axis in the direction of the stress applied at the surface. The fluid velocity clearly lies in a horizontal plane everywhere, with components $(u, v, 0)$, and the boundary conditions to be satisfied are

$$\frac{\partial u}{\partial z} = S, \quad \frac{\partial v}{\partial z} = 0, \quad \text{at } z = 0, \tag{4.4.3}$$

and
$$u, v \to 0 \quad \text{as} \quad z \to -\infty.$$

The pressure (modified to incorporate effects of gravity and centrifugal force) is also uniform in a horizontal plane, so that the equations of steady motion in the horizontal plane are

$$-2v\Omega \sin \theta = v\frac{d^2u}{dz^2}, \tag{4.4.4}$$

$$2u\Omega \sin \theta = v\frac{d^2v}{dz^2}, \tag{4.4.5}$$

where θ is the angle between the vertical z-axis and the angular velocity Ω of the frame of reference ($\frac{1}{2}\pi - \theta$ being the angle of latitude in the case of motion at the earth's surface, and regarded as uniform over the flow field). The component of the Coriolis force in the z-direction is balanced by a pressure gradient of no dynamical interest.

Equations (4.4.4) and (4.4.5) are sufficient for the determination of u and v as functions of z. Multiplication of (4.4.5) by i ($= (-1)^{\frac{1}{2}}$) and addition to (4.4.4) gives

$$v\frac{d^2(u+iv)}{dz^2} = 2i\Omega \sin \theta(u+iv), \tag{4.4.6}$$

and the solution satisfying the condition at infinity is

$$u+iv = A \exp\{k(1+i)z\}, \tag{4.4.7}$$

where $k = (\Omega \sin \theta/v)^{\frac{1}{2}}$, θ being supposed positive, corresponding to the northern hemisphere. The complex constant A is found from the free surface condition (4.4.3) to be

$$A = \frac{S(1-i)}{2k}, \tag{4.4.8}$$

giving the solution as
$$u = \frac{S}{k\sqrt{2}} e^{kz} \cos(kz - \tfrac{1}{4}\pi), \qquad (4.4.9)$$

$$v = \frac{S}{k\sqrt{2}} e^{kz} \sin(kz - \tfrac{1}{4}\pi). \qquad (4.4.10)$$

A steady flow in a layer near the surface with thickness of order k^{-1} is thus possible. At the surface the fluid velocity has its maximum magnitude $S/k\sqrt{2}$ and has a direction $45°$ in a clockwise sense (viewed from above) from the applied stress. Rather surprisingly, this angle is independent of the rate of rotation Ω (which makes one wonder what happens when $\Omega \to 0$; in that case, the time required for the steady state to be set up from rest increases indefinitely and so also does the magnitude of the velocity at the surface in the steady state). As the depth below the free surface increases, the direction

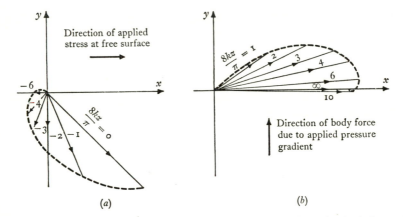

Figure 4.4.1. The velocity vector in a rotating fluid, (*a*) at various depths below a free surface at which a tangential stress is applied, and (*b*) at various heights above a rigid plane with an applied pressure gradient.

of the velocity rotates uniformly in a clockwise sense (for $\theta > 0$) and the magnitude falls off exponentially; at what might be called the penetration depth, equal to π/k, the direction is opposite to that at the surface and the magnitude has fallen to a fraction $e^{-\pi}$ (≈ 0.04) of its surface value. Figure 4.4.1 (*a*) shows the projection of the velocity vector at a number of equidistant depths on to a horizontal plane, the curve traced out by the endpoints of the vector being a logarithmic spiral.

This steady flow exhibiting a balance between the frictional and Coriolis forces was first noticed by Ekman (1905),† and was used in a discussion of wind-generated ocean currents on a rotating earth. The weakness in any such application to the ocean lies in the assumption that the tangential stress exerted across horizontal planes is due to molecular viscosity; as remarked

† See also *Physical Oceanography*, vol. 1, p. 400, by A. Defant (Pergamon Press, 1961).

above, the transport of momentum across horizontal planes in both the ocean and atmosphere is usually due primarily to irregular fluctuations in the fluid velocity arising from various sources. One possible adjustment of the above solution to take account of this fact is to regard ν, as it appears in (4.4.9) and (4.4.10), as representing an effective kinematic viscosity due to velocity fluctuations (and if information about the variation of this effective viscosity with depth is available, one can return to (4.4.4) and (4.4.5) and carry out the integration afresh). The penetration depth of the spiral, $\pi(\Omega \sin \theta / \nu)^{-\frac{1}{2}}$, increases with ν, and is 36 cm at the equator when ν represents the molecular kinematic viscosity of water; the effective viscosity due to turbulent mixing in the surface layers of the sea varies widely with the circumstances, but is almost always much larger than 0.01 cm^2/sec, sometimes by as much as a factor of 10^5, and the penetration depth is correspondingly greater.

One of the flow parameters of interest in oceanography is the net flux of water volume in the surface layer across vertical planes. This can be obtained from the integral

$$\int_{-\infty}^{0} (u+iv)\,dz = -\frac{iS}{2k^2} = -\frac{i\mu S}{2\rho\Omega\sin\theta}, \qquad (4.4.11)$$

and is independent of ν when the applied stress (μS) is given; thus ignorance of the effective value of ν is here of no consequence. It will be noticed that the net flux in the direction of the applied stress is zero, as would be expected from the fact that a net motion in that direction would give rise to a net Coriolis force in an orthogonal direction which could not be balanced by any other external force.

The layer at a rigid plane boundary

Suppose now that a large body of fluid at rest relative to uniformly rotating axes is set into motion by a uniform gradient of modified pressure which is then balanced by the Coriolis force. If the uniform pressure gradient lies in the horizontal (x, y)-plane and has components $(0, -G)$, the rotation vector being as before, the uniform velocity in the steady state has components $(U, 0)$, where

$$2U\Omega\sin\theta = G/\rho. \qquad (4.4.12)$$

If in addition the fluid is bounded by a horizontal rigid plane at rest relative to the rotating axes, the departure from a uniform stream in the 'Ekman layer' near the rigid plane involves viscous and Coriolis forces in the same manner as in the previous case, and again it happens that a steady flow exists.

The equations giving the velocity components (u, v) as functions of z in the steady flow near the rigid plane (at $z = 0$) differ from (4.4.4) and (4.4.5) only by the addition of the imposed uniform pressure gradient, so that

$$-2v\Omega\sin\theta = \nu\frac{d^2u}{dz^2}, \qquad 2u\Omega\sin\theta = \frac{G}{\rho}+\nu\frac{d^2v}{dz^2}.$$

The second equation can be rewritten as

$$2(u-U)\Omega\sin\theta = \nu\frac{d^2v}{dz^2}. \tag{4.4.13}$$

Thus the solution corresponding to a uniform stream at sufficient distance from the rigid plane (with the fluid on this occasion lying *above* the plane $x = 0$, to suit the application to atmospheric flow near the earth's surface) is

$$u - U + iv = A\exp\{-k(1+i)z\}, \tag{4.4.14}$$

where $k = (\Omega\sin\theta/\nu)^{\frac{1}{2}}$ as before. At the plane $z = 0$ we require $u = v = 0$, so that

$$A = -U$$

and the velocity components are obtained as

$$u = U(1 - e^{-kz}\cos kz), \tag{4.4.15}$$

$$v = Ue^{-kz}\sin kz. \tag{4.4.16}$$

The general features of this spiral distribution of $(u-U, v)$ near a rigid surface are much the same as in the previous case. The velocity vectors at different heights above the rigid plane are shown in figure 4.4.1 (b). Here the velocity near $z = 0$ is linear in z and is inclined at $45°$ in a clockwise direction from the direction of the body force due to the applied pressure gradient. The net flux of fluid volume in the Ekman layer in the direction normal to the uniform stream outside the layer is

$$\int_0^\infty v\,dz, \quad = U/2k = \tfrac{1}{2}U(\nu/\Omega\sin\theta)^{\frac{1}{2}}, \tag{4.4.17}$$

per unit width in the x-direction.

Horizontal pressure gradients which are approximately uniform over distances of order many kilometres arise naturally in the atmosphere from large-scale cyclonic or anticyclonic disturbances and also from horizontal variations of temperature due to differential heating of the atmosphere, and the resulting flow will normally be accompanied by something like an Ekman spiral near the earth's surface. Again the simple theoretical distributions of velocity can be applied only if ν is interpreted as an effective kinematic viscosity due to irregular turbulent mixing of different horizontal layers of fluid. With a value of ν corresponding to the molecular viscosity of air, the penetration depth π/k is 144 cm, whereas the observed penetration depth in the atmosphere may be from 500 to 1,000 m, depending on the conditions. Observations of the variation of direction and magnitude of the wind with height above the ground have been used, in conjunction with (4.4.15) and (4.4.16), as a means of obtaining the value of the effective viscosity due to turbulent mixing (Taylor 1915). A similar application of the analysis to the friction layer near the bottom of the sea can be made, although few observations are available in this case.

4.5. Flow with circular streamlines

Another simple type of fluid motion is one in which all the streamlines are circles centred on a common axis of symmetry. Such motions may be steady or unsteady, and are usually generated by the rotation of an exterior or an interior rigid boundary in the form of a circular cylinder. If the motion is to remain purely rotatory, with the axial component of velocity zero, the axial pressure gradient must be zero, and the equation of motion shows that for this to be so the motion must be two-dimensional. The velocity then depends only on the distance from the axis of symmetry. Since the component of acceleration normal to the streamlines plays a passive role in flows of this type, and changes in velocity occur wholly as a result of frictional forces between neighbouring cylindrical shells of fluid, they are equivalent to unidirectional flow in mechanical respects.

The equation of (two-dimensional) motion expressed in polar co-ordinates (r, θ) is given in appendix 2, and, on assuming that the velocity component v in the direction of the θ-co-ordinate line is a function of r and t only, and that $u = 0$, we find

$$\frac{\rho v^2}{r} = \frac{\partial p}{\partial r},$$

$$\frac{\partial v}{\partial t} = \nu \left(\frac{\partial^2 v}{\partial r^2} + \frac{1}{r} \frac{\partial v}{\partial r} - \frac{v}{r^2} \right). \tag{4.5.1}$$

The first of these equations shows that the radial variation of pressure simply supplies the force necessary to keep the fluid elements moving in a circular path. The second equation is essentially a relation for the rate of increase of angular momentum of a cylindrical shell of fluid under the action of the couples exerted by friction at its inner and outer faces. We may see this by noting that the tangential stress on an element of the surface of a cylinder of radius r (see appendix 2) is

$$\sigma_{r\theta} = \mu \left(\frac{\partial v}{\partial r} - \frac{v}{r} \right),$$

so that the couple exerted on the fluid inside a cylindrical surface of radius r by the fluid outside it is

$$2\pi \mu r^2 \left(\frac{\partial v}{\partial r} - \frac{v}{r} \right) \tag{4.5.2}$$

per unit length of cylinder. Equating the rate of change of angular momentum of the fluid in a cylindrical shell, per unit length and per unit thickness, to the couple acting on it then gives

$$\frac{\partial (2\pi \rho r^2 v)}{\partial t} = \frac{\partial}{\partial r} \left\{ 2\pi \mu r^2 \left(\frac{\partial v}{\partial r} - \frac{v}{r} \right) \right\}, \tag{4.5.3}$$

from which (4.5.1) follows.

A slightly simpler form of (4.5.1) or (4.5.3) is obtained by using as dependent variable the angular velocity $\Omega(r, t)$ of a material cylindrical shell of radius r. On putting $v = \Omega r$ in (4.5.3) we have

$$\frac{\partial \Omega}{\partial t} = \frac{\nu}{r^3} \frac{\partial}{\partial r}\left(r^3 \frac{\partial \Omega}{\partial r}\right). \tag{4.5.4}$$

Yet another form, with the vorticity magnitude

$$\omega = \frac{\partial v}{\partial r} + \frac{v}{r} = \frac{1}{r} \frac{\partial(r^2\Omega)}{\partial r} \tag{4.5.5}$$

as the dependent variable, is

$$\frac{\partial \omega}{\partial t} = \nu\left(\frac{\partial^2 \omega}{\partial r^2} + \frac{1}{r} \frac{\partial \omega}{\partial r}\right), \tag{4.5.6}$$

which is of the same form as the equation of conduction of heat in two dimensions (with circular symmetry). It will be observed from (4.2.2) that in unidirectional flow the two components of vorticity perpendicular to the streamlines also satisfy the heat conduction equation. Thus problems of unidirectional flow and of flow with circular streamlines can be described wholly in terms of diffusion of the lateral components of vorticity across the streamlines. In any particular case, a choice from the above alternative equations will usually be determined by the dependent variable that is involved in the boundary conditions.

Steady motions with circular streamlines must be maintained by the motion of rigid boundaries, and we can represent all the common cases by supposing the fluid to lie between rigid cylinders of radii r_1 and r_2 ($> r_1$) which rotate steadily with angular velocities Ω_1 and Ω_2. It is then readily found that the solution of either (4.5.3) or (4.5.4), with the left-hand side put equal to zero, that satisfies the condition of no slip at the two boundaries is

$$v(r) = \frac{1}{r}\left(\frac{\Omega_1 - \Omega_2}{r_1^{-2} - r_2^{-2}}\right) + r\left(\frac{\Omega_1 r_1^2 - \Omega_2 r_2^2}{r_1^2 - r_2^2}\right). \tag{4.5.7}$$

A flow of this kind can be set up in the laboratory with cylinders whose common length is large compared with their radii, and the velocity distribution (4.5.7) has been confirmed for various choices of r_1, r_2, Ω_1 and Ω_2. The viscosity does not appear in (4.5.7), because the net frictional couple on every cylindrical shell of fluid is zero; in this respect, steady flow with circular streamlines is the circular analogue of flow between parallel rigid planes which are in relative sliding motion (and the latter case is in fact recovered from (4.5.7) when $r_2 - r_1 \ll r_1$). From (4.5.2) and (4.5.7) we find that the frictional couple exerted across unit length of a cylindrical surface of radius r is

$$-4\pi\mu\left(\frac{\Omega_1 - \Omega_2}{r_1^{-2} - r_2^{-2}}\right), \tag{4.5.8}$$

and is independent of r as expected; in particular, this is the couple exerted on the inner rigid cylinder and is minus the couple on the outer cylinder.

The velocity distributions in various special cases of steady flow can be obtained from (4.5.7). On putting $r_1 = 0$ (with Ω_1 not so large as to make $\Omega_1 r_1^2$ non-zero), in order to represent the case of flow *inside* a single rotating cylinder we obtain

$$v = \Omega_2 r, \tag{4.5.9}$$

which is a rigid-body rotation in which the tangential stresses are everywhere zero. The other extreme case of flow in infinite fluid *outside* a single rotating cylinder is obtained by putting $r_2 \to \infty$ and $\Omega_2 = 0$, whence

$$v = \frac{r_1^2 \Omega_1}{r}. \tag{4.5.10}$$

This is an irrotational velocity distribution, in which the circulation round all closed curves linking the cylinder once is $2\pi r_1^2 \Omega_1$. The frictional couple exerted on the fluid by unit length of the cylinder is $4\pi\mu r_1^2 \Omega_1$, which implies a continual increase in the total angular momentum of the fluid; this is compatible with the assumed steadiness of the motion, since the total angular momentum associated with the distribution (4.5.10) is infinite, and the couple exerted steadily by the cylinder presumably serves to generate this velocity distribution at continually increasing distances from the cylinder.

Equation (4.5.1), or one of its equivalents, may be used to investigate the flow changes produced by starting or stopping the rotation of the two circular cylinders and the subsequent approach to a steady state. As an example, consider the motion generated from rest in the fluid contained within a single circular cylinder of radius a, the cylinder being rotated with steady angular velocity Ω. The conditions to be satisfied by $v(r, t)$ are here

$$v(r, 0) = 0 \quad \text{for} \quad 0 \leqslant r < a,$$
$$v(a, t) = \Omega_0 a \quad \text{for} \quad t > 0.$$

The method of solution is that used several times in §4.3, the form of (4.5.1) suggesting that v should be expanded in a series of Bessel functions of the first order. Again it is more convenient to consider the function

$$w(r, t) = \Omega_0 r - v(r, t),$$

since $w = 0$ for all t at $r = a$ and, in the absence of any singularities at the axis of rotation, at $r = 0$. The equation for w is the same as that for v, and a solution which satisfies this equation and the conditions at $r = 0$ and $r = a$ identically is the Fourier–Bessel series

$$w(r, t) = \sum_{n=1}^{\infty} A_n J_1\left(\lambda_n \frac{r}{a}\right) \exp\left(-\lambda_n^2 \frac{vt}{a^2}\right), \tag{4.5.11}$$

where J_1 is the Bessel function of the first kind of order one and the numbers λ_n are the positive values of λ at which $J_1(\lambda) = 0$. This expression also satisfies the condition at $t = 0$ provided

$$\sum_{n=1}^{\infty} A_n J_1\left(\lambda_n \frac{r}{a}\right) = \Omega_0 r$$

for $0 \leqslant r \leqslant a$, whence, with the aid of standard formulae,†

$$A_n = \frac{2a}{J_0^2(\lambda_n)} \int_0^1 \Omega_0 \, x^2 J_1(\lambda_n x) \, dx$$

$$= -\frac{2\Omega_0 \, a}{\lambda_n J_0(\lambda_n)}.$$

Thus the velocity distribution is given by

$$v(r,t) = \Omega_0 r + 2\Omega_0 \, a \sum_{n=1}^{\infty} \frac{J_1\!\left(\lambda_n \dfrac{r}{a}\right)}{\lambda_n J_0(\lambda_n)} \exp\left(-\lambda_n^2 \frac{\nu t}{a^2}\right). \qquad (4.5.12)$$

The term of the series that survives longest is that for $n = 1$, and the departure from solid-body rotation very soon decays exponentially, with a 'half-life' $a^2/(\lambda_1^2 \nu)$ (where $\lambda_1 = 3 \cdot 83$).

Similar, although more complicated, solutions can be obtained for the flow generated from rest in fluid between two circular cylinders (when the velocity is expressed as a Fourier–Bessel series involving Bessel functions of both the first and the second kinds) and in fluid outside a single circular cylinder (when the velocity is expressed as a Fourier–Bessel integral involving both kinds of Bessel function).‡

Finally, as a case for which (4.5.6) is the more useful equation, we consider the flow in which initially the vorticity is zero everywhere, except on the axis $r = 0$ where there is a line vortex (see §2.6) of strength C. Initially the circulation round all circles centred on the axis has the same value C, and so $v = C/2\pi r$. Vorticity is here diffused radially away from an initial concentration on a line, whereas, in the case discussed early in §4.3, the diffusion was from an initial concentration on a plane. Mathematically this problem of a spreading line vortex is identical with the two-dimensional conduction of heat, in a uniform solid, from a point at which initially there is concentrated a finite amount of heat C. The solution is obtained directly from (4.3.2):

$$\omega(r,t) = \frac{C}{4\pi\nu t} \exp\left(-\frac{r^2}{4\nu t}\right). \qquad (4.5.13)$$

The velocity distribution is then

$$v(r,t) = \frac{1}{r} \int_0^r \omega r \, dr$$

$$= \frac{C}{2\pi r}\left\{1 - \exp\left(-\frac{r^2}{4\nu t}\right)\right\}, \qquad (4.5.14)$$

which is sketched in figure 4.5.1 for several values of t. At small values of r $(\ll (4\nu t)^{\frac{1}{2}})$ the motion is a rigid-body rotation, with angular velocity

† See *Theory of Bessel Functions*, by G. N. Watson (Cambridge, 1958).
‡ See *Eigenfunction Expansions*, by E. C. Titchmarsh, §§1.11 and 4.10 (Oxford, 1962).

$C/(8\pi\nu t)$, while at large values of r ($\gg (4\nu t)^{\frac{1}{2}}$) the motion is irrotational and as it was at the initial instant. It will be observed that the circulation round a circular path centred on the origin, i.e. $2\pi r v$, has a distribution of the same form at all t. This might have been predicted on dimensional grounds, from the fact that rv/C is a dimensionless dependent variable which can depend only on ν, r and t, and so must be a function of $r^2/\nu t$ alone.

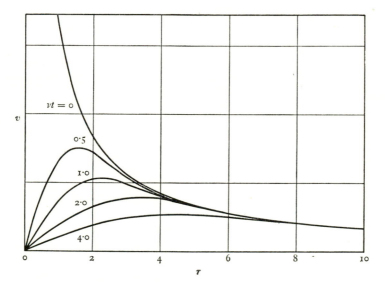

Figure 4.5.1. Velocity distribution associated with a spreading line vortex. The units for r and for νt are consistent.

4.6. The steady jet from a point source of momentum

We turn now to less simple flow fields, and consider one of the few known exact solutions of the equation of motion (4.1.8) outside unidirectional flow.

Faced with the difficulty of solving a non-linear partial differential equation, it is a sound plan to look for particular solutions in which all but one of the independent variables occur either not at all or in some simple way determined on dimensional grounds, the dependence on the remaining variable being then given by an ordinary differential equation. To take a trivial example, we could eliminate the independent variables t and angular position by choosing a steady flow with spherical symmetry about the origin of co-ordinates, leaving only the radial distance r as independent variable. Only the radial component of velocity u can then be non-zero, and the equation of mass-conservation shows at once that $u \propto r^{-2}$; in this case the equation of motion serves only to determine the pressure. This simple solution has a singularity at the origin which may be interpreted physically as a steady source of mass. In a similar way, we can consider steady flow with symmetry about an axis, leaving r and θ (the angle subtended by the radius vector and the axis of symmetry) as the only independent variables, and proceed to impose more restrictions so that the dependence on either r or θ is made evident. The flow field to be discussed in this section may be obtained by postulating that the fluid velocity varies as r^{-1}, the dependence on θ then being given by an ordinary differential equation. This kind of procedure is indirect, in that we do not know what kind of flow field we have, or whether it is physically significant, until the mathematical solution has been interpreted, but it can be quite purposeful in experienced hands.

We suppose that there is no rotation of the fluid about the axis of symmetry of the flow. It is appropriate to introduce the Stokes stream function ψ whence the velocity components (u, v, o) in a system of spherical polar co-ordinates (r, θ, ϕ) are given by

$$u = \frac{1}{r^2 \sin\theta} \frac{\partial \psi}{\partial \theta}, \quad v = -\frac{1}{r \sin\theta} \frac{\partial \psi}{\partial r}, \tag{4.6.1}$$

the equation of mass conservation thereby being satisfied identically (see §2.2). The additional restriction to be imposed, in the hope that the equation of motion can then be solved, is that u and v vary as r^{-1} and thence that $\psi \propto r$. We therefore write

$$\psi(r, \theta) = r\nu f(\theta), \tag{4.6.2}$$

the factor ν being inserted in order to make the unknown function f dimensionless.

Now the equation of motion in spherical polar co-ordinates takes the following form for steady axisymmetric flow without azimuthal 'swirl' (see appendix 2):

$$u\frac{\partial u}{\partial r} + \frac{v}{r}\frac{\partial u}{\partial \theta} - \frac{v^2}{r} = -\frac{1}{\rho}\frac{\partial p}{\partial r} + \nu\left(\nabla^2 u - \frac{2u}{r^2} - \frac{2}{r^2}\frac{\partial v}{\partial \theta} - \frac{2v\cot\theta}{r^2}\right), \tag{4.6.3}$$

$$u\frac{\partial v}{\partial r} + \frac{v}{r}\frac{\partial v}{\partial \theta} + \frac{uv}{r} = -\frac{1}{\rho r}\frac{\partial p}{\partial \theta} + \nu\left(\nabla^2 v + \frac{2}{r^2}\frac{\partial u}{\partial \theta} - \frac{v}{r^2\sin^2\theta}\right), \tag{4.6.4}$$

where

$$\nabla^2 = \frac{1}{r^2}\frac{\partial}{\partial r}\left(r^2\frac{\partial}{\partial r}\right) + \frac{1}{r^2\sin\theta}\frac{\partial}{\partial \theta}\left(\sin\theta\frac{\partial}{\partial \theta}\right).$$

When the expressions (4.6.1) and (4.6.2) are substituted into these two scalar equations, all terms except those containing p are found to be multiplied by the same power of r; this is the distinctive property of the relation (4.6.2), on which the choice of the *first* power of r on the right-hand side is based. We may remove r from the equation of motion completely by writing

$$\frac{p - p_0}{\rho} = \frac{\nu^2}{r^2} g(\theta), \tag{4.6.5}$$

where p_0 is the pressure at large distances from the origin, whence (4.6.3) and (4.6.4) become

$$g = -\frac{f^2}{2(1-\xi^2)} - \frac{1}{2}\frac{d}{d\xi}\{ff' - (1-\xi^2)f''\}, \qquad (4.6.6)$$

$$g' = -f'' - \frac{1}{2}\frac{d}{d\xi}\left(\frac{f^2}{1-\xi^2}\right), \qquad (4.6.7)$$

where $\xi = \cos\theta$ and the prime denotes differentiation with respect to ξ.

On eliminating g from equations (4.6.6) and (4.6.7) and integrating three times, we find

$$f^2 - 2(1-\xi^2)f' - 4\xi f = c_1\xi^2 + c_2\xi + c_3, \qquad (4.6.8)$$

where c_1, c_2 and c_3 are arbitrary constants of integration.

Our procedure has thus led to a solution of the governing equations (assuming (4.6.8) can be solved, by numerical methods if necessary), and we must now consider its interpretation. The three constants c_1, c_2 and c_3 are still at our disposal for the purpose of obtaining particular flow fields which are significant physically. Now if the flow is to be free from singularities on the axis of symmetry, except at $r = 0$ where (4.6.1) and (4.6.2) show a singularity to be unavoidable, v must be zero there and f must behave like $(1-\xi)$ near $\xi = 1$ $(\theta = 0)$ and like $(1+\xi)$ near $\xi = -1$ $(\theta = \pi)$; as a consequence, the expression on the left-hand side of (4.6.8) varies as $(1-\xi)^2$ near $\xi = 1$ and like $(1+\xi)^2$ near $\xi = -1$, which is impossible for the expression on the right-hand side unless $c_1 = c_2 = c_3 = 0$. Thus the flow with the minimum number of singularities on the axis, and in all probability with the greatest simplicity, is that for which

$$f^2 - 2(1-\xi^2)f' - 4\xi f = 0. \qquad (4.6.9)$$

The transformation $f = (1-\xi^2)h(\xi)$ shows that

$$h^2 - 2h' = 0,$$

whence the solution of (4.6.9) is

$$f(\xi) = \frac{2(1-\xi^2)}{1+c-\xi}, \qquad (4.6.10)$$

where c is an arbitrary constant.

The character of the flow field described by (4.6.1), (4.6.2) and (4.6.10) (found first by Landau (1944), and independently by Squire (1951)) is evident from the shape of the family of streamlines $\psi = $ constant, those shown in figure 4.6.1 being calculated for the case $c = 0.1$. The solution evidently represents a jet of fluid moving rapidly away from the origin and entraining slow-moving fluid from outside the jet. The edge of the jet can be defined conveniently as the place where the streamlines are at their minimum distance from the axis, and it is readily seen from (4.6.10) that this edge occurs at $\theta = \theta_0$, where

$$\cos\theta_0 = (1+c)^{-1};$$

and $\theta_0 = 24° 37'$ for the streamlines shown in figure 4.6.1. When the value of c is given, ψ/r is a function of θ alone, so that all the streamlines in figure 4.6.1 have the same shape, one being obtained from another by a change of the scale of r. Thus, in order to show the flow fields corresponding to

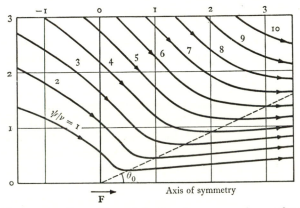

Figure 4.6.1. Streamlines of the flow for $c = 0\cdot1$, $\theta_0 = 24° 37'$. (The units for ψ/ν and r are consistent.)

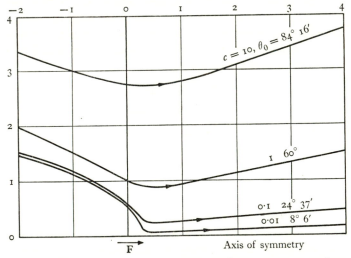

Figure 4.6.2. The streamlines $\psi/\nu = rf(\theta) = 1$ for $c = 0\cdot01$, $0\cdot1$ and $1\cdot0$, and $\psi/\nu = rf(\theta) = \frac{1}{2}$ for $c = 10$. (The units for ψ/ν and r are consistent.)

different values of c, it is sufficient to sketch one streamline for each value of c, as in figure 4.6.2. As $c \to 0$, the jet becomes more strongly concentrated near the axis of symmetry.

There are no singularities in the velocity distribution except at the origin, and the pressure is uniform at large distances from the origin. Thus the singularity at $r = 0$ is evidently the agency responsible for the whole motion, and we must examine it more closely. A jet can be produced only if momentum is supplied to the fluid continually, from some external agency, and evidence that this is exactly what the singularity at $r = 0$ does is provided by the fact that the fluid velocity is proportional to r^{-1}. To see this in detail, we use the momentum equation in integral form, in the manner described in §3.2. The fluid instantaneously outside a closed surface surrounding the origin exerts on the fluid inside the surface a force $\int \sigma_{ij} n_j \, dA$, and momentum flows outwards across this surface at a rate $\int \rho u_i u_j n_j \, dA$; the difference

$$F_i = \int (\rho u_i u_j - \sigma_{ij}) n_j \, dA \qquad (4.6.11)$$

is the steady force exerted on the contained fluid at some interior boundary. An application of the divergence theorem shows that since the flow is steady F_i has the same value for any two closed surfaces surrounding the origin whatever their shape, as would be expected.

In order to calculate F_i we choose the closed surface to be a sphere of radius r centred on the origin. Owing to the symmetry, only the axial component F of the force is non-zero and for this component we have

$$F = \int_0^{\pi} \{\rho u(u \cos\theta - v \sin\theta) - (\sigma_{rr} \cos\theta - \sigma_{r\theta} \sin\theta)\} \, 2\pi r^2 \sin\theta \, d\theta,$$

where the suffices r and θ refer to components in the positive directions of the r- and θ-co-ordinate lines. On making use of the formulae for σ_{rr} and $\sigma_{r\theta}$ given in appendix 2, and substituting from (4.6.1), (4.6.2) and (4.6.5), we have

$$\frac{F}{2\pi\rho\nu^2} = \int_{-1}^{1} \{f'(\xi f' - f) + \xi(g - 2f') + 2f + (1 - \xi^2) f''\} \, d\xi$$

and, after use of (4.6.6) and the solution (4.6.10) and some elementary calculations,

$$= \frac{32}{3} \frac{1+c}{c(2+c)} + 4(1+c)^2 \log\left(\frac{c}{2+c}\right) + 8(1+c). \qquad (4.6.12)$$

Thus the force F exerted on the fluid at the origin and the constant c in (4.6.10) are uniquely related, and it follows that the whole effect of the singularity at the origin is represented by this force.

It will be noted that, since $f(0) = f(\pi)$, there is no net flux of mass across any closed surface surrounding the origin; the singularity at $r = 0$ represents a generation of momentum alone, and not of mass.

An alternative version of the relation (4.6.12) in terms of the semi-angle θ_0 of the conical boundary of the jet is

$$\frac{F}{2\pi\rho\nu^2} = \frac{32}{3} \frac{\cos\theta_0}{\sin^2\theta_0} + \frac{4}{\cos^2\theta_0} \log\left(\frac{1 - \cos\theta_0}{1 + \cos\theta_0}\right) + \frac{8}{\cos\theta_0}, \qquad (4.6.13)$$

which is shown graphically in figure 4.6.3. For large values of F, the jet becomes very fast and narrow and (4.6.13) becomes

$$\frac{F}{2\pi\rho\nu^2} \sim \tfrac{3}{3}\theta_0^{-2}. \tag{4.6.14}$$

Under these same conditions, the flow within the jet, where $\theta \leqslant \theta_0$, is given by the following asymptotic form of (4.6.10),

$$f(\theta) \sim \frac{4\theta^2}{\theta^2+\theta_0^2}, \tag{4.6.15}$$

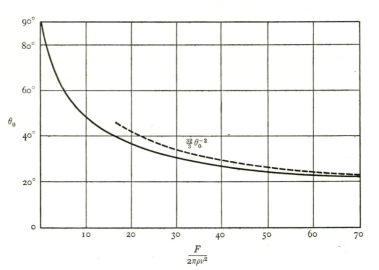

Figure 4.6.3. Relation between the force applied at the origin and the semi-angle of the jet produced by it.

and well outside the jet, where $\theta \gg \theta_0$, by

$$f(\theta) \sim 2(1+\cos\theta). \tag{4.6.16}$$

The radial velocity corresponding to (4.6.16) is $u = -2\nu/r$, this being the inward flow of fluid required to replace the fluid entrained in the jet.

It is readily seen from the calculation leading to (4.6.12) that, when θ_0 is small and $c \ll 1$, the dominant contribution to F (viz. the first term in (4.6.12)) comes from the flux of momentum. This fact makes it possible to estimate the value of F for a real jet discharged from a small orifice. If fluid is discharged from an orifice of area A with uniform velocity V, we have

$$\frac{F}{2\pi\rho\nu^2} \approx \frac{1}{2\pi}\frac{AV^2}{\nu^2}, \tag{4.6.17}$$

which must be large compared with unity if the estimate of F is to be self-consistent. The mass flux through the orifice is $\rho AV = F/V$, and the solution (4.6.15) and (4.6.16) evidently becomes a better approximation to the flow produced by the orifice as V is increased (or, equivalently, as A is decreased) for a given (large) value of F. The mass flux in the jet represented by (4.6.15) at distance r from the origin is of order $8\pi\rho\nu r$, and it therefore seems likely that the jet with a velocity profile given by (4.6.15) which develops at some distance downstream from a real orifice will appear to have come from an origin at a distance of order $AV/8\pi\nu$ upstream from the orifice.

The dimensionless parameter $(F/2\pi\rho\nu^2)^{\frac{1}{2}}$ is shown by (4.6.17) to be the 'Reynolds number' of the flow at the orifice, the interpretation of which in §4.7 makes it understandable that viscous forces should be unable to retard and diffuse the concentrated jet from the orifice when $F/2\pi\rho\nu^2 \gg 1$.

At the opposite extreme, we have $c \gg 1$ and θ_0 is close to 90°. The corresponding asymptotic form for $f(\theta)$, obtained from (4.6.10), is

$$f(\theta) \sim \frac{2}{c}\sin^2\theta,$$

and the relation between c and the force applied at the origin is found from (4.6.12) to be

$$\frac{F}{2\pi\rho\nu^2} \sim \frac{8}{c}, \quad \ll 1.$$

Consequently we may write the stream function in this asymptotic case as

$$\psi = \frac{F}{8\pi\mu} r\sin^2\theta. \tag{4.6.18}$$

A body moving through fluid exerts a force on it, and provided the speed of the body is sufficiently small for the point of application of the force to be effectively stationary we might expect (4.6.18) to have some relation to the flow generated by the body. It will be shown in §4.9 that (4.6.18) does in fact describe the resulting flow pattern at large distances (where details of the body shape are irrelevant) from a moving body exerting a force F on the fluid, under conditions such that F is small compared with $2\pi\rho\nu^2$.

4.7. Dynamical similarity and the Reynolds number

The motion of a fluid which is effectively incompressible and of uniform density ρ is governed by the equations

$$\rho\left(\frac{\partial u_i}{\partial t} + u_j\frac{\partial u_i}{\partial x_j}\right) = -\frac{\partial p}{\partial x_i} + \mu\frac{\partial^2 u_i}{\partial x_j\,\partial x_j}, \tag{4.7.1}$$

$$\frac{\partial u_i}{\partial x_i} = 0, \tag{4.7.2}$$

where p is the modified pressure. We propose to consider the effect on the flow of changes in the (uniform) values of the parameters ρ and μ. To this end it is useful to write these equations in terms of dimensionless variables, so that the effect of changing the values of ρ and μ is dissociated from the effect of mere changes of units. No parameters with the dimensions of length and velocity occur in the above equations, so that we must look to the boundary and initial conditions for dimensional quantities with which to make $\mathbf{x}$ and $\mathbf{u}$ dimensionless.

Let us suppose that the specification of the boundary and initial conditions for a particular flow involves some representative length L (which might be the maximum diameter of an interior boundary, or the distance between enclosing boundaries) and some representative velocity U (which might be the steady speed of a rigid boundary), in such a way that these conditions can be expressed in the non-dimensional form

$$\mathbf{u}' = \text{given function of } t' \text{ at some given } \mathbf{x}',$$

$$\mathbf{u}' = \text{given function of } \mathbf{x}' \text{ at some given } t',$$

where
$$\mathbf{u}' = \frac{\mathbf{u}}{U}, \quad t' = \frac{tU}{L}, \quad \mathbf{x}' = \frac{\mathbf{x}}{L}.$$

Then with these same new variables, and

$$p' = \frac{p - p_0}{\rho U^2},$$

where p_0 is some representative value of the (modified) pressure in the fluid, the governing equations become

$$\frac{\partial u_i'}{\partial t'} + u_j' \frac{\partial u_i'}{\partial x_j'} = -\frac{\partial p'}{\partial x_i'} + \frac{1}{R} \frac{\partial^2 u_i'}{\partial x_j' \partial x_j'}, \tag{4.7.3}$$

$$\frac{\partial u_i'}{\partial x_i'} = 0, \tag{4.7.4}$$

in which
$$R = \frac{\rho L U}{\mu}.$$

The equations now contain explicitly only the dimensionless parameter R, and the solution for the dependent variables $\mathbf{u}'$ and p' that satisfies the boundary conditions can depend only on

(a) the independent variables $\mathbf{x}'$ and t',

(b) the parameter R, and

(c) the dimensionless ratios needed to specify the boundary and initial conditions (e.g. the ratio of the two axes of an elliptic cylinder bounding the fluid—all these ratios can be described as specifying the 'geometry' of the boundary and initial conditions).[†]

---

† When the velocity distribution is affected by gravity, perhaps through the presence of a free surface to the fluid, and the body force due to gravity must be left explicitly in the equation of motion (see § 4.1), the dimensionless parameter U^2/gL, termed the Froude number, must also be included.

This change to dimensionless variables, a seemingly superficial step, is very revealing. It shows firstly that, once the solution for a particular flow field is known and is expressed in dimensionless form, a triply-infinite family of solutions may be obtained from it by choosing values of ρ, L, U and μ in such a way that the value of R is unchanged. All those flows that satisfy the same boundary and initial conditions when expressed in non-dimensional form, and for which the corresponding values of ρ, L, U and μ differ without the values of the combination $\rho L U/\mu$ being different, are described by one and the same non-dimensional solution; and all such flows are said to be *dynamically similar*, since the magnitudes of the various terms in the equation of motion representing the forces (viscous, pressure, and 'inertia') acting at a given non-dimensional position and instant in the fluid are in the same ratio in all such flows.

This principle of dynamical similarity is used widely as a means of obtaining information about an unknown flow field from 'model tests', that is, from experiments carried out under physical conditions more convenient than those of the unknown flow field. For instance, hydraulic and chemical engineers often wish to predict the velocity with which small solid particles will settle from a suspension in water, and, as a beginning, they need to know the terminal velocity of an isolated small particle of known size and density and of simplified shape—spherical—falling through water. Direct measurement of the fall of one particle is difficult, because the very small size of, say, silt particles makes them awkward to handle and to observe. Dynamical similarity can now be used to argue that the flow about the falling spherical particle is the same, when expressed non-dimensionally in terms of the sphere velocity U and diameter L, as that about a much larger sphere moving at such a speed and through such a fluid that $\rho L U/\mu$ has the same value in the two cases. The value of μ/ρ for lubricating oil is about 400 times that for water, and that for glycerine about 680 times; thus a dynamically similar flow field can be obtained in one of these liquids with a sphere of larger and more convenient size, and the retarding or 'drag' force D on the sphere due to the fluid can be observed for a number of values of L and U. The relation

$$D = -\int m_i\, \sigma_{ij}\, n_j\, dA$$

$$= -\rho U^2 L^2 \int m_i \left\{ -p'\, \delta_{ij} + \frac{1}{R}\left(\frac{\partial u_i'}{\partial x_j'} + \frac{\partial u_j'}{\partial x_i'}\right) \right\} n_j\, dA', \qquad (4.7.5)$$

where the unit vector **m** specifies the direction of the sphere's motion, the integration is over the area A of the sphere, and $\delta A' = \delta A/L^2$, shows that the dimensionless 'drag coefficient' $D/\rho U^2 L^2$ is the same for all the dynamically similar flow fields corresponding to the given value of R, and the model test provides values of this drag coefficient for a range of values of R covering the value appropriate to a falling silt particle. The terminal velocity of the silt particle can then be computed from its known size and density.

The equations (4.7.3) and (4.7.4) in dimensionless variables reveal secondly that, for given geometry of the boundary and initial conditions, there is no more than a singly-infinite family of different solutions in dimensionless form, the different members of the family corresponding to different values of R. In other words, for given geometry of the boundary and initial conditions, the effect on a flow field of changing ρ, L, U or μ, or of changing several of these parameters together, can be described uniquely by the consequent change of R alone. The fact that R is the parameter that determines a flow field for boundaries of given form was first recognized by Stokes (1851), although later work by Reynolds (1883) on the onset of turbulence in flow through tubes led to it being termed the *Reynolds number*.

The relation (4.7.5) for the drag on a moving body may thus be expressed in the general form

$$\frac{D}{\rho U^2 L^2} = \text{function of } R \text{ alone}, \tag{4.7.6}$$

which is valid for the set of flow fields with geometrically similar boundary and initial conditions. All other dimensionless parameters of the flow are likewise functions of R alone. A practical problem in the dynamics of a viscous fluid frequently reduces to the theoretical or experimental determination of the form of the relevant unknown function of R over a certain range of values of R.

The magnitude of the Reynolds number R may be regarded as providing an estimate of the relative importance of the non-viscous and viscous forces acting on unit volume of the fluid. The equation of motion (4.7.1) contains on the right-hand side the pressure force $-\partial p/\partial x_i$ and the viscous force $\mu\, \partial^2 u_i/\partial x_j\, \partial x_j$, and the sum of the two equals minus the so-called inertia force $-\rho \dfrac{Du_i}{Dt}$. These three forces together are in equilibrium, and the balance between them can be indicated by the ratio of any two. Since the pressure force usually plays a passive role, being set up in the fluid as a consequence of motions of a rigid boundary or of the existence of frictional stresses (although this is not so in cases of flow due to an applied pressure gradient, such as Poiseuille flow), it is customary to characterize the flow by the ratio of the magnitudes of the inertia and viscous forces. At any point in the fluid this ratio is

$$\frac{|\rho\, Du_i/Dt|}{|\mu\, \partial^2 u_k/\partial x_j\, \partial x_j|} = R\, \frac{|Du_i'/Dt'|}{|\partial^2 u_k'/\partial x_j'\, \partial x_j'|}.$$

Thus if Du_i'/Dt' and $\partial^2 u_k'/\partial x_j'\, \partial x_j'$ are each of order unity, as is likely if the flow field is a simple one and L and U are truly representative parameters (although there will usually be special places in the fluid where these dimensionless quantities are very small), R measures the relative magnitude of the inertia and viscous forces. For given geometry of the boundary and initial conditions, changes in R correspond to changes in the relative magnitude of inertia and viscous forces—although again the argument is loose, because Du'_i/Dt' and $\partial^2 u'_k/\partial x'_j\, \partial x'_j$ are themselves dependent on R and we must assume that they remain of order unity. In particular, the effect of making $R \ll 1$ is to make the inertia force much smaller than the viscous force, so that pressure and viscous forces are dominant in the flow field, while the effect of making $R \gg 1$ is to make the inertia force much greater than the viscous force, so that inertia and pressure forces are dominant. At values of R of order unity, all three forces presumably play an important part in the equation of motion.

None of the flow fields investigated hitherto in this chapter illustrates these general remarks very well, because they are all specially simple (and indeed were chosen for just this reason) in one way or another. In the case of some of them (e.g. motion generated from rest by a single moving plane boundary), only a velocity, and no length, is needed to specify the boundary conditions, so that no Reynolds number enters the problem; and in the case of the jet examined in §4.6, an effective Reynolds number $(F/\rho\nu^2)^{\frac{1}{2}}$ can be formed from parameters supplied by the boundary conditions, but no length is available with which the position co-ordinate can be made dimensionless. In the case of steady unidirectional flow, the inertia force is everywhere identically zero, so that there is no possibility here of affecting the balance of forces by a change of any of the boundary parameters; nor is there in cases of unsteady unidirectional motion due to moving boundaries, because the pressure force is everywhere zero. In the case of motion generated from rest due to steady motion of a plane boundary with velocity U, with the fluid lying between this plane and a stationary plane distance d away, the velocity distribution (see (4.3.14)) can be written as

$$\frac{u(y,t)}{U} = 1 - \frac{y}{d} - \frac{2}{\pi}\sum_{n=1}^{\infty}\frac{1}{n}\exp\left(-\pi^2 n^2 \frac{tU}{d}\frac{1}{R}\right)\sin\left(\pi n \frac{y}{d}\right),$$

where $R = \rho\, dU/\mu$. This has the general form expected in a case in which only one length and only one velocity are needed to specify the boundary conditions. As already remarked, the pressure forces are everywhere zero, so that the inertia and viscous forces are of equal magnitude everywhere, whatever the value of R. In these circumstances, change of R is entirely equivalent in its effect to change of the time scale, larger values of R corresponding to a slower approach to the ultimate steady state.

Other dimensionless parameters having dynamical significance

In the above discussion the Reynolds number emerged as the sole dimensionless parameter needed for the specification of the dynamical state of flow fields with geometrically similar boundary and initial conditions, because L, U, ρ and μ were the only parameters regarded as varying from one flow field to another. If the effect of other physical factors on the flow system is considered, other dimensionless parameters enter the analysis and provide criteria for dynamical similarity in the same general way as the Reynolds number. An extensive collection of such numbers, many of them named after the person who first made use of the number, may be found in the literature of fluid mechanics. However, nearly all the flow fields considered in this book require only the Reynolds number for their dynamical specification, this being the dimensionless number which when changed leads to the greatest variety of flow forms.

An additional dimensionless parameter arises, even though the only forces acting on the fluid are inertia, pressure and viscous forces, when the specification of the boundary and initial conditions involves three dimensional factors, L, U, and a frequency say, which may vary independently. An example of such a flow field is provided by a flat plate of length L (in two dimensions) which is advancing into still air at speed U and oscillating about some mean attitude with frequency n. In such cases, dynamical similarity of two flow fields requires that both the Reynolds number and the *Strouhal number*

$$S = nL/U$$

have the same values for the two fields. It should be noted that the Strouhal number is an independent quantity, on which the dynamical state of the flow field depends, only when independent variation of all three factors L, U and n is imposed on the flow. There are cases in which flow oscillations arise spontaneously (as in flow past a circular cylinder at Reynolds numbers in the approximate range 40 to 4×10^5), and here the frequency of the oscillation is a property of the flow system and consequently can be written as

$$n = \frac{U}{L} \times \text{function of } R;$$

it is convenient still to speak of nL/U as the Strouhal number of the flow, but it is here only a particular property of the flow and not a defining parameter.

4.8. Flow fields in which inertia forces are negligible

As already remarked, the presence of the non-linear term $\mathbf{u}.\nabla\mathbf{u}$ in the expression for the acceleration makes solution of the equation of motion very difficult for any but the simplest flow fields. It happens that in some circumstances of practical interest the non-linear term, although not identically zero, is small, and may be neglected as an approximation. If in addition the flow is steady, or so nearly so that $|\partial\mathbf{u}/\partial t|$ is not appreciably larger than $|\mathbf{u}.\nabla\mathbf{u}|$, the whole inertia force is everywhere small in magnitude compared with either the pressure or viscous force. This and the next section will be concerned with cases in which inertia forces are negligibly small, under conditions to be specified; and the more difficult cases of flow with appreciable inertia forces will be examined at length in the next chapter.

When the inertia force is negligible everywhere, the governing equations become

$$\nabla p = \mu \nabla^2 \mathbf{u}, \tag{4.8.1}$$

$$\nabla . \mathbf{u} = 0. \tag{4.8.2}$$

If the principal boundary conditions (meaning those that specify the agency generating the motion of the fluid) involve $\mathbf{u}$ alone, the problem is to find the appropriate solution of

$$\nabla^2(\nabla \times \mathbf{u}) = 0, \quad \nabla . \mathbf{u} = 0, \tag{4.8.3}$$

the pressure then being found from (4.8.1). If, on the other hand, the principal boundary conditions are given in terms of p alone, the equation to be solved is

$$\nabla^2 p = 0, \tag{4.8.4}$$

with the velocity then following from (4.8.1) with (4.8.2). In either event, the distributions of p and $\mathbf{u}$ do not depend on μ, and μ determines only the relative magnitudes of $\mathbf{u}$ and p (or of $\mathbf{u}$ and a relative pressure $p - p_0$, to be more precise). The statements made in the preceding section about the general form of the solution evidently reduce here to

$$\left. \begin{array}{l} \dfrac{\mathbf{u}}{U} = \text{function}\left(\dfrac{\mathbf{x}}{L}, \text{geometry of the boundary conditions}\right), \\[2ex] \dfrac{p - p_0}{\rho U^2} = \dfrac{\mu}{\rho L U} \times \text{function}\left(\dfrac{\mathbf{x}}{L}, \text{geometry}\right), \end{array} \right\} \tag{4.8.5}$$

when the principal boundary conditions involve $\mathbf{u}$ alone, and to

$$\left. \begin{array}{l} \dfrac{\mathbf{u}}{U} = \dfrac{\rho L U}{\mu} \times \text{function}\left(\dfrac{\mathbf{x}}{L}, \text{geometry}\right), \\[2ex] \dfrac{p - p_0}{\rho U^2} = \text{function}\left(\dfrac{\mathbf{x}}{L}, \text{geometry}\right), \end{array} \right\} \tag{4.8.6}$$

when the principal boundary conditions involve p alone.

Flow in slowly-varying channels

In this first example the non-linear term in the equation of motion is small everywhere because, for essentially geometrical reasons, the velocity $\mathbf{u}$ varies only slowly along streamlines. In the case of flow in a cylindrical tube due to an applied pressure difference at two distant ends, $\mathbf{u} . \nabla \mathbf{u}$ is identically zero; if the tube cross-section now varies along its length, $\mathbf{u} . \nabla \mathbf{u} \neq 0$, but since the viscous force is non-zero for a cylindrical tube we can always make the ratio of $|\mathbf{u} . \nabla \mathbf{u}|$ to the viscous force negligibly small by choosing a sufficiently slow rate of variation of the cross-section.

Consider first the simple case of steady flow along a circular tube whose radius a varies slowly with distance x along the centre-line. A constant difference is maintained between the pressures at two distant ends, and the resulting axial pressure gradient $-G$ will also vary slowly with x. In the neighbourhood of any station x, say within several tube radii upstream and downstream, the tube radius and the axial pressure gradient are approximately uniform with values $a(x)$ and $-G(x)$, and the approximate expression for the axial velocity obtained by neglecting the inertia force (see (4.2.5)) is

$$u(x,r) = \frac{G}{4\mu}(a^2 - r^2),\qquad (4.8.7)$$

where r denotes radial distance from the centre-line. Then if Q is the (constant) volume flux along the tube,

$$Q = \frac{\pi a^4 G}{8\mu},\qquad (4.8.8)$$

and (4.8.7) can be written as

$$u(x,r) = U\left(1 - \frac{r^2}{a^2}\right),\qquad U = \frac{2Q}{\pi a^2}.\qquad (4.8.9)$$

The streamlines are not exactly unidirectional but are inclined to the axis at a small angle whose magnitude is of order da/dx, $= \alpha(x)$ say, so that in addition to the axial component of velocity u there is a radial component v of order αu. It follows from (4.8.8) that

$$p_1 - p_2 = \frac{8\mu Q}{\pi}\int_{x_1}^{x_2} a^{-4}\,dx,$$

so that if the pressures p_1, p_2 at the two end stations $x = x_1$, $x = x_2$ are given, and the tube geometry is known, Q can be calculated and thence $G(x)$.

The expression (4.8.9) is evidently a valid approximation for sufficiently small values of α, and we can find a specific condition for its validity by using (4.8.9) itself to estimate the magnitude of the neglected inertia force. It follows from (4.8.9) that a representative value of the magnitude of each of the terms $\rho u\,\partial u/\partial x$ and $\rho v\,\partial u/\partial r$ is $\alpha\rho U^2/a$. On the other hand, for the viscous force $\mu\nabla^2 u$ retained in the equation of motion we have the representative magnitude $\mu U/a^2$, showing that the solution (4.8.9) is consistent with neglect of the inertia force provided

$$\alpha\frac{\rho a U}{\mu} \ll 1.\qquad (4.8.10)$$

Similar relations apply to steady flow in the region between two inclined planes, and an exact solution of the governing equations for this case, to be described in §5.6, shows directly how the approximate solution analogous to (4.8.7) or (4.8.9) is recovered when the condition (4.8.10) is satisfied.

The kind of approximation used in obtaining (4.8.7), viz. that the channel width and the pressure gradient are uniform in a local sense, is useful in many different circumstances. When the fluid is confined between two separate boundaries, the relative velocity of adjacent points on the two boundaries is a third parameter which may vary slowly with position, as for instance in the case of flow between two disks close together, one of which rotates in its own plane (and also as in the lubrication layer described below). Some cases in which the channel conditions are slowly-varying with respect to time, as when two plane disks are pressed close together and the fluid between them is squeezed out radially, can also be treated with the same kind of approximation. In all cases, the viscous force is dominated by velocity gradients determined by the lateral dimensions of the flow region, while the inertia force involves rates of change of the velocity determined by a relatively large length (the distance along a streamline over which the channel parameters change appreciably) or by a large time.

Lubrication theory

It is a matter of common experience that two solid bodies can slide over one another very easily when there is a thin layer of fluid between them and that under certain conditions a high positive pressure is set up in the fluid layer. For instance, a sheet of paper dropped on to a smooth floor will often 'float' on a film of air between it and the floor and thereby will be able to glide horizontally for some distance before coming to rest. The existence of this high pressure in the fluid layer between the surfaces is used widely in engineering practice as a means of substituting fluid–solid friction for the much larger friction between two solid bodies in contact; once set up, the fluid layer offers great resistance to being squeezed out and remains as a 'lubricating' film between the two surfaces. In some cases the fluid layer may be used to support a useful load, and is then called a lubrication bearing.

The essence of the phenomenon is that, as a result of the thickness of the fluid layer between the two solid boundaries being so small, the rate of strain and the stress due to viscosity in the fluid layer are very large, this large stress then being used, by suitable choice of the configuration of the fluid layer, to develop a large pressure. To see how this may be done, we shall consider the simple case of one solid body with a plane surface gliding steadily over another, the surface of the gliding block being of finite length l in the direction of motion and of great width so that the motion may be regarded as two-dimensional; this case was analysed first by Reynolds in 1886. Experience shows that it is necessary for the plane surfaces to be slightly inclined to one another, as indeed we shall see, so that the fluid boundaries will be assumed to be disposed as in figure 4.8.1. The axes will be chosen as fixed relative to the upper solid body; the lower solid surface then moves in its own plane, at speed U, and the whole flow field is steady relative to these axes.

The thickness d of the fluid layer is everywhere small compared with l, and we therefore examine the possibility that the velocity distribution near any section of the layer where the thickness and pressure gradient have certain values is approximately the same as in a uniform layer with thickness and pressure gradient having those same values everywhere. Provided the fluid velocity is everywhere of the same order of magnitude as U, as will later be seen to be so, the argument leading to (4.8.10) is applicable here. Thus the condition for the suggested approximation to be valid is

$$\alpha \frac{\rho d U}{\mu} \ll 1.$$

This requirement is usually satisfied under practical conditions of lubrication. We can therefore proceed to make use of the solutions found for uniform channels in the manner described above.

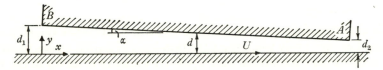

Figure 4.8.1. The lubrication layer between two plane surfaces in relative motion.

In the neighbourhood of any station where the thickness of the fluid layer is d and the pressure gradient is $-G$, we have, in accordance with (4.2.10),

$$u = \frac{G}{2\mu} y(d-y) + U\left(\frac{d-y}{d}\right). \tag{4.8.11}$$

The volume flux, per unit width of the fluid layer, is

$$Q = \int_0^d u \, dy = \frac{G d^3}{12\mu} + \tfrac{1}{2} U d, \tag{4.8.12}$$

and Q must be independent of x. This requires the pressure gradient to vary with d according to the relation

$$\frac{dp}{dx} = -G = 6\mu \left(\frac{U}{d^2} - \frac{2Q}{d^3}\right) \tag{4.8.13}$$

in which $d = d_1 - \alpha x$. Integration of (4.8.13) gives

$$p - p_0 = \frac{6\mu}{\alpha}\left\{ U\left(\frac{1}{d} - \frac{1}{d_1}\right) - Q\left(\frac{1}{d^2} - \frac{1}{d_1^2}\right)\right\}, \tag{4.8.14}$$

where p_0 is the pressure at $x = 0$ and $d = d_1$. Now the sliding block may be supposed to be completely immersed in the fluid, with narrow passages for the fluid on one side of the block only, so that the pressures at the two end-points A and B are approximately the same. This condition, that $p = p_0$ when $d = d_2$, enables us to determine Q from (4.8.14):

$$Q = U \frac{d_1 d_2}{d_1 + d_2}, \tag{4.8.15}$$

and then the expression for the pressure becomes

$$p - p_0 = \frac{6\mu U}{\alpha} \frac{(d_1 - d)(d - d_2)}{d^2(d_1 + d_2)}. \tag{4.8.16}$$

The solution represented by (4.8.11) and (4.8.16) is of the general form (4.8.5), corresponding to the fact that the principal boundary conditions are given here in terms of **u** alone.

The volume flux and the pressure distribution in the lubrication layer can now be calculated when the sliding velocity U and the inclination of the sliding block are known. The pressure increment $p - p_0$ is one-signed throughout the layer, and is positive only when $d_2 < d_1$, as anticipated in figure 4.8.1. Thus a lubrication layer will generate a positive pressure, and will be able to support a load normal to the layer, only when the layer is so arranged that the relative motion of the two surfaces tends to drag fluid (by viscous stresses) from the wider to the narrower end of the layer. The pressure increment has a single maximum in the layer, and its value there is of order $\mu l U / d_1^2$ (assuming $(d_1 - d_2)/d_1$ to be of order unity), showing that very high pressures can be set up in very thin films.

The total normal force exerted on either of the two boundaries by the fluid layer is

$$\int_0^l (p - p_0) \, dx = \frac{6\mu U}{\alpha^2} \left\{ \log \frac{d_1}{d_2} - 2 \left(\frac{d_1 - d_2}{d_1 + d_2} \right) \right\}. \tag{4.8.17}$$

The total tangential force exerted by the fluid on the lower plane is

$$\int_0^l \mu \left(\frac{\partial u}{\partial y} \right)_{y=0} dx = \frac{2\mu U}{\alpha} \left\{ 3 \left(\frac{d_1 - d_2}{d_1 + d_2} \right) - 2 \log \frac{d_1}{d_2} \right\},$$

and the tangential force on the upper boundary is

$$-\int_0^l \mu \left(\frac{\partial u}{\partial y} \right)_{y=d} dx = \frac{2\mu U}{\alpha} \left\{ 3 \left(\frac{d_1 - d_2}{d_1 + d_2} \right) - \log \frac{d_1}{d_2} \right\};$$

the two tangential forces are not equal and opposite because the normal force on one plane has a (small) component parallel to the other plane. Thus

$$\frac{\text{tangential force on block}}{\text{normal force on block}} = \alpha \times \text{function} \left(\frac{d_1}{d_2} \right), \tag{4.8.18}$$

and, if $(d_1 - d_2)/d_1$ is of order unity, the order of magnitude of this 'coefficient of friction' is d_1/l. Somewhat surprisingly, the ratio of the two components of force is independent of the viscosity of the fluid, and can be made indefinitely small by reduction of d_1 with d_2/d_1 held constant.

We have regarded α as a given quantity in the above analysis, although in any case in which the sliding block is free to move under a given load α may be a variable; the attitude and position of the block at which the given load is supported by the pressure in the lubrication layer must then be a state of stable equilibrium. Consideration of these practical matters, and of the important case of a lubrication layer between a rotating circular shaft (or 'journal') in a circular bearing of slightly larger radius, is beyond our scope here.†

The Hele Shaw cell

A related example of a flow field in which inertia forces are negligible is provided by the Hele Shaw cell. This is an arrangement of two parallel plates very close together, the space between the plates being occupied partly by fluid and partly by 'obstacles' in the form of cylinders with generators normal to the plates. The fluid is made to flow from one end of the layer to the other by a steady pressure difference applied between the ends. The layer thickness d is small compared with the linear dimension L of an obstacle (measured in the plane of the plates), so that we again examine the possibility that the flow is everywhere approximately the same as if the local pressure gradient extended to infinity. When this approximation is valid, the relation between the local velocity and the local pressure gradient is

$$u \approx -\frac{1}{2\mu}\frac{\partial p}{\partial x}z(d-z), \quad v \approx -\frac{1}{2\mu}\frac{\partial p}{\partial y}z(d-z), \tag{4.8.19}$$

where the co-ordinate z is normal to the plates.

The velocity of the fluid changes appreciably, with respect to distance along a streamline, over a distance of order L, and the inertia force is therefore of order $\rho(u^2+v^2)_{\text{max.}}/L$. Hence the condition that inertia forces should be small compared with viscous forces (at any representative value of z) is

$$\frac{d}{L}\left(\frac{\rho\,d^3\,|\nabla p|}{\mu^2}\right) \ll 1. \tag{4.8.20}$$

This condition, which is of effectively the same form as (4.8.10), the ratio of the two lengths determining the viscous and inertia forces being α in one case and d/L in the other, is readily satisfied with laboratory equipment.

Hele Shaw (1898) pointed out that the values of u and v at some constant value of z (or averaged with respect to z) given by (4.8.19) define a two-dimensional velocity field which is irrotational and which satisfies the condition of zero normal component at a rigid boundary in the (x, y)-plane, although not the condition of no slip at such a boundary. Thus the stream-lines of the steady flow past obstacles in the Hele Shaw cell are identical in shape with those in a hypothetical two-dimensional flow of inviscid fluid with zero vorticity past obstacles of the same form. By introducing colouring matter at a few points at the entry end (the choice of the location of the source normal to the plates being immaterial), the Hele Shaw cell can be

---

† Reference may be made to *Lubrication: Its Principles and Practice*, by A. G. M. Michell (Blackie, 1950).

used as a visual demonstration of the shape of the streamlines in this hypothetical flow.†

Relations of a similar kind apply to the flow in a very shallow layer of liquid (perhaps a few millimetres in thickness) in an open horizontal dish, the level of the surface at one end being kept higher at one end by the introduction of liquid there; this arrangement may be more convenient for demonstration purposes.

Percolation through porous media

When subterranean water is forced by a pressure gradient through soil, each material element of the water traces out a devious path as it passes through the irregularly arranged interstices between the soil particles. If d is a length representative of the dimensions of the interstices, and U is representative of the velocity of the fluid in the interstices, it is to be expected that in a steady motion the inertia forces will be of order $\rho U^2/d$ while the viscous forces will be order $\mu U/d^2$. The former will be small compared with the latter when

$$\rho d U/\mu \ll 1,$$

and the governing equations then reduce to (4.8.1) and (4.8.2). This condition is satisfied in most cases of soil water movement and in many other cases of percolation of fluid through a porous solid, and the corresponding laws of flow with negligible inertia forces have considerable practical importance.

Since a detailed knowledge of the shape of the interstices occupied by the fluid is not available, and in any case would be unusable owing to its complexity, it is customary to introduce dependent variables which are effectively averages over a large number of interstices. We may define the 'local' velocity $\bar{\mathbf{u}}$, in a steady flow through a porous medium, as having components equal to the flux of fluid per unit area across three plane surface elements, with linear dimensions large compared with d, normal to the co-ordinate axes. Such a definition will have meaning only when it is possible to find a length which is both large compared with d and small compared with the linear dimensions of the exterior boundaries. In a similar way we may define a pressure $\bar{p}$ which is the average value of p over a volume of fluid large enough to fill many interstices but small in linear dimensions compared with the field as a whole.

The equations governing the actual flow in and out the interstices are linear, and the flux of fluid through a given piece of porous medium large enough to contain many interstices may be expected to be proportional to the pressure gradient applied across it, and inversely proportional to μ, just as if the medium consisted of a number of tubes of small diameter in each of which the flow is of Poiseuille type. If the porous medium has a

† Whether fluids of small viscosity really flow even approximately in an irrotational way, and with a single-valued and continuous velocity potential as demanded by (4.8.19), depends critically on the shape of the obstacles, as we shall see in the next chapter.

structure which is statistically isotropic, so that pressure gradients applied in different directions produce the same flux, we may write

$$\nabla \bar{p} = \mu \bar{\mathbf{u}}/k, \qquad (4.8.21)$$

where k is a constant, called the *permeability*, which depends on the size and shape of the interstices (being proportional to the square of their linear dimensions, for a given shape). The relation (4.8.21) is known as Darcy's law (Darcy 1856), and has a long history of use in soil mechanics for a wide variety of porous media. Its justification rests partly on the above theoretical argument, and partly on its agreement with measurements of the flow produced by an applied pressure gradient in homogeneous media like sand.

The relation (4.8.21) implies that when the porous medium is statistically uniform and k is independent of position, the velocity $\bar{\mathbf{u}}$ is irrotational, with the velocity potential ϕ proportional, as in the case of the Hele Shaw cell, to the pressure. The equation of mass conservation, when averaged in the above way, then leads to

$$\nabla^2 \bar{p} = (\mu/k)\,\nabla^2 \phi = 0. \qquad (4.8.22)$$

This equation is to be solved subject to the conditions of zero normal derivative of ϕ at an impermeable surface and a given constant value of ϕ (or $\bar{p}$) at a free surface of the fluid. (In the case of water percolating through soil, a free surface for the water in the form of an air–water boundary, or 'water table', may occur in the interior of the soil). In this way many practical problems concerned with seepage from dams, variations in the level of the water table near a well, movement of ground water near a coast due to tidal variations of pressure, etc., have been solved.

Two-dimensional flow in a corner

Suppose that one rigid plane is sliding steadily over another, with constant inclination θ_0, as sketched in figure 4.8.2 for the case $\theta_0 = \frac{1}{2}\pi$. Fluid in the region between the planes is set in motion, as might happen in a cylinder with a moving piston, or at the edge of a blade used to scrape up liquid on a table. Near the intersection O, the gradients of velocity become very large, because the velocity has different values at the two rigid boundaries, and it is a reasonable presumption that viscous forces are dominant. The velocity distribution in the neighbourhood of O will be determined on this assumption, which will then be checked *a posteriori*.

The problem can be made one of steady motion by choosing an origin of co-ordinates at (and moving with) O. In cases of two-dimensional motion with negligible inertia forces, it is convenient to introduce the stream function ψ (§2.2) so that the equation of mass conservation is satisfied identically and the one non-zero component of vorticity becomes $-\nabla^2 \psi$. The first of equations (4.8.3) is then

$$\nabla^2(\nabla^2 \psi) = 0, \qquad (4.8.23)$$

and the boundary conditions, in terms of polar co-ordinates (r, θ), are

$$\frac{\partial \psi}{\partial r} = 0, \quad \frac{1}{r}\frac{\partial \psi}{\partial \theta} = -U, \quad \text{at } \theta = 0,$$

$$\frac{\partial \psi}{\partial r} = 0, \quad \frac{1}{r}\frac{\partial \psi}{\partial \theta} = 0, \quad \text{at } \theta = \theta_0.$$

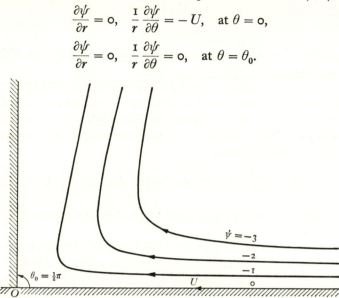

Figure 4.8.2. Two-dimensional flow in a corner due to one rigid plane sliding on another (arbitrary units for ψ).

The form of these boundary conditions is such that ψ could be proportional to r everywhere, and it is worthwhile to enquire if the differential equation allows such a possibility. We therefore write

$$\psi(r, \theta) = rf(\theta), \tag{4.8.24}$$

and substitute in (4.8.23) to find

$$\nabla^2 \left\{ \frac{1}{r}(f+f'') \right\} = \frac{1}{r^3}(f+2f''+f^{\mathrm{iv}}) = 0.$$

The solution of this equation for f is

$$f(\theta) = A\sin\theta + B\cos\theta + C\theta\sin\theta + D\theta\cos\theta, \tag{4.8.25}$$

and we need to choose values of A, B, C, D such that

$$f(0) = 0, \quad f'(0) = -U, \quad f(\theta_0) = 0, \quad f'(\theta_0) = 0.$$

The required values are

$$A, B, C, D = (-\theta_0^2, 0, \theta_0 - \sin\theta_0 \cos\theta_0, \sin^2\theta_0) \times \frac{U}{\theta_0^2 - \sin^2\theta_0}. \tag{4.8.26}$$

Thus we have a solution which satisfies the boundary conditions and the equations of motion with inertia forces neglected. The components of acceleration of the fluid at any point, evaluated according to this solution, are proportional to U^2/r, with a constant of proportionality which varies with θ and is of order unity. The viscous forces, also evaluated according to this solution, are of order $\mu U/r^2$, so that the assumption of negligible inertia forces is self-consistent if $\rho r U/\mu \ll 1$; that is, the solution obtained is valid within a neighbourhood of the intersection O defined by

$$r \ll \mu/\rho U.$$

For lubricating oil at normal temperatures, and $U = 10 \text{ cm/sec}$, this condition is $r \ll 0.4 \text{ cm}$.

For any flow of the form (4.8.24), the velocity components are independent of r and streamlines with equal increments in ψ cut any radial line at equidistant points. The motion is reversible, since the governing equations and boundary conditions are linear and homogeneous. In the particular case $\theta_0 = \frac{1}{2}\pi$, the solution becomes

$$\psi = \frac{rU}{\frac{1}{4}\pi^2 - 1}(-\tfrac{1}{4}\pi^2 \sin\theta + \tfrac{1}{2}\pi\theta \sin\theta + \theta\cos\theta), \qquad (4.8.27)$$

and the streamlines (of the motion relative to O) for this case are sketched in figure 4.8.2.

It will be noticed that both the normal and tangential components of stress in the fluid vary as r^{-1}, so that the total force exerted on the planes $\theta = 0$ and $\theta = \theta_0$ by the fluid is logarithmically infinite. In practice two plane rigid boundaries do not make perfect geometrical contact, and the maximum stress is finite; what we learn from the above solution is that the total force on the plane boundaries depends on the precise shape of the two boundaries very close to their intersection and that the force increases as the clearance is diminished.

An analogous solution of the equation (4.8.23) of the form $\psi = r^2 f(\theta)$ can be used to describe the two-dimensional flow in the neighbourhood of the point of intersection of two straight rigid boundaries which are in relative rotation about that point. Yet another, of the form $\psi = r^3 f(\theta)$, can be used to describe the flow in the neighbourhood of a point of zero tangential stress at a straight rigid boundary; and since this latter flow has some relevance to the considerations of § 5.10 the solution will be described briefly.

The velocity gradient $\partial u/\partial y$ (the y-axis being normal to the boundary) at the wall changes sign at a point of zero friction O, and there must exist a streamline OP which divides flow coming towards O from the right and from the left, as sketched in figure 4.8.3. If the inclination of this dividing streamline to the boundary is taken as θ_0, it is readily found that a solution of (4.8.23) which satisfies the no-slip condition at the wall and satisfies $\psi = 0$ at $\theta = \theta_0$ is

$$\psi(r, \theta) = Ar^3 \sin^2\theta \sin(\theta_0 - \theta), \qquad (4.8.28)$$

where A is an arbitrary constant. The streamlines corresponding to this solution are shown in the figure for $\theta_0 = \frac{1}{3}\pi$ and $A < 0$, and again the velocity may equally be taken to be everywhere in the opposite direction. The region near O in which the solution (4.8.28) is self-consistent is given by $r^3 \ll \mu/\rho |A|$. The values of A and θ_0 in this solution are evidently determined by circumstances outside the region in which the solution is valid. However, it can be deduced readily that, whatever the signs of A and $\cos\theta_0$, the directions of the pressure gradient (which, according to the solution (4.8.28), is uniform) and the velocity at $\theta = \theta_0$ lie in the same quadrant.

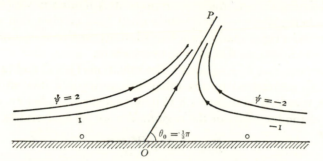

Figure 4.8.3. Two-dimensional flow near a point of zero friction at a plane rigid boundary, according to (4.8.28), with $\theta_0 = \frac{1}{3}\pi$ and $A < 0$ (arbitrary units for ψ).

Uniqueness and minimum dissipation theorems

We shall show that there cannot be more than one solution for the velocity distribution for flow in a given region with negligible inertia forces and consistent with prescribed values of the velocity vector at the boundary of the region (including a hypothetical boundary at infinity when the fluid is of infinite extent). The argument is very similar in form to that used in §2.7 to establish uniqueness of the solution for the potential ϕ of a solenoidal irrotational velocity field with certain prescribed boundary conditions. An interesting related result is that flow with negligible inertia forces has a smaller total rate of dissipation than any other incompressible flow in the same region with the same values of the velocity vector everywhere on the boundary of the region. (Both results were found first by Helmholtz (1868a).)

First let us suppose that u_i, p, e_{ij} and u_i^*, p^*, e_{ij}^* are two sets of distributions of the velocity, pressure and rate-of-strain tensor in a certain region, both of which satisfy the equations (4.8.1) and (4.8.2); and suppose further that $u_i = u_i^*$ at all points on the boundary of the region. Then,

$$\int (e_{ij}^* - e_{ij})(e_{ij}^* - e_{ij})\,d\tau = \int \frac{\partial(u_i^* - u_i)}{\partial x_j}(e_{ij}^* - e_{ij})\,dV$$

$$= \int (u_i^* - u_i)(e_{ij}^* - e_{ij})n_j\,dA$$
$$\qquad - \tfrac{1}{2}\int (u_i^* - u_i)(\nabla^2 u_i^* - \nabla^2 u_i)\,dV$$

$$= -\frac{1}{2\mu}\int (u_i^* - u_i)\frac{\partial(p^* - p)}{\partial x_i}\,dV$$

$$= -\frac{1}{2\mu}\int (p^* - p)(u_i^* - u_i)n_i\,dA$$

$$= 0,$$

where the volume integrals have been taken over the region of fluid concerned and the surface integrals over the boundary of the region. The integrand of the integral on the left-hand side is non-negative everywhere, so that evidently the rates of strain e_{ij}^* and e_{ij} are identical everywhere. The velocity difference $u_i^* - u_i$ therefore represents a motion in which no fluid element is being deformed and which must be a combination of a rigid translation and a rigid rotation; such a difference motion is excluded by the boundary conditions, so that $u_i^* = u_i$ everywhere.

Now suppose that u_i, p, e_{ij} satisfy the equations (4.8.1) and (4.8.2), but that u_i', p', e_{ij}' correspond to any other incompressible flow in the same region (i.e. that $\nabla . \mathbf{u}' = 0$, but (4.8.1) is not satisfied); as before, $u_i = u_i'$ at all points of the boundary of the region. We see by analysis like that used above that

$$\int (e_{ij}' - e_{ij}) e_{ij} \, dV = 0.$$

The total rate of dissipation of mechanical energy by the action of viscosity over the whole region in the flow represented by u_i' is then

$$2\mu \int e_{ij}' e_{ij}' \, dV = 2\mu \int \{ e_{ij} e_{ij} + (e_{ij}' - e_{ij})(e_{ij}' - e_{ij}) \} \, dV,$$

which is the sum of the total rate of dissipation corresponding to the flow represented by u_i and a non-negative term which is zero only if $e_{ij}' = e_{ij}$. Thus the rate of dissipation corresponding to the flow in a given region with negligible inertia forces is less than that in any of those solenoidal velocity distributions in the same region (including ones which satisfy the complete equation of motion) with the same values of the velocity at all points of the boundary of the region.

Exercises

1. A circular disk of radius a is parallel to and at distance h from a rigid plane, and the space between them is occupied by fluid. The pressure at the edge of the disk is atmospheric. Show that motion of the disk in the direction normal to the plane gives rise to a force on the disk in that direction equal to

$$-\frac{3\pi}{2} \frac{\mu a^4}{h^3} \frac{dh}{dt}, \quad \text{provided} \quad h \ll a, \quad \frac{\rho h}{\mu} \left| \frac{dh}{dt} \right| \ll 1.$$

Hence show that a constant force F applied to the disk will pull it well away from the plane in a time $\frac{3}{4}\pi\mu a^4/h^2 F$. (The fact that this time is large when h is small is the basis of the phenomenon of viscous adhesion, used in adhesives such as 'Scotch tape' and in the 'wringing' together of accurately ground metal surfaces.)

2. Fluid in the region between two rigid planes at $\theta = \mp \alpha$ is in steady two-dimensional motion due to some agency far from the corner. Examine the motion near the corner for which the stream function is of the form $r^\lambda f(\theta)$, where $f(\theta)$ is an even function of θ and λ is a number to be determined. Show that real solutions for λ exist only when $\alpha < 73°$ (approximately). The flow corresponding to the complex solutions obtained when $\alpha > 73°$ consists of a sequence of eddies of decreasing size and intensity as the corner is approached (Moffatt 1964).

4.9. Flow due to a moving body at small Reynolds number

When a body with representative linear dimension d is in steady translational motion, with speed U, through fluid which is otherwise undisturbed, d and U are a representative length and velocity for the flow field as a whole. The inertia forces on the fluid are therefore likely to be of order $\rho U^2/d$ and the viscous forces of order $\mu U/d^2$. The ratio of these two estimates is $\rho \, dU/\mu, = R$, so that when $R \ll 1$ the inertia forces may be negligible. We propose to examine the flow field with this assumption, on the understanding that the solution so obtained must be tested for consistency with the initial assumption. Motion of a body through fluid with a value of R which is small, usually because of the very small size of the body, is a flow problem which is important in a variety of physical contexts, such as the settling of sediment in liquid, and the fall of mist droplets in air. The quantity of greatest practical interest is the drag force exerted by the fluid on the body, since from this the terminal velocity for free fall under the action of gravity can be calculated. The velocity of the body is not always steady in these practical problems, but unless either the body or the ambient fluid is caused to move with an acceleration much greater than U^2/d (as might happen if a sound wave of high frequency passes through the fluid) the above estimate of the relative magnitude of inertia and viscous forces will stand.

The equations to be solved are (4.8.1) and (4.8.2), which we rewrite as

$$\nabla \left(\frac{p - p_0}{\mu} \right) = \nabla^2 \mathbf{u} = -\nabla \times \boldsymbol{\omega}, \tag{4.9.1}$$

$$\nabla \cdot \mathbf{u} = 0, \tag{4.9.2}$$

where p_0 is the uniform pressure far from the sphere. It is a consequence of these equations that
$$\nabla^2 p = 0 \quad \text{and} \quad \nabla^2 \boldsymbol{\omega} = 0.$$

We choose a co-ordinate system relative to which the fluid at infinity is stationary. The boundary conditions for a *rigid* body moving with velocity $\mathbf{U}$ are then

$$\left. \begin{array}{l} \mathbf{u} = \mathbf{U} \quad \text{at the body surface,} \\ \mathbf{u} \to 0 \quad \text{and} \quad p - p_0 \to 0 \quad \text{as} \quad |\mathbf{x}| \to \infty. \end{array} \right\} \tag{4.9.3}$$

We recognize, from the general result obtained at the end of the preceding section, that not more than one solution of (4.9.1) and (4.9.2) can satisfy the boundary conditions (4.9.3).

We shall make explicit use here of the fact that the equations (4.9.1) and (4.9.2), and the boundary conditions (4.9.3), are linear and homogeneous in $\mathbf{u}$, $(p - p_0)/\mu$ and $\mathbf{U}$. The expressions for $\mathbf{u}$ and $(p - p_0)/\mu$ must therefore be linear and homogeneous in $\mathbf{U}$. (A similar argument was used for irrotational flow in §2.9—see (2.9.23).)

A rigid sphere

The case of a spherical body is important, and is one of the few that are tractable. The flow field due to a rigid sphere in translational motion was first determined by Stokes (1851).

We choose the origin of the co-ordinate system to be at the instantaneous position of the centre of the sphere, which has radius a. The distributions of $\mathbf{u}$ and $(p-p_0)/\mu$ must be symmetrical about the axis passing through the centre of the sphere and parallel to $\mathbf{U}$, and the vector $\mathbf{u}$ lies in a plane through that axis. The differential operators in (4.9.1) and (4.9.2) are independent of the choice of co-ordinate system, so that $(p-p_0)/\mu$ and $\mathbf{u}$ depend on the vector $\mathbf{x}$ and not on any other combination of the components of $\mathbf{x}$. The parameters $\mathbf{U}$ and a complete the list of quantities on which $(p-p_0)/\mu$ and $\mathbf{u}$ can depend (although if the body had been of any shape other than spherical, vectors specifying orientation of the body and scalar shape parameters would have had to be included).

It follows that $(p-p_0)/\mu$ must be of the form $\mathbf{U}.\mathbf{x}F$, where $a^2 F$ is a dimensionless function of $\mathbf{x}.\mathbf{x}/a^2 (= r^2/a^2)$ alone. Since $p-p_0$ satisfies Laplace's equation, and vanishes at infinity, it can be represented as a series of spherical solid harmonics of negative degree in r (see (2.9.19)); and the only term of the series which is compatible with this form is the one of degree -2 (the 'dipole' term). Thus

$$\frac{p-p_0}{\mu} = \frac{C\mathbf{U}.\mathbf{x}}{r^3}, \tag{4.9.4}$$

where C is a constant.

Exactly the same kind of argument applies to the harmonic function $\boldsymbol{\omega}$, which is a vector in the azimuthal direction and must be proportional to $\mathbf{U} \times \mathbf{x}/r^3$. The constant of proportionality is found from (4.9.1) to be C, so that

$$\boldsymbol{\omega} = \frac{C\mathbf{U} \times \mathbf{x}}{r^3}. \tag{4.9.5}$$

The velocity corresponding to this vorticity distribution is most conveniently found in terms of the stream function ψ. With a spherical polar co-ordinate system (and $\theta = 0$ in the direction of $\mathbf{U}$), the azimuthal (or ϕ-) component of $\boldsymbol{\omega}$ is defined as

$$\frac{1}{r}\frac{\partial(ru_\theta)}{\partial r} - \frac{1}{r}\frac{\partial u_r}{\partial \theta},$$

and on replacing u_r, u_θ by the expressions (2.2.14) we find from (4.9.5) that

$$\frac{\partial^2\psi}{\partial r^2} + \frac{\sin\theta}{r^2}\frac{\partial}{\partial\theta}\left(\frac{1}{\sin\theta}\frac{\partial\psi}{\partial\theta}\right) = -\frac{CU\sin^2\theta}{r}.$$

The particular integral for ψ is clearly proportional to $\sin^2\theta$; and the inner boundary condition also requires ψ to depend on θ in this way at $r = a$. We therefore put

$$\psi = U \sin^2 \theta f(r), \qquad (4.9.6)$$

which may be seen to be equivalent to a velocity vector of the form

$$\mathbf{u} = \mathbf{U} \left(\frac{1}{r} \frac{df}{dr} \right) + \mathbf{x} \frac{\mathbf{x}.\mathbf{U}}{r^2} \left(\frac{2f}{r^2} - \frac{1}{r} \frac{df}{dr} \right). \qquad (4.9.7)$$

The equation for the unknown function f is

$$\frac{d^2 f}{dr^2} - \frac{2f}{r^2} = -\frac{C}{r}, \qquad (4.9.8)$$

of which the general solution is

$$f(r) = \tfrac{1}{2} Cr + Lr^{-1} + Mr^2. \qquad (4.9.9)$$

The terms containing the new constants L and M represent an irrotational motion.

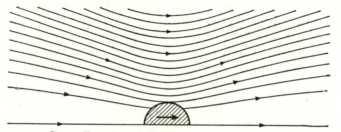

Figure 4.9.1. Streamlines, in an axial plane, for flow due to a moving sphere at $R \ll 1$ (with complete neglect of inertia forces).

Now the outer boundary condition demands that $f/r^2 \to 0$ as $r \to \infty$; and the kinematical condition $u_r = U \cos \theta$ at the surface of the sphere requires $f(a) = \tfrac{1}{2} a^2$. Hence

$$M = 0, \quad L = \tfrac{1}{4} a^3 - \tfrac{1}{4} Ca^2. \qquad (4.9.10)$$

There remains the no-slip condition at the surface of the sphere, viz.

$$u_\theta = -\frac{1}{r \sin \theta} \frac{\partial \psi}{\partial r} = -U \sin \theta \quad \text{at } r = a,$$

which is satisfied if

$$C = \tfrac{3}{2} a. \qquad (4.9.11)$$

The stream function representing the motion is thus

$$\psi = Ur^2 \sin^2 \theta \left(\frac{3}{4} \frac{a}{r} - \frac{1}{4} \frac{a^3}{r^3} \right). \qquad (4.9.12)$$

A sketch of the streamlines is shown in figure 4.9.1. The streamlines are symmetrical about a plane normal to $\mathbf{U}$, as is of course implied by the linearity of $\mathbf{u}$ in $\mathbf{U}$; reversing the direction of $\mathbf{U}$ merely leads to a change of the sign of **u** everywhere. It will also be noticed that the disturbance due to the sphere extends to a considerable distance from the sphere, the velocity approaching zero as r^{-1} at large values of r. As a consequence, the presence of an outer rigid boundary, for example in the form of a cylinder with generators parallel to **U**, can modify the fluid motion appreciably, even when it is at a distance of many diameters from the sphere; likewise the interaction between two moving spheres many diameters apart can be appreciable.

These features of the solution are consequences of the neglect of the inertia term in the equation of motion. The equation for the vorticity, viz. $\nabla^2 \boldsymbol{\omega} = 0$, shows that the flow represented by (4.9.12) is effectively due solely to steady molecular diffusion of vorticity to infinity in all directions, the sphere being a source of vorticity as a consequence of the no-slip condition. The term $\partial \boldsymbol{\omega}/\partial t$ which is present in the full equation for $\boldsymbol{\omega}$, and which represents the effect of the continual change in the position of the sphere relative to the axes, has been neglected here, and molecular diffusion spreads the vorticity as far ahead of the sphere as behind it; it is as if the sphere were stationary and acted purely as a source of vorticity. The vorticity distribution shows the decrease as r^{-2} to be expected for the diffusion of each component of $\boldsymbol{\omega}$ from a stationary steady source of dipole character (equal positive and negative quantities of each component of $\boldsymbol{\omega}$ being generated at the surface of the sphere).

It remains for us to verify that the solution found on the assumption that inertia forces can be neglected is actually consistent with that assumption. According to the solution (4.9.12), an estimate of the magnitude of the viscous force $\mu \nabla^2 \mathbf{u}$ is $\mu U a/r^3$. If the sphere velocity is exactly steady, and the rate of change of **u** at a fixed point is due simply to the sphere changing its position relative to the point concerned, the operator $\partial/\partial t$ is equivalent to $-\mathbf{U} . \nabla$ and the inertia force is

$$\rho(-\mathbf{U} . \nabla \mathbf{u} + \mathbf{u} . \nabla \mathbf{u}). \tag{4.9.13}$$

For the first of these two terms the order-of-magnitude estimate using (4.9.12) is $\rho U^2 a/r^2$, whereas for the second it is $\rho U^2 a^2/r^3$. These two terms are of the same order near the sphere, but the first is dominant far from the sphere. Thus the ratio of the order of magnitude of the neglected inertia forces to that of the retained viscous forces is

$$\frac{\rho U^2 a}{r^2} \bigg/ \frac{\mu U a}{r^3} = \frac{\rho a U}{\mu} \frac{r}{a} = \tfrac{1}{2} R \frac{r}{a}. \tag{4.9.14}$$

At positions near the sphere our solution is indeed self-consistent when $R \ll 1$, but it seems that the inertia forces corresponding to the solution become comparable with viscous forces at distances from the sphere of order a/R. The solution (4.9.12) is evidently not valid at these large distances from the sphere, although this by itself may not be of consequence since the fluid velocity and the inertia and viscous forces are all small there. We shall in fact see in §4.10 that it is possible to find a velocity distribution which is a valid approximation to the solution of the complete equation of motion everywhere in the fluid when $R \ll 1$, and which coincides with the above solution, to a consistent approximation, when r/a is of order unity.

In order to find the force exerted by the fluid on the sphere, we now evaluate the stress tensor at $r = a$. The i-component of the force per unit area exerted on the sphere at a position denoted by $\mathbf{x} = a\mathbf{n}$ is

$$n_j(\sigma_{ij})_{r=a} = n_j\left\{-p\,\delta_{ij} + \mu\left(\frac{\partial u_i}{\partial x_j} + \frac{\partial u_j}{\partial x_i}\right)\right\}_{r=a},$$

and for a velocity of the form (4.9.7) this may be found with a little working to become

$$= \left\{-pn_i + \mu n_i \mathbf{U}.\mathbf{n}\left(-\frac{f''}{r} + \frac{6f'}{r^2} - \frac{10f}{r^3}\right) + \mu U_i\left(\frac{f''}{r} - \frac{2f'}{r^2} + \frac{2f}{r^3}\right)\right\}_{r=a},$$

$$(4.9.15)$$

where f' denotes df/dr. Substitution for p and f from (4.9.4) and (4.9.9), and the use of (4.9.10), then gives

$$n_j(\sigma_{ij})_{r=a} = n_i\left\{-p_0 + \frac{3\mu\mathbf{U}.\mathbf{n}}{a}\left(\frac{2C}{a} - 3\right)\right\} + \frac{3\mu U_i}{a}\left(1 - \frac{C}{a}\right), \quad (4.9.16)$$

and, with the value of C required by the no-slip condition,

$$= -p_0 n_i - \frac{3\mu U_i}{2a}. \qquad (4.9.17)\dagger$$

It seems that the force per unit area on the sphere due to the motion has the same vectorial value $-3\mu\mathbf{U}/2a$ at all points on the sphere—a striking result, which however is not true of bodies of different shape nor of a sphere with a non-rigid surface. The first term on the right-hand side of (4.9.17) is simply the same uniform normal stress as in the fluid at infinity, and makes no contribution to the total force on the sphere, which is a retarding or drag force parallel to $\mathbf{U}$ of magnitude

$$D = 6\pi a\mu U. \qquad (4.9.18)$$

The expression (4.9.18) is usually known as Stokes's law for the resistance to a moving sphere. It is common practice to express the forces exerted on moving bodies by the fluid in terms of a dimensionless coefficient obtained by dividing the force by $\frac{1}{2}\rho U^2$ and by the area of the body projected on to a plane normal to $\mathbf{U}$; thus the drag coefficient is here

$$C_D = \frac{D}{\frac{1}{2}\rho U^2 \pi a^2} = \frac{24}{R} \quad \left(\text{where } R = \frac{2aU\rho}{\mu}\right). \qquad (4.9.19)$$

† This result can also be obtained readily from the expression for the stress at a rigid boundary given in the second exercise at the end of §4.1.

It is now a simple matter to calculate the terminal velocity which a sphere would have when falling freely under gravity through fluid, according to Stokes's law. On taking into account the buoyancy force exerted on the sphere (§ 4.1), we find for the terminal velocity $\mathbf{V}$ of a sphere of density $\bar{\rho}$

$$6\pi a\mu\mathbf{V} = \tfrac{4}{3}\pi a^3(\bar{\rho}-\rho)\mathbf{g},$$

that is,

$$\mathbf{V} = \frac{2}{9}\frac{a^2\mathbf{g}}{\nu}\left(\frac{\bar{\rho}}{\rho}-1\right), \tag{4.9.20}$$

where $\nu = \mu/\rho$. The corresponding value of the Reynolds number for a sphere falling with its terminal velocity is

$$\frac{2aV\rho}{\mu} = \frac{4}{9}\frac{a^3 g}{\nu^2}\left(\frac{\bar{\rho}}{\rho}-1\right). \tag{4.9.21}$$

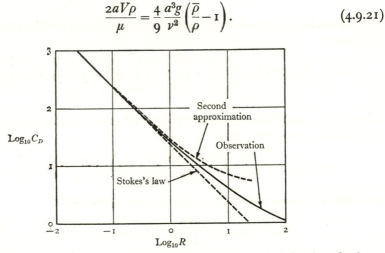

Figure 4.9.2. Comparison of measured values of the drag on a sphere (taken from Castleman 1925) and two theoretical estimates, Stokes's law $C_D = 24/R$, and a second approximation $C_D = 24R^{-1}(1+\tfrac{3}{16}R)$, where $R = 2a\rho U/\mu$.

For a particle of sand falling through water at 20 °C, we have $\bar{\rho}/\rho \approx 2$ and $\nu = 0.010$ cm²/sec, making the Reynolds number $4.4 \times 10^6 a^3$, a being in centimetres; and for a water droplet (assumed to be rigid) falling through air, we have $\bar{\rho}/\rho \approx 780$ and $\nu = 0.15$ cm²/sec, making the Reynolds number $1.5 \times 10^7 a^3$. The assumption on which neglect of the inertia force was based, namely, that $R \ll 1$, is satisfied in the case of the sand particle in water provided $a \ll 0.006$ cm and in the case of the water droplet in air provided $a \ll 0.004$ cm. The conditions under which the analysis may be applied are thus restricted to extremely small spheres. However, it seems, from a comparison of the observed and calculated terminal velocities of spheres of known size (see figure 4.9.2), that Stokes's law for the drag is tolerably accurate for most purposes when $R < 1$; and there is no detectable error when $R < 0.5$. Thus the theoretical requirement 'small compared with' used above may usually in practice be replaced, so far as the drag force is concerned, by simply 'smaller than'.

It will be noticed from figure 4.9.2 that the curve representing Stokes's law lies below the measured values of the drag and below the other theoretical estimate (which will be referred to in the next section). This was to be expected from the general result established at the end of §4.8; the velocity field obtained by neglecting inertia forces is accompanied by a smaller total rate of dissipation than that for any other solenoidal velocity distribution with the same value of the velocity vector everywhere on the boundary of the fluid, and hence is accompanied by a smaller rate of working by the sphere against fluid forces at a given speed **U**.

A spherical drop of a different fluid

In a number of cases of practical interest, the sphere in translational motion at small Reynolds number is itself composed of fluid in which differential motion may occur, and it is desirable to see if this internal circulation affects the drag significantly (Hadamard 1911). We shall suppose that the two fluids are immiscible, and that surface tension at the interface is sufficiently strong to keep the 'drop' approximately spherical against any deforming effect of viscous forces. The condition for this is that γ/a (where γ is the coefficient of surface tension) should be large compared with the normal stress due to the motion, of order $\mu U/a$, that is, that

$$\gamma \gg \mu U; \tag{4.9.22}$$

we shall refer again to this requirement at the end of the section. It will also be assumed that the Reynolds number of the motion within the drop is small compared with unity, like that of the motion outside the drop.

The argument used to determine the velocity and pressure distributions for the case of a rigid sphere can be modified without difficulty. The motions both inside and outside the sphere are axisymmetric and satisfy the equations (4.9.1) and (4.9.2) (although with different values of the viscosity). **u** and $p - p_0$ must vanish at infinity, as before, and $\bar{\mathbf{u}}$ and $\bar{p} - \bar{p}_0$ (where the overbar indicates a quantity relating to the internal fluid and its motion) are finite everywhere within the sphere. The common kinematical condition at the interface is

$$\mathbf{n}.\mathbf{u} = \mathbf{n}.\bar{\mathbf{u}} = \mathbf{n}.\mathbf{U} \quad \text{at } r = a. \tag{4.9.23}$$

In place of the no-slip condition at the surface of a rigid sphere there are certain dynamical matching conditions. No relative motion of the two fluids can occur at the interface, and the tangential stress exerted at the interface by the external fluid must be equal and opposite to that exerted by the internal fluid.† No information can be obtained from considerations of the normal stress at the interface, since we have supposed that any discontinuity in the normal stress there which cannot be eliminated by an

---

† We are assuming here that the only mechanical property of the interface is a uniform surface tension; in practice it appears that contaminant molecules may collect at the interface and give rise to other properties (§1.9).

appropriate choice of $\bar{p}_0$ is balanced by surface tension acting at a slightly deformed interface. Thus

$$\mathbf{x} \times \mathbf{u} = \mathbf{x} \times \bar{\mathbf{u}} \quad \text{at } r = a, \tag{4.9.24}$$

$$\epsilon_{mki} n_k n_j (\sigma_{ij} - \bar{\sigma}_{ij}) = 0 \quad \text{at } r = a. \tag{4.9.25}$$

The equations and boundary conditions are linear and homogeneous in $\mathbf{u}, p - p_0, \bar{\mathbf{u}}, \bar{p} - \bar{p}_0$ and $\mathbf{U}$, so that the relations (4.9.4) to (4.9.10) still stand, and are supplemented by analogous relations for the internal motion. $\bar{p}$ satisfies Laplace's equation, like p, and the appropriate solution, analogous to (4.9.4), is

$$(\bar{p} - \bar{p}_0)/\bar{\mu} = \bar{C}\mathbf{U} . \mathbf{x},$$

where $\bar{C}$ is a constant. The stream function and velocity within the sphere have the forms (4.9.6) and (4.9.7), but the internal vorticity is

$$\bar{\boldsymbol{\omega}} = -\tfrac{1}{2}\bar{C}\mathbf{U} \times \mathbf{x}$$

and so the right-hand side of the differential equation for $\bar{f}$, analogous to (4.9.8), is $\frac{1}{2}\bar{C}r^2$. Hence

$$\bar{f}(r) = \tfrac{1}{20}\bar{C}r^4 + \bar{L}r^{-1} + \bar{M}r^2. \tag{4.9.26}$$

The need to avoid a singularity at $r = 0$ and the kinematical condition at $r = a$ require

$$\bar{L} = 0, \quad \bar{M} = \tfrac{1}{2} - \tfrac{1}{20}\bar{C}a^2.$$

The velocity within the sphere is thus

$$\bar{\mathbf{u}} = \mathbf{U} - \tfrac{1}{10}\bar{C}\{\mathbf{U}(a^2 - 2r^2) + \mathbf{x}\mathbf{U} . \mathbf{x}\}. \tag{4.9.27}$$

It remains to determine C and $\bar{C}$ from the dynamical matching conditions. From (4.9.24) we have

$$C - \tfrac{1}{2}a = \tfrac{1}{10}\bar{C}a^3 + a.$$

Only the term containing U_i in the general expression (4.9.15) for the stress across the interface contributes to the tangential component, and matching of this tangential component gives

$$\frac{3\mu}{a^2}(a - C) = \tfrac{3}{10}\bar{\mu}a\bar{C}.$$

Hence

$$C = \tfrac{1}{2}a\frac{2\mu + 3\bar{\mu}}{\mu + \bar{\mu}}, \quad \bar{C} = -\frac{5}{a^2}\frac{\mu}{\mu + \bar{\mu}}. \tag{4.9.28}$$

The resultant force exerted on the interface by the external fluid is now obtained by integrating the force per unit area (4.9.16) over the interface A:

$$\int n_j(\sigma_{ij})_{r=a}\, dA = -4\pi\mu U_i C$$

$$= -4\pi a\mu U_i \frac{\mu + \tfrac{3}{2}\bar{\mu}}{\mu + \bar{\mu}}. \tag{4.9.29}$$

The terminal velocity $\mathbf{V}$ of a fluid sphere of density $\bar{\rho}$ and viscosity $\bar{\mu}$ moving freely under gravity is thus

$$\mathbf{V} = \frac{1}{3}\frac{a^2\mathbf{g}}{\nu}\left(\frac{\bar{\rho}}{\rho} - 1\right)\frac{\mu + \bar{\mu}}{\mu + \tfrac{3}{2}\bar{\mu}}. \tag{4.9.30}$$

The case of a rigid sphere is recovered by putting $\bar{\mu}/\mu \to \infty$. The case of a spherical gas bubble moving through liquid corresponds (approximately) to the other extreme, $\bar{\mu}/\mu = 0$, together with $\bar{\rho}/\rho = 0$. The speed of a spherical gas bubble rising steadily under gravity is thus given as $\frac{1}{3}a^2g/\nu$. However, observation of the terminal speed V of very small gas bubbles suggests that the drag is often closer to the value $6\pi a\mu V$ than to the expected value $4\pi a\mu V$; this is believed to be because any surface-active impurities present in the liquid are likely to form a mesh of large molecules at the bubble surface and to cause the interface to act partially like a rigid surface.†

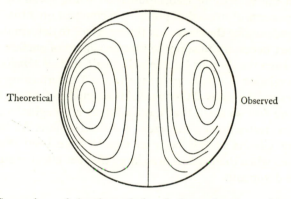

Theoretical   Observed

Figure 4.9.3. Comparison of the theoretical and observed pattern of streamlines in a spherical drop of glycerine falling through castor oil (from Spells 1952).

Observations of the general form of the flow inside liquid drops falling under gravity through a second liquid have been made, although measurements of the velocity distribution are difficult. Figure 4.9.3 shows a sketch of the streamlines observed in a spherical liquid drop, relative to axes moving with the drop. The theoretical streamlines corresponding to (4.9.27), relative to these same axes, are lines on which

$$\psi \propto \tfrac{1}{2} U r^2 (a^2 - r^2) \sin^2 \theta \qquad (4.9.31)$$

is constant, and these are also shown; the agreement is satisfactory.

Finally we note an interesting point about the normal component of stress at the surface of the fluid sphere, on which no restrictions have been placed. It will be recalled that the pressure represented in the equations of this section is the modified pressure, and that to obtain the absolute pressure (or a quantity differing from it only by a constant) we should add to the modified pressure a term $\rho \mathbf{g} . \mathbf{x}$ for the flow field outside the sphere and a term $\bar{\rho} \mathbf{g} . \mathbf{x}$ for the flow field within the sphere. The difference between the values of the normal component of the absolute stress at the surface of the

† A general discussion of the effect of adsorbed material at the surface of a small gas bubble rising through liquid will be found in *Physico-chemical Hydrodynamics*, by V. G. Levich (Prentice-Hall, 1962).

sphere as approached from the outer and inner sides is then found from the general expression (4.9.15) to be

$$n_i n_j (\sigma_{ij} - \overline{\sigma}_{ij})_{r=a} = \overline{p}_0 - p_0 - \mathbf{n} \cdot \mathbf{g} a(\rho - \overline{\rho}) + \mathbf{n} \cdot \mathbf{U} \left\{ \frac{3\mu}{a^2} (C - 2a) + \tfrac{3}{5} a \overline{\mu} \overline{C} \right\}$$

$$= \overline{p}_0 - p_0 - \mathbf{n} \cdot \mathbf{g} a(\rho - \overline{\rho}) - \mathbf{n} \cdot \mathbf{U} \frac{3\mu}{a} \frac{\mu + \tfrac{3}{2}\overline{\mu}}{\mu + \overline{\mu}}, \qquad (4.9.32)$$

apart from any contribution due to surface tension. The notable feature of the expression (4.9.32) is that when the sphere is moving steadily under gravity, with the velocity of translation then being given by (4.9.30), the normal components of stress differ only by a constant quantity $\overline{p}_0 - p_0$. Thus there is no tendency for the stresses at the interface to deform the sphere and it is not in fact necessary to suppose that the effect of surface tension is so strong as to keep the drop or bubble spherical; surface tension enters only through the relation $\overline{p}_0 - p_0 = 2\gamma/a$ (see (1.9.2)) determining $\overline{p}_0$. Provided the viscosities and densities of the two fluids are such as to make the Reynolds number of the flow so small that inertia forces are negligible, we now see that there is no restriction on the size of the fluid sphere. Air bubbles rising through very viscous liquids such as treacle have been observed to be spherical, even when their radius is so large that the effect of surface tension could not be dominant.

A body of arbitrary shape

Although it is difficult to work out the details of the flow due to a moving body at small Reynolds number for any shape other than spherical,[†] some general results are available. The following remarks refer only to circumstances in which inertia forces may be neglected completely.

Arguments like those used at the beginning of this section show that, for a body of arbitrary shape in translational motion with velocity $\mathbf{U}$, both $\mathbf{u}$ and $(p - p_0)/\mu$ are linear and homogeneous in $\mathbf{U}$. Furthermore, a change of size of the body without change of its shape simply changes the length scale of the whole flow field, so that for a body of given shape $\mathbf{u}/U$ and $(p - p_0) d/\mu U$ are (dimensionless) functions of $\mathbf{x}/d$, where d is a representative linear dimension of the body.

Both the tangential and excess normal stresses in the fluid are linear in $\mathbf{U}$, so that the resultant (vector) force exerted by the body, given by the integral

$$F_i = - \int \sigma_{ij} n_j \, dA \qquad (4.9.33)$$

taken over the body surface, is proportional to $\mu U d$. The equation (4.9.1) governing flow with negligible inertia forces is equivalent to

$$\partial \sigma_{ij} / \partial x_j = 0,$$

and it follows from an application of the divergence theorem that the integral

---

† The solution for the case of a rigid ellipsoid is given in *Hydrodynamics*, by H. Lamb, 6th ed. (Cambridge University Press, 1932).

in (4.9.33) has the same value for any surface in the fluid enclosing the body and in particular for a sphere of large radius centred on the origin. Hence

$$F_i = - \int \lim_{r \to \infty} (r \sigma_{ij} x_j) \, d\Omega(\mathbf{x}), \qquad (4.9.34)$$

where $\delta\Omega(\mathbf{x})$ is an element of solid angle at the direction of $\mathbf{x}$. This relation shows that in the case of flow due to a moving body exerting a finite force on the fluid $p - p_0$ and the rate-of-strain tensor must both decrease at least as rapidly as r^{-2} as $r \to \infty$.

We know also that $p - p_0$ is a harmonic function and can be represented as a series like (2.9.19). The first non-zero term of this series is evidently of degree -2 in r, so that

$$\frac{p - p_0}{\mu} \sim \frac{P_{ij} U_j \, dx_i}{r^3} \qquad (4.9.35)$$

is the asymptotic form as $r \to \infty$, where P_{ij} is a numerical tensor coefficient dependent only on the body shape. The vorticity $\boldsymbol{\omega}$ also satisfies Laplace's equation, and may be written as a similar series (with allowance for its axial vectorial character). Terms of the same degree in the series for $(p - p_0)/\mu$ and for $\boldsymbol{\omega}$ are related by the governing equation (4.9.1), and it may be seen that if the leading term of the series for $(p - p_0)/\mu$ is $\boldsymbol{\alpha} . \nabla r^{-1}$, that for $\boldsymbol{\omega}$ is $\boldsymbol{\alpha} \times \nabla r^{-1}$. Consequently we have

$$\omega_l \sim \epsilon_{ikl} \frac{P_{ij} U_j \, dx_k}{r^3}, \qquad (4.9.36)$$

as $r \to \infty$. Finally we may obtain the asymptotic form for the velocity, which is determined by (4.9.36) (apart from an irrotational contribution which cannot be of larger magnitude than r^{-3} when the flux of volume across the body surface is zero) and the requirement that $\mathbf{u}$ is solenoidal. We find

$$u_k \sim \tfrac{1}{2} P_{ij} U_j \left(\frac{d}{r} \delta_{ik} + \frac{d}{r^3} x_i x_k \right), \qquad (4.9.37)$$

as $r \to \infty$.

It is now possible to relate the coefficient P_{ij} to the force $\mathbf{F}$ by evaluating the stress at a spherical surface of large radius (see (4.9.34)). The working is straight-forward, and leads to the result

$$F_i = 4\pi\mu P_{ij} U_j d. \qquad (4.9.38)$$

It appears that the single numerical tensor P_{ij} is sufficient for the specification of the total force on the fluid and the asymptotic expressions for the pressure and velocity, when a body of given shape moves with translational velocity $\mathbf{U}$; and in the case of a spherical body of radius $\tfrac{1}{2}d$ composed of fluid of viscosity $\bar{\mu}$ we know from the preceding calculation that

$$P_{ij} = \tfrac{1}{2} \delta_{ij} \frac{\mu + \tfrac{3}{2}\bar{\mu}}{\mu + \bar{\mu}}.$$

The flow at large distances from the body has axisymmetry about the direction of the vector $P_{ij} U_j$. Consequently we may represent the flow in this region in terms of a stream function. With spherical polar co-ordinates (r, θ, ϕ) and the axis $\theta = 0$ in the direction of the vector $P_{ij} U_j$—which is also the direction of the force **F**—we find from (4.9.37) and (4.9.38) that

$$\psi = \frac{F}{8\pi\mu} r \sin^2 \theta \qquad (4.9.39)$$

in this region, where F is the magnitude of **F**. Now (4.9.38) shows that $F/\rho v^2$ is of the same order as Ud/v, which has been assumed to be small compared with unity. It is therefore not surprising that we should have recovered the flow field (4.6.18) due to a force of magnitude small compared with ρv^2 applied to the fluid at the origin. When a body of arbitrary shape moves through fluid at small Reynolds number, the distant flow field depends only on the resultant force exerted on the fluid and is not affected by the continual change of position of the body.

These general results have a convenient form for application to the case of a small particle, either solid or fluid, falling freely under gravity. If the volume τ and density $\bar{\rho}$ of the particle are known, the distributions of velocity and pressure far from the particle are immediately obtained from the above formulae by putting $\qquad \mathbf{F} = (\bar{\rho} - \rho)\tau\mathbf{g};$

details of the shape of the particle are irrelevant, and it presumably also does not matter whether the particle continually turns over and changes its orientation relative to the direction of gravity or whether it moves on a path which is not vertical.

The flow field represented by (4.9.39) is sometimes referred to as being due to the existence of a 'Stokeslet' at the origin.

Exercise

A rigid sphere of radius a is rotating with angular velocity Ω in fluid which is at rest at infinity. Show that when $a^2 \Omega \rho / \mu \ll 1$ the retarding couple exerted on the sphere by the fluid is of magnitude $8\pi\mu a^3 \Omega$.

4.10. Oseen's improvement of the equation for flow due to moving bodies at small Reynolds number

It has been seen that complete neglect of the inertia forces in the flow due to a moving sphere leads to some inconsistency, inasmuch as, for the solution found in this way, the inertia forces are in fact comparable with viscous forces at distances from the sphere of order a/R. This situation is not peculiar to bodies of spherical form. Since the first term in the expression (4.9.13) for the inertia force involves a first-order spatial derivative, whereas the viscous force involves a second-order derivative, it follows that for a wide class of solutions, including those for which $|\mathbf{u}| \sim r^{-n}$ as $r \to \infty$, inertia forces become comparable with viscous forces at sufficiently large distances from the origin.

This criticism of the use of equation (4.9.1) to represent flow due to bodies moving through fluid of infinite extent which is otherwise undisturbed was made by Oseen (1910), who also showed how it is possible to improve the equation and thereby to remove the inconsistency. Oseen's improvement applies to cases in which the body is moving with steady velocity $\mathbf{U}$ and in which the flow relative to the body is steady, in which event the local inertia force is as given in (4.9.13), namely,

$$\rho(-\mathbf{U}.\nabla\mathbf{u}+\mathbf{u}.\nabla\mathbf{u}),\qquad(4.10.1)$$

where $\mathbf{u}$ is the fluid velocity relative to a co-ordinate system fixed in the fluid at infinity as before. Since the first of these two terms becomes dominant at large r, and is responsible for inertia forces being comparable with viscous forces at sufficiently large r, Oseen suggested that it, alone of the two contributions to the inertia force, be retained in the equation of motion. The second term, which presents the greater mathematical difficulty in view of its non-linearity in $\mathbf{u}$, is again neglected on the assumption that $R \ll 1$; provided $|\mathbf{u}|$ falls off at least as rapidly as r^{-1} as r increases, this second term remains small relative to the viscous force however large r may be. Near the body the two terms in (4.10.1) are of the same order and will both be small compared with the viscous force, provided $R \ll 1$, so that in this region the suggested equation is neither more nor less accurate than (4.9.1).

The Oseen equations for flow due to a moving body at small Reynolds number are therefore

$$\left.\begin{aligned}\rho\frac{\partial\mathbf{u}}{\partial t}=-\rho\mathbf{U}.\nabla\mathbf{u}=-\nabla p+\mu\nabla^2\mathbf{u},\\[2mm]\nabla.\mathbf{u}=0,\end{aligned}\right\}\qquad(4.10.2)$$

with the boundary conditions, for a rigid body,

$$\mathbf{u}=\mathbf{U}\quad\text{at the surface of the body,}$$

$$\mathbf{u}\to 0\quad\text{and}\quad p-p_0\to 0\quad\text{as}\quad r\to\infty.$$

Although these equations are still linear in the dependent variables $\mathbf{u}$ and p, they are no longer linear in $\mathbf{u}$, p and $\mathbf{U}$, and are more difficult to solve than (4.9.1) and (4.9.2).

A rigid sphere

The solution of these new equations for the case of a moving sphere is not known in closed form, but an approximate solution which is consistent with the degree of approximation used in the equations themselves has been found (Lamb 1911). In terms of the stream function, this approximate solution, which will simply be quoted here, is

$$\psi = Ua^2\left[-\frac{1}{4}\frac{a}{r}\sin^2\theta+3(1-\cos\theta)\frac{1-\exp\{-\frac{1}{4}R(1+\cos\theta)\,r/a\}}{R}\right]\quad(4.10.3)$$

16

at the instant at which the centre of the sphere coincides with the origin, where $R = 2aU\rho/\mu$ as before. This expression is readily seen to satisfy the equations (4.10.2) exactly, and it also makes $\mathbf{u} \to 0$ as $r \to \infty$. Near the sphere, where r/a is of order unity and $Rr/a \ll 1$, it becomes

$$\psi = Ua^2 \sin^2\theta \left\{ -\frac{1}{4}\frac{a}{r} + \frac{3}{4}\frac{r}{a} + O\left(R\frac{r}{a}\right) \right\}, \qquad (4.10.4)$$

and therefore coincides with Stokes's solution (4.9.12)—and in particular satisfies the inner boundary condition—with a relative error of order R. This is just the degree of approximation to which (4.10.2) represents the equations of motion, so that (4.10.3) is as accurate a solution of (4.10.2) as is wanted.

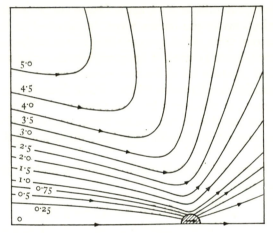

Figure 4.10.1. Streamlines in an axial plane for the outer part of the flow field due to a moving sphere, according to the Oseen equations. ψ is equal to some constant times the numbers shown on the streamlines.

Figure 4.10.1 shows the streamlines corresponding to the solution (4.10.3), with neglect of the first term within square brackets which is significant only near the sphere when $R \ll 1$. The qualitative differences between the Oseen and Stokes solutions in the outer part of the flow field are evident. The streamlines are no longer symmetrical about the plane $\theta = \frac{1}{2}\pi$, as was to be expected from the fact that the governing equation does not remain satisfied after a change in the signs of $\mathbf{u}$ and $\mathbf{U}$. Far from the sphere the flow tends to become radial, as if from a source of fluid at the sphere, except within a 'wake' directly behind the sphere. Analytically, we see from (4.10.3) that when $Rr/a \gg 1$ the flow has different forms according to whether $1 + \cos\theta$ is small compared with unity. At positions where $1 + \cos\theta$ is not small, the stream function becomes

$$\psi \sim Ua^2 \frac{3}{R}(1 - \cos\theta), \qquad (4.10.5)$$

which describes the outward radial flow from a source at the origin emitting

$12\pi a^2 U/R$ units of volume per second. On the other hand, within the wake, where $1 + \cos\theta$ is of the same order of magnitude as $4a/rR$ (that is, where $\pi - \theta$ is small and of order $(8a/rR)^{\frac{1}{2}}$), we have

$$\psi \sim Ua^2 \frac{6}{R}\left[1 - \exp\left\{-\frac{R}{8}\frac{r}{a}(\pi-\theta)^2\right\}\right], \qquad (4.10.6)$$

which describes a compensating flow towards the sphere, the inflow velocity being $3Ua/2r$ on the axis $\theta = \pi$.

Far from the sphere, the vorticity is zero in the source-flow region and is confined to the wake, which may be regarded as bounded by a paraboloid of revolution on which $(\pi-\theta)^2 r/a$ is of order R^{-1}. Whereas in the Stokes approximation the vorticity diffuses out in all directions from an effectively stationary sphere, here the motion of the sphere is allowed for, as may be seen from the equation for $\boldsymbol{\omega}$ obtained from (4.10.2):

$$\frac{\partial\boldsymbol{\omega}}{\partial t} = -\mathbf{U}.\nabla\boldsymbol{\omega} = \nu\nabla^2\boldsymbol{\omega}. \qquad (4.10.7)$$

This equation for each component of $\boldsymbol{\omega}$ is of the same form as that satisfied by temperature in a stationary conducting medium through which a steady source (which in this case has a dipole character) of heat is moving with steady velocity $\mathbf{U}$. The vorticity generated at the sphere is left behind as the sphere moves on, in a wake which becomes narrower as R increases.

We may now confirm that the solution (4.10.3) is self-consistent in the way that Stokes's solution was not; that is, we show that the neglected term $\rho\mathbf{u}.\nabla\mathbf{u}$, evaluated by means of (4.10.3), is small compared with any term retained in the equation of motion, when $R \ll 1$. In the region near the sphere, where r/a is of order unity, (4.10.3) reduces to Stokes's solution (with an error of order R), for which $\rho|\mathbf{u}.\nabla\mathbf{u}|$ is already known to be small compared with $\mu|\nabla^2\mathbf{u}|$, the ratio of these terms being of order R. Far from the sphere, in the region where Rr/a is of order unity, which is where (4.10.3) first differs significantly from the Stokes solution, the magnitude of $\mathbf{u}$ as given by (4.10.3) is of order Ua/r, or UR; hence the ratio of the neglected term $\rho|\mathbf{u}.\nabla\mathbf{u}|$ to the retained term $\rho|\mathbf{U}.\nabla\mathbf{u}|$ is of order R and is again small. Still further from the sphere, where $r/a \gg R^{-1}$, $|\mathbf{u}|$ is even smaller by comparison with U.

It appears then that the approximate form of the equation of motion suggested by Oseen has a solution such that the approximation is self-consistent over the whole of the flow field when $R \ll 1$. Near the sphere this solution has the same form as Stokes's solution—and so leads to the same expression, $6\pi a\mu U$, for the resistance experienced by the sphere†—with a

---

† It will be noticed that the resistance is ρU times the inward flux of volume in the wake far downstream. This relation follows from general considerations of momentum (see §5.12, on wakes), and holds for any body, at any Reynolds number, provided the body moves steadily and leaves behind it a wake of non-zero vorticity whose width increases less rapidly than its length.

relative error of order R, which is the degree of error involved in the replacement of the equation of motion by the Oseen equation. Since (4.10.3) is evidently an approximation to the solution of the complete equations of motion which is valid for $R \ll 1$ over the whole of the flow field, it is natural to consider making (4.10.3) the starting point of a process of successive approximation to the solution of these equations. This has been done (Kaplun and Lagerstrom 1957; Proudman and Pearson 1957), and the second approximation to the drag coefficient has been found to be

$$C_D = \frac{24}{R}(1 + \tfrac{3}{16}R). \qquad (4.10.8)$$

(This expression for C_D to order R^0 also follows from the Oseen equations, which at first sight is surprising; the explanation is that the term of order R in the difference between the solution of the Oseen equations for $\mathbf{u}$ and the second approximation to the solution of the complete equations makes zero contribution to the drag for bodies with fore-and-aft symmetry.) As indicated in figure 4.9.2, the formula (4.10.8) agrees with the measured drag for a slightly larger range of Reynolds number than does Stokes's law.

A rigid circular cylinder

The difficulties associated with the use of the equations (4.9.1) and (4.9.2), and the overcoming of these difficulties by the use of the Oseen equations (4.10.2), have been explored for a few other cases of bodies moving steadily through fluid. We shall mention here the case of a circular cylinder of radius a moving with velocity $\mathbf{U}$ normal to its axis, since this case exhibits marked differences, typical of two-dimensional flow at small Reynolds number, from the case of a sphere.

A solution of the equations (4.9.1) and (4.9.2) may be sought in exactly the same way as for a moving sphere, making use of the linearity of the solution in $\mathbf{U}$ and of the dependence on $\mathbf{x}$, $\mathbf{U}$ and a alone. In place of the relations (4.9.4) and (4.9.5) we find

$$\frac{p - p_0}{\mu} = \frac{C\mathbf{U} \cdot \mathbf{x}}{r^2}, \quad \boldsymbol{\omega} = \frac{C\mathbf{U} \times \mathbf{x}}{r^2}, \qquad (4.10.9)$$

where C is a constant and (r, θ) are the polar co-ordinates of the two-dimensional vector $\mathbf{x}$. The vorticity may also be expressed in terms of a stream function ψ. The analogue of (4.9.6) is

$$\psi = U \sin\theta f(r), \qquad (4.10.10)$$

and the equation satisfied by the function f is

$$\frac{d^2f}{dr^2} + \frac{1}{r}\frac{df}{dr} - \frac{f}{r^2} = -\frac{C}{r}.$$

The general solution is

$$f(r) = -\tfrac{1}{2}Cr \log r + Lr + Mr^{-1}, \qquad (4.10.11)$$

and there arises the difficulty that the particular integral associated with the vorticity distribution gives a divergent velocity at infinity. If for the moment we ignore the outer boundary condition on $\mathbf{u}$, we find that the required conditions at the inner boundary, viz.

$$f/r = 1, \quad df/dr = 1 \quad \text{at } r = a,$$

are satisfied when

$$L = 1 + \tfrac{1}{4}C + \tfrac{1}{2}C \log a, \quad M = -\tfrac{1}{4}a^2 C.$$

The velocity distribution is then

$$\mathbf{u} = \mathbf{U} + C\mathbf{U}\left(-\tfrac{1}{2}\log\frac{r}{a} - \frac{1}{4} + \frac{1}{4}\frac{a^2}{r^2}\right) + C\mathbf{x}\frac{\mathbf{U}.\mathbf{x}}{r^2}\left(\frac{1}{2} - \frac{1}{2}\frac{a^2}{r^2}\right). \quad (4.10.12)$$

The normal and tangential stresses at the surface of the cylinder, as derived from the expressions (4.10.9) and (4.10.12), exert a force on the cylinder which is found to be a drag of magnitude

$$D = 2\pi\mu UC \qquad (4.10.13)$$

per unit length of the cylinder.

The expressions (4.10.9) and (4.10.12) for $(p - p_0)/\mu$ and $\mathbf{u}$ satisfy the equations (4.9.1) and (4.9.2), the inner boundary condition, and the conditions of linearity in $\mathbf{U}$ and symmetry about $\theta = 0$, but the expression for $\mathbf{u}$ diverges as $\log r$ when r is large and no choice of the remaining arbitrary constant C will make $\mathbf{u} \to 0$ as $r \to \infty$. However, the solution (4.10.12) is not useless. According to (4.10.12), the two contributions to the neglected inertia force (see (4.9.13)) have the following magnitudes when r is large:

$$\left|\rho\frac{\partial\mathbf{u}}{\partial t}\right| \sim \frac{\rho U^2 C}{r}, \quad |\rho\mathbf{u}.\nabla\mathbf{u}| \sim \frac{\rho U^2 C^2}{r}\log\frac{r}{a}. \qquad (4.10.14)$$

On the other hand, the retained viscous force has the magnitude

$$|\mu\nabla^2\mathbf{u}| \sim \frac{\mu UC}{r^2}.$$

Both contributions to the inertia force become comparable with the viscous force at sufficiently large distances from the cylinder, the first when r/a is of order R^{-1} (where $R = 2a\rho U/\mu$) and the second when $(Cr/a)\log(r/a)$ is of order R^{-1}. The solution (4.10.12) is thus in any case not a self-consistent approximation to the flow field at large values of r, and its failure to satisfy the outer boundary condition might not therefore be a fatal defect in itself. Evidently some other approximation to the equation of motion at large r is needed, and (4.10.12) must match with the solution of this approximate equation as $r \to \infty$.

Detailed calculation shows that the Oseen approximate form of the equation of motion does in fact have a solution (Lamb 1911) which is self-consistent over the whole field in the sense that the neglected term $\rho\mathbf{u}.\nabla\mathbf{u}$, evaluated according to the solution obtained, proves to be small everywhere by comparison with terms retained in the equation when $R \ll 1$. Near the cylinder this solution for $\mathbf{u}/U$ approximates, with an absolute error of order R, to the form (4.10.12) provided the constant in (4.10.12) is chosen as

$$C = \frac{2}{\log(7 \cdot 4/R)}. \tag{4.10.15}$$

We note that, with this value of C, the magnitude of $\rho \mathbf{u} . \nabla \mathbf{u}$, according to the the 'inner' solution (4.10.12), does not become comparable with $|\mu \nabla^2 \mathbf{u}|$ until r/a is of order R^{-1}, which is also the value of r/a at which the Oseen improvement to equation (4.9.1) is necessary and at which the solution of the Oseen equation has begun to differ from (4.10.12).

The general features of the flow far from the cylinder as obtained from the Oseen equation are similar to those for a sphere, and in particular there is a parabolic wake of finite vorticity behind the cylinder.

Since the solution of the Oseen equation is given approximately by (4.10.12) near the cylinder, with an error of the same order as is involved in the replacement of the equation of motion by the Oseen equation (viz. $O(R)$), the estimate (4.10.13) for the drag is still appropriate. Substituting from (4.10.15), we have for the drag coefficient of unit length of the cylinder

$$C_D = \frac{D}{\frac{1}{2}\rho U^2 2a} = \frac{8\pi}{R \log(7 \cdot 4/R)}. \tag{4.10.16}$$

It is more difficult to make measurements of the drag on a cylinder at low Reynolds number than for a sphere, owing largely to the unwanted effect of the ends of a cylinder of finite length, but the relation (4.10.16) gives values near $R = 0 \cdot 5$ which are consistent with observation (see figure 4.12.7).

In some recent research a procedure for obtaining higher-order approximations to the flow past a circular cylinder and to the drag coefficient has been devised.† It appears from these investigations that (4.10.12) (with (4.10.15)) represents the true (non-dimensional) velocity distribution in the neighbourhood of the cylinder with an absolute error of order $(\log R)^{-2}$.

4.11. The viscosity of a dilute suspension of small particles

Mixtures consisting of one material in the form of small particles, either solid, liquid, or gaseous, dispersed randomly throughout another fluid material are quite common in nature and in industry. The term 'suspension' usually refers to a system of small solid particles in liquid, but the nature of the two media is not of particular significance from the dynamical point of view and our use of the word here will include also systems of solid particles in a gas, systems of drops of one liquid dispersed either in another liquid

---

† For a general account of the procedure, which may be applied to some other problems in fluid dynamics, see *Perturbation Methods in Fluid Mechanics*, by M. D. Van Dyke (Academic Press, 1964).

(an emulsion) or in a gas, and systems of gas bubbles in a liquid. It is of interest to know how such suspensions behave in response to applied forces and moving boundaries. Provided the characteristic length scale of the motion of the suspension is large compared with the average distance between the particles, as we shall assume to be the case, we can regard the suspension as a homogeneous fluid, with mechanical properties different from those of the ambient fluid in which the particles are suspended. A random distribution of spherical particles does not confer any directional properties on the medium (elongated rod-like particles might do so by tending to set themselves in a certain direction relative to the local velocity distribution, although Brownian motion of the suspended particles tends to eliminate any directional preferences), so that provided the ambient fluid is 'Newtonian' the homogeneous fluid equivalent to a suspension of roughly spherical particles is likewise 'Newtonian' and is characterized by a shear viscosity (and perhaps also by an expansion viscosity).

Our objective in this section is to calculate the effective viscosity of an incompressible fluid containing suspended particles of such small linear dimensions that (*a*) the effects of gravity and inertia on the motion of a particle are negligible, so that a particle moves with the ambient fluid locally, and (*b*) the Reynolds number of the disturbance motion resulting from the presence of one particle is small compared with unity. The particles will be assumed to be exactly spherical, for simplicity; in the case of liquid or gaseous particles of small radius, surface tension tends to keep the particles spherical against the deforming effect of the motion, so that an assumption about shape is needed only for solid particles. Finally, the suspension will be assumed to be dilute, with the average distance between particles large compared with their linear dimensions.

Under these conditions the background motion of the ambient fluid on which the disturbance flow due to the presence of one particular particle is superposed consists approximately of a uniform translation, a uniform rotation and a uniform pure straining motion. The particle translates and rotates with the surrounding fluid, so that only the background pure straining motion gives rise to a disturbance flow. It seems inevitable that the change in the straining motion due to the presence of one particle is accompanied by an increase in the total rate of dissipation and that the effective viscosity (either shear or expansion) of the suspension is *greater* than the viscosity of the ambient fluid; we shall find this to be so.

In the first instance we assume the particles to be incompressible, so that the suspension is likewise incompressible and only the effective value of the shear viscosity is to be determined. Explicit knowledge of the disturbance flow due to a single incompressible particle is needed in a determination of the effective viscosity of the suspension, and we therefore consider first the following problem of flow with negligible inertia forces.

The flow due to a sphere embedded in a pure straining motion

Fluid of viscosity μ and density ρ occupies the space outside a sphere of radius a, and far from the sphere the fluid is in pure straining motion specified by the rate-of-strain tensor e_{ij}, where $e_{ii} = 0$. The velocity and pressure in the fluid may be written as

$$u_i = u_i' + e_{ij} x_j, \quad p = p' + P, \tag{4.11.1}$$

where P is the pressure in the pure straining motion represented by e_{ij} in the absence of the sphere; u_i' and p' represent the changes due to the presence of the sphere and

$$u_i' \to 0, \quad p' \to 0, \tag{4.11.2}$$

as $r(=|\mathbf{x}|) \to \infty$. We choose the centre of the sphere to be at the origin, so that by symmetry there is no tendency for the sphere to translate and the surface of the sphere is given permanently by $r = a$; thus

$$\mathbf{n} \cdot \mathbf{u} = 0 \quad \text{at } r = a. \tag{4.11.3}$$

There are further conditions at the surface of the sphere which depend on the nature of the particle. We may include the various kinds of particle within the scope of the analysis by supposing the sphere to contain incompressible fluid of viscosity $\bar{\mu}$ (the case of a rigid particle corresponding, as in §4.9, to $\bar{\mu}/\mu \to \infty$). The velocity must be continuous across the interface, and so also is the tangential component of stress if we suppose, as in §4.9, that the interface has no mechanical properties other than a uniform surface tension. Thus

$$\left.\begin{aligned} u_i &= u_i' + e_{ij} x_j = \bar{u}_i \\ \epsilon_{kli} n_l n_j (\sigma_{ij} - \bar{\sigma}_{ij}) &= 0 \end{aligned}\right\} \quad \text{at } r = a, \tag{4.11.4}$$

where the over-bar indicates a quantity referring to the motion within the sphere and $\mathbf{n}$ is the normal to the interface. Also $\bar{p}$ and $\bar{u}_i$ are finite at $r = 0$.

The velocities $\mathbf{u}$ and $\bar{\mathbf{u}}$ satisfy the Navier–Stokes equation (with different values for the viscosity), but for a *small* spherical particle we may evidently use an approximate form of this equation as in the case of flow due to a small sphere in translational motion (§4.9). For the flow outside the sphere, the Navier–Stokes equation becomes, after substitution from (4.11.1),

$$\rho \left\{ \frac{\partial u_i'}{\partial t} + (u_j' + e_{jk} x_k) \frac{\partial u_i'}{\partial x_j} + e_{ij} u_j' \right\} = -\frac{\partial p'}{\partial x_i} + \mu \nabla^2 u_i'. \tag{4.11.5}$$

Now the variation of the undisturbed velocity over the region occupied by the particle is of magnitude $|e_{ij}| a$, and it is evident that the disturbance velocity $\mathbf{u}'$ also has this magnitude in the region not far from the particle. Hence, when the radius a satisfies the condition

$$|e_{ij}| a^2 \rho/\mu \ll 1 \tag{4.11.6}$$

(and provided the imposed rate of strain is not changing rapidly), the flow near the particle is governed by the approximate equation

$$\nabla p' = \mu \nabla^2 \mathbf{u}'. \tag{4.11.7}$$

Under the same conditions the velocity $\bar{\mathbf{u}}$ and pressure $\bar{p}$ inside the sphere also satisfy the equation of motion with neglect of inertia forces, that is,

$$\nabla \bar{p} = \bar{\mu} \nabla^2 \bar{\mathbf{u}}. \tag{4.11.8}$$

Finally, mass conservation gives the two equations

$$\nabla . \mathbf{u}' = \nabla . \mathbf{u} = 0, \quad \nabla . \bar{\mathbf{u}} = 0. \tag{4.11.9}$$

The equations (4.11.7), (4.11.8) and (4.11.9) and the boundary conditions (4.11.2), (4.11.3) and (4.11.4) governing the disturbance motion are linear and homogeneous in $\mathbf{u}'$, p', $\bar{\mathbf{u}}$, $\bar{p}$ and e_{ij}. No vector occurs in the description of the interface, and, with an argument like that used in §4.9, the pressures (which are harmonic functions) and velocities are seen to be of the form

$$\left.\begin{aligned} p' &= C\mu\, e_{ij}\, x_i x_j / r^5, \quad \bar{p} - \bar{p}_0 = \bar{C}\bar{\mu}\, e_{ij}\, x_i x_j, \\ u'_i &= e_{ij}\, x_j\, M + e_{jk}\, x_i x_j x_k\, Q, \\ \bar{u}_i &= e_{ij}\, x_j\, \bar{M} + e_{jk}\, x_i x_j x_k\, \bar{Q}, \end{aligned}\right\} \tag{4.11.10}$$

where M, Q, $\bar{M}$ and $\bar{Q}$ are functions of r alone, and C, $\bar{C}$ and $\bar{p}_0$ are constants. The forms of the functions that satisfy the governing equations and the conditions far from the particle and at $r = 0$ are readily found to be

$$\left.\begin{aligned} M &= \frac{D}{r^5}, \quad & Q &= \frac{C}{2r^5} - \frac{5D}{2r^7}, \\ \bar{M} &= \bar{D} + \tfrac{5}{21}\bar{C}r^2, \quad & \bar{Q} &= -\tfrac{2}{21}\bar{C}, \end{aligned}\right\} \tag{4.11.11}$$

and the conditions at the interface $r = a$ are then satisfied provided

$$\frac{C}{(2\mu + 5\bar{\mu})\, a^3} = \frac{D}{\bar{\mu}a^5} = -\frac{2\bar{C}a^2}{21\mu} = \frac{2\bar{D}}{3\mu} = -\frac{1}{\mu + \bar{\mu}}. \tag{4.11.12}$$

We observe in passing that at large distances from the particle

$$u'_i = \tfrac{1}{2}C\, e_{jk}\, \frac{x_i x_j x_k}{r^5} + O(r^{-4}), \tag{4.11.13}$$

showing that the disturbance velocity is one order smaller than in the case of flow due to a sphere in translational motion, as might have been expected from the 'dipole' nature of the condition on $\mathbf{u}'$ at the surface of the sphere (see (4.11.3) with (4.11.1)) in the present case.

The above solution has been obtained by neglecting the terms representing the inertia force in the equation of motion (4.11.5), and the self-consistency of the solution may be examined by using it to evaluate the order of magnitude of the neglected terms. We find in this way that the ratio of the magnitude of the neglected inertia force to the retained viscous force is of order $|e_{ij}| \, r^2\rho/\mu$. This ratio is small compared with unity in the neighbourhood of the sphere when the condition (4.11.6) is satisfied, as had been supposed, but the ratio is not small in the outer region of the flow field where r/a is of order $(|e_{ij}| \, a^2\rho/\mu)^{-\frac{1}{2}}$. Thus (4.11.7) is not a valid approximation to the complete equation of motion (4.11.5) in the outer field, although again we are reassured by the observation that all terms in the complete equation of motion are small (after being made non-dimensional with the parameters a and $|e_{ij}|$) in this same region. An improved and completely self-consistent approximation to the velocity distribution could presumably be obtained (for a steady imposed straining motion) from the equation

$$\rho\left(e_{jk}\,x_k\frac{\partial u_i'}{\partial x_j}+e_{ij}\,u_j'\right)=-\frac{\partial p'}{\partial x_i}+\mu\nabla^2 u_i',$$

which is still linear in $\mathbf{u}'$, but we shall take it for granted that this improved approximation would not differ significantly from the above solution in the neighbourhood of the particle.

The increased rate of dissipation in an incompressible suspension

We proceed to use the foregoing results in a calculation of the effective shear viscosity of a suspension of small incompressible spherical particles undergoing a prescribed bulk motion. The precise manner in which the effective viscosity should be specified and determined is not obvious, and must be described with care.

A volume V_1 of the suspension will be supposed to be bounded by a flexible surface A_1 on which the velocity is specified as a linear function of position; and in order to have a completely defined system we take this velocity to be *exactly* a linear function of $\mathbf{x}$. The rotational part of the motion at the boundary does not enter into the analysis so for convenience we choose the (i-component of the) velocity at the boundary as $e_{ij}x_j$, where e_{ij} is a symmetrical tensor with $e_{ii} = 0$. The suspension takes up a motion compatible with that of the boundary, and if the suspension were a homogeneous fluid the velocity would be $e_{ij}x_j$ everywhere in the volume V_1; owing to the presence of the particles the velocity of the ambient fluid has this value only in an average sense, and will be written as

$$e_{ij}\,x_j+u_i'.$$

Likewise the pressure in the suspension would have a certain value, P say, if the suspension were a homogeneous fluid of the same average density, whereas in reality the pressure in the ambient fluid has a more complicated dependence on position to be represented as

$$P+p'.$$

If the particles are far apart, each particle is embedded in a pure straining motion characterized by a rate-of-strain tensor e_{ij} and near one particle u_i'

and p' have the forms given in (4.11.10) *et seq.*; but there is no need for substitution yet.

The stress tensor at any point in the ambient fluid of viscosity μ is thus

$$\sigma_{ij} = -P\delta_{ij} + 2\mu e_{ij} + \sigma'_{ij},$$

where

$$\sigma'_{ij} = -p'\delta_{ij} + \mu\left(\frac{\partial u'_j}{\partial x_i} + \frac{\partial u'_i}{\partial x_j}\right). \qquad (4.11.14)$$

On the other hand, if the suspension were a homogeneous fluid of the same average density and of viscosity μ^*, the stress tensor would be

$$-P\delta_{ij} + 2\mu^* e_{ij}.$$

We wish to choose the value of the viscosity μ^* so that in a physically significant way it represents the net effect of the disturbance flows due to the presence of all the particles in the suspension. An appropriate quantity which when calculated in two different ways yields a value of the effective viscosity of the suspension is the rate of dissipation of mechanical energy in V_1; this rate of dissipation is a direct consequence of internal friction and it has the desired property of including the effect of the disturbance flows due to all the particles within the volume V_1.

Now the additional rate of dissipation per unit volume in the fluid at distance r from a single particle varies asymptotically as r^{-3} (for see (4.11.1) and (4.11.13)), and this frustrates direct evaluation of the total additional rate of dissipation by integration over the fluid in V_1; the integral is not actually divergent but is found to have a value dependent on the shape of the distant outer boundary of the region of integration for one particle. We must therefore proceed in a different way. The rate at which work is being done by forces at the boundary A_1 is

$$\int_{A_1} e_{ik} x_k \sigma_{ij} n_j \, dA, \quad = e_{ik}\int_{A_1} (-P\delta_{ij} + 2\mu e_{ij} + \sigma'_{ij}) x_k n_j \, dA;$$

and if the suspension were a homogeneous fluid of the same average density and of viscosity μ^* the rate of working at the boundary would be

$$e_{ik}\int_{A_1} (-P\delta_{ij} + 2\mu^* e_{ij}) x_k n_j \, dA.$$

The term involving P is common to the two expressions and accounts for any increase in kinetic energy associated with the linear velocity field. The remaining parts of the two expressions represent the rates of dissipation within V_1, and the effective viscosity μ^* will be defined to have a value such that they are equal; that is, after use of the divergence theorem,

$$2\mu^* e_{ij} e_{ij} V_1 = 2\mu e_{ij} e_{ij} V_1 + e_{ik}\int_{A_1} \sigma'_{ij} x_k n_j \, dA. \qquad (4.11.15)$$

The last term of (4.11.15), representing the additional rate of dissipation in V_1 due to the presence of particles, can be transformed to an integral over the surfaces of the particles. For we have

$$e_{ik} \int_{A_1} \sigma'_{ij} x_k n_j \, dA = e_{ik} \int_{V_1 - \Sigma V_0} \left(\frac{\partial \sigma'_{ij}}{\partial x_j} x_k + \sigma'_{ik} \right) dV + e_{ik} \Sigma \int_{A_0} \sigma'_{ij} x_k n_j \, dA,$$

where A_0 and V_0 are the surface and volume of one particle, with $\mathbf{n}$ as the outward normal to A_0, and Σ denotes a summation over all the particles in V_1. The disturbance motion due to the presence of each particle is governed by equation (4.11.7) when the condition (4.11.6) is satisfied, as we shall assume to be the case, so that $\partial \sigma'_{ij}/\partial x_j = 0$. Also

$$e_{ik} \int_{V_1 - \Sigma V_0} \sigma'_{ik} \, dV = e_{ik} \int_{V_1 - \Sigma V_0} 2\mu \frac{\partial u'_i}{\partial x_k} \, dV,$$

$$= -e_{ik} \Sigma \int_{A_0} 2\mu u'_i n_k \, dA$$

since $\mathbf{u'} = 0$ on the boundary A_1. Thus the relation (4.11.15) becomes

$$2(\mu^* - \mu) e_{ij} e_{ij} = \frac{e_{ik}}{V_1} \Sigma \int_{A_0} (\sigma'_{ij} x_k n_j - 2\mu u'_i n_k) \, dA, \qquad (4.11.16)$$

the right-hand side of which represents the average additional rate of dissipation per unit volume due to the presence of the particles. This expression for the effective viscosity of an isotropic suspension is valid for any spacing of the particles in V_1.

If now we assume the suspension to contain spherical particles at distances from each other large compared with their diameters, the disturbance flow near one particle is approximately independent of the existence of other particles and we may use the results obtained earlier to evaluate the integral in (4.11.16). It follows readily from (4.11.14), (4.11.10) and (4.11.11) that

$$\sigma'_{ij} x_j x_k - 2\mu u'_i x_k = \mu e_{ij} x_j x_k \left(\frac{C}{r^3} - \frac{10D}{r^5} \right) + \mu e_{jl} x_i x_j x_k x_l \left(-\frac{5C}{r^5} + \frac{25D}{r^7} \right).$$

$$(4.11.17)$$

The surface A_0 is a sphere of radius a, and with the use of the well-known identities

$$\int n_j n_k \, d\Omega = \tfrac{4}{3} \pi \, \delta_{jk}, \quad \int n_i n_j n_k n_l \, d\Omega = \tfrac{4}{15} \pi (\delta_{ij} \delta_{kl} + \delta_{ik} \delta_{jl} + \delta_{il} \delta_{jk}),$$

$$(4.11.18)$$

in which the integration is over the complete solid angle subtended at the centre of the sphere, we find

$$\int_{A_0} (\sigma'_{ij} x_k n_j - 2\mu u'_i n_k) \, dA = -\tfrac{4}{3} \pi \mu C \, e_{ik}.$$

Thus
$$\frac{\mu^*}{\mu} = 1 - \frac{2\pi}{3V_1} \Sigma C,$$

and, from (4.11.12),
$$= 1 + \frac{1}{V_1} \Sigma \left(\frac{\mu + \frac{5}{2}\bar{\mu}}{\mu + \bar{\mu}}\right) V_0. \qquad (4.11.19)$$

The summation in (4.11.19) is over the various particles, which have been assumed all to be spherical although not otherwise similar. If all the particles have the same internal viscosity,

$$\frac{\mu^*}{\mu} = 1 + \alpha \left(\frac{\mu + \frac{5}{2}\bar{\mu}}{\mu + \bar{\mu}}\right), \qquad (4.11.20)$$

where $\alpha = \Sigma V_0/V_1$ is the concentration of particles by volume. For a suspension of rigid particles the effective viscosity is greater than the viscosity of the ambient fluid by a fraction $\frac{5}{2}\alpha$ (a result first obtained by Einstein (1906, 1911)), while for a suspension of gas bubbles the corresponding fraction is α.

The formulae (4.11.19) and (4.11.20) are subject to the restriction $\alpha \ll 1$; when the concentration is not small compared with unity, the presence of neighbouring particles affects the disturbance flow due to one particle and the expression (4.11.17) then needs modification. The experimental evidence for the validity of the relation (4.11.20) appears not to be decisive, but the viscosity of a suspension of small rigid spheres is believed to be represented by the 'Einstein formula' $\mu(1 + \frac{5}{2}\alpha)$ for values of α less than about 0·02.†
The formula corresponding to (4.11.20) for rigid particles of ellipsoidal shape (with the assumption that Brownian motion is sufficiently strong to make all orientations of the ellipsoids equally probable) has been obtained by Jeffery (1922), and shows that the effective viscosity increases with the departure from sphericity, because velocity gradients in the ambient fluid are larger in magnitude for 'sharper' particles; however, the variation of the coefficient of α is found to be small until the ratio of the maximum to the minimum diameter of the particle is about 3.

The effective expansion viscosity of a liquid containing gas bubbles

As an interesting supplement to the preceding analysis, we consider now the nature of the response of a suspension to a prescribed bulk motion when the particles, but not the ambient fluid, are allowed compressibility.

Let us suppose that a volume V of liquid containing a large number of small gas bubbles in suspension is subjected to a pure straining motion of the boundaries specified as before by the rate-of-strain tensor e_{ij} although now $e_{ii} = \Delta \neq 0$. If the suspension were a homogeneous fluid, the rate of expansion would be Δ everywhere in V, but in reality the expansion occurs wholly in the gas bubbles. Near any one (spherical) bubble of radius a the

---

† Information concerning the viscosity of more concentrated suspensions is given in chapter 9 of *Low Reynolds Number Hydrodynamics*, by J. Happel and H. Brenner (Prentice-Hall, 1965).

relative motion in the liquid due to expansion of that bubble is a spherically symmetrical expansion with radial velocity

$$u(r) = \frac{a^2}{r^2}\frac{da}{dt}.$$ (4.11.21)

Associated with this motion there are viscous stresses in the liquid, and in particular there is a normal stress at the bubble surface equal to $2\mu(\partial u/\partial r)_{r=a}$ which opposes the expansive motion. As a consequence, the pressure inside the bubble differs from the pressure in the liquid at some distance from the bubble by an amount which depends on the rate of expansion. The mechanical pressure applied at the boundary of the suspension therefore differs from the pressure which would be obtained from the equilibrium equation of state for the suspension, whatever its form may be, at the given instantaneous value of the density of the suspension. This is exactly the kind of response which is described by the expansion viscosity of the suspension (§ 3.4).

An estimation of the effective value of the expansion viscosity may be made by the same procedure of equating the total dissipation as it would be for a homogeneous fluid and the total dissipation due to the ordinary shear viscosity at work in the liquid surrounding the bubbles. The calculation is much simpler in this case because the dissipation in the liquid is more strongly concentrated in the neighbourhood of the liquid and can be obtained by direct integration.

At a point distance r from the centre of one bubble, where the radial velocity of the liquid is $u(r)$, one principal rate of strain is du/dr, and each of the other two equal principal rates of strain must be $-\frac{1}{2}du/dr$. The local rate of dissipation per unit volume of the liquid is therefore

$$3\mu\left(\frac{du}{dr}\right)^2, \quad = \frac{12\mu a^4\dot{a}^2}{r^6},$$ (4.11.22)

where $\dot{a}$ stands for da/dt. The total rate of dissipation in the liquid due to one bubble is

$$\int_a^\infty \frac{12\mu a^4\dot{a}^2}{r^6} 4\pi r^2 \, dr = 16\pi\mu a\dot{a}^2,$$

and the rate of dissipation in volume V of the suspension containing a number of spherical bubbles is $16\pi\mu\Sigma a\dot{a}^2$.

On the other hand, if we imagine the suspension to be a homogeneous fluid undergoing a symmetrical expansion with uniform rate of expansion Δ, there would be no dissipation due to shear viscosity and, according to (3.4.9), the rate of dissipation per unit volume due to the existence of an expansion viscosity κ^* would be $\kappa^*\Delta^2$. The effective expansion viscosity of the suspension is thus given by

$$\kappa^* = 16\pi\mu\frac{\Sigma a\dot{a}^2}{V\Delta^2}.$$ (4.11.23)

But the expansion of the suspension is due to the change in volume of each bubble, and

$$\Delta = \frac{1}{V}\Sigma \frac{d}{dt}(\tfrac{4}{3}\pi a^3) = \frac{4\pi}{V}\Sigma a^2 \dot{a}, \tag{4.11.24}$$

the summation again being over the bubbles in the volume V. Hence we have

$$\kappa^* = \frac{\mu}{\pi}\frac{V\Sigma a\dot{a}^2}{(\Sigma a^2 \dot{a})^2}. \tag{4.11.25}$$

In the simple case of n gas bubbles of equal size and similar constitution (so that they expand at the same rate) per unit volume of the suspension

$$\kappa^* = \frac{\mu}{\pi n a^3} = \frac{4\mu}{3\alpha}, \tag{4.11.26}$$

where α is again the (instantaneous) concentration of particles by volume. The unreal feature of the model that $\kappa^* \to \infty$ as $\alpha \to 0$ is modified if allowance is made for the compressibility of the liquid.

As well as giving results of some practical value, the analysis in this section shows the kind of connection that exists between the microscopic properties of the medium, meaning here the existence of particles of known shape and constitution, and the macroscopic or bulk properties of the equivalent homogeneous fluid. Observed macroscopic properties of a medium often become more intelligible if one can imagine a microscopic structure which would give rise to a given kind of macroscopic behaviour, and the devising of microscopic models is a common exercise in the science of rheology, especially in considerations of the properties of non-Newtonian fluids.

4.12. Changes in the flow due to moving bodies as R increases from 1 to about 100

We have seen (§4.7) that when a rigid body of given shape is in steady translational motion with speed U through an infinite body of fluid which is otherwise undisturbed, the non-dimensional quantities describing the flow field depend only on the Reynolds number $R = \rho L U/\mu$ (where L is a length representing the linear dimensions of the body) and not on ρ, L, U, μ separately. The value of R thus characterizes the flow field, once the shape of the body and its attitude relative to the direction of motion have been specified. One of the most important problems of fluid mechanics is to determine the properties of the flow due to moving bodies of simple shape, over the entire range of values of R, and more especially for the large values of R corresponding to bodies of ordinary size moving through air or water, both of which are fluids of small kinematic viscosity. An account of the flow at very small Reynolds numbers has been given for a sphere and a circular cylinder (§§4.9, 4.10), and we shall now describe briefly the changes that take place in these flow fields as R is increased. The analysis of those two sections was based on the approximation that when $R \ll 1$ inertia forces are negligible compared with viscous forces over most of the flow field. Bearing in mind the interpretation of R as a measure of the ratio of these two forces, we may expect that when $R \gg 1$ the converse will hold and that viscous forces may in some sense be negligible; this approximation is the subject of the next chapter. In the intermediate range in which neither of these approximations is valid, a range which it appears may be specified very roughly as $1 < R < 100$, analytical work meets with great difficulties and our knowledge of the flow has been obtained mostly by observation and partly by numerical integration of the complete equation of motion.

At Reynolds numbers small compared with unity, the dominant process in the flow is the diffusion of vorticity away from the body. The fluid in immediate contact with the body is dragged along at the same speed as the body and this leads to the generation of vorticity at the body. The Stokes approximation of §4.9 is that the effect of motion of the body is confined to this generation of vorticity, and that vorticity diffuses in all directions away from an effectively stationary source. This leads to a flow with fore-and-aft symmetry in the flow far from the body (and also near the body if its surface has that same symmetry). The Oseen improved approximation of §4.10 takes account of inertia forces partially, and the vorticity here diffuses from a source which is moving steadily. Vorticity diffuses over a distance l from the source in a time of order l^2/ν (ignoring, for the purposes of this qualitative argument, any differences between scalar and vector quantities subject to molecular diffusion), and during this time the body moves forward a distance $l(lU/\nu)$. For diffusion distances l of the order of the body dimension L, and for $LU/\nu \ll 1$, the diffusion is dominant and the vorticity distribution has approximate fore-and-aft symmetry near the body to the extent allowed by the body shape. On the other hand, for diffusion distances $l \gg L$, the motion of the body has the effect of leaving the vorticity behind in a region which becomes more nearly parabolic (with the body at the focus) with increasing distance downstream from the body. Thus, for $R \ll 1$, when the Stokes and Oseen approximations are applicable, the flow has fore-and-aft symmetry near a body with that same symmetry, but distinct asymmetry further out.

The argument suggests that at larger values of R the asymmetry is more pronounced and affects the flow near the body (which is not a situation which could be described by the Oseen approximation). Observation confirms this, and shows the interesting form taken by the flow, in particular at the rear of the body where the vorticity is concentrated. The sequence of flow fields corresponding to different values of R will be illustrated here first by the case of a moving circular cylinder, which lends itself to observation of streamlines better than does a sphere. It is easier to appreciate the character of the flow, in these circumstances in which attention is focused on the neighbourhood of the body, if the motion relative to the body is described; the flow field may then be termed 'flow past a body', rather than 'flow due to a moving body'. The fluid at infinity now has a uniform speed U, and the flow will be supposed in this section to be from left to right.

Figure 4.12.1 (plate 1) shows the short paths executed, while a photographic plate is exposed, by small solid particles in fluid moving past a circular cylinder of diameter $2a$ at different values of the Reynolds number $R(= 2aU/\nu)$. The flow relative to the cylinder is steady, so that these particle paths are portions of streamlines, and it is possible to see the shape of whole connected streamlines, except possibly in the region where the fluid velocity is low and the distances moved by the particles are small. At $R = 0.25$, the fore-and-aft asymmetry is hardly discernible, but it is quite evident at $R = 3.64$. At $R = 9.10$ there is a region of slowly circulating fluid immediately behind the cylinder, and as R is increased further it seems that this region becomes longer and within it a definite pattern of motion intensifies. The details of the flow just behind the cylinder are not entirely clear from these photographs, but numerical integration of the complete equation of motion reveals that the streamlines there are closed and are in two groups arranged symmetrically, each group comprising a 'standing eddy' with the sense of circulation such as to harmonize with the streamlines flowing past the cylinder.

Several numerical calculations of the flow past a circular cylinder have also been made: for $R = 10$ and 20 by Thom (1933), for $R = 40$ by Kawaguti (1953), for $R = 40$ and 44 by Apelt (1961), and for $R = 2, 4, 10$ and 15 by Keller and Takami (1966). All these authors used finite difference approximations to the complete governing equations, which, in a case of two-dimensional flow like this, can be reduced to a single differential equation with the stream function ψ as the dependent variable. The calculation becomes very difficult as R increases, but the agreement between these calculated flow patterns and those observed is good, as may be seen by comparing the calculated streamlines for $R = 4$ and $R = 40$ shown in figures 4.12.2 and 4.12.3 with those observed at $R = 3.64$ and $R = 39.0$ (figure 4.12.1, plate 1). In figure 4.12.4 the calculated pressure distribution at the cylinder surface for $R = 40$ is compared with the distributions observed at $R = 36$ and $R = 45$ by Thom (1933). It will be noticed that the minimum pressure occurs at the side of the cylinder, as would be expected from Bernoulli's theorem if the fluid were inviscid (since it is evident from figure 4.12.3 that the speed is a maximum there), whereas when $R \ll 1$ it occurs at the rear stagnation point (see (4.10.9)). Figures 4.12.2 and 4.12.3 also give the calculated lines on which the vorticity ω is constant, for the cases $R = 4$ and $R = 40$. It is clear that as R increases convection becomes more effective in sweeping the vorticity downstream as it diffuses away from the body.

There appears to be a definite value of R at which closed streamlines

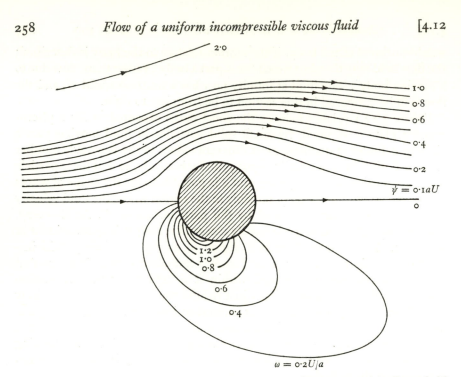

Figure 4.12.2. Streamlines (upper half of figure) and lines of constant vorticity (lower half) in flow past a circular cylinder at $R = 4$, calculated by Keller and Takami (1966).

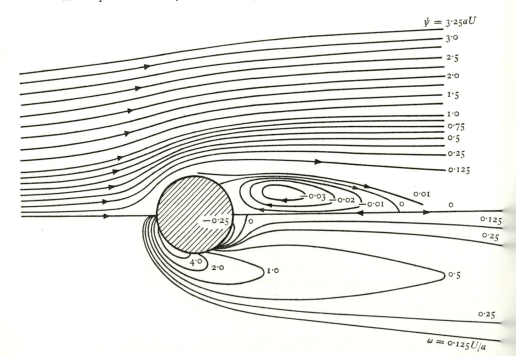

Figure 4.12.3. Streamlines (upper half of figure) and lines of constant vorticity (lower half) in flow past a circular cylinder at $R = 40$, calculated by Apelt (1961).

appear at the rear of the cylinder.† Taneda (1956 *a*) measured the length of the standing eddies from a number of photographs of streamlines like those shown in figure 4.12.1 (plate 1), and his results (see figure 4.12.5) suggest that they first appear near $R = 6$. It is not easy to give a straight-forward explanation of the formation of the standing eddies, despite the fact that they are present in the flow past most bodies, whether two-dimensional or

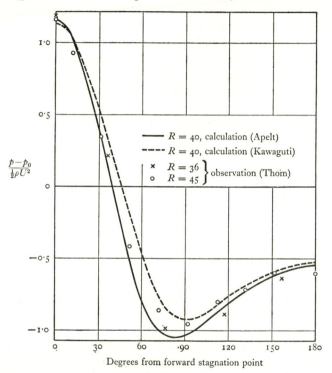

Figure 4.12.4. Pressure distribution at the surface of a circular cylinder.
p_0 = pressure at infinity.

three-dimensional (exceptions being slender bodies whose lateral dimensions are small compared with their stream-wise dimensions), and at all Reynolds numbers above some value of order 10 which depends on the body shape. Putting the matter crudely, we may say that, as R increases and the convection of vorticity becomes more effective than diffusion of vorticity,

† It is generally believed that this critical value of the Reynolds number depends on the curvature of the body at the sides. In the case of an elliptic cylinder with its major axis normal to the stream, the critical value is found to decrease as the minor axis of the ellipse is decreased in length, and becomes close to zero when the ellipse degenerates to a flat plate broadside-on to the stream. Some calculations made by Dean (1944) are interesting in this connection. Dean showed that when a (two-dimensional) solid protruberance of simple shape is added to a plane wall over which there is a simple shearing motion of fluid, back-flow occurs in the lee of the protruberance for Reynolds numbers above a certain value, and that as the tip of the protruberance is made sharper the critical value decreases, becoming zero in the limit of a pointed tip.

more and more vorticity is carried round to the rear of the cylinder, the vorticity being of negative sign (or clockwise rotation) near the upper surface and of positive sign near the lower surface. Ultimately there is more vorticity of each sign at the rear of the cylinder than is needed for the satisfaction of the no-slip condition there, and a backward flow is induced near the surface. The backward flow counters the forward-moving fluid and deflects it away from the rear of the cylinder, and this tends in turn to strengthen the rotational motion in the standing eddy.

At a value of R between 30 and 40, the steady flow appears to become unstable to small disturbances; this is a phenomenon which, as already noted, affects almost all steady flows when the Reynolds number is large enough (the dissipative or damping action of viscosity then being relatively

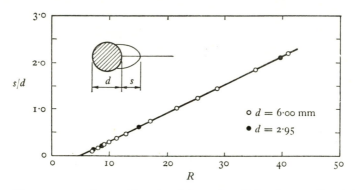

Figure 4.12.5. Observed lengths of the region of closed streamlines behind a circular cylinder (from Taneda 1956a).

weak). In the present case, the instability first affects the wake, at some distance downstream from the cylinder, and gives rise to a slow oscillation of the wake, approximately sinusoidal in both time and stream-wise distance, with an amplitude which increases with distance downstream. Figure 4.12.6 (plate 2) shows these oscillations clearly in the streak lines traced out by colouring matter released at the cylinder and carried downstream in the wake. As R increases, beyond the critical value at which instability first appears, the oscillation of the wake moves closer to the cylinder and, when R is near 60, begins to affect the two standing eddies immediately behind the cylinder. The two standing eddies oscillate together in lateral position, and appear to shed some rotating fluid at the end of every half-period, on each side of the cylinder alternately. The behaviour of the wake at this stage is very striking, as the last three photographs show. Most of the fluid passing close to the cylinder appears to gather itself into discrete lumps, arranged in two regular staggered rows on either side of the stream-wise line through the cylinder axis, and other observations have shown beyond any doubt that most of the vorticity in the wake is concentrated in these lumps, all the lumps in each row having vorticity of the same sign. This regular array of discrete fluid elements with vorticity (loosely called 'vortices', the whole array being a 'vortex street') moves downstream with a velocity less than U and persists much further downstream than figure 4.12.6 reveals, although with a slow increase of the distances between the two rows and between neighbouring vortices. The two eddies immediately behind the cylinder are not clearly recognizable for values of R much above 100, although a vortex street continues to form in the wake up to much larger Reynolds numbers.

Accompanying all these changes in the flow pattern as R is increased are changes in the total drag force D on the cylinder. Figure 4.12.7 shows the

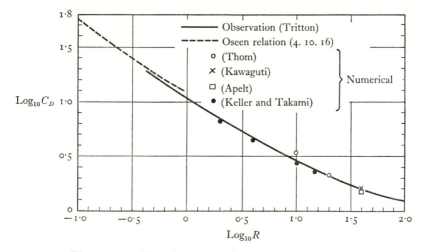

Figure 4.12.7. Drag force exerted on a circular cylinder of radius a;
$$C_D = D/\tfrac{1}{2}\rho U^2 2a, \ R = 2aU/\nu.$$

extensive set of measurements of the drag as a function of R made by Tritton (1959), together with the theoretical relation (4.10.16) obtained from the Oseen equation and the available computed values. It appears from curves showing the separate contributions to C_D from the normal pressure force (obtained from measurements of the time-mean of p at the cylinder surface (Thom 1929), or calculated) and the tangential viscous, or 'skin-friction', force (obtained as the difference between the measured total and resultant pressure forces) that the former contribution is an increasing function of R when R lies in the approximate range 50 to 110. This anomalous behaviour is presumably associated with the growing oscillation of the two standing eddies behind the cylinder. The oscillation of the standing eddies is known to be accompanied also by a considerable side-force (of zero time-mean), normal to the flow direction and the cylinder axis.

A similar sequence of changes as R is increased from values near unity occurs in the flow past most other bodies. Figure 4.12.8 (plate 3) shows a number of photographs of the streamlines for flow past a sphere, the stream-lines lying in a plane through the centre of the sphere and containing the flow direction (again $R = 2aU/\nu$, $2a$ being the sphere diameter). The region of closed streamlines behind the sphere, which forms, according to figure 4.12.9, at about $R = 24$, here contains a standing ring-eddy, the direction of circulation of the eddy again being such as to give flow in the same direction as the external flow at their common boundary. Again there is instability of the flow pattern above a critical value of the Reynolds number, possibly originating in the wake, and Taneda (1956 *b*) found that the ring-eddy first began to oscillate gently at about $R = 130$. At higher Reynolds numbers the ring-eddy oscillates with a larger amplitude and some of the fluid in the

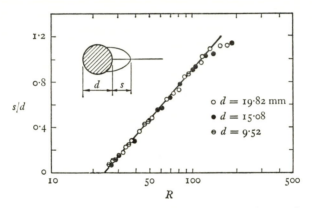

Figure 4.12.9. Observed lengths of the region of closed streamlines behind a sphere (from Taneda 1956 *b*).

region of closed streamlines breaks away and is carried downstream. No regular pattern of motion like the vortex street appears to form in the wake of a sphere (or of any three-dimensional body), although there is a general impression that vorticity is shed from the standing ring-eddy in something like a succession of distorted vortex loops not symmetrical about the central axis. The way in which the drag on a sphere varies with Reynolds number was shown earlier in figure 4.9.2.

Further evidence of the occurrence of flow patterns of the above kind for almost all shapes of bodies is provided by the photographs in figure 4.12.10 (plate 4) of streamlines in steady flow past a thin rectangular plate of length l and very great width (normal to the photograph) held broadside-on to the stream. Then, as an example of the exceptional class of slender bodies aligned with the stream, at the rear of which standing eddies do not form, the same figure shows the streamlines, at $lU/\nu = 3$, for a thin plate of length l held edge-on to the stream. There is very little change in the form of the stream-lines of this latter flow as R is increased further by as much as a factor of about 10^5. The change is confined to a continual decrease in the thickness of the retarded layer of fluid which may be seen close to the plate in figure 4.12.10, and the plate causes less and less visible disturbance to the uniform stream of fluid. We may note in passing that a thin plate edge-on to the stream is a body whose longitudinal and lateral dimensions are quite different in magnitude, and that the corresponding Reynolds numbers formed with these lengths may suggest different properties of the flow pattern; care must then be taken to choose the Reynolds number of greatest significance. So far as fore-and-aft symmetry of the flow is concerned, the Reynolds number based on length of the plate is more significant, because the plate length is a length representative of the dimensions of the 'flow as a whole'. So far as certain local features of the flow are concerned, such as the distribution of pressure near the leading edge of the plate, a Reynolds number based on the plate thickness may have more significance.

Exercises for chapter 4

1. Show that in the case of a fluid enclosed entirely by stationary rigid boundaries the rate of dissipation due to viscosity is

$$\mu \int \omega^2 \, dV,$$

where the integral is taken over the whole volume of the fluid.

2. A plane rigid surface wetted with a thin layer of liquid of uniform thickness h_0 is held vertically and the liquid drains off. Show that the layer thickness h at distance x from the upper edge of the plate satisfies the approximate equation

$$\frac{\partial h}{\partial t} + V \frac{h^2}{h_0^2} \frac{\partial h}{\partial x} = 0,$$

where $V = \rho g h_0^2 / \mu$, and that at time t after the draining begins

$$h = h_0(x/Vt)^{\frac{1}{2}} \quad \text{for} \quad x \leqslant Vt, \qquad h = h_0 \quad \text{for} \quad x \geqslant Vt.$$

3. A long circular tube has a cylindrical layer of liquid of uniform thickness adhering to its inner surface. In order to remove the liquid air is blown through the tube by the application of a difference between the pressures in the air at the two ends. Determine the ratio of the steady fluxes of volume of the air and the liquid from the end of the tube.

4. A thin layer of viscous fluid lies between two parallel rigid planes, one of which is stationary and the other of which is in oscillatory translational motion with frequency n in its own plane. Determine the ratio of the magnitudes of the (oscillatory) frictional forces on the two planes, and examine the cases of large and small values of n.

5. A slender rigid axisymmetric body with axial length l and maximum diameter $d(\ll l)$ is in translational motion with speed U through fluid, and $\rho l U / \mu \ll 1$. Show that the drag exerted on the body is approximately twice as large when it is moving normal to its length as when it is moving parallel to its length.

6. What change occurs in the spacing of the vortices in a 'vortex street' behind a body in a stream of fluid when the body size and the stream speed but not their product (nor the body shape) are varied?

5

FLOW AT LARGE REYNOLDS NUMBER: EFFECTS OF VISCOSITY

5.1. Introduction

In this chapter the discussion of the flow of a viscous incompressible fluid of uniform density will be continued.

The values of the kinematic viscosity for air and water are so small that the Reynolds numbers for most of the flow systems of importance, whether in nature or in technology or in the laboratory, are very much larger than unity. A Reynolds number of 10^3 is attained in air at 20 °C when UL has the very modest value 150 cm²/sec, and in water when UL is only 10 cm²/sec, where U and L are representative values for the velocity variations and the distances over which they occur in the flow system concerned. Such small values of the product UL are so readily and so often exceeded that flow at large Reynolds number must be regarded as the standard case.

The largeness of $R = UL/\nu$ has its implications for the relative importance of the various terms in the equation of motion, as was seen in § 4.7. Provided the non-dimensional quantities $|D\mathbf{u}'/Dt'|$ and $|\nabla^2\mathbf{u}'|$ are both of order unity over most of the flow field (which would of course exclude some simple flow fields, such as steady unidirectional flow in a tube, in which the fluid acceleration is zero everywhere), R is a measure of the ratio of the magnitudes of inertia and viscous forces acting on the fluid; and a flow field for which $R \gg 1$ is presumably one in which inertia forces are much greater than viscous forces over most of the field. It is natural then to proceed with the investigation of flow at large Reynolds number on the assumption that viscous forces may be discarded from the equation of motion altogether. This was in fact the accepted plan for a long time in the history of fluid mechanics, and the theory of flow of an inviscid fluid is highly developed as a consequence.

However, the assumption that flow at large Reynolds number is approximately the same as flow of an inviscid fluid did not prove to be very successful in accounting for what was observed. In particular, the theory of inviscid fluid flow was completely unable to explain the existence of backward-moving fluid at the rear of stationary bodies in a stream, which is such a prominent feature of flow at all except quite small Reynolds numbers and for all kinds of bodies except very slender ones aligned with the stream. We know now that a hypothetical fluid of zero viscosity would behave quite differently from a fluid of small but non-zero viscosity, and that flow of a real fluid at very large Reynolds number cannot be regarded as a slightly perturbed form of flow of a fluid without viscosity, except in rather special circumstances. Usually the reason for this difference in the character of the two flows lies in the different behaviour of the real and hypothetical fluids near a solid boundary. A real fluid satisfies the no-slip condition at a solid boundary, however small its viscosity may be, whereas an inviscid fluid does not. We cannot get round this physical difference by retaining the no-slip condition in a mathematical analysis of inviscid fluid flow, because dropping the viscous force from the equation of motion (4.1.6) reduces the order of this differential equation by one, and one of the boundary conditions consequently becomes redundant.

We shall see in this chapter how maintenance of the no-slip condition at a rigid boundary in a flow at very large Reynolds number can have an important effect on the flow as a whole, thereby rendering inappropriate the assumption that the fluid is inviscid, except under certain circumstances. We shall be particularly interested in establishing the conditions under which a real fluid *does* behave approximately as if it were inviscid, because it happens that some of the properties of inviscid flow, notably zero drag on bodies in a stream, would offer important practical advantages in aeronautics and other fields concerned with propulsion through fluids. Moreover, there exists an extensive set of mathematical results about flow of an inviscid fluid which we should like to be allowed to make use of in investigations of flow of real fluids.

In this chapter we shall have occasion to use Bernoulli's theorem (§ 3.5) for steady entropy-preserving inviscid flow, and the formula and its derivation (in the present context of uniform incompressible fluid) will be recalled here for convenience. In view of the vector identity

$$\mathbf{u} \times (\nabla \times \mathbf{u}) = \tfrac{1}{2}\nabla(\mathbf{u}.\mathbf{u}) - \mathbf{u}.\nabla\mathbf{u},$$

the equation of motion for a fluid of uniform density and viscosity (see (4.1.6)) can be written as

$$\frac{\partial \mathbf{u}}{\partial t} - \mathbf{u} \times \boldsymbol{\omega} = \mathbf{F} - \nabla\left(\frac{p}{\rho} + \tfrac{1}{2}q^2\right) + \nu\nabla^2\mathbf{u}, \tag{5.1.1}$$

where $\boldsymbol{\omega} = \nabla \times \mathbf{u}$ is the vorticity and $q^2 = \mathbf{u}.\mathbf{u}$. Then if $\mathbf{F}$ is of the form $-\nabla\Psi$, as is indeed the case when $\mathbf{F}$ represents the force of gravity, and if the motion is steady, (5.1.1) becomes

$$\nabla H = \mathbf{u} \times \boldsymbol{\omega} + \nu\nabla^2\mathbf{u}, \tag{5.1.2}$$

where
$$H = \tfrac{1}{2}q^2 + \Psi + p/\rho.$$

Bernoulli's theorem follows from (5.1.2); in the absence of viscous forces, H is constant along any streamline (and also along any vortex-line). If we introduce the modified pressure $P = p + \rho\Psi$ described in § 4.1, the variation of which is due entirely to the motion of the fluid, and represent it by the symbol p for convenience, the theorem states that the quantity

$$H = \tfrac{1}{2}q^2 + p/\rho \qquad\qquad (5.1.3)$$

is constant along any streamline.

When viscous forces are not negligible, $(5.1.2)$ gives

$$\mathbf{u}.\nabla H = \nu \mathbf{u}.(\nabla^2 \mathbf{u}). \qquad\qquad (5.1.4)$$

H decreases in the flow direction when $\mathbf{u}.(\nabla^2\mathbf{u}) < \mathrm{o}$, that is, when the local net viscous force per unit volume tends to decelerate the fluid and work is done against viscous forces as an element of fluid moves along a stream-tube; likewise H increases along the streamline when the net viscous force tends to accelerate the fluid. This variation of H is in keeping with the interpretation of Bernoulli's theorem (§ 3.5) as a statement of the energy balance for material elements moving along a stream-tube. In the particular case of steady flow past a solid body placed in a stream which is uniform at infinity, H has the same value on all streamlines far upstream, and, since the fluid elements passing close to the body usually experience a decelerating net viscous force, the value of H on such streamlines will fall in the flow direction. Thus, far downstream, where the flow is again unidirectional and the pressure is again uniform, H (and consequently q also) will be uniform except on those streamlines which have passed close to the body and which have been subjected to viscous action; on these latter streamlines, making up the wake of the body, both H and q will in general have smaller values.

5.2. Vorticity dynamics

It proves to be useful and instructive for many different purposes to describe flow at large Reynolds number primarily in terms of the distribution of vorticity. The essential reason for this is that, as we shall see, vorticity cannot be created or destroyed in the interior of a homogeneous fluid under normal conditions, and is produced only at boundaries. Rules about the way in which the vorticity associated with a material element of fluid changes as the element moves are available; and it is often possible to form a qualitative view of the distribution of vorticity throughout the fluid from inspection of the boundary conditions. Moreover, we know, from § 3.3, that the net viscous force on an element of (incompressible) fluid is determined by the local gradients of vorticity. When the fluid viscosity is small, the net viscous force is significant only at places where the vorticity gradients are large; and if for some reason the vorticity is zero over a region of the flow, viscous stresses make no contribution to the net force on elements of fluid and may be ignored for most purposes. As a preliminary to this chapter, then, we shall examine the direct consequences of the equation of motion for the vorticity.

The 'vorticity equation' is obtained by taking the curl of both sides of the equation of motion in the form (5.1.1):

$$\frac{\partial \boldsymbol{\omega}}{\partial t} = \nabla \times (\mathbf{u} \times \boldsymbol{\omega}) + \nu \nabla^2 \boldsymbol{\omega}$$

$$= -\mathbf{u}.\nabla\boldsymbol{\omega} + \boldsymbol{\omega}.\nabla\mathbf{u} + \nu\nabla^2\boldsymbol{\omega}, \tag{5.2.1}$$

in which use has been made of the auxiliary relations

$$\nabla.\mathbf{u} = 0, \quad \nabla.\boldsymbol{\omega} = 0,$$

and $\mathbf{F}$ has been assumed to be of the form $-\nabla\Psi$. Equation (5.2.1) can also be written as

$$\frac{D\boldsymbol{\omega}}{Dt} = \boldsymbol{\omega}.\nabla\mathbf{u} + \nu\nabla^2\boldsymbol{\omega}. \tag{5.2.2}$$

One of the advantages of a description of flow changes in terms of vorticity lies in the absence of the pressure from (5.2.1) or (5.2.2). The angular momentum of a material element of fluid which is instantaneously spherical changes at a rate which is determined by tangential viscous stresses alone; this angular momentum is $\frac{1}{2}\boldsymbol{\omega}I$ (see (2.3.12)), where I is the moment of inertia of the element about any axis, and (5.2.2) is in fact equivalent to a statement that the rate of change of $\frac{1}{2}\boldsymbol{\omega}I$ for the material element is equal to the couple exerted by tangential stresses, the first term on the right side expressing the rate of change of I due to change of shape of the element.

We proceed now to interpret the contributions to the rate of change of $\boldsymbol{\omega}$ at a given point in space represented by the three terms on the right-hand side of (5.2.1). The first of these terms is the familiar rate of change due to convection of fluid in which the vorticity is non-uniform past the given point. The term $\nu\nabla^2\boldsymbol{\omega}$ likewise calls for little new comment, since it represents the rate of change of $\boldsymbol{\omega}$ due to molecular diffusion of vorticity in exactly the way that $\nu\nabla^2\mathbf{u}$ represents the contribution to the acceleration from the diffusion of velocity (or momentum). Vorticity, or angular velocity of the fluid, does not seem at first sight to be a transportable quantity, capable of being conveyed from one part of the fluid to another by molecular migration, but inasmuch as all components of $\mathbf{u}$, at all points in the fluid, are transportable quantities, so too are spatial derivatives of $\mathbf{u}$ and so too, in effect, is vorticity.

The term $\boldsymbol{\omega}.\nabla\mathbf{u}$ in (5.2.1) and (5.2.2) is the one that has no counterpart in the equation of momentum and that gives vorticity changes a distinctive character. Its meaning becomes evident if we write it in the form

$$\boldsymbol{\omega}.\nabla\mathbf{u} = |\boldsymbol{\omega}| \lim_{PQ\to 0} \frac{\delta\mathbf{u}}{PQ}, \tag{5.2.3}$$

where P and Q are two neighbouring points on the local vortex-line (figure 5.2.1) and $\delta\mathbf{u}$ is the velocity of the fluid at Q relative to that at P. The corresponding contribution to the fractional rate of change of the vorticity, i.e. to either

$$\frac{1}{|\boldsymbol{\omega}|}\frac{\partial \boldsymbol{\omega}}{\partial t} \quad \text{or} \quad \frac{1}{|\boldsymbol{\omega}|}\frac{D\boldsymbol{\omega}}{Dt},$$

is thus identical with the fractional rate of change of the material line element vector extending from P to Q, P and Q now being regarded as material points (cf. (3.1.3)). It is as if $\boldsymbol{\omega}$ behaved like a material line element coinciding instantaneously with a portion of the vortex-line, part of the change in $\boldsymbol{\omega}$ coming from rigid rotation of the line element (due to the component of $\delta\mathbf{u}$ normal to $\boldsymbol{\omega}$) and part being a change in $\boldsymbol{\omega}$ coming from extension or contraction of the line element (due to the component of $\delta\mathbf{u}$ parallel to $\boldsymbol{\omega}$).

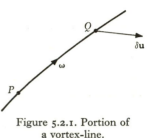

Figure 5.2.1. Portion of a vortex-line.

It is worth noticing that, in the case of two-dimensional motion, $\boldsymbol{\omega}$ is everywhere normal to the plane of flow and $\boldsymbol{\omega}.\nabla\mathbf{u} = 0$; (5.2.2) then reduces to the scalar equation

$$D\omega/Dt = \nu\nabla^2\omega, \tag{5.2.4}$$

which is of the same form as the equation satisfied by the density of some conserved substance which is convected with the fluid and diffused within it (see (3.1.17)). Another case in which $\boldsymbol{\omega}.\nabla\mathbf{u} = 0$ is unidirectional flow, since, for a velocity $(u, 0, 0)$ with u dependent on rectilinear co-ordinates y and z in the lateral plane, the vectors

$$\boldsymbol{\omega} = (0, \partial u/\partial z, -\partial u/\partial y) \quad \text{and} \quad \nabla u = (0, \partial u/\partial y, \partial u/\partial z)$$

are orthogonal.

The fact that $\boldsymbol{\omega}$ changes, so far as the first term on the right-hand side of (5.2.2) is concerned, like the vector representing a material line element which coincides instantaneously with a portion of the local vortex-line may also be interpreted in terms of the behaviour of the flux of vorticity across a material surface element. For a material surface element $\delta\mathbf{S}(t)$ which is located instantaneously at position $\mathbf{x}$ this flux of vorticity is $\boldsymbol{\omega}.\delta\mathbf{S}$, and is of course equal to the circulation $\delta C(t)$ round the closed curve bounding the surface element. The rate of change of the flux of vorticity across the material element $\delta\mathbf{S}$ is

$$\frac{d\,\delta C}{dt} = \frac{D\boldsymbol{\omega}}{Dt}.\delta\mathbf{S} + \boldsymbol{\omega}.\frac{d\,\delta\mathbf{S}}{dt}$$

which we see from (3.1.6) (with ρ constant) and (5.2.2) to reduce to

$$= \nu\,\delta\mathbf{S}.(\nabla^2\boldsymbol{\omega}) + o(\delta S)$$

$$= -\nu\,\delta\mathbf{S}.\{\nabla\times(\nabla\times\boldsymbol{\omega})\} + o(\delta S). \tag{5.2.5}$$

It seems therefore that the flux of vorticity across a material surface element changes only as a consequence of molecular diffusion; changes in the magnitude and direction of $\delta\mathbf{S}$ have an effect on the flux which is exactly balanced by the changes in $\boldsymbol{\omega}$ due to the first term on the right-hand side of (5.2.2).

We can integrate (5.2.5) over an arbitrary open material surface and thereby find the rate of change of circulation round the bounding curve. However, it proves to be more useful to begin afresh. The circulation round a closed material curve drawn in the fluid is

$$C(t) = \oint \mathbf{u}.d\mathbf{l},$$

and the typical element of integration can be regarded as a material line element $\delta\mathbf{l}$ which changes at a rate $\delta\mathbf{l}.\nabla\mathbf{u}$, in accordance with the procedure described in §3.1. Then

$$\frac{dC}{dt} = \oint\left(\frac{D\mathbf{u}}{Dt}\right).d\mathbf{l} + \oint \mathbf{u}.(d\mathbf{l}.\nabla\mathbf{u}),$$

$$= \oint \mathbf{F}.d\mathbf{l} + \oint d\mathbf{l}.\nabla\left(-\frac{p}{\rho} + \tfrac{1}{2}q^2\right) + \nu\oint(\nabla^2\mathbf{u}).d\mathbf{l} \qquad (5.2.6)$$

since ρ is uniform. Provided $F = -\nabla\Psi$ *and* provided that Ψ is a single-valued function of position (as it is when $\mathbf{F}$ represents gravity) like p, ρ and q, we have

$$\frac{dC}{dt} = \nu\oint(\nabla^2\mathbf{u}).d\mathbf{l} = -\nu\oint(\nabla\times\boldsymbol{\omega}).d\mathbf{l}. \qquad (5.2.7)$$

This relation is clearly consistent with (5.2.5). In the case of a material closed curve for which it is possible to find an open surface bounded by the curve and lying entirely in the fluid (such a closed curve usually being *reducible*, in the terminology of §2.6), integration of (5.2.5) over such a material surface and application of Stokes's theorem recovers (5.2.7) exactly. The relation (5.2.7) is actually a little stronger than (5.2.5), because, in the case of material irreducible closed curves (say those which link a rigid cylindrical body of infinite length), integration of (5.2.5) over the open surface bounded by any *two* irreducible curves which are reconcilable with a *one-to-one* correspondence leads to the statement that

$$\frac{dC}{dt} + \nu\oint(\nabla\times\boldsymbol{\omega}).d\mathbf{l}$$

has the same value for all such reconcilable closed curves, whereas (5.2.7) provides the further result that this common value is zero. This difference arises because (5.2.7) makes use of the assumption that the force potential Ψ is a single-valued function of position, whereas (5.2.1) and (5.2.5) do not and thereby they allow the possibility of generation of circulation (but not of vorticity) by a more general kind of body force than gravity acting on fluid in a multiply-connected region. Electromagnetic body forces (under certain conditions) are a case in point, and it is possible, for instance, to generate an irrotational motion with circular streamlines in a flat dish of mercury by maintaining a radial electric field between inner and outer cylindrical bounding walls in the presence of a magnetic field normal to the surface of the mercury. In this chapter we shall assume that only a body force derived from a single-valued potential acts on the fluid.

The interesting and significant feature of (5.2.7) lies in the fact that the circulation C is not affected at all by conditions at points not near the material closed curve. Neither gravity nor pressure forces have any direct effect on C, and only the viscous forces acting in the neighbourhood of the material curve can change C. These facts were encountered in a simplified form in the discussion (§ 4.5) of flow with circular streamlines, in which case radial molecular diffusion of vorticity clearly corresponds to angular acceleration of material rings of fluid under the action of a frictional couple. In more general circumstances, the process of diffusion of vorticity seems to be more complex owing to the vector character of $\boldsymbol{\omega}$. However, each component of $\boldsymbol{\omega}$ (relative to rectilinear axes) diffuses as a scalar quantity like temperature, and the complexity is largely apparent. In the case of changes in the flux of vorticity across a material surface element, described by (5.2.5), only the diffusion of the component of $\boldsymbol{\omega}$ instantaneously parallel to $\delta \mathbf{S}$ is capable of affecting the flux, as the formula shows. In the case of changes in the circulation round a closed material curve, described by (5.2.7), we can see the diffusion process more clearly by choosing the (rectilinear) x_1-axis to be parallel to the line element $\delta \mathbf{l}$ at one point of the closed curve. The contribution to the right-hand side of (5.2.7) from this line element then becomes

$$-\nu \left(\frac{\partial \omega_3}{\partial x_2} - \frac{\partial \omega_2}{\partial x_3} \right) \delta l,$$

showing explicitly the separate contributions from diffusion of ω_3 down the gradient in the x_2-direction (this being the direction that will carry ω_3 across the line element) and from diffusion of ω_2 down the gradient in the x_3-direction, with due regard for the signs of these contributions to the flux of vorticity enclosed by the whole curve.

Finally, it is worth recalling that both (5.2.1) and (5.2.7) were obtained on the assumption that the fluid is of uniform density. If the fluid is not uniform, although still incompressible, there are contributions to both $D\boldsymbol{\omega}/Dt$ and dC/dt arising, in effect, from the differential action of both gravity and pressure gradients on elements of the fluid of different density. The generation of vorticity and circulation by the action of gravity on a fluid of non-uniform density is an important process in the atmosphere, where variations in density are produced by thermal effects.

The intensification of vorticity by extension of vortex-lines

The fact that the vorticity of a material element increases, so far as non-viscous effects are concerned, when the element is extended in the direction of the local vortex-line implies the possibility of a net increase in the total 'amount' of vorticity in a body of fluid. A suitable measure of the total amount of vorticity in a material volume τ is provided by the integral $\int \frac{1}{2}\omega^2 \, d\tau$, and then (5.2.2) gives

$$\frac{d}{dt}\int \frac{1}{2}\omega^2 \, d\tau = \int \omega \cdot \frac{D\omega}{Dt} \, d\tau$$

$$= \int \omega \cdot (\omega \cdot \nabla \mathbf{u} + \nu \nabla^2 \omega) \, d\tau.$$

This relation becomes more intelligible when rewritten in suffix notation, and with use of the divergence theorem, as

$$\frac{d}{dt}\int \frac{1}{2}\omega_i \omega_i \, d\tau = \int \omega_i \omega_j \frac{\partial u_i}{\partial x_j} \, d\tau - \nu \int \left(\frac{\partial w_i}{\partial x_j}\right)^2 d\tau + \frac{1}{2}\nu \int \frac{\partial(\omega_i \omega_i)}{\partial x_j} \, n_j \, dS,$$

$$(5.2.8)$$

where S is the material surface bounding τ. As expected, the smoothing effect of viscosity can lead only to a decrease in the total amount of vorticity, apart from the change due to diffusive flux of ω^2 across the boundary of the body of fluid (represented by the last term of (5.2.8)). On the other hand, $\omega_i \omega_j \, \partial u_i/\partial x_j$ is positive at places in the fluid where the rate of extension of the fluid in the direction of ω is positive. In general, some material line elements in the fluid are being extended, and some compressed, at any moment, and clearly a suitable arrangement of the vortex-lines (not in two-dimensional flow!) can lead to a positive value of $\int \omega_i \omega_j \, \partial u_i/\partial x_j \, d\tau$ and to a positive value of the whole right-hand side of (5.2.8).

There are many flow fields in which the total amount of vorticity increases, and continues to increase, in a remarkable way, until through change of the vorticity distribution the loss by viscosity is able to balance the gain by extension of vortex-lines. This is one of the most striking properties of turbulent flow, in which the integral of ω^2 over unit volume of fluid reaches a large value (proportional to a positive power of the Reynolds number of the flow) before the loss by viscous action equals or exceeds the gain by extension of vortex-lines.

A general discussion of vorticity intensification is beyond the scope of this book, and all we shall do here is note a simple example of a steady flow field in which vortex-lines are being extended. In this example the vorticity vector is unidirectional, and has components $(\omega, 0, 0)$ relative to cylindrical co-ordinates (x, σ, ϕ), where ω is dependent only on σ and t. The velocity distribution is likewise axisymmetric, with components (u_x, u_σ, u_ϕ). The vorticity can remain unidirectional only if $\omega \cdot \nabla \mathbf{u}$ is a vector parallel to the x-axis, so that u_σ and u_ϕ must be assumed to be independent of x. The motion in an axial plane is thus of the form

$$u_x = \alpha x, \quad u_\sigma = -\tfrac{1}{2}\alpha\sigma,$$

representing axisymmetric irrotational flow in the neighbourhood of a stagnation point (§2.7) on which the azimuthal motion with vorticity ω is superposed. α is an arbitrary constant, to be taken as positive.

The vorticity equation (5.2.1) then reduces to the single scalar equation

$$\frac{\partial \omega}{\partial t} = \frac{\alpha}{2\sigma} \frac{\partial(\omega\sigma^2)}{\partial \sigma} + \nu \left(\frac{\partial^2 \omega}{\partial \sigma^2} + \frac{1}{\sigma} \frac{\partial \omega}{\partial \sigma} \right). \tag{5.2.9}$$

We are especially interested in the possible existence of a steady flow of this kind, for which ω is a function of σ alone and evidently satisfies the equation

$$\tfrac{1}{2}\alpha\omega\sigma^2 + \nu\sigma \frac{d\omega}{d\sigma} = \text{const.} \tag{5.2.10}$$

If the constant is put equal to zero to avoid a singularity in ω at $\sigma = 0$, the solution is

$$\omega(\sigma) = \omega_1 \exp \left(-\frac{\alpha\sigma^2}{4\nu} \right). \tag{5.2.11}$$

It can be shown that (5.2.11) is in fact the distribution to which ω tends, as $t \to \infty$, for an arbitrary initial distribution of ω with respect to σ subject only to the conditions that $\omega \to 0$ faster than σ^{-2} as $\sigma \to \infty$ and that $\int_0^\infty \omega 2\pi\sigma \, d\sigma$ is finite and non-zero. This latter integral represents the flux of vorticity across a plane normal to the x-axis and is invariant, thereby determining the constant ω_1 in terms of the initial conditions.

The solution (5.2.11) represents a steady flow in which the vorticity is concentrated within a radial distance of order $(\nu/\alpha)^{\frac{1}{2}}$ from the axis of symmetry and in which the intensification of vorticity due to the extension of vortex-lines is ultimately balanced by the rate of decrease due to lateral spreading by viscous diffusion. This steady distribution of vorticity is the same as the instantaneous distribution of ω with respect to σ in a spreading line vortex (see (4.5.13)), and the corresponding distributions of azimuthal velocity with respect to σ are consequently both of the form (4.5.14) (see also figure 4.5.1). The interesting aspect of (5.2.11) is the result that, however dispersed the vorticity may be initially, it is ultimately concentrated within a distance from the axis which may be very small. If initially ω is roughly uniform with magnitude ω_0 within a distance σ_0 from the axis and zero elsewhere, the condition of constant flux across planes normal to the x-axis gives

$$\omega_1 \approx \omega_0 \frac{\alpha\sigma_0^2}{4\nu}. \tag{5.2.12}$$

The conditions assumed in the analysis leading to (5.2.11), viz. uni-directional vorticity with axial symmetry and an imposed straining motion leading to uniform extension of the vortex-lines, are seemingly rather special, but an approximation to these conditions can be expected to occur locally quite often. When the assumed conditions are realized over a considerable region, the outcome can be startling. A tornado may be an approximately steady flow of this kind, with the extension of the vortex-lines arising from thermal up-currents. When a jet engine is being run on a stationary aircraft, an intense vortex is often observed to appear between the ground and the engine intake. On a more homely scale, the familiar bath-plug vortex is a steady concentrated distribution of vorticity resulting from extension, due to the draining motion, of the vortex-lines associated with casual initial motion in the bath. In all these examples, the extension of vortex-lines is caused by a roughly axially symmetrical motion away from a plane boundary.

A solution similar to (5.2.11) can be found for the case in which the contraction in the (y, z)-plane normal to the vortex-lines is wholly in one direction, say the y-direction. The vorticity is here independent of z and has the form

$$\omega(y) = \omega_1 \exp\left(-\frac{\alpha y^2}{2\nu}\right), \tag{5.2.13}$$

corresponding to a sheet vortex smeared over a layer of thickness of order $(\nu/\alpha)^{\frac{1}{2}}$.

Exercise

The vortex-lines in an axisymmetric steady flow field are helices about the x-axis, and the stream function for the motion in an axial plane is of the form

$$\psi = x\sigma^2 f(\sigma).$$

Examine the streamlines in both axial and azimuthal planes for the two-cell vortex given by

$$f(\sigma) = \tfrac{1}{2}\alpha - 6\nu(1 - e^{-\alpha\sigma^2/4\nu})/\sigma^2,$$

where α is a positive constant (Sullivan 1959). Note that some vortex-lines are here being contracted.

5.3. Kelvin's circulation theorem and vorticity laws for an inviscid fluid

Since there are conditions under which the motion of a fluid at large Reynolds number approximates to that of a completely inviscid fluid, it is useful to examine the form taken by the results of the preceding section when $\nu = 0$. Some of them become strikingly simple and powerful. As a beginning, (5.2.7) becomes

$$\frac{dC(t)}{dt} = 0; \tag{5.3.1}$$

the circulation round any closed material curve is invariant in an inviscid fluid. This is *Kelvin's circulation theorem* (Kelvin 1869), and has been established here for an inviscid incompressible fluid of uniform density acted on by a body force per unit mass which can be written as the gradient of a single-valued scalar function of position.

The vorticity equation (5.2.2) reduces here to

$$\frac{D\boldsymbol{\omega}}{Dt} = \boldsymbol{\omega}.\nabla\mathbf{u}. \tag{5.3.2}$$

In the case of a reducible closed material curve which bounds an open material surface S lying in the fluid, (5.3.1) is equivalent to

$$\frac{d}{dt}\int \boldsymbol{\omega}.d\mathbf{S} = 0. \tag{5.3.3}$$

Thus the flux of vorticity across an open material surface is invariant. This result suggests that vortex-tubes are in some sense permanent, as indeed is demonstrated by the following argument.

Consider a material tube which at an initial instant coincides with a vortex-tube of arbitrary cross-section. Initially no vortex-line passes through the tube surface, and the circulation round any closed curve lying on the tube surface and passing round it p times is equal to p multiplied by the strength of the vortex-tube. Now if these closed curves be regarded as material curves, the circulation round each of them remains constant, by Kelvin's circulation theorem. In particular, the circulation round all those material closed curves of small linear dimensions that initially lay on the surface of the vortex-tube without passing round it remains zero. That is to say, the flux of vorticity across any open surface bounded by one of these small closed curves remains zero, and this is possible only if the material closed curves continue to lie on the surface of a vortex-tube without passing round it. Moreover, the invariance of the circulation round those material curves that initially passed round the vortex-tube shows that the strength of the vortex-tube defined by the set of material curves is invariant.

It appears that a vortex-tube specified at an initial instant by drawing the vortex-lines that intersect a given closed curve in the fluid has a continuing identity. We may say that in an inviscid fluid of uniform density *a vortex-tube moves with the fluid and its strength remains constant*. This statement summarizes the very important vorticity laws first put forward by Helmholtz (1858).

If the cross-section of a vortex-tube is contracted to zero, we obtain in the limit a vortex-line. The above result shows that a material line which initially coincides with a vortex-line continues to do so. It is thus possible and convenient to regard a vortex-line as having a continuing identity and as moving with the fluid. (In a viscous fluid it is of course possible to draw the pattern of vortex-lines at any instant, but there is no way in which a particular vortex-line can be identified at different instants.) The fact that the strength of a vortex-tube of small cross-section remains constant as it moves with the fluid also has implications for a vortex-line. For if a material line coinciding with a vortex-line is extended over a portion of its length, the small cross-section of the associated vortex-tube must decrease in conformity with conservation of mass and so the magnitude of the vorticity must increase. It is clear that the length of an element of a material line coinciding with a vortex-line and the local vorticity magnitude *remain in the same ratio*, both being inversely proportional to the infinitesimal cross-section of the associated vortex-tube. Thus the direction and the magnitude of $\boldsymbol{\omega}$ in a material element have separately been shown to change with time in the same way as the direction and magnitude of the vector $\delta\mathbf{l}$ representing a material line element which at some initial instant, t_0 say, was chosen to be parallel to the local vorticity. We therefore have, for the vorticity in a material element,

$$\frac{\boldsymbol{\omega}(t)}{|\boldsymbol{\omega}(t_0)|} = \frac{\delta\mathbf{l}(t)}{|\delta\mathbf{l}(t_0)|}; \qquad (5.3.4)$$

it is implied in the argument here that $(5.3.4)$ becomes an exact relation as $|\delta\mathbf{l}(t_0)| \to 0$.

There is an obvious temptation to interpret some of these results as a consequence of conservation of angular momentum of the fluid contained in the vortex-tube. Such an interpretation is possible when the cross-section of the vortex-tube is small and circular, and remains circular, since the (inviscid) stress at the boundary of the vortex-tube then exerts no couple on the fluid in the vortex-tube. In more general circumstances the strength of a vortex-tube is not simply proportional to the angular momentum of the fluid in unit length of the tube, and more is involved in the above results than a simple conservation of angular momentum.

The above results are so basic that it is worth while to point out here their applicability for compressible fluids. The effect of compressibility on the laws of vortex motion for an inviscid fluid can be shown to be slight, provided we confine attention to homentropic flow fields (§3.4).

In a homentropic flow field the density at any point is a function of (absolute) pressure alone, and we may therefore write

$$\frac{1}{\rho}\nabla p = \nabla\left(\int \frac{1}{\rho}\,dp\right).$$

The equation of motion becomes, in the absence of viscosity, and with a body force $\mathbf{F}$ in the form $-\nabla\Psi$,

$$\frac{D\mathbf{u}}{Dt} = -\nabla\left(\Psi + \int \frac{1}{\rho}\,dp\right). \qquad (5.3.5)$$

The argument which led to $(5.3.1)$ is thus unaffected, and Kelvin's circulation theorem is obtained. Again it follows that a vortex-tube moves with the fluid, with constant strength. The vorticity equation, obtained by taking the curl of both sides of $(5.3.5)$ (after use of the identity that led to $(5.1.1)$), is

$$\frac{\partial\boldsymbol{\omega}}{\partial t} + \mathbf{u}.\nabla\boldsymbol{\omega} + \boldsymbol{\omega}\nabla.\mathbf{u} - \boldsymbol{\omega}.\nabla\mathbf{u} = 0,$$

which may be combined with the equation of mass conservation to give

$$\frac{D}{Dt}\left(\frac{\boldsymbol{\omega}}{\rho}\right) = \left(\frac{\boldsymbol{\omega}}{\rho}\right).\nabla\mathbf{u}. \tag{5.3.6}$$

The flux equation (5.3.3) is unchanged; the effects of compressibility on the rates of change of $\boldsymbol{\omega}$ and δS here cancel each other. Again it follows that vortex-lines move with the fluid, and that changes in $\boldsymbol{\omega}$ for a material element are closely linked with changes in a material line element which initially is parallel to $\boldsymbol{\omega}$ locally. It is readily seen that the counterpart of (5.3.4) is

$$\frac{(\boldsymbol{\omega}/\rho)_t}{|\boldsymbol{\omega}/\rho|_{t_0}} = \frac{\delta\mathbf{l}(t)}{|\delta\mathbf{l}(t_0)|}. \tag{5.3.7}$$

We obtain an alternative Lagrangian form of the relations (5.3.4) and (5.3.7) by noting that, if $\mathbf{a}$ and $\mathbf{X}(\mathbf{a}, t)$ are the position vectors of one end of the material line element at times t_0 and t respectively, $\delta\mathbf{l}(t)/|\delta\mathbf{l}(t_0)|$ is equal to the derivative of $\mathbf{X}$ with respect to $\mathbf{a}$ in the direction of $\delta\mathbf{l}(t_0)$. Thus we have the purely geometrical relationship

$$\frac{\delta\mathbf{l}(t)}{|\delta\mathbf{l}(t_0)|} = \frac{\omega_i(t_0)}{|\boldsymbol{\omega}(t_0)|}\frac{\partial\mathbf{X}}{\partial a_i} \tag{5.3.8}$$

when $\delta\mathbf{l}(t_0)$ is parallel to the local vorticity vector, and substitution in (5.3.7) gives

$$\left(\frac{\boldsymbol{\omega}}{\rho}\right)_t = \left(\frac{\omega_i}{\rho}\right)_{t_0}\frac{\partial\mathbf{X}}{\partial a_i}. \tag{5.3.9}$$

This equation (for the case of incompressible fluid) was obtained first by Cauchy.

The persistence of irrotationality

The particular case in which the circulation round all reducible closed curves in the fluid is zero at some initial instant is of great importance. It has been shown (§ 2.7) that a motion for which the circulation round all reducible closed curves within a certain region is zero is irrotational within this region. The result (5.3.1) shows that the circulation round all these closed curves, considered as material curves, remains zero at all subsequent times. Thus the motion in this same body of fluid remains irrotational at all subsequent times; *a body of inviscid fluid in irrotational motion continues to move irrotationally.*[†] This general property of irrotational motions, above all others, is responsible for their common occurrence (perhaps in approximate form, since real fluids are not entirely inviscid), and their great importance in fluid mechanics. We can infer, for instance, that when a motion is generated from rest (as so many motions are, in practice) in an inviscid fluid, it will necessarily be an irrotational motion, because the initial state of motion was irrotational (albeit in a trivial way).

[†] This result was obtained by Lagrange, although a rigorous proof was first given by Cauchy, in 1815.

The conditions under which irrotational motion remains irrotational are the same as those under which Kelvin's circulation theorem is valid. It should be noted in particular that the property of persistence of irrotationality, like Kelvin's circulation theorem, refers to a *material* body of fluid, and not to the fluid occupying a region fixed in space.

We can also show from (5.3.2) that the vorticity associated with a material element of fluid remains zero if it is initially zero (and thus that a body of fluid in irrotational motion continues to move irrotationally). It is not quite sufficient to assert that, if $\boldsymbol{\omega} = 0$ initially, then from (5.3.2) $D\boldsymbol{\omega}/Dt = 0$ at the initial instant and so $\boldsymbol{\omega}$ remains zero; as Stokes (1845; *Papers* **1**, 106) pointed out, a more complete argument of the following kind is needed. From (5.3.2) we have

$$\frac{D\omega^2}{Dt} = 2\omega_i\,\omega_j\,\frac{\partial u_i}{\partial x_j} = 2\omega^2\lambda_i\,\lambda_j\,\frac{\partial u_i}{\partial x_j}, \qquad (5.3.10)$$

where ω is the magnitude of $\boldsymbol{\omega}$ and $\boldsymbol{\omega} = \omega\boldsymbol{\lambda}$. Then, if K is the greatest positive value of $\lambda_i\,\lambda_j\,\partial u_i/\partial x_j$, at the position of the material element under consideration, during the time interval from t_0 to t, the solution of (5.3.10) is such as to satisfy

$$\omega^2(t) \leqslant \omega^2(t_0)\,e^{2K(t-t_0)}.$$

If now we can put $\omega(t_0) = 0$, it follows that $\omega(t) = 0$ also, provided $K(t-t_0)$ is finite. There is a similar restriction on the magnitude of velocity gradients in the proof of persistence of irrotationality from (5.3.1), as noted in the discussion of rates of change of material integrals in §3.1. The relation (5.3.4) (or (5.3.7) or (5.3.9)) also shows that the magnification of the vorticity of a material element over a certain interval of time is finite, and hence that the vorticity remains zero if it is initially zero, provided the linear extension of the element remains finite.

5.4. The source of vorticity in motions generated from rest

The discussion in §5.2 established that changes in the flux of vorticity across a material surface element take place solely as a consequence of local diffusion of vorticity, by viscosity. And when the body force per unit mass is derived from a single-valued potential, changes in the circulation round a material closed curve take place solely by viscous diffusion of vorticity across this curve, irrespective of whether the curve bounds an open surface lying entirely in the fluid. Vorticity flux or circulation cannot be created in the interior of the fluid but once there it is spread by the action of viscosity.

This raises the important question of the ultimate source of vorticity in motions which have been generated from rest in a fluid of uniform density. Initially the vorticity is everywhere zero, and the motion must remain wholly irrotational unless vorticity diffuses across the surface bounding the fluid. Real fluid motions which can be seen to possess vorticity over at least part of the field are common (for instance, rotating elements are clearly visible at the surface of a dish of water through which a knife blade is moved from rest), so that we are led to expect that some mechanism exists for the generation of vorticity at the boundary of the fluid.

When the fluid is bounded wholly or partly by a solid, any remaining part of the boundary being at infinity where the fluid is at rest, such a mechanism is provided by the no-slip condition. Mechanisms for the generation of vorticity do exist at other types of boundary, such as a 'free' surface at which the pressure is constant and the tangential stress is zero (§5.14), but the case of a solid boundary is by far the most common and it alone need be examined in detail here. An irrotational motion of the fluid is determined completely by the condition of zero flux of mass across each element of the solid boundary,† and this unique irrotational motion almost inevitably has a non-zero tangential component of relative velocity of the fluid at the solid boundary (there being no reason except fortuitous circumstances why it should not do so). Thus, the motion that would be generated from rest in the absence of diffusion of vorticity across the boundary of the fluid is accompanied by a non-zero tangential relative velocity at the boundary. Since the no-slip condition requires the tangential component of relative velocity to be zero at each point of the solid boundary, however small the viscosity may be, the vorticity in this flow is infinite at the boundary. This sheet of infinite vorticity at the boundary is the source from which—once viscosity is allowed to act—vorticity diffuses into the interior of the fluid.

The development of a flow with vorticity in the interior of the fluid can be understood readily by considering the special case of a fluid which, being initially at rest, is set in motion by a solid body whose velocity rises suddenly at $t = 0$ to some finite value and remains steady thereafter. The development of the final steady motion of the fluid relative to the body can be regarded as occurring in three stages. The first stage is the instantaneous generation of a motion in the fluid which satisfies the condition of no flux of mass across each element of the body surface. The body is given a finite velocity suddenly or 'impulsively', and the fluid must move suddenly in conjunction with it.‡ This fluid motion at $t = 0$ is necessarily irrotational at interior points of the fluid, because the vorticity was zero at $t < 0$. As already

† Provided the region occupied by the fluid is singly-connected (see §§2.7, 2.9). When the region is doubly-connected, as in the case of flow past a cylinder of infinite length, the statement made is valid when the circulation round irreducible closed curves in the fluid is given (§§2.8, 2.10), as indeed it is for a wholly irrotational motion generated from rest.

‡ In reality, only the fluid immediately in contact with the solid is set in motion instantaneously, and other parts of the fluid are later brought into motion by the action of compression waves which travel away from the body with finite speed. However, when the flow speeds are small compared with the minimum speed of compression waves (which is the velocity of sound waves in the fluid), the fluid may be regarded as effectively incompressible and compression waves as having infinite speed.

The details of the way in which sudden motion of a boundary sets up an impulsive pressure gradient throughout an incompressible fluid and thereby sets the fluid in motion suddenly are considered in §6.10.

remarked, the initial irrotational flow is fully determined by the known motion of the solid boundary, and the unique irrotational flow set up at $t = 0$ is practically certain to have a velocity at the boundary whose tangential component is different from that of the solid. Thus, at $t = 0$, there is a discontinuity in tangential velocity at the boundary, which is equivalent to a sheet vortex at the body surface. The line integral of $\boldsymbol{\omega}$ along the normal to the boundary at any point is equal in magnitude to the local jump in tangential velocity and is thus finite.

In stage two of the development, the vorticity which was concentrated at the boundary at $t = 0$ diffuses into the fluid by the action of viscosity. If the changes in vorticity at a fixed point were due to viscous diffusion alone, each component of $\boldsymbol{\omega}$ relative to rectilinear axes would satisfy the heat conduction equation and, as amply illustrated in chapter 4 by examples in which other changes in $\boldsymbol{\omega}$ are absent for one reason or another, the distance† from the body to which vorticity would diffuse in time t would be of order of magnitude $(\nu t)^{\frac{1}{2}}$. In fact, vorticity is also convected with a material element (producing a contribution to $\partial \boldsymbol{\omega} / \partial t$ at a fixed point represented by the first term on the right-hand side of (5.2.1)) and it is changed by local distortion and rotation of the fluid (the second term on the right-hand side of (5.2.1)). The second of these additional effects does not influence the extent of the vorticity distribution, but the first may do. However, the velocity of the fluid, relative to that of the body, at positions close to the body has only a small normal component, so that at small values of t, when the diffusion distance $(\nu t)^{\frac{1}{2}}$ is small, the main effect of convection is to carry vorticity parallel to the body surface rather than away from it. Thus, at small values of t, the vorticity of the fluid is non-zero within a layer of thickness of order $(\nu t)^{\frac{1}{2}}$ surrounding the body. Within this layer the vorticity is finite, because a finite velocity jump is now spread over a layer of non-zero thickness.

During the third stage, $(\nu t)^{\frac{1}{2}}$ is no longer a small distance (compared with whatever linear dimension of the boundary is relevant) and convection is able to transport vorticity toward or away from the boundary. As $t \to \infty$, a steady motion of the fluid relative to the body will (usually) be established, the changes in $\boldsymbol{\omega}$ at any fixed point relative to the body due to convection with the fluid, to local distortion and rotation of the fluid, and to viscous diffusion, then having zero resultant. Of these three effects, that due to local distortion and rotation of the fluid modifies the local vorticity and has only a secondary effect on the broad features of the whole vorticity distribution. The other two effects, convection and diffusion, evidently determine whether the vorticity spreads to all parts of the fluid in the steady state. Some motions have such a form that convection does not oppose the spreading of vorticity to all parts of the fluid by viscous diffusion. Others do not; and in the case of

---

† This is not a definite distance of course, since for $t > 0$ there is no point in the fluid at which $\boldsymbol{\omega}$, as given by the theoretical relations, is identically zero. The distance referred to is a 'penetration depth' in the sense made clear by examples like those in §4.3.

a fluid set in motion by a solid body moving through it, it is clear that the relative velocity of fluid in front of the body is towards it and that, since transfer by diffusion becomes weaker with increasing distance from the source, vorticity will spread ahead of the body to a finite distance only. In cases in which convection is strong (through the speed of the body being large), or in which diffusion is weak (through the fluid viscosity being small), there will be a large region of the fluid, ahead of and to the side of the body, in which the flow is approximately irrotational in the ultimate steady state.

Most of these general remarks about the development of a steady flow with distributed vorticity apply in analogous form to the development of a steady distribution of temperature in fluid streaming past a heated body, and some readers may find it easier to comprehend the latter, more familiar, situation. The analogy is particularly close in the case of two-dimensional fields, when the equation for the one non-zero component of $\boldsymbol{\omega}$ is identical (see (5.2.4)) with that for the distribution of temperature in a moving medium with thermal diffusivity ν. Consider, for the purpose of comparison of the two situations, the case of a solid body in the form of a flat plate of negligible thickness, infinite breadth and length l, which at the initial instant $t = 0$ is suddenly given velocity U in its own plane and is also suddenly heated to a temperature above that of the surrounding fluid. The initial irrotational motion of the fluid, relative to that of the body, is here one of uniform velocity $-U$. In stage two of the development, both vorticity and temperature diffuse away from the body and penetrate a distance of order $(\nu t)^{\frac{1}{2}}$ into the fluid after a (small) time interval t. The analogy is here not complete, because firstly the velocity and vorticity distributions are related by $\boldsymbol{\omega} = \nabla \times \mathbf{u}$ whereas temperature is an independent quantity, and secondly the boundary condition on vorticity is provided in effect by the no-slip condition and is not the same as that for temperature; however, the analogy retains qualitative validity if the temperature boundary condition ensures continued flux of temperature away from the body, as it would if, say, the body temperature were held steadily at some value above that of the fluid at infinity. In stage three, convection carries both vorticity and temperature excess to points far from the body on the downstream side; and the extent to which vorticity and temperature excess penetrate ahead of the body, and out sideways from it, will depend on the relative importance of the convection and diffusion processes, as indicated in figure 5.4.1.

The fluid velocity in the steady state will be everywhere of order U and nearly parallel to the plate, so that the time spent by a material element of fluid in the neighbourhood of the plate is of order l/U. During this time interval diffusion spreads vorticity and temperature rise laterally, across the streamlines, through a distance of order $(\nu l/U)^{\frac{1}{2}}$. Thus the lateral extent of the region of vorticity and temperature rise in the steady state will be of the same order as the plate length if Ul/ν is of order unity. If $Ul/\nu \gg 1$, the vorticity and heat are 'blown' away from the plate before they can spread far laterally, giving rise to a narrow 'wake' of vorticity and heat directly down-stream of the body (the magnitudes of vorticity and temperature rise in the wake being greater, the narrower the wake); whereas, if $Ul/\nu \ll 1$, the effect of convection is negligible and both vorticity and temperature rise spread more or less equally in all directions away from the body. So far as the distribution directly upstream of the body is concerned, the need for the convection and diffusion terms to balance shows that vorticity and tempera-ture rise will in general reach a distance of order ν/U ahead of the body; the above qualitative statements about the effect of choosing different values of Ul/ν therefore hold here too, although the quantitative estimates are different.

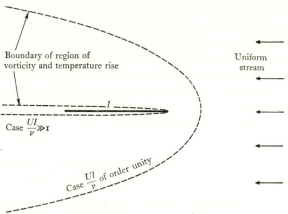

Figure 5.4.1. Diffusion and convection of both vorticity and heat from a plate in a stream.

We may here refer back to a number of particular flow systems examined in chapter 4, in order to see how they exemplify various aspects of the general statements made in this section. Several cases of development of unidirec-tional flow from rest were analysed in § 4.3, and of development of flow with circular streamlines in § 4.5. In all these cases there is no variation of vorticity along streamlines, and there is no turning or extension of vortex-lines, so that changes in vorticity are here due solely to diffusion. However, the fact that a solid boundary is a source of vorticity and that vorticity gradually diffuses from the boundary into an initially irrotational flow appears clearly and in analytical form. In most of these cases the final steady state is such that the vorticity is non-zero throughout the fluid, the steady flux of vorticity out of one solid boundary being balanced by an equal steady flux into another boundary. An interesting exception is the flow generated from rest outside a single circular cylinder which rotates steadily. In this case the vorticity diffused across the cylinder surface in the early stages of the motion all spreads to infinity, and the flow becomes steady and irrotational (see (4.5.10)) over an increasingly large region centred on the cylinder. It happens here that there is a steady state in which the solid boundary does not continue to act as a source of vorticity.

More instructive examples of the way in which the value of the Reynolds number can influence the general features of the vorticity distribution in a steady flow are provided by investigations in §§ 4.9, 4.10 and 4.12 of the flows due to a sphere and a circular cylinder moving steadily through fluid at rest at infinity. These investigations showed that as the Reynolds number increases from some value small compared with unity, the vorticity distribution takes on a more asymmetric form and tends to be confined within a parabola extending downstream with the body at the focus, in agreement with the general conclusions reached in this section about the development of such steady flows from rest. The point of over-riding importance, towards which we have gradually been working, is that, in cases in which both convection and diffusion are operative processes, the effect of increasing the Reynolds number to values large compared with unity is to confine the vorticity diffused from a solid boundary to a layer of relatively small thickness on at least the forward and part of the side portions of bodies moving through fluid, the magnitude of this vorticity being greater as the layer becomes thinner with increasing Reynolds number.

5.5. Steady flows in which vorticity generated at a solid surface is prevented by convection from diffusing far away from it

There are a few solutions of the governing equations which show analytically how the vorticity in a steady flow can be confined, over part of the field, to the neighbourhood of a solid boundary. This confinement is achieved by convection of the vorticity *toward* the boundary, in opposition to viscous diffusion which spreads the vorticity away from its source at the boundary. It is thus necessary for such solutions that, over the whole of the region of flow under consideration, the fluid velocity should have a component toward the solid boundary. This is consistent with the equation of mass conservation only if the fluid accumulating near the boundary is removed either through the wall itself or by continual increase of the tangential component of velocity with increase of distance along the boundary. One example of the former type of flow, and two of the latter, are given below.

(a) Flow along plane and circular walls with suction through the wall

Some solid materials, such as sintered bronze or metal sheet perforated with numerous small holes, are porous and are at the same time rigid and capable of being cut to a specified shape. If a layer of such a solid material is used as the boundary of a region of fluid flow, and if a low pressure is maintained on the side of the layer away from the flow, fluid will be sucked through the boundary. An appropriate boundary condition for the region of flow is then that the normal component of the relative velocity of fluid and solid at the boundary should be equal to some value determined by the porosity and the applied pressure drop; for simplicity we shall take this prescribed normal relative velocity to have the same value $-V$ (V then being positive for *suction* at the boundary) at all points of the boundary. With regard to the condition to be satisfied by the tangential component of the relative velocity at the surface, experimental investigation suggests that the no-slip condition is still appropriate; and since in any event the values of the 'suction' velocity V ordinarily used in practice are very much smaller than the main flow velocities (owing to the high resistance to flow through the porous solid), it seems reasonable to retain the no-slip condition, at least as a very good approximation.

The solution to be given here first represents steady two-dimensional flow over a plane solid boundary at which there is a uniform suction velocity V. The upstream history of the flow may be ignored for the moment, because we shall postulate (in the 'try and see' manner common in fluid mechanics) that the flow variables are independent of position in the plane parallel to the boundary. Thus, if (x, y, z) are rectilinear co-ordinates with y normal to the boundary, the fluid velocity is $(u, -V, 0)$ and the vorticity is $(0, 0, \omega)$, where $\omega = -du/dy$. The vorticity equation (5.2.1) becomes

$$-V\frac{d\omega}{dy} = \nu\frac{d^2\omega}{dy^2},$$

or, on integration, $\qquad V(\omega_0 - \omega) = \nu\frac{d\omega}{dy}, \qquad\qquad (5.5.1)$

which says simply that the rate at which (excess) vorticity is transported across unit area of the (x, z)-plane by convection with velocity $-V$ exactly cancels the rate of transport by viscous diffusion. One more integration gives

$$\omega - \omega_0 = -\frac{du}{dy} - \omega_0 = A e^{-Vy/\nu}, \qquad\qquad (5.5.2)$$

in which A and ω_0 are constants. The region of non-uniform vorticity extends to a distance of order ν/V from the boundary, which was to be expected from general arguments about the balance between convection toward the boundary and viscous diffusion away from it.

It seems, then, that a sensible solution of the postulated form exists. To make it fit circumstances likely to occur in practice we can now choose special values of ω_0 and A. The constant ω_0 evidently represents the (uniform) vorticity far from the boundary, and one obvious choice is $\omega_0 = 0$, corresponding to irrotational flow in the region out of reach of diffusion from the boundary. With $\omega_0 = 0$, we must have $u = U$ (const.) outside the region of non-uniform vorticity near the boundary. Integration of (5.5.2), and use of the boundary conditions

$$u = 0 \quad \text{at } y = 0, \qquad u \to U \quad \text{as } y \to \infty,$$

then gives for the velocity distribution

$$u = U(1 - e^{-Vy/\nu}). \tag{5.5.3}$$

To complete the solution, note from the equation of motion that the pressure is uniform throughout the region of flow concerned. The solution (5.5.3) takes on added importance from the fact that it is believed to be an *asymptotic* solution in two senses; asymptotic with respect to time for a variety of initial conditions in which the flow variables are independent of x (as may readily be established by solving the above linear equation for ω with inclusion of the term $\partial \omega / \partial t$), and asymptotic with respect to x for a variety of steady upstream conditions; provided that the vorticity is zero far from the boundary in all cases.

The case in which ω_0 is non-zero is worth a note in passing. Integration of (5.5.2) gives

$$u = B - \omega_0 y + \frac{A\nu}{V} e^{-Vy/\nu},$$

and an interpretation of the constant ω_0 may be obtained from the equation of motion, which in the present circumstances reduces to

$$-\frac{1}{\rho} \frac{dp}{dx} = V \frac{du}{dy} - \nu \frac{d^2u}{dy^2},$$

$$= V\omega_0.$$

A non-zero and uniform pressure gradient (equal to $-\rho G$, say) is a practical possibility when the fluid is confined in a tube or channel, and we therefore take, as boundary conditions for u,

$$u = 0 \quad \text{at } y = 0 \quad \text{and} \quad y = d;$$

the normal velocity at the channel wall at $y = d$ is $-V$, so that this must be a wall through which fluid is forced into the channel. With these boundary conditions we have

$$u = \frac{G}{V} \left(-y + d \frac{1 - e^{-Vy/\nu}}{1 - e^{-Vd/\nu}} \right). \tag{5.5.4}$$

When $Vd/\nu \ll 1$, diffusion is dominant over convection in the y-direction, and the velocity distribution (5.5.4) reduces to the expected parabolic form. At the other extreme, $Vd/\nu \gg 1$, we find $u \approx G(d-y)/V$, except near $y = 0$ where u falls sharply to zero. The steep gradient of u near $y = 0$ is due to the confinement of vorticity generated at that boundary to a thin layer, and the uniform, relatively small, vorticity outside that layer is generated at the boundary $y = d$ and convected across the channel. From a momentum point of view, the uniform gradient of u over most of the channel is due to continual (and uniform) acceleration of fluid elements, under the action of the pressure gradient, after they emerge through the wall at $y = d$ and until they enter the region of viscous action near $y = 0$.

A solution may also be found for the flow outside a rotating circular cylinder, at the surface of which there is an inward radial velocity V due to suction through the wall of the cylinder. It was found in §4.5 that the flow generated from rest by a rigid cylinder rotating steadily (without suction) ultimately becomes irrotational, because the vorticity created at the solid surface all diffuses to infinity. We may expect that, when suction is applied, the vorticity is prevented from diffusing to infinity and that a steady state with non-zero vorticity in the neighbourhood of the cylinder will be set up. Proceeding on the postulate that a steady state does exist, we have, as the counterpart of (5.5.1),

$$\frac{Vr_1}{r}(\omega_0 - \omega) = \nu \frac{d\omega}{dr},$$

where ω is now the axial vorticity component and r_1 is the radius of the cylinder. The solution is of the form

$$\omega - \omega_0 = A\left(\frac{r_1}{r}\right)^R, \tag{5.5.5}$$

where $R = r_1 V/\nu$, and again ω_0 may be put equal to zero to correspond with the establishment of a steady state from some initial condition in which $\omega = 0$ at large values of r. The solution has the expected feature of a maximum of the vorticity at the cylinder, but it will be noticed that ω decreases only slowly with increase of r when the Reynolds number R is small. Further integration of (5.5.5) (with $\omega_0 = 0$) gives

$$rv = r_1^2 \Omega_1 - \frac{Ar_1^2}{R-2}\left(\frac{r_1}{r}\right)^{R-2}, \tag{5.5.6}$$

where Ω_1 is a constant, which shows that the circumferential velocity v is finite at infinity only if $R \geqslant 1$, and the circulation $2\pi rv$ is finite only if $R > 2$, unless $A = 0$. If we had obtained this steady state as the asymptotic form of a time-dependent solution with initial conditions such that the circulation at infinity is finite (and perhaps zero), we should have found that it remains finite, and that consequently $A = 0$ when $R < 2$. Thus when $R < 2$ the steady state is irrotational, just as in the case of no suction, indicating that the vorticity created at the cylinder surface is not prevented by convection from spreading to infinity by diffusion. This is made possible here by the decrease of the inward radial convection velocity Vr_1/r with increase of distance from the cylinder.

(b) Flow toward a 'stagnation point' at a rigid boundary

We shall investigate here the (steady) distribution of vorticity in the immediate neighbourhood of the point on a solid surface at which fluid approaching the surface divides into streams proceeding away from the point in question. Without the no-slip condition, this dividing point would be distinguished by the fact that the fluid velocity, relative to that of the solid, is zero there; and it is common practice to refer to the flow in the neighbourhood as 'stagnation point flow', even though in fact all points on the solid boundary are equally points of zero relative velocity of a real fluid.

By confining attention to a sufficiently small neighbourhood of this dividing point, we may regard the solid boundary as being plane (unless the boundary happens to have a discontinuity of slope there).

This is clearly a situation in which the component of velocity normal to the boundary is toward it everywhere in the region concerned, so that the vorticity created at the boundary will be convected toward it, in opposition to viscous diffusion away from it. We may therefore reasonably assume that in the steady state the vorticity generated at the boundary is confined to a layer adjacent to the boundary, whose thickness decreases as the effect of convection becomes relatively stronger, and that outside this layer the vorticity has a value determined by conditions in the flow far from the boundary; the latter value will be taken to be zero, as would be so in the case, for example, of the stagnation point on the forward portion of a solid body fixed in a stream with steady uniform velocity at infinity.

Consider first two-dimensional flow near a stagnation point at a rigid surface. It proves to be convenient, for the solution of this kind of problem, to determine first the flow in the outer irrotational region (usually an easier task, in view of the severe restrictions imposed by irrotationality) and then to use this flow as an outer boundary condition for the flow in the layer of non-zero vorticity. Now when the thickness of the layer of vorticity is very small (compared with the linear dimensions of the region under discussion), its presence can make little difference to the irrotational flow. We shall therefore proceed to find the approximate form of the irrotational flow by ignoring the layer of non-zero vorticity altogether (and by ignoring the no-slip condition which gives rise to this layer). On this basis, the motion in the outer region is simply irrotational flow near a stagnation point at a *plane* boundary. We shall see that a way of improving the approximation, to take account of the effect of presence of the layer on the irrotational flow, is suggested by the form of the resulting solution.

The flow in the irrotational region is known (see (2.7.10)) to be described by the stream function
$$\psi = kxy, \tag{5.5.7}$$

where x and y are rectlinear co-ordinates parallel and normal to the boundary (see figure 5.5.1), with the corresponding velocity distribution
$$u = kx, \quad v = -ky. \tag{5.5.8}$$

k is a positive constant which, in the case of a stagnation point on a body fixed in a stream, must be proportional to the speed of the body on dimensional grounds and is found also to depend on the shape of the body as a whole.

The next step is to determine the distribution of vorticity in the thin layer near the boundary from the equation
$$u\frac{\partial \omega}{\partial x} + v\frac{\partial \omega}{\partial y} = \nu\left(\frac{\partial^2 \omega}{\partial x^2} + \frac{\partial^2 \omega}{\partial y^2}\right), \tag{5.5.9}$$

together with boundary conditions that $u = 0$ and $v = 0$ at $y = 0$ and that the flow tends to the form (5.5.8) at the outer edge of the layer. Now the existence of the no-slip condition will certainly change the dependence of the velocity components on y, but it is not evident that it will change their dependence on x; it is therefore worth while to see if there exists a solution such that $u \propto x$ throughout the vorticity layer. For such a solution we may write

$$\psi = xf(y), \tag{5.5.10}$$

corresponding to

$$u = xf'(y), \quad v = -f(y),$$

and

$$\omega = \frac{\partial v}{\partial x} - \frac{\partial u}{\partial y} = -xf''(y),$$

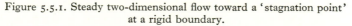

Figure 5.5.1. Steady two-dimensional flow toward a 'stagnation point' at a rigid boundary.

where $f(y)$ is an unknown function and primes denote differentiation with respect to y. Substitution in (5.5.9) shows that, provided we can find a form for $f(y)$ which satisfies

$$-f'f'' + ff''' + \nu f^{iv} = 0 \tag{5.5.11}$$

and the boundary conditions

$$f = f' = 0 \quad \text{at} \quad y = 0,$$

$$f \to ky \quad \text{as} \quad y \to \infty,$$

our guess is successful.

One integration of (5.5.11) and use of the outer boundary condition gives

$$f'^2 - ff'' - \nu f''' = k^2. \tag{5.5.12}$$

The coefficients in this equation can be made pure numbers by the transformation

$$y = \left(\frac{\nu}{k}\right)^{\frac{1}{2}} \eta, \quad f(y) = (\nu k)^{\frac{1}{2}} F(\eta),$$

whence it becomes $$F'^2 - FF'' - F''' = 1, \qquad (5.5.13)$$

with $$F = F' = 0 \quad \text{at} \quad \eta = 0,$$

$$F \to \eta \quad \text{as} \quad \eta \to \infty.$$

It was shown numerically by Hiemenz (1911) that a solution of this equation satisfying all the boundary conditions could be found, the results being given in figure 5.5.2. The corresponding streamlines and distribution of u along one co-ordinate line are sketched in figure 5.5.1. A proof of uniqueness of a solution of the flow equations, for the boundary conditions specified above, is not available, but our specification of the problem seems on

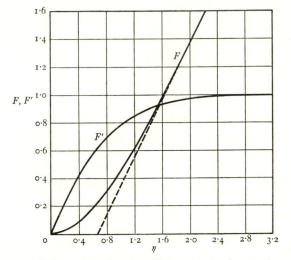

Figure 5.5.2. Values of the function $F(\eta)$ giving the flow in the vorticity layer.

physical grounds to be complete and we may reasonably accept the solution found above as a description of the flow occurring in practice.

The thickness of the layer of non-zero vorticity, defined for convenience as the value of y at which $u = 0 \cdot 99kx$, is found from the numerical solution (as improved by Howarth (1935)) to be

$$\delta = 2 \cdot 4(\nu/k)^{\frac{1}{2}}. \qquad (5.5.14)$$

We see that this thickness is independent of distance along the boundary and that, as assumed earlier, it approaches zero as the effect of convection (represented by k) becomes dominant over the effect of diffusion (represented by ν). The fact that the thickness δ is independent of x suggests that the effect of the layer of vorticity on the irrotational flow (which effect had to be neglected altogether at the beginning of our calculation) is approximately to displace it in the y-direction in the same way as a simple shift of the boundary; confirmation of this is obtained by noting from figure 5.5.2 that an improved asymptotic estimate of F, as $\eta \to \infty$, is

$$F \to \eta - 0.65,$$

so that the corresponding asymptotic forms of u and v are

$$u \to kx, \quad v \to -k(y - \delta_1),$$

in which the 'displacement thickness' δ_1 of the layer is given by

$$\delta_1 = 0.65(\nu/k)^{\frac{1}{2}}. \tag{5.5.15}$$

This simple shift of the whole field of irrotational flow does not affect the velocity distribution within that field.

It is clear now that the restriction on the value of k/ν, for the above solution for the flow in the vorticity layer to be valid, is that the region near the stagnation point within which the flow would have the form (5.5.8) in the absence of the no-slip condition should extend further from the boundary than the edge of the vorticity layer. In the case of a stagnation point at the front of a body in a uniform stream, this will usually amount to a condition of the form

$$(\nu/k)^{\frac{1}{2}} \ll \text{radius of curvature of body at the stagnation point.}$$

It is worth remarking, in conclusion, that the uniformity of thickness of the vorticity layer here is due evidently to the uniformity, with respect to x, of the velocity toward the boundary at the outer edge of the layer—or, equivalently, to the convection of the vorticity parallel to the boundary and away from the stagnation point at a speed which increases linearly with x. It will be seen later (§ 5.9) that when the tangential component u at the outer edge of the vorticity layer varies as x^m, the thickness of the layer increases with x when $m < 1$, convection then not being strong enough to prevent thickening of the layer by diffusion, and decreases as x increases when $m > 1$.

A similar solution can be found for steady axisymmetric flow (without azimuthal motion) toward a 'stagnation point' at a plane solid boundary (Homann 1936b), as would occur in approximate form at the front of a body of revolution moving parallel to its axis of symmetry through fluid at rest at infinity. The irrotational flow in the region outside the vorticity layer is here described by the relations (2.7.11), with the boundary at $x = 0$, and the calculation of the flow in the vorticity layer proceeds with only numerical differences. Both the two-dimensional and axisymmetric solutions are special (and simple) cases of the general stagnation point flow (Howarth 1951) in which the velocity in the irrotational region has components of the form (2.7.9).

(c) Centrifugal flow due to a rotating disk

In the first of the two preceding examples the convection of vorticity toward the boundary is due to suction through the boundary, and in the second it is due to the external imposition of a flow toward the boundary; in this third example, the motion toward the boundary is induced by centrifugal action on the vorticity layer. We consider a plane disk of large diameter which is made to rotate in its own plane with a steady angular velocity Ω in fluid which is initially at rest everywhere. The relative motion of the disk and the fluid sets up viscous stresses, which tend to drag the fluid round with the disk. An exactly circular motion of fluid near the disk is not possible, since there is no imposed radial pressure gradient to provide the inward radial acceleration, and the fluid near the disk therefore spirals outwards. This outward radial motion near the disk must be accompanied by an axial motion towards the disk in order that conservation of mass be satisfied, and in this way the vorticity generated at the boundary is prevented from spreading far from it. The disk acts as a centrifugal fan, throwing fluid out radially and drawing other fluid toward it to be thrown out in turn.

The resulting steady motion seems at first sight to be analytically complicated, but it happens that the linear dependence of the disk speed on radial distance r leads to a similar dependence of the radial velocity of the fluid on r and that as a consequence, just as in 'stagnation point flow', the vorticity layer has uniform thickness. Von Kármán (1921) was the first to notice that the governing equations and appropriate boundary conditions allow a solution such that u/r, v/r and w are all functions of z alone, where (u, v, w) are velocity components parallel to the (r, ϕ, z)-co-ordinate lines in a cylindrical co-ordinate system with $r = 0$ on the axis of the disk. With velocity components of this form, it follows from the equation of (steady) motion in the direction of the z-co-ordinate line that the pressure must be of the form

$$\frac{p}{\rho} = \nu \frac{dw}{dz} - \tfrac{1}{2}w^2 + F, \tag{5.5.16}$$

where F is a function of r alone. Since there is no rotation of the fluid far from the disk, and presumably no radial motion there, p must be independent of r when z is large; hence $F = $ const. The equations of motion in the directions of the r- and ϕ-co-ordinate lines (see appendix 2) then become

$$\left(\frac{u}{r}\right)^2 + w\frac{d(u/r)}{dz} - \left(\frac{v}{r}\right)^2 = \nu \frac{d^2(u/r)}{dz^2}, \tag{5.5.17}$$

$$\frac{2uv}{r^2} + w\frac{d(v/r)}{dz} = \nu \frac{d^2(v/r)}{dz^2}. \tag{5.5.18}$$

In addition we have the equation of mass conservation

$$\frac{2u}{r} + \frac{dw}{dz} = 0,$$

which enables u to be eliminated from (5.5.17) and (5.5.18).

The boundary conditions to be imposed on the solution of these equations are

$$u = w = 0, \quad v = \Omega r, \quad \text{at} \quad z = 0,$$

representing the no-slip condition, and

$$u \to 0, \quad v \to 0, \quad \text{as} \quad z \to \infty.$$

We refrain from imposing any condition on w as $z \to \infty$, since we expect the axial motion far from the disk to be an inflow which is induced by the centrifugal action near the disk; in confirmation of this expectation it turns out that the above equations and boundary conditions do in fact determine w completely.

The dimensional factors ν and Ω are the only ones occurring in the problem, and between them they determine the velocity and length scales of the flow. We put

$$z = \left(\frac{\nu}{\Omega}\right)^{\frac{1}{2}} \zeta, \quad \frac{v}{r} = \Omega g(\zeta), \quad w = (\nu\Omega)^{\frac{1}{2}} h(\zeta), \qquad (5.5.19)$$

giving the equations (5.5.17) and (5.5.18) in dimensionless form as

$$\tfrac{1}{4}h'^2 - \tfrac{1}{2}hh'' - g^2 = -\tfrac{1}{2}h''', \qquad (5.5.20)$$

$$-gh' + g'h = g'', \qquad (5.5.21)$$

with the boundary conditions

$$h = h' = 0, \quad g = 1, \quad \text{at} \quad \zeta = 0,$$

$$h' \to 0, \quad g \to 0, \quad \text{as} \quad \zeta \to \infty.$$

An accurate numerical solution satisfying these equations and boundary conditions has been obtained by Cochran (1934), and the values of g, $-h$ and $-\tfrac{1}{2}h'$ (to which u is proportional) are shown in figure 5.5.3. These values show the expected action of the disk as a centrifugal fan, with an induced axial motion toward the disk which prevents the vorticity from spreading far from the disk. If for convenience we define the edge of the vorticity layer as the place where $v/\Omega r = 0.01$, the thickness of the layer is uniform and equal to $5.4(\nu/\Omega)^{\frac{1}{2}}$. Outside the vorticity layer the axial velocity is uniform and equal to $-0.89(\nu\Omega)^{\frac{1}{2}}$; this inflow velocity decreases as ν decreases, because the vorticity layer is then thinner and less fluid is required to move in to take the place of that thrown out radially. The total volume flux outward across a cylindrical surface of radius r is $0.89\pi r^2(\nu\Omega)^{\frac{1}{2}}$.

One way of testing the correspondence between the above solution and an experimental flow system is to measure the torque exerted on (two sides of) a rotating thin disk of finite radius a. The above solution applies strictly to an infinite disk, but provided the vorticity layer thickness is small compared with the disk radius, i.e., provided $a\Omega^{\frac{1}{2}}/\nu^{\frac{1}{2}} \gg 1$, it is reasonable to suppose that the effect of the edge of the disk is small. The tangential stress at the disk is

$$\sigma_{z\phi} = \mu \left(\frac{\partial v}{\partial z}\right)_{z=0} = \rho \nu^{\frac{1}{2}}\Omega^{\frac{3}{2}}rg'(\mathrm{o}),$$

and the torque exerted by the fluid on both sides of a disk of radius a is therefore

$$2\int_0^a \sigma_{z\phi} 2\pi r^2\, dr = \pi a^4 \rho \nu^{\frac{1}{2}}\Omega^{\frac{3}{2}}g'(\mathrm{o}). \qquad (5.5.22)$$

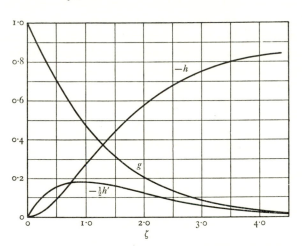

Figure 5.5.3. Dimensionless functions giving the components of velocity in flow due to a rotating disk.

Cochran's numerical solution shows that $g'(\mathrm{o}) = -\mathrm{o}\cdot 616$. This value of the torque is found to agree well with the measured value, provided $a^2\Omega/\nu$ is less than about $\mathrm{10}^5$ (and also large compared with unity); at larger values of $a^2\Omega/\nu$ the flow is unstable and a steady motion cannot be achieved in practice.

A steady axisymmetric flow of the 'similarity' form (5.5.19) also appears to exist when the rotating disk bounds fluid which is in rigid rotation about the same axis, with angular velocity Γ say, at large distances from the disk, although the flow field has not been determined in detail for all values of the ratio Γ/Ω. Analytically the problem is not very different from that described above. At large distances from the disk the pressure is now equal to $\frac{1}{2}\rho\Gamma^2 r^2$, so that in (5.5.16) we must put $F = \frac{1}{2}\Gamma r^2$ and a term $-\Gamma^2$ must be added to the right-hand side of (5.5.17). The only other change required is the replacement of the boundary condition $g \to \mathrm{o}$ as $\zeta \to \infty$ by $g \to \Gamma/\Omega$ as $\zeta \to \infty$.

However, numerical integration proves now to be more difficult, especially when Γ and Ω have opposite signs.

In the case in which the disk and the fluid at infinity are rotating rigidly with nearly the same angular velocity, the equations may be solved explicitly. This explicit solution is not an illustration of our theme of confinement of vorticity by convection toward the boundary but will be described briefly in view of its rather unexpected connection with previous work. We evidently have here

$$g = 1 + g_1, \quad |g_1| \ll 1, \tag{5.5.23}$$

and it appears from equation (5.5.21) that $|h| \ll 1$ also. To the first order in the small quantities $|g_1|$ and $|h|$ the governing equations (5.5.20) (with the inclusion of a pressure term as explained) and (5.5.21) become

$$1 + 2g_1 = (\Gamma/\Omega)^2 + \tfrac{1}{2}h''',$$

$$-h' = g_1''.$$

The solution that satisfies all the boundary conditions is readily shown to be

$$g_1(\zeta) = \frac{\Gamma - \Omega}{\Omega}(1 - e^{-\zeta}\cos\zeta), \tag{5.5.24}$$

$$h'(\zeta) = 2\frac{\Gamma - \Omega}{\Omega}e^{-\zeta}\sin\zeta. \tag{5.5.25}$$

It will be seen that our solution of the approximate equations is identical with the Ekman spiral velocity distribution near a rigid boundary over which a stream is passing in a rotating fluid (see §4.4). The radial and azimuthal velocity components here are u and v, where

$$u = -\tfrac{1}{2}r\frac{dw}{dz} = -\tfrac{1}{2}r\Omega h', \quad v = r\Omega(1 + g_1),$$

and these take the place of the velocity components $-v$ and u in the plane of the boundary given in (4.4.16) and (4.4.15). When the flow due to a disk and adjoining fluid rotating at nearly equal speeds is referred to axes which rotate with the angular velocity Ω of the disk, the radial and azimuthal components of velocity have magnitudes small compared with Ωr and the vorticity component parallel to the axis of rotation has magnitude small compared with Ω. These are just the conditions for Coriolis forces to be much larger in magnitude than inertia forces (giving so-called geostrophic flow, about which more will be said in §§7.6, 7.7), and our approximation (5.5.23) is equivalent to the assumption that variations of velocity along a streamline are without local consequence. Uniformity of the velocity along a streamline in a rotating system was the assumption on which the analysis of §4.4 was based, and so identity of the two solutions is understandable. It will be noticed that the net drift in the Ekman layer is in the direction opposite to the gradient of modified pressure, and in the present context this implies a net drift in the radial direction, inwards when $\Gamma > \Omega$ and outwards when $\Gamma < \Omega$, of magnitude proportional to r. A drift of this kind can occur in the friction layer on a rotating disk if there is a compensating *uniform* axial component of velocity, taking fluid out of the layer when $\Gamma > \Omega$ and into it when $\Gamma < \Omega$, which is what has been found above. It may be shown readily that the uniform inflow or outflow velocity outside the friction layer, i.e. the limit of w as $z \to \infty$ given by integration of (5.5.25), has exactly the value required for consistency with mass conservation and the volume flux (4.4.17) in the direction opposite to the pressure gradient in an Ekman layer.

5.6. Steady two-dimensional flow in a converging or diverging channel

Another example of the combined effects of convection and diffusion of vorticity generated at a rigid boundary is provided by two-dimensional flow in the region between two intersecting plane walls. The walls are stationary, and a steady flow is caused by the presence of a source or sink of fluid volume at the point of intersection of the walls; in practice such a point source or sink in the plane of flow could be approximated by a small hole near the point of intersection through which fluid is discharged or withdrawn, or by connection of the narrow end of the channel to a parallel-sided channel. A source at the point of intersection gives flow in a diverging channel, and a sink there gives flow in a converging channel. A set of similarity solutions of the equation of motion for this kind of flow field (first explored by Jeffrey (1915) and Hamel (1917)) is known, covering the whole range of values of the angle between the walls and of the effective Reynolds number of the flow. The solutions resemble mathematically that given in §4.6 for the steady jet from a point source of momentum, inasmuch as the velocity components are proportional to r^{-1}, where r is the distance from the point singularity in the two cases. The solutions are useful in that, like all similarity solutions, they show velocity distributions which are dynamically possible. In practice the velocity distribution would no doubt depend on the detailed conditions upstream. It may be that in some circumstances the solutions described below are asymptotic solutions, valid at sufficiently large distances downstream from the place where conditions are actually prescribed, although this is not yet known.

We shall use polar co-ordinates (r, θ), with $\theta = \mp \alpha$ at the two plane walls; (u, v) are the corresponding velocity components. We look for a solution such that the flow is purely radial and, as a consequence of the mass-conservation equation,

$$u = r^{-1}F(\theta). \tag{5.6.1}$$

Substitution of this expression for u, with $v = 0$, in the two equations of motion (which are given in terms of polar co-ordinates in appendix 2) and elimination of the pressure then gives

$$2FF' + \nu F''' + 4\nu F' = 0, \tag{5.6.2}$$

where the primes denote differentiation with respect to θ. Since the vorticity of the fluid is $-F'/r$, the three terms in this equation represent the contributions to (minus) the rate of change of vorticity at a point from, respectively, convection, diffusion in the circumferential direction, and radial diffusion. This equation is to be solved subject to the no-slip boundary conditions

$$F = o \quad \text{at } \theta = \mp \alpha. \tag{5.6.3}$$

Some condition specifying the intensity of the flow must also be imposed. One way of doing this is to specify the net volume flux into the channel from the source at the origin:

$$Q = \int_{-\alpha}^{\alpha} ur \, d\theta = \int_{-\alpha}^{\alpha} F \, d\theta. \tag{5.6.4}$$

Since some of the flow fields to be found below show some fluid moving radially outwards and some radially inwards, a more direct measure of the intensity of the flow is provided by the value of F, F_0 say ($= u_0 r$), at one of the local maxima of $|F|$; if there is only one stationary value of F in the range $-\alpha \leqslant \theta \leqslant \alpha$, $|F_0|/r$ is the maximum speed of the fluid at distance r from the origin. $|Q|/\nu$ may be regarded as a Reynolds number of the flow, and since αr is a measure of the width of the channel so also may $\alpha |F_0|/\nu$. We shall put

$$R = \alpha F_0/\nu,$$

the sign of R indicating the direction of the flow at the chosen maximum of $|F|$.

It is now convenient to introduce the dimensionless variables

$$\eta = \theta/\alpha, \quad f = F/F_0,$$

whence (5.6.2) becomes

$$2\alpha R f f' + f''' + 4\alpha^2 f' = o, \tag{5.6.5}$$

with primes now denoting differentiation with respect to η. The conditions to be satisfied by f are

$$f = o \quad \text{at } \eta = \mp 1, \tag{5.6.6}$$

$$f' = o \quad \text{at } f = 1. \tag{5.6.7}$$

Equation (5.6.5) can be integrated once as it stands, and then again after multiplication by f', to give

$$f'^2 = (1-f)\{\tfrac{2}{3}\alpha R(f^2+f) + 4\alpha^2 f + c\}; \tag{5.6.8}$$

c is one constant of integration and the other has been determined from (5.6.7). The result of one further integration of (5.6.8) can be written down in terms of elliptic functions, although we shall not need to introduce these functions for the limited purposes of this discussion.† Both c and the constant resulting from the further integration are determined by the

† Further details of the analysis may be found in papers by Rosenhead (1940), Millsaps and Pohlhausen (1953), and Fraenkel (1962).

conditions (5.6.6) and thus depend on α and R. c is evidently real and non-negative, since

$$f'^2 = c \quad \text{at } \eta = \mp 1.$$

The form of the solution of (5.6.8) depends on the location of the zeros of the expression within curly brackets on the right-hand side ($= P(f)$ say). When $R \geqslant 0$ it is clear that there is no zero of $P(f)$ at any positive value of f. Now for any R there is a local maximum of $f(\eta)$ at $f = 1$, so that $P > 0$ there; and $P = c \geqslant 0$ at $f = 0$. Since $P \to -\infty$ as $f \to \mp\infty$ when $R < 0$, it follows that the quadratic function P cannot vanish in the range $0 < f \leqslant 1$ when $R < 0$. The possible forms of $P(f)$ are sketched in figure 5.6.1. Hence f varies monotonically between $f = 0$ and $f = 1$ for all values of R. We see that only one local maximum of $f(\eta)$ can occur in a region of outflow or inflow. And when there is only one extremum of f in the channel, it must occur at $\eta = 0$ in view of the symmetry of the solution about the maximum.

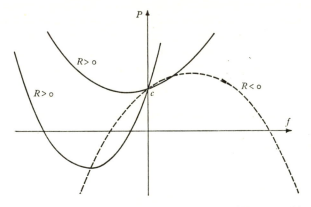

Figure 5.6.1. Sketch of the function $P(f) = \frac{2}{3}\alpha R(f^2+f) \div 4\alpha^2 f + c$ for different values of R.

It is worth while to observe in passing that we may obtain some confirmation here of the hypothesis of unidirectional flow made in § 4.8 in connection with flow between nearly-parallel boundaries of arbitrary shape. For when

$$\alpha \ll 1, \quad \alpha|R| \ll 1,$$

the approximate solution of (5.6.8) that satisfies the boundary conditions is

$$c = 4, \quad u/u_0 = f = 1 - \eta^2,$$

that is, a parabolic variation of the velocity across the channel, as obtained from the hypothesis. Moreover, the restriction $\alpha|R| (= \alpha^2 r|u_0|/\nu) \ll 1$ is identical with that found in § 4.8 to be needed for the approximation of unidirectional flow in a slowly-varying channel to be valid.

In the context of this chapter we are interested particularly in the case of large Reynolds number. Since f is of order unity equation (5.6.5) takes the approximate form

$$2\alpha R f f' + f''' = 0 \qquad (5.6.9)$$

when $|R| \gg 1$, the term representing diffusion of vorticity in the flow direction then being negligible. Equivalently, (5.6.8) takes the approximate form

$$f'^2 = \tfrac{2}{3}\alpha R f (1 - f^2) + c(1 - f). \qquad (5.6.10)$$

The solution then depends only on the single parameter αR, and not on α and R separately.

We proceed to examine the form of the solution of (5.6.8) in each of the two cases $R < 0$ and $R > 0$, with particular attention to large values of $|R|$ in both cases.

Purely convergent flow

Here there is inflow, with converging streamlines, everywhere in the channel, and $F_0 < 0$ and $R < 0$. Only one maximum of $|F|$ occurs in the channel, and the velocity distribution is symmetrical about $\theta = 0$. The boundary condition (5.6.6) then requires

$$1 = \int_0^1 \frac{df}{(1-f)^{\frac{1}{2}} \{\tfrac{2}{3}\alpha R(f^2 + f) + 4\alpha^2 f + c\}^{\frac{1}{2}}}. \qquad (5.6.11)$$

If now we allow $|R|$ to become large, with α fixed, we see that this relation can be satisfied only if one of the zeros of the expression within curly brackets tends to $f = 1$ (so that the integral tends to become divergent). Hence, as $R \to -\infty$ it is necessary that

$$c \to -\tfrac{4}{3}\alpha R,$$

and we may use this asymptotic value of c as an approximation, valid for large $|R|$, in any integral which is not then divergent. Thus, when $|R| \gg 1$, we have from (5.6.8)

$$1 - \eta \approx (-\tfrac{2}{3}\alpha R)^{-\frac{1}{2}} \int_0^f \frac{df}{(1-f)(f+2)^{\frac{1}{2}}}$$

$$= (-\tfrac{1}{2}\alpha R)^{-\frac{1}{2}} \left\{ \tanh^{-1}\left(\frac{f+2}{3}\right)^{\frac{1}{2}} - \tanh^{-1}(\tfrac{2}{3})^{\frac{1}{2}} \right\},$$

and the approximate expression for the velocity in the range $0 \leqslant \theta \leqslant \alpha$ is

$$\frac{u}{u_0} = \frac{F}{F_0} = f = 3 \tanh^2 \left\{ (-\tfrac{1}{2}\alpha R)^{\frac{1}{2}} \left(1 - \frac{\theta}{\alpha}\right) + \tanh^{-1}(\tfrac{2}{3})^{\frac{1}{2}} \right\} - 2. \quad (5.6.12)$$

It appears that a purely convergent flow at large Reynolds number is possible in which the radial velocity is approximately independent of θ, and equal to its value u_0 at the centre, over the whole channel except in layers so close to the walls that

$$O(\alpha - |\theta|) = \left(-\frac{R}{\alpha}\right)^{-\frac{1}{2}} = \left(\frac{r|u_0|}{\nu}\right)^{-\frac{1}{2}}. \qquad (5.6.13)$$

Outside these two layers the flow is irrotational, all the vorticity generated at the walls being confined within the layers. Since the velocity is everywhere in the radial direction, the component of velocity normal to the nearer wall is towards it, so that the effect of convection here is to oppose the diffusion of vorticity away from the wall and is so strong as to cause a decrease of the layer thickness with increase of distance in the flow direction. The velocity distribution across the channel is shown in figure 5.6.2. The relation between the parameters Q, ν and R is here

$$Q \approx 2\alpha u_0 r = 2\alpha F_0 = 2\nu R.$$

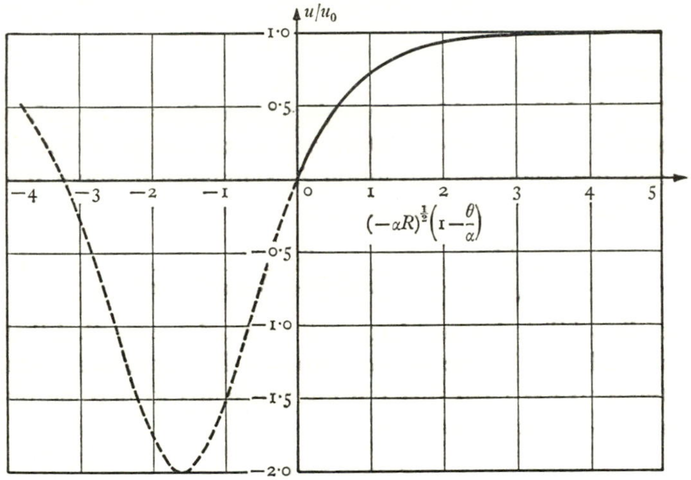

Figure 5.6.2. Flow in a converging channel at large Reynolds number.

It will be noticed that the profile represented by (5.6.12) may be continued to values of θ greater than α, into a region of outflow, with the velocity returning again to zero at

$$\theta/\alpha = 1 + 2(-\tfrac{1}{2}\alpha R)^{-\frac{1}{2}} \tanh^{-1} (\tfrac{2}{3})^{\frac{1}{2}},$$

as shown by the broken curve in figure 5.6.2. This second zero of u is another possible position of a channel boundary, corresponding to a different value of α. Apparently the approximately uniform inflow in the wide central region of the channel can be bounded by a narrow region of outflow at either wall. And further continuation of the profile leads to another region of convergent flow with a velocity distribution which is a repetition of that in the range $0 \leqslant \theta \leqslant \alpha$.

Purely divergent flow

Consider now the case of outflow, with diverging streamlines, for $|\theta| < \alpha$, with $F_0 > 0$, $R > 0$. Again there is only one maximum of $|F|$, and the velocity distribution is symmetrical about $\theta = 0$. The relation (5.6.11) must again be satisfied, but the consequences are quite different. All terms in the expression within curly brackets in (5.6.11) are now positive, and it is clearly not possible for the equation to be satisfied for all choices of α and R. The constant c is at our disposal, subject to the restriction $c \geqslant 0$; and the maximum value of R, say R_m, for which the relation can be satisfied for given α evidently occurs when $c = 0$ and is given by

$$(\tfrac{2}{3}\alpha R_m)^{\frac{1}{2}} = \int_0^1 \frac{df}{\{f(1-f)(f+1+6\alpha R_m^{-1})\}^{\frac{1}{2}}}. \tag{5.6.14}$$

This integral is related to the 'complete elliptic integral of the first kind', and the numerical value for given α and R_m can be found from available

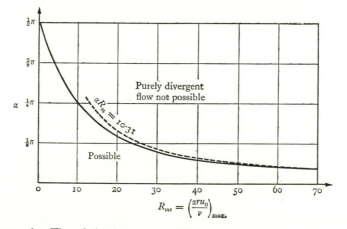

Figure 5.6.3. The relation between α and the maximum value of $R (= \alpha r u_0/\nu)$ for which purely divergent flow is possible.

tables. Figure 5.6.3 shows the relation between α and R_m represented by (5.6.14), which is in effect a restriction on the intensity of pure outflow in a channel of given angle.

When $R_m \gg 1$, the relation (5.6.14) becomes

$$\alpha R_m \approx \frac{3}{2}\left\{\int_0^1 \frac{df}{f^{\frac{1}{2}}(1-f^2)^{\frac{1}{2}}}\right\}^2$$

$$= 10\cdot31, \tag{5.6.15}$$

which involves α and R only in the combination αR as expected. Since αR is a measure of the ratio of the magnitudes of inertia and viscous forces in a flow with nearly parallel streamlines (§4.8), (5.6.15) implies that in purely divergent flow viscous forces do not decrease indefinitely in relative magnitude as the Reynolds number R tends to infinity and at no value of R are viscous forces negligible. At the other extreme, as $R_m \to 0$,

$$\alpha \to \frac{1}{2} \int_0^1 \frac{df}{f^{\frac{1}{2}}(1-f)^{\frac{1}{2}}}$$
$$= \tfrac{1}{2}\pi,$$

showing that in no circumstances is purely divergent flow possible when the angle between the channel walls exceeds π.

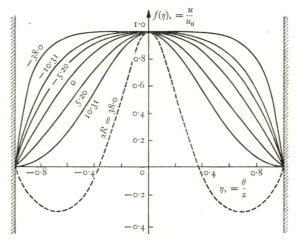

Figure 5.6.4. Symmetrical distributions of radial velocity in a divergent channel for various values of $\alpha R (= \alpha^2 r u_0/\nu)$. Positive values of αR represent outflow at the centre of the channel.

The change in character of the velocity distribution in purely divergent flow as αR increases from values satisfying $\alpha R \ll 1$ (when the velocity profile is parabolic) to values near the limit allowed by (5.6.14) is revealed by the velocity profiles shown in figure 5.6.4. These profiles have been calculated (Millsaps and Pohlhausen 1953) by direct integration of (5.6.8). Profiles for purely convergent flow with the same values of $|\alpha R|$ are shown for comparison. Whereas for purely convergent flow the effect of increasing the Reynolds number is to produce a flatter profile at the centre with steep gradients near the walls, the effect in purely divergent flow is to concentrate the volume flux at the centre of the channel with smaller gradients at the walls. The limiting situation represented by $c = 0$ and $R = R_m$ corresponds with zero wall stress. The volume flux Q can be calculated from these profiles using the relation
$$Q = \int_{-\alpha}^{\alpha} F \, d\theta = \nu R \int_{-1}^{1} f \, d\eta,$$
and it is evident that $\alpha Q/\nu$ is approximately a linear function of αR over the range $-\infty < \alpha R < 10\cdot31$, increasing less rapidly than linearly near the upper end of the range.

Solutions showing both outflow and inflow

Information about the nature of the velocity distribution when

$$\alpha R > 10\cdot31,$$

given that the profile is symmetrical about the plane $\theta = 0$, is contained in the above discussions of pure outflow and pure inflow. All the solutions shown in figure 5.6.4 can be continued into the region $\eta > 1$, and those that have a second zero of f' and thus of f can be interpreted as a combination of inflow and outflow in a channel of width chosen to suit the position of the second zero of f. One example of such a combination flow, showing a wide inflow region with nearly uniform velocity adjoining a narrow outflow region, was found simply by continuing the profile (5.6.12) to values of θ beyond the 'wall' value $\theta = \alpha$ (figure 5.6.2). Another is shown as a broken curve in figure 5.6.4; this has been obtained by continuing the solution for $\alpha R = 5\cdot20$ into the region $\eta > 1$ and then by contracting the abscissa scale so that the second zero of f occurs at $\eta = 1$, the associated value of αR being found from the required change in α. (In this way we see how the sequence of solutions should be continued to values of αR above the critical value $10\cdot31$.) Beyond the second zero of f the solutions repeat themselves, giving alternate regions of outflow and inflow, all the outflow profiles being identical and all the inflow profiles identical. Each region of either outflow or inflow is identical with the purely divergent or purely convergent flow, at an appropriate value of αR (provided $|R| \gg 1$), investigated above.

It is evident that the possibility of finding compound flows with zero values of f at $\eta = \mp 1$ increases as αR increases, and that several symmetrical solutions exist for a given (large) value of αR. For example, when $\alpha R = 114$ there are found to be three possible symmetrical distributions with outflow at the centre, the composition of these distributions being (i) one region of outflow and two of inflow, (ii) three of outflow and two of inflow, and (iii) three of outflow and four of inflow. The number of possible distributions increases with αR, although not in a way which can be specified simply. Similar remarks may be made about unsymmetrical distributions with an odd number of zeros of the velocity.

An interesting practical question is: what happens when fluid is discharged into a channel such that the angle between the (curved) walls increases very gradually from some small value near the inlet end satisfying the condition $\alpha R \ll 1$? At the inlet end of the channel there is a parabolic velocity distribution, and it is to be expected that as the effective value of αQ and thence of αR increases with distance downstream the profile passes through a sequence of shapes like those given in figure 5.6.4 for the range $0 \leqslant \alpha R \leqslant 10\cdot31$. When the local value of αR reaches and exceeds $10\cdot31$, purely divergent flow becomes impossible, and a region of inflow may be expected to appear near one or both walls. Experiment shows that something like this does occur, although diverging flow in a channel appears to be unstable and it is difficult to establish a steady state with regions of inflow near the walls.

From the point of view of the general theme of this chapter, the following aspects of the above family of similarity solutions are important. It is quite clear that there is a significant difference between the two cases of purely convergent and purely divergent flow in a channel, or, equivalently, between flow nearly parallel to a rigid wall with continual acceleration of all material elements and that with deceleration. In the flow with continual acceleration, vorticity is convected towards the wall and the vorticity generated at the wall is permanently confined to a layer adjoining the wall whose thickness tends to zero as the Reynolds number tends to infinity. In the extensive region outside this layer the velocity distribution has the form that would be predicted for an inviscid fluid. In the flow with continual deceleration, however, no similarity solution in which the fluid velocity is everywhere directed away from the source is possible at Reynolds numbers above a critical value; instead, we find similarity solutions in which there are regions of backward flow. This is a typical and practically important property of all flows with diverging streamlines, and it is also typical (judging by numerical investigations of flow in divergent channels in three dimensions) that the criterion for divergent flow without regions of backward moving fluid is of the form

$$\alpha \times \text{Reynolds number} < \text{number of order 10},$$

where the Reynolds number is based on the local maximum velocity and the local width of the region concerned. If the intensity of outflow in a channel is made very large, the possible similarity solutions contain many identical regions of outflow and of inflow. The width of each region of outflow is so small that effects of viscosity are significant everywhere in this region. Thus here the velocity distribution does not in any sense have the form that would be predicted for an inviscid fluid.

5.7. Boundary layers

The notion of a thin layer close to a solid boundary within which vorticity varies rapidly as a result of the combined effects of viscous diffusion and convection, and outside which the vorticity is zero (or is non-zero and varies only slowly), has been made clear in the preceding sections. We can now proceed to introduce the more general idea of a *boundary layer*, as being a thin layer in which the effect of viscosity is important however high the Reynolds number of the flow may be.

We have traced the development from rest of the flow due to a body moving through infinite fluid at a speed which ultimately is steady; and have noted how the solid boundary acts as a source of vorticity which is then diffused away by viscosity and convected downstream with the fluid (and which also in general undergoes changes due to rotation and stretching of vortex-lines, although these changes may be disregarded for the present purpose). As the Reynolds number of such a flow increases, the effect of convection at any point becomes relatively more important. We also saw that, in the cases of some flow systems involving a rigid boundary (for example, the flow due to an oscillating plane wall in § 4.3, or flow near a stagnation point at a plane wall in § 5.5, or convergent flow in a channel in § 5.6), the region in which viscosity has any effect on the flow shrinks to a thin layer at the boundary as $\nu \to 0$. These and many other particular cases suggest the important hypothesis, first advanced by Prandtl (1905), that, under rather broad conditions, viscosity effects (viz. stresses and forces due to viscosity, diffusion of vorticity, etc.) are significant, and comparable in magnitude with convection and other manifestations of 'inertia forces', in layers adjoining solid boundaries and in certain other layers, the thicknesses of which approach zero as the Reynolds number of the flow approaches infinity, and are small outside these layers. This hypothesis has been applied to very many different kinds of flow field since it was first advanced. No general mathematical proof of the boundary-layer hypothesis is available, but it is supported by many observations of particular flow systems, as well as by several of the known particular solutions of the complete equation of motion. The case of divergent flow in a channel examined in § 5.6 provides a useful warning that the boundary-layer hypothesis is not applicable to all flow systems. A simple statement of the class of flow field to which the hypothesis does apply cannot be given, but working rules governing use of the hypothesis will emerge from the discussion in the remainder of this chapter.

The boundary-layer hypothesis helps to reconcile the intuitive expectation that the effects of viscosity on the flow are unimportant, at any rate over most of the flow field, when the viscosity ν is small, with the fact that the no-slip condition must be satisfied at a solid boundary however small ν may be; indeed this reconciliation was Prandtl's main objective and was a land-mark in the development of fluid mechanics. The boundary layer is in effect the layer in which the fluid velocity makes a transition from the required value zero (relative to the solid) at the boundary to a finite value which is appropriate, in some sense to be examined more closely later, to an inviscid fluid.

The fact that a boundary layer is thin, compared with linear dimensions of the boundary, makes possible certain approximations in the equations of motion, also due to Prandtl, and thereby the flow in the boundary layer may be determined in certain cases. For the purpose of explaining these approximations, we take the boundary to be a plane wall (at $y = 0$) and the flow to be two-dimensional. The boundary-layer thickness (defined in any convenient manner) is supposed everywhere to be small compared with distances parallel to the boundary over which the flow velocity changes appreciably. Across the boundary layer the flow velocity changes from the value zero at the boundary to some finite value characteristic of an inviscid fluid, and derivatives with respect to y of any flow quantity are in general much larger than those with respect to x. Thus at points within the boundary layer we may use the approximations

$$\left|\frac{\partial u}{\partial x}\right| \ll \left|\frac{\partial u}{\partial y}\right|, \quad \left|\frac{\partial^2 u}{\partial x^2}\right| \ll \left|\frac{\partial^2 u}{\partial y^2}\right|,$$

whence the equation of motion in the x-direction becomes

$$\frac{\partial u}{\partial t} + u \frac{\partial u}{\partial x} + v \frac{\partial u}{\partial y} = -\frac{1}{\rho}\frac{\partial p}{\partial x} + \nu \frac{\partial^2 u}{\partial y^2}. \tag{5.7.1}$$

The velocity component v normal to the boundary must also be small, and the mass-conservation equation, viz.

$$\frac{\partial u}{\partial x} + \frac{\partial v}{\partial y} = 0, \tag{5.7.2}$$

suggests that v and the boundary-layer thickness are of equal order of smallness;† as a consequence none of the terms on the left-hand side of (5.7.1) may safely be neglected.

The difference between the boundary-layer equation (5.7.1) and the corresponding equation of motion for the inviscid flow region outside the boundary layer lies in the retention in (5.7.1) of the term $\nu \, \partial^2 u/\partial y^2$, which represents viscous diffusion across the boundary layer. The boundary layer is, by definition, the region in which viscous diffusion of vorticity is significant, so that we expect the term $\nu \, \partial^2 u/\partial y^2$ in (5.7.1) to be comparable in magnitude with the 'inertia terms' on the left-hand side at positions within the boundary layer. Provided the magnitude of $u \, \partial u/\partial x$ is representative of the group of inertia terms on the left-hand side of (5.7.1), we can therefore regard the boundary-layer region as being characterized by the order-of-magnitude relation

$$O\left(u \frac{\partial u}{\partial x} \middle/ \nu \frac{\partial^2 u}{\partial y^2}\right) = 1,$$

when the Reynolds number of the flow as a whole is large.

Now if U_0 is representative of the magnitude of u in the flow field as a whole and L represents a distance in the x-direction over which u changes appreciably, U_0^2/L is a measure of the magnitude of $u \, \partial u/\partial x$. And if δ_0 is a small length representative of the boundary thickness,‡ $\nu U_0/\delta_0^2$ is a measure of the magnitude of $\nu \, \partial^2 u/\partial y^2$. The above relation can then be written as

$$O\left(\frac{\delta_0^2}{L^2} R\right) = 1, \quad \text{where } R = \frac{U_0 L}{\nu}. \tag{5.7.3}$$

R is a Reynolds number representative of the flow as a whole, and since the

---

† A vague statement, in view of their different dimensions, but it will be made more precise shortly.

‡ The thickness of the boundary layer in general varies with position on the boundary, and δ_0 should be regarded as an average thickness.

approximations on which the boundary-layer concept is based improve as $R \to \infty$, we may evidently write (5.7.3) as

$$\delta_0/L \sim R^{-\frac{1}{2}} \quad \text{as} \quad R \to \infty. \tag{5.7.4}$$

The fact that the boundary-layer thickness varies as $\nu^{\frac{1}{2}}$ when ν is small is already familiar from its appearance in a number of special cases. The underlying reason is essentially a dimensional one—which gives the result great generality—but we have seen that in some circumstances it can also be thought of as due to the relation:

diffusion distance (for vorticity or velocity) $\propto (\nu t)^{\frac{1}{2}}$

for a layer which has developed in a time interval t (which is equivalent to L/U_0 in the above argument).

With this estimate of the order of magnitude of δ_0, and thus of derivatives with respect to y, (5.7.2) shows that v is of order $U_0 R^{-\frac{1}{2}}$. We are now in a position to examine the equation of motion in the y-direction. All terms except one are evidently small, and we are left with the approximate relation

$$\partial p/\partial y = 0; \tag{5.7.5}$$

more precisely, $\partial p/\partial y$ is of the same order of smallness as δ_0. The pressure is thus approximately uniform across the boundary layer; and if it happens that the variation of p with x just outside the boundary layer is known— perhaps from a consideration of the inviscid flow equations in the region outside the boundary layer, perhaps from measurements—the pressure term in (5.7.1) can be regarded as given. Equations (5.7.1) and (5.7.2) are then to be used to determine u and v throughout the boundary layer.

The boundary conditions are, first, that

$$u = v = 0 \quad \text{at} \quad y = 0; \tag{5.7.6}$$

and, second, that the boundary layer must join smoothly with the region of inviscid flow outside it. If U is the x-component of velocity just outside the boundary layer (and U is not a rapidly varying function of y, so that the impossibility of locating exactly the 'edge' of the boundary layer is of no consequence) we can express this second condition as

$$u(x, y, t) \to U(x, t) \quad \text{as} \quad y/\delta_0 \to \infty. \tag{5.7.7}$$

Like p, U must be regarded as given, in a consideration of the boundary layer alone, and the two are related by the approximate equation

$$\frac{\partial U}{\partial t} + U \frac{\partial U}{\partial x} = -\frac{1}{\rho} \frac{\partial p}{\partial x} \tag{5.7.8}$$

describing inviscid flow in the x-direction just outside the boundary layer (where v is small and the term $v \, \partial U/\partial y$ may be neglected); when the flow is steady, (5.7.8) is equivalent to Bernoulli's theorem, that $p/\rho + \frac{1}{2}U^2$ is constant along the streamline at the outer edge of the boundary layer. A third boundary condition is needed to describe the way in which vorticity is convected into any portion of the boundary layer from further upstream; that is, $u(y)$ must be specified at some value of x. Finally, if the motion is not steady, $u(x, y)$ must be specified at $t = 0$.

The asymptotic variation of δ_0 given in (5.7.4) may be used to transform the boundary-layer equations so that they do not involve the Reynolds number (or the viscosity). In order to obtain a more natural co-ordinate system in which lateral distances and velocities are measured with the (representative) boundary-layer thickness as the unit of length, we define dimensionless quantities

$$\left. \begin{aligned} x' &= \frac{x}{L}, \quad y' = R^{\frac{1}{2}}\frac{y}{L}, \quad t' = \frac{tU_0}{L}, \\ u' &= \frac{u}{U_0}, \quad v' = R^{\frac{1}{2}}\frac{v}{U_0}, \quad p' = \frac{p-p_0}{\rho U_0^2}, \end{aligned} \right\} \tag{5.7.9}$$

where p_0 is the value of p at some convenient reference point in the fluid. In terms of these new variables, the *complete* equations of motion in the x- and y-directions and the equation of mass conservation become

$$\left. \begin{aligned} \frac{\partial u'}{\partial t'} + u'\frac{\partial u'}{\partial x'} + v'\frac{\partial u'}{\partial y'} &= -\frac{\partial p'}{\partial x'} + \frac{1}{R}\frac{\partial^2 u'}{\partial x'^2} + \frac{\partial^2 u'}{\partial y'^2}, \\ \frac{1}{R}\left(\frac{\partial v'}{\partial t'} + u'\frac{\partial v'}{\partial x'} + v'\frac{\partial v'}{\partial y'}\right) &= -\frac{\partial p'}{\partial y'} + \frac{1}{R^2}\frac{\partial^2 v'}{\partial x'^2} + \frac{1}{R}\frac{\partial^2 v'}{\partial y'^2}, \\ \frac{\partial u'}{\partial x'} + \frac{\partial v'}{\partial y'} &= 0. \end{aligned} \right\} \tag{5.7.10}$$

If we now suppose that R is large, and that the dimensionless quantities u', v', p' and their derivatives with respect to x', y', t' at given values of x', y', t' remain finite and non-zero as $R \to \infty$ (as is implied by the boundary-layer hypothesis), an approximate form of the equations, which becomes exact in the limit $R \to \infty$, is

$$\left. \begin{aligned} \frac{\partial u'}{\partial t'} + u'\frac{\partial u'}{\partial x'} + v'\frac{\partial u'}{\partial y'} &= -\frac{\partial p'}{\partial x'} + \frac{\partial^2 u'}{\partial y'^2}, \\ 0 &= -\frac{\partial p'}{\partial y'}, \\ \frac{\partial u'}{\partial x'} + \frac{\partial v'}{\partial y'} &= 0. \end{aligned} \right\} \tag{5.7.11}$$

These are simply the transformed versions of (5.7.1), (5.7.5) and (5.7.2). The equations (5.7.11) do not contain R explicitly; nor in most common cases do the boundary conditions when expressed in terms of the above dimensionless variables; nor, as a consequence, does the solution. The role of the Reynolds number is solely to determine the thickness of the boundary layer, and the set of boundary layers corresponding to different values of R but the same (non-dimensional) boundary conditions are identical when scaled to a common thickness.

For simplicity of explanation, the boundary-layer flow has been assumed in this discussion to be two-dimensional and to adjoin a plane, rigid, wall. None of these restrictions is essential. When the flow field as a whole is three-dimensional, boundary layers form near rigid walls and in general the velocity vector in such boundary layers changes direction along the normal to the wall, while remaining nearly parallel to the wall. Again the equations describing flow in the boundary layer can be so transformed as to eliminate the Reynolds number. When the boundary layer adjoins a curved wall, it is natural, in the case of a two-dimensional system, to replace the rectilinear co-ordinates x and y by orthogonal curvilinear co-ordinates x and y such that the line $y = 0$ coincides with the curved wall. The curvature of the wall then enters into the complete equations of motion, but it can be shown (and is fairly evident) that the only effect of the wall curvature κ on the approximate equations for two-dimensional flow is to modify (5.7.5), which becomes

$$\partial p / \partial y = \rho \kappa u^2. \tag{5.7.12}$$

Provided κ is finite, the total change in pressure across the boundary layer is thus of order δ_0 and is still negligible, so that the relation (5.7.8) between the pressure in the boundary layer and the velocity U just outside the boundary layer still stands.

Moreover, the rigidity of the boundary is not necessary for the existence of a boundary layer (although a rigid wall is the commonest cause of formation of a boundary layer, and is the cause which Prandtl had in mind when forming the boundary-layer hypothesis), and affects the above argument only through the boundary condition (5.7.6). In general, a boundary layer will occur at any boundary at which the conditions required by the nature of the boundary are not satisfied exactly by a velocity distribution derived from the inviscid-fluid equations. A boundary layer may exist at a 'free' surface, at which the tangential stress must be zero (see §5.14). A thin layer in which viscous effects are significant may also exist between two regions in each of which the flow is approximately as for an inviscid fluid, in which case boundary conditions of the type (5.7.7) apply on both sides of the layer. The transition layer between two parallel uniform streams of different speeds, discussed in §4.3, is such a detached or free 'boundary' layer, although it happens in that case that the terms neglected in the above boundary-layer equations are identically zero and no appeal to approximation was necessary. Under certain conditions, which amount to a requirement that the appropriate Reynolds numbers be large, jets and wakes may also be regarded as free 'boundary' layers. It is evident that there must be at least one detached vorticity layer extending downstream from a solid body moving through fluid, because the vorticity generated at the boundary is swept backwards and is eventually carried off the rear of the body; and if the lateral gradient of vorticity is large in the layer adjoining the body boundary, it will also be large in the detached layer or wake downstream of the body, so that viscous diffusion of vorticity will be important there, at any rate for some distance downstream until the detached layer has widened considerably.

In the remaining sections of this chapter, the properties of boundary layers will be described briefly and the leading part played by boundary layers in flow at large Reynolds number will be demonstrated by discussion of some particular flow fields. The subject of boundary layers is as extensive as it is important, and we can give here only an introduction. For the sake of simplicity, only cases of either two-dimensional or axisymmetric flow will be considered; in such cases no rotation of vortex-lines occurs, and the extension of vortex-lines (in axisymmetric flow) has a specially simple character. Readers should not get the impression that these are the only cases that are of interest or that are analytically tractable.

5.8. The boundary layer on a flat plate

A simple case of steady two-dimensional boundary-layer flow occurs when a flat plate of very small thickness, length l and much larger breadth, is placed in a steady uniform stream of fluid (meaning a stream whose velocity would be uniform in the absence of the plate), with the stream parallel to the length l and normal to the edge of the plate. This case is important because it provides a standard of comparison for the skin friction on flat slender bodies, such as aeroplane wings, aligned edge-on to a stream. As a convenient idealization of real conditions, we shall regard the flat plate as being of zero thickness. In the absence of any effects of viscosity, then, the plate causes no disturbance to the stream and the fluid velocity is uniform, with magnitude U say. For a real fluid, for which the no-slip condition must be satisfied, the fluid near the plate is retarded—or, equivalently, vorticity is diffused away from the plate—and a layer in which the velocity is different from U and in which the vorticity is non-zero forms near the plate. In the resulting steady state the boundary-layer thickness will be small compared with l everywhere provided $lU/\nu \gg 1$. As a consequence of the retardation of fluid near the plate, streamlines outside the boundary layer are deflected laterally; so far as this region of inviscid flow is concerned, it is as if the plate had become endowed with a certain thickness round which the streamlines must pass. However, provided the boundary-layer thickness is everywhere small, the disturbance to the distribution of velocity in the region of inviscid flow is small and may be neglected as a first approximation.

With this approximation, the velocity just outside the boundary layer is uniform and equal to U. The pressure is likewise uniform just outside the boundary layer and is thus approximately uniform throughout the boundary layer, whence the boundary-layer equations (5.7.1) and (5.7.2) reduce (with the further assumption of steady flow) to

$$u\frac{\partial u}{\partial x} + v\frac{\partial u}{\partial y} = \nu\frac{\partial^2 u}{\partial y^2}, \tag{5.8.1}$$

$$\frac{\partial u}{\partial x} + \frac{\partial v}{\partial y} = 0. \tag{5.8.2}$$

The second of these two equations is satisfied identically by writing $u = \partial\psi/\partial y$, $v = -\partial\psi/\partial x$, and then equation (5.8.1) contains just one dependent variable (ψ). If we place the origin of co-ordinates at the leading edge of the plate, so that the trailing edge is at $x = l$, $y = 0$, the boundary conditions on the two sides of the layer are

$$\left.\begin{array}{l} u = v = 0 \quad \text{at} \quad y = 0 \\ u \to U \quad \text{as} \quad y/\delta_0 \to \infty \end{array}\right\} \quad \text{for } 0 \leqslant x \leqslant l,$$

and the boundary condition describing the upstream conditions is simply

$$u = U \quad \text{at} \quad x = 0 \quad \text{for all } y.$$

This is the complete set of equations and boundary conditions.

The local boundary-layer thickness, δ say, again defined in any convenient way, is here a function of x. It is evident that δ must increase with distance x from the leading edge of the plate, since the frictional force exerted by each additional portion of the plate surface contributes to the loss of momentum of the fluid passing over it. Inasmuch as the time spent by a fluid particle in proximity to the plate while moving at (constant) speed U is x/U, the usual diffusion argument leads us to expect that the local boundary-layer thickness increases as $(\nu x/U)^{\frac{1}{2}}$. Another way of obtaining this important conclusion is to note that since the distribution of velocity in the boundary layer at station x is determined entirely by viscous diffusion in the y-direction and convection of vorticity from *upstream* positions, and cannot depend on the existence of the solid boundary further downstream, except through its influence on the distribution of velocity just outside the boundary layer (there being no such influence here), $\delta(x)$ cannot depend on l. On the other hand, we know from (5.7.4) that the length δ_0 representative of the thickness of the boundary layer over the whole plate is proportional to $(\nu l/U)^{\frac{1}{2}}$, and for consistency we must have

$$\delta \propto \left(\frac{\nu x}{U}\right)^{\frac{1}{2}}. \tag{5.8.3}$$

The approximate equation of motion (5.8.1) will now be transformed in much the same way that led to the dimensionless equation (5.7.11). Since the velocity distribution across the boundary layer at station x is here independent of l, it is more natural to use the (variable) local boundary-layer thickness rather than a representative thickness as the scaling unit for y and to leave the co-ordinate x unchanged. We write

$$\eta = \left(\frac{U}{\nu x}\right)^{\frac{1}{2}} y, \quad u' = \frac{u}{U}, \quad v' = \left(\frac{x}{\nu U}\right)^{\frac{1}{2}} v, \tag{5.8.4}$$

and the equivalent transformation of the stream function ψ is

$$f(x, \eta) = (\nu U x)^{-\frac{1}{2}} \psi(x, y).$$

Equation (5.8.1) then becomes

$$x\left(\frac{\partial f}{\partial \eta} \frac{\partial^2 f}{\partial x \partial \eta} - \frac{\partial f}{\partial x} \frac{\partial^2 f}{\partial \eta^2}\right) - \left(\tfrac{1}{2} f \frac{\partial^2 f}{\partial \eta^2} + \frac{\partial^3 f}{\partial \eta^3}\right) = 0, \tag{5.8.5}$$

the partial differentiation with respect to x now being with η held constant.

Evidently there exists a solution for f which is independent of x, provided we can find a solution of

$$\tfrac{1}{2} f f'' + f''' = 0 \tag{5.8.6}$$

(where the prime now denotes differentiation with respect to η) which fits the boundary conditions

$$f = \frac{df}{d\eta} = 0 \quad \text{at} \quad \eta = 0,$$

and

$$\frac{df}{d\eta} \to 1 \quad \text{as} \quad \eta \to \infty.$$

The condition that $u = U$ at $x = 0$ is already satisfied by a solution such that f (and hence also $u = \partial\psi/\partial y$) is a function of $y/x^{\frac{1}{2}}$ only. Numerical integration of (5.8.6) does in fact show that a solution satisfying the above boundary conditions can be found,[†] and the resulting velocity distribution is shown in figure 5.8.1. No other solution of (5.8.5) which satisfies the boundary conditions is known, and in the absence of further information we presume that the above solution which makes u/U a function of $(U/\nu x)^{\frac{1}{2}} y$ alone is in fact the only solution. Many measurements have been made of the distribution of the velocity u in the boundary layer on a smooth flat plate of small thickness carefully aligned with the stream, and these show good agreement with the distribution shown in figure 5.8.1. The parabolic growth of boundary-layer thickness represented by (5.8.3) has also been confirmed adequately by measurements.

One of the useful features of the solution is the estimate which it gives for the tangential force exerted on the plate by the fluid. The frictional force per unit area of the plate at distance x from the leading edge is

$$\mu\left(\frac{\partial u}{\partial y}\right)_{y=0} = \rho U^2 \left(\frac{Ux}{\nu}\right)^{-\frac{1}{2}} f''(0),$$

$$= 0 \cdot 33 \rho U^2 \left(\frac{Ux}{\nu}\right)^{-\frac{1}{2}} \tag{5.8.7}$$

[†] The solution was first found by Blasius (1908) in series form, following Prandtl's earlier work on boundary layers, and has been improved numerically by later writers.

according to the numerical solution; the variation as $x^{-\frac{1}{2}}$ is of course simply a reflection of the increase of boundary-layer thickness as $x^{\frac{1}{2}}$, since the shape of the velocity profile is independent of x. The drag exerted on the two sides of unit width of a plate of length l is then

$$D = 2 \int_0^l \mu \left(\frac{\partial \mu}{\partial y}\right)_{y=0} dx = 1 \cdot 33 \rho U^2 l \left(\frac{Ul}{\nu}\right)^{-\frac{1}{2}}. \qquad (5.8.8)$$

This estimate of the total frictional drag applies approximately to any slender two-dimensional body of length l aligned with the stream (§ 5.11).

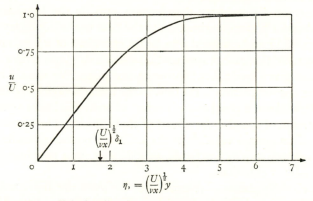

Figure 5.8.1. Velocity distribution in the boundary layer on a flat plate.

The numerical value of the boundary-layer thickness may also be found. Figure 5.8.1 shows that the value of u/U reaches the value $0 \cdot 99$ near $y = 4 \cdot 9(\nu x/U)^{\frac{1}{2}}$. A less arbitrary measure of thickness of the boundary layer is provided by the *displacement thickness*, defined as

$$\delta_1 = \int_0^\infty \left(1 - \frac{u}{U}\right) dy, \qquad (5.8.9)$$

which can be thought of as the distance through which streamlines just outside the boundary layer are displaced laterally by the retardation of fluid in the boundary layer. The numerical solution shows that

$$\delta_1 = 1 \cdot 72 \left(\frac{\nu x}{U}\right)^{\frac{1}{2}}, \qquad (5.8.10)$$

which is indicated on figure 5.8.1. With $U = 100$ cm/sec and $x = 10$ cm, this formula gives δ_1 equal to $0 \cdot 21$ cm and $0 \cdot 06$ cm for air and water respectively at normal temperature.

It will be recalled that the approximate boundary-layer equations hold only when the Reynolds number based on linear dimensions of the rigid boundary is large and when the magnitude of $\partial u/\partial x$ is small compared with that of $\partial u/\partial y$. These conditions are satisfied for a flat plate of length l when

$Ul/\nu \gg 1$, and with increasing accuracy as $x \to \infty$, everywhere except in the neighbourhood of the leading edge at $x = 0$. Within a small region here, the Reynolds number Ux/ν is of order unity, δ (as estimated by (5.8.3)) is of the same order as x, and variations with respect to x are not slow compared with those with respect to y. For this reason we cannot expect the above description of the flow to be valid within a distance of order ν/U from the leading edge of the plate. Improved approximations to the flow in this region are described in more specialized books on the subject.†

It is also possible to improve the description of the flow at values of x for which $Ux/\nu \gg 1$ by taking some account of the effect of the presence of the boundary layer on the velocity distribution outside the boundary layer. In order to obtain a first approximation to the flow in the boundary layer we neglected this effect and assumed u to be independent of x just outside the boundary layer. On this basis it appears that a boundary layer with a displacement thickness given by (5.8.10) forms at the plate. We should evidently obtain a better approximation to the flow in the inviscid region by finding the irrotational flow about a parabolic cylinder, of semi-thickness $1.72(\nu x/U)^{\frac{1}{2}}$, placed in a stream of speed U far ahead of the cylinder. The corresponding distribution of tangential velocity at the cylinder boundary could then be used as the velocity to which u tends, as $y/\delta \to \infty$, in a new integration of the boundary-layer equations.

There are obvious points of resemblance between the above steady flow along a flat plate, in which the boundary layer thickens as $(\nu x/U)^{\frac{1}{2}}$ while maintaining the same velocity profile, and the unsteady flow due to an infinite plate which is suddenly brought to velocity U in fluid initially at rest. In this latter flow, described in §4.3, the 'boundary layer' thickens as $(\nu t)^{\frac{1}{2}}$ and the velocity distribution maintains the same shape at all t. Rayleigh suggested that the time-dependent flow with an infinite plate could be regarded as an approximation to steady flow past a semi-infinite plate when t is replaced by x/U (and with axes fixed relative to the plate in both cases), and this same approximation has sometimes been adopted (for lack of more accurate solutions) in other problems of steady flow past semi-infinite bodies for which the velocity outside the boundary layer is uniform. The analogy between the two kinds of flow is only qualitative, as may be seen from a comparison of the values obtained for the frictional force per unit area of the flat plate, viz. $0.56\rho U^2(U^2 t/\nu)^{-\frac{1}{2}}$ for the infinite plate and $0.33\rho U^2(Ux/\nu)^{-\frac{1}{2}}$ for the semi-infinite plate. The difference between the two flows may be stated in terms of the equation of motion for a flow with nearly uniform pressure and slow variations in the x-direction, viz.

$$\frac{\partial u}{\partial t} + u\frac{\partial u}{\partial x} + v\frac{\partial u}{\partial y} = \nu\frac{\partial^2 u}{\partial y^2}.$$

† See *Laminar Boundary Layers*, edited by L. Rosenhead (Oxford University Press, 1963), which may also be consulted for further general information about the subject of this chapter.

For steady flow over a semi-infinite plate, the left-hand side becomes $u\,\partial u/\partial x + v\,\partial u/\partial y$, whereas for time-dependent flow over an infinite plate it becomes $\partial u/\partial t$, or $U\,\partial u/\partial \tau$ if we put $\tau = Ut$. The terms $u\,\partial u/\partial x + v\,\partial u/\partial y$ would be approximated well by $U\,\partial u/\partial \tau$ (with $\tau = x$) in the case of flow in which the velocity (u, v) is close to the uniform stream velocity $(U, 0)$ (and this was of course the basis of the Oseen equations (4.10.2), which were intended specifically to represent flow in the region far away from a body placed in a uniform stream), but not otherwise, because the replacement of u by U and of $v\,\partial u/\partial y$ by zero are changes usually in the same direction.

The above calculation of velocity distribution near the flat plate, and the associated estimates of frictional force and displacement thickness, are valid only if the flow in the boundary layer is steady, or 'laminar', over the whole surface of the flat plate. In fact, the flow in the boundary layer becomes unstable when the local Reynolds number $\delta_1 U/\nu$ exceeds a value near 600. In these circumstances, disturbances in the boundary layer grow and a transition to a different type of flow occurs at some distance downstream. Observation shows that the flow in this new régime is characterized by permanent and random unsteadiness, although the distribution of steady mean velocity exhibits the same general boundary-layer character. The frictional force at the wall in such a turbulent boundary layer is considerably larger than that in a laminar boundary layer with the same external stream speed, because the random cross-currents in the boundary layer carry the fast-moving fluid in the outer layers into the neighbourhood of the wall and are more effective in promoting lateral transport than molecular diffusion.

The subject of turbulent flow lies outside the scope of this book, but in view of the important role of the total frictional force on a flat plate as a standard of comparison for all two-dimensional slender bodies it is desirable to take brief note of some of the data. It follows from the above criterion for stability and the formula (5.8.10) that laminar flow will occur everywhere in the boundary layer on a flat plate of length l when lU/ν is less than about $1\cdot 2 \times 10^5$ (corresponding to flat plates of length 180 cm and 13 cm in a stream of speed 100 cm/sec for air and water respectively at normal temperature). As the Reynolds number is increased above this value, transition to turbulent flow occurs first near the trailing edge and gradually moves upstream, with an accompanying rise in the total frictional force on the plate. Figure 5.8.2 shows the kind of variation of total drag with Reynolds number that is observed in a wind tunnel or a water channel. At very large Reynolds number, most of the boundary layer on the plate is turbulent and the (non-dimensional) total frictional force again decreases with Reynolds number although not as rapidly as for a wholly laminar layer. Variations in the degree of steadiness of the oncoming stream at a fixed Reynolds number can change appreciably the position of the transition from laminar to turbulent flow in the boundary layer. The position of the curved line joining the two lines representing the total drag for wholly laminar and wholly turbulent layers in figure 5.8.2 thus varies according to the conditions; for a rather disturbed stream the first departure of the measured resistance from the line for a wholly laminar layer may occur near $lU/\nu = 10^5$, whereas for a very smooth stream obtained with modern wind tunnel techniques it may not occur until $lU/\nu = 4 \times 10^6$. The smoothness of the surface of the flat plate and the detailed shape of the leading edge may also be relevant to the location of the transition from laminar to turbulent flow in the boundary layer, especially in a stream with a very low level of disturbances.

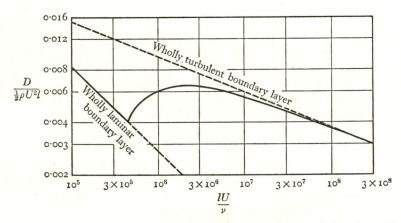

Figure 5.8.2. The total frictional force D on unit width of a smooth flat plate of length l in a stream of speed U. The curved transition line has a shape typical of observations, but its position depends on the experimental conditions.

5.9. The effects of acceleration and deceleration of the external stream

The steady two-dimensional flow near a 'stagnation point' considered in § 5.5 is an example of a flow field in which the vorticity generated at a solid surface is prevented from diffusing far from the surface by the action of convection. The resulting layer of non-zero vorticity can be regarded as a boundary layer with an external stream velocity of the form $U = kx$; the velocity distribution in the layer satisfies the approximate boundary-layer equation (5.7.1), as well as the complete equation of motion, because the neglected term $\partial^2 u/\partial x^2$ is identically zero in this case. As was seen, the thickness of the layer of non-zero vorticity, as determined by the opposing actions of convection towards the wall and viscous diffusion away from it, is uniform, and the frictional force per unit area at the wall increases with x. In the case of purely convergent flow in a channel at large Reynolds number (§5.6), the effect of convection of vorticity towards the wall is even more powerful and the thickness of the boundary layer decreases with increase of distance in the flow direction. Another case of a boundary layer on a plane wall was analysed in § 5.8, the velocity of the external stream being uniform;

here there is no convection of vorticity in the direction normal to the wall (apart from the small lateral velocity caused by the presence of the boundary layer itself), the boundary-layer thickness increases with x under the action of diffusion alone, and the wall friction decreases as x increases.

A third possible type of boundary layer, of which no example has yet been given, is one in which the speed of the external stream diminishes as x increases. It follows from the mass-conservation relation

$$v(y) = -\int_0^y \frac{\partial u}{\partial x} dy \qquad (5.9.1)$$

that in such cases the (small) normal component of velocity is directed *away* from the wall in the outer part of the boundary layer (where $\partial u/\partial x$ is certain to have the same sign as $\partial U/\partial x$) at least and probably throughout it. Convection and diffusion now combine to transport vorticity away from the wall and it is to be expected that the boundary-layer thickness increases rapidly with x; in some circumstances the rate of thickening is apparently so great as to prevent the formation of a boundary layer at all (as in flow in a divergent channel—see § 5.6) and in others, as we shall see, to cause the phenomenon of separation of the boundary layer in which event the external stream ceases to flow approximately parallel to the boundary beyond a certain point. The possibility of separation, with a consequent radical effect on the character of the whole flow field, makes the case of a decelerating external stream of particular importance.

The effect of acceleration or deceleration of the external stream on a boundary layer can also be stated in dynamical terms. The pressure is approximately uniform across a boundary layer, so that the pressure gradient which produces the acceleration of the external stream acts equally on the fluid within the boundary layer. Now the equation of motion for inviscid fluid shows that the rate of change of speed q with respect to distance s along a streamline in steady flow is given by

$$\frac{\partial q}{\partial s} = -\frac{1}{\rho q}\frac{\partial p}{\partial s},$$

and is thus numerically greater, for a given pressure gradient, for the slower-moving layers of fluid near the wall than for the external stream. In this way, a negative (or accelerating) pressure gradient tends to diminish the variation of velocity across the boundary layer and to decrease the thickness of the layer, while a positive pressure gradient has the opposite tendency. To this effect of the pressure gradient must be added the effect of viscosity, and of wall friction in particular, which continually takes momentum from the boundary layer and tends to thicken it. These two effects balance exactly, so far as boundary-layer thickness is concerned, for an external stream whose velocity increases in proportion to x.

The similarity solution for an external stream velocity proportional to x^m

The way in which acceleration or deceleration of the external stream affects the variation of boundary-layer thickness and skin friction with distance along the boundary is shown clearly by a family of solutions for steady two-dimensional flow given by Falkner and Skan (1930). These authors noticed that in the case of the external stream velocity

$$U = cx^m, \qquad (5.9.2)$$

where $c\,(>0)$ and m are constants, it is possible to obtain a solution of the boundary-layer equations of the form

$$\psi = (\nu Ux)^{\frac{1}{2}} f(\eta), \quad \eta = (U/\nu x)^{\frac{1}{2}} y. \qquad (5.9.3)$$

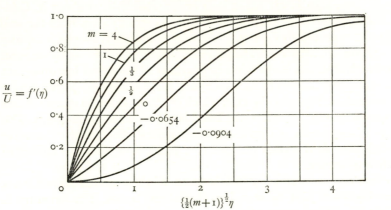

Figure 5.9.1. The distribution of velocity across the boundary layer for an external stream velocity $U = cx^m$.

With this transformation, and after elimination of the pressure using the relation (5.7.8) between U and p, the boundary-layer equation (5.7.1) becomes

$$mf'^2 - \tfrac{1}{2}(m+1)ff'' = m + f'''. \qquad (5.9.4)$$

This equation reduces to (5.5.13) when $m = 1$, corresponding to flow toward a 'stagnation point' at a plane wall, and to (5.8.6) when $m = 0$, corresponding to flow over a flat plate aligned with a uniform stream. As will be seen in §6.5, the velocity variation (5.9.2) occurs on the surface of a wedge of semi-angle $\pi m/(m+1)$ placed symmetrically in an irrotational stream of inviscid fluid (x being measured from the vertex); thus (5.9.4) can be regarded as governing the flow in the boundary layer at the surface of such a wedge, with negative values of m corresponding to flow over a flat plate which is inclined away from the stream (although the infinite value of U at $x = 0$ when $m < 0$ would not be realized in practice). However, that possible application of the results is not the point of immediate interest.

The boundary conditions to be satisfied at the inner and outer edges of the layer are

$$f(0) = f'(0) = 0, \quad f'(\eta) \to 1 \text{ as } \eta \to \infty,$$

and these are sufficient to determine the solution of (5.9.4). No conditions at $x = 0$ can be imposed, because the assumed form of solution (5.9.3) is such that the variation of u with y has the same shape at all values of x. The solution of (5.9.4) with these boundary conditions has been obtained numerically for many values of m by Hartree (1937), and the velocity profiles for several of these values are shown in figure 5.9.1; the abscissa for this graph is $\{\frac{1}{2}(m+1)\}^{\frac{1}{2}} \eta$ because that happened to be a more convenient variable than η for the numerical integration.

Two important parameters of the solution for each value of m are the displacement thickness (δ_1) and the frictional stress at the wall (τ_0 say). We find from (5.9.3) that

$$\delta_1 = \int_0^\infty \left(1 - \frac{u}{U}\right) dy = \left(\frac{\nu x}{U}\right)^{\frac{1}{2}} \int_0^\infty (1 - f') \, d\eta,$$

$$\propto x^{\frac{1}{2}(1-m)}, \tag{5.9.5}$$

and

$$\tau_0 = \mu \left(\frac{\partial u}{\partial y}\right)_{y=0} = \rho \left(\frac{\nu U^3}{x}\right)^{\frac{1}{2}} f_0'',$$

$$\propto x^{\frac{1}{2}(3m-1)}, \tag{5.9.6}$$

and these expressions show explicitly how the dependence of layer thickness and wall stress on x changes with the degree of acceleration or deceleration of the external stream. When $m = \frac{1}{3}$, it seems that the wall stress is uniform; the tendency for the velocity gradient at the wall to increase as a result of acceleration of the external stream is here balanced exactly by the opposite tendency due to diffusive thickening of the boundary layer.

No drastic change in the behaviour of δ_1 and τ_0 at any negative value of m is indicated by (5.9.5) and (5.9.6), but there are signs of such a change in the profiles shown in figure 5.9.1. It will be observed that the profiles for a decelerated external stream have a point of inflexion. This property is in fact general, as may be seen from the form taken by the equation of motion at the boundary, viz.

$$\nu \left(\frac{\partial^2 u}{\partial y^2}\right)_{y=0} = \frac{1}{\rho} \frac{\partial p}{\partial x}, \quad \approx -\frac{1}{2} \frac{dU^2}{dx}; \tag{5.9.7}$$

$\partial^2 u / \partial y^2$ is thus positive at the wall for a decelerating external stream, and, since $u \to U$ as $y \to \infty$, $\partial^2 u / \partial y^2$ must change sign for some value of y. In the case $U = \text{const.}$ (i.e. $m = 0$), the point of inflexion occurs at the boundary. As m decreases from zero, the positive value of $\partial^2 u / \partial y^2$ at the wall increases, and this, together with the constraint $u \to U$ as $y \to \infty$, leads to a continual decrease of $(\partial u / \partial y)_{y=0}$. At $m = -0.0904$ (which is the smallest value of m for which a complete numerical solution is available) the velocity gradient at the wall vanishes.† At even smaller values of m, there is a small region near the wall in which $u < 0$, i.e. where back-flow is occurring. It is found in these latter cases that $u > U$ further from the wall, and that, as $y \to \infty, u \to U$ from above; presumably the meaning of this is that such a profile cannot maintain its shape for all x unless there is something like a jet in the outer part of the layer to supply forward momentum to the boundary layer. Boundary conditions (at some initial value of x) which provide a jet in the outer part of the layer would be rather artificial, and are unlikely to occur as a consequence of past history of a boundary layer with some given variation of U (not necessarily of power-law form), so that solutions for $m < -0.0904$ may not have practical significance.

An important conclusion from the Falkner–Skan family of solutions is that, for an external stream velocity of the form x^m, the largest deceleration of the external stream for which back-flow does not occur in the boundary layer is that corresponding to $m = -0.0904$. For this value of m the forward frictional force exerted by the outer layers of fluid is just able to prevent the layer of fluid near the boundary from being moved backward by the positive ('adverse' in the conventional terminology) pressure gradient. The closeness of this critical value of m to zero is noteworthy.

Calculation of the steady boundary layer on a body moving through fluid

It is now possible to see in a general way how the boundary-layer flow will vary with position on the surface of a body moving steadily through fluid at rest at infinity with a large value of the Reynolds number based on body speed U_0 and linear dimensions. The boundary layer may be regarded as 'beginning' at the point where the dividing‡ streamline coming from far upstream intersects the body surface, and the nature of the initial development will depend on the local geometry of the surface. If it is a rounded surface which may be regarded as plane locally, the initial development of the boundary layer is like the 'stagnation point' flow analysed in § 5.5; and if it is a sharp forward facing edge, with the dividing streamline meeting the apex, the boundary layer will develop initially like one member of the family of solutions of (5.9.4) with $m \geqslant 0$, the flat plate of negligible thickness aligned with the stream corresponding to $m = 0$. Further downstream of this origin of the boundary layer, the velocity of the external stream varies in a different

† Apparently there are two solutions of (5.9.4) when $-0.0904 < m \leqslant 0$ (Stewartson 1954), the two solutions coalescing when $m = -0.0904$. All members of the second set of solutions show a region of reversed flow near the wall. As $m \to 0$, the intensity of this reversed flow in the second solution tends to zero, but the limiting solution is not the same as the Blasius solution for a boundary layer on a flat plate. The significance of this second solution is not clear, although it indicates the possibility of backflow in a boundary layer however small the magnitude of the deceleration in the external stream may be.

‡ The flow divides into two parts passing round different sides of the body only when the body is two-dimensional, which is the easiest case to describe in general terms. Boundary layers on three-dimensional bodies are considerably more complicated unless the flow is axisymmetric.

manner, determined by the body shape as a whole, and there may be both accelerations and decelerations, with consequent variation of the thickness of the boundary layer and of the velocity distribution within it. Over a limited portion of the body surface the velocity distribution in the boundary layer will be roughly like that given by the Falkner–Skan equation with a value of m chosen to represent the local conditions, except when the deceleration of the external stream is more than slight.

Many numerical methods of finding the development of the boundary layer in response to a given steady distribution of external stream speed have been devised,† all of which make use of the fact that the boundary-layer equation (5.7.1) is a second-order partial differential equation of parabolic type (unlike the full equation of motion which is of elliptic type in x and y) and allows forward integration with respect to x, conditions at any value of x being determined in general by the upstream history of the boundary layer. Important practical objectives of such a calculation are the values of the boundary-layer thickness and the tangential stress at the wall, the latter being needed in any estimate of the total force exerted on the body by the fluid. The dimensionless forms of the boundary-layer equations, (5.7.9) and (5.7.11), show that, for a two-dimensional body in a stream of speed U_0, the displacement thickness of the boundary layer at a distance x along the body surface from the forward stagnation point is

$$\delta_1(x) = \int_0^\infty \left(1 - \frac{u}{U}\right) dy = LR^{-\frac{1}{2}} \int_0^\infty \left(1 - u'\frac{U_0}{U}\right) dy', \qquad (5.9.8)$$

where L might be taken as the length of the body. Likewise the local tangential stress at the wall is

$$\mu\left(\frac{\partial u}{\partial y}\right)_{y=0} = \rho U_0^2 R^{-\frac{1}{2}} \left(\frac{\partial u'}{\partial y'}\right)_{y'=0}. \qquad (5.9.9)$$

The quantities with primes in (5.9.8) and (5.9.9) are dimensionless and independent of Reynolds number, and a numerical integration of the boundary-layer equations would yield the numerical values of

$$\int_0^\infty (1 - u' U_0/U)\, dy' \quad \text{and} \quad (\partial u'/\partial y')_{y'=0}$$

as functions of x'.

Although the various numerical methods available for calculation of the development of the boundary layer will not be described here, we shall note an integral relation, first given by von Kármán (1921), which has been made the basis of a number of approximate procedures. This relation is a special case of the momentum equation in integral form given in §3.2, and is conveniently obtained by integrating the boundary-layer equation (5.7.1) with respect to y across the layer. Some of the terms in this equation do not

---

† See *Laminar Boundary Layers*, edited by L. Rosenhead.

vanish outside the boundary layer, but we can avoid a dependence of the results on the exact range of integration by first subtracting from (5.7.1) the relation (5.7.8), which is the form taken by (5.7.1) just outside the boundary layer. The equation to be integrated is thus

$$\frac{\partial(u-U)}{\partial t} + u\frac{\partial u}{\partial x} + v\frac{\partial u}{\partial y} - U\frac{\partial U}{\partial x} = \nu\frac{\partial^2 u}{\partial y^2}, \tag{5.9.10}$$

and limits of integration for y can be taken as 0 and ∞, on the understanding that u is regarded as independent of y and equal to U outside the boundary layer (which may be seen to become an increasingly accurate assumption as $R \to \infty$, when stated in terms of the more natural lateral co-ordinate $y' = R^{\frac{1}{2}}y/L$ introduced in §5.7).

The result of the integration is

$$\nu\left(\frac{\partial u}{\partial y}\right)_{y=0} = \frac{\partial}{\partial t}\int_0^\infty (U-u)\,dy + \int_0^\infty \left\{(U-u)\frac{\partial U}{\partial x} + u\frac{\partial(U-u)}{\partial x} + v\frac{\partial(U-u)}{\partial y}\right\}dy$$

$$= \frac{\partial(U\delta_1)}{\partial t} + \frac{\partial U}{\partial x}U\delta_1 + \int_0^\infty \left\{u\frac{\partial(U-u)}{\partial x} + (U-u)\frac{\partial u}{\partial x}\right\}dy, \tag{5.9.11}$$

in which the last step has used the mass-conservation equation. The quantity

$$U^{-2}\int_0^\infty u(U-u)\,dy$$

is a length analogous to the displacement thickness, and is usually termed the momentum thickness and denoted by θ. The relation (5.9.11) may then be written as

$$\nu\left(\frac{\partial u}{\partial y}\right)_{y=0} = \frac{\partial(U\delta_1)}{\partial t} + \frac{\partial U}{\partial x}U\delta_1 + \frac{\partial(U^2\theta)}{\partial x}. \tag{5.9.12}$$

A simple procedure which often gives sufficiently accurate results is now to assume that the velocity distribution with respect to y has the same shape for all relevant values of x, this shape being chosen so as to satisfy as many of the boundary conditions as possible, and then to regard (5.9.12) as a differential equation for one of the parameters defining this shape. For instance, we could choose the profile

$$u = \begin{cases} U\sin\alpha y & \text{for} \quad 0 \leqslant y \leqslant \pi/2\alpha \\ U & \text{for} \quad y \geqslant \pi/2\alpha, \end{cases}$$

which is qualitatively reasonable for the case of a steady boundary layer with an external stream velocity which is uniform or increasing not too rapidly. With this profile we have

$$\delta_1 = \int_0^{\pi/2\alpha} (1-\sin\alpha y)\,dy = (\tfrac{1}{2}\pi - 1)/\alpha,$$

$$\theta = \int_0^{\pi/2\alpha} \sin\alpha y(1-\sin\alpha y)\,dy = (1-\tfrac{1}{4}\pi)/\alpha.$$

As a test we may take the case of uniform external stream speed, for which (5.9.12) gives

$$\nu\alpha = U(1 - \tfrac{1}{4}\pi)\frac{d\alpha^{-1}}{dx},$$

from which α can be determined. Then

$$\delta_1 = \frac{0\cdot571}{\alpha} = 0\cdot571\left(\frac{8}{4-\pi}\right)^{\frac{1}{2}}\left(\frac{\nu x}{U}\right)^{\frac{1}{2}} = 1\cdot74\left(\frac{\nu x}{U}\right)^{\frac{1}{2}},$$

which is remarkably close to the value (5.8.10).

All the available methods of calculation of steady flow in a boundary layer are frustrated if the deceleration of the external stream is more than slight, because the rate of increase of boundary-layer thickness usually then becomes so great as to render the boundary-layer approximations invalid.

Growth of the boundary layer in initially irrotational flow

As another way of examining the peculiar consequences of deceleration of the fluid at the outer edge of a boundary layer, we may consider the temporal development of boundary-layer flow from an initial state of zero vorticity everywhere, and with a given distribution of the external stream velocity U. Let us suppose, for definiteness, that a solid body immersed in infinite fluid at rest is made to move in translation. For an initial interval of time the motion of the fluid is irrotational everywhere except within a thin layer adjoining the body surface, and the relative velocity of the fluid at the surface associated with this irrotational flow field can be determined from a knowledge of the shape of the body and its instantaneous velocity. The velocity U at the outer edge of the vorticity layer is thus known as a function of distance x along the body surface (assuming, for simplicity, that it is two-dimensional) and time t. The motion in the boundary layer, relative to the body, is then governed by equations (5.7.1) and (5.7.2), with (5.7.8) to give the pressure gradient, and the conditions to be satisfied by the solution are (5.7.6), (5.7.7) and the new requirement

$$u(x, y, t) = U(x, t) \quad \text{at} \quad t = 0 \quad \text{for} \quad y > 0.$$

In the simple case in which the body speed is brought suddenly from zero to a value U_0 and subsequently is constant, U is independent of t and equations (5.7.1) and (5.7.8) reduce to

$$\frac{\partial u}{\partial t} + u\frac{\partial u}{\partial x} + v\frac{\partial u}{\partial y} = U\frac{dU}{dx} + \nu\frac{\partial^2 u}{\partial y^2}. \tag{5.9.13}$$

It seems evident that over the portion of the body surface between the forward stagnation point and the position of the first maximum of $U(x)$ the solution will exhibit a smooth approach to a steady boundary layer without unusual features. We should like to know what happens over the rear portion of the body where $dU/dx < 0$.

A method of finding the required solution of (5.9.13) at small values of t by a process of reiteration (first used by Blasius 1908) has been devised, the physical basis being that initially the boundary layer is extremely thin and the main contribution to $\partial u/\partial t$ comes from viscous diffusion. Of the four terms in the expression for $\partial u/\partial t$ given by (5.9.13), only $\nu\,\partial^2 u/\partial y^2$ becomes infinitely large as $t \to 0$ from above. (The term $v\,\partial u/\partial y$ is of the same order of magnitude as $u\,\partial u/\partial x$, as for boundary layers generally, and thus remains finite, the largeness of $\partial/\partial y$ being exactly balanced by the smallness of v.) Thus the first approximation to u, to be denoted by $u_1(x, y, t)$, satisfies the equation

$$\frac{\partial u_1}{\partial t} = \nu \frac{\partial^2 u_1}{\partial y^2},$$

of which the solution satisfying the above boundary conditions is

$$u_1(x, y, t) = U(x)\frac{2}{\sqrt{\pi}}\int_0^\eta e^{-\eta^2}\,d\eta = U(x)\,\mathrm{erf}\,\eta, \qquad (5.9.14)$$

where $\eta = \tfrac{1}{2}y/(\nu t)^{\frac{1}{2}}$. This distribution of relative velocity of fluid and boundary at position x is exactly the same as that generated in a semi-infinite fluid bounded by a plane rigid wall whose velocity in its own plane is suddenly brought to and held constant at $-U(x)$, the effects of convection and pressure gradient being identically zero in this latter case (§4.3).

We may now use this first approximation to estimate the convection terms in (5.9.13). Thus, the second approximation to the component of velocity parallel to the boundary locally is $u = u_1 + u_2$, where

$$\frac{\partial u_2}{\partial t} - \nu \frac{\partial^2 u_2}{\partial y^2} = U\frac{dU}{dx} - u_1\frac{\partial u_1}{\partial x} - v_1\frac{\partial u_1}{\partial y}, \qquad (5.9.15)$$

the normal velocity v_1 of the first approximation being obtained from (5.9.14) and the mass-conservation relation (5.9.1). The boundary conditions to be satisfied by $u_2(x, y, t)$ are

$$u_2(x, y, 0) = 0, \quad u_2(x, 0, t) = 0, \quad u_2(x, y, t) \to 0 \quad \text{as} \quad y \to \infty.$$

The right-hand side of (5.9.15) is of the form

$$U\frac{dU}{dx} \times \text{function of } \eta,$$

so that the particular integral of (5.9.15) can be written as $tU(dU/dx)f(\eta)$. The determination of the function $f(\eta)$ to satisfy (5.9.15) and the above boundary conditions is a straight-forward matter,† and we then have, as the second approximation,

$$u = U\,\mathrm{erf}\,\eta + tU\frac{dU}{dx}f(\eta). \qquad (5.9.16)$$

† See, for instance, *Laminar Boundary Layers*, edited by L. Rosenhead, §VII. 7.

The procedure can be continued to improve the approximation, the outcome of the nth stage being the addition of a term of the form

$$t^{n-1} \times \text{function of } x \times \text{function of } \eta$$

to the expression for u. The approximation (5.9.16) is sufficient for our purpose, which is to note the development of back-flow within the boundary layer in a region of deceleration of the external stream. The two functions erf η and $f(\eta)$ are everywhere non-negative, and the ratio of $f(\eta)$ to erf η is found to be greatest at $\eta = 0$. Consequently, back-flow can occur only when $dU/dx < 0$ and it does so first at $\eta = 0$, i.e. at $y = 0$—conclusions which we have seen apply also to steady boundary layers. The interval of time before back-flow occurs at any position x is the value of t which makes $(\partial u/\partial y)_{y=0}$ zero, and, according to (5.9.16) and the known solution for $f(\eta)$, is

$$-\frac{1}{dU/dx}\left\{\frac{d(\text{erf }\eta)/d\eta}{df(\eta)/d\eta}\right\}_{\eta=0}, \quad = -\frac{0\cdot 70}{dU/dx}.$$

The exact time and place on the body surface at which back-flow begins depend on the function $dU(x)/dx$, which is determined completely by the shape of the body. A simple illustrative case is a circular cylinder of radius a, for which the relative fluid velocity at the surface in a completely irrotational flow with zero circulation round the cylinder (see (2.10.12)) is

$$U(x) = 2U_0 \sin\frac{x}{a},$$

where x is measured along the surface from the forward stagnation point and U_0 is the speed of translation of the cylinder relative to the fluid at infinity. The maximum value of $-dU/dx$ in this case occurs at the rear stagnation point $x = \pi a$, and back-flow begins there after a time $0\cdot 35a/U_0$, i.e. after the cylinder has moved a distance $0\cdot 35a$. (At the third approximation to u, this number $0\cdot 35$ becomes $0\cdot 32$.) At subsequent times back-flow occurs over a finite part of the surface at the rear of the cylinder, for example, over the range $\frac{3}{4}\pi a < x \leqslant \pi a$ at time $0\cdot 50 \, a/U_0$, and it creeps forward to $x = \frac{1}{2}\pi a$ as $t \to \infty$. However, it is doubtful whether the first few terms of the series in powers of t provide an accurate estimate of u for values of t greater than about $0\cdot 5a/U_0$. The value of the calculation lies in the finding that back-flow occurs in the boundary layer at the rear of the body after a time which is inversely proportional to the maximum value of $-dU/dx$ and which does not depend on the viscosity. Steady boundary layers in which there is back-flow (and in which $u/U < 1$ for all y) have never been found, either mathematically or experimentally, so there is an indication here that steady flow past a bluff body with a thin boundary layer adjoining the whole surface of the body is not possible.

It is possible also to estimate the time needed for the vorticity to be carried far away from the cylinder. Near the rear stagnation point of the irrotational flow the component of velocity normal to the surface (and away from it) is equal to ky, where $k = 2U_0/a$ for a circular cylinder; and so the thickness of the layer of non-zero vorticity increases as $\exp(kt)$ as soon as the layer is so thick that the transfer of vorticity is dominated by convection. But viscosity is needed to provide the initial transfer from the surface of the cylinder and a dimensional argument indicates that the complete expression for the thickness of the vorticity layer near the rear stagnation (where the external stream is determined by the parameter k alone) is

$$\text{const.} \times (\nu/k)^{\frac{1}{2}} e^{kt}.$$

Thus this thickness might be expected to be comparable with the cylinder radius after a time of order

$$k^{-1}\log\frac{a^2 k}{\nu}, \quad \text{or} \quad \frac{a}{U_0}\log\frac{aU_0}{\nu}$$

for a circular cylinder, which is not very different from a/U_0 in practical terms.

Calculations of the development of the flow past a circular cylinder from initially irrotational motion have also been made using a finite difference approximation to space and time derivatives in the complete equations of motion and integrating them step by step in time over the whole flow field with the aid of a high-speed automatic computer. Payne (1958) has integrated the vorticity equation (which has the advantage that the calculations can be confined to the limited region of non-zero vorticity near the cylinder) numerically in this way, using steps separated by a time interval $0 \cdot 1a/U_0$, and the resulting distributions of vorticity ω and stream function ψ at times $2a/U_0$ and $6a/U_0$ after the circular cylinder is brought suddenly to its steady speed U_0 with a Reynolds number $2aU_0/\nu = 100$ are shown in figure 5.9.2. This Reynolds number is not so large that a thin boundary layer forms (and the finite difference method would meet great difficulties if it did), but the way in which vorticity is first carried round to the rear of the cylinder and then downstream from it is probably much the same at higher Reynolds numbers. It seems that growth of the thickness of the layer of non-zero vorticity is rapid in the region of closed streamlines associated with back-flow near the cylinder surface.

It is possible to observe this rapid growth of the vorticity layer at the rear of a bluff body which moves from rest. Figure 5.9.3 (plate 5) shows the flow at the rear of a blunt body at successive intervals of time after the body has begun to move, the first having been taken so soon after starting that the boundary layer is not discernible. Subsequent photographs show the occurrence and intensification of back-flow in the boundary layer. (They also show a typical tendency for the vorticity which accumulates in the thickening boundary layer to form roughly circular eddies or 'vortices', evidently as an outcome of local instability of the flow. These vortices are associated with quite vigorous back-flow, as may be inferred from the secondary counter-vortices on the upstream side of the main vortex in the last two photographs.) It is quite clear that the growth of the vorticity layer at the rear soon renders the boundary-layer approximations invalid there, and that the ultimate steady state is not one in which the fluid in the boundary layer on the forward surface of the body flows round the surface to the rear stagnation point.

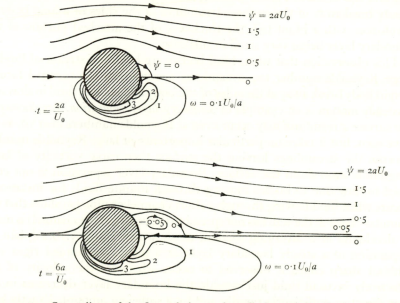

Figure 5.9.2. Streamlines of the flow relative to the cylinder and distribution of vorticity ω at times $2a/U_0$ and $6a/U_0$ after a circular cylinder is suddenly made to move at steady speed U_0; $2aU_0/\nu = 100$. (From Payne 1958.)

5.10. Separation of the boundary layer

We come now to the aspect of laminar flow at high Reynolds number which presents the greatest theoretical difficulty.

It is useful first to recall the demonstration in §4.12 that, at Reynolds numbers of the steady flow past bodies in the range 1 to about 100, back-flow at the rear of the body is a typical feature and one which becomes more pronounced as the Reynolds number increases in this range. Figure 4.12.1 (plate 1), figure 4.12.8 (plate 3), and figure 4.12.10 (plate 4) all show the appearance of large circulating eddies stationary relative to the body. The streamlines close to the forward face of the body leave the body at some point and pass round the large standing eddies. The Reynolds numbers of the flow fields described in §4.12 are not quite large enough for a boundary layer to be clearly recognizable at the body surface (with the flat plate formula (5.8.10), we have $\delta_1/l = 0.17$ when $Ul/\nu = 100$).

Observations of steady flow past bodies at higher Reynolds numbers at which the boundary-layer thickness on the forward face is small compared with the body size show the same breakaway of the surface streamlines somewhere near the sides of the body, although the flow in the wake at the rear of the body is usually unsteady (as a consequence of instability of the steady flow) and attached standing eddies with a regular closed flow become less and less evident. Figure 5.10.1 (plate 6) provides a detailed picture of steady breakaway of the surface streamlines at the sides of a long body of revolution with a blunt tail, the thickness of the attached portion of the boundary layer being very small in this case.

This observation that when bluff bodies are held in a stream of fluid at large Reynolds number the streamlines passing near the forward face of a rigid body break away at the sides of the body and enclose fluid in slow and unsteady motion is an example of *boundary-layer separation*. Breakaway of the surface streamlines may occur even at Reynolds numbers near 10, as we have seen, but it takes on particular importance at large Reynolds number because the streamlines leaving the surface then carry vorticity of large magnitude away from the surface. Boundary-layer separation is not confined to flow due to bodies moving through infinite fluid; for instance it occurs when fluid flowing through a short channel or pipe (so short that the boundary layer at the wall has not spread to the centre of the channel) is made to decelerate in a diverging section, as shown in figure 5.10.2 (plate 6). Separation of a steady boundary layer at a plane or rounded rigid wall without sharp edges is observed to occur whenever the velocity of the effectively inviscid fluid just outside the boundary layer decreases in the flow direction sufficiently rapidly and by a sufficient amount. Exactly how rapidly and by how much the external stream velocity must decrease cannot be stated, since the past history of the boundary layer also appears to be relevant, but, in practical terms, a steady boundary layer usually separates after very little retardation of the external stream. It is true that some members of the Falkner–Skan family of solutions (§ 5.9) represent boundary layers which continue at a plane wall despite an indefinite decrease of external stream velocity, but the deceleration is here very slow indeed and the initial profile of the boundary must have the prescribed form if it is to persist.

A graphic demonstration that separation is due essentially to the presence of a solid boundary, at which the no-slip condition is satisfied and vorticity is generated, is provided by the two photographs in figure 5.10.3 (plate 7). The first shows two-dimensional flow toward a 'stagnation point' at a plane solid wall (of the kind analysed in § 5.5), and the second shows how this flow is changed by the addition of a thin rigid plate at the plane of symmetry. The plate coincides with a streamline of the first flow, so that if the fluid were completely inviscid, and slip were allowed at the plate, the flow would show no change. In fact, the no-slip condition holds, and a boundary layer must form on the plate. Moreover, the deceleration of fluid on the plane of symmetry of the original flow is too severe for a steady thin boundary layer on the plate to be possible; more and more vorticity accumulates in the ever-thickening boundary layer and finally produces separation close to the forward edge of the plate, so that the oncoming irrotational flow no longer passes close to the original wall.

Despite the ubiquity of boundary-layer separation, and the great importance of its consequences for the flow as a whole, a proper understanding of the phenomenon and an analytical description of it have not yet been achieved. Separation is undoubtedly associated with the empirical fact that a steady state of the boundary layer adjoining a solid boundary is impossible with an appreciable fall in velocity of the external stream. Apparently, in cases in which the fall in velocity of the external stream is appreciable in some initial phase of establishment of the flow, the transfer of vorticity away from the solid boundary is so rapid that vorticity is soon not confined to a thin layer. The whole character of the flow then changes, and takes a form such that the boundary layer remains attached to the wall only where there is little or no deceleration of the fluid just outside the boundary layer. The difficulty is that we lack a means of seeing what that ultimate form will be.

Attempts have been made to relate mathematically the occurrence of separation to the upstream history of the boundary layer in the steady state, with the pressure distribution over the boundary regarded as known. As remarked in § 5.9, the usual plan in a numerical integration of the boundary-layer equations is to proceed from some initial value of x, where the distribution of velocity with respect to y is given (which initial station might be where the boundary layer literally begins), to larger values of x with a given distribution of pressure, or, equivalently, of external stream velocity. It has been found that all such methods encounter difficulties in regions of sustained deceleration of the external stream, usually failures of the numerical procedure such as a lack of convergence of an iteration process, and that the computational difficulties increase rapidly as the position of zero frictional stress at the wall is approached. The results of careful numerical investigation of the boundary layer in the particular case of an external stream velocity $U = U_0 - \alpha x$, where U_0 and α are constants (Hartree 1949; Leigh 1955), suggest that the solution of the boundary-layer equations has an algebraic singularity at this point. We know of one regular solution in which $(\partial u/\partial y)_{y=0} = 0$, viz. the Falkner–Skan similarity solution with $m = -0.0904$ (§ 5.9), although the fact that $(\partial u/\partial y)_{y=0} = 0$ at all values of x in this solution probably renders it unrepresentative.

If in some flow field $(\partial u/\partial y)_{y=0}$ changes smoothly from positive to negative values through zero as x increases, the position of zero wall friction is also the position at which back-flow begins.[†] The equation of motion (in either

---

† The two-dimensional flow within a very small neighbourhood (well within the boundary layer) of the point on a wall where $\partial u/\partial y$ changes sign has been described as a case of flow with negligible inertia forces in § 4.8.

the exact or the approximate boundary-layer form) in two dimensions may be regarded as representing the change in vorticity of the fluid at any point as the resultant of contributions from diffusion of vorticity and from convection with the fluid. It is thus to be expected that a numerical forward integration of the boundary-layer equations, in which conditions at some value of x are determined from those at smaller values of x, could hope to succeed in general only if the fluid velocity is in the direction of x increasing, over the whole of the region of integration; for if back-flow occurs at any point x_1, a contribution to the vorticity balance there will be made by convection from positions *at larger x*, which contribution is not determined by conditions at $x < x_1$, contrary to the assumed basis of the computation.

The position is, then, that first, we know from observation that a steady boundary layer 'separates' or breaks away from a solid boundary of smooth shape at, or a small distance downstream from, the position of maximum velocity of the external stream, and second, there are real difficulties in making a forward integration, in the direction of the external stream, of the boundary-layer equations as soon as the integration reaches the neighbourhood of the point of zero frictional stress on the wall. It is commonly taken for granted in the literature, although the assumption seems to lack proper justification, that the point of separation coincides with the point of zero wall friction. The most that can be said from observation is that the two points are usually close; experiments are hampered by the instability of the flow on the down-stream side of the separation point, and sometimes also of the boundary layer adjoining the wall. A cautious view, in the present state of knowledge, is that zero wall friction is a necessary preliminary or concomitant to separation of a steady boundary layer, and may occur at or upstream from the position of separation. In any event, an important conclusion from boundary-layer theory is that, provided the distribution of the external stream velocity remains unchanged, the position of the point of zero wall friction is independent of the Reynolds number; thus the occurrence of zero skin friction at some point and presumably also the related phenomenon of separation persist in the limit $\nu \to 0$.

Quite apart from the possibility that there is a singularity in the solution of the steady boundary-layer equations at a point of zero wall friction, *a priori* analysis of the boundary layer alone is not sufficient for the determination of the position of separation. Separation of the boundary layer, as defined above, refers to the departure, from the neighbourhood of the body surface, of those streamlines which lay within the boundary layer on the forward portion of the body, and is a feature of the flow as a whole which is determined by considerations of the flow both inside and outside the boundary layer and the flow on both sides of the separated boundary layer. The flow outside the boundary layer is governed by a partial differential equation of elliptic type (being simply Laplace's equation when this flow is irrotational), and is influenced everywhere by the shape of the whole

boundary of this region of flow, including that part of the boundary provided by the separated boundary layer. Thus the occurrence of separation
affects the distribution of pressure over the attached portion of the boundary
layer. One of the major unsolved problems of fluid mechanics is the determination of the total flow, both within and outside the boundary layer, in
cases of steady flow with boundary-layer separation.

In cases of separation from bodies of smooth geometrical form, the surface
streamlines are observed to leave the surface more or less tangentially, as
indicated in figure 5.10.4 a, which is drawn with a scale such that the body
dimensions are of order unity and the boundary layer appears as a line

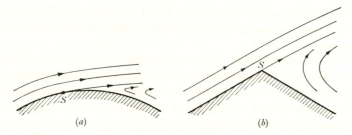

Figure 5.10.4. Separation from a rigid wall (a) without and
(b) with a salient edge.

without thickness. If the streamline leaving the separation point S made an
angle with the upstream wall less than $180°$, it is readily shown (as already
remarked in connection with the interpretation of (5.9.2) as the velocity
distribution on a wedge in irrotational flow) that the external stream velocity,
assuming it to be part of an irrotational flow, would be zero at S; there would
thus be appreciable deceleration of the external stream before reaching S
and, as has been seen, this would lead to back-flow and too great an accumulation of vorticity to be confined within a thin layer. In other words, tangential departure of the boundary-layer streamlines is the only self-consistent
possibility in a steady flow.

The case of a wall or a body surface with a salient edge possesses some
characteristic features and merits separate discussion. Here it is found that
the boundary layer always separates at the salient edge (as in the flow at the
two highest Reynolds numbers shown in figure 4.12.10, plate 4), and that it
leaves the edge tangentially to the upstream face, as sketched in figure
5.10.4 b. It seems that there is no possible steady state, for the flows inside
and outside the boundary layer, other than that in which the streamlines
near the wall leave the edge tangentially; and the flow as a whole adjusts
itself so that the separation takes place in this way. It is also found that,
unlike the case of a rigid boundary of smooth shape, the velocity of the
inviscid flow just outside the boundary layer does not decrease as the
separation point is approached. (Note that a forward integration of the

boundary-layer equations from some point upstream, with the observed distribution of external stream speed, would here give no prior indication of separation at the edge.)

The analysis in § 5.9 of the growth of the boundary layer immediately after a body begins to move through fluid at rest is relevant here also. In wholly irrotational flow past a body with a salient edge, the magnitude of the fluid velocity at the edge is very large (being theoretically infinite if the radius of curvature at the salient edge is zero, as will be seen in § 6.5). Thus the value of dU/dx to be substituted into (5.9.16) is negative with very large magnitude just downstream from the salient edge. Back-flow therefore occurs there almost immediately after the body begins to move, and moreover is strong. A standing eddy forms in the lee of the edge and the effect of this eddy is to throw oncoming fluid off the salient edge in the manner indicated in figure 5.10.4 b.

Detailed evidence about the development of separation almost immediately after a body with a salient edge begins to move is given in figure 5.10.5 (plate 8); here the air passing close to the forward surface of the body has been made visible by the release of vapour of different density from the body surface which causes a shadow in the light from an intense spark. The shedding of vorticity from the salient edge is so rapid that some fluid elements with vorticity remain approximately at the initial position of the body. These pictures reveal the characteristic spiral-like form of the line of fluid particles which have come from the salient edge, associated with the induced velocity of the vorticity shed from the edge. They also show the marked instability of a sheet vortex and the formation of undulations and, later, gathering of the vapour (and presumably of vorticity) at regular intervals in place of a uniform sheet vortex. The last four pictures of the sequence were taken when the body was moving at its ultimate steady speed, but the large vortex shed from the salient edge has not yet moved far enough downstream for the flow near the body to have attained its (statistically) steady form completely, except perhaps in the forward half of the flow field.

A less detailed illustration is provided by figure 5.10.6 (plate 9), which shows various stages in the development of flow past a model of a house with a pitched roof from rest. The existence of a small closed eddy just behind the roof peak can be observed in figure 5.10.6 b and this eddy grows in size until its internal structure is lost in the irregular fluctuations of figure 5.10.6 e. An interesting feature of figure 5.10.6 d is the appearance of smaller closed vortices at the two junctions of the house with the ground; the closed eddy on the upstream side of the house is associated with separation of the boundary layer from the ground in a region of external stream deceleration, and that on the downstream side is presumably associated in a similar way with the deceleration of the flow towards the house due to the large standing eddy (although the Reynolds number of this secondary flow is not so large that distinct boundary layers form there). It is clear from figure 5.10.6 e that when the rigid boundary has a salient edge the direction of the tangent to the boundary on the upstream side of the edge can have a dominant influence on the form of the final statistically steady flow.

The necessity of separation of the boundary layer at a salient edge and the strong influence of that separation on the form of the flow field as a whole are important factors in the theory of lift on bodies in translational motion. We shall see in §6.7 that the tendency for boundary-layer separation to occur in such a way as nearly to eliminate deceleration of the external stream over the attached portion of the boundary layer in the steady state can be exploited for useful purposes in aeronautics.

5.11. The flow due to bodies moving steadily through fluid

Determination of the high-Reynolds-number flow produced by a body moving steadily through fluid at rest at infinity, or, equivalently, the flow past a fixed body in a stream which is steady and uniform at infinity, is a basic problem in fluid dynamics of great practical importance in several engineering fields. As indicated in the preceding section, theory is not yet able to cope with the whole field of steady flow when separation occurs; and it is rendered even less effective by the inevitable turbulence which results from instability of boundary layers and wakes at large Reynolds number. In consequence, knowledge of flow past bluff bodies is drawn mostly from observation, and concerns the dependence of gross properties of the flow field on the Reynolds number rather than details of the velocity distribution.

The feature of flow due to a rigid body in steady translational motion through fluid at rest at infinity which is of most practical significance is the total force exerted on the body by the fluid. Contributions to this total force are made by the tangential stress at the body surface, integrated over that surface, and by the normal stress. The total force due to the tangential stress will usually be approximately opposite in direction to the velocity of the body, and is termed the *friction drag* since it is wholly and directly a consequence of viscosity or internal friction in the fluid. The total force due to the normal stress at the surface of a body in steady motion has a more complex origin, and it is useful to recognize the following contributions, which are additional to the buoyancy force due to the action of gravity on the fluid (§4.1).

(*a*) The *lift* on the body. This is the component of the total force normal to the direction of motion of the body, and is large for bodies of certain shape. The lift force owes its existence to the generation of vorticity at a rigid surface, in a way which will be considered in §6.7.

(*b*) The *induced drag*. An essential accompaniment to the lift on a three-dimensional body is the existence of vortices which trail downstream from the body. The body continually supplies kinetic energy to the fluid as these trailing vortices increase in length, this energy being the manifestation of work done by the body against that part of the total force known as induced drag. This drag contribution will be discussed in §7.8.

(*c*) The *form drag*. This is the component of the resultant pressure force parallel (and opposite in direction) to the velocity of the body, after the induced drag has been subtracted. The form drag depends strongly on the shape and attitude of the body, and, unlike the induced drag, offers scope for drag reduction by appropriate design in situations where reduction is desirable.

As a preliminary to the consideration of the friction and form drags acting on bodies of different shape, we note that, when the flow due to a body in steady translational motion through an inviscid fluid is wholly irrotational, the total drag (excluding any buoyancy force) on the body is zero. This important result follows from the fact that the whole irrotational flow field is determined uniquely by the instantaneous body velocity $\mathbf{U}$ in the case of a three-dimensional body of finite size (§2.9) and by $\mathbf{U}$ and the circulation round the body in the case of a cylinder, or two-dimensional body (§2.10). When $\mathbf{U}$ is constant (the circulation round a cylinder is constant in any event, by Kelvin's circulation theorem), the whole flow pattern simply moves with the body without change of the distribution, relative to the instantaneous position of the body, of the fluid velocity. Consequently the total kinetic energy of the fluid remains constant.† No means of dissipating energy exists in an inviscid fluid, and no radiation of energy to infinity can occur by sound waves in an incompressible fluid or by gravity waves in the absence of density variations or a free surface to the fluid; hence work done by the body against a non-zero drag can appear only as an increase of kinetic energy of the fluid, and, since this kinetic energy is constant when the body moves steadily, the drag exerted by the fluid on the body must be zero under these conditions. (This argument of course places no restriction on the component of force on the body normal to $\mathbf{U}$.)

The result that an inviscid fluid offers no resistance to steady translational motion of a rigid body when the flow is everywhere irrotational is sometimes referred to as d'Alembert's paradox, since rigid bodies do experience a resistance to motion through real fluid. The result is in serious disagreement with observation in the case of bluff bodies, which is hardly surprising, since the flow at the rear of a bluff body is far from having the assumed irrotational form. However, the result is quite relevant to flow due to the motion of slender bodies through a real fluid at large Reynolds number.

Flow without separation

Consider first the relatively simple case of bodies whose shape and attitude are such that no separation of the boundary layer occurs in the steady state.

† In the case of a two-dimensional body round which the circulation is non-zero, the velocity diminishes at large distances from the body as r^{-1} so that the total kinetic energy of the fluid is theoretically infinite. The above argument is then not adequate, although the result remains valid, as will be seen from the investigation in §6.4.

The various streamlines lying close to the body surface here proceed from an attachment point or points on the forward side of the body to some detachment point or points on the downstream side, where the boundary layer leaves the body and becomes a 'wake' whose thickness, at positions close to the body, is of the same order as the boundary-layer thickness. Such an absence of boundary-layer separation can occur only when the total fall in velocity of the fluid just outside the boundary layer is small, as the preceding discussion has shown. In particular, there must not be a stagnation point of the external stream at the rear of the body, and for this to be possible the body surface (whether it be two- or three-dimensional) must have a cusped edge at the rear. The body must also be slender and aligned roughly with the flow at infinity, for otherwise there would be a large velocity maximum at the sides of the body with a consequent large total reduction of external stream speed from there to the rear of the body. Figure 5.11.1 a (plate 7) shows a photograph, with an exposure for a short time, of small particles carried along with the fluid in the flow past an aerofoil (that is, a two-dimensional body of the general shape commonly used for aircraft wings), the boundary layer being just discernible as a collection of very short streaks on the rear half of the body. (A steady boundary layer subjected to a decelerating external stream like that on the upper surface of the aerofoil in figure 5.11.1 a is usually unstable and a turbulent boundary layer takes its place; turbulent boundary layers are less prone to separation, since the fluctuating cross-currents are able to transfer forward momentum from the outer layers to the slow-moving layers of fluid near the wall, but the same general rules about when separation will occur are qualitatively applicable.)

When boundary-layer separation does not occur, the vorticity generated at the surface of the body remains confined to a thin layer adjoining the surface and to a thin wake in which the vorticity is convected far downstream. The thickness (relative to the body length) of either a laminar or a turbulent† boundary layer at the surface of the body decreases to zero as the Reynolds number of the flow approaches infinity, the approach being faster for a laminar layer. Hence the flow in the limit of infinite Reynolds number is everywhere effectively inviscid and irrotational, except at certain stream surfaces which we shall refer to as singular because in general the tangential component of velocity changes discontinuously across each such surface. One singular stream surface is the surface of the body (the limiting form of the boundary layer) and the other is the sheet containing all the streamlines which extend far downstream from the detachment point or points at the rear of the body (the limiting form of the wake). Only in the neighbourhood of these singular stream surfaces is the gradient of vorticity large enough to make viscous effects significant when the Reynolds number is large.

† The existence of small roughness elements at the surface of the body may cause the thickness of a turbulent boundary layer to tend to a small but non-zero limit as the Reynolds number approaches infinity.

In the case of a two-dimensional body, or cylinder, the direction of the fluid velocity must be continuous across the singular surface coinciding with the streamline passing downstream from the point of detachment at the rear of the body; the pressure is necessarily continuous across this surface, and from Bernoulli's theorem (the flow being steady) so too is the velocity magnitude since streamlines on the two sides of this surface come from a region far upstream with uniform conditions and have the same Bernoulli constant. The singular surface extending downstream from the body is here a degenerate one across which all properties of the flow are continuous; the Bernoulli constant and the velocity magnitude right at the singular surface are smaller than elsewhere, this being all that is left of the wake at infinite Reynolds number, but this singularity is without effect on the flow. Hence we can proceed to calculate the irrotational flow without regard for the existence of singular surfaces in the fluid. The irrotational flow is of course not uniquely determined until the circulation round the cylinder is specified (§ 2.10). Observation shows that there is only one value of this circulation for which steady flow (or statistically steady flow, in the case of a turbulent boundary layer) in the boundary layer is possible, viz. that value for which a rear stagnation point does not occur on either the upper or lower surface of the (two-dimensional) body and for which, as a consequence, the fluid streams passing round the two sides of the body both leave the cusped trailing edge tangentially; more will be said about this important matter of determination of the cyclic constant in connection with lift on aerofoils in § 6.7.

In the case of three-dimensional bodies, the singular stream surface consisting of all the streamlines which pass downstream from the detachment points on the body may not be so innocuous. In this case it may convect vorticity with a non-zero component in the local flow direction which has been generated at the body surface—and always does so when a lift, or side force, acts on the body—in which event there is a jump in direction of the velocity across the singular surface. The occurrence of a lift force on the body needs special consideration, to be given briefly in § 7.8, and all that need be said here is that a knowledge of the shape of the body allows the nature of the jump, and thence the whole irrotational flow, to be determined in principle (although the practical difficulties may be insuperable). When no lift force acts on the body, streamwise vorticity is usually not generated at the body surface, the direction of the fluid velocity is continuous across the singular surface, and the singular surface has the same degenerate character as for a two-dimensional body.

Thus in all cases of bodies on which no separation occurs, the irrotational flow outside the body and thin boundary layer and wake is determinable approximately from the shape of the body (with an accuracy which increases with Reynolds number, because the boundary layer is ignored when applying the condition of zero normal velocity at the inner boundary of the region of irrotational flow). Some analytical methods of finding this irrotational flow explicitly will be described in chapter 6.

When the distribution of velocity in the irrotational flow just outside the boundary layer is known, the tangential stress $\mu(\partial u/\partial y)_{y=0}$ at each point of the body surface can be calculated by the kind of numerical integration of the boundary-layer equation mentioned in § 5.9. On introducing dimension-less boundary-layer variables, as in (5.9.9), and integrating the component of this surface force in the stream direction over the surface of the body, we find that for a two-dimensional body of length L in a stream of speed U_0

$$\text{total friction drag} = k\rho U_0^2 L R^{-\frac{1}{2}} \qquad (5.11.1)$$

per unit width normal to the plane of flow, where k is a number depending only on the shape of the body and $R = LU_0/\nu$. For a three-dimensional body there is a similar formula with some measure of the area of the body surface replacing L in (5.11.1).

On a streamlined body the changes in external stream velocity are fairly gentle (except near the forward stagnation point, where there is rapid acceleration), so that the development of the boundary layer on a two-dimensional body is not greatly different from that on a flat plate edge-on to the stream at the same value of the Reynolds number based on streamwise length of the solid surface; and for a flat plate the formula (5.8.8) gives the value of the coefficient k in (5.11.1) as 1·33. Observation shows that k is close to 1·33 for two-dimensional bodies of very small thickness, provided the Reynolds number does not exceed a certain value, and that it increases with the thickness of the body, partly because the velocity just outside the boundary layer exceeds U_0 over most of the surface by an amount which increases with the thickness. As remarked in § 5.8, the flow in the boundary layer becomes turbulent, and the frictional stress at the wall increases con-siderably, when the Reynolds number based on local boundary-layer thickness exceeds a value which for a flat plate boundary-layer profile is near 600. The kind of velocity distribution in the boundary layer that occurs over the rear portion of a body, where the external stream is decelerating, is more favourable to instability. Hence an increase in the thickness of a slender body at a given Reynolds number, and an accompanying magnification of the deceleration of the external stream, may cause an earlier transition to turbulent flow and consequently again a greater frictional drag.

In the limit of infinite Reynolds number, when the boundary layer and wake have zero thickness, the form drag on the body has the value zero corresponding to wholly irrotational flow. At a finite Reynolds number the existence of the thin boundary layer and wake has a small effect on the form of the surrounding irrotational flow and a corresponding small effect on the distribution of pressure at the body surface. The streamlines of the irrotational flow are displaced laterally by both the solid body and the boundary layer, and the thickness of the boundary layer in general increases from the leading to the trailing edge of the body; consequently the rise in velocity and fall in pressure at the sides of the body where the streamlines are crowded together may be expected to be more marked at the rear of the body than in the absence of the boundary layer, and this greater decrease of pressure at the rear causes the total drag due to normal stresses to be non-zero and positive. The pressure at any point on the body surface differs from that in wholly irrotational flow by an amount which is evidently proportional to the displacement thickness of the boundary layer, and so the form drag is proportional to $R^{-\frac{1}{2}}$, like the friction drag. The magnitude of the form drag depends on the shape of the body, being zero for a flat plate at zero incidence and generally larger the fatter the body, and is usually much smaller than the friction drag under conditions in which no separation of the boundary layer occurs.

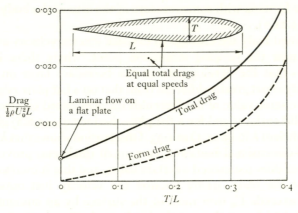

Figure 5.11.2. Observed values of the total drag and form drag for a family of symmetrical aerofoils like the one shown, at zero incidence; $U_0 L/\nu = 4 \times 10^5$. (From Fage, Falkner and Walker 1929.)

Figure 5.11.2 shows how the measured total drag and form drag vary with thickness for a typical family of symmetrical two-dimensional aerofoils or wing-sections (which for practical reasons have trailing edges in the form of thin wedges) at zero incidence to the stream, some of which are so thick that separation of the boundary layer undoubtedly occurred. It is also probable that the flow in the boundary layer was turbulent over a portion of the surface near the trailing edge in the case of the thicker aerofoils.

From an energy-balance point of view, the small drag on a slender body in steady flow must be accompanied by a corresponding rate of dissipation of kinetic energy in the fluid by viscous stresses. At large values of R, the boundary layer on the body of length L has a thickness of order $LR^{-\frac{1}{2}}$, so the velocity gradients in the boundary layer are of order $U_0 R^{\frac{1}{2}}/L$. Thus the rate of dissipation of energy per unit volume in the boundary layer is of order $\rho\nu U_0^2 R/L^2$, and the total rate of dissipation in the boundary layer, per unit area of surface of the body, is of order

$$\frac{\rho v U_0^2 R}{L^2} \times \delta, \quad \text{or} \quad \rho U_0^3 R^{-\frac{1}{2}}. \tag{5.11.2}$$

(There is also a contribution to total dissipation from the wake, but this is smaller since the velocity gradients quickly become small when the no-slip condition need not be satisfied; and the contribution from the region of irrotational flow is negligible because it is linear in μ and hence varies as R^{-1}.) This estimate is identical with that for the rate at which the body does work on the fluid by moving at speed U_0 against a total drag of order $\rho U_0^2 R^{-\frac{1}{2}}$ per unit area of the body surface (cf. (5.11.1)).

Flow with separation

The character of the flow is quite different when separation of the boundary layer occurs, as it does when the body is not sufficiently slender (as in figure 5.10.1, plate 6) or when the body is slender but not aligned with the stream (as in figure 5.11.1 b, plate 7). Attachment of the boundary layer over the whole surface of the body is impossible for such bodies because there would be considerable retardation of the external stream over the rear portion of the body and this is incompatible with a steady state of the boundary layer. In the case of the inclined aerofoil in figure 5.11.1 b, a large maximum of the velocity just outside the boundary layer occurs on the upper surface quite close to the leading edge and boundary-layer separation occurs not far downstream from this point; the boundary layer here adjoins very little of the upper surface and the aerofoil is said to be 'stalled', the term being associated with the accompanying severe reduction in lift on the aerofoil.

The photographs in figure 5.11.3 (plate 10) of the flow at different times after a circular cylinder has been made to move impulsively show the initial stages described in §5.9 and also the later stages which cannot be followed analytically and in which growth of the vorticity layer has changed radically the boundary of the region of irrotational flow. Figure 5.11.3 b seems to have been taken close to the instant at which back-flow begins in the boundary layer. In figure 5.11.3 c the boundary layer has separated, and vorticity is being convected away from the rear half of the cylinder. The region enclosed by the two separating streamlines grows larger and in figure 5.11.3 e is even larger than the cylinder itself. Instability then interrupts the approach to a steady state. The two standing vortices behind the cylinder develop asymmetric oscillations (which are probably forced on the standing eddies by the oscillations which develop at an even earlier stage in the wake, as explained in §4.12), and some of the rotating fluid in the bigger eddy ultimately breaks loose from the cylinder and moves downstream. The departure of so much vorticity from the neighbourhood of the cylinder affects the flow near the cylinder in such a way that there is then a tendency for the standing eddy of the opposite rotation to become bigger and shed some rotating fluid, and so on. At Reynolds numbers $(R = 2aU/\nu)$ not above about 2,500 these shed eddies or vortices are recognizable for some distance downstream and are observed to form a regular 'vortex-street' beyond four or five diameters from the cylinder, with vortices of the same sign on each of two nearly straight and parallel rows. A vortex-street was seen to form in the wake of a circular cylinder at Reynolds numbers above about 70 (figure 4.12.6, plate 2), and figure 5.11.4 (plate 11) shows its appearance in a similar flow field at much higher Reynolds number. A definite periodic oscillation of the flow near the cylinder is detectable until R reaches a value near 4×10^5, when the boundary layer adjoining the cylinder surface becomes turbulent.

Analysis of the flow due to a bluff body moving steadily through fluid is effectively prohibited by the large-scale unsteadiness of the flow behind the body revealed in figure 5.11.1b (plate 7) and figure 5.11.3f (plate 10). Whereas in the case of a 'streamlined' or slender body the turbulence resulting from instability is confined to the (attached) boundary layer and to the thin wake, here the resulting turbulence contains large eddies and co-ordinated velocity fluctuations which extend across the broad wake between the separating streamlines. These large eddies have an appreciable effect on the mean (i.e. time-averaged) flow properties, in a way which is difficult to represent analytically; in addition, measurements of the flow become hard to get and uncertain of interpretation. Knowledge of this kind of flow field is mostly empirical.† It remains true that on the upstream side of the field the vorticity generated at the surface of the body is confined to a thin boundary layer and that outside this layer the flow is irrotational; but the part of the boundary of this irrotational region formed by the separating streamlines is of complex, fluctuating and unknown shape and as a consequence the irrotational flow cannot be determined.

Although the flow behind a bluff body is unsteady in practice, there is no reason to doubt that a steady (unstable) solution of the equations of motion does exist. Despite its intrinsic interest, the form of this theoretical steady flow at large Reynolds number is not known. There is a general belief that, like a number of other steady flows at large Reynolds number, it consists of extensive regions of effectively inviscid flow separated by thin layers which, in the limit $R \to \infty$, become singular stream surfaces and which may enclose a region of (inviscid) flow with vorticity. The primary unknown element of the flow is the shape of the singular stream surface extending downstream from the points on the body at which the boundary layer separates; it is not even known confidently whether the region enclosed by the separating streamlines grows indefinitely long in the stream direction as $R \to \infty$ or approaches some finite form. One suggestion about the limit flow, put

† Some of the experimental data is described in *Modern Developments in Fluid Dynamics*, edited by S. Goldstein (Oxford University Press, 1938).

forward originally by Kirchhoff (1869) and Rayleigh (1876), is that the fluid within the broad wake enclosed by the streamlines passing through the separation points is everywhere at rest and at a uniform pressure equal to that at infinity far upstream. The speed of the fluid on the irrotational flow side of these same streamlines must then be uniform and equal to the free stream speed (from Bernoulli's theorem), and the wake width may be shown to increase indefinitely with increase of distance downstream. This model due to Kirchhoff will be described in more detail later (§6.13) in connection with flow of water about gaseous or vacuous cavities, a context in which there is less doubt about the validity of the assumptions of the model.

Whereas friction drag makes the major contribution to the drag on a slender body, form drag is responsible for most of the drag on a bluff body. The friction drag on a bluff body has much the same magnitude, per unit area of surface, as that on a slender body, but the form drag for a bluff body is greater than that for a slender body by a large factor. The results shown in figure 5.11.2 demonstrate the change in drag accompanying a large increase in the thickness of a slender body.

When the boundary layer separates from the side of a bluff body, the streamlines passing downstream from the separation points enclose a wide region in which the pressure does not vary greatly, since the velocities there are considerably smaller than U_0. The value of this approximately uniform pressure is roughly the same as that in the irrotational flow just outside the separated boundary layer. Thus over most of the rear of the body surface the pressure has the same low value as that at the side of the body, where the velocity exceeds the free-stream speed. On part of the forward face of the body, viz. near the stagnation point, the pressure is large, and the consequence of this fore-and-aft asymmetry of the pressure distribution is an appreciable form drag on the body. Also, since the pressure variations round the body surface are essentially a consequence of velocity variations of the same order as U_0, the link being made (over the forward part at least) by Bernoulli's theorem, we may expect that the total form drag will be of the same order of magnitude as $\frac{1}{2}\rho U_0^2$ (which is the excess of pressure at a stagnation point over that at infinity) times the frontal area of the body. Acting on this expectation, it is common practice to quote data about drag of bluff bodies in the form of values of the drag coefficient defined as

$$C_D = \frac{D}{\frac{1}{2}\rho U_0^2 A},$$

where D is the total drag in a stream of speed U_0 at infinity, and A is the area of the projection of the body on a plane normal to the stream at infinity. In the case of two-dimensional bodies, D and A refer to unit width in the direction normal to the plane of flow. The dimensionless coefficient C_D is in any event a function of Reynolds number alone (§4.7) (leaving aside effects of roughness of the body surface and of velocity fluctuations in the ambient fluid), and in the case of bluff bodies at Reynolds numbers above about 100 it has the convenient property of being of order unity.

These general arguments are supported by the measurements of the pressure distribution at the surface of a circular cylinder shown in figure 5.11.5; we note in particular the nearly uniform pressure over a large part of the rear of the cylinder, in contrast to the distribution of pressure in wholly irrotational flow. Upsteam of this region of nearly uniform pressure the boundary layer adjoins the surface of the cylinder, and the velocity U

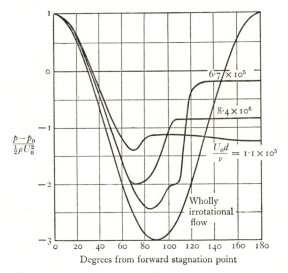

Figure 5.11.5. The measured pressure distribution at the surface of a circular cylinder in a stream of speed U_0, at different Reynolds numbers; p_0 = pressure at infinity.

in the flow just outside the boundary layer may be found from the measured pressure by using Bernoulli's theorem:

$$\frac{p - p_0}{\frac{1}{2}\rho U_0^2} = 1 - \left(\frac{U}{U_0}\right)^2. \qquad (5.11.3)$$

Figure 5.11.6 shows the measured drag of a circular cylinder over a wide range of Reynolds numbers which overlaps a little with the range covered by figure 4.12.7. A recognizable boundary layer forms on the forward side of the cylinder at Reynolds numbers above about 100, and we see that the drag coefficient is of order unity at larger Reynolds numbers, as suggested above. A striking indication of the reduction in drag that can be obtained by 'streamlining' a body, that is, by shaping it so as to avoid separation of the boundary layer, is provided by a comparison of the drag forces on a two-dimensional aerofoil and on a circular cylinder; the small black circle in figure 5.11.2 represents the circular cylinder on which the total drag is equal to that on the aerofoil drawn in the figure at the same speed (and at an aerofoil Reynolds number of about 4×10^5), despite the very much greater volume and surface area of the latter body.

Generally similar arguments and observations apply to three-dimensional bodies. Figure 5.11.6 shows the observed variation of the drag on a sphere with Reynolds number, and also that on a flat circular disk placed at right angles to the stream. The salient edge in this latter case fixes the position of boundary-layer separation, for all Reynolds numbers, and variation of the

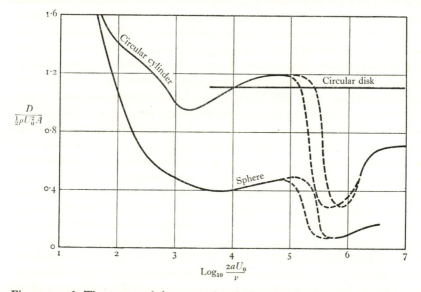

Figure 5.11.6. The measured drag on unit axial length of a circular cylinder ($A = 2a$), on a sphere ($A = \pi a^2$), and on a circular disk normal to the stream ($A = \pi a^2$), all of radius a. The broken curves represent results obtained in different wind tunnels.

Reynolds number has little effect on the drag. The drag coefficient of a disk is seen to be close to what it would be (unity) if the pressure had the stagnation-point value over the whole of the front face and the free-stream value over the whole of the rear face; in fact, the pressure falls continuously on the front face from the stagnation point at the centre to the edge, but the pressure defect (relative to the free-stream value) on the rear face more than compensates for this.

An interesting phenomenon is revealed by figures 5.11.5 and 5.11.6. It will be observed in figure 5.11.5 that increase of the Reynolds number above 10^5 leads to a marked increase in the approximately uniform pressure in the broad wake region at the back of the cylinder. The measurements of drag coefficient for a circular cylinder show a corresponding large fall as the Reynolds number increases through a value which normally lies somewhere in the range 10^5 to 4×10^5 and which depends on the wind tunnel used for the measurements. A similar fall in the drag coefficient of a sphere occurs at about the same Reynolds number, and likewise in the drag coefficients of most bluff bodies on which the position of separation is not fixed by the existence of a salient edge. In all these cases, the rate of decrease of drag coefficient as the Reynolds number increases through the critical value is so great that the absolute value of the drag is a decreasing function of U_0.

Prandtl (1914) suggested that the explanation lies in the behaviour of the boundary layer at the surface of the cylinder. When the Reynolds number for the body as a whole exceeds a certain value, steady or laminar flow in the boundary layer is unstable and may be replaced by a turbulent fluctuating flow. The rate of exchange of momentum between different layers of fluid in a turbulent boundary layer is much greater owing to the random lateral movement of elements of fluid, and so a turbulent boundary layer resists the development of zero wall stress and of separation, when the external stream speed is decreasing, more effectively than a laminar boundary layer. Thus, when transition to turbulent flow in the boundary layer occurs, the position of separation of the boundary layer moves back. Judging by the measured pressure distribution on a circular cylinder (figure 5.11.5), the separation point here moves back from about 80° to 120° from the forward stagnation point as R increases from 10^5 to 7×10^5. For reasons related to the geometry of the body, any retreat of the separation point usually implies a narrower wake and consequently a smaller form drag. The fact that the critical Reynolds number varies a little from one wind tunnel to another is a consequence of the different degrees of steadiness of the wind tunnel streams; the more disturbed the stream, the lower the Reynolds number at which the unstable steady flow in the boundary layer develops such large oscillations that turbulent flow sets in before separation occurs. A demonstration of the effect of deliberately disturbing the boundary layer can be made by fastening a wire or rough strip to the front of the body; figure 5.11.7 (plate 11) shows how a wire leads to delayed separation and a less broad wake for a sphere. The purpose of the corrugations on the surface of a golf ball is to make the boundary layer turbulent and thereby to reduce the drag.

Recent measurements on a circular cylinder at Reynolds numbers in the range 10^6 to 10^7 (Roshko 1961), included in figures 5.11.5 and 5.11.6, have shown a fall in the pressure at the rear and a corresponding increase of the drag coefficient up to a limiting value of about 0.7. In the light of these measurements it seems more likely that the change that occurs between $R = 10^5$ and $R = 7 \times 10^5$ is due to transition to a turbulent state of flow in the *detached* boundary layer just downstream from the point of separation, with reattachment of the boundary layer, now turbulent, to the cylinder surface as a result of the increased rate of spreading due to the turbulence. On this view the further change in the range 10^6 to 10^7 is to be interpreted as being due to transition to turbulent flow in the attached portion of the boundary layer, the delay in final separation being then less. The process of separation of a laminar boundary layer and transition to turbulent flow in the detached layer, followed by reattachment, is known to occur in some cases of flow past aerofoils which are inclined so much to the stream that the boundary layer on the upper surface separates quite close to the leading edge (as in the example shown in figure 5.11.1 *b*, plate 7). There may be a second separation of the boundary layer further back, but in any event the effect of the reattachment is to prevent a large increase in the drag (and, more importantly, a large decrease in the lift on the aerofoil).

These latter examples show how strongly the flow as a whole depends on the development of the boundary layer, and in particular on the position of separation. Even though the theoretical steady flow past a body tends to a limiting form as $R \to \infty$ and probably approximates closely to this asymptotic form when R is about 10^3 or 10^4, the instability of the wake, and of detached and attached boundary layers, leads to a number of important changes in the real flow at larger values of R. The replacement of laminar by turbulent flow in various parts of the field affects the flow as a whole, and when a change in the position of separation is involved the effect on total drag and lift can be dramatic.

5.12. Jets, free shear layers and wakes

Solid walls are the most common source of vorticity and give rise to adjoining boundary layers when the Reynolds number of the flow field is large. However, the near presence of a solid wall is not necessary for the ideas and approximations of boundary-layer theory to be applicable. Three different kinds of steady flow in which lateral gradients of vorticity may be relatively large, despite the absence of a solid wall, are (*a*) narrow jets, in which the steep gradients of vorticity originate at an orifice, the total change of velocity across the layer being zero; (*b*) free shear layers which are transition layers at the common boundary of two streams of different speeds; and (*c*) wakes, which are formed by vorticity being carried downstream from a body placed in a stream, the total change of velocity across the layer again being zero. These three boundary-layer-type flow systems will be described briefly in turn.

Narrow jets

We obtained in §4.6 an exact solution of the complete equation of motion which describes the steady jet due to a point singularity at which fluid momentum, but not mass, is generated steadily. This flow can be regarded as an idealization of that due to the discharge of fluid through a small orifice with high velocity. There is no need in this particular case to use an approximate form of the equation of motion, but it is instructive to notice briefly how this solution of the complete equation is related to the solution which may be obtained from the boundary-layer equations.

An appropriate Reynolds number for the steady axisymmetric jet flow described by (4.6.1), (4.6.2) and (4.6.10), at a given radial distance r from the orifice, is

(maximum radial velocity) × (semi-width of jet)$/\nu$

$$= \left(\frac{df/d\theta}{\sin\theta}\right)_{\theta=0} \sin\theta_0 = \frac{4\sin\theta_0\cos\theta_0}{1-\cos\theta_0},$$

where $\theta = \theta_0$ defines the semi-angle of the conical boundary of the jet. This Reynolds number is independent of r (and is effectively the same as the Reynolds number $(F/\rho\nu^2)^{\frac{1}{2}}$ found in §4.6 to apply to the flow right at the orifice when F is large), so that any consequences of the Reynolds number being large are equally applicable at all distances from the orifice. The Reynolds number is large and of order θ_0^{-1} in the case of a narrow jet, and these are the circumstances in which the boundary-layer approximations may be expected to be valid. The fluid surrounding the narrow jet is approximately at rest and at uniform pressure, so that the appropriate boundary-layer form of the equation of motion is simply an axisymmetric version of (5.7.1) together with $\partial u/\partial t = 0$, $\partial p/\partial x = 0$. This equation was in fact shown by Schlichting (1933) to yield the solution (4.6.15) before it was known that (4.6.15) is the asymptotic form (as $\theta_0 \to 0$) of a solution of the complete equation of motion.

No solution of the complete equation of motion is available for a steady two-dimensional jet discharged from an orifice in the form of a long slit, so that here it is necessary to turn to the approximate form of the equations. As with an axisymmetric jet, the pressure in the surrounding, nearly stationary, fluid is uniform and the boundary-layer equation (5.7.1) becomes

$$u\frac{\partial u}{\partial x} + v\frac{\partial u}{\partial y} = \nu\frac{\partial^2 u}{\partial y^2}, \tag{5.12.1}$$

the positive x-axis being the direction of the total force acting on the fluid at the origin. For fast narrow jets, the force acting on the fluid at the origin is manifested mainly as a flux of momentum across a surface surrounding the origin. On choosing a surface whose intercept in the (x,y)-plane consists of the line $x = $ const. (>0), $-\infty < y < \infty$, and a semi-circle of large radius lying mostly in the region of negative x, we have, for the force F exerted on the fluid per unit length of the slit,

$$F \approx \rho \int_{-\infty}^{\infty} u^2\,dy. \tag{5.12.2}$$

Since the choice of surface surrounding the origin is arbitrary, this integral must be independent of x.

It is convenient to introduce the stream function ψ, where $u = \partial\psi/\partial y$, $v = -\partial\psi/\partial x$, so that the mass-conservation equation is satisfied identically. In view of the difficulty of solving partial differential equations like (5.12.1), we enquire whether there exist solutions which depend on a certain combination of the two independent variables. A suitable hypothesis here is that the velocity profiles over sections at different values of x have the same shape, that is, that the solution has a 'similarity' form

$$\psi(x, y) \propto x^p f(y/x^q),$$

where p and q are unknown numbers. It follows immediately from the form of (5.12.1) that $p + q = 1$, and, from the fact that (5.12.2) is independent of x, that $2p - q = 0$, so that we must choose

$$p = \tfrac{1}{3}, \quad q = \tfrac{2}{3}.$$

Our hypothesis is, therefore,

$$\psi = 6\nu x^{\frac{1}{3}} f(\eta), \quad \eta = y/x^{\frac{2}{3}}, \\ u = 6\nu x^{-\frac{1}{3}} f', \quad v = 2\nu x^{-\frac{2}{3}}(2\eta f' - f), \bigg\} \tag{5.12.3}$$

where the factor 6ν has been inserted for later convenience. Equation (5.12.1) now reduces to

$$f''' + 2ff'' + 2f'^2 = 0. \tag{5.12.4}$$

The test of the hypothesis lies in being able to find a solution of this equation which satisfies

$$f'(\eta) \to 0 \quad \text{as} \quad \eta \to \mp \infty,$$

and the symmetry condition $f'(\eta) = f'(-\eta)$. Such a solution is

$$f(\eta) = \alpha \tanh \alpha \eta, \tag{5.12.5}$$

where α is a constant which (5.12.2) shows to be given by

$$F = 36\rho\nu^2\alpha^4 \int_{-\infty}^{\infty} \operatorname{sech}^4 \alpha\eta \, d\eta = 48\rho\nu^2\alpha^3. \tag{5.12.6}$$

The mass flux from the singularity at the origin is found to be zero, as for the round jet.

Thus a solution representing a two-dimensional narrow jet has been found, with a velocity profile of the form $\operatorname{sech}^2 \alpha\eta$, and width increasing as $x^{\frac{2}{3}}$, and its only limitation is that, having the assumed similarity form, it cannot be used to determine the development of a jet with a given velocity profile at some initial value of x. However, this is not a serious handicap if, as is frequently true of similarity forms, the above solution is approached asymptotically, as $x \to \infty$, with an arbitrary velocity profile at some initial value of x. Measurements made by Andrade (1939) suggest that this is so, since a velocity distribution like that given by (5.12.5) was found in a jet formed by discharging fluid under pressure through a long narrow slit.

The Reynolds number for the jet at a distance x from the origin, defined again in terms of the maximum velocity and the width of the jet, is

$$\frac{(6\nu x^{-\frac{1}{3}}\alpha^2) \times (x^{\frac{2}{3}}/\alpha)}{\nu} = 6\alpha x^{\frac{1}{3}} = \left(\frac{9}{2}\frac{Fx}{\rho\nu^2}\right)^{\frac{1}{3}}, \tag{5.12.7}$$

apart from a numerical factor of order unity which depends on exactly how the width is defined. The criterion for the approximate boundary-layer equations to be applicable, and equally for the jet to have a small angle of spread, is therefore

$$\left(\frac{Fx}{\rho\nu^2}\right)^{\frac{1}{3}} \gg 1.$$

The solution thus becomes more accurate as x increases, for a given value of F, although there is always a region near the origin where the boundary-layer equations are not valid. The lack of validity of the solution near $x = 0$ is not important, since the velocity profile imposed at the orifice in a real jet is in any event unlikely to have the form (5.12.5). When the velocity profile settles down to the form (5.12.5), at sufficiently large values of x, further development of the jet with distance downstream will be as if the jet had developed from some virtual origin according to the boundary-layer equations and with the similarity profile right from the origin.

Both two- and three-dimensional jets are unstable at Reynolds numbers above certain critical values, and under these conditions the steady laminar flow is replaced by a turbulent flow which has the same jet-like character although with a greater rate of spreading. Inasmuch as the Reynolds number of a portion of the two-dimensional jet continually increases with x, this jet always becomes turbulent at some distance from the orifice; indeed, experience suggests that only over a very short range of values of x is the two-dimensional jet both stable and describable by the boundary-layer equations.

Free shear layers

The simplest example of a transition layer between two uniform streams is the diffusing sheet vortex considered in §4.3. The fluid velocity is here uniform over planes parallel to the sheet, and development occurs with respect to time t, rather than with respect to distance x in the flow direction as in steady boundary-layer flows. The term $\nu \, \partial^2 u/\partial x^2$, which is neglected in the boundary-layer form of the equation of motion, is identically zero, and the pressure is exactly constant across the layer; thus the solution given in §4.3 is also a solution of the boundary-layer equations, and a particularly simple one because the non-linear terms vanish.

A free shear layer which is steady necessarily varies with respect to x, and the boundary-layer equations are then useful. A case of some generality is that in which two uniform streams of identical fluid, both moving in the direction of x increasing but with different speeds U_1 and $U_2 (< U_1)$, come into contact at $x = 0$, $y = 0$, $-\infty < z < \infty$ (figure 5.12.1). When $U_2 = 0$, this can be regarded as a specification of the two-dimensional flow at the edge of a wide slit through which an initially uniform stream is being discharged. The boundary-layer equation for this general case is (5.12.1), since the pressure is uniform outside, and consequently also inside, the layer, and the boundary conditions are

$$u \to U_1 \quad \text{as} \quad y \to \infty, \qquad u \to U_2 \quad \text{as} \quad y \to -\infty.$$

It can be shown readily that a similarity solution satisfying these conditions exists, with a transition layer of thickness proportional to $(x\nu/U_1)^{\frac{1}{2}}$, although

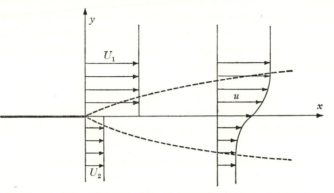

Figure 5.12.1. The steady transition layer between two parallel streams coming into contact at $x = 0$.

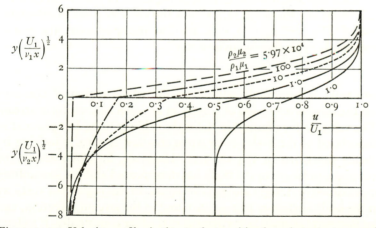

Figure 5.12.2. Velocity profiles in the steady transition layer between two parallel streams of fluids of different densities and viscosities. (From Lock 1951.)

the resulting ordinary differential equation for the velocity profile has to be solved numerically. The velocity profiles depend on the ratio U_2/U_1, and are shown in figure 5.12.2 for the cases $U_2/U_1 = 0$ and 0.5. No part of the upper region of fluid can be accelerated by contact with the lower stream, and no part of the lower region can be decelerated; thus the acceleration is always zero at the surface of contact between the two regions of fluid, which is the streamline through the origin, and the profile has a point of inflexion there.

It is also possible to obtain a similarity solution for two parallel streams of different fluids, of different densities and viscosities, which come into contact in the above manner (Lock 1951). The independent variable for the upper fluid is now $\eta_1 = y(U_1/\nu_1 x)^{\frac{1}{2}}$, and for the lower fluid $\eta_2 = y(U_1/\nu_2 x)^{\frac{1}{2}}$ (not $y(U_2/\nu_2 x)^{\frac{1}{2}}$ because U_2 might be zero), and an equation like (5.12.1) must be solved for each of the upper and lower regions, with the conditions of continuity of the velocity and the stress to be satisfied at the contact surface. Continuity of the tangential stress $\mu \, \partial u/\partial y$ implies a discontinuity of $\partial u/\partial y$ at the contact surface when $\mu_1 \neq \mu_2$, although $\partial^2 u/\partial y^2$ approaches zero there, both from above and from below, as before. In terms of similarity variables the stress condition is

$$(\rho_1 \mu_1)^{\frac{1}{2}} \left(\frac{du}{d\eta_1} \right)_{\eta_1 \downarrow 0} = (\rho_2 \mu_2)^{\frac{1}{2}} \left(\frac{du}{d\eta_2} \right)_{\eta_2 \uparrow 0},$$

which shows that the solution depends on the ratio $(\rho_2 \mu_2/\rho_1 \mu_1)^{\frac{1}{2}}$ as well as on U_2/U_1.

This solution may be used to represent the flow of air over water, subject to the limitations implied by the similarity form of the velocity distribution right from the line of first contact of the streams. A free shear layer in a uniform fluid shows marked instability, but a shear layer at an air–water interface is made less unstable by the restraining effect of gravity on disturbances to the interface. Figure 5.12.2 shows the velocity profiles calculated for the cases $U_2/U_1 = 0$, $\rho_2 \mu_2/\rho_1 \mu_1 = 10$, 100 and $5 \cdot 97 \times 10^4$, the last of which corresponds to air flowing over water.

Wakes

The term wake is commonly applied to the whole region of non-zero vorticity on the downstream side of a body in an otherwise uniform stream of fluid. The velocity distribution in the wake is likely to be complicated in the neighbourhood of the body, even when the flow is steady, judging by the flow fields described in §§4.12, 5.11. However, far downstream the direct effect of the presence of the body has disappeared and the streamlines become approximately straight and parallel again. In this region the vorticity shed from the body surface is being convected in the stream direction, and diffused by viscosity; and since vorticity is spreading continually, it follows that ultimately convection is more important than streamwise diffusion and that the streamwise gradient is small compared with that in the lateral plane. Thus the boundary-layer approximations are applicable asymptotically, irrespective of the Reynolds number of the flow based on the body size, although the distance downstream at which they begin to apply naturally depends on the circumstances, being greater for smaller Reynolds numbers and for bluff bodies.

A further effect accompanying the continual spreading of the wake is the tendency for frictional forces to make the velocity uniform. Thus, in addition to the boundary-layer approximations we may assume that the departure from the free-stream velocity is small at positions sufficiently far downstream. Some analytical deductions about steady flow in this asymptotic region will now be described for the simple case of a body on which only a drag force acts.

The free-stream velocity has magnitude U and direction parallel to the x-axis say. The velocity and pressure in the irrotational flow outside the wake are approximately uniform (for x sufficiently large), and so the pressure is approximately uniform also in the wake. Moreover $|U-u| \ll U$, and the component of acceleration in the x-direction is given approximately by $U \, \partial u/\partial x$. The equation of motion far downstream in a steady wake therefore reduces to the linear equation

$$U\frac{\partial u}{\partial x} = \nu\left(\frac{\partial^2 u}{\partial y^2} + \frac{\partial^2 u}{\partial z^2}\right), \tag{5.12.8}$$

and the boundary condition at the edge of the wake is

$$u \to U \quad \text{as} \quad (y^2 + z^2)^{\frac{1}{2}} \to \infty.$$

Equation (5.12.8) is of the same form as the equation for conduction of heat in the (y, z)-plane in a solid (with x/U in (5.12.8) corresponding to time in the conduction equation), for which it is known that the solution in an infinite region tends to an asymptotic form independent, apart from a multiplicative factor, of the initial conditions. (See the remarks in §4.3 following (4.3.6).) This asymptotic solution is

$$U - u \to \frac{QU}{4\pi\nu x}\exp\left\{-\frac{U(y^2+z^2)}{4\nu x}\right\} \tag{5.12.9}$$

as $x \to \infty$. Here Q is a constant determined by the conditions at some initial value of x through the fact that, as may be seen by integrating both sides of (5.12.8), the integral

$$\int\int_{-\infty}^{\infty} (U-u)\,dy\,dz = Q \tag{5.12.10}$$

is independent of x. The corresponding formulae for a two-dimensional wake differ only in that the operations and terms involving z in (5.12.8), (5.12.9) and (5.12.10) are absent and the factor $(U/4\pi\nu x)$ in (5.12.9) appears to the power $\frac{1}{2}$. The width of the wake, defined as the region in which the velocity deficiency $U-u$ is greater than some fraction of its maximum value, thus increases parabolically with distance downstream in both two and three dimensions.

It proves to be possible to obtain a relation between the constant Q and the total drag D on the body giving rise to the wake, despite the fact that the relation between Q and conditions in the wake at positions closer to the body than those where (5.12.8) applies remains unknown. We shall use the momentum equation in integral form in the manner explained in §3.2 (and illustrated in §5.15). For the 'control surface' we choose a cylinder with generators parallel to the undisturbed stream and plane faces of area A normal to the stream, as sketched in figure 5.12.3, the curved surface S of the cylinder being far enough from the body to lie entirely outside the wake. The fluid instantaneously within this control surface is acted on by forces at the control surface, and by forces at the body surface whose resultant in the

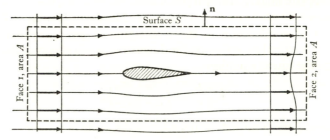

Figure 5.12.3. Control surface (broken line) surrounding fluid whose momentum balance is to be calculated.

x-direction is $-D$. On equating the sum of the x-components of these forces to the net outward flux of x-momentum across the control surface, we have

$$D = \int (p_1 + \rho u_1^2 - p_2 - \rho u_2^2)\, dA - \rho \int u\mathbf{u}.\mathbf{n}\, dS$$

$$+ \text{viscous forces acting at the control surface,} \quad (5.12.11)$$

where u_1, p_1 and u_2, p_2 are the values of the x-component of $\mathbf{u}$ and the pressure at the upstream and downstream faces respectively. The mass flux across the control surface must be exactly zero, so that an additional relation is

$$\int \mathbf{u}.\mathbf{n}\, dS + \int (u_2 - u_1)\, dA = 0. \quad (5.12.12)$$

We may assume that all parts of the control surface are at large distances from the body, to ensure that viscous forces at the control surface are relatively small and in order to be able to approximate to the values of u, u_1, u_2, p_1 and p_2 in (5.12.11). On the curved cylindrical surface, the departure from free-stream conditions is small and we may put $u = U$ in the second integral in (5.12.11). With the aid of (5.12.12), we then have

$$D = \int \{p_1 + \rho u_1(u_1 - U) - p_2 - \rho u_2(u_2 - U)\}\, dA. \quad (5.12.13)$$

Now the existence of a velocity defect in the wake is equivalent to the super-position of 'in-flow' towards the body on the uniform stream. The volume flux of this in-flow is the quantity Q defined in (5.12.10); relative to axes fixed in fluid at infinity, Q is the rate at which fluid volume is dragged across a stationary plane behind the body. This in-flow in the wake must be com-pensated, as (5.12.12) shows in effect, by an equal flux of volume away from the body in the region of irrotational flow outside the wake. Thus the presence of the wake is associated with a source-like contribution to the irrotational flow field, the strength of the effective source being Q, and it is known (§§ 2.9, 2.10) that far from the body there cannot be a larger contribution to the departure from a uniform stream. The flow at large distances r from the body is evidently a superposition of a uniform stream and the motion sketched in figure 5.12.4, and in the region outside the wake the departure of the velocity from the uniform stream value falls off as r^{-2} in three dimensions and as r^{-1} in two dimensions. In this same region Bernoulli's theorem is applicable and the integrand of (5.12.13) becomes

$$\tfrac{1}{2}\rho\{(u_1 - U)^2 - v_1^2 - w_1^2 - (u_2 - U)^2 + v_2^2 + w_2^2\}, \qquad (5.12.14)$$

showing that the integral over the part of the area A lying outside the wake tends to zero as the distance of the ends of the cylinder from the body increases.

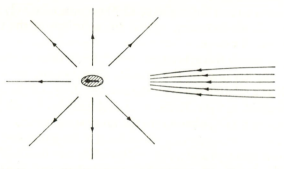

Figure 5.12.4. The flow at large distances from a body moving (from right to left) through fluid at rest at infinity, showing the wake and the compensating source-like flow away from the body.

Far from the body the streamlines are nearly parallel, and the pressure variation across the wake is very small. The limiting form of the integral in (5.12.13), as the cylinder becomes indefinitely long and u_1, u_2, p_1 and p_2 all approach the free-stream values, but with u_2 making the slowest approach, is thus

$$D = \rho U \!\int (U - u_2)\, dA = \rho U Q; \qquad (5.12.15)$$

and Q may be evaluated by integrating only over the wake cross-section. This determines the constant occurring in the asymptotic wake profile (5.12.9). The relation (5.12.15) between the drag on a body and the rate of in-flow in the wake has already been met as a consequence of the Oseen approximation to the equations for flow past a sphere at small Reynolds number (§4.10). The general momentum argument given here shows that (5.12.15) holds for any non-lifting body moving steadily through fluid at rest at infinity, and for any Reynolds number.

The velocity profile (5.12.9) also has links with the distribution of velocity about a steadily moving body as calculated from the Oseen equations in

§4.10. It is easily seen that the stream function (4.10.6) obtained for the flow far downstream $(r/a \gg R^{-1})$ from a moving sphere, and within a paraboloid of revolution on which $(\pi - \theta)^2 r/a$ is of order $8/R$, gives the same distribution of longitudinal velocity as (5.12.9), provided that Q is determined through (5.12.15) and the relation $D = 6\pi a \mu U$ which is appropriate for a sphere at small Reynolds number. This identity of the two velocity distributions in the wake, which were obtained for apparently different sets of conditions, although both with equations in which (for a fluid velocity $\mathbf{u}$ relative to the body) the non-linear term $\mathbf{u}.\nabla\mathbf{u}$ is replaced by $\mathbf{U}.\nabla\mathbf{u}$, can be explained by two comments. First, the asymptotic $(r/a \gg R^{-1})$ stream functions (4.10.5) and (4.10.6), representing the outward source flow from the sphere and the in-flow in the wake respectively, depend on the body shape *only* through its effect on the magnitude of the in-flow and this, as (5.12.15) shows, is determined by the total drag. Second, terms which are retained in the first of the Oseen equations (4.10.2) but not in (5.12.8) become relatively smaller as the distance downstream increases, so that the solutions of the two equations become identical as $x \rightarrow \infty$.

Instability of steady wake flows sets severe limitations on the applicability of the above expressions for the velocity distribution, although (5.12.15) remains valid as a relation between the average drag and the average rate of in-flow in a fluctuating wake. Reference has already been made (figure 4.12.6, plate 2, and figure 5.11.4, plate 11) to the interesting regular array of eddies, or 'vortex-street', which forms in the wake of a circular cylinder at values of $2aU_0/\nu$ between about 70 and 2500. At Reynolds numbers higher than 2,500 the wake behind a circular cylinder is turbulent and contains irregular fluctuating velocities, and even at lower Reynolds numbers the flow is usually found to be turbulent far downstream beyond a vortex-street. The width of a turbulent wake behind a cylindrical body is known to increase as $x^{\frac{1}{2}}$, just as for steady laminar flow, although the constants of proportionality have different values. The property of a wake at a given distance downstream that determines its stability and its ability to remain turbulent is the Reynolds number based on wake width and a representative relative velocity within the wake. Such a Reynolds number is independent of x for both steady laminar and turbulent wakes behind two-dimensional bodies, so that the wakes remain either laminar or turbulent indefinitely. The application of the two-dimensional version of (5.12.9) is therefore limited to body Reynolds numbers such that the steady wake is stable, which for a circular cylinder requires $2aU_0/\nu$ to be less than about 40.

The conditions for instability of the wake behind a three-dimensional body are not as well known, although it is probable that the critical value of the body Reynolds number is higher than for a cylinder. The width of a turbulent wake behind a three-dimensional body increases as $x^{\frac{1}{3}}$, and the maximum value of the mean velocity deficiency decreases as $x^{-\frac{2}{3}}$; thus the effective Reynolds number of a portion of the wake diminishes as $x^{-\frac{1}{3}}$ and

PLATE I

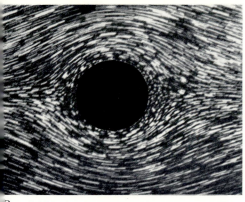

R = 0·25

R = 13·05

R = 3·64

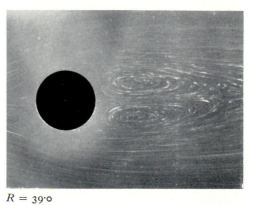

R = 39·0

= 9·10

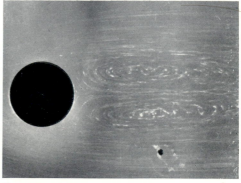

R = 57·7

Figure 4.12.1. Streamlines of steady flow (from left to right) past a circular cylinder of radius a; $R = 2aU/\nu$. The photograph at $R = 0·25$ (from Prandtl and Tietjens 1934) shows the movement of solid particles at a free surface, and all the others (from Taneda 1956 a) show particles illuminated over an interior plane normal to the cylinder axis.

PLATE 2

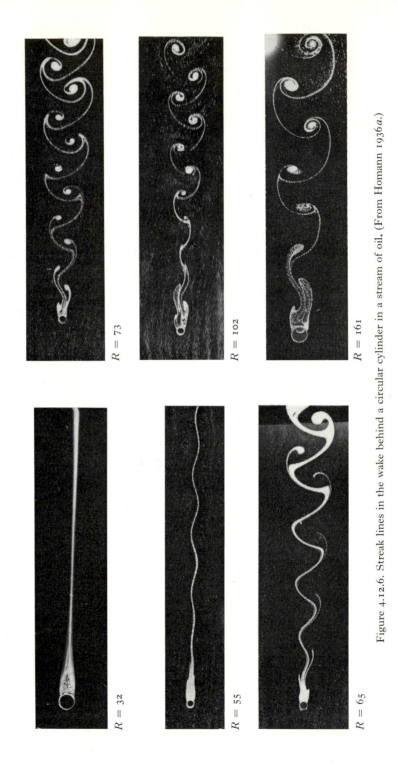

R = 73

R = 102

R = 161

R = 32

R = 55

R = 65

Figure 4.12.6. Streak lines in the wake behind a circular cylinder in a stream of oil. (From Homann 1936a.)

PLATE 3

$R = 9 \cdot 15$

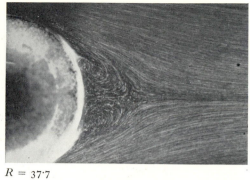

$R = 37 \cdot 7$

$R = 17 \cdot 9$

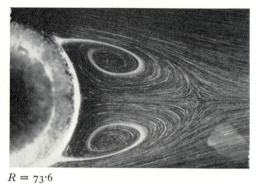

$R = 73 \cdot 6$

$R = 25 \cdot 5$

$R = 118$

$R = 26 \cdot 8$

$R = 133$

Figure 4.12.8. Streamlines, in an axial plane, of steady flow (from left to right) past a sphere of radius a (from Taneda 1956b); $R = 2aU/\nu$.

PLATE 4

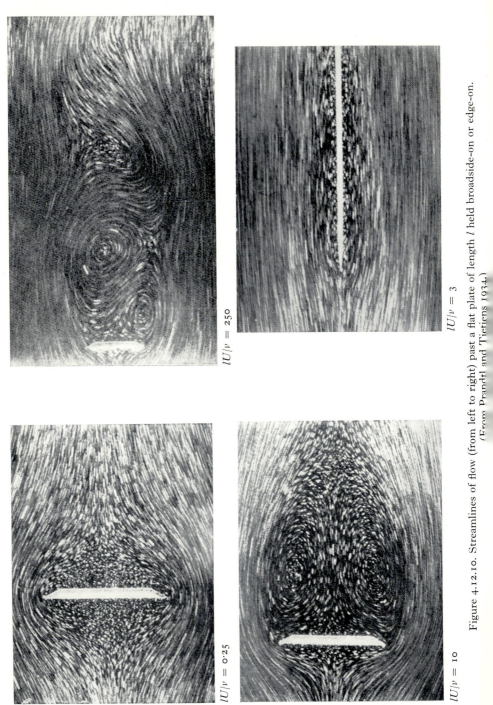

$lU/v = 0.25$

$lU/v = 250$

$lU/v = 10$

$lU/v = 3$

Figure 4.12.10. Streamlines of flow (from left to right) past a flat plate of length l held broadside-on or edge-on. (From Prandtl and Tietjens 1934.)

PLATE 5

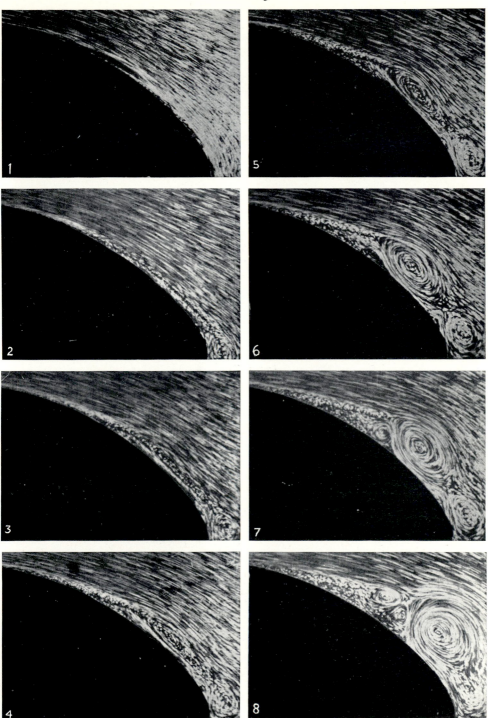

Figure 5.9.3. Stages in the growth of the boundary layer at the rear of a bluff body in motion after being initially at rest. The flow is viewed relative to the body and is from left to right. (From Prandtl and Tietjens 1934.)

PLATE 6

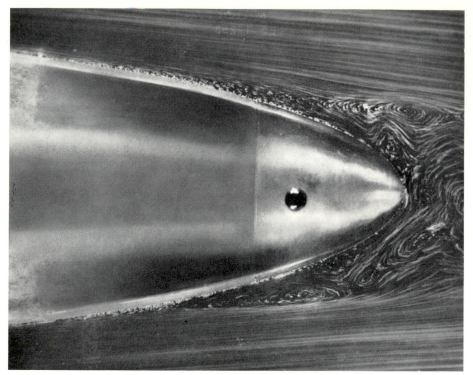

Figure 5.10.1. Separation of the steady boundary layer at the rear surface of a body of revolution in a stream of water. Reynolds number based on body length $= 1 \cdot 3 \times 10^5$. The motion was made visible by placing aluminium particles on the water surface which was also a plane of symmetry of the body. (From Clutter, Smith and Brazier 1959.)

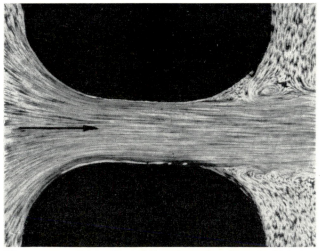

Figure 5.10.2. Streamlines of flow through a channel which first converges and then diverges. (From Föttinger 1939.)

PLATE 7

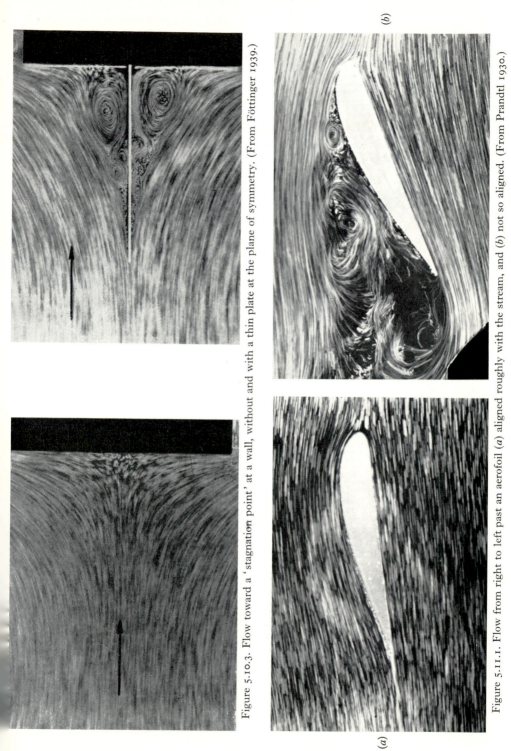

Figure 5.10.3. Flow toward a 'stagnation point' at a wall, without and with a thin plate at the plane of symmetry. (From Föttinger 1939.)

Figure 5.11.1. Flow from right to left past an aerofoil (a) aligned roughly with the stream, and (b) not so aligned. (From Prandtl 1930.)

PLATE 8

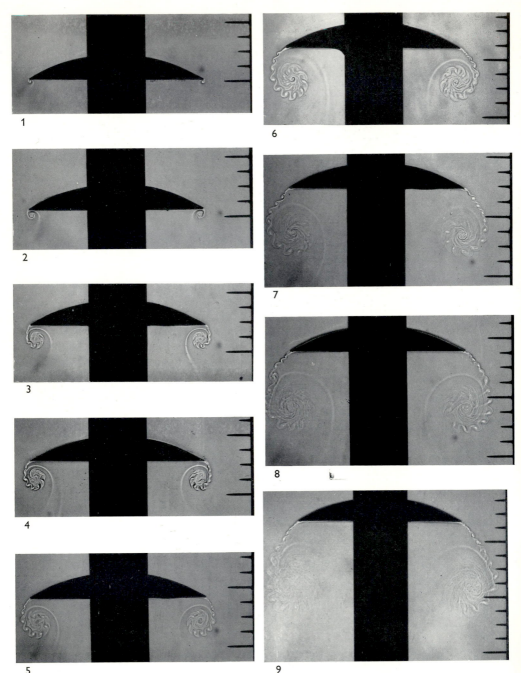

Figure 5.10.5. Sequence of shadowgraphs, taken at intervals of about $1 \cdot 1 \times 10^{-3}$ sec (with twice that interval between the last two), showing the position of vapour released from the surface of an axisymmetric body with a salient edge accelerating rapidly upwards from rest to a final steady speed of 731 cm/sec in air. The fixed scale marks on the right-hand side of each picture are at intervals of 0·63 cm. (From Pierce 1961. Crown copyright.)

PLATE 9

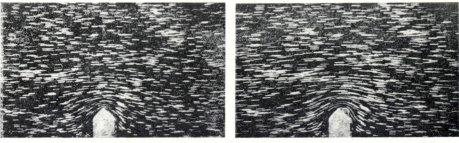

(a)   (b)

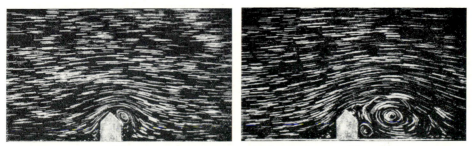

(c)   (d)

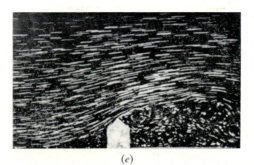

(e)

Figure 5.10.6. Stages in the development of flow past a model of a house from rest. Flow from left to right. (From Nøkkentved 1932.)

PLATE 10

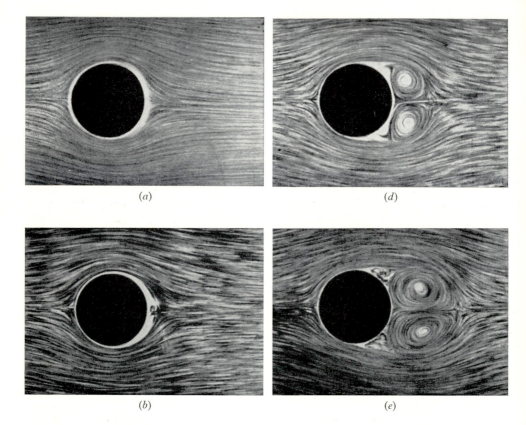

Figure 5.11.3. Stages in the development of flow (from left to right) past a circular cylinder from rest. The speed of the stream has been increased rapidly and then kept constant. (From Prandtl 1927.)

PLATE II

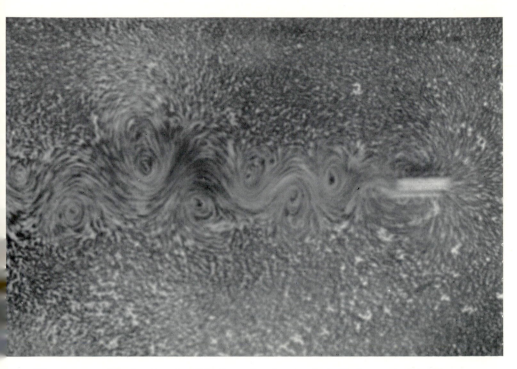

Figure 5.11.4. The 'vortex street' in the wake of a circular cylinder moving steadily, at $R = 1.93 \times 10^3$. The motion was made visible by placing aluminium particles on a water surface, and the cylinder was moving relative to the camera (so that it appears as elongated) from left to right. (From Clutter, Smith and Brazier 1959.)

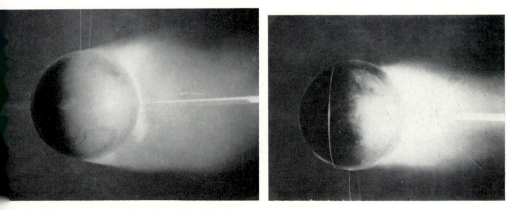

Figure 5.11.7. Smoke released at the rear of a sphere in a stream flowing from left to right. The second photograph shows the effect of disturbing the boundary layer with a wire. (From Wieselsberger 1914.)

PLATE 12

$a\Omega/U = 0$

$a\Omega/U = 3$

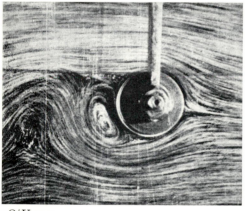

$a\Omega/U = 1$

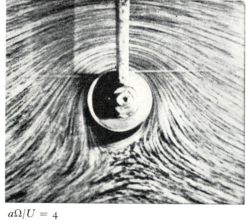

$a\Omega/U = 4$

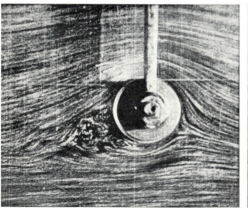

$a\Omega/U = 2$

$a\Omega/U = 6$

Figure 6.6.2. Photographs of the streamlines in flow due to a rigid cylinder rotating with anti-clockwise angular velocity Ω in a stream (from right to left) of uniform speed U at infinity. (From Prandtl and Tietjens 1934.)

PLATE 13

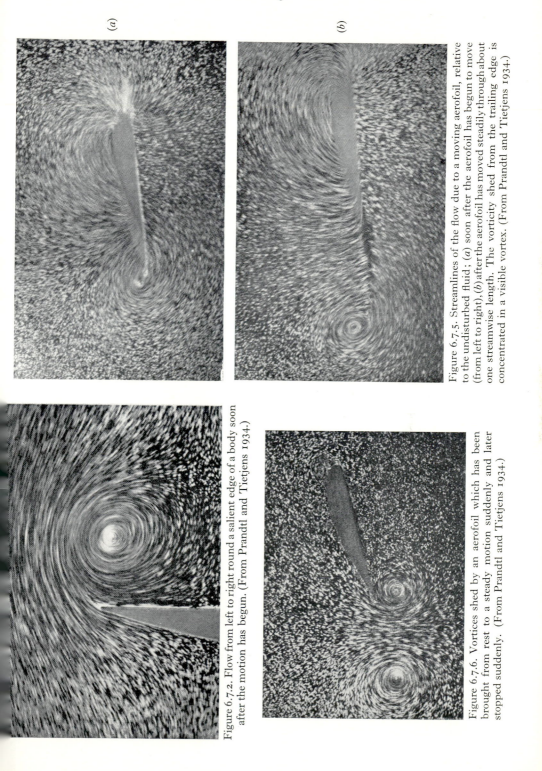

(a)

(b)

Figure 6.7.5. Streamlines of the flow due to a moving aerofoil, relative to the undisturbed fluid; (a) soon after the aerofoil has begun to move (from left to right), (b) after the aerofoil has moved steadily through about one streamwise length. The vorticity shed from the trailing edge is concentrated in a visible vortex. (From Prandtl and Tietjens 1934.)

Figure 6.7.2. Flow from left to right round a salient edge of a body soon after the motion has begun. (From Prandtl and Tietjens 1934.)

Figure 6.7.6. Vortices shed by an aerofoil which has been brought from rest to a steady motion suddenly and later stopped suddenly. (From Prandtl and Tietjens 1934.)

PLATE 14

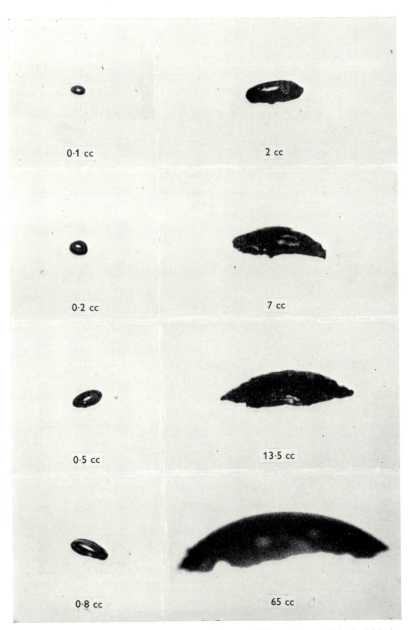

Figure 6.11.1. Air bubbles rising through water. The volume of the bubble is shown beneath each photograph. (From Jones 1965.)

PLATE 15

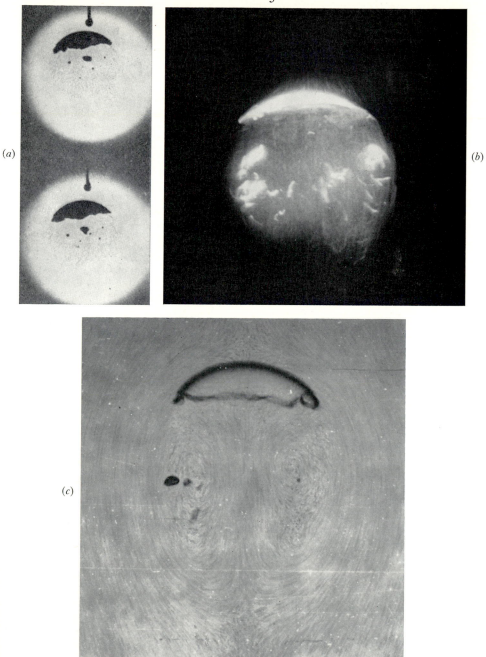

Figure 6.11.2. Photographs of large gas bubbles showing the wake structure. (a) Spark photographs of an air bubble in nitrobenzene; $R \approx 3$ cm, $U \approx 37$ cm/sec (from Davies and Taylor 1950). (b) Air bubble in water; $R \approx 5$ cm, $U \approx 45$ cm/sec (unpublished photograph by R. Collins). (c) A 'two-dimensional' air bubble in water between parallel plates 6 mm apart; $R \approx 7$ cm, $U \approx 40$ cm/sec (from Collins 1965). Photographs (b) and (c) were taken with the camera stationary relative to the bubble and with solid particles as tracers. In (a) and (b) the bubble and liquid moving with it together occupy roughly a sphere, and in (c) a circle.

PLATE 16

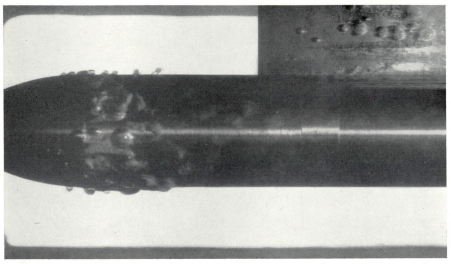

Figure 6.12.1. Formation of cavities in flow (from left to right) past an axisymmetric slender body in a water tunnel at $K = 0·26$. (From Knapp 1952.)

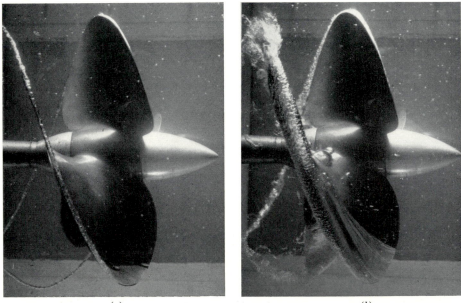

(a)   (b)

Figure 6.12.2. Cavitation due to a propeller rotating in a stream of water flowing from right to left. In (a) there is a cavity at the centre of the 'tip vortex' from each of the three blades; in (b), at higher rotation and stream speeds, there is also a cavity on the suction side of each blade of sufficiently large volume to affect the pressure distribution on the blade.

PLATE 17

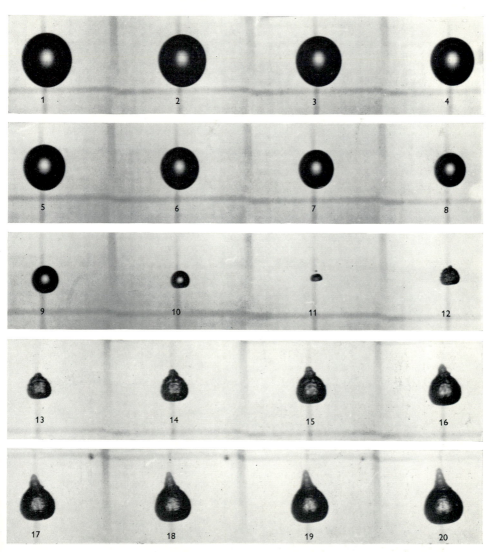

Figure 6.12.3. Photographs of a collapsing cavity in a stationary vessel of water at intervals of 2×10^{-4} sec. The cavity was formed by putting the water in tension for a brief period, with a minute bubble of gas released by electrolysis as the nucleus. During the period covered by the photographs the pressure difference $p_0 - p_v$ was maintained at 0·051 atm. The mean radius of the cavity in the first photograph is 0·69 cm, and the minimum volume occurs between the 10th and 11th photographs. Photographs 11 to 20 show a rebound with characteristic non-spherical form. (Unpublished photographs by T. B. Benjamin and A. T. Ellis.)

PLATE 18

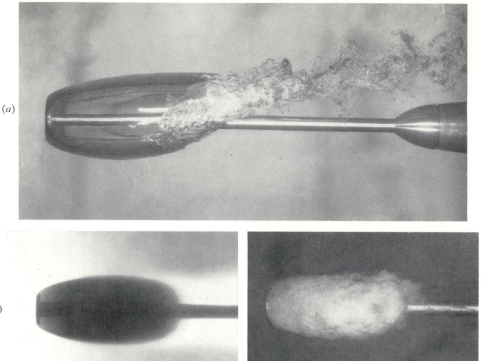

Figure 6.12.5. Steady-state cavities behind a circular disk held normal to a stream of water; $K = 0.19$. (a) Pressure in cavity maintained by continual supply of air to cavity; (b) and (c) no supply of air. (b) Exposure time 2 sec, (c) exposure time 10^{-4} sec. The asymmetry at the rear of the cavity is due to buoyancy. (Official Photographs U.S. Navy.)

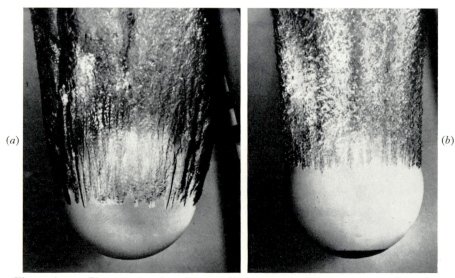

Figure 6.12.7. Photographs of a sphere of diameter 22 cm entering water at 64 cm/sec; (a) smooth surface, (b) with a rough patch of sand near the nose, presumably causing turbulent flow in the boundary layer on the sphere. (Official Photographs U.S. Navy.) See figure 5.11.7 (plate 11) for a similar effect on flow without a cavity.

PLATE 19

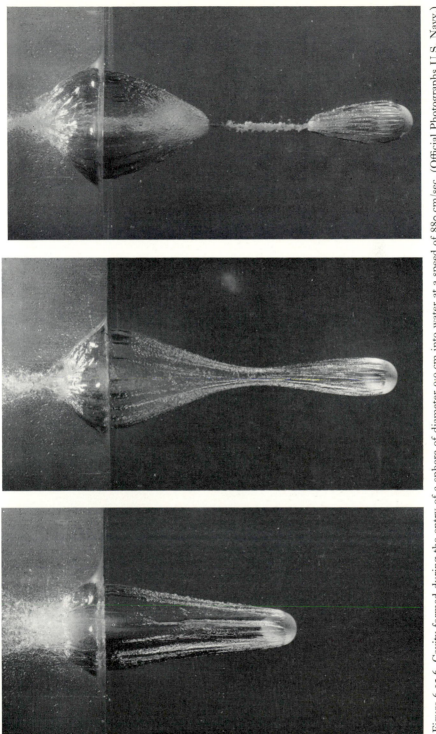

Figure 6.12.6. Cavity formed during the entry of a sphere of diameter 9·9 cm into water at a speed of 880 cm/sec. (Official Photographs U.S. Navy.)

PLATE 20

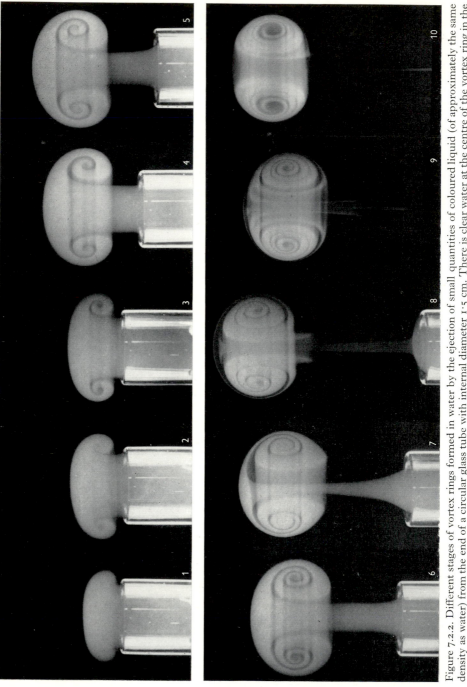

Figure 7.2.2. Different stages of vortex rings formed in water by the ejection of small quantities of coloured liquid (of approximately the same density as water) from the end of a circular glass tube with internal diameter 1·5 cm. There is clear water at the centre of the vortex ring in the last photograph, although it is not apparent from a side-view. (From Okabe and Inoue 1960.)

PLATE 21

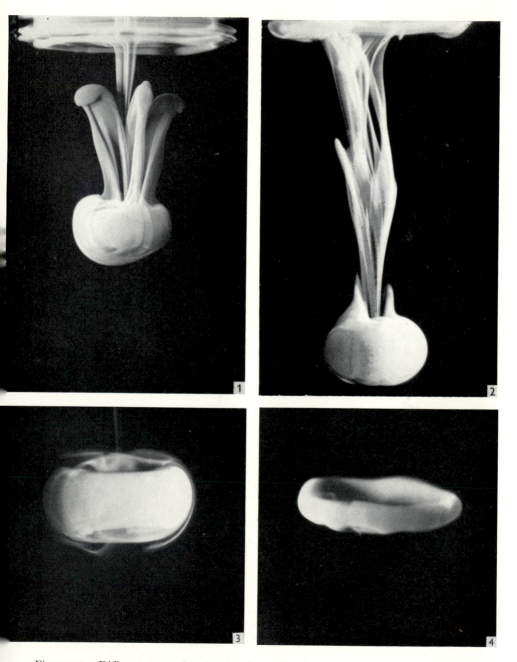

Figure 7.2.3. Different stages of vortex rings formed in water by allowing drops of coloured water to fall vertically from the end of a pipette 1 cm above a free surface. (From Okabe and Inoue 1961.)

PLATE 22

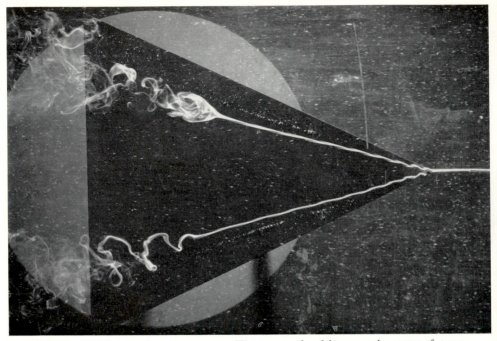

Figure 7.5.7. 'Bursting' vortices in water. The two streaks of dye are at the centre of strong vortices shed from the sides of a triangular wing. (From Lambourne and Bryer 1962.)

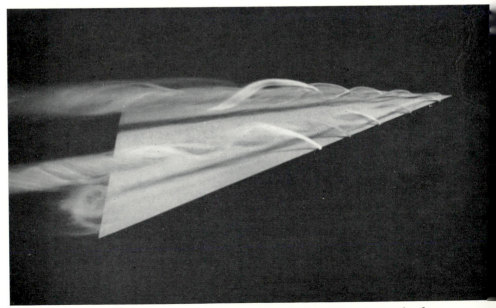

Figure 7.8.6. Flow near the upper surface of a delta wing at an angle of incidence of 12°. (From Werlé 1961.)

PLATE 23

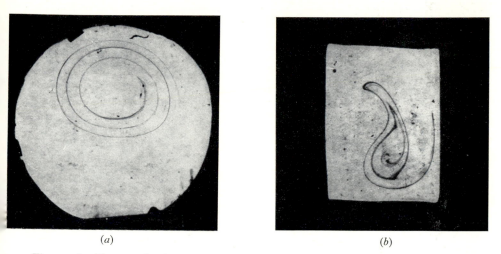

(a)   (b)

Figure 7.6.2. Photographs, from a point on the axis of rotation, of a drop of coloured liquid drawn out into very thin sheets parallel to the axis, showing the two-dimensionality of the motion. A slow motion relative to rotating axes was produced by changing slightly the speed of rotation of (a) a circular dish, and (b) a rectangular dish. (From Taylor 1921.)

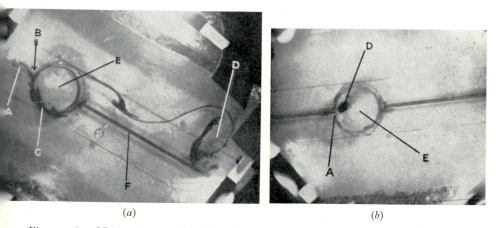

(a)   (b)

Figure 7.6.3. Motion in a rotating dish of water 4 in. deep due to slow translation of a circular cylinder (E) of height 1 in. from right to left across the bottom of the dish, viewed from above. In (a) the dye has been released at a (moving) point A above the top of the cylinder and directly ahead of it, and B (a dividing point), C and D are subsequent positions of the dye; in (b) the point of release (A) is within the upward projection of the cylinder and the dye remains in the blob D. The flow evidently has a two-dimensional character. (From Taylor 1923.)

PLATE 24

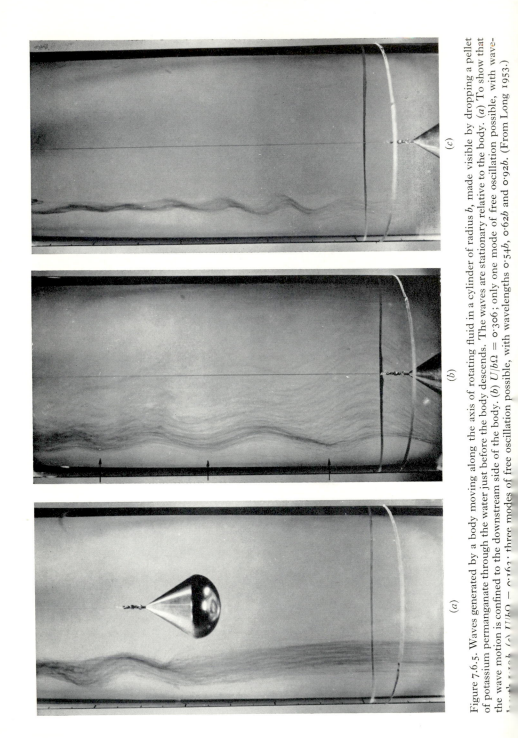

Figure 7.6.5. Waves generated by a body moving along the axis of rotating fluid in a cylinder of radius b, made visible by dropping a pellet of potassium permanganate through the water just before the body descends. The waves are stationary relative to the body. (a) To show that the wave motion is confined to the downstream side of the body. (b) $U/b\Omega = 0.306$; only one mode of free oscillation possible, with wave-length 1.13b. (c) $U/b\Omega = 0.163$; three modes of free oscillation possible, with wavelengths $0.54b$, $0.62b$ and $0.92b$. (From Long 1953.)

the flow in the wake ceases to be turbulent beyond some distance from the body. Thus the velocity profile (5.12.9) may be expected to apply to the steady wake behind a three-dimensional body at Reynolds numbers small enough for stability, and also very far downstream from a body whose wake is initially turbulent.

5.13. Oscillatory boundary layers

In a number of practical problems of flow at large Reynolds number the velocity varies periodically with respect to time, often as a consequence of an enforced oscillation of a solid boundary. It happens that the governing equations can be linearized, in certain cases of periodic flow, and a number of interesting results can be derived. The basic approximation to be adopted here, for cases in which no mean flow is imposed on the fluid, is that

$$|\partial \mathbf{u}/\partial t| \gg |\mathbf{u}.\nabla\mathbf{u}|. \qquad (5.13.1)$$

If the velocity varies periodically everywhere with frequency n and with an amplitude represented by U_0, and if L is representative of the distance over which $\mathbf{u}$ changes appreciably along a streamline, $|\mathbf{u}.\nabla\mathbf{u}|$ is of order U_0^2/L (the large gradients of $\mathbf{u}$ in a boundary layer being across the streamlines), and (5.13.1) is satisfied when

$$nL/U_0 \gg 1. \qquad (5.13.2)\dagger$$

In cases in which the periodic variation of $\mathbf{u}$ is forced on the fluid by oscillation of a solid boundary over distances of order ϵ in a direction normal to the boundary, U_0 is of order $n\epsilon$ and (5.13.2) is then equivalent to

$$\epsilon \ll L. \qquad (5.13.3)$$

(It will also be recalled from the discussion in §3.6 that the condition $nL/c \ll 1$ must be satisfied, where c is the speed of sound waves through the fluid, if, as will be assumed here, the velocity distribution is to remain uninfluenced by compressibility of the fluid.)

Vorticity arises wholly from the boundaries and, if the relative fluid motion at the boundaries is purely periodic, the rate of generation of vorticity there is alternately positive and negative. In these circumstances it is reasonable to assume, at any rate as an approximation, that no net vorticity is generated in one cycle and that the vorticity is zero outside a narrow region near the boundary within which the alternate layers of positive and negative vorticity are diffusing together and cancelling.‡ The time available

---

† The dimensionless parameter nL/U_0 is the Strouhal number, mentioned in §4.7.

‡ There will always be parts of the boundary where convection carries vorticity in a direction away from the boundary, and we are assuming that vorticity does not move far from the surface before the convection velocity is reversed. This is permissible if separation of the boundary layer does not occur. Bodies with salient edges are thus excluded, because separation occurs almost immediately such a body moves. As the frequency n becomes larger, edges of smaller radii of curvature can be admitted. We

for diffusion of vorticity of one sign from the boundary is $2\pi/n$, and so the thickness of the layer of non-zero vorticity, δ say, is of order $(\nu/n)^{\frac{1}{2}}$. We shall assume that

$$\delta \ll L, \tag{5.13.4}$$

which is equivalent to an assumption that the Reynolds number L^2n/ν is large compared with unity. A variation of the penetration depth δ as $(\nu/n)^{\frac{1}{2}}$ was obtained explicitly in §4.3 for the case of a plane rigid boundary oscillating in its own plane.

When $\delta \ll L$, the flow is irrotational over nearly all the field and the velocity potential may be determined from the instantaneous speed and position of the boundary. Corresponding to this irrotational flow there will be a non-zero tangential component of velocity of fluid at the boundary and relative to it, which, for a sinusoidal oscillation, may be written as the real part of $U e^{int}$ say, where the complex quantity U varies with position on the boundary; this tangential velocity is now an 'external stream velocity' for the boundary layer of non-zero vorticity. Thus, bearing in mind that the approximation (5.13.1) holds both inside and outside the boundary layer, we have for the flow in the boundary layer (see (5.7.1) and (5.7.8))

$$\frac{\partial u}{\partial t} = \frac{\partial}{\partial t}(U e^{int}) + \nu \frac{\partial^2 u}{\partial y^2}, \tag{5.13.5}$$

where the real part of the complex quantity u is the component of fluid velocity parallel to the boundary (in the same direction as the 'external stream') and relative to it, and y denotes distance normal to the boundary. The boundary conditions to be satisfied by u are

$$u \to U e^{int} \quad \text{as} \quad y \to \infty,$$

and, assuming the boundary to be rigid,

$$u = 0 \quad \text{at} \quad y = 0.$$

Equation (5.13.5) and the accompanying boundary conditions show that the distribution of $U e^{int} - u$ with respect to y and t at a given position on the body surface is the same as the distribution of velocity in fluid at rest at infinity and bounded by an infinite rigid plane wall whose velocity (parallel to itself) is the real part of $U e^{int}$; and so, on making use of (4.3.16), we have

$$u(y, t) = U e^{int}\{1 - e^{-(1+i)y/\delta}\}, \tag{5.13.6}$$

in which for numerical convenience we have put

$$\delta = \left(\frac{2\nu}{n}\right)^{\frac{1}{2}}. \tag{5.13.7}$$

saw in §5.9 that if a circular cylinder is suddenly brought to a steady finite velocity, back-flow does not occur anywhere in the boundary layer until the cylinder has moved a distance of about one-third of a radius, and the onset of separation is probably even later; and so separation is unlikely to occur on an oscillating circular cylinder which reverses its motion after moving a distance small compared with its radius.

The fact that U varies with position on the boundary does not affect the local distribution of u, because we have assumed the boundary-layer thickness to be small compared with the distance L over which U changes appreciably. The external irrotational flow and the associated boundary-layer flow (5.13.6) together represent an approximation to the complete velocity distribution under the assumed conditions (5.13.2) and (5.13.4).

The damping force on an oscillating body

It follows from (5.13.6) that the frictional stress at the boundary is the real part of

$$\mu\left(\frac{\partial u}{\partial y}\right)_{y=0} = \mu(1+i)\frac{U}{\delta}e^{int}. \tag{5.13.8}$$

The skin friction has a phase lead of $\frac{1}{4}\pi$ (i.e. it goes through its cycle one-eighth of a period in advance) over the external stream speed $U e^{int}$, because the imposed oscillating pressure gradient acts on all layers of the fluid equally and produces a velocity in the same direction as the acceleration more rapidly in the slower-moving layers near the boundary than in the outer layers. The fact that this phase difference is not $\frac{1}{2}\pi$ implies that, in the case of a rigid body oscillating about a fixed position, a non-zero amount of work is done by the body against frictional stresses during each cycle and that a *damping* force acts on the oscillating body. The force acting on a body oscillating with translational velocity equal to the real part of $U_0 \mathbf{k} e^{int}$ (where U_0 is real and $\mathbf{k}$ is a constant unit vector) in fluid at rest at infinity, due to the tangential component of stress at the body surface given by (5.13.8), is readily found from a knowledge of the body shape and the speed $U e^{int}$ at the boundary due to the outer irrotational flow.

However, there is also a contribution to the force on the body from the normal stress at the surface. The pressure at the body surface is approximately the same as if the fluid were completely inviscid and the flow were irrotational everywhere; and this approximation becomes increasingly accurate as $\delta/L \to 0$, i.e. as $Ln^{\frac{1}{2}}/\nu^{\frac{1}{2}} \to \infty$. The first approximation to the net force due to pressure at the body surface is therefore equal to the total force acting on the same body oscillating in an inviscid flow. But the irrotational flow generated by movement of the body in an inviscid fluid is determined uniquely by its instantaneous speed, and must vary periodically with the same frequency n as the body speed; the kinetic energy of the fluid likewise has a purely periodic value, and, since no means of absorbing energy exists, the net work done by the body against normal stresses in one cycle in wholly irrotational flow is zero. (See also §6.4, where it is shown directly that the resistive force on an accelerating body in inviscid fluid has a phase different from that of the body speed by $\frac{1}{2}\pi$.) It is thus necessary to find a better approximation to the pressure at the surface. The (dimensionless) contribution to the damping force due to tangential stress at the body surface is of the order of some Reynolds number to the power $-\frac{1}{2}$, as may be seen from the way the viscosity enters the expression (5.13.8), and the contribution from the normal stress must also be evaluated to this order. This is not easy. Stokes (1851) was able to calculate the whole flow field for both a sphere and a circular cylinder making oscillations of small amplitude without making any assumption about the magnitude of the Reynolds number, and it can be seen from his results that when $\delta \ll L$ there is a contribution to the pressure at the body surface due to the existence of the boundary layer; the correction to the irrotational flow value of the surface pressure is here the same as if the sphere or cylinder were endowed with a new radius larger by an amount of order δ (which gives rise to a small increase in the instantaneous resultant force due to normal stresses), and as if the centre of the effectively enlarged body lagged in position behind the centre of the real body by an amount of order δ (which gives rise to a contribution to the pressure force which is in phase with the body speed, and thus to a contribution to the damping force).

We shall employ instead a different and considerably simpler method of determining the damping force. As already noted, the irrotational flow and the associated boundary-layer flow (5.13.6) together give a good approximation to the velocity distribution over the whole flow field. We are therefore in a position to estimate the total rate of dissipation in the fluid, and thereby to determine the average rate of working of the body. Now the total rate of dissipation in the boundary layer (of thickness δ and external stream speed of order U_0) is of order $\mu U_0^2/\delta^2$ per unit volume and $\mu U_0^2 A/\delta$ for the whole body of surface area A, whereas the contribution from the region of irrotational flow is of order $\mu U_0^2/L^2$ per unit volume and $\mu U_0^2 A/L$ for the whole field, showing that this latter contribution may be neglected. Within the boundary layer the velocity is approximately parallel to the boundary locally, so that the rate of dissipation per unit volume at a point in the boundary layer follows from (5.13.6) as

$$\mu \left\{ \mathscr{R} \left(\frac{\partial u}{\partial y} \right) \right\}^2 = \frac{2\mu U^2}{\delta^2} e^{-2y/\delta} \cos^2 \left(nt + \frac{\pi}{4} - \frac{y}{\delta} \right), \qquad (5.13.9)$$

where $\mathscr{R}$ denotes 'the real part of'; U is real here because all parts of the outer irrotational flow oscillate in the same phase. The average, over one cycle, of the rate of dissipation in the boundary layer per unit area of the body surface at the point where the irrotational flow velocity oscillates with amplitude U is thus

$$\int_0^\infty \frac{\mu U^2}{\delta^2} e^{-2y/\delta} \, dy = \frac{\mu U^2}{2\delta}, \qquad (5.13.10)$$

and the average total rate of dissipation is obtained by integrating over the body surface A. But if $F\mathbf{k}\,e^{int}$ is the damping force on the body in phase with its velocity $U_0\,\mathbf{k}\,e^{int}$, the average rate at which the body does work against forces exerted on it by the fluid is $\frac{1}{2}U_0 F$; hence

$$\tfrac{1}{2}U_0 F = \frac{\mu}{2\delta}\int U^2 dA. \qquad (5.13.11)$$

To proceed further, we must know the shape of the body and thence U. For a sphere it follows from (2.9.28) (with appropriate changes of notation) that
$$U = \tfrac{3}{2}U_0\sin\theta, \quad \int U^2 dA = 6\pi a^2 U_0^2,$$
and for (unit length of) a circular cylinder, from (2.10.12), that
$$U = 2U_0\sin\theta, \quad \int U^2 dA = 4\pi a U_0^2,$$
where a is the radius and θ is the polar angle with $\theta = 0$ in the direction of $\mathbf{k}$ in both cases. Hence we have

$$F = 6\pi a^2 \mu U_0/\delta \quad \text{for a sphere,}$$
$$4\pi a\mu U_0/\delta \quad \text{for unit length of a circular cylinder.}$$

It can readily be found from (5.13.8) that the contribution to F from tangential stresses at the body surface is $\frac{2}{3}$ of the above total value for a sphere, and $\frac{1}{2}$ in the case of a circular cylinder, the remaining parts being evidently accounted for by the normal stresses as affected by the presence of the boundary layer. Expressed as dimensionless coefficients in the usual way, the above damping forces become

$$\frac{F}{\pi a^2 \frac{1}{2}\rho U_0^2} = 12\frac{\mu}{\rho U_0 \delta} = 6\sqrt{2}\left(\frac{\nu}{U_0 \epsilon}\right)^{\frac{1}{2}} \quad \text{for a sphere,} \qquad (5.13.12)$$

and $$\frac{F}{2a\frac{1}{2}\rho U_0^2} = 4\pi\frac{\mu}{\rho U_0 \delta} = 2\pi\sqrt{2}\left(\frac{\nu}{U_0 \epsilon}\right)^{\frac{1}{2}} \quad \text{for a circular cylinder,} \quad (5.13.13)$$

where $\epsilon = U_0/n$ is half the range over which the centre of the body oscillates. Variation as the $(-\frac{1}{2})$-power of a Reynolds number was to be expected for a boundary-layer flow with no separation, although the fact that the Reynolds number is formed from the oscillation amplitude ϵ rather than from a dimension of the body is a novel feature.

For a freely oscillating elastic circular cylinder such as a piano string, of density ρ_s, the average total energy of the cylinder is $\frac{1}{2}\pi a^2 \rho_s U_0^2$ per unit length, and the equation for the decay of the oscillation due to the viscosity of the fluid† is
$$\frac{d}{dt}(\tfrac{1}{2}\pi a^2 \rho_s U_0^2) = -\tfrac{1}{2}U_0 F = -2\pi\frac{a\mu U_0^2}{\delta}.$$

Thus $U_0^2 \propto e^{-\beta nt}$, and the fractional decrease in energy of the cylinder in one cycle is approximately

$$2\pi\beta = 8\pi\frac{\mu}{a\rho_s n\delta} = 4\pi\frac{\rho}{\rho_s}\left(\frac{2\nu}{na^2}\right)^{\frac{1}{2}}. \qquad (5.13.14)$$

† There is also a loss of energy of a vibrating cylinder due to the direct generation of sound waves in the fluid, but this is usually much smaller; however, the indirect loss of energy to sound radiation *via* a sounding board may not be negligible.

Measurements of the rate at which a freely oscillating circular cylinder loses energy due to fluid damping at large values of a^2n/ν have been found to yield a damping force in good agreement with the above estimate for oscillation amplitudes ϵ less than about 0·1a.

Steady streaming due to an oscillatory boundary layer

Another feature of the velocity distribution (5.13.6) with interesting consequences is the rapid variation of the amplitude of the oscillation across the boundary layer. Now when either the amplitude or the phase of the 'external stream' velocity $U e^{int}$ in a two-dimensional flow varies with co-ordinate x in the direction of this external stream, the mass-conservation relation shows there must exist a non-zero component of velocity normal to the boundary throughout the boundary layer, of order $\delta \, dU/dx$, being given explicitly by the real part of

$$v = -\int_0^y \frac{\partial u}{\partial x} \, dy = -e^{int} \frac{dU}{dx} \int_0^y \{1 - e^{-(1+i)\,y/\delta}\} \, dy$$

$$= -e^{int} \frac{dU}{dx} \left\{ y - \frac{\delta}{1+i} + \frac{\delta}{1+i} e^{-(1+i)\,y/\delta} \right\}. \quad (5.13.15)$$

If it happens that the real parts of u and v at some point oscillate with a phase difference other than $\frac{1}{2}\pi$, so that the average of their product is non-zero, there will be a net transfer of x-momentum across a surface element with normal in the y-direction during one cycle of the oscillation. As a result of the increase of the amplitude of u with distance from the boundary, this effective stress will vary across the boundary layer and so will give rise to a non-zero average force on the fluid. The consequent steady motion of the fluid may be weak, but, inasmuch as it leads to extensive migration of fluid elements in an apparently purely oscillatory system, its effect is sometimes important. Such a drift motion, or 'steady streaming', is likely to be generated in any oscillatory flow in which there is a non-zero mean flux of momentum across surfaces in the fluid, and we expect it to be more significant when there exists a boundary layer in which the lateral gradient of amplitude of oscillation of the velocity (and so also the stress gradient) is large.

To determine the drift motion, we must return to the boundary-layer equation without neglect of the non-linear terms. The flow will be assumed to be two-dimensional for the purpose of exposition. Suppose that u_1 and U_1 are the real (i.e. not complex) sinusoidally-varying velocities relative to the boundary, inside and just outside the boundary layer respectively, as determined by the above linear theory, and that $u_1 + u_2$ and $U_1 + U_2$ satisfy the complete non-linear equations which apply inside and just outside the boundary layer respectively. Then, since $|u_2|$ and $|U_2|$ are presumably small compared with $|u_1|$ and $|U_1|$, the equation for u_2 is approximately

$$\frac{\partial u_2}{\partial t} - \nu\frac{\partial^2 u_2}{\partial y^2} - \frac{\partial U_2}{\partial t} = U_1\frac{\partial U_1}{\partial x} - u_1\frac{\partial u_1}{\partial x} - v_1\frac{\partial u_1}{\partial y}. \qquad (5.13.16)$$

The expressions for u_2 and U_2 evidently consist of terms proportional to $\sin 2nt$ and to $\cos 2nt$ and of constant terms. The latter represent the steady streaming under consideration, and we therefore average all terms in (5.13.16) over one cycle to obtain

$$-\nu\frac{\overline{\partial^2 u_2}}{\partial y^2} = \overline{U_1\frac{\partial U_1}{\partial x}} - \overline{u_1\frac{\partial u_1}{\partial x}} - \overline{v_1\frac{\partial u_1}{\partial y}}. \qquad (5.13.17)$$

Now U_1 is the real part of $U(x)\,e^{int}$ (where U is complex in general), and may be written as
$$U_1 = \tfrac{1}{2}(U\,e^{int} + U^*\,e^{-int}),$$

where the asterisk denotes a complex conjugate; and

$$\overline{U_1^2} = \tfrac{1}{2}UU^*.$$

u_1 and v_1 are the real parts of the quantities u and v given in (5.13.6) and (5.13.15), and may be written similarly. Evaluation of the three terms on the right-hand side of (5.13.17) in this way gives

$$-\nu\frac{\overline{\partial^2 u_2}}{\partial y^2} = \frac{1}{4}\frac{d(UU^*)}{dx}\{1 - (1 - e^{-\alpha y})(1 - e^{-\alpha^* y})\}$$

$$+ \frac{1}{2}\mathscr{R}\left\{U^*\frac{dU}{dx}\frac{\alpha^*}{\alpha}\,e^{-\alpha^* y}(\alpha y - 1 + e^{-\alpha y})\right\}, \qquad (5.13.18)$$

$$= G(x,y) \quad \text{say,}$$

in which $\alpha = (1+i)/\delta$. The steady streaming velocity $\overline{u_2}$ is determined by this equation, together with the boundary condition

$$\overline{u_2} = 0 \quad \text{at} \quad y = 0$$

when the boundary is rigid, and a condition expressing the fact that $\overline{u_2}$ tends to a constant value as $y/\delta \to \infty$.

A formal interpretation of equation (5.13.18) is that $\overline{u_2}$ is the velocity in a steady, effectively unidirectional, flow generated by a body force G per unit mass of fluid in the x-direction, this force being simply the net gain of x-momentum in unit time arising from the small oscillatory movements of fluid in the y-direction. $G(x,y)$ varies slowly with x but rapidly with y, and vanishes outside the boundary layer, showing that the whole field of steady streaming motion is driven by the effective tangential force acting on the fluid in a thin layer near the boundary. Inasmuch as G varies rapidly across the boundary layer, $\overline{u_2}$ will also vary rapidly with y, and this gives the distribution of $\overline{u_2}$ an apparent boundary-layer character. It should be noted that we have not assumed that the Reynolds number of the secondary streaming motion is large and that viscous diffusion of the vorticity associated with the streaming motion is negligible except near the boundary. The distribution of u_1 shows rapid variations across a thin layer near the boundary for conventional reasons, whereas the rapid variation of $\overline{u_2}$ across the same layer is due entirely to the nature of the distribution of the effective body force G.

In the case of the first approximation to the velocity, represented by u_1, the distribution within the boundary layer is determined by that outside the boundary layer. However, the converse is true of the 'correction' term $\overline{u_2}$. In the integration of (5.13.18), $\overline{U_2}$ is unknown, and we are therefore obliged to take, as the outer boundary condition,

$$\frac{\partial \overline{u_2}}{\partial y} \to 0 \quad \text{as} \quad y \to \infty,$$

corresponding to the fact that there is no source of *rapid* variations of the streaming velocity outside the boundary layer.

The solution of (5.13.18) satisfying these boundary conditions has been obtained (Schlichting 1932) and exhibits a complicated dependence on y. The most interesting feature of the solution is the non-zero value of $\overline{u_2}$ approached as $y \to \infty$, and this alone will be given here. From (5.13.18) and the above boundary conditions we have

$$\overline{vu_2} = \int_0^y \left\{ \int_{y'}^\infty G(y')\,dy' \right\} dy'',$$

showing that $\overline{u_2} \to \overline{U_2}$ as $y \to \infty$, where

$$\overline{U_2} = \frac{2}{\delta^2 n} \int_0^\infty y\,G(y)\,dy. \tag{5.13.19}$$

Substitution of the expression (5.13.18) for $G(y)$ then gives, after a few lines of working,

$$\overline{U_2} = \frac{3}{8n} \left\{ -\frac{d(UU^*)}{dx} + i\left(U^* \frac{dU}{dx} - U \frac{dU^*}{dx} \right) \right\},$$

$$= -\frac{3}{8n} \left(\frac{dA^2}{dx} + 2A^2 \frac{d\gamma}{dx} \right) \tag{5.13.20}$$

if we write $U = A\,e^{i\gamma}$ and allow both the amplitude A and the phase angle γ to be functions of x. (5.13.20) gives the value of the mean velocity at points just outside the boundary layer that is imposed on the fluid there by the balance between the effective body force G exerted on fluid near the boundary and viscous forces arising from the lateral variation of $\overline{u_2}$. The ratio of $\overline{U_2}$ to U_0 (the representative amplitude of the velocity fluctuations) is of order U_0/nL, which is small compared with unity. A remarkable property of $\overline{U_2}$ is its lack of dependence on the viscosity.

The distribution of the mean velocity over the whole region outside the boundary layer can now be determined, in principle, using (5.13.20) as one of the boundary conditions for this distribution. If the flow outside the boundary layer were exactly irrotational, no steady streaming could exist there, because an irrotational motion is fully determined by the instantaneous normal velocity of the boundaries and must have the same oscillatory character as the motion of the boundary. However, there is a slow transfer of vorticity away from the boundary as a consequence of the steady flow near the boundary, and ultimately, some time after the motion has started from rest, there will be a second-order steady vorticity throughout the fluid. The form of the equation satisfied by the mean velocity in the region outside the boundary layer depends on the effective Reynolds number $\overline{U_2}L/\nu$ of the steady flow. In view of (5.13.20), this Reynolds number may be written as

$$U_0^2/n\nu \quad \text{or} \quad \epsilon^2/\delta^2$$

(where ϵ is, as before, the amplitude of the oscillation of position of the solid body in cases in which that is the source of the flow), which may be either small or large compared with unity, consistently with the assumed conditions (5.13.4) and (5.13.2) or (5.13.3). Transfer of vorticity associated with the steady streaming takes place mainly by viscous diffusion in the former case, whereas in the latter case convection is dominant everywhere except in certain thin layers, one of which adjoins the rigid boundary with a thickness much greater than that of the boundary layer of the primary oscillatory flow. Corresponding approximate forms of the governing equation in these two cases, and their solution, are beyond our scope here.†

Applications of the theory of steady streaming

As remarked earlier, the steady streaming motion is of practical interest chiefly because it leads to extensive migration of elements of fluid. This drift is connected with the mean velocity of a given fluid element, which may not be the same as the mean velocity at a point, and it is necessary to relate these two velocities. Suppose that $\mathbf{w}(\mathbf{x}_0, t)$ is the velocity at time t of a material element which at a previous instant t_0 was at the point $\mathbf{x}_0$, and that $\mathbf{u}(\mathbf{x}, t)$ is the velocity at time t at the point $\mathbf{x}$. Then, without approximation, we have

$$\mathbf{w}(\mathbf{x}_0, t) = \mathbf{u}\left(\mathbf{x}_0 + \int_{t_0}^{t} \mathbf{w}\,dt, t\right).$$

For values of $t - t_0$ of the order of a period of the oscillation, the displacement of the element is small compared with L, so that

$$\mathbf{w}(\mathbf{x}_0, t) \approx \mathbf{u}(\mathbf{x}_0, t) + \left\{\left(\int_{t_0}^{t} \mathbf{w}\,dt\right).\nabla\mathbf{u}(\mathbf{x}, t)\right\}_{\mathbf{x}=\mathbf{x}_0}, \qquad (5.13.21)$$

† The problem is pursued in papers by Longuet-Higgins (1953) and Stuart (1966).

and on the right-hand side **w** may be replaced by **u** with a consistent approximation. The second-order drift may now be obtained by using the first-order motion to evaluate the non-linear term, in the manner that led to (5.13.16). On taking an average over one cycle, we find for the mean velocity of fluid elements just outside the boundary layer (where both **w** and **u** are nearly parallel to the boundary),

$$\overline{W_2} = \overline{U_2} + \frac{i}{4n}\left(U^*\frac{dU}{dx} - U\frac{dU^*}{dx}\right),$$

$$= \overline{U_2} - \frac{A^2}{2n}\frac{d\gamma}{dx} \qquad (5.13.22)$$

if we write $U = A\,e^{i\gamma}$ again. Thus the mean velocity of an element initially at a point $\mathbf{x}_0$ and the mean velocity at $\mathbf{x}_0$ are different, for kinematical reasons, whenever the phase of the velocity fluctuation varies with distance in the direction of the velocity, and their difference, which has nothing directly to do with viscosity, is of the same form as one of the two terms in (5.13.20), both of which do arise from the effect of viscosity at the rigid boundary.

Many oscillating flows are of either the 'standing wave' type, in which the phase γ of the oscillation is independent of position, or the 'progressive wave' type, in which the amplitude A is independent of position. A rigid body oscillating about a fixed mean position gives rise to flow oscillations of the standing wave type, for which, according to (5.13.20) and (5.13.22),

$$\overline{U_2} = \overline{W_2} = -\frac{3}{8n}\frac{dA^2}{dx} \qquad (5.13.23)$$

at points just outside the boundary layer, showing that fluid elements drift towards the positions where the amplitude A is a minimum. In the case of a circular cylinder of radius a, the velocity of whose centre is (the real part of) $U_0\,e^{int}$ in the direction $\theta = 0$ normal to the axis, we have

$$U = A = 2U_0\sin\theta,$$

so that the drift velocity just outside the boundary layer is here

$$\overline{U_2} = \overline{W_2} = -\frac{3U_0^2}{2na}\sin 2\theta \qquad (5.13.24)$$

(a positive value signifying a velocity in the direction of θ increasing). This drift round the surface towards the front and rear stagnation points sets up a circulation within each quadrant of the fluid as a whole, which can be observed qualitatively by oscillating a circular cylinder in a tank of water on the surface of which small visible particles have been sprinkled.

Flow oscillations of the standing wave type are also set up when a stationary rigid body is immersed in fluid through which a plane sound wave is progressing, provided the sound wave-length is large compared with the body dimensions;† the steady secondary motion is often referred to as 'acoustic streaming' in this case. Another acoustic phenomenon which may be explained by the effects discussed above is the tendency for fine dust particles to gather at the nodal points (of velocity) on the wall of a tube in which there is a standing sound wave (the so-called Kundt's dust tube). Again analysis based on the assumption of incompressible fluid is applicable, provided λ (the sound wave-length) $\gg \delta$. Since $A = A_0 \sin(2\pi x/\lambda)$ here, (5.13.23) becomes

$$\overline{U_2} = \overline{W_2} = -\frac{3A_0^2}{8c} \sin \frac{4\pi x}{\lambda}, \qquad (5.13.25)$$

where $c = n\lambda/2\pi$ is the velocity of propagation of the sound wave. Accompanying this steady flow towards the nodal points $x = 0, \frac{1}{2}\lambda, \lambda, \ldots$ just outside the boundary layer, there is a circulation in the interior of the fluid, with flow towards the tube wall over the anti-nodes and away from it over the nodes; the whole field of steady streaming in this case was calculated by Rayleigh (1883), without the aid of boundary-layer approximations to the first-order motion near the wall. The fine dust particles are able to settle at places on the wall where they are not disturbed by the periodic first-order oscillation, i.e. at the velocity nodes, and they are carried towards such points by the second-order drift.

The flow oscillations produced in water of uniform depth by the passage of a progressive surface wave provides an example of the other type, in which γ varies with position. The accompanying drift of fluid elements here may have consequences of practical importance, such as transport of sediment near the bottom. The amplitude of the (sinusoidal) flow oscillations just outside the boundary layer on the horizontal rigid boundary is independent of position on the boundary, so that we have from (5.13.20) and (5.13.22)

$$\overline{U_2} = -\frac{3A^2}{4n} \frac{d\gamma}{dx}, \qquad \overline{W_2} = -\frac{5A^2}{4n} \frac{d\gamma}{dx}. \qquad (5.13.26)$$

We also have here $\gamma = -2\pi x/\lambda$ (the direction of progression of the surface wave being that of x increasing), where λ is the wavelength. Thus the mean speeds at points, and of fluid elements, just outside the boundary layer near the rigid bottom of the body of water become

$$\overline{U_2} = \frac{3A^2}{4c}, \qquad \overline{W_2} = \frac{5A^2}{4c}, \qquad (5.13.27)$$

respectively, where c is the speed of propagation of the surface wave. The oscillation amplitude A is known to decrease exponentially with depth, and the additional drift speed due to viscosity is appreciable only for bottom

---

† It is a general result in acoustics that under this same condition the relative flow in the neighbourhood of the body is approximately the same as if the fluid were incompressible, so that the above results are applicable.

depths less than about one wavelength. Observations have shown the existence of a steady drift of small solid particles near the bottom, in the direction of progression of the wave, and at about the predicted speed $5A^2/4c$.

There is a 'boundary layer' at free surfaces, as we shall see in the next section, and this too leads to a drift motion of fluid near the surface of water over which a wave is advancing (Longuet-Higgins 1953), although the matter will not be pursued here.

5.14. The boundary layer at a free surface

Although a rigid boundary is the commonest source of vorticity and consequently of boundary layers in flow at large Reynolds number, the case of a boundary at which the tangential stress vanishes exhibits interesting differences and deserves to be considered briefly. At a 'free' surface of the fluid (§ 3.3), the conditions to be satisfied are that the normal component of stress is the sum of a constant term and any contribution from surface tension and that the tangential component is zero.

Following the plan of previous sections, it is useful to imagine a flow which is set up from rest by bringing all the given boundaries to their prescribed velocities. Immediately after the boundaries are made to move, an irrotational motion exists everywhere in the fluid. We need to enquire whether such a motion is capable of satisfying the complete boundary conditions; if so, the irrotationality persists and the steady state is one of irrotational flow everywhere. Now if the shape of the free boundary were prescribed in advance, the irrotational motion would be fully determined and the conditions on both normal and tangential components of stress at the boundary would in general remain unsatisfied. In fact, the shape of a free boundary is itself affected by the fluid motion and adjusts itself in accordance with the boundary conditions. Of the two boundary conditions, the one which must be satisfied by the initial irrotational flow is the condition on the normal component of stress; for any departure of the normal stress on the fluid side of the boundary from the specified value implies the existence of infinite acceleration of the fluid at the boundary in a direction normal to the boundary, and thus a change of boundary shape would occur rapidly. We may therefore assume that the shape of the free boundary at all times is such that the normal stress at the boundary is equal to a constant plus any jump in pressure due to surface tension. This in general determines fully the initial irrotational motion and the boundary shape.

There remains the condition of zero tangential stress at the boundary, which in general cannot also be satisfied by the initial irrotational motion. Any departure of the tangential stress in the irrotational flow at points close to the boundary from zero implies the existence of infinite acceleration of the fluid at the boundary in a direction parallel to the boundary and in such a sense as to bring the tangential stress in the fluid closer to zero. This acceleration of the fluid by viscous forces generates vorticity at the boundary which then diffuses into the fluid in the familiar manner. However, whereas a solid boundary requires a non-zero jump in velocity at the boundary of a region of irrotational flow, and consequently generates a sheet of (initially) infinite vorticity, a free surface requires a non-zero jump in velocity *derivatives* and generates a finite vorticity. The exact amount of vorticity generated at the free surface can be calculated readily in the following way.

We choose orthogonal curvilinear co-ordinates (ξ, η, ζ) such that the free surface coincides instantaneously with one of the surfaces on which ζ is constant. Then with (u, v, w) representing corresponding velocity components, and the usual meanings for h_1, h_2, h_3, the definition of the ξ-component of vorticity (see appendix 2) is

$$\omega_\xi = \frac{1}{h_2 h_3} \left\{ \frac{\partial(h_3 w)}{\partial \eta} - \frac{\partial(h_2 v)}{\partial \zeta} \right\},$$

which may be rearranged as

$$-\left\{ \frac{h_2}{h_3} \frac{\partial(v/h_2)}{\partial \zeta} + \frac{h_3}{h_2} \frac{\partial(w/h_3)}{\partial \eta} \right\} + \frac{2}{h_2} \frac{\partial w}{\partial \eta} - \frac{2v}{h_2 h_3} \frac{\partial h_2}{\partial \zeta}. \tag{5.14.1}$$

The quantity within braces is equal to twice one of the non-diagonal elements of the rate-of-strain tensor (appendix 2), which is zero right at the free surface from the need for the corresponding tangential component of the stress tensor to vanish there. The normal velocity component w is necessarily continuous across the thin layer at the free surface, and so too is $\partial w/\partial \eta$; and no jump in the tangential component of velocity v is required so that the last term in (5.14.1) may also be taken as continuous. Thus the jump in ω_ξ across the thin boundary layer which forms at the free surface is simply the value of the quantity within braces in (5.14.1) at the boundary of the region of irrotational flow:

$$\Delta \omega_\xi = -\left\{ \frac{h_2}{h_3} \frac{\partial \left(\frac{1}{h_2^2} \frac{\partial \phi}{\partial \eta} \right)}{\partial \zeta} + \frac{h_3}{h_2} \frac{\partial \left(\frac{1}{h_3^2} \frac{\partial \phi}{\partial \zeta} \right)}{\partial \eta} \right\}_{\text{boun.}}, \tag{5.14.2}$$

where ϕ is the velocity potential for the irrotational flow. There is a corresponding expression for the jump in ω_η, while that in ω_ζ is zero.

In cases in which the free surface is stationary, or can be made so by suitable choice of the translational and rotational speeds of the co-ordinate system (rates of strain being independent of such movements of the co-ordinate system), we have $\partial \phi/\partial \zeta = 0$ at all points on the boundary, and (5.14.2) reduces to

$$\Delta \omega_\xi = \left(\frac{2}{h_2^2 h_3} \frac{\partial h_2}{\partial \zeta} \frac{\partial \phi}{\partial \eta} \right)_{\text{boun.}} = \left(\frac{2\kappa_\eta}{h_2} \frac{\partial \phi}{\partial \eta} \right)_{\text{boun.}}, \tag{5.14.3}$$

where κ_η is the curvature of the intersection of the free surface with the plane normal to the ξ-co-ordinate line. We see incidentally that there is no jump in vorticity at a plane free surface. This is because, with rectilinear co-ordinates, $\partial v/\partial \zeta = \partial w/\partial \eta$ in the irrotational region and, if $w = 0$ at all points of the free surface, $\partial v/\partial \zeta = 0$ there also, showing that the tangential stress in the irrotational flow vanishes at the free surface. Thus, in this case of a stationary plane free surface, such as the surface of a tank or a pond in which the water motions are too gentle to deform the surface, irrotational motion in the water is capable of satisfying *all* the boundary conditions at the free surface, no vorticity is generated at the free surface and no boundary layer forms there. At a stationary curved free surface, we may choose one of the co-ordinate lines to be parallel locally to $\nabla\phi$, whence the jump in vorticity is seen to be a vector lying in the tangential plane of the free surface and orthogonal to the local streamline, with magnitude

$$\Delta\omega = (2\kappa q)_{\text{boun.}}, \qquad (5.14.4)$$

where κ is the curvature of the intersection of the free surface with the plane normal to it and parallel to $\nabla\phi$, and $q = |\nabla\phi|$.

Thus the boundary layer that forms at a free surface is a layer in which vorticity is diffused by viscosity and convected by the flow (as well as being changed by rotation or extension in three-dimensional fields), the vorticity at the free surface being always greater than that just outside the boundary layer by the amount represented by (5.14.2), or by (5.14.4) where it is applicable (or by an obvious modification of (5.14.2) if the motion outside the boundary layer is not irrotational). The jump in velocity across the boundary layer is evidently of order $\delta\Delta\omega$, where δ is the boundary layer thickness, and thus varies as $R^{-\frac{1}{2}}$ in the many cases in which diffusion of vorticity works in such a fashion as to make δ vary in this way.

The smallness of this variation of velocity across the boundary layer has three noteworthy implications. (*a*) The equation of motion in the boundary layer may be linearized in the departure of the velocity from the value just outside the boundary layer. For instance, for a two-dimensional boundary layer and with the notation of §5.7, the equation of mass conservation (5.7.2) gives

$$v \approx -y\frac{\partial U}{\partial x} \qquad (5.14.5)$$

and (5.7.1) and (5.7.8) then yield, with consistent approximation,

$$\frac{\partial u'}{\partial t}+u'\frac{\partial U}{\partial x}+U\frac{\partial u'}{\partial x}-y\frac{\partial U}{\partial x}\frac{\partial u'}{\partial y} = \nu\frac{\partial^2 u'}{\partial y^2}, \qquad (5.14.6)$$

where $u' = u - U$. If the flow is steady, and U is known as a function of x, it may be possible to solve this linear equation for u' by standard methods. (*b*) The tendency for back-flow to develop in the boundary layer when the external stream is decelerating is very much weaker at a free surface than at a rigid wall, and separation is unlikely to occur unless there is large curvature of the boundary at some point. (c) Since velocity gradients are not of larger order of magnitude within the boundary layer than outside it, the rate of dissipation of energy per unit volume is of the same order throughout the fluid. Thus the total rate of dissipation is dominated by the contribution from the more extensive region of irrotational flow—by contrast with the case of flow with a boundary layer at a rigid wall, for which the contribution to total dissipation from the region of irrotational flow is small.

Two applications of these results concerning flow at large Reynolds number in the presence of a free surface are as follows.

The drag on a spherical gas bubble rising steadily through liquid

The flow due to a gas bubble rising freely through liquid, with the diameter so small that viscous forces are dominant, was investigated in § 4.9. The Reynolds number of the flow increases rapidly with increase of bubble size, and it is useful to consider the flow at such large Reynolds numbers (based on the rate of rise and diameter of the bubble) that boundary-layer ideas are applicable. We shall assume that the bubble is nevertheless so small that it remains approximately spherical under the action of surface tension. Variations of pressure in the liquid over the bubble boundary due to the motion tend to distort the bubble, but observation shows that the distortion is small for bubbles of radii up to about 0·05 cm (or volume up to 6×10^{-4} c.c.) in pure water; for bubbles near the upper end of this range the Reynolds number is certainly much larger than unity. We shall also assume, on the basis of the above discussion, that the boundary layer does not separate from the bubble surface. Qualitative observation of the flow near rising bubbles in the relevant size range does in fact suggest that back-flow does not occur, at any rate if the liquid is pure (Hartunian and Sears 1957). (It is known that bubbles of volume larger than about 5 c.c. in water take a shape like a slice off a sphere and that the boundary layer separates at the sharp rim of the spherical cap (see §6.11); such bubbles certainly lie outside the scope of the present discussion.) Finally, we suppose that the internal motion of the gas has no effect on the liquid motion.

Under the assumed conditions, the vorticity is confined to a thin boundary layer at the bubble surface and a narrow axisymmetric wake, and the irrotational flow outside this region is approximately the same as if the liquid were wholly inviscid. Thus for a spherical bubble of radius a moving with speed U through fluid at rest at infinity the flow outside the boundary layer and wake is given approximately (see (2.9.26)) by the velocity potential

$$\phi = -\tfrac{1}{2}Ua^3 \frac{\cos\theta}{r^2}, \tag{5.14.7}$$

where r, θ are spherical polar co-ordinates with origin at the instantaneous position of the centre of the sphere.

Now for the purpose of estimating the drag D on a bubble in steady motion it is not necessary to analyse the flow in the boundary layer, because the rate at which buoyancy forces on the bubble do work, viz. UD, must be equal to the total rate of dissipation in the liquid, and this, as we have seen, can be determined approximately from the irrotational flow alone. The general expression for the rate of dissipation in incompressible fluid is given in (4.1.5), and in the present circumstances of irrotational flow the rate of dissipation in a volume of fluid V is

$$2\mu \int \frac{\partial^2 \phi}{\partial x_i \partial x_j} \frac{\partial^2 \phi}{\partial x_i \partial x_j} \, dV, \quad = \mu \int \frac{\partial^2 q^2}{\partial x_i \partial x_i} \, dV$$

$$= \mu \int \mathbf{n} . \nabla q^2 \, dA, \tag{5.14.8}$$

where $q^2 = (\partial \phi / \partial x_i)^2$ and the latter integral is taken over the surface A bounding the volume V (with $\mathbf{n}$ directed out of V). Hence the drag on the gas bubble is given by

$$UD = -\mu \int_0^\pi \left(\frac{\partial q^2}{\partial r} \right)_{r=a} 2\pi a^2 \sin \theta \, d\theta,$$

and evaluation of this integral for the motion specified by (5.14.7) gives

$$D = 12\pi \mu a U. \tag{5.14.9}$$

The corresponding drag coefficient is

$$C_D = \frac{D}{\pi a^2 \frac{1}{2} \rho U^2} = \frac{48}{R}, \tag{5.14.10}$$

where $R = 2aU\rho/\mu$. The drag coefficient of a rigid body from which the boundary layer does not separate is proportional to $R^{-\frac{1}{2}}$ (§5.11), and that for a 'body' with a free surface is of smaller order because the free surface does not retard the fluid in the boundary layer as effectively as a solid surface.

It is possible to obtain an improved estimate of the drag on the bubble by calculating also the dissipation of energy in the boundary layer at the bubble surface and in the wake, the new result being (Moore 1963)

$$C_D = \frac{48}{R} \left(1 - \frac{2 \cdot 2}{R^{\frac{1}{2}}} \right). \tag{5.14.11}$$

The terminal velocity V of a bubble moving under the action of gravity alone may now be determined by equating the drag to the buoyancy force on a bubble of volume $\frac{4}{3}\pi a^3$. With the first approximation to the drag, given by (5.14.9), we have

$$V = \frac{1}{9} \frac{ga^2}{\nu}. \tag{5.14.12}$$

Observations have been made of the rates of rise of bubbles of different sizes through various liquids free from impurities, and the inferred values of the drag coefficient for two liquids are shown in figure 5.14.1 as a function of the Reynolds number (in which the length a is taken as the radius of a sphere of the same volume as the bubble). The agreement with (5.14.10) is fair, and with (5.4.11) is quite good, for Reynolds numbers above about 20 and less than a critical value at which the drag coefficient begins to rise rapidly. This critical value of the Reynolds number varies from one liquid to another, and appears to mark the development of a non-spherical shape of the bubble. The variation of pressure in the water at the surface of a bubble rising with steady speed U is approximately ρU^2, and the bubble may be

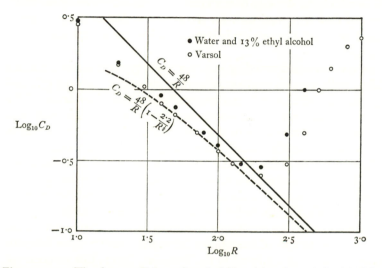

Figure 5.14.1. The drag coefficient of gas bubbles rising through liquids. The points for the two particular liquids are taken from experimental curves given by Haberman and Morton (1953).

expected to remain spherical under the action of surface tension γ only when $\rho U^2 \ll \gamma/a$, that is, assuming U to be the terminal speed given by (5.14.12), only when

$$\frac{\rho g^2 a^5}{\nu^2 \gamma} \ll 81.$$

For pure water this restriction on the radius of the bubble is

$$a \ll (6 \cdot 1 \times 10^{-7})^{\frac{1}{5}} \, \text{cm}, \quad = 0 \cdot 06 \, \text{cm},$$

which is in fact close to the value of the radius of air bubbles in pure water at which non-sphericity is first observed.

When pressure variations first become comparable with the surface tension stress, the bubble is an oblate ellipsoid, being flattened at the fore-and-aft points by the stagnation pressure there. It is possible to calculate the total dissipation in the irrotational flow due to a moving oblate ellipsoid, and to obtain a new estimate of the terminal velocity, but the result is probably of limited value since, when the departure from a spherical shape is appreciable, the bubble motion is unstable and the bubble zig-zags from side to side or rises in a spiral path.

The attenuation of gravity waves due to the boundary layer at a free surface

Gravity waves at the free surface of a body of liquid is a topic outside the scope of this book, but a few remarks about the features associated with the resulting boundary layers are appropriate here. We shall consider only the simple case in which the velocity of material elements oscillates sinusoidally (apart from any slow attenuation) with frequency n, due to the passage of the wave, and in which the motion is irrotational to the extent allowed by viscosity.

If a surface wave of either standing or progressive type occurs in a channel with rigid side-walls, or with depth smaller than about one wave-length, and if the frequency is sufficiently large, an oscillatory boundary layer of the kind considered in the previous section is established at each rigid wall; and estimates of both the attenuation of the wave due to the dissipation within the boundary layer and the steady streaming velocity just outside the boundary layer can be made by the methods described (provided the frequency and amplitude of the wave motion satisfy the condition (5.13.2)). If on the other hand the side-walls are effectively absent and the depth is larger than a wave-length, the rather weak effects of the boundary layer at the free surface act alone. An oscillatory boundary layer at a free surface can be shown to give rise to a steady second-order motion, and it again happens that the steady flow is not confined to the boundary layer; but whereas for a rigid surface the steady streaming velocity at the outer edge of the boundary layer tends to a definite value, dependent only on local conditions, the corresponding result for a free surface is that the normal gradient of the steady velocity tends to a definite value (Longuet-Higgins 1953, 1960). The existence of a boundary layer at a free surface also causes a small change in the phase of the normal stress acting at the boundary of the region of irrotational flow, so that the work done by this normal stress over one cycle of rise and fall of the free surface is non-zero, and negative. Thus the amplitude of the wave motion slowly decreases, at a rate which can be determined in the following simple way for a progressive wave.

The loss of total energy (kinetic plus potential) of the liquid over one cycle is necessarily equal to the rate of viscous dissipation of energy per cycle, provided the net flux of energy into the volume of liquid concerned is zero. This dissipation occurs mainly in the region of irrotational flow and can thus be obtained from a knowledge of the velocity potential there; in this way the attenuation of the wave can be calculated without the need for explicit consideration of the flow in the boundary layer at the free surface. Suppose that the velocity in the liquid varies sinusoidally with respect to one horizontal position co-ordinate x (with wave-number k) and lies in the

(z, x)-plane, where z is vertical depth below the mean position of the free surface. Then, in the region of irrotational flow we have

$$\phi = A(t) \sin(kx - nt) e^{-kz}, \qquad (5.14.13)$$

the dependence on z being a simple consequence of the need for ϕ to satisfy Laplace's equation; the amplitude A varies slowly as the wave loses energy. We may again use the expression (5.14.8) for the rate of dissipation, and find that the rate of dissipation of the whole motion, per unit area of a horizontal plane, is $2\mu k^3 A^2(t)$, provided the wave amplitude is small compared with the wave-length so that the integral (5.14.8) can be evaluated at the plane of the undisturbed surface ($z = 0$).

Now the total kinetic energy of the liquid, per unit area of a horizontal plane, and averaged over one cycle, is $\tfrac{1}{4}\rho k A^2(t)$. The liquid also has potential energy, and since all material elements are executing small oscillations under the action of gravity and mutual reactions, the average total potential energy is equal to the average total kinetic energy. Hence the equation for the gradual loss of energy is

$$\frac{d(\tfrac{1}{2}\rho k A^2)}{dt} = -2\mu k^3 A^2,$$

showing that A decreases as $e^{-\beta nt}$, where

$$\beta = 2\nu k^2/n. \qquad (5.14.14)$$

The frequency of a deep-water wave of wave-length $2\pi/k$ is known from the theory of surface waves to be given by

$$n = (gk)^{\frac{1}{2}},$$

so that the damping of the waves reduces their amplitude to a fraction $\exp\{-4\pi\nu(k^3/g)^{\frac{1}{2}}\}$ of the initial value after one period.

The length $(\nu/n)^{\frac{1}{2}}$ will be recognized, from the argument of the preceding section, as a measure of the thickness of the oscillatory boundary layer. The above analysis is valid only if this thickness is small compared with the length $2\pi/k$ representative of the flow field as a whole, that is, if

$$\left(\frac{\nu k^{\frac{3}{2}}}{4\pi^2 g^{\frac{1}{2}}}\right)^{\frac{1}{2}} \ll 1;$$

satisfaction of this condition thus ensures that the change of amplitude of the wave during one period is small.

As a numerical example, suppose that the wave-length is 10 cm; the period is then about 0·25 sec, the length $(\nu/n)^{\frac{1}{2}}$ representing the boundary-layer thickness is about 0·02 cm for water, and the amplitude is reduced by a fraction 0·0022 after one period. It will not happen often in practice that the boundary layer at the free surface is the primary cause of the attenuation of surface waves and that the above estimate of β is applicable, for in the case of waves produced in the laboratory the dissipation in the boundary layer at the rigid walls of the container will normally be dominant and in the case of waves on lakes or the sea casual disturbances due to wind will usually be more effective as a means of dissipating the energy of the wave motion.

5.15. Examples of use of the momentum theorem

As mentioned in § 3.2, there are circumstances in which it is possible to obtain useful and quite strong results about a steady flow system by application of the 'momentum theorem', that is, of the momentum equation in the integral form (3.2.4). Success depends on being able to choose the control surface in such a way that the flux of momentum and the stress can be evaluated at all points of the control surface directly from the available information. The two examples of the use of the momentum theorem given in this section are concerned with a transition from one uniform stream to another, this being a type of flow for use on which the theorem is obviously well suited.

The force on a regular array of bodies in a stream

In the first example, a very simple one, a stream of fluid, of uniform velocity U far upstream, impinges on an array of similar rigid bodies distributed regularly over a plane normal to the stream (figure 5.15.1); the rigid bodies have small linear dimensions, and are close together, as in the case of wire gauze for example, and the stream is bounded laterally by walls parallel to the stream. It is found that in circumstances like these the fluid velocity again becomes uniform at some distance downstream from the array of rigid bodies, and with the same speed U as required for conservation of mass (in the absence of any appreciable variation of density of the fluid). The fluid exerts a force on the array which is directed downstream when the rigid bodies are individually symmetrical about the normal to the array, and we enquire whether the average force per unit area of the plane of the array can be determined from observation of the conditions far upstream and far downstream.

We choose the control surface shown as a broken line in figure 5.15.1, consisting of an inner boundary A_1, and an outer boundary A_2 in the form of a cylinder, the cross-section of the cylinder being sufficiently large to include a large number of elements of the array, and apply (3.2.4), ignoring for the present purposes the term arising from the gravitational body force since it can be absorbed in the modified pressure. The net flux of momentum out of the region between A_1 and A_2 is zero, and so we have the result

i-component of total force on portion of array enclosed by A_1

$$= -\int \sigma_{ij} n_j \, dA_1 = \int \sigma_{ij} n_j \, dA_2,$$

the normal **n** being directed away from the region at all parts of the boundary. The stress σ_{ij} is a purely normal stress at the end faces of the cylinder A_2, where the velocity is uniform and the pressure is p_1 far upstream and p_2 far downstream, and it is also purely normal over the side of the cylinder except in the neighbourhood of the array. Hence, if the contribution to the integral from tangential stress at the side of the cylinder is made small (relatively) by choosing the cylinder cross-section to be large, we have

average normal force exerted on unit area of array $= p_1 - p_2$.　　(5.15.1)

Thus the normal force exerted by the fluid on the array can be found without either theoretical or experimental examination of the flow near the bodies making up the array.

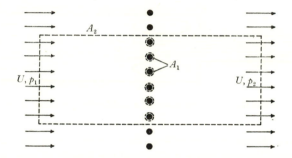

Figure 5.15.1. Use of the momentum theorem to determine the force on a regular array of rigid bodies.

The result (5.15.1) clearly remains valid when the incident stream is not normal to the plane of the array or when the individual bodies are not symmetrical, since the incident and emergent streams are again uniform far from the array and the above argument is still applicable in terms of the normal component of the stream velocity on the two sides of the array and the normal component of the force on the bodies. In each of these new cases the fluid exerts on the array of bodies a force which has a non-zero component in the plane of the array and there is a corresponding deflection of the stream at the array.

The effect of sudden enlargement of a pipe

The second example concerns the effect of a sudden increase in the cross-sectional area of a short length of pipe through which fluid is flowing with approximately uniform velocity U_1 (figure 5.15.2). In normal circumstances, when $U_1 d/\nu \gg 1$, where d is a measure of the linear dimensions of the pipe cross-section, the stream discharges into the wide section in the form of a straight jet (as in the flow shown in figure 5.10.2, plate 6), irregular eddying motion develops at the side of the jet, the surrounding fluid is gradually entrained and mixed with the jet, and finally the stream again has an approximately uniform speed, U_2 say. The details of this unsteady mixing process are complicated and beyond the scope of calculation; can anything be said about conditions far downstream without knowledge of the flow near the sudden expansion? The speed U_2 follows from conservation of mass (again on the assumption that the density is uniform), and we should like to be able to determine also the pressure p_2 far downstream, given that the pressure is p_1 upstream of the sudden enlargement.

The momentum integral theorem can be employed to this end, with the help of the inference, which is readily confirmed by direct observation, that the modified pressure is approximately uniform across the pipe at the sudden enlargement in view of the absence of appreciable lateral velocity of the

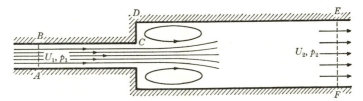

Figure 5.15.2. Uniform flow in a pipe towards a sudden enlargement of the cross-section.

fluid there. We choose the control surface consisting of the cross-sections shown as broken lines in figure 5.15.2 and the length of pipe wall between them, and denote the up- and downstream cross-sections of the pipe by S_1 and S_2 respectively. Just downstream of the pipe enlargement, the fluid velocity and pressure fluctuate, as a result of the unsteadiness of the mixing process there, but the fluctuations are about steady average values and we could suppose the relation (3.2.3) to be averaged over a long time, yielding a relation like (3.2.4) between averaged quantities. However, it is more convenient to choose the cross-sections AB and EF to be outside the region of fluctuation, and the existence of fluctuations will have no consequences in the present example.

The flux of momentum in the downstream direction outwards across the control surface is

$$\rho U_2^2 S_2 - \rho U_1^2 S_1, \quad = \rho U_2 S_2 (U_2 - U_1).$$

The stresses at the portions of the control surface represented by AB and CD are purely normal with pressure p_1, and that at EF is purely normal with pressure p_2. Boundary layers are formed at the sections BC and DE of the pipe wall, and it will be assumed that the Reynolds number of the flow is so large that the corresponding tangential components of stress are negligibly small when made non-dimensional with the quantities ρ, d and U_1 (which is equivalent to the assumption already made that the velocity distribution across the pipe remains approximately uniform over the section BC). The momentum balance is then represented by

$$\rho U_2 S_2 (U_2 - U_1) = p_1 S_1 + p_1 (S_2 - S_1) - p_2 S_2,$$

giving the pressure in the downstream region of uniform flow as

$$p_2 = p_1 + \rho U_2(U_1 - U_2). \tag{5.15.2}$$

This pressure rise due to a sudden enlargement may be compared with that associated with a gradual increase of the cross-sectional area from S_1 to S_2. In the latter circumstances, the flow is everywhere steady and viscosity has a negligible effect except near the pipe wall, so that we can use Bernoulli's theorem for an effectively incompressible fluid (see (3.5.16)) to find

$$p_2 = p_1 + \tfrac{1}{2}\rho(U_1^2 - U_2^2), \tag{5.15.3}$$

p_1 and p_2 being modified pressures. The final pressure is thus less for a sudden enlargement than for a gradual one, by an amount

$$\tfrac{1}{2}\rho(U_1 - U_2)^2, \quad = \tfrac{1}{2}\rho U_1^2\left(1 - \frac{S_1}{S_2}\right)^2. \tag{5.15.4}$$

Alternatively, we may say that whereas there is no change in the Bernoulli constant due to a gradual enlargement, a sudden enlargement causes an eddying mixing flow with an accompanying fall in the Bernoulli constant of the amount (5.15.4). We may infer, from the fact that the Bernoulli constant measures the total mechanical energy per unit mass of fluid, that the eddying entrainment by the jet emerging at the sudden enlargement is associated with a dissipation of mechanical energy (by internal friction). How the dissipation occurs is not evident in detail, but the momentum theorem shows that the total loss of energy by dissipation for each unit mass of fluid is determined by the overall conditions.

The result (5.15.4) allows a useful application to flow problems of the kind considered in the first example, inasmuch as certain kinds of array of rigid bodies can be regarded as a regular arrangement of narrow passages through which the fluid must pass before meeting a sudden enlargement on the downstream face. Consider, for example, the case of a flat rigid plate through which holes of diameter comparable with the thickness have been drilled at regularly distributed positions (figure 5.15.3). When such a plate is placed at right angles to a steady stream of speed U, the fluid discharges from the rear of the plate as a number of jets which ultimately mix with surrounding fluid, with an irregular eddying motion, and form a uniform stream of speed U again. The analysis for the sudden enlargement in a pipe thus applies here, and shows that the process of emergence of fluid from the holes in the plate is accompanied by a total fall in the Bernoulli constant of amount

$$\tfrac{1}{2}\rho U^2\left(\frac{1-\alpha}{\alpha}\right)^2, \tag{5.15.5}$$

where α is the fraction of the plate area coinciding with holes. On the other hand, the flow on the upstream side of the plate and into the holes is one in which, as we have seen in this chapter, viscosity does not play an important part (provided the hole diameter is not too small) and to most of which

Bernoulli's theorem applies. Thus (5.15.5) is the difference between the values of the Bernoulli constant at positions far upstream and far down-stream from the plate, and, since the stream speeds at these positions are both equal to U, (5.15.5) is also the difference between the pressures at these positions. Hence, in view of (5.15.1), the average drag force on unit area of the plate is also given by (5.15.5). This theoretical result is found to agree with observations of the overall pressure drop across such a plate.

Similar simple theories may be devised to give the force exerted on other kinds of perforated plates, and on arrays of cylinders or other bodies. Such theories are more accurate when the cross-sectional area of each jet emerging from the plate or array is well defined by the geometry of the boundary, as in the above case of a plate with drilled holes.

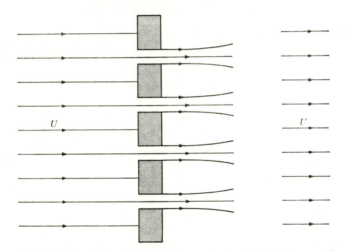

Figure 5.15.3. Calculation of the drag exerted on a perforated plate in a uniform stream.

Further applications of the momentum equation in integral form will be described in §§ 6.3 and 6.8 in the discussion of flow fields in which effects of viscosity can be neglected. In these latter cases, the momentum theorem may provide a short cut, or a neat way of obtaining useful results, whereas in cases like the two examples given above effects of viscosity play an essential part (although they may not appear explicitly in analysis using the momentum theorem—this is its great merit), direct calculation of the flow field is usually impossible, and use of the momentum theorem is essential if results are to be obtained.

It will be evident that integral relations representing the overall balance of quantities other than linear momentum will occasionally be useful; energy and angular momentum are obvious candidates, the latter being especially relevant in considerations of the action of pumps, turbines and other machines with rotating components.

Further reading relevant to chapter 5

Applied Hydro- and Aeromechanics, by L. Prandtl and O. G. Tietjens (McGraw-Hill, 1934; also Dover Publications, 1957).
Modern Developments in Fluid Dynamics, edited by S. Goldstein (Oxford University Press, 1938).
The Essentials of Fluid Dynamics, by L. Prandtl (Blackie, 1952).
Laminar Boundary Layers, edited by L. Rosenhead (Oxford University Press, 1963).

Exercises for chapter 5

1. A long rigid cylinder is placed in a steady uniform stream of fluid with its generators inclined at angle α to the stream (a so-called 'yawed' cylinder). The cross-section is streamlined and no separation of the boundary layer occurs. Show that the velocity component $\mathbf{u}$ in the plane normal to the generators has a distribution of the same form for all values of α (except those near o or π), both inside and outside the boundary layer, and that the equation for the component w parallel to the generators in the boundary layer is

$$\mathbf{u} . \nabla w = \nu \, \partial^2 w / \partial y^2,$$

where y is the normal distance from the boundary.

2. A steady narrow two-dimensional jet of fluid adjoins a plane rigid wall and the fluid is at rest far from the wall. Use the boundary-layer equations to show that the quantity

$$\int_0^\infty u \left\{ \int_y^\infty u^2 \, dy \right\} dy, \quad = P \text{ say,}$$

is independent of distance x along the wall. Show that the asymptotic form of the velocity distribution, provided it depends only on P and ν, is given by

$$\psi = (P\nu x)^{\frac{1}{4}} f(\eta), \quad \eta = (P/\nu^3 x^3)^{\frac{1}{4}} y, \quad \text{where} \quad f''' + ff'' + 4f'^2 = \text{o}.$$

Show also that in the analogous axisymmetric wall jet spreading out radially the velocity and thickness of the jet vary as $x^{-\frac{3}{2}}$ and $x^{\frac{5}{4}}$ and the velocity distribution across the jet is of the same form as in the plane case. (Glauert 1956.)

3. A three-dimensional body is held in a stream of uniform speed U in the x-direction, and a drag force D and a lift force L in the y-direction act on the body. Show that far downstream the approximate expressions for the velocity components are

$$u = U - \frac{D}{4\pi\mu x} e^{-\eta^2}, \quad v = \frac{\partial\phi}{\partial y} - \frac{L}{4\pi\mu x} e^{-\eta^2}, \quad w = \frac{\partial\phi}{\partial z},$$

where

$$\phi = \frac{L}{8\pi\mu x} \frac{y}{\eta^2} (\mathrm{1} - e^{-\eta^2}), \quad \eta^2 = \frac{(y^2 + z^2) \, U}{4\nu x}.$$

Note that there is now stream-wise vorticity in the wake, and that the irrotational motion in the (y, z)-plane well outside the wake is the same as that associated with a point vortex doublet of strength $L/\rho U$ parallel to the z-axis and located at the origin of that plane.

6

IRROTATIONAL FLOW THEORY
AND ITS APPLICATIONS

6.1. The role of the theory of flow of an inviscid fluid

We have completed a study of the general effects of the viscosity of the fluid, and are now in a position to take advantage of the fact that the viscosities of the common fluids air and water are quite small. The Reynolds number $\rho L U/\mu$ (in the notation of §4.7) is usually a measure of the ratio of the representative magnitude of inertia forces to that of viscous forces; and, when this Reynolds number is large compared with unity, viscous forces frequently play a negligible part in the equation of motion over *nearly* all the flow field. In many cases in which separation of the boundary layer from a rigid boundary does not occur, the flow field tends to the form appropriate to an inviscid fluid, as $\rho L U/\mu \to \infty$, over the whole of the region occupied by the fluid, and the fact that viscous forces remain significant in certain thin layers in the fluid, however large the Reynolds number may be, is of little consequence for many purposes. However, in cases in which the boundary layer does separate from a rigid boundary, the limit is a singular one and, although the region of fluid in which viscous forces are significant may decrease in size to zero as $\rho L U/\mu \to \infty$, the limiting form of the flow field is not the same as that appropriate to a completely inviscid fluid. This singular limiting behaviour is made possible, mathematically speaking, by the facts that the viscosity occurs in the equation of motion as the coefficient of the highest-order derivative and that a viscous fluid must satisfy the additional condition of no slip at a rigid boundary however small the viscosity may be.

Theoretical results concerning the flow of an inviscid fluid may thus be applied directly to the former class of cases in which boundary-layer separation does not occur;† here inviscid-fluid theory provides us with a good approximation to the flow of a real fluid at large Reynolds number (with the same initial and boundary conditions) over the whole region of fluid, except in thin layers whose thickness approaches zero as $\rho L U/\mu \to \infty$ and whose position is known from the inviscid-fluid solution. Analysis of the flow of an inviscid fluid is much simpler than that of a viscous fluid, and it is therefore of the greatest importance that we should be able to predict— by inspection and the use of general rules, not by detailed calculation—

---

† Absence of boundary-layer separation is a necessary but possibly not a completely sufficient condition for the limiting form of the flow field to be the same as for an inviscid fluid. It is difficult to make general statements which are both useful and reliable.

whether inviscid flow theory is applicable. More specifically, it is necessary that we should be able to tell by inspection if the flow of a real fluid in the circumstances of a given set of boundary and initial conditions would be accompanied by boundary-layer separation. If separation would not occur, we can apply the extensive results of inviscid-fluid theory. This need to make decisions about how a fluid with non-zero viscosity would behave in given circumstances demands an understanding of the flow of a viscous fluid before the relevance and usefulness of inviscid-fluid theory can be appreciated; for this reason, inviscid-fluid theory takes its place in this book after a study of viscous flow, notwithstanding its greater simplicity.

Even in cases of flow of a real fluid in which separation of the boundary layer does occur, there are large parts of the flow which locally are not affected significantly by the viscosity of the fluid and to which inviscid-fluid theory may be applied. However there is here the difficulty that the position and shape of the separated boundary layer in the real fluid flow are usually unknown, so that although the effect of viscosity is negligible locally everywhere except near certain singular surfaces (some of which lie in the interior of the fluid) the shape of the boundary of the domain of effectively inviscid flow is unknown and in general cannot be found from considerations of a wholly inviscid fluid. The scope for application of inviscid-fluid theory is therefore restricted in these cases. Moreover, as already mentioned on several occasions, separated boundary layers are invariably unstable to small disturbances and a fluctuating turbulent flow is set up. The influence of the turbulent fluctuations of velocity may not extend directly over the whole flow field (for instance, the flow near the forward stagnation point on a bluff body moving steadily through fluid and generating a turbulent wake behind it may be fairly steady), but the non-zero average flux of momentum due to the turbulence does have an effect on the form of the flow as a whole; it is still true that certain regions of the flow are steady and locally unaffected by the fluid viscosity, but determination of the shape of the boundary of one of these regions and of the conditions at that boundary is certainly beyond the reach of rigorous calculation, and we must often fall back on plausible speculation or empiricism.

In this and the following chapter, various aspects of the flow of a fluid regarded as entirely inviscid† (and incompressible) will be considered. The results presented are significant only inasmuch as they represent an approximation to the flow of a real fluid at large values of the Reynolds number, and the limitations of each result must be regarded as information as important as the result itself.

This chapter is devoted to the particular case of irrotational flow. Highly

† In a number of cases the existence of certain features of the flow having their origin in the effect of viscosity will nevertheless be assumed as a part of the formulation of the problem. For instance, it would be appropriate to assume that 'inviscid' fluid passing through a hole in a thin plate emerged as a concentrated jet with a sheet vortex at its boundary.

special though the property of irrotationality may seem to be, it is given great practical importance by the consequence of Kelvin's circulation theorem (§ 5.3) that material elements of a uniform fluid set into motion from rest remain without rotation unless they move into a region where viscous forces are significant. A proper understanding of irrotational flow theory and an appreciation of its numerous applications is essential in all branches of fluid dynamics. Chapter 7 takes up the more general situation in which either localized or distributed vorticity plays an essential part in the flow. Examples of flow fields without rigid boundaries are prominent in the two chapters, since they provide wide scope for the application of inviscid-fluid theory. Cases of irrotational flow of liquid with a free surface are considered, although the subject of gravity waves requires separate treatment and is omitted from this volume. The theory of lift generated by slender bodies moving through fluid—one of the scientific foundations of aeronautics—is developed in §§ 6.7 and 7.8; here the application of inviscid-fluid theory is made possibly by the use of simple rules, based on the considerations of chapter 5, about the occurrence and consequence of separation of the stream-lines at the surface of the body. The peculiar dynamical properties of fluid with a general rotation are described in chapter 7, together with some of their geophysical manifestations.

Finally, we recall the relations which govern the motion of an inviscid fluid and on which the work of chapters 6 and 7 will be based.

We shall continue to regard the fluid as incompressible (the conditions for validity of this assumption being as described in § 3.6), so that the equation of mass conservation is

$$\nabla . \mathbf{u} = 0. \tag{6.1.1}$$

We shall also continue to assume that the density ρ is uniform throughout the fluid.

The body force acting on the fluid will be assumed to be due to gravity, so that $\mathbf{F} = \mathbf{g}$. In some of the flow fields to be considered, the fluid has a free surface, and gravity here affects the distribution of velocity in the fluid.

In the absence of viscosity, tangential stresses in the fluid are zero every-where, the stress tensor reduces to $-p\,\delta_{ij}$, and the equation of motion becomes

$$\frac{D\mathbf{u}}{Dt} = \mathbf{g} - \frac{1}{\rho}\nabla p. \tag{6.1.2}$$

When the uniform value of ρ is given, the two variables $\mathbf{u}$ and p are to be found as functions of $\mathbf{x}$ and t from the equations (6.1.1) and (6.1.2).

6.2. General properties of irrotational flow

Many of the general kinematical properties of irrotational flow of an incompressible fluid have already been given in §§ 2.7 to 2.10 in the course of the discussion of that part of an arbitrary velocity distribution that is solenoidal and irrotational. The results described below supplement that discussion.

When the velocity $\mathbf{u}$ is irrotational, we may introduce a velocity potential ϕ given by

$$\mathbf{u} = \nabla\phi, \tag{6.2.1}$$

as explained in § 2.7, in which case the mass-conservation equation for an incompressible fluid becomes

$$\nabla^2\phi = 0. \tag{6.2.2}$$

Although the equation of motion (6.1.2) is non-linear in $\mathbf{u}$, the velocity distribution is here determined completely by a linear equation derived from the restrictive condition of irrotationality and the mass-conservation equation. This linearity is the distinctive property of irrotational flow which allows the employment of many powerful mathematical techniques. The non-linear equation of motion is needed here only for the calculation of the pressure after the velocity distribution has been determined; and we shall see that the equation of motion can be integrated to give an explicit expression for the pressure.

Since equation (6.2.2) is linear, different solutions for the velocity potential ϕ can be superposed to form a new solution. The corresponding velocity distributions can likewise be superposed, although not the pressure distributions in view of the non-linear dependence of p on $\mathbf{u}$. In particular, new irrotational flow fields may be constructed by superposing the velocity potentials that were shown in §§ 2.5 and 2.6 to be associated with certain singular distributions of the expansion Δ and vorticity $\boldsymbol{\omega}$ (Δ or $\boldsymbol{\omega}$ being zero everywhere except at a point or on a line or surface, where it has infinite magnitude). In the case of a point or line singularity in the distribution of Δ or $\boldsymbol{\omega}$, the 'induced' velocity at position $\mathbf{x}$ was found to increase indefinitely as $\mathbf{x}$ approaches the point or line concerned; clearly this is also true of the total induced velocity associated with several superposed singular distributions of Δ and $\boldsymbol{\omega}$. For instance, the irrotational velocity distribution associated with a point source of strength m' at point $\mathbf{x}'$ and one of strength m'' at point $\mathbf{x}''$ is (see (2.5.2))

$$\mathbf{u}(\mathbf{x}) = \frac{m'}{4\pi}\frac{\mathbf{x}-\mathbf{x}'}{s'^3} + \frac{m''}{4\pi}\frac{\mathbf{x}-\mathbf{x}''}{s''^3},$$

where $s'^2 = (\mathbf{x}-\mathbf{x}')^2$ and $s''^2 = (\mathbf{x}-\mathbf{x}'')^2$; this velocity field is dominated by the contribution from the source m' when $\mathbf{x}$ is near $\mathbf{x}'$ and by that from m'' when $\mathbf{x}$ is near $\mathbf{x}''$.

The point or line singularities in the distributions of Δ or $\boldsymbol{\omega}$ considered in §§ 2.5 and 2.6 impose a singularity in ϕ at the boundary of the region of solenoidal irrotational flow. For example, when a point source of strength m is located at the point $\mathbf{x}'$, the region of solenoidal irrotational flow does not include the point $\mathbf{x}'$, which must be regarded as surrounded by a closed surface across which the flux of volume is prescribed as having the value m;

and since the velocity near $\mathbf{x}'$ is dominated by the induced velocity due to the point source at $\mathbf{x}'$, the precise boundary condition is that

$$\mathbf{u} \sim \frac{m'}{4\pi} \frac{\mathbf{s}}{s^3}, \quad \text{or} \quad \phi \sim -\frac{m'}{4\pi s},$$

at all points $\mathbf{x}$ on a closed surface of infinitesimal linear dimensions surrounding $\mathbf{x}'$, where $\mathbf{s} = \mathbf{x} - \mathbf{x}'$. Likewise, for a source doublet of strength $\boldsymbol{\mu}$ at $\mathbf{x}'$, the appropriate condition at the boundary of the region of solenoidal irrotational flow is (see (2.5.3))

$$\phi \sim -\frac{\boldsymbol{\mu} . \mathbf{s}}{4\pi s^3}$$

at all points $\mathbf{x}$ on a closed surface of infinitesimal linear dimensions surrounding $\mathbf{x}'$. However, it is a known general property of the differential equation (6.2.2) that ϕ and all its derivatives with respect to $\mathbf{x}$ are finite and continuous at all *interior* points of the region of solenoidal irrotational flow.

Conditions under which at most one solution for the velocity $\nabla\phi$ exists have been obtained in §§ 2.7 to 2.10. The most important of the results found there is that the solution for $\nabla\phi$ in a singly-connected region of fluid, which may extend to infinity in all directions provided the fluid is at rest there, is determined uniquely when the value of the normal component of $\mathbf{u}$ is prescribed at all points of the boundaries. Uniqueness is also ensured in these circumstances if the value of ϕ at all points of the boundaries is prescribed and, in cases in which the fluid extends to infinity and is at rest there, the net flux of volume across the inner boundary is prescribed. When the region of solenoidal irrotational flow is not singly-connected, specification of the cyclic constants of the flow must be added to the above conditions for uniqueness.

Integration of the equation of motion

The vector identity $\quad \frac{1}{2}\nabla(\mathbf{u}.\mathbf{u}) = \mathbf{u}.\nabla\mathbf{u} + \mathbf{u}\times\boldsymbol{\omega}$

enables (6.1.2) to be written in the alternative form

$$\frac{\partial\mathbf{u}}{\partial t} - \mathbf{u}\times\boldsymbol{\omega} = -\nabla\left(\tfrac{1}{2}q^2 + \frac{p}{\rho} - \mathbf{g}.\mathbf{x}\right), \tag{6.2.3}$$

where $q^2 = \mathbf{u}.\mathbf{u}$. When $\mathbf{u} = \nabla\phi$ and $\boldsymbol{\omega} = 0$, this becomes

$$\nabla\left(\frac{\partial\phi}{\partial t} + \tfrac{1}{2}q^2 + \frac{p}{\rho} - \mathbf{g}.\mathbf{x}\right) = 0, \tag{6.2.4}$$

showing that the quantity within brackets must be a function of t alone, $F(t)$ say. The form of this unknown function is without significance, because we could define a new velocity potential ϕ' such that

$$\phi' = \phi - \int F(t)\,dt, \quad \nabla\phi' = \nabla\phi,$$

and thereby remove the function of t without affecting the velocity distribution. It is customary to ignore the arbitrary function of t and to write the integral of (6.2.4) as

$$\frac{\partial \phi}{\partial t} + \tfrac{1}{2}q^2 + \frac{p}{\rho} - \mathbf{g} \cdot \mathbf{x} = \text{const.} \tag{6.2.5}$$

throughout the fluid.

It will be observed that when irrotational flow is also steady, the left-hand side of (6.2.5) reduces to the quantity previously denoted by H and is constant throughout the fluid. This is also the result that would have been expected from the proof of Bernoulli's theorem (see §5.1); in steady flow H is constant along any streamline and along any vortex-line, and when in addition $\boldsymbol{\omega} = 0$ everywhere H must be constant throughout the fluid.

The relation (6.2.5) provides an explicit expression for the pressure when the velocity distribution is known. It is particularly useful in this way, because ϕ satisfies Laplace's equation and is determined uniquely by certain types of boundary conditions on ϕ or $\nabla \phi$, and can therefore be determined without regard for the pressure.

Expressions for the kinetic energy in terms of surface integrals

Here again most of the relevant analysis has already been given in §§2.7 to 2.10. For flow in a singly-connected region bounded internally and externally, we see from (2.7.6) that the total kinetic energy of the fluid is

$$T = \tfrac{1}{2}\rho \int \phi \mathbf{u} \cdot \mathbf{n}_2 \, dA_2 - \tfrac{1}{2}\rho \int \phi \mathbf{u} \cdot \mathbf{n}_1 \, dA_1, \tag{6.2.6}$$

where the integrals are taken over the whole of the internal boundary A_1 and the external boundary A_2, and the unit normals $\mathbf{n}_1$ and $\mathbf{n}_2$ are both outward relative to the closed surfaces to which they refer. If the fluid is not bounded externally, but extends to infinity in all directions and is at rest there, we see from (2.9.17) that

$$T = \tfrac{1}{2}\rho \int (C - \phi) \mathbf{u} \cdot \mathbf{n} \, dA, \tag{6.2.7}$$

where A is the internal boundary and C is the constant value to which ϕ tends at infinity. If the flux of volume across the internal boundary is zero, (6.2.7) reduces to

$$T = -\tfrac{1}{2}\rho \int \phi \mathbf{u} \cdot \mathbf{n} \, dA. \tag{6.2.8}$$

If the region of flow is doubly-connected, and ϕ has the cyclic constant κ, the formula (6.2.6) for flow bounded internally and externally must be supplemented by a term

$$\tfrac{1}{2}\rho\kappa \int \mathbf{u} \cdot \mathbf{n} \, dS, \tag{6.2.9}$$

as shown by (2.8.8), the integral being taken over the whole of the (topological) barrier S. Alternatively, the formula (2.8.10) involving two contributions to ϕ, one single-valued and one many-valued with the appropriate cyclic constant, may be employed in cases in which the normal component of $\mathbf{u}$ is specified at all points of the boundary. If the fluid is not bounded externally, but extends to infinity in all directions in three-dimensional space and is at rest there, the kind of modification represented by (6.2.7) or (6.2.8) again applies. In the case of a fluid extending to infinity in two-dimensional space, however, we need to proceed more cautiously, because the magnitude of the velocity is in general of order $|\mathbf{x}|^{-1}$ at large values of $|\mathbf{x}|$ and the integral expression for the kinetic energy is then not convergent. The discussion of this case is taken up again in §6.4.

Kelvin's minimum energy theorem

The uniqueness of the solution for a single-valued velocity potential with a prescribed value of the normal component of $\nabla\phi$ at each point of the boundary is associated with a minimum value of the total kinetic energy, as is shown by the following result obtained first by Kelvin (1849).

Let $\mathbf{u}(\mathbf{x})$ and $\mathbf{u}_1(\mathbf{x})$ be two solenoidal velocity distributions in a given region occupied by fluid, with equal values of their normal components at each point of the boundary of the region (and, if the fluid extends to infinity, with zero values there); and suppose that $\mathbf{u}$ is irrotational, with a single-valued potential ϕ. Then the difference between the total kinetic energies corresponding to these two velocity distributions is

$$T_1 - T = \tfrac{1}{2}\rho\!\int(\mathbf{u}_1^2 - \mathbf{u}^2)\,dV$$
$$= \tfrac{1}{2}\rho\!\int(\mathbf{u}_1 - \mathbf{u})^2\,dV + \rho\!\int(\mathbf{u}_1 - \mathbf{u})\,.\,\mathbf{u}\,dV. \qquad (6.2.10)$$

For the second volume integral we have

$$\int(\mathbf{u}_1 - \mathbf{u})\,.\,\nabla\phi\,dV = \int\nabla\,.\,\{(\mathbf{u}_1 - \mathbf{u})\,\phi\}\,dV - \int\phi\nabla\,.\,(\mathbf{u}_1 - \mathbf{u})\,dV$$
$$= \int\phi(\mathbf{u}_1 - \mathbf{u})\,.\,\mathbf{n}\,dA,$$

where the surface integral is taken over the whole of the boundary of the fluid (the contribution from a hypothetical boundary at infinity, in cases in which the fluid extends to infinity, being zero) and so is zero. Thus $T_1 - T > 0$ if $\mathbf{u}_1 \neq \mathbf{u}$, showing that no motion compatible with given values of the normal component of the velocity at the boundary can have as small a total kinetic energy as the one possible irrotational motion.

In the case of a many-valued potential with a prescribed value of the normal component of velocity at each point of the boundary, it is evident that the above theorem applies to the single-valued part ϕ_1 described at the end of §2.8.

Positions of a maximum of q and a minimum of p

We show first that ϕ cannot have a maximum or a minimum value at an interior point of the fluid. It follows from (6.2.2) that

$$\int\mathbf{n}\,.\,\nabla\phi\,dA = 0$$

for any closed surface A enclosing a region wholly occupied by fluid in solenoidal irrotational motion. Hence $\mathbf{n}\,.\,\nabla\phi$ cannot be one-signed over any such closed surface, and an extremum of ϕ at some point with some different and uniform value of ϕ over a small closed surface surrounding this point is impossible.

In this argument the only relevant property of ϕ is that it satisfies Laplace's equation. The same conclusion thus holds for $\partial\phi/\partial x$, and it follows that near any interior point P of the fluid it is possible to find another point P' such that

$$|\partial\phi/\partial x|_{P'} > |\partial\phi/\partial x|_P.$$

We are free to choose the direction of the rectilinear co-ordinate x to be parallel to $\nabla\phi$ at the point P, in which case, *a fortiori*,

$$(\nabla\phi)^2_{P'} > (\nabla\phi)^2_P, \quad \text{and} \quad q_{P'} > q_P.$$

Hence a maximum of the velocity magnitude q can occur only at a point on the boundary. The occurrence of a minimum of q at an interior point is not excluded; indeed, stagnation points at which q has the smallest possible value do occur in the interior of the fluid.

A related result can be obtained for the pressure p. It follows from (6.2.5) that

$$\nabla^2 p = -\tfrac{1}{2}\rho\nabla^2 q^2 = -\rho\frac{\partial u_i}{\partial x_j}\frac{\partial u_i}{\partial x_j}, \qquad (6.2.11)$$

and consequently

$$\int \mathbf{n}.\nabla p \, dA = -\rho \int \frac{\partial u_i}{\partial x_j}\frac{\partial u_i}{\partial x_j} dV, \quad < 0, \qquad (6.2.12)$$

for any closed surface A enclosing a region wholly occupied by fluid. Now if p had a minimum value at some interior point of the fluid, $\mathbf{n}.\nabla p$ would be positive at all points of a small closed surface surrounding that point and the value of $\int \mathbf{n}.\nabla p \, dA$ taken over the same surface would be positive; which is impossible, according to (6.2.12). Hence any point at which p has a minimum value must lie on the boundary, although a maximum may occur at an interior point. The position of a minimum of p does not in general coincide with the position of a maximum of q, although there is such a coincidence when the flow is steady and the variation of $\mathbf{g}.\mathbf{x}$ is negligible.

These results have qualitative application to situations in which certain physical phenomena occur at that place in the fluid where the pressure is a minimum or the speed is a maximum. For instance, cavitation occurs in water when the absolute pressure falls below a critical value, and we infer that, for given boundaries to the (irrotational) flow, cavitation will occur first at some point on the boundary as the pressure everywhere is decreased. Likewise the occurrence of shock waves is known to be associated with local fluid velocities in excess of the speed of sound waves. For given boundaries to the flow, the maximum speed occurs at some point on the boundary while the fluid speeds everywhere are small enough for the fluid to be effectively incompressible and as the speed everywhere is increased shock waves will occur first (in the absence of any tendency for compressibility effects to change the location of the maximum) in the neighbourhood of the boundary.

Local variation of the velocity magnitude

Some simple but useful results follow immediately from the expression for local vorticity in terms of rectangular co-ordinates with axes parallel to the local directions of **u**, of the principal normal to the streamline (directed towards the centre of curvature), and of the binormal to the streamline. If (s, n, b) represent co-ordinates in these three directions respectively, and (u, v, w) are the corresponding velocity components, we have

$$v = w = 0, \quad u = q, \quad \frac{\partial v}{\partial s} = \frac{q}{R}, \quad \frac{\partial w}{\partial s} = 0$$

locally, where R is the local radius of curvature of the streamline. The components of the vorticity locally are then

$$\omega_s = \frac{\partial w}{\partial n} - \frac{\partial v}{\partial b}, \quad \omega_n = \frac{\partial u}{\partial b}, \quad \omega_b = \frac{u}{R} - \frac{\partial u}{\partial n}.$$

Moreover,
$$\left(\frac{\partial}{\partial s}, \frac{\partial}{\partial n}, \frac{\partial}{\partial b} \right) u = \left(\frac{\partial}{\partial s}, \frac{\partial}{\partial n}, \frac{\partial}{\partial b} \right) q$$

locally. Hence, in irrotational flow we have

$$\frac{\partial q}{\partial n} = \frac{q}{R}, \quad \frac{\partial q}{\partial b} = 0. \tag{6.2.13}$$

The first of the relations (6.2.13) shows that, when the streamlines are curved, the speed q is greater on the inside of the bend than on the outside. When water flowing along a straight length of pipe with approximately uniform velocity over the cross-section meets a bend in the pipe, the maximum speed, and consequently the minimum pressure and the first appearance of cavitation, occur at the wall of the pipe on the inside of the bend. Likewise, when fluid flows over a protruding salient edge, the maximum speed occurs at the edge itself provided the flow near the edge is approximately irrotational; the fluid at the boundary on the downstream side of the edge is therefore being decelerated, which leads to separation of the boundary layer (and in due course to a new flow régime without a maximum of the speed at the salient edge), as remarked already in § 5.10.

6.3. Steady flow: some applications of Bernoulli's theorem and the momentum theorem

In the particular case of steady flow of an inviscid fluid, we have available Bernoulli's theorem (§§ 3.5 and 5.1) and the momentum equation in integral form (§ 3.2). When the general character of the flow field is known from other considerations, as it is in the cases of irrotational flow to be described below, these relations are often sufficient for the purpose of determining the flow properties that are important in practice. Neither of the two relations involves a knowledge of the details of the flow, and it is appropriate to consider a few simple examples of their use before turning to more elaborate methods of investigation.

Bernoulli's theorem for an incompressible inviscid fluid of uniform density, on which gravity acts, states that the quantity

$$H = \tfrac{1}{2}q^2 + \frac{p}{\rho} - \mathbf{g}.\mathbf{x} \tag{6.3.1}$$

is constant on any streamline of a steady flow, where $q^2 = \mathbf{u}.\mathbf{u}$ as before. In the present context of *irrotational* steady flow the value of H is uniform throughout the fluid, as remarked in §6.2. In cases in which the absolute pressure does not enter into the conditions to be satisfied at the boundary of the fluid (i.e. when no free surface of liquid is present), the combination $p - \rho\mathbf{g}.\mathbf{x}$ will be defined as a 'modified pressure', as explained in §4.1, and represented by the symbol p, so that the gravity term is then suppressed.

The integral form of the momentum equation for an incompressible uniform inviscid fluid in steady motion (see (3.2.4)) is

$$\rho\!\int\!\mathbf{u}\mathbf{u}.\mathbf{n}\,dA = \int(\rho\mathbf{g}.\mathbf{x} - p)\,\mathbf{n}\,dA, \tag{6.3.2}$$

where $\mathbf{n}\,\delta A$ is an element of a freely chosen control surface which bounds a volume V entirely occupied with fluid and $\mathbf{n}$ is directed out of the volume V. The term containing $\mathbf{g}$ may be transformed back to a volume integral, giving

$$\rho\!\int\!\mathbf{u}\mathbf{u}.\mathbf{n}\,dA = \rho V\mathbf{g} - \int p\mathbf{n}\,dA \tag{6.3.3}$$

and showing explicitly that gravity may be ignored when components of the momentum flux and resultant force in a horizontal plane are under consideration.

The examples of the use of the momentum theorem given in §5.15 concerned steady flow fields in which viscous forces play an essential part, although it was found possible to choose control surfaces on which viscous stresses are small and thereby to avoid detailed consideration of the effects of viscosity. Even when viscous forces do not play an essential part and can safely be neglected everywhere, as in the examples given below, it is sometimes convenient to use the integral form of the momentum equation. The same results can usually be obtained by detailed solution of the flow field when the fluid is inviscid, although, if the integral approach succeeds at all, it will usually do so with greater speed and economy.

Efflux from a circular orifice in an open vessel

When a vessel containing water has a small orifice in one of its walls, the water is observed to flow out steadily in the form of a smooth jet. In the case of a *circular* orifice the jet becomes cylindrical at a short distance from the orifice, and remains so until it is deflected or accelerated by the action of gravity.† The flux of fluid volume through the orifice is balanced by a corresponding slow fall of the surface of the water in the vessel. All the streamlines passing through the orifice must begin at the free surface, where the velocity is negligibly small and the pressure is uniform and equal to the atmospheric pressure p_0 (for an open vessel); and the Bernoulli parameter H has the same value for all the emerging streamlines, except those coming from the boundary layer at the vessel wall which we shall ignore.

We may now use Bernoulli's theorem to determine the velocity q_0 in the cylindrical section of the jet where the pressure is necessarily uniform (the accelerations and viscous forces there being negligible) and equal to p_0. Evaluation of H at a point on the free surface and at a point in the jet gives

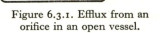

$$\frac{p_0}{\rho} = \frac{p_0}{\rho} + \tfrac{1}{2}q_0^2 - gh,$$

where h is the vertical distance between these two points (figure 6.3.1), and so

Figure 6.3.1. Efflux from an orifice in an open vessel.

$$q_0 = (2gh)^{\frac{1}{2}}. \tag{6.3.4}$$

This is also the speed that the elements of water would attain in free fall from a height h, as indeed was to be expected from energy considerations; the part played by fluid pressure here is to make the jet emerge in a direction perpendicular to the vessel wall, without influencing its speed. The result (6.3.4) is often described as Torricelli's theorem, having been found by him long before Bernoulli's work.

In practical problems one would like to use the above expression for q_0 to determine the mass flux through a circular orifice. There arises then the further problem, which cannot be resolved by Bernoulli's theorem alone, of finding the cross-section of the jet in the region in which it is cylindrical and the water velocity is q_0. Examination of a jet issuing from a circular orifice reveals that the convergence of the streamlines on the upstream side of the orifice persists for a few diameters downstream and that the cross-sectional area of the jet when it is cylindrical is smaller, by a factor α say, than that of the orifice. The coefficient of contraction α depends only on the exact shape of the rigid boundary near the orifice, and is observed to be from 0·61 to 0·64 in the simple case of a circular hole in a thin plane wall. We shall see

† When the orifice is not circular, the convergence of the streamlines leads to complicated changes of cross-sectional form; for example, the convergence at the corners of a square orifice is greater than at the midpoints of the sides and the cross-section of the jet develops the form of a cross as the corners come inwards. If the effect of surface tension is important the cross-section of the jet from a non-circular orifice oscillates periodically with distance downstream.

from the momentum theorem that α lies between $\frac{1}{2}$ and 1 for all except some peculiar boundary shapes. The value of α and the shape of the issuing jet can be worked out in detail for some simple boundary shapes in a two-dimensional flow field, by methods to be described in §6.13.

We turn now to the use of the momentum theorem. The discharge of water through the orifice of the vessel is accompanied by a flux of momentum in the direction of the jet axis (assumed to be horizontal), and this implies the existence of a horizontal reaction on the vessel. The resultant force exerted on the vessel by fluid in contact with it, either by water at the wetted surface A', where pressure is p, or by air at the remaining unwetted surface A'' where the pressure has the uniform value p_0, is

$$\mathbf{R} = \int p \mathbf{n} \, dA' + \int p_0 \mathbf{n} \, dA'',$$

where $\mathbf{n}$ is always directed away from the fluid. Since $A' + A''$ is a closed surface and the integral of a constant over the whole of a closed surface is zero, we have

$$\mathbf{R} = \int (p - p_0) \mathbf{n} \, dA'. \tag{6.3.5}$$

The vertical component of $\mathbf{R}$ represents the weight of the fluid supported in the vessel and does not interest us here; the horizontal component is the jet reaction, which we shall determine with the help of the momentum theorem.

We choose a control surface A consisting of (1) the free surface of the water in the vessel, (2) the wetted surface of the vessel A', (3) the surface bounding the portion of the jet between the orifice and any station where the jet is cylindrical, and (4) the cross-section of the jet at this latter station. Over the parts (1), (3) and (4) of the control surface, the pressure is p_0, so that

$$\int p \mathbf{n} \, dA = \int (p - p_0) \mathbf{n} \, dA', \quad = \mathbf{R}. \tag{6.3.6}$$

Consideration of the components of terms of (6.3.3) in the direction $\mathbf{k}$ of the jet axis then shows that

$$\mathbf{k} . \mathbf{R} = -\rho q_0^2 \alpha S$$

$$= -2\rho g h \alpha S, \tag{6.3.7}$$

where S is the area of the circular orifice and αS that of the cross-section of the jet in the cylindrical region.

At points in the water in the vessel, on the same horizontal plane as the jet axis, and not near the orifice, the velocity is negligibly small and the pressure is $p_0 + \rho g h$. The part of the reaction (6.3.7) to be attributed to the fact that on one side of the vessel there is a hole of area S, and on the other side, directly opposite, an equal area of wall on which an excess pressure acts, is thus $-\rho g h S$. A further contribution to the reaction arises from the fall in pressure (relative to the static-fluid value $p_0 + \rho \mathbf{g} . \mathbf{x}$) at the vessel wall, in the neighbourhood of the orifice, accompanying the increase of velocity as the water moves towards the orifice. The magnitude of this latter contribution depends on the exact shape of the vessel wall at the orifice, and cannot in general be determined by integral arguments of the above kind. In view of (6.3.7), a calculation of the second contribution to the reaction is effectively a calculation of the contraction coefficient α.

For two particular kinds of orifice it happens that the value of α can be obtained immediately. One is the case of a long slowly converging mouthpiece in which the streamlines become straight and parallel before emerging

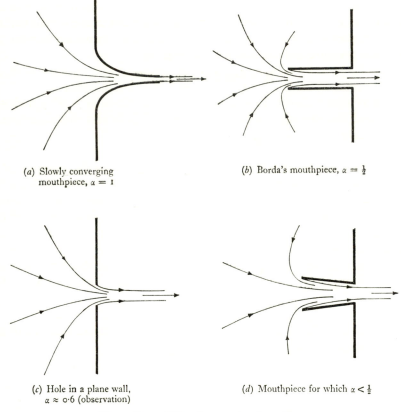

(a) Slowly converging
mouthpiece, $\alpha = 1$

(b) Borda's mouthpiece, $\alpha = \frac{1}{2}$

(c) Hole in a plane wall,
$\alpha \approx 0.6$ (observation)

(d) Mouthpiece for which $\alpha < \frac{1}{2}$

Figure 6.3.2. Efflux from a circular orifice.

(figure 6.3.2 (a)). Evidently $\alpha = 1$ here, and (6.3.7) then shows that the contribution to the reaction on the vessel from the pressure drop at the wall near the orifice is of the same magnitude as that from the uncompensated pressure on the area S of the wall opposite to the orifice. The other special case is an orifice consisting of a cylindrical tube (of internal cross-sectional area S) projecting into the vessel (figure 6.3.2 (b)), called Borda's mouthpiece. The region in which the water velocity is appreciable is here near the entrance to the tube, and the pressure has approximately the static-fluid value $p_0 + \rho \mathbf{g} \cdot \mathbf{x}$ at all points of the vessel wall except on the tube. No contribution to the component of **R** in the direction of the tube axis arises from the pressure on the tube,† so that direct evaluation of the integral in (6.3.5) gives

$$\mathbf{k.R} = -\rho ghS. \tag{6.3.8}$$

Comparison with (6.3.7) shows that $\alpha = \frac{1}{2}$ in this case.

For most other shapes of orifice α lies between $\frac{1}{2}$ and 1, since the departure of the pressure from the static-fluid value at the vessel wall near the orifice cannot be other than a suction and the rest is a matter of geometry. For the unusual orifice shown in figure 6.3.2 (d) (for which the area S is taken at the inner, or wider, end of the conical mouthpiece), the relative suction on the wetted side of the mouthpiece makes a contribution to the reaction on the vessel which is in the same direction as the jet, instead of opposite to it, so that here we see from (6.3.7) that α should be slightly less than $\frac{1}{2}$.

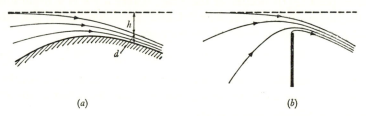

(a)   (b)

Figure 6.3.3. Steady flow over a weir, (a) broad-crested and (b) sharp-crested.

Flow over a weir

Hydraulic engineers often need to know the quantity of water flowing steadily along an open channel or through a gate of a reservoir. A simple means of obtaining this information approximately is to obstruct the flow of water, at some point in the channel or at the gate of the reservoir, by a submerged obstacle or 'weir', and to observe the level of the surface of the dammed slowly-moving water upstream of the obstacle.

A common type of weir is 'broad-crested', as sketched in figure 6.3.3 (a). The slopes of both the weir and the free water surface are here small, and we may assume that the water velocity q is approximately uniform across the stream above any point of the weir. Then if d is the depth of this stream, the rate of discharge of water volume, per unit width of the weir in a direction normal to the figure, is

$$Q = qd. \tag{6.3.9}$$

We also know from Bernoulli's theorem for the streamline at the surface that

$$\tfrac{1}{2}q^2 - gh = 0, \tag{6.3.10}$$

---

† Curiously enough, in irrotational flow round a sharp edge of exterior angle 2π, the infinitely large velocity and infinitely large suction at the edge *does* yield a non-zero force on the boundary (see §6.5); however, here the angular extent of the region occupied by fluid near the edge is supposed to be less than 2π, owing to the separation of the jet from the sharp edge, and the force on the boundary in the direction of the tube axis is zero.

where h is the fall in water level below that some distance upstream where the velocity is negligibly small. Thus the discharge is

$$Q = (2gh)^{\frac{1}{2}} d, \qquad (6.3.11)$$

and may be calculated from observations of both h and d at any point.

A further piece of information can be deduced by noticing that the vertical distance of any point of the weir from the level of the dammed water, viz.

$$d+h, \quad = \frac{Q}{q} + \frac{q^2}{2g}, \qquad (6.3.12)$$

has a minimum with respect to q. Consequently, if the upstream and down-stream conditions are such that the speed of a material element increases from zero in the reservoir to a value greater than $(gQ)^{\frac{1}{3}}$ as it passes over the weir, the speed $(gQ)^{\frac{1}{3}}$ occurs at the place where $d+h$ is a minimum, i.e. above the highest point of the weir.† Under these conditions, the values of h, d and q at the highest point of the weir are

$$h_1 = \tfrac{1}{2}(Q^2/g)^{\frac{1}{3}}, \quad d_1 = (Q^2/g)^{\frac{1}{3}}, \quad q_1 = (gQ)^{\frac{1}{3}}. \qquad (6.3.13)$$

Thus a measurement of either h_1 or d_1 (or of h_1+d_1 more conveniently, since this quantity can be measured at a position where the water is almost still) suffices for the determination of Q.

Formulae of the kind (6.3.13) are also valid for weirs which are not broad-crested (as indeed seems likely on dimensional grounds), although the numerical coefficients are different. Weirs with a sharp edge, as shown in figure 6.3.3 (*b*), are sometimes used and here it is found by observation that

$$Q = cg^{\frac{1}{2}}(h_1+d_1)^{\frac{3}{2}}, \qquad (6.3.14)$$

where c is approximately 5% greater than the theoretical value $(\frac{2}{3})^{\frac{3}{2}}$ for a broad-crested weir. It is necessary here to allow air to have free access to the region beneath the emerging jet, for if this region is confined the air in it is gradually entrained and removed by the jet and the jet is sucked downwards.

Jet of liquid impinging on a plane wall

If a steady cylindrical jet of water surrounded by air is directed against an inclined plane rigid wall, the jet is converted into a sheet of water adjoining the wall in which the flow is everywhere away from the point of impact. We suppose that the velocity of the water in the approaching jet is uniform and of sufficiently large magnitude U for the effect of gravity to be negligible. The speed everywhere on the free surface is then equal to U, by Bernoulli's theorem. The speed within the sheet must likewise be approximately

† The significance of the speed $(gQ)^{\frac{1}{3}}$ lies in the fact that it is also the maximum speed of propagation of surface waves of small amplitude when the depth is d_1; disturbances can be propagated into the reservoir from any point upstream of the highest point of the weir—and in this way the presence of the weir dams up the water—but not from any point downstream of it since the water speed is here greater than $(gQ)^{\frac{1}{3}}$.

uniform (except in the thin boundary layer near the wall) at some distance from the point of impact, because the velocity is there approximately unidirectional, and the pressure consequently uniform, across the sheet, again by Bernoulli's theorem. Thus all that remains to be determined about the sheet at some distance from the point of impact is the distribution of sheet thickness with respect to direction of the flow away from the point of impact. The total mass flux away from the point of impact is of course equal to that in the jet, but its directional distribution remains unknown.

Jets with circular cross-section are of special interest, since they can be produced easily in a laboratory. In laboratory work the plane rigid wall can be replaced by a plane of symmetry, that is, by directing two similar circular jets so that their two axes intersect. The resulting sheet of water is then free from the effect of viscosity at a rigid wall, and spreads out radially until its thickness (which varies as the reciprocal of the radial distance, in order to satisfy conservation of mass) is so small that it breaks up into discrete drops under the action of surface tension.

The momentum equation clearly imposes a constraint on the way in which the jet is diverted by the wall. Since there is no net force exerted by boundaries on the water in directions parallel to the wall, the momentum of the sheet is equal to the component of the jet momentum in the plane of the wall. This further relation does not enable the directional distribution of sheet thickness to be determined in general, but it is sufficient in the case of a two-dimensional jet, which produces a sheet whose thickness distribution is specified by just two values, one for each of the two streams moving away from the point of impact. We therefore proceed with the two-dimensional case, despite its limited physical importance, as a further illustration of use of the momentum equation in integral form.

Figure 6.3.4 shows the two-dimensional jet of width b impinging at angle α to the normal to the wall and dividing into two streams whose widths are ultimately uniform and equal to b_1 and b_2. We take the control surface shown as a broken line in the figure, on which the velocity is equal to U, and the pressure is equal to that in the surrounding air (p_0), except at points on the wall near the central impact point O. Then the component of the vector equation (6.3.3) parallel to the wall reduces to

$$\rho U^2(-b\sin\alpha + b_1 - b_2) = 0. \tag{6.3.15}$$

Conservation of mass of the water requires

$$b_1 + b_2 = b, \tag{6.3.16}$$

and these two relations together give

$$b_1 = \tfrac{1}{2}b(1 + \sin\alpha), \quad b_2 = \tfrac{1}{2}b(1 - \sin\alpha). \tag{6.3.17}$$

It is also possible to obtain some information about the distribution of pressure on the wall near O. The component of equation (6.3.3) normal to the wall becomes

$$\rho U^2 b \cos \alpha = \int (p - p_0)\, dA_w = F, \qquad (6.3.18)$$

where the integral is taken over the surface of the wall, and F is the magnitude of the normal force exerted on the wall by the jet (both per unit depth, normal to the plane of the figure). Furthermore, consideration of the moment about O of the momentum entering and leaving this same control region and of the forces acting on the fluid within the region shows that the centre of pressure C (that is, the point at which a concentrated normal force F on the wall has an anti-clockwise moment about O equal to that of the

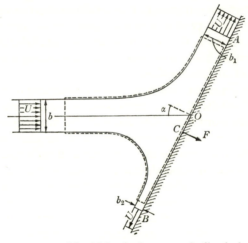

Figure 6.3.4. Jet of liquid impinging on an inclined plane wall (two-dimensional case).

pressure distribution on the wall) is a point on the anti-clockwise side of the normal and at a distance from O given by

$$OC \times F = \tfrac{1}{2}\rho U^2 b_1^2 - \tfrac{1}{2}\rho U^2 b_2^2.$$

Hence

$$OC = \tfrac{1}{2} b \tan \alpha. \qquad (6.3.19)$$

Thus if the rigid wall were hinged about O it would tend to set itself at right angles to the jet.

This latter qualitative result holds also for a rigid flat plate of finite breadth hinged about one of the bisecting lines and immersed in a stream of great width; and rectangular plates falling through infinite fluid tend to fall broadside-on. The explanation is to be found in the location of the stagnation point, where the pressure is a maximum, on the wall or plate. For two reasons the stagnation point moves away from the central point O, as α increases from zero, towards the point B in figure 6.3.4. The first is that more of the fluid in the oncoming jet flows towards the point A, so that the dividing streamline in the jet (which later must intersect the plate at the stagnation point) lies on the *B*-side of the jet axis. The second is that the streamline intersecting the plate at the stagnation point must intersect it at right angles (since any streamline passing through a stagnation point in irrotational flow must be parallel to a principal axis of the local rate-of-strain tensor, and the stream-line at the plate surface is one such principal axis), and this requires the streamline to turn towards *B* as it approaches the plate.

A special case of an axially symmetric jet for which the momentum equation does provide most of the information required arises in the theory of 'shaped charges'.† A typical shaped charge consists of a hollow metal

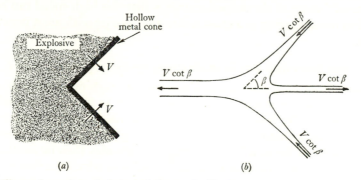

Figure 6.3.5. A conical shaped charge. In (*b*) the velocities are relative to axes moving with the apex of the cone.

cone with an open base and with explosive packed round the outside of the cone, as sketched in figure 6.3.5 (*a*). When detonation occurs, more or less simultaneously throughout the explosive, the metal wall of the cone is forced inwards under enormous pressure, and becomes plastic and capable of flow like a liquid under the action of the very large stresses. Each portion of the cone wall moves initially in the direction of the inward normal, so that the metal layer continues to have conical form (although with increasing wall thickness), except near the moving apex where there is an accumulation of metal. The situation becomes clearer when we consider the flow relative to axes which move with the apex of the cone. The metal layer now appears as a jet in the form of a conical sheet moving towards the apex, and the outcome must be the formation of two jets of circular cross-section moving along the cone axis away from the apex, as in figure 6.3.5 (*b*) (and in a two-dimensional case, with a hollow metal wedge, the flow diagram is as in figure 6.3.4, the wall *AB* there being the axis of symmetry of the wedge).

If we imagine for the sake of simplicity that the original metal layer had a thickness equal to $A/2\pi r$, where r is distance from the cone axis and A is a constant, and that the cone walls move inwards with a constant normal

† Developed during the last war for use in anti-tank and demolition weapons designed to make holes in thick metal sheets.

velocity V, the flow relative to axes moving with the cone apex is steady. The speed in all the jets is then $V \cot \beta$ at some distance from the apex, where β is the semi-angle of the cone, and the cross-sectional areas A_1, A_2 of the axial jets are found from conservation of mass and (axial) momentum to be

$$A_1 = \tfrac{1}{2}A(1 + \cos \beta), \quad A_2 = \tfrac{1}{2}A(1 - \cos \beta). \tag{6.3.20}$$

Thus a fraction $\tfrac{1}{2}(1 - \cos \beta)$ of the mass of the conical metal layer is ejected through the open base of the cone as a narrow jet with speed $V \cot \beta$ relative to the cone apex, that is, with speed $V(\cot \beta + \operatorname{cosec} \beta)$ relative to the un-detonated charge. Very large values of V can be achieved with explosive, and the special feature of the shaped charge is that this high speed can be conferred on metal moving in a chosen direction as a jet with great penetrating power.

Irrotational flow which may be made steady by choice of rotating axes

It sometimes happens, in machines such as turbines and pumps, that fluid passes from a stationary tube or channel to an enclosure which is rotating as a whole. If the fluid moves irrotationally through the stationary tube, it will be without vorticity in the rotating portion of the system. The determination of the flow in the rotating section may then be facilitated by the use of a rotating frame or reference, relative to which the flow is steady. Relative to these new axes the vorticity is not zero, but it was noted in § 3.5 that the Bernoulli quantity H, modified to include a contribution from centrifugal force, is nevertheless constant over the whole region of steady flow.

When all quantities representing a flow field are referred to axes rotating with the steady angular velocity $\mathbf{\Omega}$, the equation of motion (6.2.3), with inclusion of the fictitious body force (3.2.10), becomes

$$\frac{\partial \mathbf{u}}{\partial t} - \mathbf{u} \times (\mathbf{\omega} + 2\mathbf{\Omega}) = -\nabla \left\{ \tfrac{1}{2}q^2 + \frac{p}{\rho} - \mathbf{g} \cdot \mathbf{x} - \tfrac{1}{2}(\mathbf{\Omega} \times \mathbf{x})^2 \right\}.$$

Consequently, when the flow relative to these axes is steady and has vorticity $-2\mathbf{\Omega}$ (corresponding to zero vorticity relative to an absolute frame), the quantity

$$H = \tfrac{1}{2}q^2 + \frac{p}{\rho} - \mathbf{g} \cdot \mathbf{x} - \tfrac{1}{2}(\mathbf{\Omega} \times \mathbf{x})^2 \tag{6.3.21}$$

is constant throughout the flow field. A simple explicit expression for the pressure in terms of the velocity is thus available, and may be employed in many different ways.

As an illustrative example, consider the system of pipes shown in figure 6.3.6(a). Fluid is forced up the vertical pipe and divides into two streams in the horizontal pipe which rotates about the vertical z-axis with steady angular velocity Ω. We assume that the velocity is uniform across the vertical pipe, so that the flow is irrotational everywhere relative to an absolute frame.

Relative to x- and y-axes which rotate with the horizontal pipe, the vorticity is of magnitude -2Ω and is parallel to the z-axis, so that when the velocity in the horizontal pipe has become unidirectional, at some distance from the junction, the x-component of velocity is

$$u = U + 2\Omega y;$$

the fluid moves more quickly on the side of the rotating pipe which leads. According to Bernoulli's equation in the above form, the pressure in the rotating pipe is then

$$\frac{p}{\rho} = H - \tfrac{1}{2}(U+2\Omega y)^2 + \tfrac{1}{2}\Omega^2(x^2+y^2), \qquad (6.3.22)$$

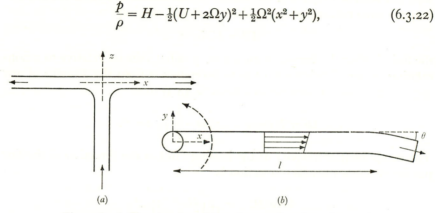

(a)   (b)

Figure 6.3.6. Flow in a branching pipe rotating about the z-axis.

the gravity term being ignored. As expected intuitively, there is an excess pressure on the side of the pipe which lags, and a torque must be exerted on the pipe to keep it rotating and to keep providing moment of momentum about the z-axis for the stream of fluid. If the rotating pipe has a circular section of radius a, the torque which must be exerted on a length $2l$ of straight pipe to keep it rotating is easily found from (6.3.22) to be

$$2\int_0^l 2\pi\rho U\Omega a^2 x\,dx, \quad = 2\pi\rho U\Omega a^2 l^2.$$

The two ends of the rotating pipe might be open to the atmosphere, where the pressure is p_0 say. If the pipe length is large compared with its radius, the average axial speed with which the fluid is flung out of the two open ends of the rotating pipe is then found from (6.3.22) to be given by

$$U^2 = 2\left(H - \frac{p_0}{\rho}\right) + \Omega^2 l^2, \qquad (6.3.23)$$

where H, the total energy per unit mass of the water coming up the vertical pipe, can be regarded as given. Even when the fluid comes from a reservoir where the speed is very small and the pressure is less than p_0 (aside from the contribution due to gravity), so that $H < p_0/\rho$, fluid may still be 'pumped' from the reservoir. This is essentially the action of a centrifugal pump.

If the axis of the rotating pipe near the two open ends does not pass through the axis of rotation, the reaction of the issuing jets may be used to make the pipe rotate, as in a water sprinkler or a 'catherine' wheel fire-work. In the simple case shown in figure 6.3.6 (b), the two jets are discharged at angle θ to the x-axis at an average velocity U (relative to the rotating axes), and the moment about the axis of rotation of the force exerted on the pipe elbow is approximately $lU\sin\theta$ multiplied by the mass of fluid discharged from the pipe per second ($2\pi\rho a^2 U$). On equating this moment to that required to keep the pipe rotating with angular velocity Ω, we find

$$\Omega l = U\sin\theta;$$

the fluid is thus discharged with a purely radial velocity relative to a non-rotating frame of reference, as is to be expected from the absence of any applied moment. It then follows from (6.3.23) that

$$\Omega l = U\sin\theta = \left\{2\left(H - \frac{p_0}{\rho}\right)\right\}^{\frac{1}{2}}\tan\theta.$$

The component of velocity of the issuing fluid in the direction of the line passing through the axis of rotation is

$$U\cos\theta, \quad = \left\{2\left(H - \frac{p_0}{\rho}\right)\right\}^{\frac{1}{2}},$$

showing, again as expected on other grounds, that the area of ground which can be watered by such a sprinkler is independent of θ.

Exercise

A cylindrical column of liquid of length l and density ρ is moving in a direction parallel to the generators, and impinges against a solid wall at a speed high enough to cause the material of the wall to behave as a liquid of density ρ_s locally. Show that the depth of penetration of the jet into the wall is approximately $l(\rho/\rho_s)^{\frac{1}{2}}$.

6.4. General features of irrotational flow due to a moving rigid body

Solenoidal irrotational flow associated with a rigid body moving in fluid extending to infinity in all directions and at rest there arises often in both theoretical and practical contexts and is particularly important. The corresponding mathematical results may be applied directly to cases of flow at large Reynolds number in which boundary-layer separation does not occur (which would include slender bodies moving parallel to their length, and bodies of arbitrary shape accelerating from rest or executing translational or rotational oscillations of small amplitude about a fixed position), and are also valuable in a number of indirect ways. The mathematical theory is highly developed, and a great variety of both analytical and numerical methods is available for the determination of the flow field.† In this and the next few sections we shall consider the main features of the problem and some illuminating special cases.

In this section we present some results concerning the asymptotic form of the velocity distribution far from the body, the total energy of the fluid, and the net force exerted on a body in translational motion, without explicit regard for the body shape except that it will be assumed to be singly-connected. The generality of these results gives them considerable value.

There are significant differences between the properties of two- and three-dimensional flow fields due to moving bodies, arising from the fact that the region occupied by fluid is necessarily doubly-connected in one case and singly-connected in the other, and the arguments and results mostly need statement for the two cases in turn. The three-dimensional flow field usually has simpler properties and will be taken first.

The velocity at large distances from the body

As a preliminary, we recall from § 2.9 that, in a three-dimensional field in which $\nabla\phi$ is zero at infinity, the velocity potential may be written as an infinite series of spherical solid harmonics of negative degree in the region external to a spherical surface which encloses the inner boundary. More specifically, in this region we have

$$\phi(\mathbf{x}) - C = \frac{c}{r} + c_i \frac{\partial}{\partial x_i}\left(\frac{1}{r}\right) + c_{ij} \frac{\partial^2}{\partial x_i \, \partial x_j}\left(\frac{1}{r}\right) + \dots, \qquad (6.4.1)$$

where $r = |\mathbf{x}|$ and the tensor coefficients $c, c_i, c_{ij}, \dots$ are given by (2.9.20) in terms of integrals of ϕ and $\nabla\phi$ over the inner boundary of the region of irrotational flow. In the case of interest here, the inner boundary is the body surface, and the net flux of volume across the inner boundary is zero; as remarked in § 2.9, $c = 0$ as a consequence. Thus, at large distances from the body we have, in general,

$$\phi(\mathbf{x}) - C \sim \mathbf{c}.\nabla\left(\frac{1}{r}\right), \qquad = -\frac{\mathbf{c}.\mathbf{x}}{r^3}, \qquad (6.4.2)$$

where

$$4\pi\mathbf{c} = \int(\mathbf{x}\,\mathbf{n}.\nabla\phi - \mathbf{n}\phi)\,dA, \qquad (6.4.3)$$

the integral being taken over the whole of the body surface A, with $\mathbf{n}$ as the outward normal. At large distances from the body, the velocity distribution is of the same form as for a source doublet of strength $4\pi\mathbf{c}$ located at the origin and the magnitude of the velocity is of order r^{-3}. In the particular case of a sphere with centre instantaneously at the origin, c_i is the only non-zero coefficient in the series (6.4.1) (and $c_i = \frac{1}{2}a^3U_i$, where a is the radius of the

---

† The most comprehensive account of the analytical methods and results will be found in *Hydrodynamics*, 6th ed., by H. Lamb (Cambridge University Press 1932, and Dover Publications).

sphere and $\mathbf{U}$ its instantaneous velocity), so that here the form (6.4.2) applies throughout the fluid.

Similar results are available for flow due to a rigid body moving in infinite fluid in a two-dimensional field, provided we make proper allowance for the cyclic constant of the motion, that is, for the circulation κ round the body. In place of (6.4.1) we have the series (2.10.6), viz.

$$\phi(\mathbf{x}) - C = \frac{\kappa}{2\pi}\theta + c\log r + c_i \frac{\partial}{\partial x_i}(\log r) + c_{ij}\frac{\partial^2}{\partial x_i\,\partial x_j}(\log r) + \dots,$$

$$(6.4.4)$$

in which the coefficients $c, c_i, c_{ij}, \dots$ are given as integrals, not of ϕ and $\nabla\phi$, but of $\phi - (\kappa/2\pi)\,\theta$ and $\nabla\{\phi - (\kappa/2\pi)\,\theta\}$, over the body surface. This series applies in the region external to a circle centred at the origin and enclosing the inner boundary. Since the first coefficient c is again proportional to the net flux of volume across the inner boundary (per unit depth normal to the plane of motion) and is zero here, the asymptotic form corresponding to (6.4.2) is

$$\phi(\mathbf{x}) - C - \frac{\kappa}{2\pi}\theta \sim \mathbf{c}.\nabla(\log r), \qquad = \frac{\mathbf{c}.\mathbf{x}}{r^2}, \qquad (6.4.5)$$

at large values of r, where

$$-2\pi\mathbf{c} = \int\left\{\mathbf{x}\mathbf{n}.\nabla\left(\phi - \frac{\kappa\theta}{2\pi}\right) - \mathbf{n}\left(\phi - \frac{\kappa\theta}{2\pi}\right)\right\}dA. \qquad (6.4.6)$$

At large distances from the body, the dominant contribution to the velocity distribution is of order r^{-1} and has the same form as for a 'point vortex' of strength κ located at the origin, and if $\kappa = 0$ the dominant contribution is of order r^{-2} and has the same form as for a point source doublet of strength $-2\pi\mathbf{c}$ located at the origin. In the case of a moving circular cylinder with centre instantaneously at the origin, c_i is the only non-zero coefficient (and $c_i = -a^2 U_i$) and the form (6.4.5) applies throughout the fluid.

It is a remarkable feature of the integral in (6.4.3) or (6.4.6) representing the strength of the effective source doublet to which the body corresponds at large distances that it may equally well be evaluated over any closed surface S in the fluid enclosing the body instantaneously. For we have

$$\int(\mathbf{x}\mathbf{n}.\nabla\phi - \mathbf{n}\phi)\,dA = \int\{\mathbf{n}.\nabla(\mathbf{x}\phi) - 2\mathbf{n}\phi\}\,dA$$

$$= -\int\{\nabla^2(\mathbf{x}\phi) - 2\nabla\phi\}\,dV + \int\{\mathbf{n}.\nabla(\mathbf{x}\phi) - 2\mathbf{n}\phi\}\,dS$$

$$= \int(\mathbf{x}\mathbf{n}.\nabla\phi - \mathbf{n}\phi)\,dS,$$

and likewise with $\phi - (\kappa/2\pi)\,\theta$ replacing ϕ in the case of a two-dimensional field, where $\mathbf{n}$ represents the outward normal to both A and S and the integral with respect to V is taken over the volume of fluid bounded by A and S.

The velocity distribution is uniquely determinate when we specify the value of $\mathbf{n}.\nabla\phi$ at each point of the body surface, and also the circulation round the body in a two-dimensional field. Now the instantaneous motion of the body is given generally by its angular velocity Ω and by the velocity U of some material point of the body, which for convenience we choose as the centre of volume of the body, with instantaneous position vector x_0; in these circumstances the inner boundary condition is

$$n \cdot \nabla \phi = n \cdot \{U + \Omega \times (x - x_0)\} \tag{6.4.7}$$

at all points of A. The expression for $4\pi c$ for a body in a three-dimensional field then becomes, from (6.4.3),

$$4\pi c = \int x n \cdot \{U + \Omega \times (x - x_0)\} \, dA - \int \phi n \, dA$$
$$= \int \{U + \Omega \times (x - x_0)\} \cdot \nabla x \, dV - \int \phi n \, dA,$$

where the first integral is taken over the volume V_0 of the body. Hence

$$4\pi c = V_0 U - \int \phi n \, dA, \tag{6.4.8}'$$

and in a two-dimensional field the corresponding relation is

$$-2\pi c = V_0 U - \int \left\{ n \left(\phi - \frac{\kappa\theta}{2\pi} \right) + x n \cdot \nabla \left(\frac{\kappa\theta}{2\pi} \right) \right\} dA. \tag{6.4.8}''$$

In the particular case of a body which is symmetrical about each of three orthogonal planes in a three-dimensional field and which is rotating about the line of intersection of any two of these planes (so that $U = 0$), the values of ϕ at the two points on A at the ends of a line through the centre of volume of the body are necessarily equal and the unit normal vectors are anti-parallel; thus $\int \phi n \, dA$ and c are both zero, and the velocity at large distances from the body is of order r^{-4}. For a body symmetrical about each of two orthogonal planes in a two-dimensional field, and with $U = 0$, we find by a similar symmetry argument that $c = 0$.

Further information about the dependence of ϕ, and therefore also of c, on U, Ω and κ is available in relations like (2.9.23) and (2.10.13). We can write the velocity potential as $\phi_1 + \phi_2$, where ϕ_1 is a single-valued velocity potential satisfying the specified inner boundary condition (which is (6.4.7) here) and ϕ_2 is a many-valued velocity potential with cyclic constant κ (which is non-zero only for a two-dimensional flow) satisfying $n \cdot \nabla \phi_2 = 0$ on A. ϕ_1 may be represented as the sum of two single-valued velocity potentials, one whose normal derivative on A has the value $n \cdot U$ and which is thus of the form (2.9.23), and one whose normal derivative on A has the value $n \cdot \{\Omega \times (x - x_0)\}$ and which is therefore linear in Ω. ϕ_2 does not depend on U or Ω in any way, is necessarily linear in κ, and, as noted in §2.10, can be written as

$$\phi_2(x) = \kappa \left(\frac{\theta}{2\pi} + \Psi \right), \tag{6.4.9}$$

where Ψ is a single-valued velocity potential depending only on $x - x_0$ and on the shape of the body. The complete velocity potential for a two-dimensional field, and for a three-dimensional field when κ is put equal to zero, is then

$$\phi(\mathbf{x}) = \mathbf{U}.\boldsymbol{\Phi} + \boldsymbol{\Omega}.\boldsymbol{\Theta} + \kappa\left(\frac{\theta}{2\pi} + \Psi'\right), \qquad (6.4.10)$$

in which $\boldsymbol{\Phi}$, $\boldsymbol{\Theta}$ and Ψ' are all functions of $\mathbf{x} - \mathbf{x}_0$ dependent on the shape of the body and independent of $\mathbf{U}$, $\boldsymbol{\Omega}$ and κ, and $\boldsymbol{\Theta}$ is an axial vector, like $\boldsymbol{\Omega}$, which is normal to the plane of motion in the case of a two-dimensional field. By substituting (6.4.10) in the boundary condition (6.4.7), and using the fact that these two relations are valid for all $\mathbf{U}$ and $\boldsymbol{\Omega}$, the inner boundary conditions on $\boldsymbol{\Phi}$ and $\boldsymbol{\Theta}$ are found to be

$$\mathbf{n}.\nabla\boldsymbol{\Phi} = \mathbf{n}, \quad \mathbf{n}.\nabla\boldsymbol{\Theta} = -\mathbf{n}\times(\mathbf{x} - \mathbf{x}_0) \qquad (6.4.11)$$

at all points on A.

We see from substitution of (6.4.10) in (6.4.8) that $\mathbf{c}$ is a linear function of $\mathbf{U}$, $\boldsymbol{\Omega}$ and κ; in a three-dimensional field

$$4\pi c_i = U_j(V_0\delta_{ij} - \int\Phi_j n_i\, dA) - \Omega_j\int\Theta_j n_i\, dA, \qquad (6.4.12)'$$

and in a two-dimensional field

$$-2\pi c_i = U_j(V_0\delta_{ij} - \int\Phi_j n_i\, dA) - \Omega_j\int\Theta_j n_i\, dA - \kappa\int\{\Psi' n_i + x_i\mathbf{n}.\nabla(\kappa\theta/2\pi)\}\, dA. \qquad (6.4.12)''$$

A significant property of the second-order tensor forming the coefficient of U_j here is that it is symmetrical in the indices i and j; for it follows from (6.4.11) that

$$\int\Phi_j n_i\, dA - \int\Phi_i n_j\, dA = \int\left(\Phi_j\frac{\partial\Phi_i}{\partial x_k} - \Phi_i\frac{\partial\Phi_j}{\partial x_k}\right) n_k\, dA$$

$$= \int\left(\Phi_j\frac{\partial\Phi_i}{\partial x_k} - \Phi_i\frac{\partial\Phi_j}{\partial x_k}\right) n_k\, dS - \int(\Phi_j\nabla^2\Phi_i - \Phi_i\nabla^2\Phi_j)\, dV,$$

in which the volume integral vanishes since $\boldsymbol{\Phi}$ satisfies Laplace's equation and the integral over the surface S can be seen to be zero from the choice of S as a sphere (or a circle) of indefinitely large radius.

The kinetic energy of the fluid

Expressions for the kinetic energy of the fluid in terms of surface integrals have been given in §6.2. Here we adapt these expressions to the case in which the fluid is bounded internally by a rigid body having a prescribed motion. Rather unexpectedly, a simple relation between the kinetic energy and the coefficient c_i in the series (6.4.1) or (6.4.4) can be obtained for bodies in translatory motion.

Take first the case of a simply-connected body moving in a three-dimensional field, or a body in a two-dimensional field with zero circulation round it. Then ϕ is single-valued and the general expression for the kinetic energy of the fluid (see (6.2.8)) is

$$T = -\tfrac{1}{2}\rho\int\phi\mathbf{u}.\mathbf{n}\, dA,$$

the integral being taken over the body surface. For a body moving with both translation and rotation, the value of $\mathbf{u} \cdot \mathbf{n}$ at the body surface is given by (6.4.7), and so

$$T = -\tfrac{1}{2}\rho \int \phi \mathbf{U} \cdot \mathbf{n}\, dA - \tfrac{1}{2}\rho \int \phi \{\boldsymbol{\Omega} \times (\mathbf{x} - \mathbf{x}_0)\} \cdot \mathbf{n}\, dA. \qquad (6.4.13)$$

Substitution of the general form (6.4.10) (with $\kappa = 0$) then shows that T is a quadratic function of $\mathbf{U}$ and $\boldsymbol{\Omega}$, like the kinetic energy of a rigid body, and can be written as

$$T = \tfrac{1}{2}\rho V_0 (\alpha_{ij}\, U_i\, U_j + \beta_{ij}\, U_i \Omega_j + \gamma_{ij}\, \Omega_i \Omega_j), \qquad (6.4.14)$$

where V_0 is the volume of the body as before and the tensor coefficients α_{ij}, β_{ij} and γ_{ij} depend on the shape and size of the body. α_{ij} is dimensionless, depends only on the body shape, and is given by

$$\alpha_{ij} = -\frac{1}{2V_0} \int (\Phi_j\, n_i + \Phi_i\, n_j)\, dA, \quad = -\frac{1}{V_0} \int \Phi_j\, n_i\, dA \qquad (6.4.15)$$

since we saw above that this latter integral is symmetrical in i and j.

We can go further in the simple, but nevertheless important, case of a body moving without rotation (and still with $\kappa = 0$). We have then

$$T = -\tfrac{1}{2}\rho U_i \int \phi n_i\, dA, \quad = \tfrac{1}{2}\rho V_0\, \alpha_{ij}\, U_i\, U_j. \qquad (6.4.16)$$

Thus the kinetic energy of the fluid due to translational motion of the body is equal to $\tfrac{1}{2}\rho V_0 |\mathbf{U}|^2$ multiplied by a factor $\alpha_{ij}\, U_i\, U_j / |\mathbf{U}|^2$ which depends on the shape of the body and the direction of its motion. The known expressions for $\boldsymbol{\Phi}$ for a sphere and a circular cylinder show that α_{ij} is equal to $\tfrac{1}{2}\delta_{ij}$ and δ_{ij} respectively in those cases. Other particular values will be pointed out later in this chapter. In the case of an axisymmetric body the principal axes of α_{ij} consist of the axis of symmetry of the body and any two orthogonal axes.

A further result for the case $\boldsymbol{\Omega} = 0$, $\kappa = 0$, comes from a comparison of (6.4.16) with (6.4.8); we see that in that case

$$T = \tfrac{1}{2}\rho (4\pi \mathbf{U} \cdot \mathbf{c} - V_0 \mathbf{U} \cdot \mathbf{U}) \qquad (6.4.17)$$

for a three-dimensional field, and a similar relation with $4\pi \mathbf{c}$ replaced by $-2\pi \mathbf{c}$ holds for a two-dimensional field. There is evidently a connection, although not a simple one, between the volume of the body and the magnitude of the disturbance that it causes in the fluid at large distances; the magnitude of the component, in the direction of $\mathbf{U}$, of the strength of the source doublet representing the effect of the body at large distances cannot be less than $V_0 |\mathbf{U}|$ in either a two- or a three-dimensional field. An alternative expression of this relation between T and $\mathbf{c}$ is obtained from (6.4.12) (with $\boldsymbol{\Omega} = 0$, $\kappa = 0$) and (6.4.15) as

$$4\pi c_i = V_0\, U_j (\delta_{ij} + \alpha_{ij}) \qquad (6.4.18)$$

for a three-dimensional field, $4\pi c_i$ being replaced by $-2\pi c_i$ for a two-dimensional field.

In the case of a two-dimensional field with non-zero circulation round the body, the velocity of the fluid is of order r^{-1} at large distances from the body, and theoretically the flow has infinite kinetic energy. The implication for a real flow system is that the amount of kinetic energy in the fluid is influenced by the position and form of the distant outer boundary. The separation of the velocity potential into two parts described at the end of §2.8 is useful, because the single-valued part ϕ_1 makes a finite contribution to the kinetic energy. We have from (2.8.10) (in which the normal to A is away from the fluid)

$$T = -\tfrac{1}{2}\rho\int\phi_1\,\mathbf{n}.\nabla\phi_1\,dA + \text{kinetic energy associated}$$
$$\text{with the circulation } \kappa \text{ and zero normal}$$
$$\text{component of velocity at the boundary,} \quad (6.4.19)$$

of which the second term on the right-hand side is infinite in a way which does not affect the value of the first term and which is independent of both $\mathbf{U}$ and $\mathbf{\Omega}$. The first term on the right-hand side of (6.4.19) is the kinetic energy of the motion corresponding to the given values of $\mathbf{U}$ and $\mathbf{\Omega}$ with $\kappa = 0$, and so the remarks made above apply directly to it.

The force on a body in translational motion

We consider the total force $\mathbf{F}$ exerted instantaneously by the surrounding fluid on a body moving without rotation. This force arises from the pressure at the body surface, and with the aid of (6.2.5) we have

$$\mathbf{F} = -\int p\mathbf{n}\,dA$$
$$= \rho\int\frac{\partial\phi}{\partial t}\mathbf{n}\,dA + \tfrac{1}{2}\rho\int q^2\mathbf{n}\,dA - \rho\int\mathbf{g}.\mathbf{x}\mathbf{n}\,dA, \quad (6.4.20)$$

the integrals being taken over the fixed surface A that coincides instantaneously with the body surface. The last integral in (6.4.20) represents the buoyancy force on the body (§4.1), and we shall ignore it in what follows.

Now $\partial\phi/\partial t$ is non-zero, even for a body in steady translational motion, since we are using axes fixed in the fluid at infinity and the position of the body is changing relative to these axes. The unknown functions $\mathbf{\Phi}$ and $\mathbf{\Psi}$ in (6.4.10) are functions of $\mathbf{x} - \mathbf{x}_0$, where $\mathbf{x}_0$ is the instantaneous position vector of a material point of the body and

$$d\mathbf{x}_0/dt = \mathbf{U}. \quad (6.4.21)$$

The velocity $\mathbf{U}$ may also depend on t, but variation of κ with t is excluded by Kelvin's circulation theorem in wholly irrotational flow. The rate of change of ϕ at a point fixed relative to the fluid at infinity is then found from (6.4.10) (with $\mathbf{\Omega} = 0$) to be

$$\frac{\partial\phi}{\partial t} = \dot{\mathbf{U}}.\mathbf{\Phi} + \frac{\partial(\mathbf{x} - \mathbf{x}_0)}{\partial t}.\nabla\phi$$
$$= \dot{\mathbf{U}}.\mathbf{\Phi} - \mathbf{U}.\mathbf{u}, \quad (6.4.22)$$

where $\dot{\mathbf{U}}$ stands for $d\mathbf{U}/dt$.

The relation (6.4.20) (without the contribution from buoyancy) becomes

$$F_i = \rho U_j \int \Phi_j n_i \, dA + \rho \int (\tfrac{1}{2} q^2 - U_j u_j) n_i \, dA. \qquad (6.4.23)$$

The first of these two terms on the right-hand side is non-zero only when $\mathbf{U}$ is changing, whereas the value of the second term is independent of the fact that $\mathbf{U}$ may be changing. Thus the first term represents what may be called the *acceleration reaction*, and the second term is the force on a body in steady motion. We postpone further discussion of the acceleration reaction for the moment.

To obtain definite results about the contribution to the force that remains when the velocity of translational motion is steady, we introduce a surface S in the fluid which encloses the body instantaneously, and relate integrals over the closed surfaces A and S to integrals over the volume V bounded by A and S. Then, with $\mathbf{n}$ as the outward normal to both surfaces, we have

$$\int \tfrac{1}{2} q^2 n_i \, dA = \int \tfrac{1}{2} q^2 n_i \, dS - \int \frac{\partial (\tfrac{1}{2} u_j u_j)}{\partial x_i} \, dV$$

and, since $\mathbf{u}$ is both irrotational and solenoidal,

$$= \int \tfrac{1}{2} q^2 n_i \, dS - \int \frac{\partial (u_i u_j)}{\partial x_j} \, dV$$

$$= \int (\tfrac{1}{2} q^2 n_i - u_i u_j n_j) \, dS + \int u_i u_j n_j \, dA.$$

q is at least as small as r^{-3} in a three-dimensional field, and as small as r^{-1} in a two-dimensional field, when r is large; consequently the special choice of S as a sphere or circle of indefinitely large radius shows that the integral over S on the right-hand side is identically zero. Hence

$$\int \tfrac{1}{2} q^2 n_i \, dA = U_j \int u_i n_j \, dA, \qquad (6.4.24)$$

and the force on a body in steady motion is

$$F_i = \rho U_j \int (u_i n_j - u_j n_i) \, dA = \rho U_j \int (u_i n_j - u_j n_i) \, dS, \qquad (6.4.25)$$

where S is again an arbitrary surface in the fluid enclosing the body.

We see that $U_i F_i = 0$, showing that the fluid provides no resistance to steady translational motion of the body; thus we have recovered the result (d'Alembert's paradox) obtained from an energy argument with less generality in § 5.11.

The existence of a component of $\mathbf{F}$ normal to $\mathbf{U}$ can also be excluded in the case of a rigid body of finite dimensions in a three-dimensional field, since q is of order r^{-3} at large distances from the body and the special choice of S as a sphere of large radius shows that the whole expression in (6.4.25) is zero.

On the other hand, in a two-dimensional field, q is of order r^{-1} at large distances from the body when there is a non-zero circulation round it, and the integrals in (6.4.25) may not vanish; on taking S as a circle of large radius and using the asymptotic relation

$$\nabla\phi \sim \frac{\kappa}{2\pi}\nabla\theta$$

on S, we have, for the component of $\mathbf{F}$ in the y-direction in a case in which $\mathbf{U}$ has magnitude U and direction parallel to the x-axis (figure 6.4.1),

$$F_y = \rho U \frac{\kappa}{2\pi}\int_0^{2\pi}\left(\frac{\partial\theta}{\partial y}\cos\theta - \frac{\partial\theta}{\partial x}\sin\theta\right)r\,d\theta$$

$$= \rho U \kappa. \tag{6.4.26}$$

This remarkable side-force or 'lift' on the body, which is the foundation of the theory of the lifting action of aeroplane wings, arises from the combined effect of the forward motion of the body and the circulation round it, and is

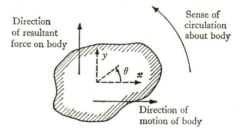

Figure 6.4.1. Definition sketch for calculation of force on a body in steady translational motion in a two-dimensional field.

independent of the size, shape and orientation of the body. The relation (6.4.26) is associated with the names of Kutta (1910) and Joukowski, two of the pioneers of the scientific study of aeronautics. It should be noted from figure 6.4.1 that the direction of the resultant force on the body is obtained by rotating the vector representing the velocity of the body relative to the fluid at infinity through 90°, in the sense of the circulation.

We obtain further insight into the mechanism of this side-force by an application of the momentum theorem, which was in fact the method used by Joukowski to establish the result (6.4.26). We suppose that the body is moving steadily and, in order to obtain a steady flow (as is convenient for use of the momentum theorem), choose axes moving with the body and with origin within the body. The velocity at any point of the fluid relative to these new axes is then $-\mathbf{U}+\mathbf{u}$, where $\mathbf{u} = \nabla\phi$ is, as before, the velocity relative to the fluid at infinity. The control 'surface' consists of the boundary of the body A and a circle S of large radius centred at the origin, with $\mathbf{n}$ as the outward normal to both closed curves. The total momentum of the fluid contained within this control surface is constant, so that the force exerted on the body is given by

$$\mathbf{F} = -\int\rho(-\mathbf{U}+\mathbf{u})(-\mathbf{U}.\mathbf{n}+\mathbf{u}.\mathbf{n})\,dS - \int p\mathbf{n}\,dS, \tag{6.4.27}$$

in which the first integral represents the flux of momentum outward across the control surface. The pressure is given by Bernoulli's theorem as

$$p = p_0 + \tfrac{1}{2}\rho\{U^2 - (-\mathbf{U} + \mathbf{u}).(-\mathbf{U} + \mathbf{u})\}.$$

Since $|\mathbf{u}|$ is of order r^{-1} at large distances from the origin, terms quadratic in $\mathbf{u}$ need not be retained in the integrands in (6.4.27); moreover, $\int \mathbf{n}\, dS = 0$ and $\int \mathbf{u}.\mathbf{n}\, dS = 0$. Hence

$$F_i = \rho U_j \int u_i n_j\, dS - \rho U_j \int u_j n_i\, dS,$$

and then (6.4.26) is recovered as before. It follows from the calculation immediately preceding (6.4.26) that the side-force exerted by the cylinder appears in the fluid far from the body half as a momentum flux and half in the form of a pressure distribution.

It should be remembered that all these results about the force on a body in steady translational motion apply equally to the second of the two contributions in (6.4.23) to the force on a body whose velocity is changing.

A similar analysis may be made of the moment exerted by the fluid on a body which is rotating with one point of the body fixed. The main result is that when $\boldsymbol{\Omega}$ is constant the component of the moment parallel to $\boldsymbol{\Omega}$, which is the only surviving component in the case of a two-dimensional field, is zero; this is the result to be expected from the fact that the kinetic energy of the fluid does not change when $\boldsymbol{\Omega}$ is constant, although this latter argument is not wholly satisfactory since the kinetic energy is theoretically infinite when there is a non-zero cyclic constant. The case of combined translation and rotation of the body is considerably more complicated owing to the continual change of the direction of $\mathbf{U}$ relative to the body.

The acceleration reaction

We return now to the first term on the right-hand side of (6.4.23), that is, to the (translational) acceleration reaction $\mathbf{G}$ given by

$$G_i = \rho \dot{U}_j \int \Phi_j n_i\, dA,$$
$$= -\rho V_0 \alpha_{ij} \dot{U}_j \qquad\qquad (6.4.28)$$

in view of the definition of the dimensionless (symmetrical) tensor α_{ij} given in (6.4.15). Since the force on a rigid body arising from translational acceleration of the body through the surrounding fluid is a linear function of the components of the acceleration, it is natural to regard the coefficient of $-\dot{U}_j$ in (6.4.28), viz. $\rho V_0 \alpha_{ij}$, as an induced or virtual tensor inertia which must be added to the real mass of the body when determining its response to a given applied force. ρV_0 is the mass of fluid displaced by the body, and α_{ij} may be termed the (tensor) *coefficient of virtual inertia*. It appears that in the case of translational motion with zero cyclic constant, the velocity at large distances from the body, the total kinetic energy of the fluid, and the virtual inertia, are all determined by V_0, $\mathbf{U}$ and the coefficient α_{ij}.

The acceleration reaction is obviously related to the fact that, when the velocity of the body changes, the total kinetic energy of the fluid changes also. The part of the kinetic energy of the fluid arising from the translational motion of the body, and not from any circulation which may exist, is a quadratic function of the components of **U** at any time (for see (6.4.14)), and is thus representable as an amount of energy arising from a certain addition to the mass of the body (mass being regarded as a second-order tensor). When the velocity of the body is changing, the rate of change of the kinetic energy of the fluid associated with the translational motion of the body is

$$\frac{d}{dt}(\tfrac{1}{2}\rho V_0 \alpha_{ij} U_i U_j), \quad = \rho V_0 \alpha_{ij} U_i \dot{U}_j$$
$$= -U_i G_i.$$

Thus the whole of the work done by the body against the acceleration reaction appears as kinetic energy of the part of the fluid motion associated with the translation of the body; the kinetic energy arising from any circulation round the body in a two-dimensional field, although infinite, can evidently be regarded as constant when the body accelerates. Quite apart from any difficulties over the case with circulation, it might be thought surprising that the kinetic energy and the acceleration reaction are determined by precisely the same tensor coefficient α_{ij}, since the rate of working against the acceleration reaction involves only the component of **G** parallel to **U**. The explanation lies in the fact that **G** is independent of **U**.

A similar interpretation of the acceleration reaction can in principle be given in terms of the total linear momentum of the fluid, which we expect to be a linear function of the components of **U**. However, direct calculation of the rate of change of total linear momentum of the fluid is frustrated by the fact that the integral $\rho \int \mathbf{u} \, dV$ is in general not absolutely convergent as the volume V approaches infinity; $|\mathbf{u}|$ is of order r^{-3} and r^{-2} for r large in three and two dimensions respectively (leaving aside any circulation), and although no logarithmic divergence occurs the value of the integral is found to depend on the shape of the outer boundary whose linear dimensions are made large. The difficulty is that the 'doublet' motion far from the body which is set up by pressure gradients in the fluid when the body is made to accelerate contains indefinitely large amounts of momentum both in the forward-moving fluid and in the backward-moving fluid. However, we can say that the rate at which the body communicates linear momentum to the fluid is $-\mathbf{G}$, and hence that the (i-component of the) total amount of momentum given to the fluid while the body velocity is brought from zero to **U** is

$$-\int G_i \, dt, \quad = \rho V_0 \alpha_{ij} U_j. \tag{6.4.29}$$

It is immaterial whether this momentum is given to the fluid gradually or instantaneously, and $\rho V_0 \alpha_{ij} U_j$ is sometimes referred to as the *fluid impulse*, meaning the impulse which must be given to the rigid body in order to generate from rest the irrotational flow due to motion of the body with translational velocity **U**.

It will be noted that when the velocity of the body is periodic in time, the mean value of **U**.**G** over one cycle is zero, showing that no net work is done by the body against the acceleration reaction in irrotational flow, as observed in § 5.13 from an energy argument.

The force on a body in accelerating fluid

Some of the preceding formulae can usefully be extended to allow for an acceleration of the fluid through which the body moves. We suppose that the mass of fluid which surrounds the body has a uniform acceleration **f** relative to a Newtonian frame of reference. It is then convenient to choose moving axes such that the velocity of the fluid far from the body (or, equivalently, the velocity of the fluid in the absence of the body) is, and remains, zero. The equation of motion of the fluid relative to these accelerating axes must include a fictitious uniform body force $-\mathbf{f}$ per unit mass. There is consequently an additional contribution $-\rho\mathbf{f}.\mathbf{x}$ to the pressure, and an additional contribution $\rho V_0\mathbf{f}$ (an effective 'buoyancy' force, analogous to the force on the body arising from the action of gravity on the fluid) to the total force on the body.

If now we suppose that the body is in translational motion with instantaneous velocity $\mathbf{U}(t)$ relative to a Newtonian frame of reference, the pattern of streamlines of flow relative to the body will be unaffected by the acceleration of the fluid but the resultant force on the body will be affected in several ways. Firstly, the expression for the side-force arising from any circulation round the body (see (6.4.26)) must be modified by the replacement of U by the (magnitude of the) velocity of the body relative to that of the fluid at infinity. Secondly, the acceleration of the body relative to the fluid at infinity is $\dot{\mathbf{U}}-\mathbf{f}$, so that the acceleration reaction (6.4.28) becomes

$$-\rho V_0\alpha_{ij}(\dot{U}_j-f_j).$$

Thirdly there is the new contribution $\rho V_0\mathbf{f}$ mentioned above. Thus, on leaving aside any side-force due to circulation and any buoyancy force due to gravity, the i-component of the force on the body becomes

$$-\rho V_0\alpha_{ij}\dot{U}_j+\rho V_0 f_j(\alpha_{ij}+\delta_{ij}). \tag{6.4.30}$$

6.5. Use of the complex potential for irrotational flow in two dimensions

It was remarked in §2.7 that the velocity potential ϕ and the stream function ψ relating to two-dimensional irrotational flow of an incompressible fluid have some striking conjugate properties. These properties are summarized in the statement that the complex potential $w, = \phi+i\psi$, is an *analytic* function of $z, = x+iy$, in the region of the z-plane occupied by the flow, meaning that w has a unique derivative with respect to z at all points of that region. Conversely, any analytic function of z can be regarded as the complex potential of a certain flow field. Thus, simply by choosing different mathematical forms of $w(z)$, we obtain possible forms of the functions ϕ and ψ, although it may not happen that the flow fields that they represent are physically interesting. A more direct way of determining irrotational flow fields is provided by the method of conformal transformation of functions of a complex variable. We shall illustrate in this section these indirect and direct procedures and other uses of the complex potential.

It will be useful as a preliminary to notice the form taken by w in the case of the following simple irrotational flow fields whose description in terms of ϕ or ψ is already known.

Uniform flow with velocity (U, V): $w = (U - iV)z$

Simple source of strength m at the point z_0 (§2.5):
$$w = \frac{m}{2\pi} \log(z - z_0)$$

Source doublet of strength μ with direction parallel to the x-axis at z_0:
$$w = -\frac{\mu}{2\pi(z - z_0)}$$

The same, with direction parallel to the y-axis:
$$w = -\frac{i\mu}{2\pi(z - z_0)}$$

Point vortex of strength κ at z_0 (§2.6):
$$w = -\frac{i\kappa}{2\pi} \log(z - z_0)$$

Vortex doublet of strength λ with direction parallel to the x-axis at z_0:
$$w = \frac{i\lambda}{2\pi(z - z_0)}$$

Flow due to a circular cylinder of radius a moving with velocity (U, V) and circulation κ round it, centre instantaneously at z_0 (§2.10):
$$w = -\frac{i\kappa}{2\pi} \log(z - z_0) - \frac{a^2(U + iV)}{z - z_0}$$

Arbitrary flow outside a circle, centre at z_0, which encloses all boundaries, in fluid at rest at infinity (Laurent series, §2.10):
$$w = \frac{m - i\kappa}{2\pi} \log(z - z_0) + \sum_{n=0}^{\infty} A_n(z - z_0)^{-n}$$

Flow near a stagnation point at the origin (§2.7):
$$w = \tfrac{1}{2}kz^2.$$

Flow fields obtained by special choice of the function $w(z)$

Perhaps the simplest mathematical form for w is

$$w(z) = Az^n, \tag{6.5.1}$$

where A and n are real constants. If r, θ are polar co-ordinates in the z-plane, we have $z = r e^{i\theta}$ and

$$\phi = Ar^n \cos n\theta, \quad \psi = Ar^n \sin n\theta. \tag{6.5.2}$$

The physical interest of a mathematical solution for irrotational flow usually depends on whether it satisfies boundary conditions likely to occur in practice. The commonest type of boundary condition is that the flux of volume across each element of a given surface is zero, either because there are certain symmetry properties across the surface (as when two similar jets of water meet at a plane of symmetry) or because the surface is the boundary of a rigid body (in which case we must check that the distribution of velocity along the rigid surface determined as a part of an irrotational flow field is indeed such as not to cause separation of the boundary layer at the surface in a real fluid). At a stationary 'zero-flux' boundary the normal component of velocity is zero; and in the present context of two-dimensional flow a

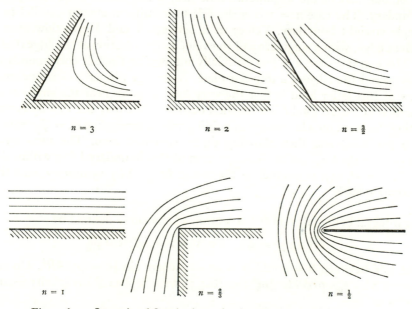

Figure 6.5.1. Irrotational flow in the region between two straight zero-flux boundaries intersecting at an angle π/n.

boundary is a curve in the (x, y)-plane. This condition is satisfied on any streamline of the flow, so that we are free to regard any one of the family of streamlines given by (6.5.2) as a stationary zero-flux boundary. Zero-flux boundaries of simple geometrical form occur more often in practice, and zero-flux plane boundaries are the most common. We are therefore led to look in particular for any straight lines among the family of streamlines.

The expression for ψ in (6.5.2) is constant, and equal to zero, for all r when $\theta = 0$ and when $\theta = \pi/n$. Consequently (6.5.1) and (6.5.2) provide a representation of irrotational flow in the region between two straight zero-flux boundaries intersecting at an angle π/n. Different choices of n give particular cases, some of which have interesting features (figure 6.5.1). There is clearly a marked change in the character of the flow near the intersection as n decreases through unity, since

$$q = \left| \frac{dw}{dz} \right| = |nA| r^{n-1} \qquad (6.5.3)$$

and, as $r \to 0$,

$$q \to 0, \ |A| \text{ or } \infty, \quad \text{according as} \quad n > 1, = 1 \text{ or } < 1.$$

For $n > 1$ the straight boundaries include an angle less than π; and for $n = 2$ we have flow in a region bounded by a right angle, with streamlines in the form of rectangular hyperbolae, which has already been seen (in §2.7) to be one-half of the irrotational flow near a stagnation point at a plane boundary. The case $n = 1$ corresponds to a uniform stream parallel to a single straight boundary. Values of n between 1 and $\frac{1}{2}$ give flow over a salient edge, with a singularity of the velocity distribution at the edge itself.

The extreme case $n = \frac{1}{2}$ is especially interesting since it corresponds to flow round the edge of a thin flat plate. A strange property of irrotational flow of this latter kind is that the very low pressure near the sharp edge exerts a non-zero total force on the boundary. We may see this by calculating the force on a boundary coinciding with a streamline $\psi = \psi_0$ (which is a parabola) and then allowing ψ_0 to approach zero. The total force exerted by the fluid on the finite portion of this boundary lying within the circle $r = R$, say, is parallel to the x-axis ($\theta = 0$) by symmetry, and the x-component is

$$F_x = \int p \, dy,$$

the integral being taken over the section of the curve defined by

$$A r^{\frac{1}{2}} \sin \tfrac{1}{2}\theta = \psi_0, \quad \text{i.e. } y = 2(\psi_0/A)^2 \cot \tfrac{1}{2}\theta,$$

that lies between $\theta = \epsilon$ and $\theta = 2\pi - \epsilon$, where $\sin \tfrac{1}{2}\epsilon = \psi_0/AR^{\frac{1}{2}}$. On replacing p by $p_0 - \rho \, \partial\phi/\partial t - \tfrac{1}{2}\rho q^2$ (see (6.2.5)), with neglect of the gravity term), we have

$$F_x = \int_\epsilon^{2\pi - \epsilon} \left(p_0 - \rho \frac{dA}{dt} \frac{\psi_0}{A} \cot \tfrac{1}{2}\theta - \rho \frac{A^4}{8\psi_0^2} \sin^2 \tfrac{1}{2}\theta \right) \frac{\psi_0^2}{A^2} \operatorname{cosec}^2 \tfrac{1}{2}\theta \, d\theta$$

and

$$F_x \to -\tfrac{1}{4}\pi\rho A^2 \qquad (6.5.4)$$

as $\psi_0 \to 0$. The limiting value of F_x is independent of R and represents a suction force concentrated at the sharp edge and parallel to the plate. This non-zero force on the sharp edge is not of direct practical significance since a real fluid in steady motion would separate from the edge and the very low pressure near the edge would not occur; however, a clear understanding of irrotational flow in the presence of sharp-edged plates is indirectly useful in view of the possibility of transforming a flow field with a sharp-edged boundary to a flow field with a boundary of different shape.

All these particular cases of irrotational flow in the region between two straight intersecting zero-flux boundaries are given greater generality by the fact that they hold in the neighbourhood of the intersection of two straight zero-flux boundaries of finite length irrespective of the form of the remainder of the flow. The proof is given later in this section. Thus, in any irrotational flow, the velocity at a point on a zero-flux boundary where there is a discontinuity in the direction of the tangent to the boundary is zero when the angle on the fluid side is less than π, and is infinite when it is greater than π. In the former case the surface streamline in a real fluid would separate (in steady flow, at any rate) from a rigid boundary *before* reaching the discontinuity in view of the deceleration preceding the stagnation point, giving rise to a standing eddy in the corner; and in the latter case it would separate *at* the discontinuity of a rigid boundary, unless the change in direction of the tangent is small.

When $\frac{1}{2} > n > -\frac{1}{2}$, the angle between the two straight intersecting streamlines on which $\psi = 0$ is larger than 2π and it is no longer possible for both of those streamlines to be regarded as zero-flux boundaries in a flow field; and when $n < 0$ both ϕ and ψ become infinite as $r \rightarrow 0$. For these values of n there is less scope for the finding of interesting flow fields by regarding certain streamlines as zero-flux boundaries.

Trigonometric functions may also be tried in an indirect search for flow fields. Suppose, for instance, that we begin with the relation

$$w(z) = A \sin kz, \tag{6.5.5}$$

where A and k are real constants. The corresponding expressions for ϕ and ψ are

$$\phi = A \sin kx \cosh ky, \quad \psi = A \cos kx \sinh ky, \tag{6.5.6}$$

representing a flow field with periodicity in the x-direction. The indefinite increase of the velocity as $y \rightarrow \mp \infty$ is a handicap to practical application of this form for $w(z)$. However, we can obtain a flow field in which the velocity tends to zero in one direction by superposing two solutions of the above form; thus, for

$$w(z) = A(\sin kz + i \cos kz),$$

we have

$$\phi = A e^{ky} \sin kx, \quad \psi = A e^{ky} \cos kx. \tag{6.5.7}$$

These expressions are familiar in the theory of surface waves, and are known to describe the instantaneous motion of a semi-infinite fluid with a free surface which has the equilibrium position $y = 0$ and over which a sinusoidal wave of wave-length $2\pi/k$ and small amplitude is progressing as a result of the action of gravity and/or surface tension.

Conformal transformation of the plane of flow

If the complex variable $\zeta, = \xi + i\eta$, is an analytic function of $z, = x + iy$, given by $\zeta = F(z)$, there is a connection between the shape of a curve in the z-plane and the shape of the curve traced out by the corresponding set of points in the ζ-plane. This connection is a consequence of the defining property of an analytic function of z, viz. that the value of the derivative $\lim_{\delta z \to 0} (\delta\zeta/\delta z)$ is independent of the way in which the increments δx and δy separately tend to zero. For suppose that $\delta z'$ and $\delta z''$ are two different small increments in z and that $\delta\zeta'$ and $\delta\zeta''$ are the corresponding increments in ζ, that is to say, $\zeta + \delta\zeta' = F(z + \delta z'), \quad \zeta + \delta\zeta'' = F(z + \delta z'').$

The two short straight lines joining the points $z + \delta z'$ and $z + \delta z''$ to the point z in the z-plane have lengths whose ratio is $|\delta z'/\delta z''|$ and intersect at an angle

$$\arg \delta z' - \arg \delta z'' = \arg(\delta z'/\delta z'')$$

(where, as is customary in complex variable theory, $\arg z$ denotes the angle whose tangent is equal to the ratio of the imaginary and real parts of z). The two short straight lines joining the corresponding points $\zeta + \delta\zeta'$ and $\zeta + \delta\zeta''$ to the point ζ in the ζ-plane have lengths whose ratio is $|\delta\zeta'/\delta\zeta''|$ and intersect at an angle $\arg(\delta\zeta'/\delta\zeta'')$. But

$$\delta\zeta' = \delta z' \frac{d\zeta}{dz} + O(\delta z'^2), \quad \delta\zeta'' = \delta z'' \frac{d\zeta}{dz} + O(\delta z''^2),$$

so that to the first order in the small increments the length ratios $|\delta z'/\delta z''|$ and $|\delta\zeta'/\delta\zeta''|$ are equal and the angles $\arg(\delta z'/\delta z'')$ and $\arg(\delta\zeta'/\delta\zeta'')$ are equal.

Thus to a closed curve of small linear dimensions in the z-plane there corresponds a closed curve of small linear dimensions *of the same shape* (to the first order in the linear dimensions) in the ζ-plane. The two infinitesimal figures in general have different orientations and different sizes, but are similar. This kind of transformation from the z-plane to the ζ-plane, by means of an analytic relation between the two complex variables, is said to be a *conformal* transformation. There may of course be a difference between the shapes of two corresponding figures in the z- and ζ-planes of finite linear dimensions, but, if we imagine one of these figures to be divided up into a large number of figures of small linear dimensions, the corresponding set of small, approximately similar, figures in the other plane will compose the corresponding figure of finite size.

The relation between the sizes of two small corresponding figures in the z- and ζ-planes depends on the form of the function F. Any short straight line in the z-plane transforms into a short straight line in the ζ-plane of length greater in the ratio $|d\zeta/dz|$, and the magnification of area of a small figure in a transformation from the z-plane to the ζ-plane is thus $|d\zeta/dz|^2$. At any points of the z-plane where $d\zeta/dz$ is either zero or infinite the above remarks clearly do not apply; these are singular points of the transformation at which the representation is not conformal.

These properties of a conformal transformation are relevant to the theory of irrotational flow in two dimensions. If $w(z)$ is the complex potential of an irrotational flow in a certain region of the z-plane, and if z is an analytic function of another complex variable given by $z = f(\zeta)$, then w can also be regarded as an analytic function of ζ; for the incremental ratio from which the derivative of w with respect to ζ is obtained can be written as

$$\frac{\delta w}{\delta \zeta} = \frac{\delta w}{\delta z} \frac{\delta z}{\delta \zeta},$$

and both of the factors on the right-hand side tend to a unique limit as δx and δy, or $\delta \xi$ and $\delta \eta$, tend independently to zero. $w\{f(\zeta)\}$ is thus the complex potential of an irrotational flow in a certain region of the ζ-plane, and the flow in the z-plane is said to have been 'transformed' into flow in the ζ-plane. The families of equipotential lines and streamlines in the z-plane given by $\phi(x, y) = $ const. and $\psi(x, y) = $ const. transform into families of curves in the ζ-plane on which ϕ and ψ are constant and which are equipotential lines and streamlines of the flow in the ζ-plane, the two families being orthogonal in the ζ-plane, as in the z-plane, except at singular points of the transformation. The velocity components at a point of the flow in the ζ-plane are given (see (2.7.13)) by

$$u_\xi - iu_\eta = \frac{dw}{d\zeta} = \frac{dw}{dz} \frac{dz}{d\zeta}. \tag{6.5.8}$$

This shows incidentally that the magnitude of the velocity is changed, in the transformation from the z-plane to the ζ-plane, by the reciprocal of the factor by which linear dimensions of small figures are changed; thus the kinetic energy of the fluid contained within a closed curve (whether of small linear dimensions or not) in the z-plane is equal to the kinetic energy of the corresponding flow in the region enclosed by the corresponding curve in the ζ-plane.

In some flow fields the motion of the fluid is 'caused' by (or, properly speaking, associated with) the presence of source or vortex point singularities of the kind described in chapter 2. We may be given the information that singularities of specified type and strength are located (instantaneously) at certain points in the z-plane, and the problem is to determine the irrotational flow that is compatible with these singularities and a boundary of given shape. It is necessary therefore to consider the relation between the flow near a singularity in the z-plane and that near the corresponding point in the ζ-plane. Now near a source or vortex singularity ϕ or ψ takes very large values and is dominated by the contribution from that singularity. That is to say, if there is a simple source singularity of strength m at the point $z = z_0$ and a simple vortex singularity of strength κ at the same point (and other more complicated singularities can be built up from a simple source and a simple vortex in the manner described in §§2.5, 2.6), we have

$$w(z) \sim \frac{m - i\kappa}{2\pi} \log(z - z_0) \tag{6.5.9}$$

near $z = z_0$, irrespective of the nature of the flow elsewhere. But, provided

$z = z_0$ is a non-singular point of the transformation represented by $\zeta = F(z)$, we also have

$$\zeta - \zeta_0 \sim (z - z_0)(d\zeta/dz)_{z_0}$$

near $z = z_0$, so that (6.5.9) can be written as

$$w(\zeta) \sim \frac{m - i\kappa}{2\pi} \log(\zeta - \zeta_0)$$

near the point $\zeta = \zeta_0$ in the ζ-plane corresponding to $z = z_0$. Thus the irrotational flow in the ζ-plane has a similar singularity, located at the corresponding point and with the same source strength and vortex strength; a point source in the z-plane may be said to transform into an identical source in the ζ-plane, and likewise for a point vortex.

Corresponding results may be obtained for more complex point singularities by noting the way they are made up of either simple sources or simple vortices. The two simple sources which together can be regarded as composing a source doublet in the z-plane transform into identical sources at corresponding neighbouring points in the ζ-plane, and, since the infinitesimal distance between the two points is magnified by the factor $|d\zeta/dz|$ as a result of the transformation, the strength of the source doublet is changed in magnitude by this factor (and may also be changed in direction). Evidently a multipole point singularity made up of 2^n simple sources or vortices transforms to a singularity of the same type at the corresponding point in the ζ-plane, with a strength changed in magnitude by a factor $|d\zeta/dz|^n$.

The result that a simple source or vortex in the z-plane corresponds to an identical source or vortex in the ζ-plane can be regarded alternatively as a consequence of the fact that, if w is many-valued at points in the z-plane, as it may be when the presence of interior boundaries or singularities renders the region of irrotational flow multiply-connected, w is similarly many-valued at corresponding points in the ζ-plane. As a point in the z-plane is taken round an irreducible closed curve, say in a doubly-connected region of irrotational flow in the z-plane, the value of ϕ varies and, when the point returns to its initial position, is greater by an amount equal to the cyclic constant κ; likewise the value of ψ increases by an amount equal to the net volume flux m across the closed curve. Exactly the same changes must take place in ϕ and ψ as the corresponding curve in the ζ-plane is traced out (provided there is a one-to-one correspondence between points in the relevant regions of the z- and ζ-planes).

The usefulness of conformal transformation as a technique in irrotational flow theory lies in the possibility of transforming a given flow field of unknown form into a flow field which is easier to determine. The difficulty of finding ϕ or ψ for a given flow field depends greatly on the geometrical form of the boundaries at which certain conditions must be satisfied. If the boundary is an infinite straight line or a circle, many standard methods for determining ϕ or ψ are available; for a complicated shape of the boundary, it may be that no direct method of solution (other than a numerical one involving use of a computing machine) is known. Thus conformal transformation can make a problem of irrotational flow tractable by converting an awkwardly shaped boundary into one of the simpler forms. The procedure depends to some extent on the nature of the given flow field, and later we shall consider two main types of transformation. The transformation may also affect the condition to be applied at the boundary, as well as the shape of the boundary. In many common cases, the condition to be applied at the boundary in the original or given flow field is that the velocity normal to it is everywhere zero, that is, that ψ is constant on the boundary. Another possibility is that a rigid body immersed in the fluid in the original flow field has a prescribed instantaneous motion, say an angular velocity Ω and velocity components (U, V) of the centre of volume of the body which is instantaneously at position (x_0, y_0). In this case the condition is (see (6.4.7))

$$\mathbf{n} \cdot \nabla\phi = \frac{\partial \psi}{\partial s} = n_1 U + n_2 V + \Omega\{(x - x_0) n_2 - (y - y_0) n_1\} \qquad (6.5.10)$$

at the boundary, where s denotes distance along the boundary curve (in the anti-clockwise sense) and (n_1, n_2) are the components of the unit outward normal to the boundary. Then since

$$n_1 = \partial y / \partial s, \quad n_2 = -\partial x / \partial s,$$

the condition may be written as

$$\psi - \psi_0 = Uy - Vx - \tfrac{1}{2}\Omega\{(x - x_0)^2 + (y - y_0)^2\} \qquad (6.5.11)$$

at the boundary. This may be converted into a relation between ψ and ξ and η on the boundary in the ζ-plane when the form of the transformation between z and ζ is known.

We conclude with two remarks of an incidental nature. The first is that the technique of conformal transformation renders solutions for irrotational flow fields associated with simple boundary shapes such as a circle and an ellipse potentially useful in a new way. One irrotational flow field may be transformed to another by means of an analytic relation between the two complex co-ordinates z and ζ, and, although the occurrence of boundary-layer separation may make it impossible to realize the first flow field in practice, the second may be quite realistic; the unrealistic first flow field may thus be useful as a mathematical stepping stone to irrotational flow fields of direct physical interest. The second is that the use of conformal transformation as a working tool has its tricks and difficulties, and demands practice on more examples than will be described here.†

† Numerous worked examples will be found in text-books concerned primarily with inviscid fluids, such as *Theoretical Hydrodynamics*, by L. M. Milne-Thomson, 5th ed. (Macmillan, 1967).

Transformation of a boundary into an infinite straight line

In cases of irrotational flow with an exterior zero-flux boundary, or with an interior zero-flux boundary and stationary fluid at infinity, it is often convenient to transform the flow field in such a way that the boundary becomes an infinite straight line and the region of flow becomes a half-plane. Suppose, for instance, that we wish to determine the irrotational flow in the region of the z-plane bounded by two straight walls intersecting at angle π/n; there may also be a boundary and a motion-producing agency at a great distance from the point of intersection, but these need not be specified. The transformation $\zeta = z^n$ 'opens up' the region $0 < \theta < \pi/n$ between the two intersecting walls of the z-plane into the upper half ($\eta > 0$) of the ζ-plane (note the singularity of the transformation at the point of intersection, where an angle π/n is transformed into an angle π), and the corresponding flow in the ζ-plane can be determined immediately. The only possible irrotational flow in the upper half of the ζ-plane due to a distant agency, with the implication that any non-uniformity of the flow near $\zeta = 0$ is due solely to non-uniformity of the boundary, is a uniform stream parallel to the boundary at $\eta = 0$ described by

$$w = A\zeta,$$

where A is a real constant; the required irrotational flow in the z-plane is then

$$w = Az^n, \tag{6.5.12}$$

as already discovered by an indirect argument.

Cases of irrotational flow in a region bounded externally by a closed polygon, of which one or more vertices may be at infinity, may always be handled by means of the Schwarz–Christoffel theorem, which states that a polygonal boundary in the z-plane with interior angles $\alpha, \beta, \gamma, \ldots$ is mapped on to the real axis $\eta = 0$ of the ζ-plane by the transformation given by

$$\frac{d\zeta}{dz} = K(\zeta - a)^{1 - \frac{\alpha}{\pi}}(\zeta - b)^{1 - \frac{\beta}{\pi}}(c - \zeta)^{1 - \frac{\gamma}{\pi}}\ldots, \tag{6.5.13}$$

where K is a constant and $a, b, c, \ldots$ are the (real) values of ζ corresponding to the vertices of the polygon. The corresponding region of flow in the ζ-plane is the half-plane $\eta > 0$, and again the expression for the complex potential in terms of ζ may be written down when information about the motion-producing agency is provided.

If one vertex of the polygon is at infinity, the associated interior angle, α say, is zero; in such a case we may, without loss of generality, take the corresponding point $\zeta = a$ in the ζ-plane to be at infinity, so that the factor $(\zeta - a)$ in (6.5.13) is effectively constant and may be regarded as being absorbed in the new constant K'. For example, a semi-infinite strip in the z-plane ($\alpha = 0, \beta = \frac{1}{2}\pi, \gamma = \frac{1}{2}\pi$) is mapped on to the upper half of the ζ-plane by the transformation

$$\frac{d\zeta}{dz} = K'(\zeta-b)^{\frac{1}{2}}(\zeta-c)^{\frac{1}{2}},$$

i.e.
$$\zeta = \tfrac{1}{2}(b+c) + \tfrac{1}{2}(b-c)\cosh\{K'(z-z_0)\}. \tag{6.5.14}$$

The points $\zeta = b$, $\zeta = c$ in the ζ-plane correspond to the vertices $z = z_0$, $z = z_0 + i\pi/K'$ respectively in the z-plane, where the constants z_0 and K' can be determined in terms of the location and width of the given semi-infinite strip; the constants b and c control the position on the real axis and the degree of magnification of a line element in the ζ-plane corresponding to an element of the boundary of the strip in the z-plane, and may be chosen freely. The general case in which two vertices of the polygon in the z-plane are at infinity is more difficult, although if these are the only vertices, giving an infinite strip, the required transformation follows either directly from

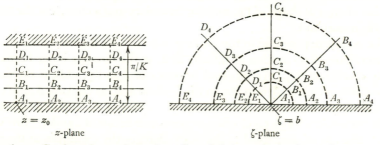

Figure 6.5.2. Conformal transformation of an infinite strip in the z-plane on to the upper half of the ζ-plane by the relation $\zeta = b + e^{K(z-z_0)}$, where K is real and positive.

(6.5.13) by putting $\alpha = \beta = 0$, and $(\zeta-a)/a \to -1$, $Ka \to -K'$ as before, or indirectly by putting

$$\mathscr{R}(K'z_0) \to -\infty, \quad b-c \to 0, \quad \tfrac{1}{4}(b-c)e^{-K'z_0} \to e^{-K'z_0'}$$

in (6.5.14); either way, we obtain

$$\zeta = b + e^{K'(z-z_0')}. \tag{6.5.15}$$

The width of the strip is $|\pi/K'|$ as before. The correspondence between certain lines in the z- and ζ-planes for this simple and useful transformation is shown in figure 6.5.2 (with omission of the primes).

    These transformations of a flow field with an exterior polygonal boundary are not of much direct practical value, since polygonal boundaries are unusual, but they are often useful as one of a sequence of transformations. The discussion of some·interesting flow fields involving 'free streamlines', to be given in §6.13, will illustrate their use in this way.

Transformation of a closed boundary into a circle

A different way of proceeding is to seek a transformation which will convert the region outside a given closed boundary curve in the z-plane into the region outside a circle in the ζ-plane. The most important application of this procedure is to cases of flow due to a rigid cylinder moving through fluid at rest at infinity, and the method will be described in terms of this application. Since the general purpose is to obtain a new flow system which can be determined easily, it is desirable to use a transformation which converts the simple motion at infinity in the z-plane into an equally simple motion in a part of the ζ-plane; the obvious plan is to choose an analytic relation $\zeta = F(z)$ of such a form that

$$\zeta \sim z \quad \text{as} \quad |z| \to \infty \qquad (6.5.16)$$

so that the fluid also extends to infinity in the ζ-plane and has the same motion there as in the z-plane. The primary consequence of the transformation is a change of shape of the interior boundary and a change of the flow in its neighbourhood.

The complete potential in the ζ-plane is then to be determined subject to the condition of no motion at infinity and, at the circular inner boundary, to the condition (6.5.11) expressed in terms of ξ and η; also, if the flow in the z-plane is cyclic, it is cyclic in the ζ-plane and, as explained above, has the same cyclic constant κ. The chance of success of the method thus turns on the form taken by (6.5.11) when expressed in terms of ξ and η. Now when $\Omega = 0$, the use of axes fixed in the body changes the condition at the inner boundary in the z-plane to $\psi = \psi_0$ (const.) and at the outer boundary to

$$w(z) \sim -(U-iV)z \quad \text{as} \quad |z| \to \infty.$$

The corresponding conditions in the ζ-plane are $\psi = \psi_0$ at the circular inner boundary and

$$w(\zeta) \sim -(U-iV)\zeta \quad \text{as} \quad |\zeta| \to \infty;$$

in other words, the flow in the ζ-plane is that due to a circular cylinder held in a stream of uniform velocity $(-U, -V)$ at infinity and with circulation κ round it, for which the complex potential is known. If we wished we could now use axes in the ζ-plane which reduce the fluid at infinity to rest, in which case the inner boundary condition would become

$$\psi - \psi_0 = U\eta - V\xi \qquad (6.5.17)$$

at the circular boundary, corresponding to translation of the circular cylinder in the ζ-plane with velocity (U, V). However, it should be noted that the two flows with a common complex potential are those relative to axes fixed in the inner boundaries and not those relative to axes fixed in the fluid at infinity. Only when the inner boundary is a streamline do the flows in the two planes correspond; the correspondence is lost when the motion is referred to other axes because a uniform stream in one plane corresponds to a flow in the other plane which is uniform only at infinity.

Thus, for any transformation satisfying (6.5.16), the flow in the ζ-plane that corresponds to a body held in a stream of uniform velocity $(-U, -V)$ at infinity in the z-plane is that due to the transformed body held in a stream of the same uniform velocity at infinity. The same argument and result do not apply to rotation of the body in the z-plane because the choice of axes fixed in the body then confers a rotational motion on the fluid. Investigation of the flow due to a rotating cylinder needs less direct and more specialized procedures,† and we shall concentrate on the case of translation alone.

The method of determining flow due to a cylinder of given shape in translational motion consequently requires a knowledge of (a) the complex potential for a circular cylinder held in a stream of specified velocity and with specified circulation, and (b) an analytic relation between z and ζ such that the cylinder boundary in the z-plane corresponds to a circle in the ζ-plane and which satisfies (6.5.16). With regard to (a), we need only recall results obtained in § 2.10. From (2.10.12), the single-valued part of ϕ for flow in the ζ-plane due to a circular cylinder of radius c held in a stream of uniform velocity $(-U, -V)$ at infinity and with centre at the point ζ_0 (a generalization which will be needed later) is

$$-\{U(\xi-\xi_0) + V(\eta-\eta_0)\}\left\{1 + \frac{c^2}{(\xi-\xi_0)^2 + (\eta-\eta_0)^2}\right\}.$$

The effect of the circulation is to add a term $(\kappa/2\pi)\tan^{-1}\left(\dfrac{\eta-\eta_0}{\xi-\xi_0}\right)$ to ϕ (see (2.10.15)). The corresponding complex potential, which is an analytic function of ζ with ϕ as its real part, is

$$w(\zeta) = -(U-iV)(\zeta-\zeta_0) - (U+iV)\frac{c^2}{\zeta-\zeta_0} - \frac{i\kappa}{2\pi}\log\frac{\zeta-\zeta_0}{c}; \quad (6.5.18)$$

this makes the value of ψ at the inner boundary zero.

With regard to (b), the details depend on the given shape of the cylinder but one general remark about the transformation may be made. Since the relation between z and ζ is analytic everywhere in the region of the ζ-plane outside the circle of radius c, and satisfies (6.5.16) at large distances from the origin, we may write z as a Laurent series

$$z = \zeta + \sum_{n=1}^{\infty} \frac{B_n}{\zeta^n}, \quad (6.5.19)$$

in which the (complex) coefficients $B_1, B_2, \ldots$ depend on the shape of the cylinder. By omitting a constant term B_0 we are choosing the origin of the ζ-plane so that, when the two planes are laid over one another with

† For which reference may be made to *Hydrodynamics*, by H. Lamb.

corresponding points at infinity coinciding, it coincides with the origin of the z-plane; in these circumstances the position of the centre of the circular boundary in the ζ-plane is not disposable. The series (6.5.19) may be inverted to give a series

$$\zeta = z - \frac{B_1}{z} + \sum_{n=2}^{\infty} \frac{B'_n}{z^n},$$ (6.5.20)

valid for sufficiently large values of $|z|$, where the coefficients $B'_n (n \geqslant 2)$ differ from B_n; this is the general form taken by the analytic relation $\zeta = F(z)$ at sufficiently large values of z. The complex potential $w(z)$ for the flow in the z-plane due to the given cylinder held in a stream is obtained by substituting $\zeta = F(z)$ in (6.5.18), and it appears that at sufficiently large distances from the cylinder $w(z)$ can be written (compare (2.10.7)) as

$$w(z) = -(U-iV)z - \frac{i\kappa}{2\pi} \log \frac{z}{c} + \sum_{n=0}^{\infty} \frac{A_n}{z^n},$$ (6.5.21)

where
$$A_0 = (U-iV)\zeta_0, \quad A_1 = B_1(U-iV) - c^2(U+iV) + \frac{i\kappa\zeta_0}{2\pi}.$$ (6.5.22)

Examples of the use of this method of conformal transformation for the determination of the flow due to moving bodies will occur in the next two sections.

The circle theorem

A knowledge of the analytic relation which transforms the region outside a closed curve of given shape into the region outside a circle is also useful, when supplemented by a general result now to be established, in cases in which flow singularities exist in the fluid outside the closed boundary.

The following result, known as the circle theorem (Milne-Thomson 1940) concerns the complex potential representing motion of fluid of infinite extent in the presence of a single internal boundary of circular form. Suppose first that in the absence of the circular cylinder the complex potential is

$$w = f(z)$$

and that $f(z)$ is free from singularities in the region $|z| \leqslant a$, where a is a real length. If now a stationary circular cylinder of radius a and centre at the origin bounds the fluid internally, the flow is modified; to every singularity of $f(z)$ there will correspond an 'image' in the circular boundary, such that the flow due to the singularity and its image together has the circle $|z| = a$ as a streamline. We obtain a general expression for the entire image system by noticing that, on the circle $|z| = a$,

$$a^2 = z\bar{z},$$

where the overbar denotes a complex conjugate, so that

$$f(z) + \bar{f}(a^2/z)$$ (6.5.23)

is purely real on $|z| = a$. A complex potential of the form (6.5.23) thus has $|z| = a$ as a streamline; and it has the same singularities outside $|z| = a$ as $f(z)$, since if z lies outside $|z| = a$, a^2/z lies in the region inside this circle where $f(z)$ is known to be free from singularities. Consequently the additional term $\bar{f}(a^2/z)$ in (6.5.23) represents fully the modification to the complex potential due to the presence of the circular cylinder. It should be noted that the complex potentials considered, both in the absence and in the presence of the circular cylinder, refer to flow relative to axes such that the cylinder is stationary.

The simplest possible application of the circle theorem is to the case of a circular cylinder held fixed in a stream whose velocity at infinity is uniform with components $(-U, -V)$. In the absence of the cylinder the complex potential is $-(U-iV)z$, and the circle theorem shows that, with the cylinder present,

$$w = -(U-iV)z - (U+iV)a^2/z,$$

as already known. Another simple case, not as readily treated in other ways, is flow due to a point vortex of strength κ at the point $z = z_0$, for which the complex potential is $-(i\kappa/2\pi)\log(z-z_0)$. In the presence of a circular cylinder of radius a $(< |z_0|)$, the complex potential becomes

$$w = -\frac{i\kappa}{2\pi}\log\left\{\frac{(z-z_0)z}{a^2 - z\bar{z}_0}\right\},$$

showing that the image system consists of a point vortex of strength κ at the origin and one of strength $-\kappa$ at the point $z = a^2/\bar{z}_0$ inverse to the position of the original vortex.

If the internal boundary of the fluid is not circular, conformal transformation of the region external to the boundary into the region external to a circle, in the manner previously described, gives a new problem of flow in the presence of a circular boundary with a new set of flow singularities. The correspondence between flow singularities in two planes related by a conformal transformation has been described earlier in this section, so that the determination of the complex potential is reduced to an application of the circle theorem.

6.6. Two-dimensional irrotational flow due to a moving cylinder with circulation

One of the purposes of this section is to examine the effect of the circulation round the cylinder on the flow field. We saw in § 6.4 that one important effect of the circulation, in combination with translation of the cylinder, is to produce a side-force on the cylinder. It will be useful to look closely into the details of one or two particular flow fields in order to see clearly the origin of this side-force. We begin with a discussion of the simple and fundamental case of a circular cylinder, for which formulae describing the flow field are already known.

A circular cylinder

The instantaneous motion of a rigid circular cylinder is specified fully by the velocity of its centre and the angular velocity of the cylinder; the latter motion has no effect on an inviscid fluid and will be ignored. We shall be concerned with the form of streamlines, the significance of which is most easily seen when the motion is referred to axes fixed in the cylinder, chiefly because the cylinder surface is itself a streamline then. Thus the velocity potential and stream function required are those describing irrotational flow due to a circular cylinder of radius a held in a stream with uniform velocity $(-U, -V)$ far from the cylinder, and with circulation κ round the cylinder. The appropriate complex potential is known (see (6.5.18)) to be

$$w(z) = -(U-iV)z - (U+iV)\frac{a^2}{z} - \frac{i\kappa}{2\pi}\log\frac{z}{a}. \tag{6.6.1}$$

In this case of a body with circular symmetry, no greater generality is obtained, so far as the instantaneous motion is concerned, by allowing V to be non-zero, and we therefore suppose the relative motion of the cylinder and the fluid at infinity to be parallel to the x-axis. The velocity potential and stream function are then

$$\phi = -U\left(r+\frac{a^2}{r}\right)\cos\theta + \frac{\kappa\theta}{2\pi}, \tag{6.6.2}$$

$$\psi = -U\left(r-\frac{a^2}{r}\right)\sin\theta - \frac{\kappa}{2\pi}\log\frac{r}{a}, \tag{6.6.3}$$

where $r^2 = x^2+y^2$ and $\theta = \tan^{-1}y/x$. There is a singly-infinite family of different flow fields corresponding to different values of κ/aU. The pattern of streamlines for the case $\kappa/aU = 0$, which is the only one for which the pattern is symmetrical about the x-axis, is shown in figure 6.6.1 (a).

We gain one view of the effect of changing the value of κ/aU by noting that the velocity of the fluid at the surface of the cylinder is

$$\left(\frac{1}{r}\frac{\partial\phi}{\partial\theta}\right)_{r=a} = 2U\sin\theta + \frac{\kappa}{2\pi a}, \tag{6.6.4}$$

and vanishes at the two points at which

$$\sin\theta = -\frac{\kappa}{4\pi aU}.$$

The two stagnation points, which are at the foremost and rearmost points of the cylinder when $\kappa = 0$, both move down as κ/aU increases and coalesce at $\theta = -\frac{1}{2}\pi$ when $\kappa/aU = 4\pi$. Streamlines for a case in which $0 < \kappa/aU < 4\pi$ are shown in figure 6.6.1 (b), and those for the particular case $\kappa/aU = 4\pi$ in figure 6.6.1 (c). At values of κ/aU larger than 4π, the velocity is non-zero, and in the direction of θ increasing, at all points of the surface of the cylinder.

The stagnation point has here moved away from the cylinder along the line $\theta = -\frac{1}{2}\pi$ and is at a radial position given by one of the roots (the other referring to a motion inside the cylinder) of

$$\left(\frac{1}{r}\frac{\partial\phi}{\partial\theta}\right)_{\theta=-\frac{1}{2}\pi} = -U\left(1+\frac{a^2}{r^2}\right)+\frac{\kappa}{2\pi r} = 0,$$

since $\partial\phi/\partial r = 0$ on this line by symmetry. The streamlines in figure 6.6.1 (*d*) show that some of the fluid simply circulates round the cylinder and remains near it.

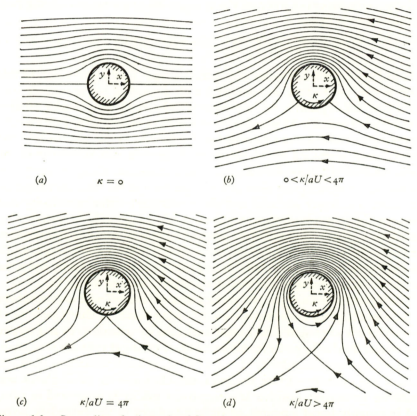

Figure 6.6.1. Streamlines for irrotational flow due to a circular cylinder held in a stream of uniform velocity $(-U, 0)$ at infinity, with circulation κ (anti-clockwise positive) round the cylinder.

The case of negative values of κ need not be considered, because the effect of changing the sign of κ is to reflect the streamlines in the x-axis.

It is evident that the effect of increasing κ/aU from zero is to cause an increasingly marked difference between the flow regions on the two sides of the x-axis, and in particular to cause the speeds on the upper surface of the cylinder to be high and those on the lower surface to be low. Remembering that $p + \frac{1}{2}\rho q^2$ is constant in a steady irrotational flow (leaving aside effects of gravity), we see plainly the greater pressure on the lower surface which leads to the non-zero side-force on the cylinder. It is possible to calculate the side-force directly and explicitly in this simple case of a circular cylinder. For we see from (6.6.4) and Bernoulli's theorem that when the flow is steady† the pressure at a point on the cylinder surface is

$$p_{r=a} = \rho H - \tfrac{1}{2}\rho \left(2U \sin\theta + \frac{\kappa}{2\pi a} \right)^2, \tag{6.6.5}$$

and the y-component of force on the cylinder (the x-component being zero by symmetry) is then

$$F_y = - \int_0^{2\pi} p_{r=a}\, a \sin\theta\, d\theta = \rho U\kappa, \tag{6.6.6}$$

as found in §6.4 for a cylinder of arbitrary shape. The positive sign of this force in the y-direction is associated with a positive (or anti-clockwise) circulation and a positive x-component of velocity of the cylinder relative to the fluid at infinity. Memorization of the signs is doubtless aided by the observation that undercut tennis and golf balls tend to rise, although in fact the two phenomena are not closely related; there is no counterpart to the cyclic constant in a three-dimensional field, and the rise of a spinning sphere is associated with different positions of separation of the boundary layer on the upper and lower surfaces and the consequent different velocity and pressure distributions on these surfaces. A better method of remembering the direction of the side-force is to visualize the general pattern of streamlines for flow relative to the cylinder and to recall that the side-force acts towards the high-velocity side, on which the contributions to the velocity from the circulation and the uniform stream are additive (see also figure 6.4.1).

The fact that the fluid velocity at the surface of the cylinder is everywhere in the same angular direction when $\kappa/aU > 4\pi$ points to the possibility of generating the corresponding irrotational velocity distribution as a steady flow of real fluid at large Reynolds number. If the rigid cylinder is made to rotate with an angular velocity $\Omega = \kappa/2\pi a^2$, the relative velocity of fluid and solid at the cylinder surface becomes anti-clockwise in some places and clockwise in others; positive and negative vorticity is generated in turn, and in roughly equal amounts, at different places in the boundary layer, so that separation may be avoided. The maximum and minimum values of the relative velocity at the cylinder surface (see (6.6.4)) are now $2U$ and $-2U$, and the relative velocity at a material point of the rigid cylinder surface varies periodically with frequency $\Omega/2\pi$, $= \kappa/(2\pi a)^2$; during one period the relative displacement of fluid and solid is of order $a^2 U/\kappa$, and provided this displacement is small compared with a (compare the results in §5.9 for a circular

---

† As remarked in §6.4, the side-force which acts when the translational velocity of the cylinder is steady continues to act when $\mathbf{U}$ is changing but it will not then be the only contribution to the force on the cylinder.

cylinder moving from rest, and see also § 5.13), separation of the boundary layer will be inhibited. We may expect therefore that in a real fluid the flow is irrotational everywhere except very near the cylinder surface provided $aU/\nu \gg 1$ and $\kappa/aU \gg 1$, and provided the cylinder rotates with the appropriate angular velocity.

The circulation round the cylinder is of course not a controllable quantity in an experiment, but is determined by the angular velocity and translational speed of the cylinder. The appropriate value of the circulation may be recognized, at any rate in principle, by supposing the motion to be set up in two stages. First the cylinder is given a steady angular velocity Ω in fluid initially at rest; as deduced in § 4.5, vorticity is generated in the fluid and diffuses to infinity, leaving a steady irrotational motion with circulation $2\pi a^2 \Omega$. The cylinder is then given a translational velocity U. If U is small enough in relation to $a\Omega$, no separation of the boundary layer occurs, vorticity is confined to a thin boundary layer at the cylinder surface, and the circulation remains constant and equal to $2\pi a^2 \Omega$. On the other hand if U is not small compared with $a\Omega$ separation occurs and a close correspondence with the irrotational flow pattern is lost. At some values of $U/a\Omega$ of order unity, it seems likely that separation does not occur but that vorticity accumulates near the cylinder, diffuses across the confining streamline, and is swept downstream, thereby leaving the cylinder with a circulation slightly different from $2\pi a^2 \Omega$ in the steady state.† Figure 6.6.2 (plate 12) shows some photographs of the streamlines of the actual flow set up by a cylinder with angular velocity Ω in a stream of steady speed U for several values of the ratio $a\Omega/U$; the circulation in each case is not known, but it seems that when $a\Omega/U \geqslant 4$ there is qualitative correspondence with flow fields like those in figure 6.6.1 (c) and (d), although not for equal values of $2\pi a^2 \Omega$ and the theoretical circulation κ used in figure 6.6.1.

It is evident from the photographs in figure 6.6.2 that the fluid velocity on the upper side of the cylinder is generally higher, and the pressure correspondingly lower, than on the lower side whenever the rotational speed of the cylinder is appreciable, irrespective of whether the rotational speed is large enough to prevent boundary-layer separation and the formation of a large region of non-zero vorticity. The existence of the side-force on a rigid circular cylinder which is both rotating and moving forward, and likewise on a sphere, is usually known as the *Magnus effect*, after the person who made the first relevant laboratory experiments (Magnus 1853).

An elliptic cylinder in translational motion

The expression for the complex potential describing flow due to an elliptic cylinder in translational motion may be obtained by conformal

---

† The circulation compatible with a steady flow in this case may be calculated from considerations of the boundary layer at the cylinder surface (Glauert 1957), and is found to be less than $2\pi a^2 \Omega$.

transformation of the region outside the ellipse in the z-plane into the region outside a circle in the ζ-plane, in the manner explained in §6.5. We require a transformation which will give linear relations between ξ and x and between η and y for points on a circle in the ζ-plane, since this 'strains' a circle into an ellipse. Bearing in mind the need also to make $\zeta \sim z$ as $|z| \to \infty$, it is evident that the required transformation is given by

$$z = \zeta + \frac{\lambda^2}{\zeta},\tag{6.6.7}$$

where λ is a real constant, so that

$$x = \xi\left(1 + \frac{\lambda^2}{|\zeta|^2}\right), \quad y = \eta\left(1 - \frac{\lambda^2}{|\zeta|^2}\right).$$

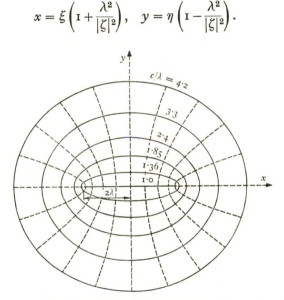

Figure 6.6.3. The family of ellipses in the z-plane corresponding to the circle $|\zeta| = c$ in the ζ-plane with the transformation $z = \zeta + \lambda^2/\zeta$ and different values of c. The broken lines show the orthogonal family of hyperbolae.

This converts a circle of radius c with centre at the origin in the ζ-plane into the ellipse

$$\frac{x^2}{a^2} + \frac{y^2}{b^2} = 1$$

in the z-plane, where

$$c = \tfrac{1}{2}(a+b), \quad \lambda = \tfrac{1}{2}(a^2 - b^2)^{\frac{1}{2}}.\tag{6.6.8}$$

Differently shaped ellipses, with different values of b/a, are thus obtained by choosing different values of c/λ. Some members of this family of confocal ellipses are shown in figure 6.6.3, including the limiting case of a flat plate ($b = 0$, $c/\lambda = 1$). The transformation (6.6.7) may also be written in the inverse form

$$\zeta = \tfrac{1}{2}z + \tfrac{1}{2}(z^2 - 4\lambda^2)^{\frac{1}{2}};\tag{6.6.9}$$

only the positive root is relevant for our purposes, since the negative root gives a relation which maps the region of the z-plane outside the ellipse on to the region *inside* the circle $|\zeta| = c$ in the ζ-plane.

The transformation represented by (6.6.7) and (6.6.9) may also be used to convert slender sharp-tailed bodies (or 'aerofoils') in the z-plane into a circle in the ζ-plane, as will be explained in § 6.7, and was first used for this purpose by Joukowski (1910).

The complex potential representing the flow in the z-plane due to an elliptic cylinder held in a stream of uniform velocity $(-U, -V)$ at infinity and with circulation κ round the cylinder is now obtained by substituting for ζ from (6.6.9) in (6.5.18), with $\zeta_0 = 0$:

$$w(z) = -\tfrac{1}{2}(U-iV)\{z+(z^2-4\lambda^2)^{\frac{1}{2}}\} - (U+iV)\frac{c^2}{2\lambda^2}\{z-(z^2-4\lambda^2)^{\frac{1}{2}}\}$$

$$-\frac{i\kappa}{2\pi}\log\left\{\frac{z}{2c}+\frac{1}{2c}(z^2-4\lambda^2)^{\frac{1}{2}}\right\}. \quad (6.6.10)$$

The flow in the z-plane relative to axes fixed in the fluid at infinity may be obtained by adding a term $(U-iV)z$ to the right-hand side of (6.6.10).

This solves the problem, formally, and we now consider the properties of the flow represented by (6.6.10). For this purpose it is convenient to introduce 'elliptic co-ordinates' (μ, ν) such that μ is constant on the elliptic boundary and on ellipses confocal with it and ν varies monotonically from 0 to 2π round any one of these ellipses, in place of the rectilinear co-ordinates (x, y). These co-ordinates, well known for their usefulness in problems involving Laplace's equation and elliptic boundaries, may be regarded as playing the same role in this problem as that played by polar co-ordinates (r, θ) in the previous discussion of the complex potential for a circular cylinder. The elliptic co-ordinates appropriate here are defined by

$$z = (a^2-b^2)^{\frac{1}{2}}\cosh\omega, \quad = 2\lambda\cosh\omega, \quad (6.6.11)$$

where $\omega = \mu + i\nu$, that is, by

$$x = (a^2-b^2)^{\frac{1}{2}}\cosh\mu\cos\nu, \quad y = (a^2-b^2)^{\frac{1}{2}}\sinh\mu\sin\nu. \quad (6.6.12)$$

Thus the co-ordinate line on which μ has the constant value μ_0 is an ellipse with semi-axes $(a^2-b^2)^{\frac{1}{2}}\cosh\mu_0$ and $(a^2-b^2)^{\frac{1}{2}}\sinh\mu_0$, which is identical with the cylinder boundary provided we choose μ_0 so that

$$e^{\mu_0} = \left(\frac{a+b}{a-b}\right)^{\frac{1}{2}}, \quad = \frac{c}{\lambda}. \quad (6.6.13)$$

The complex potential (6.6.10) can now be written as

$$w = -\tfrac{1}{2}(a+b)\{(U-iV)e^{\omega-\mu_0}+(U+iV)e^{\mu_0-\omega}\} - \frac{i\kappa}{2\pi}(\omega-\mu_0).$$

$$(6.6.14)$$

It is worth noticing, incidentally, that the linear dependence of the circulation term in (6.6.14) on ω is not an accident. The relation between z and ω is such that the real part of ω is constant on the inner boundary and the imaginary part of ω varies monotonically round the boundary. Consequently a complex potential proportional to $i\omega$ represents a flow which satisfies the zero-flux condition at the boundary and which gives a non-zero circulation round the cylinder; provided that $dw/dz \to 0$ as $z \to \infty$, as indeed it does, the outer boundary condition (zero velocity at infinity in the z-plane) is also satisfied. Thus a complex potential proportional to $i\omega$ represents completely the effect of a non-zero circulation round the cylinder and may be added to the complex potential representing the effect of the uniform stream alone. (We see in passing that a streamline in this pure circulatory flow about the elliptic cylinder is a line on which μ is constant, and so is an ellipse with semi-axes $(a^2 - b^2)^{\frac{1}{2}} \cosh \mu$ and $(a^2 - b^2)^{\frac{1}{2}} \sinh \mu$, which is confocal with the inner elliptic boundary; thus the family of streamlines is the set of ellipses shown in figure 6.6.3 lying outside whichever ellipse represents the inner boundary. In the conjugate flow field, for which $w \propto \omega$, the ellipses in figure 6.6.3 become equipotential lines and the streamlines are the orthogonal family of hyperbolae shown as broken lines. Any one of these hyperbolae could be interpreted as a boundary, and when the limiting hyperbola $\nu = 0$ is taken as a rigid boundary we have a representation of irrotational flow through a slit in a plane wall.) Allowance for the effect of a non-zero circulation can be made in this way for other shapes of cylinder.

Further convenience is obtained by writing $-\alpha$ for the angle which the direction of motion of the body makes with the x-axis,† so that

$$U + iV = (U^2 + V^2)^{\frac{1}{2}} e^{-i\alpha}$$

and (6.6.14) becomes

$$w = -(U^2 + V^2)^{\frac{1}{2}}(a+b)\cosh(\omega - \mu_0 + i\alpha) - \frac{i\kappa}{2\pi}(\omega - \mu_0). \quad (6.6.15)$$

The corresponding velocity potential and stream function are

$$\phi = -(U^2 + V^2)^{\frac{1}{2}}(a+b)\cosh(\mu - \mu_0)\cos(\nu + \alpha) + \frac{\kappa}{2\pi}\nu, \quad (6.6.16)$$

$$\psi = -(U^2 + V^2)^{\frac{1}{2}}(a+b)\sinh(\mu - \mu_0)\sin(\nu + \alpha) - \frac{\kappa}{2\pi}(\mu - \mu_0). \quad (6.6.17)$$

It is now a straightforward matter to calculate the streamlines and other flow properties, using μ and ν as parametric co-ordinates. The form of the streamlines depends on the ratio of axes of the elliptic boundary b/a, on the direction of motion of the body given by α, and on the relative magnitude of

† The reason for the negative sign in the definition of this angle is that in the theory of lifting bodies, developed in the next section, it is more natural to think of the axis of a long slender body as being inclined at a positive angle α to the direction of its motion.

the circulation measured by $\kappa(U^2+V^2)^{-\frac{1}{2}}(a+b)^{-1}$. We take first the case of zero circulation. The streamlines for one rather fat ellipse and for the limiting case of a flat plate, in both cases with $\alpha = 45°$, are shown in figure 6.6.4. The streamline which divides the parts of the stream passing round different sides of the cylinder intersects the cylinder, on which $\mu = \mu_0$, and is therefore a streamline on which $\psi = 0$. The upstream and down-stream branches of this streamline are consequently given by $\nu = -\alpha$ and $\nu = \pi - \alpha$, and are hyperbolae which are orthogonal to and confocal with the elliptic boundary and which asymptote to the line $Uy = Vx$; moreover, these branches of the dividing streamline are the same for all members of the family of elliptic boundaries shown in figure 6.6.3 (for given α) since they depend only on $a^2 - b^2$ (or λ).

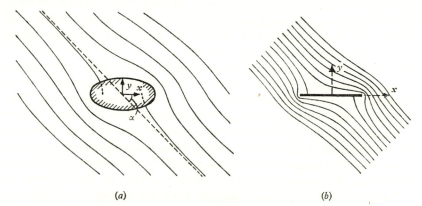

(a)   (b)

Figure 6.6.4. Streamlines of the flow due to an elliptic cylinder held in a stream of uniform velocity at infinity, with zero circulation round the cylinder and $\alpha = 45°$; (a) $b/a = 0.53$, $\mu_0 = 0.59$, (b) $b/a = 0$, $\mu_0 = 0$.

The kinetic energy (per unit length of the cylinder) of the motion of the fluid relative to axes fixed in the fluid at infinity, for $\kappa = 0$, may be obtained in several different ways, the simplest plan here being to use the formula

$$T = -\tfrac{1}{2}\rho \oint \phi \mathbf{n}.\nabla\phi \, dA = -\tfrac{1}{2}\rho \oint \left(\phi \frac{\partial\psi}{\partial\nu}\right)_{\mu=\mu_0} d\nu. \qquad (6.6.18)$$

The expressions for ϕ and ψ to be substituted in the integrand are those in (6.6.16) and (6.6.17) plus the terms $Ux + Vy$ and $Uy - Vx$ respectively arising from the change of axes; after a straight-forward calculation we find

$$T = \tfrac{1}{2}\pi\rho(a^2V^2 + b^2U^2). \qquad (6.6.19)$$

The tensor α_{ij} introduced in §6.4 and defined by (6.4.15) thus has components

$$\alpha_{11} = b/a, \quad \alpha_{22} = a/b, \quad \alpha_{12} = \alpha_{21} = 0 \qquad (6.6.20)$$

for an elliptic cylinder with major axis parallel to the x_1- or x-axis and minor axis parallel to the x_2- or y-axis. Information about the velocity far from the moving body, and about the acceleration reaction on the body, is also contained in (6.6.19), as explained in §6.4.

Further useful information comes from the velocity distribution at the cylinder surface, which (on reverting to axes fixed in the cylinder) is given by

$$(u - iv)_{\text{boun.}} = \left(\frac{dw}{d\omega}\frac{d\omega}{dz}\right)_{\mu = \mu_0}$$

$$= \frac{-i(U^2 + V^2)^{\frac{1}{2}}(a + b)\sin(\nu + \alpha) - i\kappa/2\pi}{(a^2 - b^2)^{\frac{1}{2}}\sinh(\mu_0 + i\nu)}. \quad (6.6.21)$$

When $\kappa = 0$, there are stagnation points on the cylinder at $\nu = -\alpha$ and $\nu = \pi - \alpha$, these being points on the dividing streamline. These are also points of maximum pressure in a steady flow, and their location (see figure 6.6.4) suggests that there is then a couple on the cylinder tending to turn it broadside-on to the stream; the couple on the cylinder in steady flow could be calculated from (6.6.21) and Bernoulli's theorem, but a more general method will be found later in this section.

When $\kappa \neq 0$, there is no longer any symmetry in the family of streamlines, and the displacement of points of minimum and maximum velocity on the boundary, relative to their positions for $\kappa = 0$, is not the same on the upper and lower surfaces of the cylinder. Thus whereas when $\kappa = 0$ the net force exerted on the body in steady flow is necessarily zero in view of the symmetry of the streamlines about the origin, it is non-zero when the circulation is non-zero. The force on the cylinder could be calculated directly from a knowledge of the velocity at the cylinder surface, as was done for the circular cylinder, but we know already from the general investigation of §6.4 that the force has magnitude $\rho\kappa(U^2 + V^2)^{\frac{1}{2}}$ and direction 90°, in the same sense as the circulation, from the direction of the vector (U, V) representing the velocity of the cylinder. The general effect of increasing the circulation on the two stagnation points at the cylinder surface is to make them approach one another on the side of the cylinder on which the contributions to the fluid velocity (relative to the cylinder) from the stream at infinity and the circulation are opposed. The value of κ for which the two stagnation points coalesce is that value which makes (6.6.21) zero when $\sin(\nu + \alpha)$ has its extreme value -1, viz.

$$2\pi(U^2 + V^2)^{\frac{1}{2}}(a + b).$$

With larger values of κ the two coincident stagnation points move away from the surface, leaving a volume of fluid which in a steady flow circulates round the cylinder continually, just as in the case of the circular cylinder.

The flat plate obtained by putting $b = 0$, or $\mu_0 = 0$, is an exceptional member of the family of ellipses, since the velocity is infinite (see (6.6.21)) at its two end points specified by $\nu = 0$ ($x = a, y = 0$) and $\nu = \pi$ ($x = -a, y = 0$), as expected from the results found earlier for flow round a salient edge. The properties of the flow past a flat plate will be considered in the next section in the context of sharp-tailed bodies generally. Incidentally, it may occasion surprise that when $\kappa \neq 0$ the pressure distribution on a flat plate in steady motion gives a resultant force which is not normal to the plate. The explanation lies in the fact that the infinitely low pressure which occurs at the sharp edges (except when a stagnation point is superposed—see §6.7) gives rise to a non-zero component of force parallel to the plate, as shown in §6.5.

The force and moment on a cylinder in steady translational motion

The force and moment on (unit length of) a moving cylinder of arbitrary cross-section can be determined by complex variable methods, and although results about the net force on a rigid body are available from §6.4 we shall recover them here in view of their importance and the interest in the special methods available for two-dimensional fields. We shall consider only the case of a body in steady translational motion; when the velocity of the body is changing, a second contribution to the force and moment, arising wholly from the effect of the acceleration, must be added to the expressions obtained below.

We suppose that (X, Y) are the components of force exerted on the body, and begin by forming the complex quantity

$$X - iY = -\oint_B p(dy + i\,dx), \quad = -i\oint_B p\,d\bar{z},$$

where B denotes a closed curve of integration coinciding with the surface of the body and the overbar denotes a complex conjugate. The velocity relative to the body is steady, with components (u, v), and we may use Bernoulli's theorem (with the effect of gravity ignored) to replace p by $\rho H - \frac{1}{2}\rho(u^2 + v^2)$, the first term of which makes no contribution to the integral. Now when the flow is irrotational we have

$$u^2 + v^2 = \frac{dw}{dz}\frac{\overline{dw}}{dz},$$

and since at the body surface $\overline{dw/dz}\ (= u + iv)$ and an element $\delta\bar{z}$ of the path of integration are complex numbers with equal arguments, the product $(\overline{dw/dz})\,\delta\bar{z}$ is real and can equally well be written as $(dw/dz)\,\delta z$. Hence

$$X - iY = \tfrac{1}{2}i\rho\oint\left(\frac{dw}{dz}\right)^2 dz. \tag{6.6.22}$$

Similarly the (anti-clockwise) moment about the origin of the normal stresses exerted on the cylinder is

$$M_0 = \oint_B p(x\,dx + y\,dy)$$

$$= -\tfrac{1}{2}\rho\oint_B \frac{dw}{dz}\frac{\overline{dw}}{dz}\mathscr{R}(z\,d\bar{z})$$

$$= -\tfrac{1}{2}\rho\mathscr{R}\oint\left(\frac{dw}{dz}\right)^2 z\,dz. \tag{6.6.23}$$

The path of integration for the integrals in (6.6.22) and (6.6.23) is the body surface. However, Cauchy's theorem† states that, if $f(z)$ is an analytic function of z in the region between two contours C_1 and C_2,

$$\oint_{C_1} f(z)\, dz = \oint_{C_2} f(z)\, dz, \qquad (6.6.24)$$

so that the paths of integration in (6.6.22) and (6.6.23) can equally be taken as any closed curve surrounding the body (provided of course that there are no singularities of w in the region between the body and the closed curve chosen). The special choice of a circle of large radius will clearly be useful when the form of $w(z)$ at large distances from the body is known.

The formulae (6.6.22) and (6.6.23), due to Blasius (1910), apply to any steady irrotational flow in fluid surrounding the body. We now adapt them to the case in which the fluid extends to infinity and has uniform velocity $(-U, -V)$ there. For this purpose we make use of the Laurent series (2.10.7) for the complex potential in the region external to a circle centred at the origin and enclosing the body (with the addition of a term $-(U-iV)z$, as in (6.5.21), since (2.10.7) refers to the flow relative to axes fixed in the fluid at infinity). Then (6.6.22) gives

$$X - iY = \tfrac{1}{2} i\rho \oint_C \left(-U + iV + \frac{m - i\kappa}{2\pi z} - \frac{A_1}{z^2} - \frac{2A_2}{z^3} + \dots \right)^2 dz, \qquad (6.6.25)$$

where the coefficients $A_1, A_2, \dots$ depend on U, V, m, κ and the size, shape and orientation of the body (and $m = 0$ here but is retained temporarily), and the integral is taken over any closed curve C enclosing a circle which encloses the body. The integral can be evaluated directly, either by supposing C to be a circle of indefinitely large radius, or by noting that, in the language of the theory of functions of a complex variable, the integrand has a pole at the origin; either way, the integral is equal to $2\pi i$ times the coefficient of z^{-1} in the integrand, and so

$$X = \rho m U - \rho \kappa V, \quad Y = \rho m V + \rho \kappa U. \qquad (6.6.26)$$

Thus the combination of the translational motion of the body and the circulation leads to a side-force, normal to the body velocity (U, V), as already established; and if the net flux of volume m across the body surface were non-zero and positive, this flux in combination with the translational motion would lead to a thrust, or negative drag, parallel to (U, V).

Likewise for the total couple on the body about the origin we have

$$\begin{aligned}
M_0 &= -\tfrac{1}{2} \rho \mathscr{R} \oint_C \left(-U + iV + \frac{m - i\kappa}{2\pi z} - \frac{A_1}{z^2} - \frac{2A_2}{z^3} + \dots \right)^2 z\, dz \\
&= -\tfrac{1}{2} \rho \mathscr{R} \left\{ 2\pi i \left(\frac{m - i\kappa}{2\pi} \right)^2 - 4\pi i A_1(-U + iV) \right\} \\
&= -\frac{\rho m \kappa}{2\pi} + 2\pi \rho \{ U \mathscr{I}(A_1) - V \mathscr{R}(A_1) \}. \qquad (6.6.27)
\end{aligned}$$

Unlike the force on the body, the couple depends on the body shape.

† See *Theory of Functions of a Complex Variable*, by E. T. Copson (Oxford, 1935).

In the case of a cylinder whose bounding curve in the z-plane is transformed to a circle of radius c and centre at the point ζ_0 in the ζ-plane by the general relation (6.5.19) or (6.5.20), which has the property that $z \sim \zeta$ when $|z|$ is large, the coefficient A_1 is given by (6.5.22). The couple on the cylinder is here (with $m = 0$)

$$M_0 = 2\pi\rho\left\{(-2UV)\mathscr{R}(B_1) + (U^2 - V^2)\mathscr{I}(B_1) + \frac{\kappa}{2\pi}(U\xi_0 + V\eta_0)\right\}. \quad (6.6.28)$$

In the case of an elliptic cylinder with semi-axes a and b, the appropriate transformation is (6.6.7) and

$$\zeta_0 = 0, \quad B_1 = \lambda^2 = \tfrac{1}{4}(a^2 - b^2),$$

so that the couple on the cylinder about the origin is

$$M_0 = -\pi\rho UV(a^2 - b^2). \quad (6.6.29)$$

The clockwise sense confirms the inference from the streamline pattern that the pressure distribution over the surface of the ellipse tends to turn it about the origin to a position broadside-on to the stream.

6.7. Two-dimensional aerofoils

The fact that in a two-dimensional field fluid in irrotational motion exerts a side-force, but no drag, on a steadily moving body round which there is a circulation is turned to advantage in engineering. The side-force may be used for instance to support an aircraft against gravity, or it may be used to generate axial momentum of the fluid when the body is one blade of a rotating propeller or turbine. Aeroplane wings and propeller blades are not infinitely long cylinders, and effects of finite length of the wing and of variation of the cross-section along its length play an important part in the theory of lift, as we shall see in chapter 7; nevertheless, an understanding of the operation of a lifting wing in the form of an infinitely long cylinder of appropriate cross-section moving normal to the generators—commonly termed an aerofoil, although the name is sometimes taken to embrace wings of finite length considered apart from the aeroplane—is an essential preliminary.

The practical requirements of aerofoils

The primary requirements of an aerofoil in practice are that when in motion through fluid a side-force should be exerted on it and that the drag, which would need to be balanced by some propulsive device and which would lead to the expenditure of power, should be small. These requirements are both met by a flow which is irrotational everywhere except in a thin boundary layer and wake, provided a circulation round the aerofoil can be established. Avoidance of boundary-layer separation when the aerofoil is in steady motion is thus one objective, and the establishment of circulation is another. We saw in chapter 5 that separation of the boundary layer from the body surface can be avoided only if the fluid just outside the boundary layer is not decelerated appreciably. The stagnation point at the rear face of a body in a two-dimensional field is a source of trouble, and separation would be inevitable near the rear of a body with finite curvature. The natural suggestion is to use a slender aerofoil with a sharp cusped edge at the rear and to align the aerofoil roughly parallel to the direction of its motion. The photograph of the streamlines of flow relative to the aerofoil in figure 5.11.1a (plate 7) shows that separation is then avoided. In practice it is difficult to make cusped edges, but the presence of the boundary layer and wake displaces the irrotational flow away from the aerofoil by a small distance which is non-zero near the rear edge, and the inner boundary to the region of irrotational flow is made cusp-like even when the trailing edge of the actual aerofoil is a wedge of small angle.

It is of course not inevitable that in irrotational flow past a sharp-tailed slender body the streams of fluid on the two sides should flow towards the sharp edge and join there smoothly. The analysis of two-dimensional irrotational flow due to a flat plate held in a stream of uniform velocity $(-U, -V)$ at infinity given in §6.6 makes this quite clear. In general there are two stagnation points at the plate surface (see figure 6.6.4(b)) and the fluid flows round the two sharp edges, with infinite speed at these edges. These two peaks in the velocity at the surface disappear, as do also the two stagnation points, only in the special case $\kappa = 0$, $\alpha = 0$, when the fluid has velocity $(-U, 0)$ everywhere. In this special case each stagnation point has moved to a sharp edge and has 'cancelled' the infinite velocity normally there. We therefore enquire if, for a given non-zero α (or κ) and with a special choice of κ (or α), the streams on the two sides of the plate can be made to flow smoothly off the rear (or 'trailing') edge of the plate by the rear stagnation point being placed at that edge. The relation (6.6.21) shows that the velocity at the surface of the flat plate given by $b = 0$ and $\mu_0 = 0$ is

$$u = \frac{-(U^2 + V^2)^{\frac{1}{2}} a \sin(\nu + \alpha) - \kappa/2\pi}{a \sin \nu}, \quad v = 0.$$

Thus u is finite at the trailing edge $(\nu = \pi)$ provided

$$\kappa = 2\pi a (U^2 + V^2)^{\frac{1}{2}} \sin \alpha, \tag{6.7.1}$$

and the velocity at the surface of the plate is then

$$u = -(U^2 + V^2)^{\frac{1}{2}} \frac{\sin(\frac{1}{2}\nu + \alpha)}{\sin \frac{1}{2}\nu}, \quad v = 0. \tag{6.7.2}$$

The change in the form of the streamlines in flow past the flat plate (with $\alpha = 26°$) due to the imposition of circulation with this magnitude is shown in figure 6.7.1. There is still a forward stagnation point at

$$\nu = -2\alpha, \quad \text{i.e. at } x = a\cos 2\alpha, y = -o,$$

and an infinite velocity at the forward edge ($\nu = o$), but neither need concern us since fluid at the plate surface near a forward stagnation point is accelerating and the forward velocity peak can be eliminated almost wholly by giving the plate some thickness and rounding the forward edge.

It seems then that, if a side-force is to be exerted on a sharp-tailed slender body in translational motion and if boundary-layer separation is to be avoided, not only should *some* circulation be established, but the circulation should have a particular value depending on the orientation of the body relative to the direction of its motion. The circulation should have that value

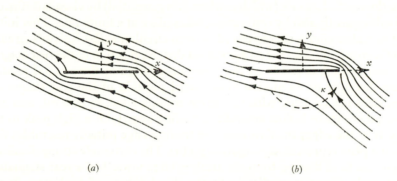

(a)   (b)

Figure 6.7.1. Streamlines of the flow due to a flat plate held in a stream of uniform velocity at infinity with $\alpha = 26°$; (a) with zero circulation, (b) with circulation such that there is smooth flow off both surfaces of the plate at the trailing edge.

for which, for the given orientation of the body, the rear stagnation point is located at the sharp trailing edge; for that value of the circulation the stagnation point and the velocity peak at the sharp edge 'cancel' each other and the velocity is finite and non-zero there. It is a remarkable fact that in practice a circulation is generated round an aerofoil, owing to the convection of a non-zero amount of vorticity from the rear edge of the aerofoil at an initial stage of the motion, and that when the aerofoil is in steady motion the circulation is established with just this special value. (See figure 5.11.1(a), plate 7, for an example.) This fortunate circumstance, that the effect of viscosity acting in the boundary layer initially is to cause the establishment of precisely the value of the circulation that enables effects of viscosity to be ignored (since no separation of the boundary layer occurs) in the subsequent steady motion, is usually given the name *Joukowski's hypothesis*. It was used as an empirical rule in the early development of aerofoil theory, but current knowledge of boundary layers enables us to account, at any rate in qualitative terms, for the establishment of the circulation with a specific value.

We shall digress briefly from the discussion of wholly irrotational flow to consider the remarkable controlling influence exerted by the sharp trailing edge of an aerofoil on the circulation.

The circulation round an aerofoil with a rounded leading edge and a sharp trailing edge in steady translational motion is observed to be independent of the past history of the flow, and for purposes of explanation we may suppose that the motion has been set up from rest and that the aerofoil has been brought rapidly to its ultimate steady velocity without change of the direction of its motion. Immediately after the aerofoil begins to move, the motion of the fluid is irrotational everywhere, because the transport of vorticity away from the aerofoil surface (which is where it is generated) by viscous diffusion, and subsequently also by convection, takes place at a finite rate. This initial irrotational motion is characterized by zero circulation (by Kelvin's circulation theorem), and there is an associated definite position of the rear stagnation point which depends on the given orientation of the aerofoil relative to the direction of its motion. The initial position of the rear stagnation point does not coincide with the sharp trailing edge, in general, and as a consequence there is flow *around* the trailing edge with a high peak in the velocity at the edge; the flow near the trailing edge initially resembles that for a flat plate represented in figure 6.7.1 (*a*). The extremely strong deceleration of the fluid flowing from the trailing edge towards the rear stagnation point leads almost immediately to the development of back-flow in the boundary layer there and to separation of the boundary layer (which at this stage is still very thin) at the sharp trailing edge.

In the next phase of the motion, the vorticity discharged from the trailing edge by the separated boundary layer affects the irrotational flow near the trailing edge, and so modifies it as to reduce the rate of discharge of vorticity. This process occurs near any sharp edge, and we can think about the flow near the trailing edge in isolation for a moment. The shape and location of the separated boundary layer shed from a salient edge almost immediately after it begins to move is shown by the sequence of photographs in figure 5.10.5 (plate 8), and further information about the streamlines on both sides of the shear layer is provided by the photograph in figure 6.7.2 (plate 13). Before any vorticity is convected from the salient edge, the irrotational flow locally has a form described by (6.5.2), with $n = \frac{1}{2}$ if the edge is a cusp (see also figure 6.5.1), and at subsequent times the vorticity shed from the edge modifies this irrotational flow over a region near the edge which increases in size. Figure 6.7.3 shows an attempt to sketch the development of the flow near the edge of a flat plate and the rolling up of the detached shear layer into a spiral under the action of its own induced velocity. The shed vorticity is carried away from the edge by the fluid, and so needs continual reinforcement by further vorticity shed from the edge in order to be able to induce a velocity near the edge which exactly cancels the velocity *round* the edge due to the background irrotational flow given by $w = Az^{\frac{1}{2}}$. It seems likely that the shapes of the detached shear layer and of the streamlines remain roughly similar, while increasing in scale, with the constant A in the specification of the original irrotational flow as the only given parameter to influence the flow, until the region of vorticity is so large that it can no longer be regarded as being embedded in an irrotational flow of the form $w = Az^{\frac{1}{2}}$.

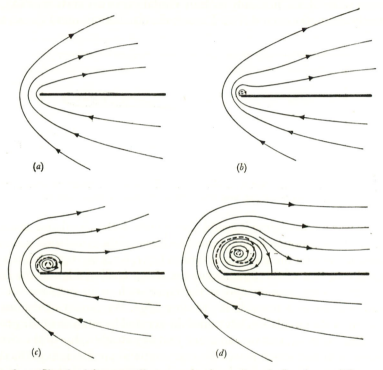

Figure 6.7.3. Sketch of the streamlines round a sharp edge of a flat plate at different stages after the motion begins. (*a*) Wholly irrotational flow specified by $w(z) = Az^{\frac{1}{2}}$; (*b*), (*c*) and (*d*) the same irrotational flow modified by the presence of a spiral vortex sheet (broken line) comprising vorticity shed from the boundary layers on the two surfaces of the plate (the negative vorticity from the lower surface being dominant).

In the third phase the intense vorticity shed from the trailing edge in the early stages of the motion is carried far downstream. The sense of the shed vorticity is the same as the sense of the motion round the trailing edge in the initial wholly irrotational flow (i.e. clockwise in figure 6.7.1 (*a*)), and it is evident that a circulation of opposite sense must be left round the aerofoil. For consider the material circuit $ABCD$ in figure 6.7.4 which is large enough to enclose both the initial position of the aerofoil (which is approximately the location of the vorticity shed initially) and its current position. The circulation round *ABCD* was initially zero, and is therefore zero at the instant considered. Thus the circulation round *ABFE* is equal and opposite to the flux of vorticity across the area *EFCD*, which includes practically all the vorticity shed from the aerofoil up to the instant considered. Photographs like those in figure 6.7.5 (plate 13) suggest that the shedding is virtually complete after the aerofoil has moved forward a distance equal to one or two times its streamwise length since the speed became steady. Thus the fluid enclosed by *ABFE* is in irrotational motion (except in the thin boundary layer and wake which, in steady motion, contain zero net vorticity flux), and the steady circulation round *ABFE* is also the circulation round the aerofoil.

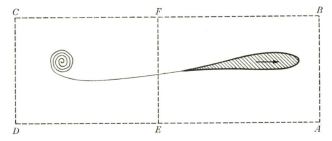

Figure 6.7.4. Definition sketch to show that the circulation round an aerofoil in steady motion is equal and opposite to the flux of shed vorticity.

In this way a flow régime is established in which the circulation round an aerofoil in steady motion is non-zero. The sense of the circulation generated —anti-clockwise for the aerofoils in figures 6.7.1, 6.7.4 and 6.7.5 (plate 13)— is opposite to that of the flow round the sharp trailing edge in the initial wholly irrotational motion and is therefore such as to displace the rear stagnation point back towards the trailing edge. We cannot determine the exact value of the circulation left round the aerofoil by analysis of the process of shedding vorticity, but we can argue that *any* steady value of the circulation other than that which places the rear stagnation point right at the sharp trailing edge would immediately set the above sequence of changes in train and cause further adjustment of the circulation, always of such a kind as to make the rear stagnation point move towards the trailing edge. The circulation prescribed by Joukowski's hypothesis is the only possible steady value for an aerofoil in steady motion.

The circulation demanded by Joukowski's hypothesis clearly depends on the steady velocity of the aerofoil, and, since the hypothesis requires cancellation of the two contributions to the velocity round the sharp edge from circulation and from the motion of the aerofoil, the circulation is *proportional* to the aerofoil speed. (Equation (6.7.1) shows this explicitly for the case of a flat plate.) It follows that vorticity must be shed from the aerofoil whenever its speed changes, and not only when it moves from rest. Figure 6.7.6 (plate 13) shows the striking effect of starting the aerofoil from rest and stopping it suddenly soon after. The vorticity shed as a result of a rapid change of speed is usually concentrated, and it is convenient to speak of a 'starting vortex', and, as in this figure, a 'stopping vortex'. The starting and stopping vortices here are of equal and opposite strength, and, if the aerofoil remains at rest, they subsequently move, each under the influence of the other, with approximately equal velocities in a direction normal to the line joining them. Starting, stopping, 'accelerating', and 'decelerating', vortices may also be demonstrated clearly by dipping a broad knife blade normally into a dish of water and moving it in a direction nearly parallel to the blade, the existence of a shed vortex being made visible by the surface depression at its centre.

If c is a length representative of the dimensions of the aerofoil, the circulation required by Joukowski's hypothesis must be of the form

$$\kappa \propto c(U^2 + V^2)^{\frac{1}{2}}, \tag{6.7.3}$$

where the constant of proportionality can depend only on the shape of the aerofoil and its orientation, represented by the angle α between the direction of its motion and some line fixed in the aerofoil. The determination of this constant of proportionality and its dependence on α (which is relevant to control of the lift force on the aerofoil by change of attitude) is now wholly a matter for irrotational flow theory.

Aerofoils obtained by transformation of a circle

The determination of the irrotational flow due to the translational motion of a slender sharp-tailed body obtained by transformation of a circle makes a nice exercise of the general method outlined at the end of §6.5. Aerofoils for which the corresponding flow properties (particularly the distribution of pressure on the wing) can be obtained analytically in this way were favoured in the early development of aeronautics, although many other ways of obtaining the required information are now available and there is no longer any reason for choosing these particular aerofoils for use. The simple procedure to be described here also has the practical disadvantage of being indirect, that is, it provides specific aerofoil shapes with known flow properties but does not enable one to calculate the properties of an aerofoil of given shape.

The distinctive feature of an aerofoil is its sharp trailing edge, which we shall take to be a cusp. The slope of the tangent to the aerofoil surface is discontinuous at this sharp edge, and a closed curve with this property in the z-plane can correspond to a circle in the ζ-plane, where $\zeta = F(z)$, only if there is a singular point of the transformation at the point $z = z_1$ at which the sharp edge lies. At this point the transformation must reduce the external angle 2π between the two sides of the trailing edge in the z-plane to an angle π at the corresponding point $\zeta = \zeta_1$ of the ζ-plane, which, as explained in §6.5 in connection with the transformation of intersecting walls into a single straight line, requires the analytic relation between ζ and z to be locally of the form

$$\zeta - \zeta_1 \propto (z - z_1)^{\frac{1}{2}}. \tag{6.7.4}$$

(The power of $z - z_1$ on the right-hand side would be $\frac{1}{2}(1 - \gamma/2\pi)^{-1}$ in the case of a sharp trailing edge in the form of a wedge of internal angle γ.) It is therefore possible to write

$$\frac{d\zeta}{dz} = \frac{f(z)}{(z - z_1)^{\frac{1}{2}}}, \tag{6.7.5}$$

where $f(z)$ is finite and non-zero at $z = z_1$, and at all points on the aerofoil surface unless there is a second sharp edge, and at all points of the z-plane outside the aerofoil (since singularities cannot occur in the interior of the fluid).

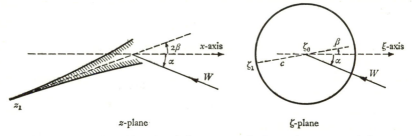

Figure 6.7.7. Definition sketch for an aerofoil obtained from a circle by conformal transformation.

This first step in the transformation is shown diagrammatically in figure 6.7.7. The cusp of the aerofoil has been drawn with an arbitrary orientation at this stage, and makes an angle $\pi + 2\beta$, say, with the x-axis. Thus the argument of $z - z_1$ for points on the upper surface of the aerofoil close to the trailing edge is 2β, and for points on the lower surface, reached by taking z on a path lying outside the aerofoil, is $2\beta + 2\pi$. The argument of $\zeta - \zeta_1$ at points on the circle in the ζ-plane corresponding to points on the upper surface of the aerofoil near the trailing edge is then seen from (6.7.5) to be $\arg\{f(z_1)\} + \beta$. We are at liberty to choose $f(z_1)$ to be pure imaginary, in which case the argument of $\zeta_1 - \zeta_0$, where ζ_0 is the centre of the circle, is $\pi + \beta$, as drawn in figure 6.7.7. The radius of the circle in the ζ-plane corresponding to the aerofoil in the z-plane is designated as c. As explained in §6.5, the relation between ζ and z must also be such that $\zeta \sim z$ as $|z| \to \infty$ so that in both planes the velocity at infinity is $(-U, -V)$.

Surprising though it may seem, enough has now been said about the aerofoil to enable us to determine the dependence of the circulation required by Joukowski's hypothesis, and thence that of the lift force on the aerofoil, on the direction of the stream of fluid at infinity (which makes an angle $\pi - \alpha$ with the x-axis, as in §6.6). The complex potential of the flow relative to the circular cylinder in the ζ-plane with an arbitrary circulation is given by

(6.5.18), and the fluid velocity in complex form at a point in the z-plane is

$$u - iv = \frac{dw}{dz} = \frac{dw}{d\zeta}\frac{d\zeta}{dz}.$$

At the trailing edge of the aerofoil ($z = z_1$), $|d\zeta/dz|$ is infinite, and so also is $|dw/dz|$ unless the circulation has such a value that $|dw/d\zeta|$ has a sufficiently small zero at $\zeta = \zeta_1$. Joukowski's hypothesis thus requires κ to be chosen so that there is a stagnation point at $\zeta = \zeta_1$ on the surface of the circular cylinder, in the flow in the ζ-plane, and the equation determining κ is

$$\left(\frac{dw}{d\zeta}\right)_{\zeta=\zeta_1} = -(U-iV)+(U+iV)\frac{c^2}{(\zeta_1-\zeta_0)^2}-\frac{i\kappa}{2\pi(\zeta_1-\zeta_0)} = 0.$$

Then, since
$$\zeta_1-\zeta_0 = c\,e^{i(\pi+\beta)}, \quad U+iV = (U^2+V^2)^{\frac{1}{2}}e^{-i\alpha}, \tag{6.7.6}$$

and with the abbreviation $(U^2+V^2)^{\frac{1}{2}} = W$, we have

$$\kappa = 4\pi Wc\sin(\alpha+\beta). \tag{6.7.7}$$

We may confirm that the velocity at the trailing edge in the z-plane is in fact finite when κ has this value by noting from (6.5.18) that, near $\zeta = \zeta_1$,

$$\frac{dw}{d\zeta} \approx (\zeta-\zeta_1)\left(\frac{d^2w}{d\zeta^2}\right)_{\zeta=\zeta_1} = \frac{2W}{c}\cos(\alpha+\beta)e^{-2i\beta}(\zeta-\zeta_1).$$

Also, near $z = z_1$,
$$\frac{d\zeta}{dz} \sim \frac{f(z_1)}{(z-z_1)^{\frac{1}{2}}}, \quad \sim \frac{2\{f(z_1)\}^2}{\zeta-\zeta_1},$$

showing that
$$\lim_{z\to z_1}\left(\frac{dw}{d\zeta}\frac{d\zeta}{dz}\right) = \frac{4W}{c}\cos(\alpha+\beta)e^{-2i\beta}\{f(z_1)\}^2, \tag{6.7.8}$$

which is finite, and non-zero. The reader may care to check that the finite velocity at the trailing edge of a flat plate given by (6.7.2) and obtained from the transformation (6.6.7) is consistent with the general expression (6.7.8).

The side-force or lift on unit length (normal to the plane of motion) of the embryonic aerofoil of figure 6.7.7 is thus

$$L = \rho W\kappa$$

$$= 4\pi\rho W^2 c\sin(\alpha+\beta). \tag{6.7.9}$$

The variation as ρW^2 was to be expected on dimensional grounds, since ρ is the only available parameter containing mass in its dimensions and W is the only one containing time (κ contains time, but is itself determined by W through the use of Joukowski's hypothesis). A length is also needed on the right-hand side of (6.7.9) to make up the dimensions of L, and c is the only remaining dimensional quantity from which to provide it. We do not yet know the relation between c and the size of the aerofoil, or between β and its shape, but (6.7.9) provides useful information about the dependence of L

on α. For values of α such that $\sin(\alpha + \beta)$ can be approximated by $\alpha + \beta$ (and in practice aerofoils do normally operate within a few degrees from the no-lift attitude), we have

$$L \propto \alpha + \beta. \tag{6.7.10}$$

The angle α between the direction of motion of the aerofoil and a line fixed in the aerofoil is termed the *angle of incidence*, and the lift is zero when $\alpha = -\beta$. The fact that the lift on an aerofoil is approximately proportional to the angle of rotation of the aerofoil from the no-lift attitude is of great importance since it makes possible a uniform control of the lift force by adjustment of the attitude of the aerofoil.

The result (6.7.10) has been established for all those aerofoils which can be obtained from a circle by conformal transformation. Measurement of the lift force on a portion of a cylindrical aerofoil between two plane parallel walls of a wind tunnel shows that it is valid for all normal aerofoil shapes for sufficiently small values of $\alpha + \beta$. When $\alpha + \beta$ exceeds some value depending on the aerofoil shape, which lies between 10° and 20° for many shapes in common use, the lift ceases to rise with further increase of $\alpha + \beta$, and may fall quite rapidly, the aerofoil then being 'stalled'. The explanation of this breakdown of the relations (6.7.9) and (6.7.10) lies in the behaviour of the boundary layer on the upper, or 'suction', side of the aerofoil. When a slender body is inclined at any but a quite small angle to an oncoming stream, there is a pronounced maximum of the velocity on the upper surface of the raised nose (which can be reduced by making the aerofoil nose fat and smoothly rounded, but only to a limited extent if the aerofoil is to remain slender) and the subsequent deceleration of the fluid outside the boundary layer causes it to separate. A photograph of the streamlines of flow past a stalled aerofoil was shown in figure 5.11.1(*b*) (plate 7), in the discussion of the effect of the boundary layer on flow due to moving bodies generally.

Joukowski aerofoils

For more specific information about the shape of an aerofoil obtained from a circle by conformal transformation and about the distribution of pressure on it, we must consider particular transformations. As an example, we shall consider briefly the Joukowski transformation already employed to determine the flow due to moving elliptic cylinders, in view of its relative simplicity and historical interest. Fuller accounts of the aerofoils obtained by this and other transformations are available elsewhere.†

The Joukowski transformation is defined by the relation

$$z = \zeta + \frac{\lambda^2}{\zeta}, \tag{6.7.11}$$

---

† See in particular *Aerofoil and Airscrew Theory*, by H. Glauert (Cambridge University Press, 1926), and, for more recent developments in aerofoil theory, *Incompressible Aerodynamics*, edited by B. Thwaites (Oxford University Press, 1960).

where λ is a real constant having the dimensions of length, and there are singular points of the transformation at

$$\zeta = \lambda \quad (z = 2\lambda) \quad \text{and} \quad \zeta = -\lambda \quad (z = -2\lambda).$$

Since (6.7.11) can be written as

$$z \mp 2\lambda = (\zeta \mp \lambda)^2/\zeta, \tag{6.7.12}$$

the transformation is of the general form (6.7.4) near both singular points, and either singular point may be used for the generation of a cusped figure in the z-plane from a curve with finite curvature in the ζ-plane. Both could be used, giving a body with two cusped edges, like the flat plate, but a practical aerofoil should have only one sharp edge. We choose the singularity at $\zeta = -\lambda$, in order to give an aerofoil with its forward edge towards the positive x-axis, and so, in the notation used earlier in this section, $\zeta_1 = -\lambda$, $z_1 = -2\lambda$. Then

$$\frac{d\zeta}{dz} = \frac{\zeta^2}{\zeta^2 - \lambda^2} = \frac{1}{2} + \frac{\frac{1}{2}z}{(z^2 - 4\lambda^2)^{\frac{1}{2}}},$$

and since the function $f(z)$ defined by (6.7.5) is $(z + 2\lambda)^{\frac{1}{2}}$ times this quantity

$$f(z_1) = \tfrac{1}{2} i \lambda^{\frac{1}{2}},$$

which is pure imaginary as assumed in the general discussion.

The circle of radius c in the ζ-plane must pass through the point

$$\zeta = \zeta_1 = -\lambda$$

corresponding to the trailing edge of the aerofoil, and must enclose the other singular point $\zeta = \lambda$ (or, as a limit, may pass through it, in which case there are two cusped edges to the aerofoil). Suppose first that the centre of the circle lies on the ξ-axis, at $\zeta = c - \lambda$, where $c \geqslant \lambda$, whence it follows from the symmetry of the transformation about the x-axis that the corresponding aerofoil is symmetrical about the x-axis and that $\beta = 0$. If $c = \lambda$, the corresponding figure in the z-plane is a flat plate of length 4λ, whereas if $c > \lambda$ we obtain an aerofoil which touches the flat plate at their common cusped trailing edge (see figure 6.7.8) and surrounds it elsewhere. Also, if

$$(c - \lambda)/\lambda \ll 1,$$

the aerofoil cannot differ appreciably from the flat plate and the aerofoil then has the required property of slenderness. It is now a simple matter to use (6.7.11) to calculate numerically the co-ordinates of points on the aerofoil corresponding to given points on the circle, for a given value of $(c - \lambda)/\lambda$; a typical symmetrical Joukowski aerofoil obtained in this way is shown in figure 6.7.8. The length of the aerofoil in the stream direction, known as the 'chord', is

$$2\lambda + (z)_{\zeta = 2c - \lambda} = \lambda + 2c + \frac{\lambda^2}{2c - \lambda},$$

$$\approx 4\lambda \left\{ 1 + \left(\frac{c - \lambda}{\lambda} \right)^2 \right\}$$

when $(c-\lambda)/\lambda \ll 1$. The general expression for the maximum thickness of the aerofoil is not as simple, but when $(c-\lambda)/\lambda \ll 1$ it follows readily that the maximum thickness occurs at approximately one-quarter of the chord from the leading edge and is approximately $3\sqrt{3}(c-\lambda)$. When $(c-\lambda)/\lambda = 0{\cdot}1$, the maximum thickness divided by the chord is about $0{\cdot}13$, and since this value is not often exceeded in practice it is evident that the chord is normally given quite accurately by 4λ. Thus the lift force on a slender symmetrical aerofoil, which is most conveniently expressed by a lift coefficient C_L analogous to the drag coefficient C_D defined in §5.11, is given (see (6.7.9)) by

$$C_L = \frac{L}{\tfrac{1}{2}\rho W^2 \times \text{chord}} \approx \frac{2\pi c}{\lambda}\sin\alpha$$

$$\approx 2\pi \sin\alpha \left(1 + 0{\cdot}77\,\frac{\text{thickness}}{\text{chord}}\right). \qquad (6.7.13)$$

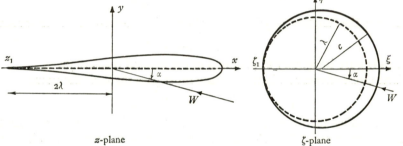

Figure 6.7.8. Transformation of a circle to a flat plate (broken line) and to a symmetrical Joukowski aerofoil (full line).

It is sometimes desirable to use an aerofoil on which the onset of stall is delayed to larger values of the angle of incidence, and for which the maximum obtainable lift coefficient is larger, than for a symmetrical aerofoil of the above type. Blunting the nose is only a partial aid, and it proves to be more effective to give the aerofoil *camber*, that is, to give the centre-line a convex-upwards curvature, so that the leading edge is pointing roughly into the stream when the bulk of the aerofoil is inclined to it. The Joukowski transformation can be used to generate aerofoils with camber, by choosing the centre of the circle in the ζ-plane to lie off the ξ-axis. A limiting case of an aerofoil with two cusped edges and zero thickness everywhere, like the flat plate, is obtained by choosing the centre of the circle to lie on the η-axis, at $\xi = 0$, $\eta = \lambda\tan\beta$, where the angle β has the same meaning as earlier. The radius of the circle is then (see figure 6.7.9) $c = \lambda\sec\beta$. Now it follows from (6.7.12) that

$$\arg(z-2\lambda) - \arg(z+2\lambda) = 2\{\arg(\zeta-\lambda) - \arg(\zeta+\lambda)\},$$

and since the right-hand side is constant and equal to $2(\tfrac{1}{2}\pi - \beta)$ for points on the arc of the circle above the ξ-axis, and constant and equal to $2(-\frac{1}{2}\pi-\beta)$ for points on the arc of the circle below the ξ-axis, the curves in the z-plane corresponding to these two arcs must be two coincident arcs of a circle whose radius is $2\lambda \operatorname{cosec} 2\beta$.

The lift force on this circular arc aerofoil when the circulation is adjusted to remove the infinite velocity at the trailing edge is given by (6.7.9). The chord of the aerofoil is 4λ exactly, so that we have

$$C_L = 2\pi \frac{\sin(\alpha+\beta)}{\cos\beta}. \tag{6.7.14}$$

It is possible now to obtain a cambered aerofoil with thickness by combining the operations described above, that is, by choosing a circle in the ζ-plane which passes through the point $\zeta = -\lambda$ and whose centre lies in the

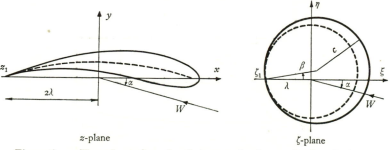

Figure 6.7.9. Transformation of a circle to a circular arc (broken line) and to a cambered Joukowski aerofoil (full line).

first quadrant. The transformation and one of the resulting aerofoils are shown in figure 6.7.9. This more general aerofoil shape may be regarded as being obtained approximately by bending the straight centre-line of a symmetrical Joukowski aerofoil into a circular arc. The shape of the aerofoil is determined by the two small parameters $(c-\lambda)/\lambda$, which controls the thickness-chord ratio, and β, which controls the camber. For these Joukowski aerofoils the lift coefficient is again $2\pi(\alpha+\beta)$ to a first approximation.

Figure 6.7.10 shows how observed values of the lift compare with the exact result from irrotational flow theory, for one cambered Joukowski aerofoil with $\beta = 8°$. The downward displacement of the experimental values is due in part to the boundary layer being thicker on the upper or suction side of the aerofoil, especially on the rear half; so far as the outer irrotational flow is concerned, the effect is to replace the aerofoil by one of slightly different shape and slightly smaller angle of incidence. The observed value of $dC_L/d\alpha$ is usually about 6 per radian, and so is close to the approximate theoretical value for slender aerofoils (2π per radian). The observed small values of the drag coefficient C_D on the same aerofoil are also shown in figure 6.7.10. The sharp change of shape of the lift curve and the rise in drag, at angles of incidence near $9°$, mark the onset of boundary-layer separation from the upper surface of the aerofoil. At higher values of α the aerofoil is stalled.

For the moment about the origin of the normal forces on a Joukowski aerofoil, we have the formula (6.6.28), in which κ is given by (6.7.7), B_1 is a complex number which here is λ^2 (for see (6.5.19) and (6.7.11)), and

$$\xi_0 = c\cos\beta - \lambda, \quad \eta_0 = c\sin\beta, \quad U = W\cos\alpha, \quad V = -W\sin\alpha.$$

Thus

$$M_0 = 2\pi\rho W^2\{\lambda^2\sin 2\alpha + c^2\sin 2(\alpha+\beta) - 2\lambda c\cos\alpha\sin(\alpha+\beta)\}, \quad (6.7.15)$$

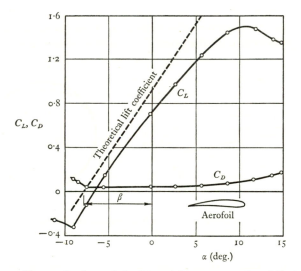

Figure 6.7.10. Observed values of the lift and drag on a cambered Joukowski aerofoil as a function of angle of incidence. (From Betz 1915.)

or, with the approximation that α, β and $(c-\lambda)/\lambda$ are all small compared with unity (in which case the circulation makes a negligible contribution),

$$M_0 \approx 4\pi\rho W^2\lambda^2\alpha. \quad (6.7.16)$$

For practical purposes a more convenient quantity is the moment about the leading edge of the aerofoil, which is given approximately by

$$M_{\text{l.e.}} = M_0 - 2\lambda L.$$

It is customary to express the moment on an aerofoil in terms of a dimensionless coefficient:

$$(C_M)_{\text{l.e.}} = \frac{M_{\text{l.e.}}}{\frac{1}{2}\rho W^2 \times (\text{chord})^2}$$

$$\approx (C_M)_0 - \tfrac{1}{2}C_L$$

$$\approx \tfrac{1}{2}\pi\alpha - \pi(\alpha+\beta), \quad = -\tfrac{1}{2}\pi(\alpha+2\beta). \quad (6.7.17)$$

This formula has been found to agree with observed values of the moment.

The lift on the aerofoil may evidently be regarded as acting at a position, known as the centre of pressure, which is a fraction of the chord back from the leading edge equal to

$$-\frac{(C_M)_{\text{l.e.}}}{C_L}, \quad \approx \frac{\alpha+2\beta}{4(\alpha+\beta)}. \tag{6.7.18}$$

The main structural support for the aerofoil should be placed near the centre of pressure, and it is desirable that the position of the centre of pressure should not vary much over the ordinary working range of values of α. For symmetrical Joukowski aerofoils, for which $\beta = 0$, the centre of pressure remains fixed at the quarter-chord point.

6.8. Axisymmetric irrotational flow due to moving bodies

Complex variable theory provides powerful general methods for the determination of the irrotational flow due to a moving body in a two-dimensional field. There is no counterpart to complex variable theory in the corresponding problem in three dimensions, and it is necessary to rely on a rather limited number of special methods. The analytical difficulties are less severe in the particular case of irrotational flow due to an axisymmetric rigid body moving in the direction of its axis, without rotation, since the whole flow field is then axisymmetric, and only this case will be considered here.

Generalities

It was remarked in §2.9 that two fundamental solutions of Laplace's equation in a three-dimensional field are $r^n S_n$ and $r^{-n-1} S_n$, where $r = |\mathbf{x}|$ and S_n is the spherical surface harmonic of order n (a positive integer) given by

$$S_n = r^{n+1} \frac{\partial^n(\mathrm{1}/r)}{\partial x_i \, \partial x_j \dots}.$$

When the flow field is axisymmetric, only those surface harmonics that are axisymmetric can occur in an expression for the velocity potential, that is, only those for which the suffixes $i, j, \dots$ all have the value corresponding to the direction of the axis of symmetry. Thus, if the x-axis is coincident with the axis of symmetry of the flow, the relevant surface harmonic of order n is

$$S_n = r^{n+1} \frac{\partial^n(\mathrm{1}/r)}{\partial x^n}. \tag{6.8.1}$$

This expression is a function only of $\mu = \cos\theta = x/r$, and is proportional to the Legendre polynomial $P_n(\mu)$ (or Legendre function which satisfies Legendre's differential equation of *integral* order n) as commonly defined.†
The following alternative expression for $P_n(\mu)$ in terms of derivatives with

† See *Methods of Mathematical Physics*, chap. 24, by H. Jeffreys & B. S. Jeffreys.

respect to μ (known as Rodrigues' formula) is also available:

$$P_n(\mu) = \frac{(-1)^n}{n!} r^{n+1} \frac{\partial^n}{\partial x^n} \left(\frac{1}{r}\right) = \frac{1}{2^n n!} \frac{d^n}{d\mu^n} (\mu^2 - 1)^n. \qquad (6.8.2)$$

As particular cases of (6.8.2) we have

$$P_0(\mu) = 1, \quad P_1(\mu) = \mu, \quad P_2(\mu) = \tfrac{1}{2}(3\mu^2 - 1). \qquad (6.8.3)$$

It is often convenient to use the Stokes stream function ψ as the dependent variable in cases of axisymmetric flow. As shown in §2.2, the velocity components in the directions of increase of the spherical polar co-ordinates r and θ are given by

$$u_r = \frac{\partial \phi}{\partial r} = \frac{1}{r^2 \sin\theta} \frac{\partial \psi}{\partial \theta}, \quad u_\theta = \frac{1}{r} \frac{\partial \phi}{\partial \theta} = -\frac{1}{r \sin\theta} \frac{\partial \psi}{\partial r}. \qquad (6.8.4)$$

Thus the expressions for ψ corresponding to the spherical solid harmonics

$$\phi = r^n P_n(\mu), \quad r^{-n-1} P_n(\mu),$$

are
$$\psi = \frac{1}{n+1} r^{n+1} (1 - \mu^2) \frac{dP_n(\mu)}{d\mu}, \quad -\frac{1}{n} r^{-n} (1 - \mu^2) \frac{dP_n(\mu)}{d\mu}. \qquad (6.8.5)$$

Another fundamental form of solution which occasionally is useful may be obtained by writing Laplace's equation for the velocity potential in terms of cylindrical co-ordinates consisting of x, $\sigma\ (=(r^2 - x^2)^{\frac{1}{2}})$ and an azimuthal angle, and putting

$$\phi = e^{\mp kx} F(\sigma).$$

The function F is seen to satisfy Bessel's equation of order zero, of which the solution that is finite at $\sigma = 0$ is $J_0(k\sigma)$ in the usual notation. It is readily seen, by comparing the two alternative expressions for the velocity components as derivatives of ϕ or ψ, that expressions for the stream function corresponding to the two solutions

$$\phi = e^{\mp kx} J_0(k\sigma) \qquad (6.8.6)$$

are
$$\psi = \pm \sigma e^{\mp kx} J_0'(k\sigma), \quad = \mp \sigma e^{\mp kx} J_1(k\sigma). \qquad (6.8.7)$$

For the relation between the solutions (6.8.6) and spherical solid harmonics, and for the use made of them in constructing irrotational flow fields of some physical interest, the reader is referred to more comprehensive texts.†

All the above expressions for ψ are of course solutions of the differential equation for ψ obtained from the condition that the flow be irrotational. Irrotationality requires

$$\frac{\partial u_r}{\partial \theta} - \frac{\partial (r u_\theta)}{\partial r} = 0,$$

and the differential equation for ψ then follows by substitution from (6.8.4):

$$\frac{\partial^2 \psi}{\partial r^2} + \frac{(1 - \mu^2)}{r^2} \frac{\partial^2 \psi}{\partial \mu^2} = 0. \qquad (6.8.8)$$

† Such as *Hydrodynamics*, by H. Lamb.

Alternatively, in terms of the cylindrical co-ordinates x, σ we find

$$\frac{\partial^2 \psi}{\partial x^2} + \frac{\partial^2 \psi}{\partial \sigma^2} - \frac{1}{\sigma}\frac{\partial \psi}{\partial \sigma} = 0. \tag{6.8.9}$$

The closeness in form of these equations to those satisfied by ϕ, viz.†

$$\frac{\partial^2 \phi}{\partial r^2} + \frac{2}{r}\frac{\partial \phi}{\partial r} + \frac{1}{r^2}\frac{\partial}{\partial \mu}\left\{(1-\mu^2)\frac{\partial \phi}{\partial \mu}\right\} = 0 \tag{6.8.10}$$

or

$$\frac{\partial^2 \phi}{\partial x^2} + \frac{\partial^2 \phi}{\partial \sigma^2} + \frac{1}{\sigma}\frac{\partial \phi}{\partial \sigma} = 0, \tag{6.8.11}$$

is striking, although it should be remembered that ϕ and ψ have quite different meanings, and also have different dimensions in an axisymmetric flow field.

The boundary conditions to be satisfied by ϕ when the fluid is at rest at infinity are, as established in earlier sections,

$$\phi \to C(\text{const.}) \quad \text{as} \quad r \to \infty,$$

and

$$\mathbf{n}.\nabla\phi = \mathbf{n}.\mathbf{U}$$

at the surface of the body, where $\mathbf{U}$ is the instantaneous velocity of the body, parallel to its axis of symmetry. When the body occupies a singly-connected region of space, these boundary conditions cannot be satisfied by more than one irrotational velocity distribution in the fluid.

We also need to specify the boundary conditions to be satisfied by ψ, for use when (6.8.8) or (6.8.9) is taken as the governing equation. The outer boundary condition, when the fluid is at rest at infinity, is seen from (6.8.4) to be, in effect,

$$\frac{1}{r}|\nabla\psi| \to 0 \quad \text{as} \quad r \to \infty.$$

At the surface of the body, the normal component of $\mathbf{u}$, expressed as a derivative of ψ, must be equal to $\mathbf{n}.\mathbf{U}$. This latter condition can be put in convenient analytical form by noticing that, relative to axes moving with a constant velocity equal to the value of $\mathbf{U}$ at the moment under consideration, the body is stationary (perhaps only instantaneously) and the intersection of the surface of the body and an axial plane is a streamline, on which ψ is equal to a constant which we may take as zero. The velocity fields relative to the two sets of axes differ only by a uniform velocity $\mathbf{U}$ parallel to the axis and the corresponding two stream functions differ by a term $\frac{1}{2}Ur^2\sin^2\theta$ or $\frac{1}{2}U\sigma^2$. Hence the inner boundary condition to be satisfied by ψ, for axes fixed in the fluid at infinity, is

$$\psi = \tfrac{1}{2}Ur^2\sin^2\theta \quad \text{or} \quad \tfrac{1}{2}U\sigma^2 \tag{6.8.12}$$

at the surface of the body.

---

† See appendix 2 for the expressions for $\nabla^2\phi$ in terms of spherical polar or cylindrical co-ordinates.

A method of constructing flow fields is suggested by the form of (6.8.12). For if we replace ψ in (6.8.12) by any function of r and θ (or x and σ) satisfying the differential equation (6.8.8) (or (6.8.9)) and also the outer boundary condition, we obtain a relation between r and θ defining the meridian curves of a family of rigid surfaces, each of which would generate a flow with the adopted stream function when it is moved with speed U parallel to its axis. However, not all solutions of (6.8.8) used in this way yield surfaces which are closed and which can therefore be regarded as rigid bodies.

A moving sphere

In this simple case of a sphere of radius a moving with velocity $\mathbf{U}$ in the direction $\theta = 0$, the inner boundary condition is

$$\frac{\partial \phi}{\partial r} = \mathbf{U}.\mathbf{n} = U\cos\theta \quad \text{at} \quad r = a.$$

It is evident that this condition can be satisfied for all θ if ϕ is proportional to the axisymmetric surface harmonic, or Legendre polynomial, of order one (see (6.8.2) and (6.8.3)) and that a solution satisfying the inner and outer boundary conditions is

$$\phi = -\tfrac{1}{2}Ua^3 \frac{\cos\theta}{r^2}, \tag{6.8.13}$$

in agreement with the solution found in §2.9 by other means. This solution applies at the instant at which the centre of the sphere is at the origin, and at any other instant, when it is at the point $\mathbf{x}_0$,

$$\phi = -\tfrac{1}{2}a^3 \frac{\mathbf{U}.(\mathbf{x}-\mathbf{x}_0)}{|\mathbf{x}-\mathbf{x}_0|^3}. \tag{6.8.14}$$

The expression for ψ corresponding to (6.8.13) is

$$\psi = \tfrac{1}{2}Ua^3 \frac{1-\mu^2}{r} \frac{dP_1(\mu)}{d\mu} = \tfrac{1}{2}Ua^3 \frac{\sin^2\theta}{r}, \tag{6.8.15}$$

which is of the same form as that for a source doublet located at the origin and directed parallel to $\theta = 0$ (see (2.5.5)). The stream function for flow relative to axes moving with the sphere is obtained by adding $-\tfrac{1}{2}Ur^2\sin^2\theta$ to the right-hand side of (6.8.15), giving

$$\psi = -\tfrac{1}{2}Ur^2\sin^2\theta \left(1 - \frac{a^3}{r^3}\right), \tag{6.8.16}$$

and the corresponding streamlines are shown in figure 6.8.1. The fore-and-aft symmetry of the pattern of streamlines is not reproduced in practice when a sphere moves steadily (compare with figure 5.11.7, plate 11), but, as previously explained, it is a realistic feature either of the flow immediately after a sphere moves from rest or of the flow due to a sphere in rapid oscillatory motion about a stationary mean position.

The kinetic energy of the fluid motion due to the moving sphere is found from (6.8.13) to be

$$T = -\tfrac{1}{2}\rho U_i \int (\phi)_{r=a}\, n_i\, dA$$

$$= \tfrac{1}{3}\pi\rho a^3 U^2. \tag{6.8.17}$$

The tensor coefficient α_{ij} defined by (6.4.15) and (6.4.16) thus has the value

$$\alpha_{ij} = \tfrac{1}{2}\delta_{ij}. \tag{6.8.18}$$

It follows from the formula (6.4.28) that the acceleration reaction on a sphere is parallel (and in the opposite sense) to $\dot{\mathbf{U}}$, irrespective of the direction of $\mathbf{U}$, and that the effect of the presence of the fluid on movement of the sphere

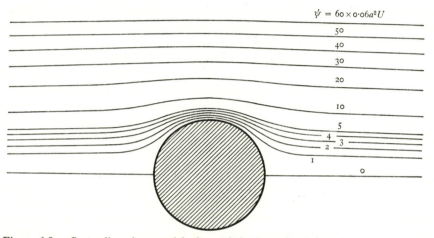

Figure 6.8.1. Streamlines in an axial plane of the irrotational flow due to a stationary sphere in a stream with uniform velocity at infinity.

under the action of given applied forces is the same, apart from buoyancy effects, as if the mass of the sphere were increased by one-half the mass of the displaced fluid.

Spheres in fluids appear in many different practical contexts, as solid or liquid spheres in a gaseous medium or as solid or gaseous spheres in a liquid medium, and the above formulae have wide application despite the limitations imposed by the assumption of irrotational flow. We shall indicate briefly the nature of some applications involving spheres moving freely.

Consider first the equation of motion of a sphere of mass M moving with velocity $\mathbf{U}$ through infinite fluid (which is set in irrotational motion), under the action of an applied force $\mathbf{X}$ say, and with allowance for the effect of gravity acting directly on the sphere and indirectly exerting a buoyancy force on the sphere by its action on the fluid:

$$M\dot{\mathbf{U}} = \mathbf{X} - \tfrac{1}{2}M_0\dot{\mathbf{U}} + M\mathbf{g} - M_0\mathbf{g}, \tag{6.8.19}$$

where $M_0 = \tfrac{4}{3}\pi a^3 \rho$ is the mass of fluid displaced by the sphere. The case of a sphere moving under gravity alone is of particular interest, and here we have

$$\dot{\mathbf{U}} = \frac{M - M_0}{M + \frac{1}{2}M_0}\,\mathbf{g}. \tag{6.8.20}$$

This formula will be accurate for a limited time after a rigid sphere accelerates from rest through fluid at rest at infinity. When $M \gg M_0$ the fluid has little effect on the initial acceleration of the sphere; but when $M \ll M_0$

$$\dot{\mathbf{U}} \approx -2\mathbf{g}. \tag{6.8.21}$$

Thus a spherical gas bubble moves from rest in water with an upward acceleration of $2\mathbf{g}$, and, since in this case boundary-layer separation seems not to occur (in a liquid free from impurities), continues to have this acceleration until either the bubble is deformed or the velocity becomes comparable with the terminal velocity considered in § 5.14.

Problems in which a sphere is set into motion relative to the fluid by the passage of a sound wave through the fluid are also of interest. Suppose that the sphere radius is small compared with the wavelength of the sound wave, and that the fluid everywhere in the neighbourhood of the sphere would have had the velocity $\mathbf{V}$ in the absence of the sphere. The acceleration of this fluid, again in the absence of the sphere, is approximately $\dot{\mathbf{V}}$, the contribution $\mathbf{V}.\nabla\mathbf{V}$ being negligible for a sound wave. We now choose axes moving with velocity $\mathbf{V}$ and acceleration $\dot{\mathbf{V}}$, recognizing that an effective force per unit mass $-\dot{\mathbf{V}}$ will appear in the equation of motion for the fluid and will lead to a 'buoyancy' force on the body of amount $M_0\dot{\mathbf{V}}$, as explained in §6.4. Provided the fluid moves irrotationally, the equation of motion for the sphere, with no force applied directly to the sphere and with neglect of gravity, is then

$$M\dot{\mathbf{U}} = -\tfrac{1}{2}M_0(\dot{\mathbf{U}} - \dot{\mathbf{V}}) + M_0\dot{\mathbf{V}}, \tag{6.8.22}$$

where $\mathbf{U}$ is the sphere velocity relative to unaccelerated axes; this is of course simply a particular case of the general formula (6.4.30). Integration gives

$$\mathbf{U} = \frac{\tfrac{3}{2}M_0}{M + \tfrac{1}{2}M_0}\,\mathbf{V}, \tag{6.8.23}$$

the constant of integration being put equal to zero on the understanding that there is no drift of the sphere through the fluid. The relation (6.8.23) is applicable to the oscillations of a small sphere suspended in a fluid through which a sound wave is passing, provided the frequency is large enough to make the thickness of the vorticity boundary layer small (see § 5.13). If the density of the sphere is greater than that of the fluid, the amplitude of the oscillations of the sphere is smaller than that of the fluid surrounding it; if the sphere is lighter, it oscillates with greater amplitude than the fluid.

A technique for rendering visible the displacement of different elements of a large tank of water with a free surface into which a projectile of some kind is fired is to distribute small air bubbles throughout the water and to

expose a photographic plate for the duration of the initial impulsive movement of the water. The air bubbles appear as streaks on the photograph, the
direction of a streak being the direction of displacement of the water locally,
and (6.8.23) shows that the length of a streak is approximately three times
the displacement of the water.

A further application of the above ideas may be made to the tendency for
gas bubbles in a liquid to approach one another and coalesce when the gas
bubbles undergo oscillations in volume in the same phase. Each oscillating
bubble produces an accelerating radial motion in the surrounding liquid,
and two neighbouring bubbles are thus able to influence each other's motion.
Since $M \ll M_0$ in this case, (6.8.22) becomes

$$\dot{U} \approx 3\left(1 - \frac{2M}{M_0}\right)\dot{V}, \qquad (6.8.24)$$

and it is evident that when both M_0 and $\dot{V}$ are periodic the average value of
$\dot{U}$ over one cycle may be non-zero. More specifically, suppose that two
spherical bubbles distance r apart displace masses of liquid equal to

$$M_1 + M_1' \sin nt \quad \text{and} \quad M_2 + M_2' \sin nt,$$

where $M_1' \ll M_1$ and $M_2' \ll M_2$, so that the second bubble produces, at the
position of the first, an approximately uniform acceleration of magnitude

$$\frac{d^2}{dt^2}\left(\frac{M_2 + M_2' \sin nt}{4\pi r^2 \rho}\right), \qquad = -\frac{n^2 M_2'}{4\pi r^2 \rho}\sin nt,$$

where ρ is the uniform density of the liquid. Then the average acceleration
of the first bubble (of mass M) along the line joining the bubbles is approximately

$$-\frac{3n^2 M}{4\pi r^2 \rho}\frac{M_1' M_2'}{M_1^2}, \qquad (6.8.25)$$

the negative sign indicating an acceleration towards the second bubble.
This effective force of attraction between the two bubbles leads to a steady
velocity of drift of each bubble as soon as viscous forces resisting the migration come into operation. The attractive force is normally very small, but
ultrasonic vibrations of a liquid can be used to clear the liquid of small gas
bubbles.   ·

Ellipsoids of revolution

The axisymmetric body that seems to be next in order of simplicity of
shape to the sphere is the ellipsoid of revolution. A first step which proves
to be useful here is to transform the independent variables (x, σ) in the
governing equation (6.8.9) or (6.8.11) to elliptic co-ordinates (ξ, η)† such
that ξ is constant on the surface of the body. The relation between the (x, σ)

† The notation (μ, ν) for elliptic co-ordinates introduced in §6.6 is unsuitable here in
  view of the use of $\mu = \cos\theta$.

and (ξ, η) 'planes' is *conformal* (see §§ 6.5, 6.6), and since, somewhat un-expectedly, the general properties of conformal transformations play a part in the analysis, we shall suppose for the moment that the relation between (x, σ) and (ξ, η) is simply of the form

$$x + i\sigma = f(\xi + i\eta). \tag{6.8.26}$$

To obtain the governing equation in terms of (ξ, η), we may use the following alternative expressions for the velocity components in an axial plane, which are consequences of the properties of ϕ and ψ:

$$u_\xi = \frac{1}{h_\xi}\frac{\partial\phi}{\partial\xi} = \frac{1}{\sigma h_\eta}\frac{\partial\psi}{\partial\eta}, \quad u_\eta = \frac{1}{h_\eta}\frac{\partial\phi}{\partial\eta} = -\frac{1}{\sigma h_\xi}\frac{\partial\psi}{\partial\xi}. \tag{6.8.27}$$

Here $h_\xi\,\delta\xi$ and $h_\eta\,\delta\eta$ are the lengths of line elements corresponding to small changes in ξ and η alone respectively, and can be found from the standard formulae

$$h_\xi^2 = \left(\frac{\partial x}{\partial\xi}\right)^2 + \left(\frac{\partial\sigma}{\partial\xi}\right)^2, \quad h_\eta^2 = \left(\frac{\partial x}{\partial\eta}\right)^2 + \left(\frac{\partial\sigma}{\partial\eta}\right)^2;$$

moreover, since $x + i\sigma$ is an analytic function of $\xi + i\eta$, it follows from the Cauchy-Riemann relations between (x, σ) and (ξ, η) that $h_\xi = h_\eta$. If now ξ, η and the azimuthal angle (for which the corresponding scale parameter h is equal to σ) are regarded as new orthogonal curvilinear co-ordinates, the expression of $\nabla \cdot \mathbf{u}$ and $\nabla \times \mathbf{u}$ in terms of these co-ordinates (see appendix 2) in conjunction with (6.8.27) will give the governing equations for ϕ and ψ respectively. ψ is a more convenient dependent variable than ϕ because it satisfies a simpler condition at the inner boundary. On equating to zero the azimuthal component of vorticity, we obtain

$$\frac{\partial(h_\eta u_\eta)}{\partial\xi} - \frac{\partial(h_\xi u_\xi)}{\partial\eta} = 0,$$

that is

$$\frac{\partial}{\partial\xi}\left(\frac{1}{\sigma}\frac{\partial\psi}{\partial\xi}\right) + \frac{\partial}{\partial\eta}\left(\frac{1}{\sigma}\frac{\partial\psi}{\partial\eta}\right) = 0, \tag{6.8.28}$$

in which σ is given in terms of ξ and η by (6.8.26).

For a *prolate* ellipsoid obtained by rotating an ellipse with major and minor semi-diameters a and b about its major axis, the appropriate transformation is

$$x + i\sigma = (a^2 - b^2)^{\frac{1}{2}}\cosh(\xi + i\eta),$$

the constant value of ξ on the ellipsoid (ξ_0) being given (see (6.6.13)) by

$$e^{\xi_0} = \left(\frac{a+b}{a-b}\right)^{\frac{1}{2}}.$$

The condition (6.8.12) to be satisfied by ψ at the inner boundary is that

$$\psi = \tfrac{1}{2}U(a^2 - b^2)\sinh^2\xi\sin^2\eta$$

when $\xi = \xi_0$. This suggests we should seek a solution of (6.8.28) of the form

$$\psi = F(\xi)\sin^2\eta. \tag{6.8.29}$$

Substitution in (6.8.28) yields a second-order ordinary differential equation for $F(\xi)$, and after integration and choice of the two constants to suit the inner and outer boundary conditions it may be shown that

$$\psi = \frac{\frac{1}{2}Ub^2(a^2-b^2)\sin^2\eta}{a(a^2-b^2)^{\frac{1}{2}}+b^2\log\left\{\dfrac{a-(a^2-b^2)^{\frac{1}{2}}}{b}\right\}}(\cosh\xi+\sinh^2\xi\log\tanh\tfrac{1}{2}\xi) \tag{6.8.30}$$

is the required solution.

To obtain the flow due to an *oblate* ellipsoid moving parallel to its axis of revolution, we need to begin with a transformation to elliptic co-ordinates defined by

$$x+i\sigma = (a^2-b^2)^{\frac{1}{2}}\sinh(\xi+i\eta).$$

ξ is now constant and equal to ξ_0 on the surface of an ellipsoid obtained by rotating an ellipse with major and minor semi-diameters a and b about its minor axis. (6.8.29) is again the form of solution appropriate (so far as the dependence on η is concerned) to the inner boundary condition, and on proceeding as before we obtain

$$\psi = \frac{\frac{1}{2}Ua^2(a^2-b^2)\sin^2\eta}{b(a^2-b^2)^{\frac{1}{2}}-a^2\cos^{-1}b/a}(\sinh\xi-\cosh^2\xi\cot^{-1}\sinh\xi) \tag{6.8.31}$$

as the required solution.

The corresponding velocity potentials may now be found without difficulty from either of the relations

$$\frac{\partial\phi}{\partial\xi} = \frac{1}{\sigma}\frac{\partial\psi}{\partial\eta}, \quad \frac{\partial\phi}{\partial\eta} = -\frac{1}{\sigma}\frac{\partial\psi}{\partial\xi},$$

but will not be written out here; clearly ϕ is proportional to $\cos\eta$.

When $(a-b)/a \to 0$ (or, equivalently, $\xi_0 \to \infty$), both ellipsoids become spheres of radius a, and it may be shown that both (6.8.30) and (6.8.31) reduce to the stream function (6.8.15) already found for a sphere. Another limiting case of (6.8.31) is obtained by putting $b = 0$ (or $\xi_0 = 0$), giving the irrotational flow due to a circular disk of radius a moving normal to its plane. The stream function here is

$$\psi = -\frac{a^2U}{\pi}(\sinh\xi-\cosh^2\xi\cot^{-1}\sinh\xi)\sin^2\eta, \tag{6.8.32}$$

and the streamlines are shown in figure 6.8.2. The velocity potential found in the above manner reduces to

$$\phi = -\frac{2aU}{\pi}\cos\eta$$

on the surface of the disk, so that the kinetic energy of the fluid is

$$T = -\tfrac{1}{2}\rho U_i \int (\phi)_{\xi=0}\, n_i\, dA$$

$$= -\pi\rho U \int_0^a \left\{ (\phi)_{\substack{\xi=0\\0<\eta<\frac{1}{2}\pi}} - (\phi)_{\substack{\xi=0\\\frac{1}{2}\pi<\eta<\pi}} \right\} \sigma\, d\sigma$$

$$= 4\rho a^3 U^2 \int_0^{\frac{1}{2}\pi} \cos^2\eta \, \sin\eta \, d\eta = \tfrac{4}{3}\rho a^3 U^2.$$

The virtual inertia for acceleration of the disk in the direction of its motion is thus $\tfrac{8}{3}\rho a^3$. The velocity of the fluid is infinite at the edge of the disk, and the consequent strong deceleration of the fluid after passing round the edge makes this irrotational flow an unlikely candidate for application to real flow systems; however irrotational flow solutions have a habit of turning up in different guises, and we shall see in § 6.10 that this particular solution applies to the motion produced by impact of a hammer with a flat circular head on a water surface.

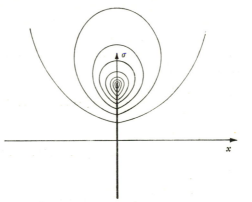

Figure 6.8.2. Streamlines in an axial plane, with equal increments in ψ, for the irrotational flow due to a circular disk moving normal to its plane.

Body shapes obtained from source singularities on the axis of symmetry

The properties of the point-source singularities considered in § 2.5 can be used in an interesting way to construct particular irrotational axisymmetric flow fields of the kind that would be generated by a moving body, although the shape of the body cannot be freely chosen (except when the body is slender—see § 6.9). The basis of the method is that, if a number of point sources and sinks (or perhaps a continuous distribution of source strength) are located on the axis of symmetry in a fluid at rest at infinity, and the net source strength is zero, all the streamlines emanating from sources terminate at the sinks. In certain circumstances (not readily specified in general terms), this will continue to be the case when a uniform flow parallel to the axis is superimposed on the motion generated by the sources and sinks. There will then be one closed streamline which surrounds the group of sources and sinks and which separates the streamlines originating at a source from those coming from infinity where the speed is uniform; this streamline can be regarded as the surface of a rigid body held stationary in a uniform stream, and the velocity distribution outside the body can be calculated as the resultant of the induced velocity fields of the various sources and sinks and the uniform stream.

If just one source of strength m and one sink of strength $-m$ are placed on the axis, it is evident that, provided the source lies upstream of the sink, none of the streamlines coming from infinity flows into the sink and the dividing

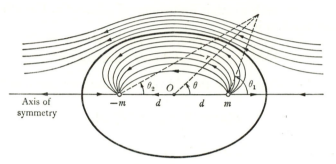

Figure 6.8.3. Streamlines in an axial plane for a combination of a point source, a point sink of equal strength, and a uniform stream.

streamline is closed. The stream function describing the flow due to a source of strength m at the origin is $-(m/4\pi)\cos\theta$. Hence, for a source of strength m at $r = d$, $\theta = 0$, a sink of strength $-m$ at $r = d, \theta = \pi$, and a uniform stream of speed U in the direction $\theta = \pi$ (so as to give a flow generated by a body moving in the direction $\theta = 0$ when axes fixed in the fluid at infinity are used), we have

$$\psi = -\frac{m}{4\pi}\cos\theta_1 + \frac{m}{4\pi}\cos\theta_2 - \tfrac{1}{2}Ur^2\sin^2\theta. \qquad (6.8.33)$$

The notation is made clear by figure 6.8.3, which also shows the streamlines in one particular case. The streamline which is closed and which may be regarded as the surface of a body has another branch on the axis of symmetry upstream and downstream of the body, where $\psi = 0$, so that the meridian curve of the body is given by

$$\cos\theta_2 - \cos\theta_1 = \frac{2\pi Ud^2}{m}\frac{r^2}{d^2}\sin^2\theta. \qquad (6.8.34)$$

An infinite family of possible body shapes, known as Rankine ovoids (Rankine 1871), is thus obtained, corresponding to different values of the dimensionless parameter Ud^2/m. The body shape changes smoothly from a long narrow cigar-shape to a slightly extended sphere as Ud^2/m is decreased from large to small values compared with unity. When $Ud^2/m \ll 1$, the body surface is approximately a sphere of radius $(md/\pi U)^{\frac{1}{2}}$, which is large compared with d; the effect of the source and sink together at points outside the body surface is here approximately the same as for a source doublet (see §2.5) of strength $2md$ at the origin directed along the axis of symmetry, as was of course already implicit in a comparison of (2.5.3) and (2.9.26).

One or two general results can be established in cases in which the presence of the body may be represented by a continuous distribution of source strength along a portion of the axis of symmetry. Suppose that the source strength over the range x to $x+\delta x$ of the axis of symmetry is $m(x)\,\delta x$, where $m(x) = 0$ outside a certain finite range of values of x. Then the velocity potential of the flow is

$$\phi(\mathbf{x}) = -\frac{1}{4\pi}\int_{-\infty}^{\infty}\frac{m(x')\,dx'}{(r^2+x'^2-2rx'\cos\theta)^{\frac{1}{2}}}, \tag{6.8.35}$$

which may be written, with the aid of a standard property of Legendre polynomials, as

$$\phi(\mathbf{x}) = \sum_{n=0}^{\infty} K_n r^{-n-1} P_n(\mu), \tag{6.8.36}$$

where the coefficients K_n are given by

$$K_n = -\frac{1}{4\pi}\int_{-\infty}^{\infty} x^n m(x)\,dx.$$

The relation (6.8.2) shows that this series is simply the special axisymmetric version of the general expansion (6.4.1) in terms of solid harmonics. The coefficient K_n is equal, as (6.8.36) has been written, to $(-1)^n n!$ times the nth-order tensor coefficient $c_{ij}...$ in (6.4.1) with $i,j,...$ all put equal to the value, 1 say, corresponding to the direction of the axis of symmetry. The first non-zero coefficient is K_1, since there is zero flux of mass across the body surface, and

$$c_1 = -K_1 = \frac{1}{4\pi}\int_{-\infty}^{\infty} xm(x)\,dx. \tag{6.8.37}$$

The expressions obtained in §6.4 for the kinetic energy of the fluid and the virtual mass due to an axisymmetric body in translational motion in terms of c_1 may now be utilized.

Semi-infinite bodies

When the source and sink in figure 6.8.3 are placed far apart (or, more precisely, when $Ud^2/m \gg 1$), the streamlines are approximately parallel to the axis of symmetry everywhere, except near either the source or the sink, and the corresponding body surface given by the relation (6.8.34) approximates to a cylinder with rounded ends. We can then determine the details of the flow near the forward half of the body (which is in any event the only part of steady flow past a body with fore-and-aft symmetry which is irrotational in practice) by imagining the sink to be located at infinity downstream and the body length to be semi-infinite. It is now more useful to put the origin at the position of the source, in which case

$$\psi = -\frac{m}{4\pi}\cos\theta - \tfrac{1}{2}Ur^2\sin^2\theta \qquad (6.8.38)$$

and the relation specifying the body surface becomes

$$\frac{m}{2\pi U}(1 - \cos\theta) = r^2\sin^2\theta.$$

Variation of m or U now simply changes the length scale of the whole field, and the one possible body shape is as shown in figure 6.8.4. The entire volume flux m from the source crosses any section of the 'body' far from the source at speed U, so that the radius of the cylindrical portion of the body is $(m/\pi U)^{\frac{1}{2}}$.

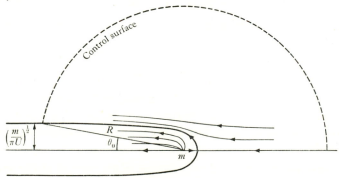

Figure 6.8.4. Flow past a semi-infinite body obtained from a point source and a uniform stream.

It is evident that the radius of the cylindrical portion of other semi-infinite bodies obtained from distributions of sources and sinks over a finite length of the axis of symmetry, with net source strength m (> 0), will be $(m/\pi U)^{\frac{1}{2}}$.

Since there is a stagnation point at the front of a semi-infinite body of the above kind, it might be thought that the fluid outside the body exerts a non-zero drag force on the body in a steady flow. This turns out not to be so, because the velocity of the fluid at the 'shoulder' of the body exceeds U and the corresponding pressure deficiency there balances the excess pressure near the stagnation point. The balance is an exact one, irrespective of the shape of the rounded nose of the body, as we shall now show by use of the momentum equation in integral form.

As control surface we choose the portion of a sphere of radius R, with centre at the centre of the distribution of source strength, that lies in the fluid (figure 6.8.4), together with the surface of the body from the nose to its intersection with the sphere. The total axial component of force exerted on the enclosed fluid by pressure at the spherical portion of the control surface is

$$-\int_0^{\pi-\theta_0} \{p_0 + \tfrac{1}{2}\rho U^2 - \tfrac{1}{2}\rho(u^2+v^2)\} \cos\theta \, 2\pi R^2 \sin\theta \, d\theta,$$

where u, v are components of velocity of the fluid on the sphere parallel and perpendicular to the axis respectively, p_0 is the pressure in the fluid far from the body, θ_0 is the semi-cone angle subtended at the centre by the intercept of the sphere with the body, and $R\sin\theta_0 = (m/\pi U)^{\frac{1}{2}}$. Now since m is the net strength of the sources on the axis that give the body surface as the dividing streamline, we have

$$u \approx -U + \frac{m}{4\pi R^2}\cos\theta, \quad v \approx \frac{m}{4\pi R^2}\sin\theta$$

when R is large, and the above force reduces, in the limit $R\to\infty$, to

$$-p_0(m/U) - \tfrac{1}{3}\rho m U.$$

Likewise the total flux of axial momentum outwards across the control surface is

$$\rho\int_0^{\pi-\theta_0} u(u\cos\theta + v\sin\theta) \, 2\pi R^2 \sin\theta \, d\theta,$$

$$\to -\tfrac{1}{3}\rho m U \quad \text{as} \quad R\to\infty.$$

Thus the force exerted on the enclosed fluid at the portion of the control surface coinciding with the body surface must be $p_0(m/U)$. The fluid exerts on the body a force in the direction of the velocity of the fluid at infinity and of magnitude

$p_0 \times$ area of projection of body on the plane normal to the axis.

This resultant force on the body is simply the effect of the uncompensated ambient pressure on one side of the body and is of no dynamical significance. So far as dynamical effects are concerned, the fluid exerts no force on a semi-infinite body. Inasmuch as the steady flow of real fluid past axisymmetric bodies at large Reynolds number is more nearly irrotational everywhere near the forward half of the body than near the rearward half, this is perhaps a more significant deduction than that the drag on a finite body in steady irrotational flow is zero.

Exercise

By assuming a form of solution in which the spherical polar co-ordinates r and θ are separated, show that axisymmetric irrotational flow near the vertex of a cone is represented by $\phi \propto r^m P_m(\mu)$, where $\mu = \cos\theta$, $P_m(\mu)$ is the Legendre function of the first kind, and m is determined by the equation

$$dP_m(\mu)/d\mu = 0 \quad \text{at} \quad \mu = -\cos\theta_0,$$

where θ_0 is the angle between the axis of the cone and one of its generators ($1 \leqslant m \leqslant 2$).

6.9. Approximate results for slender bodies

In the case of bodies whose length is large compared with their width, and which are moving through fluid at rest at infinity, there is a simple approximate method of determining the corresponding irrotational flow field. Under the name of 'slender-body theory', this method has been extended to many different kinds of problem and is widely used in aeronautics and other fields concerned with the movement of streamlined bodies through fluid. The basis of the method, in the simple version to be explained here, is that a distribution of flow singularities (such as sources and sinks) along a line can be chosen, when the body is slender, in such a way that the irrotational flow associated with these singularities in combination with a uniform stream satisfies approximately the condition of zero normal component of velocity at the surface of a body of given shape. In the case of slender bodies of revolution, the method is a natural sequel to the discussion in §6.8 of bodies represented by sources placed along the axis of symmetry, and we therefore take that case first.

Slender bodies of revolution

We take the x-axis to coincide with the axis of symmetry of the body, as in §6.8, and suppose to begin with that the body is stationary in a stream which at infinity has uniform speed U in the direction of the negative x-axis (corresponding to a body moving in the direction of the positive x-axis through fluid at rest at infinity). The area of the cross-section of the body at position x is A (figure 6.9.1). We shall suppose that the tangent to the meridian curve of the body surface is inclined at a small angle β to the axis, and that A is consequently a slowly-varying function of x; this assumption goes beyond the mere requirement of a small ratio of maximum width to overall length of the body, but is commonly regarded as implied by the term 'slender body'.

Now the streamlines of both the flow outside the body and the fictitious flow inside the body associated with the replacement singularities will be nearly parallel to the axis, like the surface streamline. Furthermore, the mass-conservation equation in cylindrical co-ordinates is

$$\frac{\partial u_x}{\partial x} + \frac{1}{\sigma}\frac{\partial(\sigma u_\sigma)}{\partial \sigma} = 0,$$

where (u_x, u_σ) is the increment to the uniform stream velocity $(-U, 0)$ due to the presence of the body, so that, provided the derivatives $\partial/\partial x$ and $\partial/\partial\sigma$ are of the same order of magnitude (as may be expected in a field governed by Laplace's equation), the changes in u_x over the flow field are of the same order as those in u_σ. Since u_σ is of order $|\beta(-U+u_x)|$ near the body surface, the conclusion is that the perturbation components u_x, u_σ are both of small order βU, and the axial component of velocity may be taken as $-U$ everywhere to a first approximation.

It follows that the flux of fluid volume across the section of the body at position x is approximately $-UA$ and that at a neighbouring position $x+\delta x$ is $-U(A+\delta x\, dA/dx)$. No fluid crosses the dividing streamline representing the body surface (figure 6.9.1), so that the flux difference

$$-U\, \delta x\, dA/dx$$

must be supplied by sources on the axis with strength $-U\, dA/dx$ per unit length. Thus, when the shape of the slender body is given, it is possible to

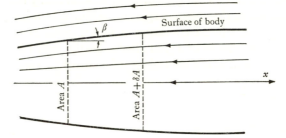

Figure 6.9.1. Axisymmetric irrotational flow due to a slender body in a stream simulated by a distribution of sources on the axis of symmetry.

specify the line density of source strength on the axis, which, together with the uniform stream, generates a flow with a dividing streamline coinciding approximately with the body surface, and from which the whole flow field may be determined. The approximation of nearly parallel flow is not accurate near the rounded nose and tail of a body of finite length, but the error is likely to be a local one only, since the total source strength on the axis between position x and the nose is UA according to the above rule, which is in fact what is required to give the correct total lateral displacement of the oncoming stream at position x.

It will be noticed from (6.8.37) that there is now a particularly simple expression for the strength of the axial source doublet which gives rise to the same asymptotic flow field (at large distances from the origin) as the body. This doublet strength is

$$4\pi c_1 = -U\int_{-\infty}^{\infty} x\frac{dA}{dx}\, dx$$

$$= U\int_{-\infty}^{\infty} A\, dx = UV_0, \tag{6.9.1}$$

showing that, for any slender body of volume V_0 moving in the direction of its axis of symmetry ($\theta = 0$) with speed U,

$$\phi \sim -\frac{UV_0}{4\pi}\frac{\cos\theta}{r^2} \tag{6.9.2}$$

for r large. Unfortunately the corresponding approximation to the kinetic energy of the fluid as obtained from (6.4.17) is zero, which is too crude to be useful!

The accuracy of this kind of slender-body theory clearly depends on the use to which it is put and on the particular flow parameters which the theory is used to estimate. As a partial guide to its general accuracy, we may compare the estimate (6.9.1) of the coefficient c_1 with the exact value, obtained from a complete solution for the velocity potential ϕ, for prolate ellipsoids of revolution moving parallel to their axis. The stream function of the flow in this case is given in (6.8.30), and our concern is with the form of this flow field at large distances from the origin. It follows from the definition of the elliptic co-ordinates used there that, as $r \to \infty$,

$$\eta \sim \theta \quad \text{and} \quad \tfrac{1}{2}(a^2 - b^2)^{\frac{1}{2}} e^{\xi} \sim r,$$

so that $\quad (\cosh \xi + \sinh^2 \xi \log \tanh \tfrac{1}{2}\xi) \sin^2 \eta \sim \tfrac{2}{3}(a^2 - b^2)^{\frac{1}{2}} \dfrac{\sin^2 \theta}{r}$.

Consequently the exact expression for the corresponding asymptotic form of ϕ is

$$\phi \sim -c_1 \frac{\cos \theta}{r^2},$$

where $\qquad c_1 = \dfrac{\tfrac{1}{3} U b^2 (a^2 - b^2)^{\frac{3}{2}}}{a(a^2 - b^2)^{\frac{1}{2}} + b^2 \log \left\{ \dfrac{a - (a^2 - b^2)^{\frac{1}{2}}}{b} \right\}}.$ \hfill (6.9.3)

When $b = a$, the value of c_1 given by (6.9.3) is $\tfrac{1}{2} U a^3$, as it should be for a sphere; the approximate expression (6.9.1) for c_1 is here $\tfrac{1}{3} U a^3$, showing that even for a body with no claims to slenderness the error is not excessive. When the thickness ratio $b/a(= \gamma$ say) is small compared with unity, (6.9.3) becomes

$$c_1 \approx \tfrac{1}{3} U a^3 \gamma^2 (1 - \gamma^2 \log \gamma),$$

whereas the approximate formula (6.9.1) is, for a prolate ellipsoid,

$$c_1 = \tfrac{1}{3} U a^3 \gamma^2.$$

Thus the relative error of the slender-body estimate is $\gamma^2 \log \gamma$.

A similar kind of approximation can be employed to take account of any 'side wind' on a slender axisymmetric body. Suppose that the velocity of the body through fluid at rest at infinity has components $(U, V, 0)$ relative to rectilinear axes of which the first is parallel to the axis of the body. The resulting irrotational flow relative to axes fixed in the body can be regarded as a superposition of the flows due to two uniform streams, one with velocity components $(-U, 0, 0)$ to which the above results apply, and the other a pure side wind with velocity components $(0, -V, 0)$. Since the cross-section of the body varies only slowly along the axis, the flow due to this side wind near position x is approximately the same as for a circular cylinder of cross-sectional area A in a stream of speed V normal to the generators (and with zero circulation round it); that is, it is approximately the same as that due to a two-dimensional source doublet† of strength $2VA$ at the centre directed against the uniform stream of speed V. Thus the presence of the body in the side wind is represented approximately by a distribution of source doublets on the axis of the body, with vectorial strength $(0, 2VA, 0)$ per unit length. The entire flow field may now be calculated when the body shape is given.

The total strength of the source doublets representing the body in a side wind has components $(0, 2VV_0, 0)$, where V_0 is again the volume of the body. On combining this result with (6.9.1), we see that the strength of the source doublet that has the same asymptotic flow field as an axisymmetric slender body moving with velocity $(U, V, 0)$ through fluid at rest at infinity is

$$4\pi\mathbf{c} = V_0(U, 2V, 0). \tag{6.9.4}$$

The corresponding approximate expression for the kinetic energy of the fluid, obtained from (6.4.17), is $\frac{1}{2}\rho V_0 V^2$, which is equal, as was to be expected from the nature of the approximation used, to the kinetic energy of the fluid between two cross-sectional planes of a circular cylinder of arbitrary radius moving with speed V normal to the axis, the distance between the planes being such that they enclose a volume V_0 of the cylinder.

Slender bodies in two dimensions

When a body in a two-dimensional field is symmetrical about a centre-line and has a small ratio of thickness to length, irrotational flow due to movement of the body in the direction of its centre-line can again be simulated approximately by a distribution of sources along the centre-line. The body surface is defined here by the curve $y = \mp \frac{1}{2}y_0(x)$, where $y_0(x)$ is again assumed to be a slowly-varying function of x. Then the argument used above shows that the curve $y = \mp \frac{1}{2}y_0(x)$ will be approximately a streamline in the flow due to a stream whose uniform velocity at infinity has components $(-U, 0)$ and a distribution of sources on the centre-line with strength $-U\,dy_0/dx$ per unit length.

However, a new approach is needed when a symmetrical body is not moving in the direction of its centre-line or when the body is not symmetrical. Provided the tangent to the body profile is still approximately parallel to the direction of its motion, the effect of the finite thickness of the body is still mainly to displace fluid elements laterally without changing appreciably their speed relative to the body, and so can be represented approximately by a distribution of sources as above, with $y_0(x)$ now denoting the thickness of the body at position x. However, we also need to find some way of representing the fact that in the flow relative to the body the streamlines at the two sides of the body are not only separated from each other by

† See the remarks following (6.4.6).

a distance $y_0(x)$ but are both inclined to the direction of the stream at infinity at small angles whose sum is non-zero.

There is no distributed flow singularity of which the local strength density imposes a certain direction on the streamlines, but there is a singularity, viz. a point vortex, which if distributed along the centre-line of the body would cause streamlines to intersect this centre-line at a non-zero angle. We saw in § 2.6 that the local strength density of a sheet vortex (which in this context of two-dimensional flow signifies a continuous distribution of vortex strength along a line in the (x, y)-plane, with the vorticity everywhere normal to this plane) is equal in magnitude to the local jump in tangential component of velocity across the sheet. This suggests we should be on the right track in choosing a sheet vortex as the appropriate singularity, since we expect that any lack of mirror-symmetry of the flow about a line drawn through the length of the body will be accompanied by a difference between the fluid velocities at two neighbouring points on different sides of the body.

These considerations of the manner of representing inclination of the surface streamlines by means of a distribution of singularities are most useful in the context of two-dimensional aerofoil theory, and we shall therefore present the analysis as an application to that case.

Thin aerofoils in two dimensions

The typical thin aerofoil has both thickness and camber, like that shown in figure 6.7.9, and the tangent to the surface makes a small angle with the direction of the incident stream everywhere except near the nose. The sharp trailing edge will be placed at the origin and the leading edge L, defined as the point of the aerofoil furthest from the trailing edge, on the x-axis at $x = c$ say (c being the chord of the aerofoil). The equations to the upper and lower surfaces of the aerofoil may then be written as

$$y = y_1(x) \pm \tfrac{1}{2} y_0(x) \quad (0 \leqslant x \leqslant c).$$

The effect of the non-zero thickness $y_0(x)$ on the flow can be simulated separately by a distribution of sources, as explained above. Our present concern is thus with the irrotational flow due to a curved plate of the form $y = y_1(x)$ held stationary in stream whose uniform velocity at infinity has components $(-W \cos \alpha, W \sin \alpha)$, where α is the (small) angle of incidence of the aerofoil (figure 6.9.2). There is no flux of mass across the curve $y = y_1(x)$, and there is in general a discontinuity in the tangential component of velocity at the curve; that is, the curved plate is exactly equivalent to a sheet vortex coinciding with the curve $y = y_1(x)$ whose strength Γ density is distributed so as to make the normal component of velocity zero at $y = y_1(x)$.

To the first order in the perturbation velocity (u, v) due to the presence of the plate, the condition of zero flux of mass across the plate can be written as

$$\left(\frac{v + \alpha W}{W} \right)_{y = y_1(x)} = -\frac{dy_1}{dx}.$$

We now make further use of the fact that $|y_1| \ll c$ to assume that, for the purpose of evaluation of the perturbation velocity (u, v), the sheet vortex lies on the x-axis, in the range $0 \leqslant x \leqslant c$, rather than on the line $y = y_1(x)$. An element δx of the x-axis acts as a point vortex of strength $\Gamma(x) \delta x$, and the approximate relation from which $\Gamma(x)$ can be found, when the aerofoil shape is given, is thus

$$\frac{1}{2\pi W} \int_0^c \frac{\Gamma(x')}{x - x'} \, dx' = -\alpha - \frac{dy_1}{dx}. \qquad (6.9.5)$$

There is the practical drawback here that the intensity of the appropriate singularity is not specified by the local aerofoil geometry but must instead be found as the solution of an integral equation involving the whole aerofoil.

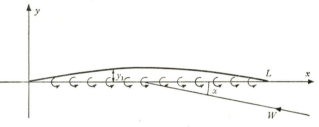

Figure 6.9.2. Representation of an aerofoil without thickness in a stream by a plane sheet vortex.

It is not to be expected that equation (6.9.5) has a unique solution for $\Gamma(x)$, because the flow about any body in a two-dimensional field is not determined unless the circulation round the body is specified. We saw in §6.7 that, in the case of sharp-tailed bodies like aerofoils which are moving steadily, the action of viscosity at the aerofoil surface causes the circulation to take up that value for which the two streams of fluid on either side of the aerofoil leave the trailing edge smoothly without passing round it (Joukowski's hypothesis). In these circumstances the fluid velocity is the same in these two streams near the trailing edge, so that the strength density of the sheet vortex replacing the aerofoil is zero there. Hence we should solve (6.9.5) for $\Gamma(x)$ subject to the condition

$$\Gamma(x) = 0 \quad \text{at} \quad x = 0.$$

On the other hand, it is not possible to apply any condition analogous to Joukowski's hypothesis at the leading edge, and in general there will be an infinite value of the fluid velocity at a sharp leading edge. In practice aerofoils are rounded at the leading edge, but our skeletonized thin aerofoil is sharp there and an infinity of the velocity is inevitable; however the usefulness of the analysis is not affected because (6.9.5) involves only the small component of velocity normal to the chord of the aerofoil. Near the sharp (and cusped) leading edge at $x = c$ the velocity at the aerofoil surface varies as $\frac{1}{2}A_0 Wc^{\frac{1}{2}}(c-x)^{-\frac{1}{2}}$ on one side and as $-\frac{1}{2}A_0 Wc^{\frac{1}{2}}(c-x)^{-\frac{1}{2}}$ on the other (see §6.5), where A_0 is a constant and the factors W and $c^{\frac{1}{2}}$ are included to make A_0 dimensionless, so that we anticipate

$$\Gamma(x) \sim A_0 W \left(\frac{c}{c-x}\right)^{\frac{1}{2}} \quad \text{near} \quad x = c.$$

Indeed we can go further and expect

$$\Gamma(x) - A_0 W \left(\frac{c}{c-x}\right)^{\frac{1}{2}} \to 0 \quad \text{as} \quad x \to c, \tag{6.9.6}$$

since the difference between the velocities on the two sides of the aerofoil near the leading edge vanishes when the local solution corresponding to flow round the leading edge is subtracted out (as may be seen in detail from the relation (6.7.2) for the velocity at the surface of a flat plate).

A solution of the integral equation (6.9.5) may be obtained, although not in closed form, by writing $\Gamma(x)$ as a Fourier series in the variable θ defined by

$$x = \tfrac{1}{2}c(1 - \cos\theta); \tag{6.9.7}$$

θ varies from o to π over the chord of the aerofoil. It is preferable to work with an unknown function which is finite everywhere, so that we consider, not $\Gamma(x)$, but the modified function

$$\Gamma(x) - A_0 W \left(\frac{x}{c-x}\right)^{\frac{1}{2}},$$

which has the convenient property of being required to be zero at $x = 0$ ($\theta = 0$) and at $x = c$ ($\theta = \pi$). We may then suppose, without any loss of generality, that this modified strength distribution is periodic in θ, with period 2π, and an odd function of θ, so that

$$\Gamma = A_0 W \tan \tfrac{1}{2}\theta + W \sum_{n=1}^{\infty} A_n \sin n\theta. \tag{6.9.8}$$

This satisfies the required condition of smooth flow at the trailing edge.

Substitution of (6.9.8) in the integral equation (6.9.5) gives

$$\alpha + \frac{dy_1}{dx} = -\frac{1}{2\pi} \int_0^{\pi} \left(A_0 \tan \tfrac{1}{2}\theta' + \sum_1^{\infty} A_n \sin n\theta' \right) \frac{\sin\theta' \, d\theta'}{\cos\theta' - \cos\theta}$$

$$= \frac{1}{2\pi} A_0(-I_0 + I_1) + \frac{1}{4\pi} \sum_{n=1}^{\infty} A_n(-I_{n-1} + I_{n+1}),$$

where $\quad I_n = \int_0^{\pi} \dfrac{\cos n\theta'}{\cos\theta' - \cos\theta} d\theta' = \pi \dfrac{\sin n\theta}{\sin\theta} \quad \text{for} \quad 0 < \theta < \pi$

is a standard definite integral.† Hence

$$\alpha + \frac{dy_1}{dx} = \tfrac{1}{2}A_0 + \tfrac{1}{2} \sum_{n=1}^{\infty} A_n \cos n\theta, \tag{6.9.9}$$

† I_n is to be regarded as the so-called principal value of the integral, defined as

$$\lim_{\epsilon \to 0} \left(\int_0^{\theta-\epsilon} + \int_{\theta+\epsilon}^{\pi} \right),$$

and the explicit relations for the coefficients $A_0, A_1, \ldots$ in terms of the aerofoil shape are

$$A_0 = 2\alpha + \frac{2}{\pi}\int_0^\pi \frac{dy_1}{dx}\,d\theta, \quad A_n = \frac{4}{\pi}\int_0^\pi \frac{dy_1}{dx}\cos n\theta\, d\theta \quad (n>0).$$

$$(6.9.10)$$

The feature of the flow due to a moving aerofoil that is of most interest is the total circulation round it and the consequent lift generated by the aerofoil. We find for the lift coefficient

$$C_L = \frac{\text{lift force}}{\tfrac{1}{2}\rho W^2 c} = \frac{2}{Wc}\int_0^c \Gamma(x)\,dx$$

$$= \pi(A_0 + \tfrac{1}{2}A_1)$$

$$= 2\pi\alpha + 2\int_0^\pi \frac{dy_1}{dx}(1+\cos\theta)\,d\theta. \qquad (6.9.11)$$

In order to calculate the moment of the forces acting on the aerofoil we may regard an element δx of the sheet vortex as generating a lift force $\rho WT\delta x$. The dimensionless coefficient giving the moment about the leading edge is then

$$(C_M)_{\text{l.e.}} = \frac{\text{moment}}{\tfrac{1}{2}\rho W^2 c^2} = -\frac{2}{Wc^2}\int_0^c (c-x)\,\Gamma(x)\,dx$$

$$= -\tfrac{1}{4}\pi(A_0 + A_1 - \tfrac{1}{2}A_2)$$

$$= -\tfrac{1}{2}\pi\alpha - \int_0^\pi \frac{dy_1}{dx}\cos\theta(1+\cos\theta)\,d\theta. \quad (6.9.12)$$

Thus for many practical purposes it may be necessary to evaluate numerically only one or two integrals involving the aerofoil shape.

The accuracy of some of these results for thin aerofoils may be tested by comparison with the exact results found in §6.7 for Joukowski aerofoils by a method of conformal transformation. For a symmetrical Joukowski aerofoil, for which the corresponding skeleton is a flat plate, the lift coefficient was found (see (6.7.13)—this result is not exact, but is close enough for the comparison) to be

$$2\pi\sin\alpha\left(1+0.77\,\frac{\text{thickness}}{\text{chord}}\right),$$

whereas the relation (6.9.11) from thin aerofoil theory gives $2\pi\alpha$ irrespective

---

so that the large values of the integrand on either side of $\theta' = \theta$ cancel. For $n = 0$ and $0 < \theta < \pi$ we have

$$I_0 = \frac{1}{\sin\theta}\lim_{\epsilon\to 0}\left[\left\{\log\frac{\sin\tfrac{1}{2}(\theta+\theta')}{\sin\tfrac{1}{2}(\theta-\theta')}\right\}_{\theta'=0}^{\theta-\epsilon} + \left\{\log\frac{\sin\tfrac{1}{2}(\theta'+\theta)}{\sin\tfrac{1}{2}(\theta'-\theta)}\right\}_{\theta'=\theta+\epsilon}^{\pi}\right]$$

$$= 0.$$

Also $\qquad\qquad\qquad\qquad I_1 = \pi + I_0\cos\theta = \pi.$

The recurrence formula $\qquad\quad I_{n+1} + I_{n-1} = 2I_n\cos\theta \quad (n \geqslant 1)$

then leads to the result stated.

of the thickness. For a circular-arc aerofoil, which is the member of the family of cambered Joukowski aerofoils considered in §6.7 that has zero thickness, the lift coefficient was found (see (6.7.14)) to be

$$2\pi \frac{\sin(\alpha+\beta)}{\cos\beta},$$

where 2β is the angle between the chord and the tangent at the trailing edge; whereas from (6.9.11) we obtain, with a little calculation,

$$C_L = 2\pi\alpha + 2\sin 2\beta \int_0^\pi \frac{\cos^2\theta\, d\theta}{(1-\sin^2 2\beta\cos^2\theta)^{\frac{1}{2}}}$$

$$\approx 2\pi(\alpha+\beta) + 3\pi\beta^3$$

for $\beta \ll 1$. Thus in each case the dominant term is given correctly.

6.10. Impulsive motion of a fluid

In some circumstances the acceleration of the boundaries and of the fluid is of large magnitude and short duration, and we may usefully consider the limiting case of an impulsive change, as in problems of rigid-body mechanics. Body forces of large magnitude do not act directly on the fluid, but a sudden change of the motion of the boundaries will establish large pressure gradients which in turn produce a sudden change in the velocity at every point of the fluid. Neither the velocity of the boundary nor that of the fluid is large during the sudden change, so that terms in the equation of motion of the fluid involving only velocities or their spatial gradients are negligible by comparison with the term $\partial\mathbf{u}/\partial t$. The approximate form of the equation of motion (without restriction on the viscosity of the fluid) during the sudden change is thus

$$\frac{\partial\mathbf{u}}{\partial t} = -\frac{1}{\rho}\nabla p. \tag{6.10.1}$$

These two surviving terms are of large magnitude for a short interval of time, and the relation between the fluid velocity $\mathbf{u}'$ just before the change begins to the velocity $\mathbf{u}''$ at the same point just after the change is

$$\mathbf{u}'' - \mathbf{u}' = -\frac{1}{\rho}\nabla\Pi, \tag{6.10.2}$$

where
$$\Pi = \int p\, dt \tag{6.10.3}$$

may be termed the *pressure impulse*. p is not zero before and after the impulsive change, but the range of integration in (6.10.3) is small (being the duration of the sudden change) and the value of the integral is presumed not to be affected significantly by the initial and final values of p.

A significant feature of (6.10.2) is that, if the velocity distribution in the fluid before the impulse is irrotational, with potential ϕ', that after the impulse is likewise irrotational (as expected from the fact that the conditions for validity of Kelvin's circulation theorem are satisfied during the sudden change), with potential

$$\phi'' = \phi' - \frac{1}{\rho}\Pi. \tag{6.10.4}$$

This relation provides us with a physical interpretation of the velocity potential. The potential ϕ of a given irrotational velocity distribution may be interpreted as $(-1/\rho)$ times the pressure impulse required to set up the given motion from rest, or, alternatively, as $(1/\rho)$ times the pressure impulse required to reduce the given motion to rest. No rotational motion can be generated from rest or reduced to rest by the action of a pressure impulse.

The same interpretation of the velocity potential is relevant to the formulae for the total kinetic energy of the fluid in terms of integrals over the boundaries. We may imagine the given irrotational motion to have been set up from rest by an impulsive motion of the boundaries, in which case the mean of the initial and final velocities of an element δA of the boundary is $\frac{1}{2}\mathbf{u}$ and the work done by that element of the boundary against the impulsive pressure exerted by the fluid is, according to the usual formula of mechanics,

$$\tfrac{1}{2}\mathbf{u}.\mathbf{n}\,\delta A \times \text{force impulse on unit area of boundary} = -\tfrac{1}{2}\rho\phi\mathbf{u}.\mathbf{n}\,\delta A$$

for a normal $\mathbf{n}$ directed into the fluid. The total kinetic energy is the sum of such contributions from all parts of the boundary of the fluid, including a hypothetical boundary at infinity when the fluid extends to infinity, yielding the formula (6.2.6).

In the particular case of a body moving through fluid at rest at infinity, the effect of the large pressure generated in the fluid at the body surface by any sudden change of velocity of the body is evidently related to the acceleration reaction $\mathbf{G}$ (§6.4). Suppose that the translational velocity of the body changes quickly from $\mathbf{U}'$ to $\mathbf{U}''$, with an accompanying change of the velocity potential of the fluid from ϕ' to ϕ''. Then the (i-component of the) force impulse on the body resulting from the change is

$$-\int\Pi n_i\,dA, \quad = -\rho\int(\phi' - \phi'')n_i\,dA$$
$$= -\rho\int(U'_j - U''_j)\Phi_j n_i\,dA$$
$$= \rho V_0\alpha_{ij}(U'_j - U''_j), \tag{6.10.5}$$

where the integration is taken over the body surface and Φ_j, V_0 and α_{ij} have the meanings stated in §6.4. It follows from (6.4.28) that the force impulse on the body is $\int G_i\,dt$, as was to be expected. As already remarked in connection with (6.4.29), $\rho V_0\alpha_{ij}U_j$ is the impulse which must be given to a rigid body in order to generate from rest the irrotational flow due to motion of the body with velocity $\mathbf{U}$.

Impact of a body on a free surface of liquid

Since an irrotational motion is determined uniquely by given values of the normal component of velocity at each point of the boundary of the fluid, a pressure impulse is generated throughout the fluid by a sudden change of the normal component of velocity anywhere at the boundary of the fluid. Problems involving a sudden change of velocity at part of the boundary arise in connection with the impact of a hammer or a projectile on the free surface of a stationary body of liquid.

Consider for example the simple case of a flat-nosed projectile which impinges normally on to the free surface of a semi-infinite region of stationary liquid at speed U, the requirement being to determine the velocity distribution in the liquid, in particular at the free surface, immediately after impact. Thus we need to determine the velocity potential ϕ of the motion immediately after impact (or, equivalently, the pressure impulse $\Pi = -\rho\phi$), given the boundary conditions

(a) $\mathbf{n}.\nabla\phi = -U$ at the part of the 'free' surface in contact with the projectile;

(b) $\phi = 0$ at the part of the free surface not in contact with the projectile (since the pressure and pressure impulse are necessarily zero at a free surface);

(c) $|\nabla\phi| = 0$ everywhere at large distances from the projectile.

This mathematical problem is equivalent to determining the irrotational flow due to a rigid flat plate moving through infinite fluid broadside-on at speed U, since the plane containing the flat plate instantaneously is here a plane of anti-symmetry in ϕ on which (except at the plate itself) $\phi = 0$. The solution to the problem in this latter form is known, for a flat plate of finite width in two dimensions (§ 6.6), for a flat circular plate in three dimensions (§ 6.8), and for some other plate shapes of less practical relevance.

For illustration we may take the case of an impacting body with a flat circular nose of radius a. The streamlines of the motion set up by the impact are those in one half ($x > 0$, say) of figure 6.8.2, and the stream function of the motion is given by (6.8.32). At the free surface $x = 0$ (where x, σ are the cylindrical co-ordinates used in §6.8), we have $\eta = \tfrac{1}{2}\pi$ and $\sigma = a\cosh\xi$, so that the velocity after impact is normal to the surface with magnitude

$$\left(\frac{1}{\sigma}\frac{\partial\psi}{\partial\sigma}\right)_{x=0} = \frac{1}{\sigma}\left(\frac{\partial\xi}{\partial\sigma}\frac{\partial\psi}{\partial\xi} + \frac{\partial\eta}{\partial\sigma}\frac{\partial\psi}{\partial\eta}\right)_{\eta=\frac{1}{2}\pi}$$

$$= -\frac{aU}{\pi}\frac{1}{\sigma}(2\coth\xi - 2\cosh\xi\cot^{-1}\sinh\xi) \quad (\xi > 0, \text{ or } \sigma > a)$$

$$= -\frac{2U}{\pi}\left\{\left(\frac{\sigma^2}{a^2}-1\right)^{-\frac{1}{2}} - \tan^{-1}\left(\frac{\sigma^2}{a^2}-1\right)^{-\frac{1}{2}}\right\} \quad (\sigma > a). \quad (6.10.6)$$

This distribution of normal velocity at the surface is shown in figure 6.10.1, and reveals the characteristic concentration of 'splash' near the sides of the impacting body. The corresponding force impulse on the body is directed upwards and has magnitude

$$\int_0^a (\Pi)_{x=0}\, 2\pi\sigma\, d\sigma = -\rho \int_0^a (\phi)_{x=0}\, 2\pi\sigma\, d\sigma,$$

$$= \tfrac{4}{3}\rho a^3 U \tag{6.10.7}$$

since $\sigma = a\sin\eta$ and $\phi = -(2aU/\pi)\cos\eta$ on the surface of the body (see §6.8). This expression also follows from (6.10.5) and the result in §6.8 that the virtual inertia for acceleration of a circular disk through infinite fluid (on *both* sides of the disk) in the direction of its axis is $\tfrac{8}{3}\rho a^3$.

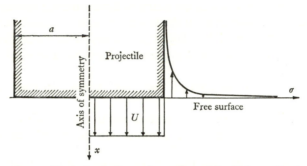

Figure 6.10.1. Vertical velocity at the surface of liquid immediately after normal impact of a body with a flat circular nose.

It should be noted that the motion produced by impact of the flat-nosed body is identical only instantaneously with that in (one half of) the flow field of a flat plate moving through infinite fluid.

6.11. Large gas bubbles in liquid

When a liquid of uniform density contains a bubble or cavity in which the density is negligibly small, the motion of the (gaseous) fluid inside the cavity has no effect on the flow of the surrounding liquid. Thus we have a problem of flow of a liquid of uniform density with a free surface of variable shape which bounds a cavity of finite volume. The mass of gas in the cavity may be large enough for the gas pressure to control the volume of the cavity, as in the case of a gas bubble rising through water under the action of gravity, or the mass may be too small for the gas pressure to be relevant, as in the case of (certain phases of) the gas bubble created by an underwater explosion. In any event, the primary mathematical difficulty usually lies in the determination of the shape of the cavity, and only in special circumstances is it possible to overcome this difficulty. We consider here 'large' gas bubbles on which the effect of surface tension is negligible and for which the Reynolds number of the liquid motion is large. Other cases of flow of liquid with a free surface of unknown shape will be described in §6.13.

A spherical-cap bubble rising through liquid under gravity

It was remarked in § 5.14 that, for small gas bubbles of volume less than about 6×10^{-4} c.c. rising through water, the effect of surface tension is sufficiently strong to keep the bubble approximately spherical. If the volume of a bubble in water is increased beyond this value, the bubble becomes oblate, owing to the variation of pressure in the water over the surface of the bubble, and it is also observed to rise in an oscillatory manner about which little is known. Further increase in bubble volume is accompanied by a progressive flattening of the rear face of the bubble, and for volumes above about 5 c.c., for which surface tension effects are negligible, the bubble is shaped like an umbrella, or a slice off a sphere, as shown by the photographs

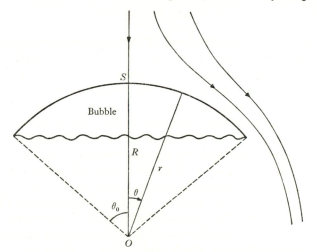

Figure 6.11.3. Definition sketch for flow due to a spherical-cap bubble rising in liquid.

in figure 6.11.1 (plate 14) and figure 6.11.2 (plate 15). The vertical motion of the bubble is now approximately steady. The rear face of the bubble is rather unsteady, and the edge of the slice is jagged and irregular, but by contrast the whole of the front appears to be steady, smooth, and closely spherical. These latter features of the bubble allow the derivation of a simple formula for the steady rate of rise, which will be reproduced here.

We consider the steady flow near the stagnation point on the forward face of the bubble, with axes fixed relative to the bubble, and employ Bernoulli's theorem for a streamline at the bubble surface. The pressure in the water must be uniform over the forward face of the bubble, so that

$$\tfrac{1}{2}q_s^2 = g(R - r\cos\theta), \qquad (6.11.1)$$

where R is the radius of curvature of the bubble surface at the stagnation point S, and r, θ are spherical polar co-ordinates, with origin at the centre of curvature O, for a point on the bubble surface (figure 6.11.3). q_s is the speed of the water at the bubble surface, and, at the high Reynolds numbers relevant to this flow system, presumably depends only on the bubble size and shape and the speed U with which the bubble rises through the water. Within a sufficiently small neighbourhood of the stagnation point, q_s varies linearly with distance from S (see (2.7.11)) and can be written as

$$q_s = \alpha U\theta, \qquad (6.11.2)$$

where α is a dimensionless constant dependent only on the bubble shape. Inspection of (6.11.1) suggests that we should now expand the right-hand side in powers of θ, putting

$$r = R + O(\theta^3), \qquad (6.11.3)$$

the terms of order θ and θ^2 being zero in view of the definition of R, whence we obtain

$$\alpha^2 U^2 = gR. \qquad (6.11.4)$$

The relation (6.11.4) is exact, but further progress demands some approximation to the shape factor α. Now a number of photographs of the flow behind large bubbles (see figure 6.11.2, plate 15) show that separation of the boundary layer on the forward face of the bubble occurs at the jagged edge (in a way which is not clear; the loss of momentum near a zero-stress surface is not normally sufficient for back-flow and separation to occur and presumably the sharpness of the edge of the spherical surface plays an important part in the phenomenon), and that the detached boundary layer lies roughly on the same sphere as the forward face of the bubble, at any rate as far back as the position of maximum cross-section, beyond which the form of the rotational wake appears to fluctuate. The fact that the whole of the inner boundary of the region of irrotational flow is approximately spherical allows us to estimate q_s as if the bubble were part of a sphere of radius R moving through an inviscid fluid, in which case

$$q_s = \tfrac{3}{2} U \sin\theta \qquad (6.11.5)$$

and $\alpha = \tfrac{3}{2}$. We then find from (6.11.4) that

$$U = \tfrac{2}{3}(gR)^{\frac{1}{2}}. \qquad (6.11.6)$$

This relation has been tested experimentally for a wide range of bubble sizes and for several different liquids, and found to be quite accurate for bubble volumes greater than about 2 c.c. The same observations (Davies and Taylor 1950) also revealed that the angle θ_0 specifying the edge of the spherical cap bubble lies between about 46° and 64°, without any evident systematic variation with volume of the bubble. Among other applications, (6.11.6) gives a rough approximation to the rate of rise of the cloud of very hot gas created by an atomic bomb explosion (after the cloud has left the ground and the rate of rise has become steady) in terms of its observed diameter.

The essential idea underlying the relation (6.11.4) is that the linear dependence of static-fluid pressure on vertical height implies a quadratic dependence (locally) on distance from the highest point S of a curved surface like that shown in figure 6.11.3, so that the pressure variation at the bubble surface due to gravity can be balanced by the dynamic term $\frac{1}{2}\rho q_s^2$ when q_s is linear in θ, as it is when there is a stagnation point at S. This idea is applicable in a wide range of circumstances, and the above formulae may be extended in various ways. For instance, we may remove (partially) the restriction that the density $\bar{\rho}$ of fluid inside the bubble is small compared with the density ρ of surrounding fluid; provided the pressure variation inside the bubble due to the dynamic term $\frac{1}{2}\bar{\rho}q_s^2$ is still small compared with that outside, the factor g in (6.11.4) is simply replaced by $g(\rho-\bar{\rho})/\rho$. The argument allows negative values of $\rho-\bar{\rho}$, as when a raindrop falls through air. If the fluid as a whole is given a uniform acceleration $\mathbf{f}$, the factor g in (6.11.4) is replaced by $|\mathbf{g}-\mathbf{f}|$ and no other change is needed. Another interesting and related extension is to the case of a bubble (or 'drop', if $\rho < \bar{\rho}$) which is accelerating relative to the surrounding fluid although remaining approximately of the same shape together with the wake. A term $\partial\phi/\partial t$ must then be added to the left-hand side of (6.11.1), where ϕ is the velocity potential of the flow relative to the bubble (and $\phi \approx \frac{1}{2}\alpha U\theta^2 R$ near $\theta = 0$), and allowance must be made for an effective body force $-dU/dt$ per unit mass on the whole fluid. A case of interest is that in which dU/dt is much larger in magnitude than g, as when a drop of water is released in a high-speed stream of air; in these circumstances, (6.11.4) is replaced by

$$\alpha R\frac{dU}{dt}+\alpha^2 U^2 = R\frac{dU}{dt}\frac{\rho-\bar{\rho}}{\rho}. \tag{6.11.7}$$

Since $\rho \ll \bar{\rho}$ for a water drop in air, the solution for the relative velocity U is

$$U \approx \frac{R\bar{\rho}}{\alpha^2\rho}\left(\frac{1}{t-t_0}\right). \tag{6.11.8}$$

Finally the photograph in figure 6.11.2(c) (plate 15) shows that relations like (6.11.5) and (6.11.6) are applicable to a two-dimensional bubble, the coefficient $\frac{2}{3}$ in (6.11.6) then being replaced by $\frac{1}{2}$.

The remarkable feature of (6.11.4) and its various extensions is that the speed of movement of the bubble is derived in terms of the bubble shape, without any need for consideration of the mechanism of the retarding force which balances the effect of the buoyancy force on a bubble in steady motion. That retarding force is evidently independent of Reynolds number, and the rate of dissipation of mechanical energy is independent of viscosity, implying that stresses due to turbulent transfer of momentum are controlling the flow pattern in the wake of the bubble.

A bubble rising in a vertical tube

The shape of the forward face of large bubbles rising in vertical tubes of circular cross-section is also steady and independent of size, provided the bubble is sufficiently large to fill the tube. As shown schematically in figure 6.11.4, in these circumstances the annular region of the tube containing water tapers down towards the rear of the bubble, which is unsteady and irregular, with a roughly flat base. Increase of the volume of the bubble leads to greater length of the bubble without change of shape of the existing portion. The limiting case corresponds to a vertical tube closed at the top and initially full of water draining from its lower end; here the irregular base of the bubble does not form at all.

Again the effect of viscosity on the flow may be ignored, except perhaps where the annular layer of water is very thin, whence (6.11.4) gives the relation between the steady rate of rise U and the radius of curvature of the bubble boundary at the nose of the bubble. α is constant, for all bubbles sufficiently large to extend nearly across the tube, and characterizes the velocity distribution in the neighbourhood of the stagnation point at the front of the bubble; however, the bubble shape is not geometrically simple and we cannot find the value of α by the method used for a bubble in infinite fluid. Observations show that U is $0.48(ga)^{\frac{1}{2}}$, where a is the radius of the circular tube containing the bubble. This value of U is less than that for a spherical-cap bubble (in infinite fluid) of cap radius R provided $R/a > 0.52$, which suggests that bubbles not quite large enough to fill the tube catch up on one which does and amalgamate with it. This is found to be so in practice, and a stream of bubbles released at the bottom

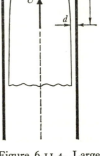

Figure 6.11.4. Large bubble rising through water in a vertical circular tube.

of a long vertical tube is observed to form ultimately into a small number of large bubbles which fill the tube and which all progress up the tube at approximately the same speed.

Far back from the nose of a large bubble which fills the tube, the bubble boundary is nearly cylindrical, with radius $a-d$ say, and the velocity of the water is approximately vertical. Provided the effect of viscosity is negligible here (which requires d to be not too small), the water velocity relative to axes moving with the bubble will then be approximately uniform across the narrow annulus, and equal to its value at the bubble boundary, which we know from Bernoulli's theorem to be $(2gx)^{\frac{1}{2}}$, where x is the axial distance from the bubble nose (figure 6.11.4). Conservation of mass gives the approximate relation

$$\pi a^2 U = \pi\{a^2 - (a-d)^2\}(2gx)^{\frac{1}{2}},$$

from which we find

$$\frac{d}{a} = 1 - \left\{1 - \frac{U}{(2gx)^{\frac{1}{2}}}\right\}^{\frac{1}{2}}, \quad \sim \left(\frac{U^2}{8ga}\frac{a}{x}\right)^{\frac{1}{2}} \qquad (6.11.9)$$

when $x/a \gg 1$. With the observed value of U^2/ga, this becomes

$$\frac{d}{a} \sim 0.17 \left(\frac{a}{x}\right)^{\frac{1}{2}}. \tag{6.11.10}$$

This estimate of the thickness of the annular sheet of water will cease to apply when d is so small as to be comparable with the thickness of the layer adjoining the tube wall within which the vorticity is non-zero. Far enough downstream the vorticity will have diffused completely across the annulus and the viscous force on an element then balances the gravitational force. The velocity profile becomes parabolic (for $d \ll a$), and, as shown in §4.2 for unidirectional flow in a layer with a free surface on a vertical plate, the volume flux relative to axes fixed in the wall is $\frac{1}{3}gd^3/\nu$ per unit width of the layer parallel to the wall. Conservation of mass then gives the relation

$$\pi a^2 U = 2\pi a(Ud + \tfrac{1}{3}gd^3/\nu), \tag{6.11.11}$$

and the approximate solution for d is

$$\frac{d}{a} = \left(\frac{3}{2}\frac{\nu}{a^2 g} U\right)^{\frac{1}{3}},$$

$$= 0.90 \left(\frac{\nu^2}{a^3 g}\right)^{\frac{1}{6}} \tag{6.11.12}$$

with the observed value of U. For water draining out of a tube of radius 10 cm this formula gives $d/a \approx 0.02$.

A spherical expanding bubble

A simple and important case of flow due to a single bubble which is changing in size is provided by a spherical cavity produced by an underwater explosion. The radial accelerations here are usually a good deal larger than g, so that as a first approximation we may neglect the effect of gravity. Provided the speed of the water is small enough for it to be regarded as incompressible, the radial velocity u in the water (assumed to depend only on r) is

$$u = R^2 \dot{R}/r^2, \tag{6.11.13}$$

where R is the radius of the bubble and $\dot{R} = dR/dt$. This is an irrotational velocity distribution, for which the velocity potential is

$$\phi = -\int_r^\infty u\,dr = -R^2 \dot{R}/r. \tag{6.11.14}$$

For the pressure in the water we find (see (6.2.5))

$$\frac{p - p_0}{\rho} = -\frac{\partial \phi}{\partial t} - \tfrac{1}{2}u^2$$

$$= \frac{2R\dot{R}^2 + R^2\ddot{R}}{r} - \frac{\tfrac{1}{2}R^4\dot{R}^2}{r^4}, \tag{6.11.15}$$

where p_0 is the uniform pressure far from the bubble. Hence, if the pressure p_b in the bubble is known as a function of t from data about the explosion, the differential equation governing the radius of the bubble is

$$R\ddot{R} + \tfrac{3}{2}\dot{R}^2 = \frac{p_b - p_0}{\rho}. \tag{6.11.16}$$

This relation may be interpreted as an energy equation. The total kinetic energy of the water is

$$\tfrac{1}{2}\rho \int_R^\infty u^2 4\pi r^2 \, dr, \quad = 2\pi\rho R^3 \dot{R}^2,$$

and the rate at which this energy is changing is equal to the rate of working of the normal force at the bubble boundary and at a material spherical surface 'at infinity', that is, equal to

$$4\pi R^2 p_b \dot{R} - \lim_{r \to \infty} (4\pi r^2 p u), \quad = (p_b - p_0) \, 4\pi R^2 \dot{R};$$

thus
$$\frac{1}{2}\frac{d}{dt}(R^3 \dot{R}^2) = \left(\frac{p_b - p_0}{\rho}\right) R^2 \dot{R}, \tag{6.11.17}$$

thereby recovering (6.11.16).

Explosive which is detonated in a large body of water generates a mass of hot gas at high pressure. As the gas bubble expands and sets the surrounding water into radial motion, the pressure in the gas decreases, approximately according to an adiabatic law which depends on the explosion products but which is usually such as to make the pressure proportional to a large (inverse) power of the bubble radius. In the later stages of the expansion of the bubble, the bubble over-shoots its equilibrium radius and the gas pressure falls well below p_0; we may then regard the bubble as effectively vacuous. The ambient pressure p_0 is constant here, so that (6.11.17) may be integrated to give

$$R^3 \dot{R}^2 = \frac{2}{3}\frac{p_0}{\rho}(R_m^3 - R^3), \tag{6.11.18}$$

where R_m is the maximum radius of the bubble attained when it ceases to expand and begins to contract. The relation between t and the radius of an effectively vacuous bubble in liquid in which the constant uniform pressure p_0 at infinity is positive is thus

$$|t - t_m| = \left(\frac{3\rho}{2p_0}\right)^{\frac{1}{2}} \int_R^{R_m} \frac{dR}{\{(R_m/R)^3 - 1\}^{\frac{1}{2}}}, \tag{6.11.19}$$

where t_m is the instant at which $R = R_m$. This formula is valid for either positive or negative values of $t - t_m$, the motion in the contraction phase $(t > t_m)$ being a simple reversal of that in the expansion phase. If none of the energy E released by the explosion is lost to the surrounding water by thermal radiation or conduction, we can equate E to the total work done against the fluid at infinity up to the instant t_m, at which instant the kinetic energy is zero, giving

$$E = \tfrac{4}{3}\pi p_0 R_m^3. \tag{6.11.20}$$

We shall use these relations again in another context in §6.12.

The relations (6.11.19) and (6.11.20) are found to agree with observations of the radius of bubbles which have expanded beyond their equilibrium radius, provided that, as assumed, the effect of gravity is negligible. Equation (6.11.16) shows that, in the absence of gravity, the radial accelerations when the bubble is near its maximum radius are of order $p_0/\rho R_m$; since a spherical gas bubble would rise through water with an initial acceleration equal to $2g$ (§6.8), the condition that gravity should have negligible effect on the bubble is evidently that

$$R_m \ll p_0/\rho g,$$

which, for an explosion under the sea, can be interpreted as a requirement that the maximum bubble radius be small compared with its depth. When this condition is not satisfied, an upward migration of the bubble is combined with the radial motion in an unsymmetrical way, the speed of rise of the centre of the bubble being greater during the contraction phase than during the expansion phase (Taylor 1963).

6.12. Cavitation in a liquid

As the volume of a given mass of gas is increased, the pressure exerted by the gas decreases; but this pressure remains positive, however large the volume may be. The same is not true of liquids, owing to the quite different form of their equations of state. Liquids have a very small coefficient of compressibility, and quite large changes in pressure are accompanied by small changes in the specific volume of a liquid. In particular, the specific volume of a typical liquid at positive pressures near zero is different from the specific volume at atmospheric pressure by less than 0·01 per cent. Thus the relation $\nabla \cdot \mathbf{u} = 0$ is satisfied accurately in flow fields in which there is a wide range of positive values of the pressure. There arises then the question: what happens when the dynamical conditions in a flowing liquid are such as to produce negative pressures at certain places in the liquid? That such negative pressures are dynamically possible in an incompressible liquid is evident from, for instance, Bernoulli's theorem for steady flow (see (6.3.1)), the pressure being negative here at places on a streamline where the speed exceeds the value $(2H + 2\mathbf{g} \cdot \mathbf{x})^{\frac{1}{2}}$.

The answer, which has important practical consequences for hydraulic machinery and underwater propulsion, is that liquid which has not been specially treated cannot withstand tension and tends to form cavities which expand and relieve the negative pressure. Continuity of the liquid is then lost, and a description of the flow involves the position and motion of the cavity boundaries. The formation and subsequent history of such cavities constitutes the phenomenon of *cavitation*.

Tests on liquids at rest show that the tendency to form cavities when the pressure is reduced nearly to zero is associated with the continual presence of nuclei which are believed to be tiny pockets of undissolved gas; some liquid vapour is certain to exist also in any small pockets, but the gas, usually air, appears to be the more essential element for cavity formation. It is not known with certainty how these pockets of gas are able to persist in a liquid under normal conditions. The inward force at the boundary of a small spherical bubble due to surface tension is very strong, much too strong to be balanced by vapour pressure, and gas subjected to this pressure would quickly pass into solution in the liquid. A common postulate is that the pockets of gas and vapour are able to persist in equilibrium under normal conditions by being trapped in crevices in small hydrophobic (non-wetting) solid particles such as dust particles, which are usually present in liquids; the liquid surface in such cracks and crevices can be concave outwards, in which case the direction of the surface tension force is outward. Then when the pressure in the surrounding liquid is reduced sufficiently below the vapour pressure (which for water at 15 °C is $1 \cdot 704 \times 10^4$ dyn/cm^2, or about 0·017 atmosphere), the gaseous pocket grows and, despite the fact that for a cavity larger than the host solid particle the surface tension force is inward, will not be able to find a new equilibrium radius. For tap water and sea water, the critical steady ambient pressure, below which cavities grow indefinitely in size, is found to be different from the vapour pressure of the water by only a small margin which is usually neglected. On the other hand, water which has been compressed for a few minutes at about 700 atmospheres and which is saturated with air can withstand *tensions* of about 25 atmospheres (Harvey, McElroy and Whiteley 1947), presumably because all except the smallest pockets of undissolved air have then been eliminated. Water which has been 'degassed' in this way likewise does not boil at atmospheric pressure until the temperature is considerably above 100 °C; the phenomena of boiling and growth of cavities in liquids at low pressures are of course mechanically similar.

When the ambient pressure in the liquid is changing rapidly, the critical pressure, below which visible cavities appear in the liquid, depends on the size of the nuclei and on the duration of the applied low pressure, and simple relations are not available. However it proves to be useful in hydraulic practice to continue to assume that the critical pressure is equal to the vapour pressure as a rough working rule.

Examples of cavity formation in steady flow

Perhaps the simplest case of realization of the low pressure needed for cavity formation is provided by the flow due to a 'streamlined' solid body moving steadily through infinite liquid. Relative to axes fixed in the body, the speed of the liquid at infinity is uniform and equal to U say, and that at any fixed point in the liquid (located relative to the body) is αU, where α is independent of U and t except at points in the boundary layer on the body or in the wake downstream from it. The flow is irrotational outside the boundary layer and wake, and so the local absolute pressure in this region is

$$p = \rho(H + \mathbf{g} \cdot \mathbf{x} - \tfrac{1}{2}\alpha^2 U^2),$$

where H is a constant. If the body were absent, the flow (referred to the same axes) would have the same Bernoulli constant H and the local absolute pressure would be

$$p_0 = \rho(H + \mathbf{g} \cdot \mathbf{x} - \tfrac{1}{2}U^2);$$

p_0 varies with position in the vertical, and differs from the static-fluid pressure by a constant quantity. Hence we can write

$$p = p_0 - \tfrac{1}{2}\rho U^2(\alpha^2 - 1). \tag{6.12.1}$$

A measure of the tendency for cavitation to occur locally is then provided by the difference between p and the vapour pressure of the water p_v, which, in an appropriate non-dimensional form, is

$$\frac{p - p_v}{\tfrac{1}{2}\rho U^2} = \frac{p_0 - p_v}{\tfrac{1}{2}\rho U^2} - (\alpha^2 - 1); \tag{6.12.2}$$

as a rough rule, cavities will form in the liquid wherever this quantity is less than zero.

The first term on the right-hand side of (6.12.2) depends only on the operating conditions, whereas the second term depends only on the shape and orientation of the body. For a body of given shape and orientation, the criterion for avoidance of cavitation locally is thus that the so-called *cavitation number*

$$K = \frac{p_0 - p_v}{\tfrac{1}{2}\rho U^2} \tag{6.12.3}$$

should not fall below some critical value. The critical value varies with position in the liquid, since both p_0 and α are functions of position. For a body of sufficiently small vertical extent, p_0 can be regarded as uniform in the neighbourhood of the body and is then the 'ambient pressure' for the body. In such a case the first occurrence of cavitation, as K is decreased, is at the point where α is a maximum and equal to α_m say; and we know from the general result of §6.2 that this maximum must occur on the boundary of the region of irrotational flow. The restriction to be placed on the operating conditions if the absolute pressure is not to fall below the vapour pressure at any point in the fluid is then

$$K > \alpha_m^2 - 1. \tag{6.12.4}$$

For a body moving horizontally in the sea, avoidance of cavitation (which normally is desirable for a variety of reasons) is favoured by increase of the depth, by decrease of U, and by 'streamlining' the body to decrease α_m. For water $\tfrac{1}{2}\rho U^2$ has a value corresponding to two atmospheres when U is about 20 m/sec, and at a depth of h metres in the sea $p_0 - p_v$ is roughly $1 + 0.1h$ atmospheres; thus cavitation will be avoided on a body moving at this speed if the depth in metres is greater than $10(2\alpha_m^2 - 3)$. Cavitation will be avoided at *all* depths at this speed in the case of slender bodies, such as a submarine or a fish, for which $\alpha_m < (\frac{3}{2})^{\frac{1}{2}}$.

Figure 6.12.1 (plate 16) shows the appearance of cavities in the flow past a slender cigar-shaped body in a water tunnel at $K = 0.26$. The cavities are evidently forming near the shoulder of the body, which is where we should expect the pressure to be a minimum, and are being carried downstream to regions of higher pressure where they disappear. Direct measurements of the pressure at the surface of this body at a given speed (from which α_m can be deduced) showed that the vapour pressure would be reached first at a value of K equal to about 0.37, although smaller values of K were needed for the cavities to grow to visible size during the short time spent in a region of negative values of $p - p_v$.

Other cases of steady flow in water in which cavitation may occur are provided by Venturi tubes and propellers. The cross-section of a Venturi tube decreases to a minimum and then increases gradually, so that water flowing along the tube has a local pressure minimum at the throat of the tube.† For certain flow conditions the pressure at the throat may be lower than the vapour pressure, in which event a foaming mixture of water and bubbles forms on the downstream side of the throat, usually with the bubbles congregating near the pipe wall. Examples of this kind of cavity formation are common in water supply systems, in which partially opened taps act like the throat of a Venturi tube. The appearance of cavities in the pipe is accompanied by characteristic hissing noises.

A prominent feature of flow due to propellers is the concentration of vorticity in a stream-tube of small cross-section passing downstream from the tip of each propeller blade in spiral form; this 'tip vortex' is associated with the thrust produced by the blade, as we shall see in §7.8. Outside the tip vortex and the boundary layer on the blade the flow is irrotational, and at a small radial distance r from the centre of the vortex, and outside it, the azimuthal component of velocity v is approximately $C/2\pi r$, where C is the circulation round the vortex (see §2.6). The complete distribution of v near a tip vortex is probably something like that shown in figure 4.5.1. Since the flow is steady, relative to axes rotating and translating with the propeller, the radial pressure gradient at points near a tip vortex is

$$\frac{\partial p}{\partial r} = \frac{\rho v^2}{r},$$

and the pressure at the centre of the vortex is therefore less than that at some

† A Venturi tube is used for the measurement of the quantity of fluid flowing along a cylindrical pipe. A short length of the pipe is replaced by the Venturi tube and the pressure at the wall of the pipe is measured at the throat and at a section just upstream of the Venturi tube. The flux of mass can then be deduced from Bernoulli's theorem if the velocity is approximately uniform across the pipe or it can be related empirically to the difference between these two pressures if the velocity is not uniform.

distance from it by an amount of order ρv_{max}^2. The maximum circumferential speed, v_{max}, depends on C, which is usually known in terms of the
characteristics and operating conditions of the propeller, and on the diameter
of the stream-tube containing the vorticity, which is not. Observation shows
that v_{max} is normally larger than the speed of the water at any other point
in the field of the propeller, and that the pressure at the centre of the tip
vortex may be less than the vapour pressure for quite modest forward speeds
and rates of rotation of propellers of conventional type; consequently the
centre of the tip vortex from each propeller blade may be the first part of the
flow due to a ship or submerged submarine to show cavity formation as the
forward speed is increased.

Figure 6.12.2(*a*) (plate 16) shows a model propeller in a water tunnel
operating under such conditions that in the absence of cavitation the pressure
at the centre of each tip vortex would be less than the vapour pressure.
A cylindrical cavity has formed at the centre of the tip vortex from each of
the three blades and, by imposing a larger minimum path length on closed
circuits round the tip vortex, has eliminated the region of very high velocity
and of pressure lower than the vapour pressure.

Examples of cavity formation in unsteady flow

The possibility of the pressure being less than the vapour pressure through
the effects of temporal variation of the velocity is illustrated well by the case
of liquid contained in a length of pipe on one side of a tightly fitting piston
with a specified motion. Initially the piston and liquid are stationary, and
then the piston is drawn away from the liquid with velocity $u(t)$. In the
absence of cavities, and with the assumption of irrotational flow (which is
valid for a certain time depending on the pipe dimensions), the velocity
everywhere in the liquid is $u(t)$ and the gradient of pressure in the flow
direction is $-\rho \, du/dt$, apart from any contribution due to gravity. Thus if the
absolute pressure is fixed and equal to p_0 at a point in the liquid distance x
from the piston, perhaps through the pipe being open to the atmosphere at
this point, the pressure in the liquid is lowest at the piston and is there
equal to

$$p_0 - x\rho \, du/dt. \qquad (6.12.5)$$

The receding piston is equivalent in its effect on the pressure distribution
to a body force of uniform magnitude directed along the pipe axis (whether
straight or slightly curved) and away from the piston. A reduction of the
pressure at the piston to one atmosphere less than p_0 would be achieved in
a water column of length about $1020n^{-1}$ cm by a piston acceleration of ng cm
sec^{-2}; thus a sharp blow to a piston (in the direction away from the liquid)
at one end of a water column of only a few tens of cm length can produce a
state of tension, accompanied, if the duration of the acceleration is sufficient,
by cavities in the water.

If we now imagine the piston and the liquid to be moving initially with speed U in the direction of the pipe axis, in the sense opposite to the subsequent acceleration of the piston, we have a simplified version of the process of suddenly turning off a tap at some distance upstream from the outlet of a water supply pipe. The time T taken to close the tap can be regarded as an estimate of the time taken by the piston to reduce its velocity from U to zero, giving U/T as a measure of the piston acceleration. The formation of a cavity on the downstream side of the tap when the tap is turned off quickly is usually made evident by the metallic thud resulting from the subsequent collapse of the cavity.

Similar calculations may be made for cyclic accelerations imposed on a fluid. If a beaker of water open at the top is clamped to a vibrating table which oscillates in the vertical direction, the pressure at the bottom of the beaker goes through a cycle with a minimum value which is $\rho h f$ below the static-fluid value, where h is the depth of water and f is the peak acceleration. The formation of visible cavities here depends on the duration of the low-pressure phase and on the size of the small air or vapour bubbles which act as nuclei.

A common method of causing cavitation deliberately in order to rid the liquid of dissolved gas is to focus a beam of acoustic (or compression) waves of ultrasonic frequency on some point in the liquid. When the intensity of the radiation is large enough in the selected region of the liquid, the liquid is in tension during part of each cycle. The tension phase is usually too brief for appreciable growth of cavities during one cycle, and one might suppose that each tiny pocket of undissolved gas simply oscillates about a small equilibrium size. However, the average size of a bubble over one cycle is observed to increase gradually, at any rate while there is still some gas dissolved in the liquid. The explanation is evidently that effects of the second order in the range of oscillation of the bubble radius cause a slightly greater gain of gas to the bubble by diffusion from the surrounding liquid during the expansion phase than the corresponding loss during the compression phase. As noted in § 6.8, there is also a tendency for neighbouring bubbles oscillating in the same phase to approach one another and coalesce, whence larger bubbles are formed and rise to the surface by gravity.

Collapse of a transient cavity

If the pressure in the neighbourhood of a cavity rises above the vapour pressure again, the cavity collapses. The walls of the cavity rush towards one another, usually with only a small amount of gas (mostly acquired during the life of the cavity by evaporation at the cavity boundary and by diffusion of dissolved gas across it) in the cavity to act as a cushion, and the impact has a hard, almost metallic, quality owing to the low degree of compressibility of the water. A very large pressure develops in the water near the cavity at the moment of impact, and is then propagated throughout the water as a compression wave. From a practical stand-point, this pressure pulse of high intensity spreading out from each collapsed cavity is an important, and usually an undesirable, feature of cavitation. It is heard as a disturbingly loud noise in main water supply systems and in hydraulic pumps. When arising from cavitation near the propeller of a ship, it may cause strong vibration of the propeller and is detectable many miles away by underwater acoustic listening devices. Most seriously, the continual collapse of cavities leads rapidly to deterioration and erosion of nearby solid surfaces. The mechanism of this damage to metal blades of propellers and turbines and concrete spillways of dams is not well understood, but local fatigue failure of the solid material due to the repeated high stressing seems to play a central part.

Estimates of the maximum pressure developed and of other properties of the collapsing cavity are consequently of interest. This is a fairly simple matter in the effectively one-dimensional case of the cavity formed by a withdrawing piston as described above. Suppose, for instance, that, at a certain time after the piston ceases to accelerate, the returning water column is about to strike the piston with relative velocity U. (In the case of a tap closed instantaneously, energy considerations show that the water on the downstream side returns to the tap with the same speed as it had just before a cavity formed, provided the mass of the column of water remains unchanged.) If the water were truly incompressible, the whole column of water would have its relative velocity reduced to zero instantly and the pressure generated impulsively in the water would be infinite; but, owing to the actual compressibility of water, pressure pulses propagate through the water with a finite velocity, c say (this being the speed of travel of sound waves if the amplitude of the pulse is not too large), and only that part of the column which can be reached by a pressure wave emanating from the piston in the interval of time since impact occurred has its relative velocity reduced to zero. Thus the rate of change of momentum of the column of water is $\rho c U$ per unit area of cross-section, and this must be the excess pressure developed in the water near the piston. The speed of sound waves in water is 1400 m/sec, so that for water flowing along a supply pipe at a speed of 1 m/sec the 'water-hammer' pressure developed, by sudden closing of a tap, first on the upstream side and subsequently on the downstream side by collapse of the resulting cavity (when the tap is at some distance from the outlet, and in the absence of energy losses) is about 1.4×10^7 dyn/cm^2, or 14 atmospheres.

In the case of a cavity bounded in all three dimensions, the collapsing motion evidently depends strongly on the shape of the cavity boundary. Photographs of cavities formed in the low-pressure region of steady flow past a body, as in figure 6.12.1 (plate 16), show that the cavities are roughly spherical when their size is a maximum. Observations of the collapse of a single cavity under more controllable conditions, like those shown in figure 6.12.3 (plate 17), confirm that a cavity which is initially spherical remains approximately spherical until the collapse is nearly complete. We therefore assume spherical symmetry of the collapsing motion, with considerable gain in simplicity of the analysis.

The differential equation governing the radius $R(t)$ of a spherical cavity for given values of the pressure p_0 'at infinity' and p_b in the cavity has already been obtained (see (6.11.16)). In the present context p_0 is the ambient pressure for the cavity and is equal to the pressure in the liquid at the position of the cavity and in the absence of the cavity; and p_b can be equated to the vapour pressure p_v at all times except when R is very small, either near the inception of the cavity (when surface tension may be important) or near the instant of complete collapse (when in addition the enclosed water vapour may be being compressed too rapidly for condensation to occur). This equation for $R(t)$ with $p_b = p_v$ can be integrated numerically when p_0 is known as a function of time from the circumstances of the water flow field. Agreement between the radius calculated in this way and the observed cavity size has been obtained in the case of cavities forming near the shoulder of the body shown in figure 6.12.1 (plate 16) and being carried downstream to places where $p_0 > p_v$ (Plesset 1949). The observations cannot follow the cavity in the last stages of collapse, since the radius is then very small and the radial velocity very large, so that it is necessary to rely mostly on calculation for information about this phase.

In the simplest case the spherical cavity collapses from rest with $p_0 - p_v$ constant. Equation (6.11.18) then applies, with a slight modification, and

$$\dot{R}^2 = \frac{2}{3}\frac{p_0 - p_v}{\rho}\left(\frac{R_m^3}{R^3} - 1\right), \qquad (6.12.6)$$

where R_m is the maximum radius of the cavity. Integration of this equation must be done numerically, and the resulting variation of cavity radius R with time t since the fluid was at rest is shown in figure 6.12.4. The instant t_0 at which $R = 0$ and the collapse is complete, according to equation (6.12.6), can be found either by direct numerical integration or analytically in terms of the Gamma function and is

$$t_0 = 0.915 R_m \left(\frac{\rho}{p_0 - p_v}\right)^{\frac{1}{2}}. \qquad (6.12.7)$$

The relation between R/R_m and t/t_0, viz.

$$\frac{t}{t_0} = 1.34 \int_{R/R_m}^{1} \frac{dx}{(x^{-3} - 1)^{\frac{1}{2}}}, \qquad (6.12.8)$$

involves no parameters, and has been found to agree well with observations of a collapsing cavity with $p_0 - p_v$ constant (see the comparison in figure 6.12.4 with the observations of figure 6.12.3, plate 17).

The violent final stage of the collapse takes place in such a short time that the variation of p_0 is likely to be negligible in all circumstances; on the other hand, the assumption that the cavity pressure remains equal to p_v may no

longer be accurate. If for simplicity we regard $p_0 - p_v$ as constant in this final
phase, when $R \ll R_m$, we have

$$\dot{R} \sim -\left(\frac{2}{3}\frac{p_0 - p_v}{\rho}\right)^{\frac{1}{2}}\left(\frac{R_m}{R}\right)^{\frac{3}{2}}. \tag{6.12.9}$$

In these last stages most of the available work $E, = \frac{4}{3}\pi R_m^3(p_0 - p_v)$, has been
converted to kinetic energy of the water, and this kinetic energy becomes
concentrated in a smaller volume of liquid as $R \to 0$. At a given radial
position in the liquid, the velocity of the liquid varies as $R^{\frac{1}{2}}$ as $R \to 0$ (for see
(6.11.13)), and there is evidently a strong deceleration of most of the liquid,
although a strong acceleration of the liquid within one or two cavity radii

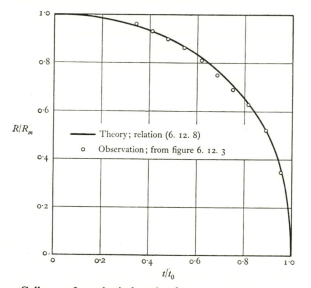

Figure 6.12.4. Collapse of a spherical cavity from rest with constant overall pressure
difference. The value of R_m for the cavity photographed in figure 6.12.3 (plate 17) was
found by extrapolation to be 0·72 cm, and the origin of time was located by the instant of
complete collapse.

from the cavity surface. This implies the existence of a maximum of the
pressure in the liquid. The explicit expression for the pressure in the liquid
is found from (6.11.15) and (6.11.16) to be

$$\frac{p - p_0}{\rho} = -\left(\frac{p_0 - p_v}{\rho}\right)\frac{R}{r} + \frac{1}{2}\dot{R}^2\left(\frac{R}{r} - \frac{R^4}{r^4}\right), \tag{6.12.10}$$

of which the dominant second term on the right-hand side is positive with
a maximum value of $\frac{3}{8}\dot{R}^2/4^{\frac{1}{3}}$ at the position $r = 4^{\frac{1}{3}}R$. It follows from (6.12.10)
and (6.12.9) that the asymptotic value of the peak pressure, p_m say, as $R \to 0$,
is given by

$$p_m - p_0 \sim 4^{-\frac{4}{3}}(p_0 - p_v)(R_m/R)^3. \tag{6.12.11}$$

For a cavity collapsing from rest under the influence of a pressure difference

$p_0 - p_v$ of about an atmosphere, the maximum pressure in the liquid thus reaches 157 atmospheres when the cavity radius is 0·1 of its initial value, at which moment the radial velocity at the cavity boundary is 260 m/sec, and both peak pressure and boundary speed are shooting up as the cavity shrinks further.

These considerations, which were first put forward by Rayleigh (1917), suffice to show that the maximum pressures which may be developed at a plane solid surface on which a hemispherical cavity repeatedly forms and collapses, or at the wall of a conical crevice in a solid surface occupied by a cavity, are large enough to account for local failure of metals. They are less conclusive about the effect of a collapsing cavity not in contact with a solid surface, because the analysis is based on the assumption of incompressibility of the liquid and provides no information about the way in which a pressure pulse is propagated away from a collapsed cavity. Compressibility effects become important when the speed at the cavity boundary becomes comparable with the speed of sound in the liquid (1400 m/sec for water), since pressure signals certainly do not then pass almost instantaneously from one part of the liquid to another as assumed implicitly in the above analysis, but a simple method of allowing for them as in the deduction of water-hammer pressure is not available. Realistic analysis of the final stages of the collapse must also take into account other neglected physical effects. Departures from sphericity may occur as a consequence of spatial variation of the ambient pressure p_0 (due either to gravity or associated with the background motion of the water) or perhaps through the influence of a neighbouring rigid boundary, and tend to become more marked near the end of the collapse phase.† There is also the cushioning of the collapse by water vapour which enters the cavity during the large-radius phase and by air which either was in the cavity nucleus initially or passes out of solution across the cavity boundary during the large-radius phase.‡ It is generally believed that a non-negligible fraction of the available energy E is ultimately converted to acoustic radiation away from the collapsed cavity (the remainder being dissipated locally by effects of viscosity and heat conductivity), but reliable estimates of the peak pressure in the radiating pulse or shock wave are difficult to obtain.

† The conical protuberance visible on the rebound cavity in figure 6.12.3 (plate 17) is characteristic of cavities which collapse in the presence of a gradient of ambient pressure (due to gravity in this case), and is believed to be a jet from the high pressure side of the cavity piercing the opposite side following severe distortion of the cavity near the instant of minimum volume (Benjamin and Ellis 1966).

‡ If the cavity contains so much air that the increase in air pressure due to shrinking of the cavity retards the collapse and prevents the speed of the cavity boundary from becoming comparable with the speed of sound in the water, analysis of the above kind can be used to follow a rebound of the cavity from some minimum radius to its original maximum and a subsequent oscillation about an equilibrium radius between these two values. Cavities which make several rebounds with diminishing maximum radius have been observed (Knapp 1952).

Steady-state cavities

It may happen in a steady flow that the pressure in a certain region is so far below the vapour pressure that numerous cavities form and reach non-negligible size (relative to the dimensions of the relevant part of the flow field) before being swept downstream. The total cavity volume is then significant, and affects the distribution of velocity in the liquid, always in such a way as to tend to bring the minimum pressure in the liquid up to the vapour pressure. Under even more extreme conditions, the cavities in the region of low pressure join to form one large permanent cavity. Figure 6.12.2(*b*) (plate 16) shows the development of such a cavity on the outer half of the low-pressure side of a propeller blade and extending downstream in the tip vortex. This steady-state or 'sheet' cavitation poses additional engineering problems, quite apart from the noise, vibration and damage resulting from the collapse of small cavities being detached continually from the irregular edges of the main cavity, because the existence of the large cavity changes the velocity and pressure distributions on the body surface in a way which is not readily predicted. In the case of a propeller blade the existence of the large cavity on the suction side prevents the realization of the intended low pressure there and the resultant force exerted by the blade on the water is correspondingly reduced. If the shape of the permanent cavity under different conditions could be predicted, the propeller could be designed to allow for the existence of the cavity, but this is not possible in general.

Just as the cavitation number defined in (6.12.3) was used to characterize the conditions for inception of cavitation, so also it may be used as a defining parameter for the form of a steady-state cavity. When a rigid body of given shape is placed in a steady stream of liquid, and cavities are absent, the non-dimensional form of the flow field is determined wholly by the Reynolds number. If now a steady-state cavity of appreciable size is admitted as a feature of the flow field, the uniform pressure in the cavity, p_c say, enters as a new relevant physical parameter. The effect of the existence of a permanent cavity is to tend to prevent the occurrence in the water of a region in which the pressure is below vapour pressure, and if this prevention is assumed to be complete it follows that the water is in compression everywhere except at the cavity boundary. In these circumstances a uniform increase in the pressure throughout the cavity and the whole body of water can have no effect on the flow field, showing that p_c is relevant not as an absolute but as a relative pressure. The approximately uniform pressure p_0 that would exist in the neighbourhood of the cavity in the absence of the body giving rise to it—that is, the ambient pressure for the cavity— again serves as a convenient reference pressure, so that a non-dimensional number representing the effect of the permanent cavity is

$$K = \frac{p_0 - p_c}{\frac{1}{2}\rho U^2};\qquad\qquad (6.12.12)$$

this cavitation number and the Reynolds number are then sufficient for the determination of the flow field in non-dimensional form and in particular for the determination of the shape of the cavity, provided that effects of gravity are negligible.

The fact that the steady cavity shape depends on $p_0 - p_c$ and not on the absolute magnitude of p_c (provided only that p_c is not less than p_v, since, if it were, further extensive cavitation would occur in the water) has useful consequences. It allows the experimenter to achieve a desired small value of K, and to obtain a flow which is dynamically similar to one in which a permanent vapour-filled cavity occurs, by increasing p_c by external means (usually by leading air at given pressure to the cavity through a tube embedded in the body to which the cavity is attached) rather than by the technically more difficult procedure of increasing U or reducing p_0. Figure 6.12.5 (plate 18) shows permanent cavities attached to the downstream side of a circular disk held normal to a stream of water with $K = 0.19$; in figure 6.12.5 (a) the cavity pressure is being maintained by an air supply at a level above vapour pressure, whereas in figures 6.12.5 (b) and (c) the cavity pressure is not controlled and is presumably close to the vapour pressure. As well as confirming the expected similarity of shape of cavities obtained at equal values of K, these photographs reveal what appears to be a characteristic difference in the surfaces of vapour-filled and air-filled cavities. The water close to the surface of the vapour-filled cavity is in a state of incipient cavitation, the roughness and continual oscillation of the surface being probably attributable to 'boiling' at the surface.

There are also circumstances in which an air-filled cavity at a pressure well above vapour pressure occurs naturally, viz. when a projectile enters water through an air–water boundary. If a solid sphere is dropped or shot vertically into water as in figure 6.12.6 (plate 19), a cavity joined temporarily to the surface forms. If the sphere speed is large enough, the cavity later detaches from the surface and continues forward with the sphere, the pressure in the cavity then being above vapour pressure. However, a close correspondence with steady flow past a fixed sphere at a definite value of the cavitation number is not to be expected, since the free sphere is decelerating, pressure varies with depth in the undisturbed water and the cavity attached to the sphere is continually losing air by entrainment at the cavity boundary. The dynamics of the air entering the cavity in the initial stages appear to influence the way in which the cavity closes and the total volume of air which continues downward with the sphere.

It appears from observations that the shape of a steady-state cavity is determined mainly by the cavitation number, in the case of a body of given shape held in an otherwise uniform stream of liquid. However, only in the case of an air-filled cavity behind a body with a salient edge is the cavity reasonably steady, and even then the rear of the cavity does not have a well-defined shape, as figure 6.12.5 (a) (plate 18) demonstrates. For all vapour-filled cavities, and for vapour- or air-filled cavities behind bodies without a salient edge, the 'shape' of the cavity has meaning, in practice, only as an average over time. In the absence of a salient edge of the body to locate and stabilize the position of both boundary-layer separation and the point of attachment of the cavity, there can also be an interaction of the properties of the boundary layer and the cavity, as the striking photographs in figure 6.12.7 (plate 18) show. It seems likely that a cavity boundary usually coincides with a separated boundary layer.

Leaving aside the instabilities and unsteadiness and the boundary-layer influences that may occur in some circumstances, the determination of the shape of a steady-state cavity behind a body in a stream at a given value of K presents an interesting problem in irrotational flow theory to be taken up in the next section.

6.13. Free-streamline theory, and steady jets and cavities

We shall consider here some problems in which steady flow at large Reynolds number is bounded partly by rigid walls and partly by 'free streamlines' of unknown shape on which the pressure is constant and has a known value. These flow systems can mostly be classified loosely as in-volving either 'unsubmerged' jets, that is, columns of liquid of finite lateral dimensions (normal to the general flow direction) surrounded by gas, or gaseous cavities of finite lateral dimensions surrounded by liquid; and there are some involving bodies moving along the free surface of a body of liquid otherwise at rest under gravity. We shall assume that the upstream con-ditions are such that the flow is irrotational everywhere except near the boundaries, with the usual qualifications about the applicability of the results.

In a number of these problems the broad features of the flow fields can be discerned by inspection, supplemented perhaps by momentum integral arguments as in §6.3, but in others, particularly those involving cavities, the character of the flow may not be at all evident; in any event detailed calculation is needed for quantitative information. The fact that the shape of the free streamlines is unknown makes the mathematical problem exceedingly difficult, except in the case of two-dimensional flow fields in which the rigid boundaries consist of straight segments.† The use of large automatic computers is enabling more numerical solutions to be obtained, at any rate in two-dimensional and axisymmetric flow fields.

As mentioned in §5.11, some of the early interest in flow fields of this type was stimulated by the notion of free streamlines as a model of the boundary of the broad wake behind a bluff body immersed in a uniform stream (without cavities) at large Reynolds number. It is true that the

† For extensive accounts of the theory of flow with free streamlines, see *Jets, Wakes and Cavities*, by Garrett Birkhoff and E. H. Zarantonello (Academic Press, 1957) and *The Theory of Jets in an Ideal Fluid*, by M. I. Gurevich (Academic Press, 1965 or Pergamon Press, 1966).

velocity in the wake region close to a bluff body is generally smaller than that in the undisturbed stream, but it is rather an over-simplification to take the pressure in the wake to be uniform and in any event the instability of the sheet vortex constituting the wake boundary leads to eddying motion and mixing of the fluid on the two sides of this boundary at short distances from the body. The application of free-streamline theory to cases in which a stream surface has liquid on one side and gas on the other is not open to the same objections, because any motion which develops in the gas has negligible effect on the liquid and the liquid–gas interface is not always dynamically unstable.

We shall assume that the static-fluid pressure is approximately uniform over the region of interest, and that as a consequence of Bernoulli's theorem the fluid velocity is uniform in magnitude on a free streamline; this will evidently be an accurate assumption when

$$gh \ll U^2,$$

where h is a measure of the vertical extent of the region concerned and U is a representative velocity.

A method of solution introduced by Kirchhoff (1869) which applies to two-dimensional fields with piece-wise straight rigid boundaries and which makes use of the complex velocity potential (§§ 2.7, 6.5) will now be described. The name of Helmholtz (1868 b) is also linked with this 'free-streamline theory', since he was the first to solve a problem involving free streamlines. The key to the method is the introduction of a new complex variable†

$$\Omega = \log\frac{dz}{dw} = \log(u - iv)^{-1}$$

$$= \log q^{-1} + i\theta, \tag{6.13.1}$$

where $z = x + iy$, $w = \phi + i\psi$, as before, and q and θ are the magnitude and direction (relative to the x-axis) of the velocity vector (u, v). This variable has the simple properties that the real part of Ω is constant on each free streamline and the imaginary part of Ω is constant on each straight portion of the rigid boundary. The whole boundary of the liquid is therefore represented in the Ω-plane by a straight-sided figure. It also is the case that the boundary of the liquid is represented by a straight-sided figure in the w-plane, namely, two straight lines parallel to the real axis corresponding to the two bounding streamlines. Now we know from the Schwarz–Christoffel theorem (§ 6.5) that it is always possible to find a conformal transformation which maps the interior or exterior of a polygon in one plane on to one half of another plane. Thus it is possible to find conformal relations between Ω and a new complex variable λ, and between w and λ, such that the flow region is mapped on to the upper half of the λ-plane in both cases. In this way a relation between Ω and w can be obtained, from which an expression for w in terms of z follows by integration.

† Kirchhoff's idea was to work with dz/dw; use of the more convenient variable $\log(dz/dw)$ was suggested a little later by Planck (1884).

The procedure will be illustrated by application to one problem involving a jet and to one involving a cavity.

Jet emerging from an orifice in two dimensions

As remarked in §6.3, it is desirable to know the degree of contraction of a jet of liquid emerging from an orifice; and momentum integral arguments are adequate for this purpose only for one or two special orifice shapes. Free-streamline theory is able to provide additional information in some two-dimensional cases (although of course these have much less practical significance).

Suppose first that the orifice is simply a hole in a plane wall of small thickness, and that the wall is part of a large vessel containing liquid. The speed of the liquid on the free streamlines separating from the edges of the orifice is uniform and equal to U, say, and this is also the speed in the interior of the jet far downstream from the orifice where (in the absence of effects of gravity) the streamlines are straight and parallel (figure 6.13.1). The two streamlines bounding the flow field, on which $\psi = \mp \psi_1$ say, are ABC and $A'B'C'$, where A, A', C, C' represent points 'at infinity', and figure 6.13.1 shows the corresponding straight-line boundaries in the Ω- and w-planes, where Ω is now defined a little more conveniently as

$$\Omega = \log\left(U\frac{dz}{dw}\right) = \log\frac{U}{q} + i\theta. \qquad (6.13.2)$$

The conformal transformation of a semi-infinite strip on to the upper half of another complex plane is achieved by the relation (6.5.14), and on adapting this relation to suit the position, width and orientation of the strip in the Ω-plane (which requires $K' = 1$, $z_0 = -\frac{1}{2}\pi i$ in (6.5.14)), we have

$$\lambda = i\sinh\Omega; \qquad (6.13.3)$$

a further choice of the constants in (6.5.14) (namely, $b = -c = 1$) has been made so as to give $\lambda = \mp 1$ as the positions of the points B, B' in the λ-plane. We must now find the relation between w and λ that maps the interior of the infinite strip in the w-plane on to the upper half of the λ-plane, with correspondence between the points A, B, C, and A', B', C' shown in the two planes in figure 6.13.1, the value of ϕ at B and B' being chosen as zero for convenience. The general form of the required transformation is (6.5.15), and again on choosing the constants appropriately ($K' = -\frac{1}{2}\pi/\psi_1, b = 0$, $z_0' = \psi_1$), we have

$$\lambda = ie^{-\frac{1}{2}\pi w/\psi_1}. \qquad (6.13.4)$$

The flow field has thus been mapped on to the upper half of the λ-plane in two coincident ways, implying that

$$\lambda = ie^{-\frac{1}{2}\pi w/\psi_1} = i\sinh\Omega$$
$$= \frac{1}{2}i\left(U\frac{dz}{dw} - \frac{1}{U}\frac{dw}{dz}\right).$$

Hence
$$U\frac{dz}{dw} = -i\lambda \mp (1-\lambda^2)^{\frac12},\qquad(6.13.5)$$

and the negative root may be discarded, since $U\,dz/dw \to 1$ as $\phi \to \infty$, $\lambda \to 0$. Then integration (with the aid of (6.13.4)) gives

$$\frac{\pi U}{2\psi_1}(z-z_0) = i(\lambda-1)-(1-\lambda^2)^{\frac12}+\tanh^{-1}(1-\lambda^2)^{\frac12},\qquad(6.13.6)$$

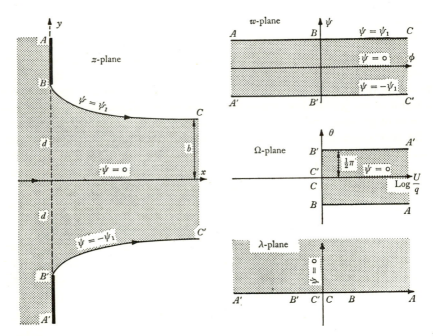

Figure 6.13.1. Conformal transformations required for the determination of the flow from an orifice in a plane wall in two dimensions.

where z_0 is a constant, and since $\lambda = 1$ at the point B where $z = id$ ($2d$ being the width of the orifice) we have
$$z_0 = id.$$

After substitution for λ from (6.13.4), (6.13.6) is the required relation between w and z.

On the free streamline BC, we have
$$\psi = \psi_1,\quad \phi = Us,\quad \Omega = i\theta,$$

and, in view of (6.13.3) and (6.13.4),
$$\lambda = -\sin\theta = e^{-\frac12\pi Us/\psi_1},\qquad(6.13.7)$$

where s denotes distance along the free streamline from B. The equation to this free streamline in parametric form is thus given by (6.13.6) as

$$x = \frac{2\psi_1}{\pi U}(\tanh^{-1}\cos\theta - \cos\theta), \quad y = d - \frac{2\psi_1}{\pi U}(1 + \sin\theta), \quad (6.13.8)$$

and the asymptotic semi-width of the jet is

$$b = \lim_{s \to \infty} y(s) = d - \frac{2\psi_1}{\pi U}.$$

The exact form of the free streamlines is shown in figure 6.13.1. The relation between w and z becomes linear as $s \to \infty$, showing that, as expected, the velocity of the liquid is uniform far downstream, so that $\psi_1 = bU$ and

$$\frac{b}{d} = \frac{\pi}{\pi + 2} = 0.61. \qquad (6.13.9)$$

This value of the contraction ratio, which is close to that found experimentally, may be compared with the values given in figure 6.3.2 for a circular orifice with various forms of the boundary near the orifice.

A similar calculation of the shape of the free streamlines may be made for the jet issuing from a two-dimensional Borda mouthpiece similar to that shown in figure 6.3.2(b), and for an orifice formed by a gap between two inclined plane walls (one of which may be infinite). For a jet from the symmetrical slit between two semi-infinite plane walls at an angle 2α, the contraction ratio is found to be

$$\left(1 + \int_0^1 \frac{\sin\alpha\beta}{\tan\frac{1}{2}\pi\beta}\,d\beta\right)^{-1}.$$

Complex variable methods of determining the flow due to impinging jets in two dimensions are also available.[†]

Two-dimensional flow past a flat plate with a cavity at ambient pressure

Again a number of different cases can be treated, but for the purpose of illustrating the general method we shall consider in detail only the simple case of a flat plate normal to a stream of infinite extent. The pressure in the cavity, which may be chosen freely so far as irrotational flow theory is concerned, will be assumed to be equal to the pressure in the undisturbed stream.[‡] Thus the speed of the liquid on the free streamlines bounding the cavity is equal to U, the uniform speed far ahead of the plate. The definition of Ω given in (6.13.2) will be used again, and we shall take $\psi = 0$ on the central streamline which divides at the stagnation point O (at which $\phi = 0$, say) and later becomes the two free streamlines.

The correspondence between various points on the streamline $\psi = 0$ in the z-, w- and Ω-planes is shown in figure 6.13.2. The region of flow occupies the whole of the w-plane except for a slit along the positive section of the real axis. As before, the procedure is to find transformations which will map

† See *Theoretical Hydrodynamics*, chap. 11, by L. M. Milne-Thomson.
‡ We shall return to the general question of cavity pressure later in this section.

the flow regions in both the w- and Ω-planes coincidentally on to the upper half of a λ-plane. The semi-infinite strip in the Ω-plane has the same width, position and orientation as that in figure 6.13.1, so that

$$\lambda = i \sinh \Omega$$

is again the appropriate relation between Ω and λ, with the positions of the points B, B' in the λ-plane again being $\lambda = \mp 1$. The appropriate relation

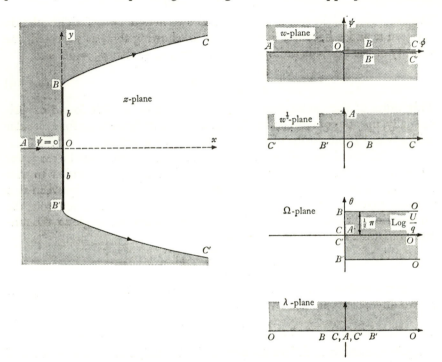

Figure 6.13.2. Conformal transformations required for the determination of the flow past a flat plate with a cavity at ambient pressure.

between w and λ may be recognized by noting first that the flow region occupies the upper half of the $w^{\frac{1}{2}}$-plane (see figure 6.13.2) and then that an inversion and a change of sign are needed to bring the corresponding points on the two real axes into coincidence; thus

$$\lambda = -(kU/w)^{\frac{1}{2}}, \tag{6.13.10}$$

where k is a constant to be determined by the correspondence between the positions of the point B in the two planes.

The required relation between w and Ω is then

$$\lambda = -(kU/w)^{\frac{1}{2}} = i \sinh \Omega$$

$$= \tfrac{1}{2}i\left(U\frac{dz}{dw} - \frac{1}{U}\frac{dw}{dz}\right).$$

Hence
$$\frac{1}{U}\frac{dw}{dz} = -i\left(\frac{kU}{w}\right)^{\frac{1}{2}} \mp \left(1 - \frac{kU}{w}\right)^{\frac{1}{2}}, \tag{6.13.11}$$

and again the negative root may be discarded because $dw/dz = 0$ at O, where $w = 0$. Integration of (6.13.11) gives

$$\frac{z - z_0}{k} = 2i\left(\frac{w}{kU}\right)^{\frac{1}{2}} + \left(\frac{w}{kU}\right)^{\frac{1}{2}}\left(\frac{w}{kU} - 1\right)^{\frac{1}{2}} + \tfrac{1}{2}\pi i - \log\left\{\left(\frac{w}{kU}\right)^{\frac{1}{2}} + \left(\frac{w}{kU} - 1\right)^{\frac{1}{2}}\right\}, \tag{6.13.12}$$

where z_0 is a constant which must be zero in view of the requirement that $w = 0$ at $z = 0$. We can now evaluate the constant k from the information that, at the point B where $\lambda = -1$,

$$z = ib, \quad w/kU = 1;$$

the result is
$$k = \frac{2b}{\pi + 4}, \tag{6.13.13}$$

where $2b$ is the breadth of the plate.

The shape of the free streamlines may now be determined. Since $w = kU$ at B, we have

$$\left.\begin{aligned} w &= \phi = U(k + s), \quad \Omega = i\theta, \\ \lambda &= -\sin\theta = -\left(\frac{k}{k + s}\right)^{\frac{1}{2}} \end{aligned}\right\} \tag{6.13.14}$$

on the free streamline BC, where s is distance along the free streamline from B. The real and imaginary parts of (6.13.12), together with (6.13.14), give the equation to the free streamline BC in parametric form as

$$x = (s^2 + sk)^{\frac{1}{2}} - k\log\left\{\left(\frac{s}{k} + 1\right)^{\frac{1}{2}} + \left(\frac{s}{k}\right)^{\frac{1}{2}}\right\}, \tag{6.13.15}$$

$$y = 2(sk + k^2)^{\frac{1}{2}} + \tfrac{1}{2}\pi k, \tag{6.13.16}$$

k being given by (6.13.13). Thus the cavity extends to infinity downstream and the cavity boundary asymptotes to the parabola

$$y^2 = 4kx = \frac{8b}{\pi + 4}x. \tag{6.13.17}$$

If the cavity were of finite extent, the drag force exerted on the body-cavity combination by the liquid would be zero, by the usual argument for steady irrotational flow (§6.4), and consequently the drag on the body alone would be zero. However, the argument is not applicable here, and the drag is clearly non-zero in view of the fact that the speed of the liquid on the forward face of the plate is everywhere smaller than U. The resultant force on the plate is in the direction of the stream and has magnitude

$$D = \int_{-b}^{b}(p - p_0)_{x=0}\,dy$$

$$= \int_{0}^{b}\rho U^2\,dy - \int_{0}^{kU}\rho(\partial\phi/\partial y)_{x=0}\,d\phi,$$

32-2

where p_0 is the pressure at infinity and in the cavity. Now $-\partial\phi/\partial y$ is the imaginary part of dw/dz, and on OB we have $w = \phi < kU$, whence it follows from (6.13.11) and (6.13.13) that

$$D = \rho U^2 b - \rho U^2 k \int_0^1 \frac{\mathrm{I} - (\mathrm{I} - \gamma)^{\frac{1}{2}}}{\gamma^{\frac{1}{2}}} \, d\gamma \quad (\gamma = \phi/kU)$$

$$= \frac{2\pi}{\pi+4}\rho U^2 b.$$

The drag coefficient is thus

$$C_D = \frac{D}{\frac{1}{2}\rho U^2 2b} = \frac{2\pi}{\pi+4} = 0\cdot88. \tag{6.13.18}$$

Observation shows that the drag coefficient for a flat plate of rectangular plan form, with one dimension very much larger than the other to reproduce the assumed two-dimensionality, set normal to a stream at large Reynolds number, and without a gaseous cavity, is about 2·0, more than twice the above figure. The general reasons why free stream-line theory is not applicable to this case have been stated; more particularly, the larger drag coefficient of the flow without a cavity arises from the development of a considerable suction (relative to the ambient pressure) in the wake immediately behind the plate.

The two-dimensional flow past a flat plate of width $2b$ which is inclined at an arbitrary angle α to the stream, and behind which there is a cavity at ambient pressure, may be calculated in much the same way (Rayleigh 1876). It is found that the free streamlines asymptote to the parabola

$$\left(\frac{y}{b}\right)^2 = \frac{8\sin^2\alpha}{\pi\sin\alpha+4}\frac{x}{b}, \tag{6.13.19}$$

and that the drag force on the plate is

$$D = \frac{2\pi\sin^2\alpha}{\pi\sin\alpha+4}\rho U^2 b. \tag{6.13.20}$$

The resultant force on the plate is necessarily normal to the plate, and there is here a 'lift' force of magnitude $D\cot\alpha$ normal to the direction of the undisturbed stream.

The relations (6.13.19) and (6.13.20) yield a relation between the drag and the single length parameter specifying the parabola to which the cavity asymptotes:

$$\frac{D}{\rho U^2} = \pi \lim_{x\to\infty} \left(\frac{y^2}{4x}\right)_{\text{cavity boundary}}. \tag{6.13.21}$$

A general method of determining the two-dimensional flow past a body with a *curved* boundary of unspecified shape with an attached cavity at ambient pressure has been devised by Levi-Civita (1907),[†] and it may be

† See also *Theoretical Hydrodynamics*, §12.4, by L. M. Milne-Thomson.

shown both that the cavity is asymptotically parabolic and that the drag is related to the asymptotic parabola by the same relation (6.13.21). Two corresponding general results have been established for axisymmetric bodies and cavities by rather difficult mathematics (Levinson 1946). The boundary of an axisymmetric cavity at ambient pressure tends asymptotically to the surface

$$\sigma^2 = 4\,lx\,(\log x)^{-\frac{1}{2}}, \qquad (6.13.22)$$

where (x, σ) are cylindrical co-ordinates with origin near the body and l is a constant length depending on the shape and size of the body, and the drag on the body is found to be $2\pi\rho U^2 l^2$.

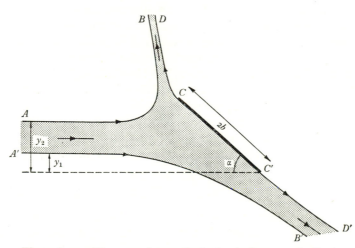

Figure 6.13.3. The general case of a jet impinging on an inclined plate with a cavity at ambient pressure.

Several other two-dimensional flow fields involving a flat plate with an attached cavity at ambient pressure may be determined by the method used above. A very general situation which comes within the scope of the method, and which includes several interesting flow fields as special cases, is shown in figure 6.13.3. The undisturbed stream is here a straight jet of uniform speed U with parallel free boundaries distance $y_2 - y_1$ apart. If $y_2 \to \infty$, $y_1 \to -\infty$, we recover the case of an inclined plate with attached cavity in an infinite stream; if $y_2 - y_1 \ll y_1$ and $0 < y_1 < 2b\sin\alpha$, we obtain the case of a jet impinging on an inclined plane wall, to which the momentum theorem was applied in §6.3; and if $y_1 \to -\infty$, we obtain the quite new case of an inclined flat plate planing on the free surface of a semi-infinite body of water (with the effect of gravity on the motion neglected). In this latter case the jet bounded by the free streamline CD, commonly termed the 'splash', is thrown away from the plate in a direction which depends on y_2, although in reality it would ultimately fall back on to the water surface under gravity and spoil the application of the mathematical solution to some extent.

One can get some idea of the form of a flow field like that sketched in figure 6.13.3, in which $q = U$ on every free streamline, by noting that the free streamlines are likely to be convex to the liquid everywhere. This follows from (6.2.13) and the other result of §6.2 that q cannot have a maximum at an interior point of the liquid; an exception is possible only when a maximum of q occurs at a point on a rigid boundary close to the point of attachment of a free streamline.

Steady-state cavities attached to bodies held in a stream of liquid

We conclude this section with a brief description of some characteristic features of steady flow past bodies with attached cavities (supplementing the remarks made at the end of the last section) and the associated irrotational flow theory. The whole problem is rendered difficult, both by the multiplicity of physical factors affecting the flow under different conditions (tensile strength of the liquid, gravity, viscosity) and by the complexity of the mathematical theory in cases other than two-dimensional flow past bodies with straight-line boundaries and attached cavities at ambient pressure, and many aspects are not yet understood.

The most important parameter governing the form of the cavity and the flow as a whole is the cavitation number which, as in (6.12.12), we may write as

$$K = \frac{p_0 - p_c}{\frac{1}{2}\rho U^2}, \quad = \frac{U_f^2}{U^2} - 1, \tag{6.13.23}$$

where p_0 is the ambient pressure for the cavity (assumed uniform), p_c is the cavity pressure, U is the speed of the undisturbed stream, and U_f is the uniform speed of the liquid on the free streamlines at the cavity boundary. Only cases of cavity flow with $K = 0$ have been considered hitherto in this section, that being mathematically the simplest condition. Cavities for which $K = 0$ are probably formed behind bodies which are projected into water, as in figure 6.12.6 (plate 19), although the correspondence with mathematical solutions of the above kind is at best a temporary one for reasons given in §6.12. If an air-filled cavity is formed artificially behind a body in a water stream by the release of air from the rear of the body, the cavitation number can be controlled, within limits, but a steady-state cavity with $K = 0$ is of infinite extent and cannot be realized in such experiments; however observations of flow past axisymmetric bodies with attached cavities with K taking values between zero and about 0·5 have been made, and these allow extrapolation to $K = 0$ of measurements of, for example, the drag on the body. Very small positive values of K may also occur in the case of an underwater projectile moving at high speeds with an attached cavity for which p_c is equal to the vapour pressure.

The mathematical and physical properties of steady cavity flows with $K \neq 0$ are not well established. If $K < 0$, the speed on the free streamlines is less than that in the undisturbed stream, and it is presumably necessary for the point of detachment of a free streamline from the body to be in the low-velocity region at the rear of the body. Two known mathematical solutions for two-dimensional cavity flows with $K < 0$ yield the cavity shapes shown in figure 6.13.4; the finiteness of the cavity length and the cusp at the end of the cavity are typical, and are consequences of the condition $U_f < U$. Such cavities have not been observed, perhaps because the boundary layer at the rigid surface would separate before reaching the low velocity region where the free streamlines begin.

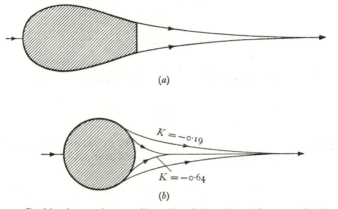

(a)

(b)

Figure 6.13.4. Cavities in steady two-dimensional irrotational flow past bodies, with the cavity pressure greater than the ambient pressure: (a) cavity behind a truncated aerofoil with $K = -0.111$, $U_f/U = 0.943$, obtained by conformal transformation (Lighthill 1949); (b) cavities behind a circular cylinder with $K = -0.19$, $U_f/U = 0.9$, and $K = -0.64$, $U_f/U = 0.6$, obtained by numerical analysis (Southwell and Vaisey 1946).

Cavities for which $K > 0$ are of real physical interest, because a steady cavity which forms in order to avoid the occurrence of regions of tension of the liquid is necessarily one in which the pressure in the cavity is a minimum over the flow field. Sketches of the cavities observed at various positive values of K formed by the emission of air behind a circular disk in a water channel are shown in figure 6.13.5; these shapes appear to depend on K alone. The cavity lengthens as $K \to 0$, and the boundary at positions far from the disk is presumably approaching the near-paraboloid represented by (6.13.22). Corresponding measurements of the drag on the disk, and on other axisymmetric bodies at different values of K, are given in figure 6.13.6. The drag coefficient for a circular disk at $K = 0$ obtained by extrapolation of the measurements is 0.80, which is quite close to the value appropriate to a normal flat plate in two dimensions (0.88), no doubt because the pressure is not far from the stagnation point value over most of the forward face in both cases.

A simple formula for the drag coefficient as a function of K which is found to fit the data for all the body shapes shown in figure 6.13.6 is

$$C_D = C_{D0}(1 + K), \qquad (6.13.24)$$

where C_{D0} is the drag coefficient at $K = 0$. This formula can also be obtained theoretically, on the basis of two assumptions. The first is that the intersection of the cavity surface and the body surface does not change with K, which is certainly valid for bodies with a salient edge. The second assump-

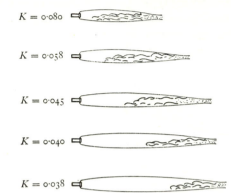

$$K = 0.080$$

$$K = 0.058$$

$$K = 0.045$$

$$K = 0.040$$

$$K = 0.038$$

Figure 6.13.5. Steady-state cavities attached to a circular disk at positive cavitation numbers. (From Reichardt 1946.)

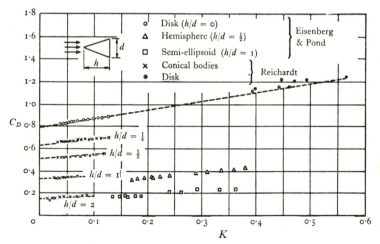

Figure 6.13.6. Measurements of the drag on different axisymmetric bodies with cavities at positive values of the cavitation number (from Reichardt 1946 and Eisenberg and Pond 1948). Each broken line is of the form $C_D = C_{D0}(1+K)$.

tion is that the velocity at any point on the wetted surface of the body is proportional to U_f, as the cavity pressure is changed, with p_0 and U fixed; this is correct for the two end points of each streamline on the body surface (one end being a stagnation point and the other the point of attachment of the cavity where the velocity is equal to U_f) and may be a reasonable approximation for intermediate points. It follows then from Bernoulli's theorem that at every point on the body surface the pressure relative to that in the cavity is proportional to U_f^2 and consequently to $1 + K$, whence the drag is given by (6.13.24).

No steady cavity flow with $K > 0$ which is free from anomalies has been found mathematically. The difficulty lies in finding an appropriate cavity shape in the region far from the body. Two ways of avoiding the difficulty, although with some sacrifice of potential correspondence with reality, have been devised for two-dimensional flow and are indicated in figure 6.13.7 for the case of flow past a flat plate. The first device, suggested by Riabouchinsky (1919), supposes that the whole flow field is symmetrical about a transverse plane, and that in effect a second or image plate exists at an arbitrarily chosen distance downstream from the first plate.† The second device is to

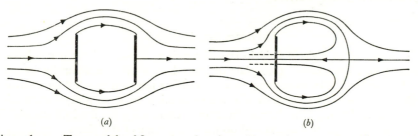

(a)   (b)

Figure 6.13.7. Two models of flow past a flat plate with a cavity at a pressure less than the ambient pressure ($K > 0$); (a) the image-plate symmetrical flow model, (b) the re-entrant jet model.

allow the free streamlines to turn inwards and to produce a jet moving towards the rear of the plate (possible in the mathematical solution, if not in reality, because the free streamlines are continued on to a second sheet of a Riemann surface). The idea underlying the use of either device is that it might be possible thereby to obtain a realistic description of the flow in the neighbourhood of the body; figure 6.12.5 (plate 18) shows that in any event the rear of a cavity at a positive value of K is ill-defined and perhaps has a definite shape only in a statistical sense. Some of the photographs of cavities attached to bodies with $K > 0$ do suggest that there is a tendency for the cavity to be filled up from the rear with a foaming mass of water and then for the contents of the cavity to be swept downstream suddenly, with repetition of the whole process.

Finally, we note that when the point at which a free streamline leaves the body is not fixed by the occurrence of a salient edge, new questions arise. It is not clear, even in principle, how the position of the attachment point on a body with a smooth shape is determined, although some restrictions on the position are evident. It may readily be seen from (6.13.14) that the

† Other forms of rigid body may be introduced some distance downstream so as to terminate the free streamlines.

curvature $(d\theta/ds)$ of the free streamline springing from the edge of a flat plate in two-dimensional flow varies as $s^{-\frac{1}{2}}$ near $s = 0$, and it is indeed a general mathematical property of cavity boundaries (in two dimensions, at least) that the curvature is infinite at the point of attachment to a rigid boundary, whether the latter be curved or straight. The sign of this curvature at the attachment point may be negative or positive, depending on the cavitation number (being convex to the liquid in the cases represented in figures 6.13.2 and 6.13.7, and concave to the liquid in that in figure 6.13.4). The free streamline necessarily leaves the rigid boundary tangentially, since the velocity at the junction would otherwise be zero or infinite, so that only for some combinations of cavitation number and the position of the attachment point will it be possible to construct free streamlines which do not cut across the body surface.

Exercises for chapter 6

1. A stream of inviscid fluid in two dimensions is bounded on one side by a plane wall, from which a thin plate of finite length projects normally into the fluid. The fluid is in irrotational motion and has uniform velocity far from the wall. Use the Schwarz–Christoffel transformation to determine the complex potential, and verify that the resultant force exerted by the fluid on the wall is equal and opposite to that exerted on the sharp edge of the plate.

2. A rigid ellipsoid with semi-diameters a, b, c is rotating with angular velocity Ω about an axis through the centre and the fluid contained in it is in irrotational motion. Show that the velocity potential is the sum of three terms like $\zeta xy(a^2 - b^2)/(a^2 + b^2)$, where ξ, η, ζ are the components of Ω in the directions of the principal axes of the ellipsoid, and that relative to the container a material element of the fluid moves on an ellipsoid which is similar to the boundary ellipsoid.

3. (a) A tongue of fluid moving under gravity down an inclined plane is immersed in lighter ambient fluid which is at rest far from the plane. The flow is two-dimensional and steady relative to axes moving with the tongue. Show that, if the effects of viscosity are negligible, the tangent to the interface between the two fluids at the foremost point of the tongue is at an angle of 60° to the plane.

(b) A progressive gravity wave of stationary form with straight crests at the free surface of water of great depth has the largest amplitude possible without breaking. The motion in the water is irrotational. It is known that under these conditions the water near the crest occupies a wedge with the vertex at the crest and with the two faces symmetrical about the vertical. Show that the angle between the two faces of this wedge is 120°.

4. Two similar but oppositely directed circular jets of water in air impinge symmetrically, and form a sheet of water spreading out radially at the plane of symmetry. What physical factor limits the outward movement in the sheet? Obtain an estimate of the radius of the circular outer boundary of the sheet. (Taylor 1959.)

7

FLOW OF EFFECTIVELY INVISCID
FLUID WITH VORTICITY

7.1. Introduction

We continue in this chapter the study of flow of a uniform incompressible fluid in circumstances in which the direct effect of viscosity may be neglected. The vorticity will be taken as non-zero, over part of the fluid at least. It will not be possible to take the theory as far, or to analyse as many different representative flow fields, as in the case of wholly irrotational flow, since in general we no longer have a linear governing equation.

We recall the kinematical result of §2.4 that the velocity of incompressible fluid associated with a vorticity distribution $\boldsymbol{\omega}(\mathbf{x})$ is

$$\mathbf{u}(\mathbf{x}) = \mathbf{v}(\mathbf{x}) - \frac{1}{4\pi} \int \frac{\mathbf{s} \times \boldsymbol{\omega}(\mathbf{x}')}{s^3} \, dV(\mathbf{x}'), \qquad (7.1.1)$$

where $\mathbf{v}$ is an irrotational solenoidal vector, $\mathbf{s} = \mathbf{x} - \mathbf{x}'$, and the volume integral is taken over the whole fluid (or over a more extensive region if $\boldsymbol{\omega} . \mathbf{n} \neq 0$ at the boundary of the fluid, as explained in §2.4). When the vorticity distribution is known, the irrotational contribution $\mathbf{v}(\mathbf{x})$ will be determined, in general, by the boundary conditions imposed on the velocity $\mathbf{u}$; equivalently we may think of the boundary conditions on $\mathbf{u}$ as being satisfied by the introduction of an 'image' of the vorticity distribution in the boundary with $\mathbf{v}$ then being zero.

The preliminary remarks in §6.1, about the role of the theory of flow of inviscid fluid, apply equally to this chapter. The governing equations set down in that section are also equally applicable here, viz. the equation of mass conservation

$$\nabla . \mathbf{u} = 0,$$

and the equation of motion, with gravity as the only body force acting on the fluid,

$$\frac{D\mathbf{u}}{Dt} = \mathbf{g} - \frac{1}{\rho} \nabla p. \qquad (7.1.2)$$

In this latter equation $\mathbf{g}$ and ρ are constant quantities. We shall also need to make use of the relation (6.2.3), which is simply an alternative form of the equation of motion. The quantity within brackets in (6.2.3) has previously been designated by the symbol H in connection with the derivation of Bernoulli's theorem for steady flow (see (3.5.16)), and it is convenient to employ this notation also in more general circumstances; that is,

$$\mathbf{u} \times \boldsymbol{\omega} - \frac{\partial \mathbf{u}}{\partial t} = \nabla H, \qquad (7.1.3)$$

where

$$H = \tfrac{1}{2}q^2 + \frac{p}{\rho} - \mathbf{g} . \mathbf{x} \qquad (7.1.4)$$

and $q^2 = \mathbf{u} . \mathbf{u}$. Cases of flow of a liquid with a free surface will not arise in this chapter, and the gravity term in (7.1.2) and (7.1.4) will usually be suppressed, p then being the modified pressure.

The vorticity equation is particularly relevant to the considerations of this chapter. It is obtained by taking the curl of both sides of (7.1.2) or (7.1.3):

$$\frac{\partial \boldsymbol{\omega}}{\partial t} = \nabla \times (\mathbf{u} \times \boldsymbol{\omega}), \quad \text{or} \quad \frac{D\boldsymbol{\omega}}{Dt} = \boldsymbol{\omega} . \nabla \mathbf{u}. \qquad (7.1.5)$$

It was shown in § 5.3 to be a consequence of this equation that vortex-tubes move with the fluid and are constant in strength.

In the case of two-dimensional flow, only the component of vorticity normal to the plane of flow (to be denoted by ω) is non-zero. The gradient of $\mathbf{u}$ in the direction of the vorticity vector is zero here, so that (7.1.5) reduces to

$$\frac{D\omega}{Dt} = 0. \qquad (7.1.6)$$

The vorticity associated with a material element is constant, because no turning or extension of vortex-lines can occur in two-dimensional flow.

The other case in which (7.1.5) takes a simple form is axisymmetric flow without 'swirl' (i.e. without azimuthal motion), when the vorticity vector at any point is normal to a plane containing that point and the axis of symmetry. In terms of cylindrical polar co-ordinates (x, σ, ϕ) with corresponding velocity components (u, v, w), and with $\mathbf{i}, \mathbf{j}, \mathbf{k}$ representing unit vectors in the directions of the x-, σ- and ϕ-co-ordinate lines respectively, we find here that

$$\boldsymbol{\omega} = \omega \mathbf{k}, \quad \frac{D\boldsymbol{\omega}}{Dt} = \frac{D\omega}{Dt}\mathbf{k}, \quad \boldsymbol{\omega} . \nabla \mathbf{u} = \frac{\omega v}{\sigma} \frac{\partial \mathbf{j}}{\partial \phi} = \frac{\omega v}{\sigma}\mathbf{k},$$

whence (7.1.5) reduces to

$$\frac{D(\omega/\sigma)}{Dt} = 0. \qquad (7.1.7)$$

This relation represents the constancy of strength of a material vortex-tube of small cross-section and length $2\pi\sigma$.

Relations like (7.1.6) and (7.1.7) are especially useful in the context of steady flow, when a zero value of a material derivative implies constancy along a streamline. Also, when the flow is steady (7.1.3) reduces to

$$\mathbf{u} \times \boldsymbol{\omega} = \nabla H, \qquad (7.1.8)$$

as found for steady homentropic flow in general (see (3.5.9)). Thus

$$\mathbf{u}.\nabla H = 0, \quad \boldsymbol{\omega}.\nabla H = 0 \qquad (7.1.9)$$

in steady flow, showing that the quantity H is constant over a surface on which intersecting families of streamlines and vortex-lines lie. A surface on which H is constant may appropriately be termed a *Bernoulli surface* (although, as in §3.5, we reserve the term Bernoulli's theorem for the result that H is constant along a streamline in steady flow).

The self-induced movement of a line vortex

In §2.6 there was introduced the notion of a line vortex, that is, a vortex-tube of infinitesimal cross-section but non-zero strength. To that kinematical discussion we are now able to add the consequences of the dynamical theorem that a vortex-tube moves with the fluid without change of strength and to deduce the way in which the shape of a line vortex changes with time. The result given below concerning the movement of a line vortex is of limited application, but is rather surprising and has important implications for the mathematical use of line vortices as an approximate representation of real vorticity distributions.

The velocity distribution in the fluid is given by (7.1.1), and we wish to use that expression to evaluate the velocity at points near a line vortex. We shall suppose that the line vortex (of strength κ) exists in infinite fluid which is at rest at infinity and which is without interior boundaries; this requires $\mathbf{v} = 0$ everywhere, so that only the integral term in (7.1.1) remains. We shall also assume that the vorticity is zero at points of the fluid not on the line vortex, in which case (7.1.1) reduces to (see (2.6.3))

$$\mathbf{u}(\mathbf{x}) = -\frac{\kappa}{4\pi}\oint\frac{\mathbf{s}\times d\mathbf{l}(\mathbf{x}')}{s^3}, \qquad (7.1.10)$$

where $\mathbf{s} = \mathbf{x} - \mathbf{x}'$ and $\delta\mathbf{l}$ is a line element of the closed curve of integration coincident with the line vortex. As was seen in §2.6, this can also be written in the form

$$\mathbf{u}(\mathbf{x}) = \frac{\kappa}{4\pi}\nabla\Omega, \qquad (7.1.11)$$

where Ω is the solid angle subtended at the point $\mathbf{x}$ by the line vortex. It is evident from both these formulae that there is a singularity of the velocity distribution at points on the line vortex. There is of course a circulatory motion round any portion of the line vortex, with a velocity which increases as the reciprocal of the distance from the line vortex as the line vortex is approached, but this circulatory motion can only rotate the infinitesimal cross-section of this local portion of the line vortex about its centre and cannot translate it. We need to examine the value of $\mathbf{u}(\mathbf{x})$ at points near the line vortex with some care in order to determine what remains when the circulatory motion is subtracted.

We consider the induced velocity in the neighbourhood of a point O on the line vortex, and choose rectilinear axes which are parallel to the tangent, principal normal, and binormal to the line vortex at O, as indicated in figure 7.1.1. With O as origin, and $\mathbf{t}, \mathbf{n}, \mathbf{b}$ unit vectors in the direction of the axes, the position vector of a point in the plane normal to the line vortex at O can be written as

$$\mathbf{x} = x_2\mathbf{n} + x_3\mathbf{b},$$

and our task is to examine the form taken by the velocity at this point as $(x_2^2 + x_3^2)^{\frac{1}{2}} (= \sigma) \to 0$. Now for a certain range of values of the distance l along the line vortex from O, say $L \geqslant l \geqslant -L$, the position vector $\mathbf{x}'$ of a point on the line vortex is given by

$$\mathbf{x}' \approx l\mathbf{t} + \tfrac{1}{2}cl^2\mathbf{n},$$

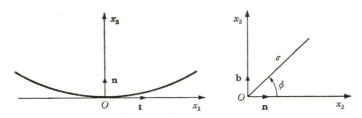

Figure 7.1.1. Definition sketch for the induced velocity near a line vortex.

where c is the curvature of the line vortex at O. Thus, near O,

$$\delta\mathbf{l}(\mathbf{x}') \approx (\mathbf{t} + cl\mathbf{n})\, \delta l$$

and

$$\frac{(\mathbf{x} - \mathbf{x}') \times \delta\mathbf{l}(\mathbf{x}')}{|\mathbf{x} - \mathbf{x}'|^3} \approx \frac{-x_3\, cl\mathbf{t} + x_3\mathbf{n} - (x_2 + \tfrac{1}{2}cl^2)\mathbf{b}}{\{x_2^2 + x_3^2 + l^2(1 - x_2 c) + \tfrac{1}{4}c^2l^4\}^{\frac{3}{2}}}\, \delta l.$$

The contribution to the velocity at the point $(0, x_2, x_3)$, or $(0, \sigma \cos\phi, \sigma \sin\phi)$, from the above nearby portion of the vortex line is then

$$\frac{\kappa}{4\pi} \int_{-L/\sigma}^{L/\sigma} \frac{(\mathbf{b}\cos\phi - \mathbf{n}\sin\phi)\,\sigma^{-1} + \tfrac{1}{2}cm^2\mathbf{b}}{\{1 + m^2(1 - c\sigma\cos\phi) + \tfrac{1}{4}c^2\sigma^2 m^4\}^{\frac{3}{2}}}\, dm, \qquad (7.1.12)$$

where m has been substituted for l/σ. As $\sigma \to 0$, the denominator of the integrand tends to $(1 + m^2)^{\frac{3}{2}}$ and the whole contribution takes the asymptotic form

$$\frac{\kappa}{2\pi\sigma}(\mathbf{b}\cos\phi - \mathbf{n}\sin\phi) + \frac{\kappa c}{4\pi}\mathbf{b}\log\frac{L}{\sigma} + \text{const.} \qquad (7.1.13)$$

The contribution to the velocity at O which arises from parts of the line vortex outside the range $L \geqslant l \geqslant -L$ is certainly bounded in magnitude and we shall not need to know more about it.

The first of the two variable terms in (7.1.13) represents the expected circulatory motion about the line vortex, and does not lead to displacement of the line vortex. The second term is novel, and shows the interesting result that there is another and weaker singularity of the velocity distribution associated with the local curvature of the line vortex. It seems that the fluid in the neighbourhood of the point O on the line vortex has a large velocity in the direction of the binormal, with the magnitude varying asymptotically as $\log \sigma^{-1}$. An ideal curved line vortex in the mathematical sense will thus move with infinite speed, and in general will change its shape with infinite speed. The implication is that the speeds of movement and of deformation of a strong vortex-tube of small cross-section have large values which depend critically on the dimensions of the cross-section of the tube when the curvature c is non-zero; this is an inconvenient conclusion since information about the cross-section is unlikely to be available.

Now that we have found that an isolated curved line vortex moves with infinite speed under the action of its own associated velocity field, it is evident that there was no loss of generality in the assumptions made earlier in reducing (7.1.1) to (7.1.10). If $\mathbf{v} \neq 0$, or if the vorticity is non-zero at points not on the line vortex, there will be additional contributions to $\mathbf{u}$ at points on the line vortex, but all such contributions are finite and hence negligible compared with the second of the two variable terms in (7.1.13) when σ is small.

In a case in which $c\mathbf{b}$ is uniform over the line vortex—so that it is circular—no deformation occurs, and the line vortex moves in the direction of the normal to its plane at infinite speed. And in the limiting case of a straight line vortex, for which $c = 0$, the speed of movement is zero and the difficulty disappears. Two other special cases in which the line vortex does not change its shape are known. One is a helical line vortex, which advances in the direction of its axis and rotates about it. The other is a line vortex in the form of the plane curve

$$x_2 = A \sin \alpha x_1, \quad x_3 = 0,$$

with $\alpha A \ll 1$. The curvature is here

$$c = \frac{-\alpha^2 x_2}{(1 + \alpha^2 A^2 \cos^2 \alpha x_1)^{\frac{3}{2}}}, \quad \approx -\alpha^2 x_2,$$

and so the line vortex rotates rigidly about the x_1-axis; but in view of the approximation a change of shape occurs in due course.

It must be concluded that the mathematical notion of a line vortex is of limited direct value in problems involving development with time, except when the line vortex is straight or circular or helical.

The instability of a sheet vortex

Another limitation on the mathematical use of singularities in the vorticity distribution arises from the intrinsic instability of a sheet vortex to small disturbances, which was first noted by Helmholtz (1868 b). Observation shows that undulations of increasing amplitude tend to form at a transition layer between two approximately uniform streams of different velocity, usually with an irregular turbulent motion as the outcome. Figure 5.10.5 (plate 8), particularly photographs 5, 6 and 7 in the sequence, demonstrates clearly this instability of transition layers; here the transition layer is formed by separation of a boundary layer at a salient edge, and the undulations of the layer increase in magnitude with increase of distance from the separation point. The instability appears to be more marked, in the sense that the undulations grow more rapidly, in the case of thinner transition layers, and there is no doubt that the mathematical sheet vortex of infinitesimal thickness introduced in §2.6 is strongly unstable. The dynamical instability of flow systems forms a large and coherent branch of fluid dynamics, and most cases of instability are best discussed within the general framework of that subject. However, an elementary analysis of instability of a sheet vortex does not require special techniques or concepts, and we shall present it here as an illustration of the rather surprising kind of dynamical instability possessed by many systems in which the flow is unidirectional.

The instability of a sheet vortex is essentially a local phenomenon, and we shall therefore suppose the sheet to be plane and of uniform strength density. With a suitable choice of (rectilinear) axes, the steady flow field whose stability is to be considered is two-dimensional, with velocity components $(-\frac{1}{2}U, 0)$ in the region $y > 0$ of the (x, y)-plane and components $(\frac{1}{2}U, 0)$ in the region $y < 0$. We propose to examine the behaviour of small disturbances superposed on this flow field. In a real fluid the vorticity in the sheet would diffuse laterally and the thickness of the transition layer would increase continually in the manner described in §4.3. It is difficult to take into account the growth in thickness of a *disturbed* transition layer, and we shall ignore it on the assumption that the thickness of the layer remains small compared with the characteristic length of the disturbance during the relevant time interval; this clearly implies a restriction on some Reynolds number of the disturbed flow field. Also, it will be assumed that the disturbance is produced by processes which satisfy the conditions of Kelvin's circulation theorem, so that the only vorticity present in the disturbed state is that in the sheet vortex as modified by the disturbance. Analysis of the disturbed flow field is now a simple matter, because it consists of two irrotational motions on either side of a sheet vortex whose shape is slightly different from a plane. On the upper side of this sheet (figure 7.1.2), the fluid velocity is derived from a potential $-\frac{1}{2}Ux + \phi_1$ and on the lower side from $\frac{1}{2}Ux + \phi_2$, ϕ_1 and ϕ_2 being the two disturbance potentials.

We write the geometrical equation of the deformed sheet vortex at any instant as
$$y \equiv \eta(x, z, t);$$
note that the disturbance is not assumed to be two-dimensional. The quantities η, ϕ_1 and ϕ_2 are related, since the sheet vortex is a material surface which continues to be a boundary of each of the two regions. By regarding

the sheet vortex as the boundary of the upper region, we find

$$\left(\frac{\partial\phi_1}{\partial y}\right)_{y=\eta} = \frac{D\eta}{Dt}$$

$$= \frac{\partial\eta}{\partial t} + \left(-\tfrac{1}{2}U + \frac{\partial\phi_1}{\partial x}\right)\frac{\partial\eta}{\partial x} + \frac{\partial\phi_1}{\partial z}\frac{\partial\eta}{\partial z},$$

or, correct to first order in the small disturbance quantities,

$$\left(\frac{\partial\phi_1}{\partial y}\right)_{y=0} = \frac{\partial\eta}{\partial t} - \tfrac{1}{2}U\frac{\partial\eta}{\partial x}. \qquad (7.1.14)$$

Similarly, by regarding the sheet as the boundary of the lower region, we have, approximately,

$$\left(\frac{\partial\phi_2}{\partial y}\right)_{y=0} = \frac{\partial\eta}{\partial t} + \tfrac{1}{2}U\frac{\partial\eta}{\partial x}. \qquad (7.1.15)$$

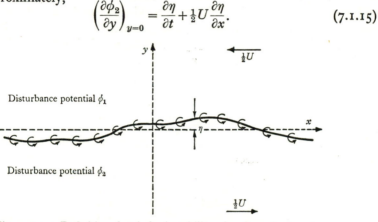

Figure 7.1.2. Definition sketch for instability of a plane sheet vortex to a small disturbance.

In addition to these kinematical matching conditions at the common boundary of the two streams there is a condition on the pressure to be satisfied. When the two streams are composed of the same fluid, no surface tension acts at the interface and the pressure must be continuous across the interface, that is,

$$(p_1 - p_2)_{y=\eta} = 0. \qquad (7.1.16)$$

Now in each of the two regions the pressure is given (see (6.2.5)) by a relation of the form

$$p = \text{const.} - \rho\left(\frac{\partial\phi}{\partial t} + gy + \tfrac{1}{2}q^2\right).$$

On substituting such an expression for p_1 and for p_2 in (7.1.16), and again assuming the two streams to have the same composition so that ρ is continuous at $y = \eta$, we find the approximate relation

$$\left(\frac{\partial\phi_2}{\partial t} - \frac{\partial\phi_1}{\partial t}\right)_{y=0} + \tfrac{1}{2}U\left(\frac{\partial\phi_2}{\partial x} + \frac{\partial\phi_1}{\partial x}\right)_{y=0} = \text{const.} \qquad (7.1.17)$$

These linear equations may be solved if we represent the sheet displacement η as a Fourier integral with respect to x and z. It is evident from (7.1.14) and (7.1.15) that a particular sinusoidal dependence of η on x or z demands a similar dependence of ϕ_1 and ϕ_2. Hence a disturbance will be a superposition of Fourier 'components' like

$$\eta, \phi_1, \phi_2 \propto e^{i(\alpha x + \gamma z)}$$

which behave independently; the complex form is convenient, because the various quantities do not all have the same phase. Here γ and α are components of a wave-number vector in the (z, x)-plane whose magnitude is

$$k = (\gamma^2 + \alpha^2)^{\frac{1}{2}};$$

that is to say, the corresponding disturbance quantities vary sinusoidally with wave-length $2\pi/k$ in a direction making an angle $\tan^{-1}(\gamma/\alpha)$ with the x-axis in the (z, x)-plane. Also, ϕ_1 and ϕ_2 satisfy Laplace's equation, which shows they depend on y as $\exp(\pm ky)$. With the restriction that the disturbance motion vanishes far from the sheet vortex, on both sides of it, we then have

$$\left.\begin{aligned}\eta &= A \\ \phi_1 &= B_1 e^{-ky} \\ \phi_2 &= B_2 e^{ky}\end{aligned}\right\} \times e^{i(\alpha x + \gamma z)}, \tag{7.1.18}$$

where A, B_1 and B_2 are functions of time t alone and only the real parts of the right-hand sides are relevant.

On substitution of these expressions in (7.1.14) and (7.1.15) we find

$$\left.\begin{aligned}-kB_1 &= \frac{dA}{dt} - \tfrac{1}{2}i\alpha UA, \\ kB_2 &= \frac{dA}{dt} + \tfrac{1}{2}i\alpha UA.\end{aligned}\right\} \tag{7.1.19}$$

From (7.1.17) we see first that the constant on the right-hand side can be non-zero only for a Fourier component independent of x and z (and also y), which is a case we can ignore; and second, with the aid of (7.1.19), that

$$\frac{d^2A}{dt^2} = \tfrac{1}{4}\alpha^2 U^2 A.$$

It follows that

$$A \propto e^{\sigma t}, \quad \sigma = \mp \tfrac{1}{2}\alpha U, \tag{7.1.20}$$

and

$$\frac{-kB_1}{\sigma - \tfrac{1}{2}i\alpha U} = \frac{kB_2}{\sigma + \tfrac{1}{2}i\alpha U} = A. \tag{7.1.21}$$

The positive root for σ corresponds to an exponentially-growing disturbance, the mathematical existence of which shows that the sheet vortex is unstable to any disturbance periodic with respect to x and z for which $\alpha \neq 0$.

It will be noticed from the expressions for η, ϕ_1 and ϕ_2 that the quantity U occurs only in combination with α. To account for this we may think of the velocity vector in each of the two undisturbed streams as being resolved into components parallel and perpendicular to the wave-number vector (γ, α) in the (z, x)-plane, with a corresponding resolution of the vorticity in the sheet. The component of vorticity in the undisturbed sheet which is parallel to the wave-number vector corresponds to stream velocities parallel to the crests of the disturbance wave, and does not interact with the disturbance; that is to say, a sinusoidal deformation of a uniform sheet vortex with the crests of the deformed sheet parallel to the stream velocities on each side of the sheet is a steady-flow configuration (compare the results obtained for sheet vortices of uniform strength density in §2.6) for any small amplitude of the deformation. The component of vorticity in the undisturbed sheet which is normal to the wave-number vector, and so parallel to the wave crests, would by itself give the sheet a strength density $\alpha U/k$; this component corresponds to streams flowing *across* the crests of the disturbance wave and alone is responsible for the interaction with, and amplification of, the disturbance wave.

It follows from these remarks that there is a gain in simplicity if we transform to new co-ordinates in the plane of the undisturbed sheet given by

$$kx' = \alpha x + \gamma z, \quad kz' = -\gamma x + \alpha z, \tag{7.1.22}$$

the new x'-axis being parallel to the wave-number vector and the z'-axis normal to it. The disturbance velocity vector lies in the (x', y)-plane everywhere, and the jump in the component of velocity in this plane across the undisturbed sheet is $\alpha U/k, = U'$ say.

The mechanism of the instability of a sheet vortex can be understood in terms of the changes in the vorticity distribution and the consequent effect on the velocity distribution. The vector obtained by integrating the vorticity across the sheet at a given point (and denoted by $\boldsymbol{\Gamma}$ in §2.6) is related to the discontinuity in velocity across the sheet according to (2.6.11), and the change in this vector due to the disturbance has components in the z' and x' directions equal to

$$-\left(\frac{\partial \phi_1}{\partial x'} - \frac{\partial \phi_2}{\partial x'}\right)_{y=0} \quad \text{and} \quad \left(\frac{\partial \phi_1}{\partial z'} - \frac{\partial \phi_2}{\partial z'}\right)_{y=0},$$

i.e., equal to $\qquad 2i\sigma\eta \quad \text{and} \quad \text{o}$

in view of (7.1.18), (7.1.21) and (7.1.22). The z'-component of the disturbance vorticity varies sinusoidally with respect to x' and has a phase different from that of η by $\frac{1}{2}\pi$ (being $\frac{1}{2}\pi$ greater in the case of a growing disturbance and $\frac{1}{2}\pi$ less if σ is negative). The relation between the displacement η and the local strength density for a *growing* disturbance to a sheet with vorticity in the z'-direction alone is sketched in figure 7.1.3. The vorticity is evidently being swept towards points like A where $\eta = 0$, $\partial \eta / \partial x' > 0$, and away from points like C where $\eta = 0$, $\partial \eta / \partial x' < 0$. This convecting motion, which can be seen explicitly from the expression for $(\partial \phi_1/\partial x' + \partial \phi_2/\partial x')_{y=0}$, arises from the fact that the vorticity in parts of the sheet displaced downwards (or upwards) induces a velocity with a negative (or positive) x'-component at any point of the sheet where $\eta > 0$ (or < 0), the inference being clearest for points like B at which the induced velocity due to *every* portion of the sheet vortex has a negative x'-component. Now when some accumulation of (positive) vorticity does exist near points like A, there will be a corresponding induced velocity distribution which tends to carry fluid round A in an anticlockwise sense and thereby to increase the amplitude of the sinusoidal displacement of the sheet vortex. A larger amplitude gives a more rapid accumulation of vorticity near points like A, and so the whole cycle accelerates. The special feature of a sinusoidal disturbance of the sheet vortex is that the two processes of accumulation of vorticity near points like A and rotation of the neighbouring portions of the sheet vortex go on simultaneously and lead to exponential growth without change of the spatial form of the disturbance.

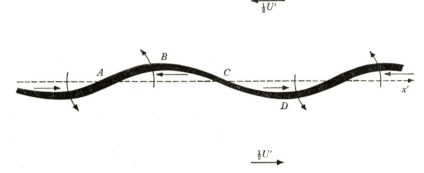

Figure 7.1.3. Growth of a sinusoidal disturbance to an initially uniform sheet vortex with positive vorticity normal to the paper. The local strength density of the sheet is represented by the thickness of the sheet. The arrows indicate the direction of the self-induced movement of the vorticity in the sheet, and show (*a*) the accumulation of vorticity at points like A and (*b*) the general rotation about points like A, which together lead to exponential growth of the disturbance.

It is now possible to see the significance of the negative root in (7.1.20). If we could distribute vorticity in the wavy sheet at some initial instant so that points like C in figure 7.1.3 were centres of accumulation (with the appropriate sinusoidal distribution), the subsequent motion would tend (*a*) to rotate portions of the sheet near C in the anti-clockwise sense about C and (*b*) to sweep vorticity toward points like A as before, thereby diminishing the disturbance exponentially. However this kind of initial condition is most unlikely to occur naturally, and the existence of a negative root for σ in this and other similar stability problems may be ignored.

The analysis shows that the *growth rate* of a sinusoidal disturbance, defined as $d(\log A)/dt$, $= \sigma$, is equal to $\frac{1}{2}\alpha U$. Thus in the case of a disturbance of more general form the Fourier components with larger wave-number (and, among those with equal wave-number magnitude, with the wave-number vector parallel to the two undisturbed streams) are magnified more rapidly and will ultimately dominate the resultant form of the disturbance. More extensive analysis shows that, although the above theory gives an accurate description of the stability of a transition layer of thickness d between two uniform streams to sinusoidal disturbances with wave-length large compared with d, disturbances with wave-length smaller than a certain length of order d do not grow and that for some wave-length, likewise of order d, the growth rate is a maximum. In this more realistic case we should therefore expect that a disturbance of arbitrary initial form is converted, by selective amplification of Fourier components, into an approximately sinusoidal form with wave-length close to that for which the growth rate is a maximum; but of course these conclusions hold only while the disturbance magnitude is small enough for the linear equations to be applicable. The large and roughly periodic oscillations visible on the sheet vortex in figure 5.10.5 (plate 8) are an outcome of the amplification of a small disturbance, and the wave-length is presumably of the same order of magnitude as the thickness of the transition layer being shed from the salient edge.

It is not difficult to extend the above analysis to a case in which the sheet vortex separates two streams of fluid with different densities, with surface tension acting at the interface and with gravity acting. The results then have application to such problems as the generation of waves at the free surface of a liquid over which gas is being blown.

Exercise

Show from (7.1.5) that in a case of axisymmetric flow with swirl

$$\frac{D}{Dt}\left(\frac{\omega_\phi}{\sigma}\right) = -\frac{2w\omega_\sigma}{\sigma^2},$$

where (x, σ, ϕ) are cylindrical polar co-ordinates and (u, v, w) are corresponding velocity components.

7.2. Flow in unbounded fluid at rest at infinity

Cases of flow of an infinite fluid without interior boundaries are of special interest in inviscid-fluid theory, since there are here none of the singular effects of viscosity associated with rigid boundaries which usually spoil the correspondence with flow of real fluids. There is no scope for inviscid-fluid theory in such cases when the flow is irrotational everywhere, because the only solution consistent with zero velocity at infinity is a state of rest everywhere. But many interesting possibilities arise when a confined region of vorticity is embedded in fluid at rest at infinity, a 'vortex ring' being perhaps the best known example. We shall consider first the more natural three-dimensional case, and leave the special features of two-dimensional flow for discussion in the next section.

Some of the kinematical results of §2.9 are relevant here. The velocity of incompressible fluid associated with a vorticity distribution $\boldsymbol{\omega}(\mathbf{x})$ is given by (7.1.1), with $\mathbf{v} = 0$ in the present context of infinite fluid without interior boundaries and which is at rest at infinity; and we saw in §2.9 that the asymptotic form for $\mathbf{u}$, as $r\,(= |\mathbf{x}|) \to \infty$, is

$$\mathbf{u}(\mathbf{x}) \sim \frac{1}{8\pi} \nabla \left\{ \left(\nabla \frac{1}{r} \right) \cdot \int \mathbf{x}' \times \boldsymbol{\omega}' \, dV(\mathbf{x}') \right\}, \qquad (7.2.1)$$

where the integral is taken over the whole fluid. Alternatively we have (see (2.9.4) and (2.9.5)) $\mathbf{u} = \nabla \times \mathbf{B}$, where

$$\mathbf{B}(\mathbf{x}) \sim \frac{1}{8\pi} \left(\nabla \frac{1}{r} \right) \times \int \mathbf{x}' \times \boldsymbol{\omega}' \, dV(\mathbf{x}') \qquad (7.2.2)$$

as $r \to \infty$. This asymptotic expression for $\mathbf{u}$ is valid provided $|\boldsymbol{\omega}|$ is of smaller order than r^{-4} when r is large,† and represents the irrotational velocity distribution associated with either a single closed line vortex of small linear dimensions or a source doublet located at the origin.

The resultant force impulse required to generate the motion

The value of the integral $\int \mathbf{x} \times \boldsymbol{\omega} \, dV$ occurring in (7.2.1) has dynamical significance, as we shall now show. The dimensions of the integral suggest a consideration of the total linear momentum in the fluid; but the velocity diminishes in magnitude as r^{-3} as $r \to \infty$ and we encounter the same difficulty as in the case of irrotational flow due to a rigid body in translational motion (see §6.4), viz. that the integral $\int \mathbf{u} \, dV$ is in general not absolutely convergent and appears to depend on the way in which the volume of integration is allowed to tend to infinity. We must therefore proceed differently. Whereas in §6.4 we determined the force impulse which must be applied to the rigid body in order to generate the given fluid motion from rest, here we calculate the resultant of the distributed force impulse that must be applied to a limited portion of the fluid in order to generate the whole of the given motion from rest. The resultant force impulse will again be called the *fluid impulse* of the flow field and will be denoted by $\mathbf{P}$.

It is immediately evident that if $\mathbf{u}'$ and $\mathbf{u}''$ are two velocity distributions for which the values of $\int \mathbf{x} \times \boldsymbol{\omega} \, dV$ are equal, the difference motion $\mathbf{u}' - \mathbf{u}''$ is one for which the total linear momentum in the fluid is zero. For if V is now a volume bounded externally by the closed surface A, to which the unit outward normal is $\mathbf{n}$, we have

$$\int (\mathbf{u}' - \mathbf{u}'') \, dV = \int \nabla \times (\mathbf{B}' - \mathbf{B}'') \, dV$$
$$= \int \mathbf{n} \times (\mathbf{B}' - \mathbf{B}'') \, dA,$$

and since the difference between $\mathbf{B}'$ and $\mathbf{B}''$ is seen from (7.2.2) to be of smaller order than r^{-2} when r is large, this surface integral tends to zero as the surface

† In this and the following section it will be assumed that $|\boldsymbol{\omega}|$ is sufficiently small when r is large to make convergent all the integrals involving vorticity which arise.

A recedes to infinity in all directions. The fluid impulse of the difference motion is also zero, since the fluid impulse and the total linear momentum are equal when the integral giving the total momentum is absolutely convergent. Now $\mathbf{P}$ is necessarily a linear functional of $\mathbf{u}$, and also of $\boldsymbol{\omega}$, and it follows first that all those flow fields for which the values of $\int \mathbf{x} \times \boldsymbol{\omega}\, dV$ are equal have the same fluid impulse and second that

$$\mathbf{P} \propto \int \mathbf{x} \times \boldsymbol{\omega}\, dV. \tag{7.2.3}$$

To find the constant of proportionality, we choose a length R so large that the velocity distribution (7.2.1) holds accurately for $r \geqslant R$, and proceed to calculate the contributions to the fluid impulse from the two parts of the flow field given by $r \leqslant R$ and $r \geqslant R$. The total linear momentum of the fluid in the region $r \leqslant R$ is

$$\rho \int_{r \leqslant R} \mathbf{u}\, dV = \rho \int_{r \leqslant R} \nabla \times \mathbf{B}\, dV$$

$$= \rho \int_{r = R} \mathbf{n} \times \mathbf{B}\, dA;$$

and since (7.2.2) holds on the spherical surface A we have

$$8\pi R^2 (\mathbf{n} \times \mathbf{B})_{r=R} = \int \mathbf{x} \times \boldsymbol{\omega}\, dV - \mathbf{n}\mathbf{n} . \int \mathbf{x} \times \boldsymbol{\omega}\, dV.$$

The total momentum in the inner region is thus $\tfrac{1}{3}\rho \int \mathbf{x} \times \boldsymbol{\omega}\, dV$. This contribution to the fluid impulse is independent of R, and we are tempted to allow R to tend to infinity and to assert that there is then no outer region to make an additional contribution. But that argument is invalidated by the lack of absolute convergence of the integral $\int \mathbf{u}\, dV$. Now the asymptotic velocity distribution (7.2.1) has the same form as in flow due to a rigid sphere of radius R moving with velocity $\mathbf{U}$ (see (6.8.13)), provided

$$4\pi R^3 \mathbf{U} = \int \mathbf{x} \times \boldsymbol{\omega}\, dV \tag{7.2.4}$$

(showing incidentally that the total momentum in the inner region is the same as if all the fluid within the spherical surface of radius R were moving with the velocity $\mathbf{U}$ given by (7.2.4)). Hence the outer region makes a contribution to the fluid impulse which is equal to the fluid impulse of the flow due to a rigid sphere of radius R moving with the velocity $\mathbf{U}$ given by (7.2.4), that is, in view of (6.4.29) and (6.8.18), equal to

$$\tfrac{2}{3}\pi R^3 \rho \mathbf{U}, \quad = \tfrac{1}{6}\rho \int \mathbf{x} \times \boldsymbol{\omega}\, dV.$$

The two contributions together give

$$\mathbf{P} = \tfrac{1}{2}\rho \int \mathbf{x} \times \boldsymbol{\omega}\, dV, \tag{7.2.5}$$

the integral being taken over the whole fluid.

We can also prove that, as would be expected, this expression for the resultant impulse required for the generation of the flow from rest is independent of time, even though the flow itself is not necessarily steady. We see from (7.1.5) that

$$\frac{d\mathbf{P}}{dt} = \tfrac{1}{2}\rho \int \mathbf{x} \times \frac{\partial \boldsymbol{\omega}}{\partial t}\, dV$$

$$= \tfrac{1}{2}\rho \int \mathbf{x} \times \{\nabla \times (\mathbf{u} \times \boldsymbol{\omega})\}\, dV.$$

Expansion of the integrand (suffix notation being simpler) shows that it is equal to $2\mathbf{u} \times \boldsymbol{\omega}$, plus several terms in the form of derivatives, and these derivatives yield surface integrals which vanish in view of the smallness of $|\boldsymbol{\omega}|$ at infinity. Thus

$$\frac{d\mathbf{P}}{dt} = \rho \int \mathbf{u} \times \boldsymbol{\omega}\, dV$$

$$= \rho \int \left\{ \tfrac{1}{2}\nabla q^2 - \frac{\partial (u_i\, \mathbf{u})}{\partial x_i} \right\} dV,$$

which also may be transformed to a surface integral which is zero in view of the known smallness of the velocity far from the origin.

It may also be shown that the resultant moment (about the origin) of the distributed force impulse which must be applied to the fluid to generate the motion from rest is

$$\tfrac{1}{3}\rho \int \mathbf{x} \times (\mathbf{x} \times \boldsymbol{\omega})\, dV, \tag{7.2.6}$$

and that this is another invariant of the motion.

The total kinetic energy of the fluid

It is also possible to find an expression for the total kinetic energy in terms of the vorticity distribution. We have

$$T = \tfrac{1}{2}\rho \int \mathbf{u} \cdot (\nabla \times \mathbf{B})\, dV$$

$$= \tfrac{1}{2}\rho \int \{\mathbf{B} \cdot (\nabla \times \mathbf{u}) - \nabla \cdot (\mathbf{u} \times \mathbf{B})\}\, dV,$$

the integrals being taken over the whole fluid. The term in the integrand in the form of a divergence gives rise to a surface integral which is zero, and on substituting for the vector potential from (2.4.10) in the remaining term we find

$$T = \tfrac{1}{2}\rho \int \mathbf{B} \cdot \boldsymbol{\omega}\, dV = \frac{\rho}{8\pi} \int \int \frac{\boldsymbol{\omega} \cdot \boldsymbol{\omega}'}{s}\, dV(\mathbf{x})\, dV(\mathbf{x}'). \tag{7.2.7}$$

An alternative expression for T follows from the identity

$$\nabla \cdot (\mathbf{u}\mathbf{x} \cdot \mathbf{u}) = \nabla \cdot (\tfrac{1}{2}q^2 \mathbf{x}) - \tfrac{1}{2}q^2 + \mathbf{u} \cdot (\mathbf{x} \times \boldsymbol{\omega}).$$

When both sides of this relation are integrated over the whole fluid, two terms give rise to integrals over surfaces at infinity which are evidently zero, and

$$T = \rho \int \mathbf{u} \cdot (\mathbf{x} \times \boldsymbol{\omega})\, dV. \tag{7.2.8}$$

This quantity may also be shown to be independent of time.

Flow with circular vortex-lines

The foregoing expressions for the velocity distribution, the fluid impulse and the total kinetic energy take a simple form in cases in which all the vortex-lines are circles centred on a common axis of symmetry.

Since the whole flow field is here axisymmetric (and there is no azimuthal motion) we may describe the velocity distribution in terms of a stream function ψ, which is related to the one non-zero component of the vector potential $\mathbf{B}$. With cylindrical co-ordinates (x, σ, ϕ), $\boldsymbol{\omega}$ and $\mathbf{B}$ are everywhere parallel to the ϕ-co-ordinate line, and (2.4.10) shows that

$$\psi(x, \sigma) = \sigma\,|\mathbf{B}| = \frac{\sigma}{4\pi}\int_{-\infty}^{\infty}\int_{0}^{\infty}\int_{0}^{2\pi}\frac{\omega(x', \sigma')}{s}\,\sigma'\cos\theta\,dx'\,d\sigma'\,d\theta, \quad (7.2.9)$$

where

$$\omega = |\boldsymbol{\omega}|, \quad \theta = \phi' - \phi, \quad s^2 = (x-x')^2 + \sigma^2 + \sigma'^2 - 2\sigma\sigma'\cos\theta.$$

The resultant impulse required to generate the flow from rest is given by (7.2.5), and is evidently here a vector in the direction of the axis of symmetry with magnitude

$$P = \pi\rho\int_{-\infty}^{\infty}\int_{0}^{\infty}\omega\sigma^2\,dx\,d\sigma. \quad (7.2.10)$$

For the total kinetic energy of the fluid we have the two general formulae (7.2.7) and (7.2.8), the former of which becomes here

$$T = \pi\rho\int_{-\infty}^{\infty}\int_{0}^{\infty}\psi\omega\,dx\,d\sigma. \quad (7.2.11)$$

These formulae are easily adapted to the case of a single circular line vortex of radius a and strength κ at $x = 0$. In the usual way, by supposing that $\omega(x', \sigma')\,\delta x'\,\delta\sigma' = \kappa$ for the element of integration in the axial plane that includes the line vortex, we find

$$\psi(x, \sigma) = \frac{\kappa a\sigma}{4\pi}\int_{0}^{2\pi}\frac{\cos\theta\,d\theta}{(x^2 + \sigma^2 + a^2 - 2a\sigma\cos\theta)^{\frac{1}{2}}}. \quad (7.2.12)$$

The integrals with respect to θ in (7.2.9) and (7.2.12) can be evaluated from available tables. For if we put

$$k^2 = \frac{4a\sigma}{x^2 + (\sigma + a)^2},$$

(7.2.12) can be written as

$$\begin{aligned}\psi(x, \sigma) &= \frac{\kappa(a\sigma)^{\frac{1}{2}}}{4\pi}\int_{0}^{\pi}\left\{\left(\frac{2}{k}-k\right)(1 - k^2\cos^2\tfrac{1}{2}\theta)^{-\frac{1}{2}} - \frac{2}{k}(1 - k^2\cos^2\tfrac{1}{2}\theta)^{\frac{1}{2}}\right\}d\theta \\ &= \frac{\kappa(a\sigma)^{\frac{1}{2}}}{2\pi}\left\{\left(\frac{2}{k}-k\right)K(k) - \frac{2}{k}E(k)\right\}, \quad (7.2.13)\end{aligned}$$

where K and E are the so-called complete elliptic integrals of the first and second kinds, for which numerical values are known.

The forms of the streamlines obtained from (7.2.13) are shown in figure 7.2.1. It may seem a little surprising that the streamlines are small closed curves near the intersection of the axial plane with the line vortex, despite the fact that we have found the axial component of velocity to tend to infinity as the line vortex is approached from any direction; but it should be remembered that this axial velocity diverges as $\log\{(\sigma-a)^2+x^2\}$, whereas the circulatory motion round the line vortex has a speed which varies as $\{(\sigma-a)^2+x^2\}^{-\frac{1}{2}}$ and is thus dominant.

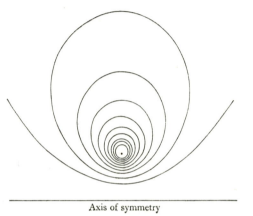

Axis of symmetry

Figure 7.2.1. Streamlines in an axial plane, with equal intervals of ψ, for the flow associated with a single circular line vortex in fluid at rest at infinity.

For the fluid impulse of the flow associated with a single circular line vortex, we have

$$P = \pi\rho\kappa a^2; \qquad (7.2.14)$$

it is noteworthy that this is finite, despite the singularity of the velocity distribution at the line vortex. The total energy is of course infinite, as it is for a flow field containing a line vortex of any shape.

Vortex rings

A familiar and intriguing example of flow with circular vortex lines is the 'smoke ring' which can be formed by ejecting a puff of smoke suddenly from the mouth through rounded lips and which travels forward steadily with a smoke-filled core. The essential requirement for the production of a vortex ring (to give it a more appropriate name) is that linear momentum should be imparted to the fluid with axial symmetry; the smoke is simply a marking agent which makes some of the fluid visible. Jerking a circular disk normal to its plane and bringing it to rest again is in principle the simplest method of producing a vortex ring; there is the practical disadvantage that the freely moving vortex ring would travel towards the disk and be obstructed by it, although this may be overcome by dipping half a circular disk into the free surface of a liquid and withdrawing it quickly when the horizontal motion ceases. A more common plan is to discharge a puff of fluid through a circular hole, either a hole in a plane rigid sheet or the open end of a tube, producing a vortex ring which travels away from the hole. Figure 7.2.2 (plate 20) shows photographs taken at different stages of the formation of vortex rings by the ejection of small quantities of fluid from the end of a tube. Rather surprisingly, it is even possible to produce a vortex ring by blowing a limited volume of air from a tube pointing upwards in a tank of water (Walters and Davidson 1963); the vortex ring, or toroidal gas bubble, here travels vertically upwards, and its radius is observed to increase, presumably because the buoyancy force is increasing the fluid impulse (see (7.2.14)). Yet another method of producing a vortex ring, illustrated in figure 7.2.3 (plate 21), is to allow drops of liquid to fall vertically on to a free surface of the same liquid.

From the theoretical point of view the striking property of all observed vortex rings in uniform fluid is the approximate steadiness of the motion relative to the ring when the ring is well clear of the generator. There is some decay of the motion always, presumably due to the action of viscosity, but the decay is less for larger rings, suggesting that the motion would be truly steady at infinite Reynolds number. We have seen that a circular line vortex has this property of steady propagation, although the speed of travel is infinite. At least some of the observed vortex rings look like approximations to a line vortex, in which the vorticity is contained in a tube of small but non-zero cross-section which propagates steadily in the direction of the axis of symmetry.

Mathematical analysis of such vortex rings is made difficult by ignorance of the exact shape of the cross-section of the tube containing the vorticity that is compatible with steady motion. But when the cross-section is small the difficulty is less severe, because the curve bounding the cross-section is a streamline of the steady flow relative to axes moving with the ring (since vortex-lines move with the fluid) and is approximately circular when the circulatory motion round the neighbouring portion of the vortex-tube is dominant. The formula (7.1.13) for the induced velocity in the neighbourhood of a curved line vortex suggests (and detailed analysis confirms) that, when the vorticity is spread roughly uniformly over a tubular core whose cross-section is a circle of small radius ϵ, the speed of movement of such a vortex-tube of uniform curvature a^{-1} is given by cutting off the logarithmic divergence at the core boundary and so is asymptotically (for $a/\epsilon \gg 1$) of the form

$$\frac{\kappa}{4\pi a} \log \frac{a}{\epsilon}. \qquad (7.2.15)†$$

† The speed thus decreases as a increases, which explains the well-known amusing game played by two similar vortex rings at some distance apart on a common axis of symmetry. The velocity field associated with the rear vortex ring has a radially outward component at the position of the front ring and so the radius of the front ring gradually increases (with κ constant). This leads to a decrease in its speed of travel, and there is

The formula (7.2.10) shows that the fluid impulse for a vortex ring whose core radius is small is approximately independent of the dimensions of the core and so is given by the same expression (7.2.14) as for a line vortex. The total kinetic energy of the fluid is no longer infinite, and, so far as the asymptotic form as $\epsilon \to 0$ is concerned, may be evaluated by noting that the kinetic energy associated with a straight line vortex of strength κ in the region between an inner circular cylinder of radius ϵ and an outer cylinder of radius b is $(\rho\kappa^2/4\pi) \log (b/\epsilon)$ per unit length of the line vortex. This gives

$$T \sim \tfrac{1}{2}\rho a\kappa^2 \log \frac{a}{\epsilon} \qquad (7.2.16)$$

for the vortex ring, as may also be seen to follow from (7.2.11) and the fact that ψ is of order $(a\kappa/2\pi) \log \epsilon$ at a small distance ϵ from a line vortex. The speed of travel of the vortex ring is thus approximately $\tfrac{1}{2}T/P$.

The flow associated with a vortex ring having a small circular core is determined approximately by the parameters a, κ and ϵ, everywhere except within the core itself where the motion depends on the actual vorticity distribution there. Alternatively, two of these parameters may be replaced by the fluid impulse P given by (7.2.14) and the total kinetic energy T given by (7.2.16). The manner in which the values of three independent parameters are determined in practice will depend on the nature of the generating mechanism, and the details are often obscure. In the case of a vortex ring produced by suddenly making a circular disk of radius R move normal to its plane, we might suppose that the disk is somehow removed from the fluid at the instant when its speed is V without immediate effect on the fluid motion. The fluid impulse and the kinetic energy of the motion associated with the vortex ring are then the same as for the circular disk moving with speed V, and the circulation for the vortex ring could be put equal to the difference between the values of the velocity potential at the central points on the two faces of the disk; that is, according to the formulae of §6.8,

$$P = \tfrac{8}{3}R^3\rho V, \quad T = \tfrac{4}{3}R^3\rho V^2, \quad \kappa = 4RV/\pi.$$

It follows then from (7.2.14) and (7.2.16) that the dimensions of the resulting vortex ring are

$$a = \sqrt{\tfrac{2}{3}}R, \quad \epsilon = a\exp\left(-\tfrac{1}{2}\pi^2/\sqrt{6}\right) = 0{\cdot}13a,$$

and, from (7.2.15), that the speed of movement of the vortex ring is $\tfrac{1}{4}V$; but the value of ϵ/a is perhaps not small enough for the expressions (7.2.15) and (7.2.16) to apply accurately.

Two of the three parameters specifying the flow associated with a vortex ring of small core radius may be regarded as defining the length and velocity

a corresponding increase in the speed of travel of the rear vortex which ultimately passes through the larger vortex and in turn becomes the front vortex. The manoeuvre is then repeated. It is possible to demonstrate in the laboratory one or two such passages of one vortex through the other before they decay.

scales of the flow system. Thus, with a reference frame which moves with the ring, we may write the fluid velocity at any point $\mathbf{x}$ (except within the core itself) in the form

$$\frac{\kappa}{a} \text{func.}\left(\frac{\mathbf{x}}{a}, \frac{\epsilon}{a}\right),$$

showing that there is a singly-infinite family of such vortex rings corresponding to different values of ϵ/a. The primary effect of changing the value of ϵ/a, leaving aside details of the flow in and near the core, is to change the speed of travel of the vortex ring. Consequently we obtain an approximation

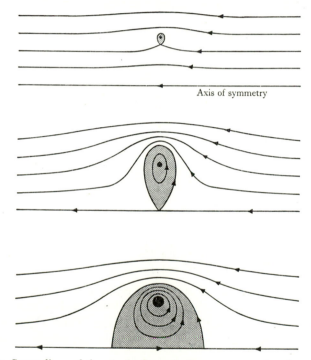

Axis of symmetry

Figure 7.2.4. Streamlines of the steady flow relative to a vortex ring for various (small) values of the ratio ϵ/a (sketch only). The inner black area marks the core of vorticity, and the shaded area represents fluid carried along with the ring.

to the steady flow patterns at different (small) values of ϵ/a by superposing a uniform axial velocity $-(\kappa/4\pi a)\log(a/\epsilon)$ on the streamlines shown in figure 7.2.1, these latter streamlines being determined by κ and a alone. In this way we find a sequence of flow patterns of the kind sketched in figure 7.2.4. There is a striking increase in the mass of fluid carried along with the vortex ring, as ϵ/a increases, and for ϵ/a larger than a value of order 0·01 the body of fluid moving with the ring extends to the axis.

It is natural to enquire if there are any steady vortex rings with cores which are not small. The distribution of vorticity is relevant here, and all that is required by inviscid-fluid theory is that ω/σ should be constant on any streamline in steady flow (for see (7.1.7)). The only available analytical evidence lies in the existence of a remarkably simple flow field known as 'Hill's spherical vortex' (Hill 1894). The vorticity here occupies a sphere of radius a, and is distributed according to the relation

$$\omega = A\sigma, \qquad (7.2.17)$$

where A has the same value for *all* the streamlines within the sphere. The corresponding stream function ψ for the flow within the sphere, relative to axes such that the sphere is stationary (so that $\psi = 0$ at $x^2 + \sigma^2 = a^2$), is readily found to be

$$\psi = \tfrac{1}{10}A\sigma^2(a^2 - x^2 - \sigma^2), \qquad (7.2.18)\dagger$$

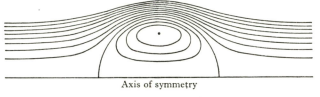

Figure 7.2.5. Streamlines of the steady flow relative to a Hill's spherical vortex, with equal intervals of ψ.

and the tangential component of velocity at the surface of the sphere, approached from within, is

$$\left\{\left(\frac{1}{\sigma}\frac{\partial\psi}{\partial\sigma}\right)^2 + \left(\frac{1}{\sigma}\frac{\partial\psi}{\partial x}\right)^2\right\}^{\frac{1}{2}}_{x^2+\sigma^2=a^2} = \tfrac{1}{5}Aa\sigma$$

in the direction away from the stagnation point at $\sigma = 0$, $x = a$. Now the velocity at the surface of a stationary sphere immersed in irrotational flow of a fluid with uniform speed U in the direction of the negative x-axis at infinity is $\tfrac{3}{2}U\sigma/a$ (see §6.8), and so the inner and outer velocity distributions match provided

$$U = \tfrac{2}{15}a^2A.$$

Streamlines of this steady flow relative to the vortex are shown in figure 7.2.5. It is evident, from a consideration of the separate contributions made by the regions inside and outside the sphere of radius a, that the fluid impulse (of the motion relative to axes fixed in the fluid at infinity) is of magnitude $2\pi a^3 \rho U$, as may also be found directly from (7.2.10).

It seems possible that a Hill's spherical vortex represents one extreme member of a family of vortex rings, with a circular line vortex as the other extreme member.

---

† This is the velocity distribution that was found also to hold inside a spherical drop of fluid in translational motion through a second fluid at small Reynolds number § 4.9.

7.3. Two-dimensional flow in unbounded fluid at rest at infinity

Two-dimensional flow fields without interior boundaries and in which the fluid extends to infinity and is at rest there differ in important respects from those considered in the previous section. There is first the possibility of an asymptotic variation of the velocity as r^{-1} at large distances from the origin, associated with the existence of a non-zero circulation round a circle of large radius, as in cases of irrotational flow with an interior boundary (see §6.4). Secondly, the self-induced velocity of movement of a straight line vortex is not infinite, and the behaviour of a vortex-tube with straight vortex-lines does not depend critically on the cross-section of the tube, so that concentrations of vorticity in two-dimensional flow can safely be approximated analytically by point vortices.

In a two-dimensional flow field without interior boundaries the irrotational contribution $\mathbf{v}$ in (7.1.1) is again zero, and we have, for the velocity at any point,

$$\mathbf{u}(\mathbf{x}) = -\frac{1}{4\pi}\int\int \frac{\mathbf{s}\times\boldsymbol{\omega}(\mathbf{x}')}{s^3} dA(\mathbf{x}')\,dz',$$

where δA is an element of area of the plane of flow, z' is a co-ordinate normal to that plane, and the integration is over all (three-dimensional) space. Since $\boldsymbol{\omega}$ is normal to the plane of flow, $\mathbf{s}\times\boldsymbol{\omega}(\mathbf{x}')$ is independent of z' and the integration with respect to z' can be carried out, giving

$$\left.\begin{aligned} u(x,y) &= -\frac{1}{2\pi}\int \frac{y-y'}{(x-x')^2+(y-y')^2}\,\omega(x',y')\,dA(x',y') \\ v(x,y) &= \frac{1}{2\pi}\int \frac{x-x'}{(x-x')^2+(y-y')^2}\,\omega(x',y')\,dA(x',y') \end{aligned}\right\}, \quad (7.3.1)$$

where (x,y) are rectilinear co-ordinates in the plane of flow, (u,v) are corresponding components of velocity, and the integration is over the whole plane of flow. It is evident that the velocity distribution is derivable from the stream function

$$\psi(x,y) = -\frac{1}{4\pi}\int \omega(x',y')\log\{(x-x')^2+(y-y')^2\}\,dA(x',y'). \quad (7.3.2)$$

This can equally be regarded as an expression for the one non-zero component of the vector potential.

Provided the vorticity is sufficiently small in magnitude at large distances from the origin, we have

$$\psi(x,y) = -\frac{1}{2\pi}\log r\int \omega\,dA + O(r^{-1}) \quad (7.3.3)$$

as $r\,(= (x^2+y^2)^{\frac{1}{2}})\to\infty$. Thus the velocity distribution far from the origin is the same as if there were a point vortex of strength equal to $\int\omega\,dA$ at the origin.

Integral invariants of the vorticity distribution

Straight-forward considerations of the total linear and angular momenta and kinetic energy of the fluid are not possible, because the integral expressions for these quantities are divergent. However, related quantities which have the expected property of invariance with respect to time do exist. It proves to be more convenient to look directly for invariant integrals of the vorticity distribution, and subsequently to consider their relation to the above physical quantities.

Both the vorticity and the area of material elements of the plane of flow are constant, so that the first and simplest of the invariant integrals is

$$\int \omega \, dA, \tag{7.3.4}$$

the integral being taken over the whole of the plane of flow; this integral is equal to the circulation round a closed curve everywhere at a large distance from the origin, so that the invariance can be regarded as a direct consequence of Kelvin's circulation theorem.

The first integral moments of the vorticity distribution are also constant. For we have

$$\frac{d}{dt} \int x\omega \, dA = - \int x \left\{ \frac{\partial(u\omega)}{\partial x} + \frac{\partial(v\omega)}{\partial y} \right\} dA$$

$$= \int u\omega \, dA,$$

and substitution of the expression for u given in (7.3.1) shows this integral to be zero; and similarly for $\int y\omega \, dA$. We may thus define the two invariant quantities

$$X = \frac{\int x\omega \, dA}{\int \omega \, dA}, \quad Y = \frac{\int y\omega \, dA}{\int \omega \, dA}, \tag{7.3.5}$$

representing the co-ordinates of the 'centre of vorticity'. If $\int \omega \, dA = 0$, this centre is at infinity.

The next integral moment which seems to warrant examination is $\int (x^2 + y^2) \omega \, dA$. The rate of change is

$$\frac{d}{dt} \int (x^2 + y^2) \omega \, dA = - \int (x^2 + y^2) \left\{ \frac{\partial(u\omega)}{\partial x} + \frac{\partial(v\omega)}{\partial y} \right\} dA$$

$$= 2 \int (xu + yv) \omega \, dA,$$

and substitution of the expressions for u and v given in (7.3.1) shows that this last integral is zero. Thus the length D defined by

$$D^2 = \frac{\int \{(x - X)^2 + (y - Y)^2\} \omega \, dA}{\int \omega \, dA} \tag{7.3.6}$$

and representing the dispersion of the vorticity distribution about its fixed centre (X, Y) is an invariant of the motion.

The dimensions of the integral quantities

$$\int x\omega\, dA, \quad \int y\omega\, dA \quad \text{and} \quad \int (x^2+y^2)\,\omega\, dA$$

suggest that they bear some relation to the linear and angular momenta of (unit depth of) the fluid. The relation is not a direct one when $\int \omega\, dA \neq 0$, because then the velocity is not small enough at infinity for the integrals representing total momenta to have meaning. Now the stream function

$$\psi(x,y)+\frac{1}{2\pi}\log r\int \omega\, dA$$

represents the difference between the given motion and a steady comparison flow with the same total amount of vorticity concentrated at the origin. The magnitude of the velocity in this difference motion diminishes as r^{-2} as $r \to \infty$, so that the integral representing the corresponding total linear momentum is still not absolutely convergent. However, it is possible to show, as in the case of three-dimensional flow considered in §7.2, that the total force impulse that must be applied to the fluid to generate this difference motion from rest has components

$$\rho\int y\omega\, dA, \quad -\rho\int x\omega\, dA. \tag{7.3.7}$$

Similarly it may be shown that the total moment, about the origin, of the force impulse required to generate the difference motion from rest is

$$-\tfrac{1}{2}\rho\int (x^2+y^2)\,\omega\, dA.$$

We have not yet found an invariant which corresponds in some way to the kinetic energy of the fluid. The use of a 'difference motion' is not profitable in the case of a non-linear quantity like kinetic energy, so we must proceed in another way. Now the kinetic energy of the fluid lying within a finite area A_1 bounded by a closed curve of which $\delta\mathbf{x}$ is an element is

$$T = \tfrac{1}{2}\rho \int_{A_1} (u^2+v^2)\, dA = \tfrac{1}{2}\rho \int_{A_1} \left(u\frac{\partial\psi}{\partial y} - v\frac{\partial\psi}{\partial x} \right) dA$$

$$= \tfrac{1}{2}\rho \int_{A_1} \left\{ \psi\omega + \frac{\partial(u\psi)}{\partial y} - \frac{\partial(v\psi)}{\partial x} \right\} dA$$

$$= \tfrac{1}{2}\rho \int_{A_1} \psi\omega\, dA - \tfrac{1}{2}\rho \oint \psi\mathbf{u}\,.\,d\mathbf{x}.$$

The first of these two integrals converges as $A_1 \to \infty$. The second does not, but its asymptotic form can be determined readily from (7.3.3). On choosing the bounding curve to be a circle of radius R centred on the origin we find that, as $R \to \infty$,

$$T - \tfrac{1}{2}\rho\int \psi\omega\, dA - \frac{1}{4\pi}\rho\log R(\int \omega\, dA)^2 \to 0, \tag{7.3.8}$$

in which the integrals are taken over the whole plane. It follows that, for some large fixed value of R, the quantity

$$W = \tfrac{1}{2}\rho \int \psi \omega \, dA,$$

$$= -\frac{1}{8\pi}\rho \iint \omega(x,y)\,\omega(x',y') \log\{(x-x')^2+(y-y')^2\}\,dA(x,y)\,dA(x',y'),$$

$$(7.3.9)$$

represents the part of the kinetic energy of the fluid that depends on the way in which the given total amount of vorticity is distributed. Since no work is being done on the fluid and no energy lost by dissipation, it is to be expected that W is independent of time; this can be confirmed by direct calculation.

We have thus found that $\int \omega \, dA$ and the quantities X, Y, D and W defined above are all constants of the motion. The material elements move with constant vorticity round a fixed centre, with constant dispersion about that centre and with a constant value of the integral in (7.3.9). These conditions to be satisfied by the changes in the vorticity distribution are quite strong, and may make possible a qualitative prediction of the development of the motion when the initial distribution of vorticity is a simple one.

Motion of a group of point vortices

The above integral invariants take a simpler form when the vorticity is concentrated at a number of points. Let us suppose that instantaneously there are point vortices of strength $\kappa_1, \kappa_2, \ldots, \kappa_n$ at the points

$$(x_1, y_1),\ (x_2, y_2),\ \ldots,\ (x_n, y_n)$$

respectively and that the vorticity is zero elsewhere. The strengths of the vortices remain constant but their positions change, the changes being such as to keep constant the values of X, Y, and D, where

$$X\sum_i \kappa_i = \sum_i \kappa_i x_i, \quad Y\sum_i \kappa_i = \sum_i \kappa_i y_i, \qquad (7.3.10)$$

$$D^2\sum_i \kappa_i = \sum_i \kappa_i\{(x_i-X)^2+(y_i-Y)^2\}, \qquad (7.3.11)$$

the summation being over all values of i from 1 to n in each case.

The expression (7.3.2) for the stream function becomes

$$\psi(x,y) = -\frac{1}{4\pi}\sum_i \kappa_i \log\{(x-x_i)^2+(y-y_i)^2\}. \qquad (7.3.12)$$

The velocity of movement of the vortex of strength κ_j is equal to the velocity of the fluid at the point (x_j, y_j) due to all the *other* vortices, since there is no self-induced movement of a point vortex. Hence

$$\frac{dx_j}{dt} = -\frac{1}{2\pi}\sum_{i(\neq j)}\frac{\kappa_i(y_j-y_i)}{r_{ij}^2}, \quad \frac{dy_j}{dt} = \frac{1}{2\pi}\sum_{i(\neq j)}\frac{\kappa_i(x_j-x_i)}{r_{ij}^2}, \qquad (7.3.13)$$

for all values of j from 1 to n, where

$$r_{ij}^2 = (x_i - x_j)^2 + (y_i - y_j)^2.$$

There is the further invariant represented by (7.3.9), but this needs modification in view of the infinite kinetic energy of an isolated point vortex. Proceeding as before, we consider the total kinetic energy T of the fluid bounded externally by a circle of large radius R and internally by circles of small radius ϵ centred on each point vortex, and find that

$$T + \frac{\rho}{4\pi} \sum_i \sum_j \kappa_i \kappa_j \log r_{ij} + \frac{\rho}{4\pi} \left(\sum_i \kappa_i^2\right) \log \epsilon - \frac{\rho}{4\pi} \left(\sum_i \kappa_i\right)^2 \log R \to 0$$
$$\scriptstyle{(i \neq j)}$$

as $R \to \infty$, $\epsilon \to 0$. Thus the expression

$$W = -\frac{\rho}{4\pi} \sum_i \sum_j \kappa_i \kappa_j \log r_{ij}, \qquad (7.3.14)$$
$$\scriptstyle{(i \neq j)}$$

in which the summation is over all values of i and j except the combinations $i = j$, is the part of the kinetic energy which, for fixed R and ϵ, depends on the positions of the vortices relative to each other. Again dW/dt may be shown to be zero.

It will be noticed that the equations (7.3.13) can also be written in the form

$$\kappa_j \frac{dx_j}{dt} = \frac{\partial W}{\partial y_j}, \quad \kappa_j \frac{dy_j}{dt} = -\frac{\partial W}{\partial x_j} \qquad (7.3.15)$$

(with no summation convention), showing the remarkable property of the flow field that, if the $2n$ quantities $\kappa_j^{\frac{1}{2}} x_j$, $\kappa_j^{\frac{1}{2}} y_j$ be regarded as generalized co-ordinates and momenta, (7.3.15) is a Hamiltonian system of differential equations.

In the case of just two vortices, the distance d between them must remain constant and the two vortices move in circular paths about the centre of vorticity with the same constant angular velocity

$$\frac{\kappa_1 + \kappa_2}{2\pi d^2}, \qquad (7.3.16)$$

as indicated in figure 7.3.1. When $\kappa_1 + \kappa_2 = 0$, the centre of vorticity is at infinity and the two vortices move in parallel straight lines with constant speed $\kappa_1/2\pi d$ (giving a two-dimensional analogue of a circular line vortex); the perpendicular bisector of the line joining the two vortices is a stream-line, and could therefore be replaced by a rigid boundary.

When $n = 3$, the details of the motion are not evident, but the above invariants suggest that all three vortices remain within a distance of order D from the centre of vorticity (except in a case in which the sum of two of κ_1, κ_2 and κ_3 is close to zero) and that the distance between any two vortices can never be much less than the smallest distance between any pair of vortices initially. The same remarks apply to larger groups of vortices, and one forms

34-2

a qualitative picture of a cloud of vortices in ceaseless motion although stationary as a whole and with constant overall size and average spacing of the vortices. Attempts have been made (Onsager 1949) to derive some of the general properties of the motion of a group of many vortices with randomly chosen initial positions, using the methods of statistical mechanics and the fact that the system is Hamiltonian, but the results are not yet conclusive.

When there are stationary interior and exterior boundaries to the fluid, the above relations must be modified to allow for the effect of the appropriate image system of point vortices. The method of conformal transformation explained in §6.5 may also be used to determine the flow due to a number of point vortices in the presence of boundaries of suitable shape.

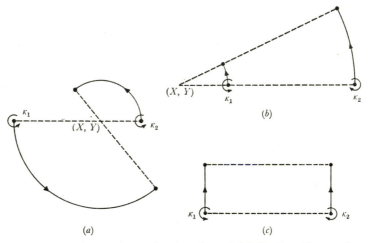

Figure 7.3.1. Motion of two point vortices in unbounded fluid. Case (*a*), κ_1 and κ_2 of same sign; (*b*) κ_1 and κ_2 of opposite sign; (*c*) $\kappa_1 + \kappa_2 = 0$.

Steady motions

As in the case of axisymmetric motion with circular vortex-lines, it is of interest to consider what distributions of vorticity in unbounded fluid in two dimensions yield steady motions.

It is immediately evident that any distribution of vorticity with circular symmetry about some point is associated with a steady motion since the streamlines are all circular. With (r, θ) as polar co-ordinates and origin at the centre of symmetry, we find, by equating the circulation round a circle to the flux of vorticity across the surface bounded by the circle,

$$u_\theta = -\frac{d\psi}{dr} = \frac{1}{r} \int r\omega(r)\, dr.$$

At positions outside the region of non-zero vorticity, the velocity is the same as if all the vorticity flux were concentrated at the origin.

If the vorticity distribution has only approximate circular symmetry, it seems likely that the departure from circular symmetry moves round the origin with the fluid although perhaps not without modification. We can see this process in detail when the vorticity has the uniform value ω_0 within the region bounded instantaneously by the curve

$$r = a + \epsilon \cos s\theta \qquad (7.3.17)$$

and is zero outside it, where s is an integer and $\epsilon \ll a$. This vorticity distribution may be regarded as a superposition of a uniform value ω_0 within the circle $r = a$ and a layer of vorticity at the circumference with strength density $\omega_0 \epsilon \cos s\theta$, as indicated in figure 7.3.2. The former contribution produces a pure rotatory motion which makes the bulges and depressions of the bound-

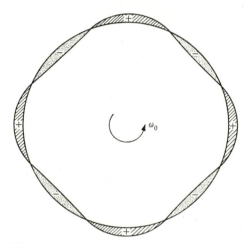

Figure 7.3.2. The non-symmetrical part of the vorticity distribution when a circular core of uniform vorticity ω_0 is perturbed ($s = 4$).

ary rotate about the origin with angular velocity $\frac{1}{2}\omega_0$. The latter deforms the boundary, by producing at angular position θ at the circumference a radial component of velocity equal to (the principal value of)

$$\frac{1}{4\pi} \epsilon \omega_0 \int_{-\pi}^{\pi} \cos s\theta' \cot \tfrac{1}{2}(\theta' - \theta) \, d\theta',$$

i.e., since

$$\int_{-\pi}^{\pi} \sin s\theta' \cot \tfrac{1}{2}\theta' \, d\theta' = 2\pi,$$

equal to $-\frac{1}{2}\epsilon\omega_0 \sin s\theta$. But this radial component of velocity at the boundary is exactly what is needed to make the boundary shape (7.3.17) rotate rigidly about the origin with angular velocity $-\frac{1}{2}\omega_0/s$. The two contributions together thus give rise to rigid rotation of the whole vorticity distribution with angular velocity

$$\tfrac{1}{2}\omega_0 \left(\frac{s-1}{s} \right), \qquad (7.3.18)$$

so that the flow field is steady relative to axes rotating with this (steady) angular velocity.

Whenever the vorticity is one-signed in a singly-connected region and zero outside it, the distribution will tend to rotate as a whole. Direct methods for the determination of those distributions that are steady relative to rotating axes are not available, but some special cases are known. It may be shown, as Kirchhoff first noticed, that a region of uniform vorticity ω_0 bounded by an ellipse $x^2/a^2 + y^2/b^2 = 1$ rotates, without change of shape, with angular velocity

$$\frac{ab}{(a+b)^2}\omega_0$$

(in agreement with the previous result when $a - b \ll a$). In the limit $b/a \to 0$, the region of non-zero vorticity becomes a sheet vortex on the x-axis of strength density

$$2b\omega_0\left(1 - \frac{x^2}{a^2}\right)^{\frac{1}{2}}, \qquad (7.3.19)$$

and this also rotates without change of form.

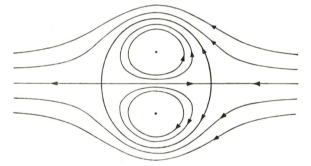

Figure 7.3.3. Streamlines of the steady flow relative to a pair of point vortices of strengths κ and $-\kappa$.

When the vorticity is positive in some parts of the fluid and negative in others, with $\int \omega \, dA = 0$, it is evidently possible for steady motions relative to translating axes to exist. As seen above, the motion due to two point vortices with strengths κ and $-\kappa$ is steady relative to axes which move with speed $\kappa/2\pi d$ in a direction normal to the line joining the vortices. The streamlines in this steady flow are shown in figure 7.3.3. It seems probable that the vorticity concentrated in each point vortex could be spread out over regions with boundaries approximating to the closed streamlines in figure 7.3.3 without violating the conditions for steady motion.

One case of steady motion with distributed vorticity of this latter kind is known, and is found by assuming that within the region of non-zero vorticity

$$\omega = k^2\psi,$$

where k is a constant, so that in polar co-ordinates

$$\frac{\partial^2 \psi}{\partial r^2} + \frac{1}{r}\frac{\partial \psi}{\partial r} + \frac{1}{r^2}\frac{\partial^2 \psi}{\partial \theta^2} = -k^2\psi. \tag{7.3.20}$$

We require a solution of this equation which will match with the stream function for the external irrotational flow, which suggests that we should try

$$\psi \propto \sin\theta,$$

as in irrotational flow past a circular cylinder. The full solution of (7.3.20) is then

$$\psi = CJ_1(kr)\sin\theta, \tag{7.3.21}$$

which makes the circle $r = a$ a streamline provided $J_1(ka) = 0$. The velocity on this streamline according to (7.3.21) then has the same value $2U\sin\theta$

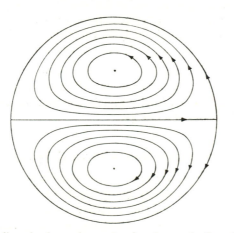

Figure 7.3.4. Streamlines in the region $r \le a$ for the steady flow due to vorticity proportional to $J_1(kr)\sin\theta$ $(r \le a,\ ka = 3\cdot83)$ and a uniform stream with suitably chosen speed at infinity.

as in irrotational flow due to a circular cylinder of radius a placed in a stream of uniform speed U in the direction $\theta = \pi$ at infinity, provided we choose

$$U = -\tfrac{1}{2}CkJ_1'(ka) = -\tfrac{1}{2}CkJ_0(ka). \tag{7.3.22}$$

Figure 7.3.4 shows the streamlines in the region $r \le a$ for this steady flow in the case $ka = 3\cdot83$, which is the smallest possible value of ka. With larger values of ka, ω and ψ change sign one or more times along a radial line before the boundary with the region of irrotational flow is reached. By considering the separate contributions from the regions inside and outside the circle $r = a$, or by use of the formula (7.3.7), we see that the fluid impulse of the flow relative to axes fixed in the fluid at infinity is a vector of magnitude $2\pi a^2 \rho U$ in the direction of the positive x-axis.

Exercises

1. Show, by conformal transformation of the plane of flow, that the path of a point vortex in the region between two straight intersecting boundaries is given by

$$r \sin n\theta = \text{const.},$$

where r, θ are polar co-ordinates with $\theta = 0$ and $\theta = \pi/n$ at the two boundaries.

2. Point vortices of equal strength κ are equidistant on a straight line extending to infinity, the distance between two consecutive vortices being a, and there is a similar parallel row of point vortices of strength $-\kappa$, the distance between the rows being b. Show that the speed with which all the vortices move along the rows is

$$\frac{\kappa}{2a} \coth \frac{\pi b}{a} \quad \text{or} \quad \frac{\kappa}{2a} \tanh \frac{\pi b}{a}$$

if a vortex in one row is opposite one in the other row or equidistant from two vortices of the other row. (Such a 'vortex-street' can be used as an approximate representation of the wake behind a body moving through fluid, in a certain range of Reynolds number; see figure 4.12.6, plate 2, and figure 5.11.4, plate 11.)

3. Show that when the effect of viscosity on flow fields of the kind considered in this section is taken into account, the total vorticity $\int \omega \, dA$ and the co-ordinates of the centre of vorticity remain constant but D^2 (where D is the dispersion length) increases at a rate 4ν.

7.4. Steady two-dimensional flow with vorticity throughout the fluid

In two-dimensional flow the mass-conservation equation may be satisfied identically by writing the velocity components (u, v) in terms of a stream function ψ. The magnitude of the vorticity vector, which is everywhere normal to the plane of flow (the (x, y)-plane), is then given by

$$\omega = \frac{\partial v}{\partial x} - \frac{\partial u}{\partial y} = -\left(\frac{\partial^2 \psi}{\partial x^2} + \frac{\partial^2 \psi}{\partial y^2}\right).$$

Now it was seen from (7.1.6) that in two-dimensional flow the vorticity associated with a material element is constant; and in steady flow the paths of material elements are streamlines. Hence ω has the same value at all points of a streamline, and can evidently be written as a function of ψ alone, as $f(\psi)$ say. We thus have

$$\frac{\partial^2 \psi}{\partial x^2} + \frac{\partial^2 \psi}{\partial y^2} = -f(\psi) \tag{7.4.1}$$

as the equation determining the velocity distribution in steady flow, once the function f is known.

The quantity H is constant along a streamline, and so is also a function of ψ alone. Thus (7.1.8) can be written as

$$\mathbf{u} \times \boldsymbol{\omega} = \frac{dH}{d\psi} \nabla \psi, \tag{7.4.2}$$

which yields the one scalar relation

$$\frac{dH}{d\psi} = -\omega = -f(\psi). \tag{7.4.3}$$

H can be found by integration when the function f is known, thereby providing an explicit expression for the pressure.

The mathematical procedure for solving a problem of steady two-dimensional flow is thus clear (although it may not be easy to determine ψ analytically from (7.4.1)), provided the distribution of vorticity over the different streamlines is known. This vorticity distribution is arbitrary, so far as inviscid-fluid theory is concerned. In practice the vorticity distribution is determined by the history of establishment of the steady flow, and this previous history will normally include a significant effect of viscosity. It will not often be possible to analyse the process of establishment of the steady state in detail, and a knowledge of the function f is likely to be available only in simple cases.

Possible solutions of the inviscid-fluid equations may be investigated by choosing specific forms for the function f in equation (7.4.1). One obviously convenient choice is $f(\psi) \propto \psi$, which yields the linear equation

$$-\omega = \frac{\partial^2\psi}{\partial x^2} + \frac{\partial^2\psi}{\partial y^2} = -\alpha^2\psi, \tag{7.4.4}†$$

which is familiar from the theory of transverse vibrations of an elastic membrane with a fixed line boundary in the (x, y)-plane, with ψ taking the place of displacement of the membrane; solutions are known for a number of shapes of boundary on which ψ is constant—circular, rectangular, triangular—but it is not known whether, and in what circumstances, the corresponding flow fields can be generated.

Another simple case, and one of practical interest, is obtained by taking the vorticity to be uniform, and equal to ω_0 say, throughout the fluid. The

---

† It happens that this form for the vorticity distribution is also capable of satisfying the complete vorticity equation for a viscous fluid, which for two-dimensional flow may be written as

$$\frac{\partial\omega}{\partial t} + \frac{\partial(\omega, \psi)}{\partial(x, y)} = \nu\left(\frac{\partial^2\omega}{\partial x^2} + \frac{\partial^2\omega}{\partial y^2}\right).$$

Any distribution of vorticity which is constant along streamlines makes the second term vanish, and with the particular distribution (7.4.4) the remaining terms in the equation balance if

$$\frac{\partial\omega}{\partial t} = -\alpha^2\nu\omega,$$

i.e. if $\omega \propto \exp(-\alpha^2\nu t)$.

Thus a solution for ψ as a function of x and y obtained from (7.4.4) represents either a steady motion of an inviscid fluid, or, when multiplied by $\exp(-\alpha^2\nu t)$, a decaying motion of a viscous fluid.

equation for ψ is then

$$\frac{\partial^2 \psi}{\partial x^2} + \frac{\partial^2 \psi}{\partial y^2} = -\omega_0, \tag{7.4.5}$$

which is Poisson's equation with a constant right-hand side, and was met also in §4.2 in a quite different context. (This equation holds for both steady and unsteady flow when ω is uniform, although the boundary conditions will not be the same in the two cases.) Equation (7.4.3) can be integrated when ω is constant, giving

$$H = \text{const.} - \omega_0 \psi,$$

or, with p representing the modified pressure which includes the effect of gravity,

$$\frac{p}{\rho} = \text{const.} - \tfrac{1}{2}q^2 - \omega_0 \psi. \tag{7.4.6}$$

The remainder of this section will discuss three different forms of this case of uniform vorticity.

Uniform vorticity in a region bounded externally

There is no need to say much about the detailed solution of flow problems of this kind, but the fact that cases of steady flow with uniform vorticity in a region bounded externally may arise naturally in at least two different ways is worthy of notice. The first and more obvious of these two ways requires an initial rotation of the fluid as a whole. Fluid which is enclosed by a rigid cylinder in steady rotation about an axis parallel to the generators comes ultimately, through the action of viscous stresses, to a state of rest relative to the rigid boundary and then has uniform vorticity. If now the rotation of the boundary suddenly ceases, the fluid in the cylinder continues to move with uniform vorticity except in a thin layer near the boundary (in the absence of any separation) where the effect of viscous diffusion of vorticity from the wall is significant. This boundary layer grows in thickness until the whole fluid is brought to rest, but for suitably large values of the Reynolds number of the flow there is a period of time in which it is of negligible thickness. During this time the flow in the bulk of the fluid is governed by (7.4.5), with ψ constant on the stationary enclosing boundary.

The other way in which regions of uniform vorticity may arise in steady two-dimensional flow also involves the action of viscosity in an initial phase of the motion. Let us suppose that in the steady state there exists a set of closed streamlines which do not enclose an interior boundary and on which the effect of viscous stresses is everywhere small (that is, none of these streamlines passes through a layer in which viscous and inertia forces are comparable). The vorticity will be approximately constant along each one of these streamlines. Now the *exact* equation satisfied by the vorticity in this case of steady two-dimensional motion is

$$\mathbf{u}.\nabla \omega = \nu \nabla^2 \omega,$$

which is simply the diffusion equation in a moving medium; and it follows that if the vorticity has different values on different streamlines there will be a diffusive flux of vorticity across streamlines, either inwards or outwards at all points of any one streamline. Since there is no source or sink of vorticity at the centre of the nest of closed streamlines, the only possible steady state (which will require a long time for its establishment, in view of the assumption of small viscous forces) is one of uniform vorticity. The argument can be given rigorous analytical form (Batchelor 1956), and the result is found to hold also when the streamlines do enclose an inner boundary.[†]

Whatever the cause of the uniformity of the vorticity, the determination of the stream function from (7.4.5) is a purely mathematical problem once ω_0 is given. The solutions of this equation mentioned in §4.2 may be employed again here, with a different interpretation. For instance, for steady flow with uniform vorticity ω_0 in fluid bounded externally by an ellipse with semi-axes a and b we have

$$\psi = -\tfrac{1}{2}\omega_0 \left(\frac{x^2}{a^2}+\frac{y^2}{b^2}\right)\bigg/\left(\frac{1}{a^2}+\frac{1}{b^2}\right), \qquad (7.4.7)$$

valid everywhere except in the neighbourhood of the boundary where viscous forces may be important; interesting features of this motion are that material elements move on similar ellipses, with equal times of orbit, and with constant moment of momentum of each material element about the centre.

Fluid in rigid rotation at infinity

When a mass of fluid which extends to infinity is initially in rigid rotation with angular velocity $\tfrac{1}{2}\omega_0$, any two-dimensional motion generated in this fluid has uniform vorticity ω_0, provided the conditions for Kelvin's circulation theorem to be valid are satisfied. We shall concern ourselves with steady motions of this kind with a stationary interior rigid boundary. It is found useful here to represent the motion as a superposition of (*a*) a rigid rotation with angular velocity $\tfrac{1}{2}\omega_0$ about the origin of the co-ordinate system, (*b*) a uniform velocity $-\mathbf{U}$ (the minus sign being included to facilitate correspondence with earlier analysis of irrotational motion due to a body moving with velocity $\mathbf{U}$ through fluid at rest at infinity) which depends on the relative positions of the origin and the actual centre of rotation, and (*c*) a disturbance motion (not necessarily of small magnitude) due to the presence of the boundary, which is irrotational with velocity potential ϕ. Then

---

† There is an analogous result for steady axisymmetric flow with closed streamlines in an axial plane. Under certain conditions the vorticity in the region of approximately inviscid flow is found then to be an azimuthal vector, with magnitude proportional to the distance from the axis of symmetry (as in a Hill's spherical vortex—see (7.2.17)). The results for both two-dimensional and axisymmetric flow may be summarized by the statement that $H \propto \psi$ in a region of steady approximately inviscid flow with closed streamlines.

in terms of polar co-ordinates (r, θ), with $\theta = 0$ in the direction of $\mathbf{U}$, we have the radial and circumferential velocity components

$$-U \cos \theta + \frac{\partial \phi}{\partial r}, \quad \tfrac{1}{2} \omega_0 r + U \sin \theta + \frac{1}{r} \frac{\partial \phi}{\partial \theta}, \tag{7.4.8}$$

where U is the magnitude of $\mathbf{U}$. The disturbance velocity $\nabla \phi$ is zero at infinity and the condition of zero flux of fluid across each portion of the inner boundary requires the normal component of $\nabla \phi$ there to take a prescribed value (which depends on the shape of the boundary and on ω_0 and $\mathbf{U}$); thus, provided the value of the cyclic constant of the irrotational motion is given, the disturbance motion is unique and the standard methods of irrotational flow theory are available for its determination.

The manner in which the flow field is affected by the vorticity of the fluid is illustrated by the simple case of a circular interior boundary of radius a. If we choose the centre of the circular cylinder as the origin, the inner boundary condition is unaffected by the rigid rotation of the fluid about the origin and ϕ has the same form as for irrotational flow due to a cylinder in a uniform stream. With κ as the cyclic constant of ϕ, we have for the disturbance motion

$$\phi = \frac{\kappa \theta}{2\pi} - \frac{U a^2 \cos \theta}{r}. \tag{7.4.9}$$

The complete velocity distribution may be represented conveniently in terms of a stream function of the form

$$\psi = -\tfrac{1}{4} \omega_0 r^2 - U r \sin \theta - \frac{\kappa}{2\pi} \log r + \frac{U a^2 \sin \theta}{r}. \tag{7.4.10}$$

When made non-dimensional ψ evidently depends on the two parameters $\omega_0 a / U$ and $\kappa / a U$. Figure 7.4.1 shows a sketch of the streamlines for $\omega_0 a / U = \tfrac{1}{2}$ (in which case the centre of rotation is four radii from the centre of the cylinder) and $\kappa / a U = 0$.

It is evident that the rotation of the whole fluid makes the flow past the circular cylinder unsymmetrical about the line $\theta = 0$, just as a non-zero circulation does in the absence of rotation, and that there is now a non-zero force exerted on the cylinder normal to this line. The resultant force exerted on the interior boundary A by the fluid is seen from (7.4.6) (on leaving aside effects of gravity) to be

$$\mathbf{F} = -\int p \mathbf{n}\, dA = \tfrac{1}{2} \rho \int q^2 \mathbf{n}\, dA, \tag{7.4.11}$$

since ψ is constant on the inner boundary. On making use of (7.4.10) we find for the non-zero component of $\mathbf{F}$ in the direction of the y-axis

$$F_y = \tfrac{1}{2} \rho a \int_0^{2\pi} \left(\tfrac{1}{2} \omega_0 a + U \sin \theta + \frac{\kappa}{2\pi a} + U \sin \theta \right)^2 \sin \theta\, d\theta$$

$$= \rho U (\pi a^2 \omega_0 + \kappa). \tag{7.4.12}$$

The nature of the relation (7.4.12) suggests that there might exist an expression for the force on a body of arbitrary shape, analogous to the Kutta-Joukowski relation to which (7.4.12) reduces when $\omega_0 = 0$. We may explore this suggestion by an application of the momentum theorem, as in §6.4. The force exerted on a body of arbitrary shape replacing the circular cylinder in figure 7.4.1 is given in terms of the conditions at a circle centred on the origin by the relation (6.4.27), provided (*a*) we replace $-\mathbf{U}$ in that relation by the undisturbed fluid velocity $-\mathbf{U} + \frac{1}{2}\boldsymbol{\omega}_0 \times \mathbf{x}$ which here is non-uniform, and (*b*) use the new expression (7.4.6) for the pressure. It will be left as an exercise for the reader to establish that the force on the body is the

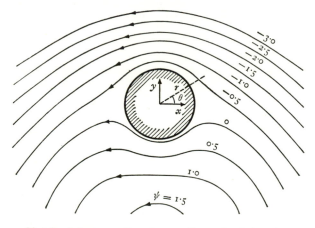

Figure 7.4.1. Sketch of the streamlines in two-dimensional flow due to a stationary circular cylinder in a rotating fluid; $\omega_0 a/U = \frac{1}{2}$, $\kappa/aU = 0$.

resultant of the usual force of magnitude $\rho U\kappa$ and direction normal to $\mathbf{U}$, and a force $-\pi\rho\boldsymbol{\omega}_0 \times \mathbf{c}$, where $\mathbf{c}\cdot\mathbf{x}/r^2$ is the leading term in the expansion of the acyclic part of ϕ in terms of circular harmonics of negative degree. It is possible to determine this coefficient $\mathbf{c}$, when the body shape is given, by conformal transformation of the boundary curve into a circle. As already remarked, the inner boundary condition satisfied by ϕ involves both $\mathbf{U}$ and ω_0, and so $\mathbf{c}$ depends on these two quantities, in general, as well as on the shape of the body.

Fluid in simple shearing motion at infinity

Generally similar remarks may be made concerning the steady flow about a stationary body in a stream of fluid whose undisturbed velocity is $(-U - \omega_0 y, 0)$ relative to a Cartesian co-ordinate system. Again the disturbance motion due to the presence of the body may be represented by a velocity potential ϕ, the acyclic part of which is determined uniquely by the condition of zero flux of volume across each portion of the boundary of the body.

The details of the flow field can again be determined readily when the body is a circular cylinder of radius a. We place the centre of the cylinder at the origin, and the inner boundary condition on ϕ is then

$$(\partial\phi/\partial r)_{r=a} = (U + \omega_0 a \sin\theta)\cos\theta.$$

The two terms on the right-hand side can be matched by separate solutions for ϕ, whence we find

$$\phi = \frac{\kappa\theta}{2\pi} - \frac{Ua^2\cos\theta}{r} - \frac{\omega_0 a^4 \sin 2\theta}{4r^2}. \qquad (7.4.13)$$

The stream function for the whole motion is thus

$$\psi = -\tfrac{1}{2}\omega_0 r^2 \sin^2\theta - Ur\sin\theta - \frac{\kappa}{2\pi}\log r + \frac{Ua^2\sin\theta}{r} - \frac{\omega_0 a^4 \cos 2\theta}{4r^2}.$$

$$(7.4.14)$$

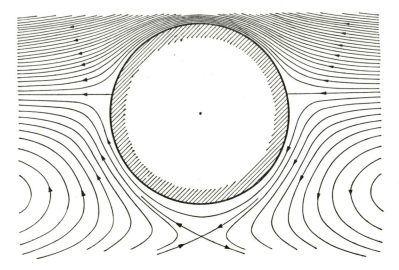

Figure 7.4.2. Streamlines of the two-dimensional flow due to a stationary circular cylinder embedded in a simple shearing motion; $\omega_0 a/U = 1$, $\kappa/aU = 0$. (From Tsien 1943.)

The streamlines are shown in figure 7.4.2 for the particular case $\omega_0 a/U = 1$, $\kappa/aU = 0$, representing a rather strong shearing motion. As was to be expected, the difference between (7.4.14) and (7.4.10) represents flow due to a circular cylinder embedded in a pure straining motion.

Again it is clear that the vorticity in the approaching fluid causes the exertion of a non-zero force on the circular cylinder. The component of force on the cylinder in the x-direction is zero, since the flow field is symmetrical about the y-axis. For the y-component we find, by straight-forward evaluation of the integral in (7.4.11),

$$F_y = \tfrac{1}{2}\rho a \int_0^{2\pi} \left(\frac{\partial\psi}{\partial r}\right)^2_{r=a} \sin\theta\, d\theta = \rho U(2\pi a^2 \omega_0 + \kappa). \qquad (7.4.15)$$

Again it is possible to generalize the relation (7.4.15) for a cylinder of arbitrary cross-section, and to find that

$$F_x = 0, \quad F_y = \rho(-2\pi\omega_0 c_x + U\kappa),$$
(7.4.16)

where the vector **c**, with components (c_x, c_y), is again the coefficient of the leading circular harmonic in the expansion of the acyclic part of ϕ. That F_x should be zero here is plausible from energy considerations; for the body is in steady motion relative to axes moving with velocity $(-U, 0)$ and any non-zero drag on the body would lead to work being done by the body and so to a change of the kinetic energy of the fluid. Again conformal transformation of the plane of flow may allow the determination of **c** for a body of given shape (as Tsien (1943) has shown for a 'Joukowski' aerofoil).

Any undisturbed motion in which the two velocity components are linear functions of x and y has uniform vorticity, but the above two particular cases—rigid rotation and simple shearing motion—seem to be the significant ones. Formulae like (7.4.12) are of some interest in considerations of flow in rotating machinery, while formulae like (7.4.15) may be relevant to the behaviour of bodies moving through fluid already in motion. Direct quantitative applications are of course restricted to bodies of streamlined shape from which no boundary-layer separation occurs. If the body has a sharp tail, like an aerofoil, the circulation κ around the body in steady flow will be determined by the Joukowski hypothesis (§ 6.7), as in the case of flow with zero vorticity.

Exercise

In one two-dimensional flow system the centre of a circular cylinder of radius a moves on a circular path of radius R with uniform angular velocity $-\tfrac{1}{2}\omega_0$ through fluid at rest at infinity, and the circulation round the cylinder is zero. In a second system the fluid at infinity is in rigid rotation with angular velocity $\tfrac{1}{2}\omega_0$ and an identical cylinder is stationary with its centre at distance R from the centre of rotation, the circulation round a path immediately enclosing the cylinder being $\pi a^2 \omega_0$. Show that the force exerted on the cylinder by the fluid is directed away from the centre of rotation in both cases and that the magnitude of the force in the first flow system is half that in the second. (Note that, according to the results of §4.1, the two velocity distributions are identical if rotating axes are used in the first system.)

7.5. Steady axisymmetric flow with swirl

We shall use cylindrical polar co-ordinates (x, σ, ϕ), with corresponding velocity components (u, v, w) and vorticity components $(\omega_x, \omega_\sigma, \omega_\phi)$ given, in axisymmetric flow, by

$$\omega_x = \frac{1}{\sigma}\frac{\partial(\sigma w)}{\partial \sigma}, \quad \omega_\sigma = -\frac{\partial w}{\partial x}, \quad \omega_\phi = \frac{\partial v}{\partial x} - \frac{\partial u}{\partial \sigma}.$$
(7.5.1)

The mass-conservation equation may be satisfied by writing the velocity components u and v in the form

$$u = \frac{1}{\sigma}\frac{\partial \psi}{\partial \sigma}, \quad v = -\frac{1}{\sigma}\frac{\partial \psi}{\partial x}, \tag{7.5.2}$$

where $\psi(x, \sigma)$ is the stream function; and the azimuthal component of vorticity is then

$$\omega_\phi = -\frac{1}{\sigma}\left(\frac{\partial^2 \psi}{\partial x^2}+\frac{\partial^2 \psi}{\partial \sigma^2}-\frac{1}{\sigma}\frac{\partial \psi}{\partial \sigma}\right). \tag{7.5.3}$$

It will be observed that the vorticity components ω_x and ω_σ are derived from the function σw in exactly the way that u and v are derived from ψ.

The dynamical equation (7.1.3) yields the three scalar relations

$$v\omega_\phi - w\omega_\sigma - \frac{\partial u}{\partial t} = \frac{\partial H}{\partial x}, \tag{7.5.4}$$

$$w\omega_x - u\omega_\phi - \frac{\partial v}{\partial t} = \frac{\partial H}{\partial \sigma}, \tag{7.5.5}$$

$$u\omega_\sigma - v\omega_x - \frac{\partial w}{\partial t} = 0, \tag{7.5.6}$$

H being a function only of x and σ in axisymmetric flow. The last of these relations may also be written as

$$\frac{D(\sigma w)}{Dt} = 0, \tag{7.5.7}$$

representing the constancy of the circulation round a material curve in the form of a circle centred on the axis of symmetry and normal to it. Problems of axisymmetric flow with swirl usually involve interesting and difficult questions concerning the interaction of the swirling component w and the motion in an axial plane with velocity components u and v.

When the motion is *steady*, a material element moves along a streamline and thus moves on the surface of revolution formed by rotating the curve in the axial plane given by $\psi = $ const. about the axis of symmetry. It follows then from Bernoulli's theorem and from (7.5.7) that

$$\tfrac{1}{2}(u^2+v^2+w^2)+\frac{p}{\rho} = H(\psi),$$

$$\sigma w = C(\psi), \tag{7.5.8}$$

H and C being arbitrary functions of ψ. Two of the relations (7.5.1) then become

$$\omega_x = u\frac{dC}{d\psi}, \quad \omega_\sigma = v\frac{dC}{d\psi}, \tag{7.5.9}$$

showing that the components of $\mathbf{u}$ and $\boldsymbol{\omega}$ in an axial plane are locally parallel and that the Bernoulli surfaces are surfaces of revolution on which ψ is constant. Either of the dynamical equations (7.5.4) or (7.5.5) may now be used to obtain an expression for ω_ϕ in terms of H and C. For (7.5.4) becomes, when $\partial u/\partial t = 0$,

$$\frac{\omega_\phi}{\sigma} = \frac{w\omega_\sigma}{\sigma v} + \frac{1}{\sigma v}\frac{dH}{d\psi}\frac{\partial\psi}{\partial x}$$

$$= \frac{C}{\sigma^2}\frac{dC}{d\psi} - \frac{dH}{d\psi}, \tag{7.5.10}$$

and this is also the form taken by (7.5.5). (In the case of zero swirl, ω_ϕ/σ is a function of ψ alone, as implied by (7.1.7).) Combination of (7.5.3) and (7.5.10) then gives

$$\frac{\partial^2\psi}{\partial x^2} + \frac{\partial^2\psi}{\partial\sigma^2} - \frac{1}{\sigma}\frac{\partial\psi}{\partial\sigma} = \sigma^2\frac{dH}{d\psi} - C\frac{dC}{d\psi}. \tag{7.5.11}$$

Cases in which all quantities are independent of x and $v = 0$ are of some importance, arising often in connection with flow along circular ducts and tubes, and may be termed *cylindrical flows* since the Bernoulli surfaces are circular cylinders. The radial equation of motion reduces here (for a steady flow) to

$$\frac{1}{\rho}\frac{dp}{d\sigma} = \frac{w^2}{\sigma} = \frac{C^2}{\sigma^3}, \tag{7.5.12}$$

so that

$$H = \tfrac{1}{2}(u^2 + w^2) + \int\frac{C^2}{\sigma^3}\,d\sigma$$

$$= \tfrac{1}{2}u^2 + \int\frac{C}{\sigma^2}\frac{dC}{d\sigma}\,d\sigma. \tag{7.5.13}$$

Equation (7.5.11) then reduces to an identity; any distributions of u and w, or alternatively of H and C, with respect to σ, give a possible steady cylindrical flow.

In cases of steady flow in which all the streamlines come from a region, perhaps at 'infinity', where the values of H and C for the different streamlines are known, the functions $H(\psi)$ and $C(\psi)$ in (7.5.11) are known and ψ can then be determined, in principle, as a function of x and σ over the whole field. Such a determination is likely to be feasible only when H and C are very simple functions of ψ. Fortunately the relatively simple case in which the fluid far upstream has uniform axial velocity U and rotates as a rigid body with angular velocity Ω is the one of most important in practice. The upstream conditions are then given by

$$\psi = \tfrac{1}{2}U\sigma^2, \quad C = \Omega\sigma^2,$$

and, since the flow is cylindrical in this upstream region and (7.5.13) applies,

$$H = \tfrac{1}{2}U^2 + \Omega^2\sigma^2.$$

We can rewrite these upstream conditions as

$$C = \frac{2\Omega}{U}\psi, \quad H = \tfrac{1}{2}U^2 + \frac{2\Omega^2}{U}\psi, \tag{7.5.14}$$

and this must be the dependence of C and H on ψ over the whole flow field. The governing equation then takes the linear form

$$\frac{\partial^2 \psi}{\partial x^2} + \frac{\partial^2 \psi}{\partial \sigma^2} - \frac{1}{\sigma} \frac{\partial \psi}{\partial \sigma} = \frac{2\Omega^2}{U} \sigma^2 - \frac{4\Omega^2}{U^2} \psi.$$

It is convenient to use as dependent variable the departure of the stream function from its upstream form, and to write

$$\psi(x, \sigma) = \tfrac{1}{2} U\sigma^2 + \sigma F(x, \sigma), \qquad (7.5.15)$$

whence the equation becomes

$$\frac{\partial^2 F}{\partial x^2} + \frac{\partial^2 F}{\partial \sigma^2} + \frac{1}{\sigma} \frac{\partial F}{\partial \sigma} + \left(k^2 - \frac{1}{\sigma^2} \right) F = 0, \qquad (7.5.16)$$

where
$$k = 2\Omega/U.$$

Some idea of the rather puzzling way in which the axial and azimuthal motions interact in steady flow will now be given by means of examples of solutions of (7.5.16). It should be borne in mind that the linearity of this equation is a consequence of the special form of the assumed upstream conditions represented by (7.5.14). Little is known about solutions of the equation (7.5.11) corresponding to other upstream conditions.

The effect of a change of cross-section of a tube on a stream of rotating fluid

We suppose that fluid in steady motion along a tube passes through a transition from one long cylindrical section to another of different cross-section, and that at some distance upstream of the transition the fluid has uniform axial velocity U and rotates as a rigid body with angular velocity Ω. The tube has an axially symmetrical boundary, and the flow is assumed axisymmetric throughout. The transition may be a simple increase or decrease of radius of the tube, and two other possibilities of practical interest are indicated in figure 7.5.1. In all the cases under consideration the flow is cylindrical both upstream and downstream of the transition, and the problem is to determine the properties of the cylindrical flow on the downstream side.

Equation (7.5.16) is applicable over the whole flow field, and in the downstream cylindrical region, where ψ and F are functions of σ alone, we have

$$\frac{d^2 F}{d\sigma^2} + \frac{1}{\sigma} \frac{dF}{d\sigma} + \left(k^2 - \frac{1}{\sigma^2} \right) F = 0. \qquad (7.5.17)$$

This is Bessel's equation of order unity, and the general solution is

$$F = AJ_1(k\sigma) + BY_1(k\sigma), \qquad (7.5.18)$$

where, in the standard terminology, J_1 and Y_1 denote Bessel functions of the first and second kind. The constants A and B are now to be determined from known values of ψ at two values of the radius.

We may include all the above kinds of transition within the scope of the formulae if we suppose that in the upstream cylindrical section the fluid lies in the annular region $a_1 \geqslant \sigma \geqslant a_2$ and in the downstream cylindrical section the fluid lies in the annular region $b_1 \geqslant \sigma \geqslant b_2$ (figure 7.5.2). The streamlines on which ψ is equal to $\frac{1}{2}Ua_1^2$ and $\frac{1}{2}Ua_2^2$ in the upstream cylindrical section are at radial positions $\sigma = b_1$ and $\sigma = b_2$ respectively in the downstream

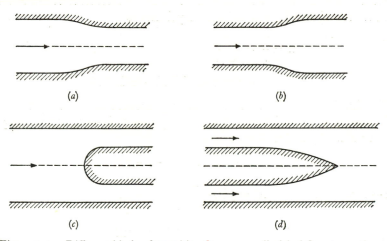

Figure 7.5.1. Different kinds of transition from one cylindrical flow to another.

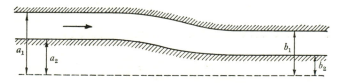

Figure 7.5.2. The general transition.

cylindrical region, and the boundary conditions to be satisfied by the solution (7.5.18) are thus

$$F = \tfrac{1}{2}U\left(\frac{a_1^2 - b_1^2}{b_1}\right) \quad \text{at} \quad \sigma = b_1,$$

$$F = \tfrac{1}{2}U\left(\frac{a_2^2 - b_2^2}{b_2}\right) \quad \text{at} \quad \sigma = b_2.$$

These conditions require

$$A = \frac{U}{2b_1 b_2}\,\frac{b_2(a_1^2 - b_1^2)\,Y_1(kb_2) - b_1(a_2^2 - b_2^2)\,Y_1(kb_1)}{J_1(kb_1)\,Y_1(kb_2) - J_1(kb_2)\,Y_1(kb_1)}, \qquad (7.5.19)$$

and similarly for B with J_1 and Y_1 exchanged.

The axial velocity in the downstream cylindrical region is then seen from (7.5.15) and (7.5.18) to be

$$u = \frac{1}{\sigma}\frac{\partial \psi}{\partial \sigma} = U + \frac{1}{\sigma}\frac{d}{d\sigma}\{A\sigma J_1(k\sigma) + B\sigma Y_1(k\sigma)\}$$

$$= U + AkJ_0(k\sigma) + BkY_0(k\sigma), \qquad (7.5.20)$$

on making use of the known relations between the Bessel functions J_0 and J_1 and between Y_0 and Y_1. The azimuthal velocity is

$$w = \frac{C}{\sigma} = \frac{2\Omega}{U}\frac{\psi}{\sigma}$$

$$= \Omega\sigma + kAJ_1(k\sigma) + kBY_1(k\sigma). \qquad (7.5.21)$$

The simple change in radius of a tube represented in figure 7.5.1 (*a*) or (*b*) is the case of greatest interest. On putting $a_2 = 0$, $b_2 \to 0$, and writing a, b in place of a_1, b_1, and making use of the limits

$$J_1(z) \to 0, \quad zY_1(z) \to -2/\pi$$

as $z \to 0$, we find

$$A = \tfrac{1}{2}U\frac{a^2 - b^2}{bJ_1(kb)}, \quad B = 0,$$

so that

$$\frac{u}{U} = 1 + \left(\frac{a^2}{b^2} - 1\right)\frac{\tfrac{1}{2}kbJ_0(k\sigma)}{J_1(kb)} \qquad (7.5.22)$$

and

$$\frac{w}{\Omega\sigma} = 1 + \left(\frac{a^2}{b^2} - 1\right)\frac{bJ_1(k\sigma)}{\sigma J_1(kb)}. \qquad (7.5.23)$$

When $kb \ll 1$, the last two of these formulae reduce to

$$\frac{u}{U} \approx \frac{a^2}{b^2}, \quad \frac{w}{\Omega\sigma} \approx \frac{a^2}{b^2},$$

corresponding to the changes expected for a combined stream-tube and vortex-tube of small cross-section over which the velocity and vorticity are uniform. For larger values of kb, the nature of the changes in the distributions of u and w with respect to σ can be envisaged from the sketch in figure 7.5.3 of the functions $J_0(z)$ and $J_1(z)$. Provided $kb < 2 \cdot 40$ (the first zero of the function $J_0(z)$), the departures of u/U and $w/\Omega\sigma$ from unity have the same sign as $a - b$ everywhere in the downstream cylindrical region (i.e. u and w/σ are increased by a contraction of the tube and decreased by an expansion), and vary monotonically across the tube, being greater in magnitude at the centre. At the centre of the downstream cylindrical flow we have

$$\left(\frac{u}{U}\right)_{\sigma=0} = \left(\frac{w}{\Omega\sigma}\right)_{\sigma=0} = 1 + \left(\frac{a^2}{b^2} - 1\right)\frac{\tfrac{1}{2}kb}{J_1(kb)}, \qquad (7.5.24)$$

the equality of u/U and $w/\Omega\sigma$ being attributable to the fact that the axis is embedded in a combined stream-tube and vortex-tube of small cross-section. The factor $\frac{1}{2}kb/J_1(kb)$ varies from $1\cdot0$ to $2\cdot32$ as kb varies from zero to $2\cdot40$, so that the changes in u/U and $w/\Omega\sigma$ across the transition at the axis may differ from those estimated on the assumption of uniform axial velocity and axial vorticity over the whole cross-section by a factor as large as $2\cdot32$. The changes in u/U and $w/\Omega\sigma$ must of course be correspondingly smaller in magnitude than $(a/b)^2-1$ at positions near the outer boundary in order to give the right total axial flux of mass and total axial flux of angular momentum.

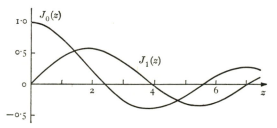

Figure 7.5.3. Bessel functions of the first kind.

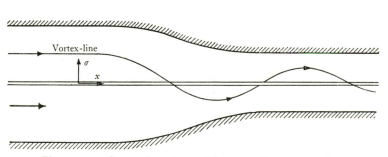

Figure 7.5.4. Conversion of a straight vortex-line into a spiral
on passage through a contraction.

The qualitative nature of these changes in u and w due to a change in radius of the tube may be explained in terms of the shape of the vortex-lines. In the upstream cylindrical region the vortex-lines are straight and parallel to the axis, and rotate about this axis with the fluid. As one end of a vortex-line passes into the transition, it moves radially inward or outward, and the azimuthal velocity of a material point on the vortex-line changes according to the rule $\sigma w = \text{const}$. Thus if a vortex-line moves radially inward on passing through the transition (figure 7.5.4), a material point on the line moves round the axis more quickly than does the vortex-line in the upstream cylindrical region, and the vortex-line is deformed into a spiral with a *positive* value (provided the axial vorticity is positive in the upstream region) of the azimuthal component of vorticity. This yields a negative value of $\partial u/\partial\sigma$ in the downstream cylindrical region, so that a contraction of the tube produces a maximum of the axial velocity at the axis, as is found from (7.5.22) (provided $kb < 3\cdot83$). Similarly an expansion produces a minimum at the axis.

An interesting feature of the formulae (7.5.22) and (7.5.23) is the occurrence of negative values of u and w for certain combinations of values of kb and a/b—roughly speaking, for sufficiently strong initial rotation. For a transition to a larger tube radius ($a < b$), the effect of increasing kb from zero is to make u and w negative first at the axis; and for a contraction of the tube u becomes negative first at the outer boundary, when kb reaches some value which must exceed $2\cdot40$. However, practical cases in which reversal of the axial velocity occurs are not likely to be described by equation (7.5.16), since it rests on the assumption that *all* the streamlines have come from a region in which there is a specified dependence of H and C on ψ and it would be difficult to contrive this same dependence for those streamlines coming from large positive values of x. Consequently the formulae should be regarded as being of practical significance only when $u \geqslant 0$ everywhere in the downstream cylindrical region.

It will also be noticed that something strange happens when kb approaches the value $3\cdot83$, where $J_1(kb) = 0$, for then the magnitudes of u and w/σ become indefinitely large everywhere in the downstream region for any value of a/b. Deeper analysis suggests that this anomaly is associated with failure of our assumption that the flow becomes cylindrical again on the downstream side of the transition. It appears that at such a large value of kb it is possible for an axisymmetric wave motion to exist in the fluid and that the effect of the change of cross-section of the tube is to set up a train of waves on the downstream side, in the way that an obstacle spanning an open channel containing a stream of water may set up a train of surface waves for certain values of the stream velocity. We shall look briefly at these axisymmetric waves in a rotating fluid in the next section.

The kind of transition represented by figure 7.5.1 (*c*) does not present any new features, apart from the fact that the coefficient B in (7.5.18) is now non-zero. In a case of disappearance of the inner boundary to the flow, represented by figure 7.5.1 (*d*), B is again non-zero (and negative) and both u and w consequently become indefinitely large and positive as $\sigma \to 0$ in the downstream region; thus here the transition produces a strong forward jet of rapidly rotating fluid near the axis.

The effect of a change of external velocity on an isolated vortex

A particularly interesting case of axisymmetric flow with swirl is provided by what may loosely be called a free or isolated vortex, that is, a vortex-tube embedded in irrotational flow. Viewed from a distance, a vortex of this kind appears simply as a line vortex (§2.6), specified by the circulation round any closed path looping it once, but a closer view will show that the vortex has a structure, with a certain distribution of vorticity within the tube. The spreading line vortex (§4.5) is perhaps the simplest example, the structure being determined in that case wholly by viscous diffusion of vorticity away from the axis. Another vortex with structure was examined near the end of §5.2; there the vorticity is everywhere parallel to the axis of the vortex, and the flow is steady as a result of a balance between radially inward convection of vorticity and outward spreading by viscous diffusion. In the present context of flow with negligible effects of viscosity, the vortex-lines move with the fluid; and we shall assume the flow to be steady. Any effects of curvature of the axis of the vortex will be ignored.

In the case of an exactly cylindrical vortex, any distributions of u and w with respect to σ are possible inside the vortex. Interest lies in the features of the velocity distribution in a vortex which are likely to be typical in practice, and we may look for these in the changes which occur in the structure when the fluid in the vortex passes through a region of non-cylindrical flow. For this purpose it is convenient to take a vortex which over some portion of its length is exactly cylindrical with simple distributions of H and C with respect to ψ, and to consider the properties of this vortex at some other section where the flow is again cylindrical. For the initial cylindrical flow, a uniform distribution of u and of the axial component of vorticity within the vortex is the obvious choice, from the point of view of mathematical convenience; this seems likely also to be a representative choice, at any rate for vortices which at some stage have been subjected to the smoothing effects of viscosity.

Our vortex is thus specified as having the velocity distribution

$$u = U_1, \quad v = 0, \quad w = \Omega\sigma, \quad \text{for } \sigma \leqslant a,$$

over some finite portion of its length, with

$$u = U_1, \quad v = 0, \quad w = \Omega a^2/\sigma, \quad \text{for } \sigma \geqslant a,$$

in the irrotational flow surrounding this portion of the vortex. We now suppose that, over some other finite portion of its length further downstream, the irrotational flow just outside the vortex is again independent of x, with velocity components

$$u = U_2, \quad v = 0, \quad w = \Omega a^2/\sigma.$$

The vortex is presumably cylindrical again (although the possibility of a wave-like flow must be kept in mind), with a different radius, b say, and a velocity distribution given by the appropriate solution of (7.5.17). Since all components of velocity remain continuous at the boundary of the vortex, the boundary condition to be satisfied by the solution of (7.5.17) is

$$u = U_2 \quad \text{at } \sigma = b, \quad \text{where} \quad \psi = \tfrac{1}{2}U_1 a^2;$$

and there is the implicit boundary condition that u is not singular at $\sigma = 0$, so that only the term $AJ_1(k\sigma)$ in the general solution (7.5.18) is to be retained. Thus the required solution is identical with that already found for flow in the downstream cylindrical region of a tube of radius b, where b is determined from the relation (see (7.5.22))

$$\frac{U_2}{U_1} = 1 + \left(\frac{a^2}{b^2} - 1\right)\frac{\frac{1}{2}kbJ_0(kb)}{J_1(kb)}, \tag{7.5.25}$$

where $k = 2\Omega/U_1$. When the radius of the vortex is known, the velocity distribution in the vortex is given by (7.5.22) and (7.5.23), with U replaced by U_1.

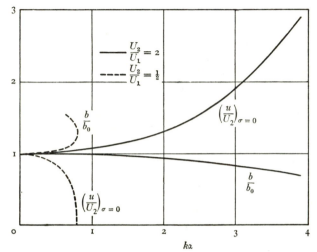

Figure 7.5.5. Properties of a vortex after increase or decrease of the external axial velocity.

In the case of a vortex of infinitesimal cross-section ($a \to 0$), we see from (7.5.25) that b also is small and that

$$\frac{b}{a} = \left(\frac{U_1}{U_2}\right)^{\frac{1}{2}}, \quad = \frac{b_0}{a} \quad \text{say.}$$

This is also the value for b/a which would be required by conservation of mass if the axial velocity of all the fluid in the vortex changed from U_1 to U_2, as would happen in the absence of any swirling motion. It is convenient now to use b_0/a as a standard of comparison for the value of b/a given by (7.5.25) which does take account of the effect of swirl. Two of the curves in figure 7.5.5 show values of b/b_0 calculated from (7.5.25) for different values of ka and for changes in the external axial velocity by factors of 2 and $\frac{1}{2}$. As an indication of the corresponding changes in the distribution of axial velocity (and, by implication, of azimuthal velocity, since both are determined by ψ) across the vortex, calculations of the ratio of the axial velocity at the centre of the vortex to that at the boundary have been made from the expression

$$\left(\frac{u}{U_2}\right)_{\sigma=0} = \frac{U_1}{U_2}\left\{1 + \left(\frac{a^2}{b^2} - 1\right)\frac{\frac{1}{2}kb}{J_1(kb)}\right\}, \quad = \frac{U_1}{U_2} + \frac{U_2 - U_1}{U_2 J_0(kb)},$$

obtained from (7.5.24) and (7.5.25), and are also shown in figure 7.5.5. Values of ka of order unity are known to be attained in the case of vortices shed from the side edge of aircraft wings (§7.8), and it appears that when the axial velocity in the irrotational flow surrounding such a vortex changes appreciably there must be significant changes in the structure of the vortex, particularly when the external fluid is decelerated. The development of a rapid variation of axial velocity across the vortex appears to be a typical feature of vortices passing through a region of non-uniform irrotational flow when $ka\,(=2\Omega a/U_1)$ is of order unity.

We can account qualitatively for the general character of the curves in figure 7.5.5 by noting the implications of the expression (7.5.12) for the radial pressure gradient. The circulation C is constant over the range $b \leqslant \sigma < \infty$ in the downstream region and decreases to zero as σ decreases from b to zero. The difference between the pressure in the vortex, say at the axis for definiteness, and that far from the vortex in the same plane normal to the axis, is consequently strongly dependent on the value of b; and an increase in b corresponds to a decrease in this pressure difference, and vice versa. Thus when the fluid outside the vortex decelerates and the vortex radius increases with increase of distance in the flow direction, there will be an additional axial pressure gradient *within the vortex* which is positive and which consequently leads to further axial deceleration and further thickening of the vortex. Acceleration or deceleration of the fluid outside the vortex thus leads to a change of axial velocity of the fluid within the vortex which is in the same direction and of greater magnitude than that outside, and to a change of vortex radius which is greater than would be expected if the axial velocity were uniform over a lateral plane.

Two other features of the curves in figure 7.5.5 relating to $U_2/U_1 = \frac{1}{2}$ call for comment. There is first the fact that the axial velocity at the centre becomes zero at a certain value of ka; as remarked earlier, continuation of the solution to combinations of values of ka and U_2/U_1 for which the axial velocity is negative is not useful, since fluid coming from the 'downstream' direction is unlikely in practice to have the assumed dependence of H and C on ψ. Secondly, and rather surprisingly, there is a critical value of ka (which is a little greater than the value at which backflow first occurs) beyond which there is no value of b satisfying (7.5.25) and flow with radial equilibrium is presumably impossible.

Similar features of the flow are revealed when we think about the changes that take place in the downstream cylindrical region as U_2/U_1 is decreased continuously from unity, with ka fixed, although we must here keep in mind the fact that the *direction* of the changes depends on the value of ka. The relation (7.5.25) shows that the signs of the quantities

$$\frac{U_1 - U_2}{b^2 - a^2} \quad \text{and} \quad \frac{J_0(kb)}{J_1(kb)}$$

are the same. Now when U_2/U_1 is close to unity, $b \approx a$ and it follows that the sign of $J_0(ka)/J_1(ka)$ determines the initial direction of the changes in vortex radius and velocity variation across the vortex. For $0 \leqslant ka < 2\cdot40$ (a range within which practical values of ka normally lie), there is the 'natural' behaviour: an increasing vortex radius as U_2/U_1 decreases, and the development of an axial velocity defect at the axis. But for $2\cdot40 < ka < 3\cdot83$ (the ends of this range being the first zeros of $J_0(ka)$ and $J_1(ka)$ respectively), axial deceleration of the external stream leads to a *smaller* vortex radius and an excess axial velocity at the axis. It may also be seen from the form of (7.5.25) that, whatever the (fixed) value of ka, it is not possible to find a value of kb which satisfies (7.5.25) when U_2/U_1 falls below a certain critical or minimum value. Figure 7.5.6 shows schematically the way in which kb

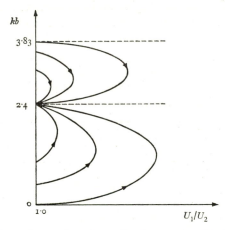

Figure 7.5.6. The dependence of vortex radius b on external axial velocity $U_2\,(\leqslant U_1)$, shown schematically. The different curves refer to different values of ka, and the arrows show the direction of change from a cylindrical vortex of radius a with rigid-body rotation.

varies with $U_1/U_2\,(\geqslant 1)$ for different given values of ka between 0 and $3\cdot83$, according to (7.5.25). For values of ka greater than $3\cdot83$ there are similar sets of curves all of which terminate at a value of kb at which $J_0(kb) = 0$.

The increased thickening of the vortex which accompanies deceleration of the external stream (when ka lies between zero and $2\cdot40$) evidently becomes catastrophic at the critical or minimum value of U_2/U_1. There is a striking phenomenon, termed 'vortex breakdown' or 'vortex bursting', which at first sight looks like a manifestation of this rapid increase in the diameter of the vortex when the external axial velocity falls to a certain value. Observations of a streak of dye inside a strong steady vortex show that in certain circumstances, which are not yet well established but which are believed to include sufficient deceleration of the external flow, the vortex may suddenly enlarge or 'burst', giving rise to a confused flow of quite a different kind. Figure 7.5.7 (plate 22) shows a photograph of two such 'bursting' vortices in water, the vortices here being a part of the trailing vortex system of a triangular plate slightly inclined to the oncoming stream. A more appealing explanation of vortex breakdown (Benjamin 1962) is that it is a finite jump from one cylindrical flow to the second cylindrical flow possible with the same external axial velocity (that two cylindrical flows are possible for given ka and U_2/U_1 is evident from figure 7.5.6), the sudden transition being analogous to the well-known hydraulic jump in a stream of water in an open channel; according to this theory vortex breakdown would occur before U_2 falls to the critical value.

7.6. Flow systems rotating as a whole

As mentioned in the preceding section, a body of fluid in steady rotation is able to sustain a wave motion propagating along the axis of rotation with axial symmetry. We now consider explicitly the 'elasticity' which rotation confers on a fluid and which provides the restoring mechanism that makes possible the propagation of waves. This effective elasticity of the fluid exists for a variety of distributions of vorticity in the fluid, but the discussion here will be confined to the case of a body of fluid which either initially or in some average sense is in steady rigid-body rotation. Such rotating flow systems exhibit a number of interesting properties which are still being studied actively.†

The restoring effect of Coriolis forces

When the motion is referred to axes which rotate steadily with the bulk of the fluid, the fictitious Coriolis and centrifugal forces (3.2.10) must be supposed to act on the fluid. The centrifugal force per unit mass may be written as $\frac{1}{2}\nabla(\mathbf{\Omega} \times \mathbf{x})^2$ and is equivalent in its effect to a contribution to the pressure (in a fluid of uniform density). The Coriolis force on the other hand gives rise to effects of a new type, among which is the elasticity of the fluid. With p now denoting a modified pressure which includes allowance for the centrifugal force as well as for gravity, the equation of motion with velocity $\mathbf{u}$ relative to axes rotating with steady angular velocity $\mathbf{\Omega}$ is

$$\frac{\partial \mathbf{u}}{\partial t} + \mathbf{u}.\nabla \mathbf{u} + 2\mathbf{\Omega} \times \mathbf{u} = -\frac{1}{\rho}\nabla p. \tag{7.6.1}$$

The Coriolis force is directed at right angles to both the axis of rotation and the local velocity vector, and is a deflecting force which does no work on a material element. Only the component of $\mathbf{u}$ in the plane normal to $\mathbf{\Omega}$ (which we shall term the lateral plane) is involved, and the Coriolis force tends to change the direction of this component. The sense of this change is opposite to that of the change of direction of motion, relative to an absolute frame, of a point with co-ordinates fixed relative to the rotating axes (figure 7.6.1); that

† For an account of these properties see *The Theory of Rotating Fluids*, by H. P. Greenspan (Cambridge University Press, 1968).

is to say, if the basic rotation in the lateral plane is anti-clockwise, the Coriolis force tends to turn the direction of motion of an element relative to the rotating frame to the right. Moreover, the Coriolis force is linear in the velocity, and will tend to change the direction of the component of **u** in a lateral plane at the same rate for all magnitudes and directions of that component. Thus a material element whose motion is dominated by the Coriolis force moves on a path whose projection on a lateral plane is a circle, the whole circle being completed in a time of order Ω^{-1}. The Coriolis force evidently tends to restore an element to its initial position in the lateral plane. Note that there is no special significance about the position of the axis of rotation, so far as the Coriolis force is concerned.

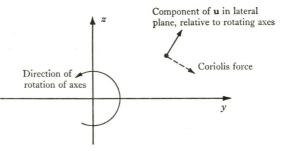

Figure 7.6.1. To show the direction of the Coriolis force which acts in a rotating reference system. The (y, z)-plane is normal to the axis of rotation.

Since the motions of different elements of the fluid normally exercise a strong mutual influence through the action of pressure gradients, it is desirable to consider also the collective effect of Coriolis forces on different elements. Suppose that relative to the rotating axes there is generated a motion which leads to a non-zero and positive value of the rate of expansion in the lateral plane, that is, to a positive value of

$$\partial v/\partial y + \partial w/\partial z$$

(in terms of the co-ordinate system of figure 7.6.1) over a certain region of the fluid. The area enclosed by the projection on a lateral plane of a closed material curve in this region will then be increasing. The effect of the Coriolis force accompanying this general outward movement of the projected curve is to generate a tangential motion of the material curve, which makes a *negative* contribution to the circulation round it. This change in the circulation of the motion relative to rotating axes is of course simply the change required to keep the circulation relative to an absolute frame constant during a movement leading to increase of the projected area on a lateral plane. Now the new tangential motion of the material curve itself leads to a Coriolis force in a direction normal to the curve; and inasmuch as the new tangential motion makes a net negative contribution to the circulation, the associated Coriolis force is directed mostly inwards and so tends to produce a *reduction* in the area enclosed by the projected curve. In other words, at places in the fluid where there is a positive value of $\partial v/\partial y + \partial w/\partial z$ the effect of Coriolis forces is to tend to produce a negative value, and vice versa. Thus the net effect of Coriolis forces is to oppose displacements of fluid elements which together lead to change of the area enclosed by the projection of a material curve on a lateral plane, that is, to a non-zero expansion in a lateral plane.

The extent to which the restoring effect of Coriolis forces restricts the displacement of fluid elements evidently depends on the relative magnitudes of Coriolis forces and other forces acting on the fluid; and in the present context these other forces are inertia forces. If U is a representative velocity magnitude (relative to rotating axes) and L is a measure of the distance over which $\mathbf{u}$ varies appreciably, the ratio of the magnitudes of the terms $\mathbf{u}.\nabla\mathbf{u}$ and $2\mathbf{\Omega} \times \mathbf{u}$ in (7.6.1) is of order

$$U/L\Omega.$$

The value of this ratio, known as the *Rossby number* in recognition of the work of the Swedish meteorologist, provides a convenient measure of the importance of Coriolis forces. When $U/L\Omega \gg 1$, Coriolis forces are likely to cause only a slight modification of the flow pattern; but when $U/L\Omega \ll 1$, the tendency for Coriolis forces to oppose any expansion in a lateral plane is likely to be dominant. And in the intermediate case when $U/L\Omega$ is of order unity, an interesting mixture of effects is to be expected, some hint of which was provided by the discussion of steady axisymmetric flow with swirl in §7.5.

Steady flow at small Rossby number

The dominance of Coriolis forces in flow at values of $U/L\Omega$ small compared with unity has strange consequences when the flow is also steady relative to rotating axes, as was first pointed out by J. Proudman (1916). In steady flow a material element of fluid moves along the same streamline, without reversal of direction, at all times. But the strong Coriolis forces oppose any displacement of fluid elements leading to a non-zero expansion in a lateral plane. It follows that, in the limit $U/L\Omega \to 0$, the form of the streamlines must be consistent with a zero rate of expansion in a lateral plane.

We can establish this result formally by noting that when $\partial\mathbf{u}/\partial t = 0$ and the term $\mathbf{u}.\nabla\mathbf{u}$ is negligible by comparison with the Coriolis force, the equation of motion (7.6.1) becomes

$$2\mathbf{\Omega} \times \mathbf{u} = -\frac{1}{\rho}\nabla p, \qquad (7.6.2)$$

that is,

$$\frac{1}{\rho}\left(\frac{\partial p}{\partial x}, \frac{\partial p}{\partial y}, \frac{\partial p}{\partial z}\right) = (0, 2\Omega w, -2\Omega v)$$

with the co-ordinates shown in figure 7.6.1. Elimination of p then gives

$$\frac{\partial v}{\partial y} + \frac{\partial w}{\partial z} = 0 \qquad (7.6.3)$$

in steady flow relative to rotating axes, and, as a consequence of the mass-

conservation equation,
$$\partial u / \partial x = 0. \tag{7.6.4}$$

The curious property of these approximate equations which hold when $U/L\Omega \ll 1$ is that the motion in the lateral or (y, z)-plane is not coupled with the motion parallel to the axis of rotation. Furthermore, none of the flow properties depends on x. Proudman's theorem is sometimes stated as being that 'slow' steady motions relative to rotating axes must be two-dimensional. Since in this book we have regarded the term two-dimensional motion as implying that the velocity vector everywhere lies in a certain plane, it would be more appropriate here to say that steady motions at small Rossby number must be a superposition of a two-dimensional motion in the lateral plane and an axial motion which is independent of x.

The value of the velocity component u parallel to the axis of rotation is evidently determined by the boundary conditions. It will often happen that every line in the fluid parallel to the axis meets a stationary boundary; in such cases the above relations require $u = 0$ everywhere, and only the two-dimensional motion remains. The photographs on plate 23 (made by G. I. Taylor many years before the subject of rotating fluids had attracted much notice) of the flow in an open flat dish of water which is rotating show that Coriolis forces do indeed make the motion two-dimensional in these circumstances. In figure 7.6.2 (plate 23) a drop of coloured liquid has been drawn out into a thin sheet by a 'slow' motion imparted to the rotating fluid, and the two photographs, taken by a camera placed on the axis of rotation of the dish, show that the sheet is everywhere parallel to the axis and that the component of velocity in a lateral plane is independent of x. The flow revealed by the streak of dye released from point A in figure 7.6.3 (plate 23) is more startling. The motion relative to rotating axes is due here to a portion of a circular cylinder E being drawn slowly across the bottom of the dish. The depth of water is 4 in. and the cylinder is 1 in. high, and in a non-rotating fluid the water would pass over the top of the moving cylinder as well as round the sides. However, the dye emerging from a point $1\frac{1}{2}$ in. above the top of the cylinder and directly ahead of it (figure 7.6.3a) divides at point B, as if it had met an upward extension of the cylinder, and passes round this imaginary cylinder in two sheets,† the sheet on one side (D) even showing separation and the formation of eddies. In figure 7.6.3b the dye is being released from a point just inside the cylindrical region vertically above the body, and collects in a blob which moves with the cylinder. It seems that the flow outside the upward projection of the cylinder is approximately the same as if the cylinder extended from the bottom to the top of the layer of water, and that vertically above the cylinder there is a cylindrical column of water which moves with it. Thus the motion is two-dimensional in the way that is consistent with translation of the cylinder, even though the height of that cylinder is only one-quarter of the depth of water.

† Another striking photograph of this phenomenon is reproduced in Greenspan's book.

When the fluid is not enclosed by stationary boundaries intersected by lines parallel to the axis of rotation, the value of the axial velocity component in the fluid will usually be determined by conditions at an inner boundary. An interesting and fundamental case is flow due to translation of a rigid body, with velocity U parallel to the axis of rotation, through fluid which is unbounded in that direction. It seems that here the above requirements for flow with $U/L\Omega \to o$ can be satisfied only if all the fluid in the cylinder circumscribing the body moves parallel to the axis with the body, the component of velocity in a lateral plane being zero everywhere. Experiments do in fact suggest that a column of fluid is pushed ahead of a body moving parallel to the axis, although the flow behind the body seems not to be wholly in accord with the above simple theory. Further reference to these experiments will be made later in this section.

In the case of bodies moving either parallel to the axis of rotation or normal to it, the above theory for flow at small Rossby number leads to the conclusion that a so-called 'Taylor column' of fluid parallel to the axis accompanies the body. At the edge of the column there are shear layers where the vorticity is large. It is to be expected that the approximate linear equation (7.6.2) is not applicable everywhere in these layers, although the consequences for the whole flow field are not well understood.

Propagation of waves in a rotating fluid

We have seen that any displacement of the elements of a fluid in rigid-body rotation which leads to a non-zero expansion in a lateral plane is accompanied by Coriolis forces which tend to eliminate this expansion. Since there is no dissipation of energy in an inviscid fluid, it follows that a displacement of this kind which is given to the fluid initially may set up an oscillation. This raises the possibility that a train of waves can propagate through a rotating fluid, with different phases of the wave being associated with positive and negative values of the expansion in a lateral plane. We can examine this possibility by seeking solutions of the equation governing departures from a state of rigid-body rotation which are periodic in time and in certain spatial co-ordinates.

We shall consider first the physically simple case of an axisymmetric wave motion with propagation in the direction of the axis of rotation. Relative to rotating axes, the wave motion is superimposed on stationary fluid and so, for a simple harmonic wave, all flow quantities vary sinusoidally in time with angular frequency β say (period $2\pi/\beta$) and sinusoidally with respect to the axial co-ordinate x with wave-number α say (wavelength $2\pi/\alpha$). The governing equation for flow relative to rotating axes is (7.6.1), in axisymmetric form, and, following the usual pattern of investigation of wave motions, we might proceed by neglecting terms in this equation of degree higher than the first in quantities representing the departure from the undisturbed state. However, there is no need to go through the details, because use can be made of the analysis of the preceding section. We saw there that, for any *steady* axisymmetric flow in which the functions C and H have the same dependence on the stream function ψ as in flow with uniform axial speed U and uniform angular velocity Ω, ψ satisfies the equations (7.5.15) and (7.5.16); and it follows that any solution of equation (7.5.16) which is periodic with respect to x can be regarded as representing a progressive wave of arbitrary amplitude propagating with phase velocity U through fluid which in the absence of the wave motion would be in rigid rotation.

We are therefore led to examine a steady flow represented by the stream function (7.5.15), in which $F(x, \sigma)$ satisfies equation (7.5.16) and is of the form (for a simple harmonic wave of wave-number α)

$$F(x, \sigma) = G(\sigma) \sin(\alpha x + \epsilon),$$

where ϵ is a constant. The equation for G is then

$$\frac{d^2 G}{d\sigma^2} + \frac{1}{\sigma}\frac{dG}{d\sigma} + \left(k^2 - \alpha^2 - \frac{1}{\sigma^2}\right) G = 0, \qquad (7.6.5)$$

where $k = 2\Omega/U$, which is of the same form as (7.5.17). A relevant solution is

$$G(\sigma) = A J_1\{(k^2 - \alpha^2)^{\frac{1}{2}} \sigma\},$$

where A is an arbitrary constant; there is a second possible solution involving a Bessel function of the second kind which we omit, since it gives a singularity of the velocity distribution at the axis. We thus have a steady flow represented by

$$\psi = \tfrac{1}{2} U\sigma^2 + A\sigma J_1\{(k^2 - \alpha^2)^{\frac{1}{2}} \sigma\} \sin(\alpha x + \epsilon) \qquad (7.6.6)$$

for a positive value of $k^2 - \alpha^2$ and an arbitrarily chosen value of A. The associated azimuthal component of velocity is seen from (7.5.14) to be

$$\frac{C}{\sigma} = \frac{2\Omega}{U}\frac{\psi}{\sigma} = \Omega\sigma + kF.$$

Far from the axis the axial and radial components of velocity become uniform with values U and zero, although the approach to these values is rather slow (as $\sigma^{-\frac{1}{2}}$). We see from this expression for C that the radial component of Coriolis force that acts (in a rotating frame) on each element of a material circle in a lateral plane has the same sign as F and always tends to restore the radius of the circle to an equilibrium value.

The same flow viewed relative to axes moving in the (positive) axial direction with speed U is given by

$$\psi = A\sigma J_1\{(k^2 - \alpha^2)^{\frac{1}{2}} \sigma\} \sin\{\alpha(x + Ut) + \epsilon\}, \qquad (7.6.7)$$

and the instantaneous streamlines have the appearance shown in figure 7.6.4; we have here a simple harmonic wave progressing in the negative x-direction with phase velocity U through fluid which is otherwise in rigid-body rotation. It will be noted, from the manner in which the solution

(7.6.7) representing a progressive wave has been obtained, that it is not restricted to small values of the amplitude A. This is a special feature of those axisymmetric oscillatory motions about a state of rigid-body rotation which can be converted to a steady motion by reference to axes in translational motion parallel to the axis of rotation, and arises from linearity of the governing equation in that case. Thus two solutions of the form (7.6.7) may be superposed, without restriction on the amplitudes, only when the two phase velocities are equal both in magnitude and direction; and when the two phase velocities differ, superposition is possible only if the two amplitudes are sufficiently small for non-linear terms in the governing equation (7.6.1) to be negligible.

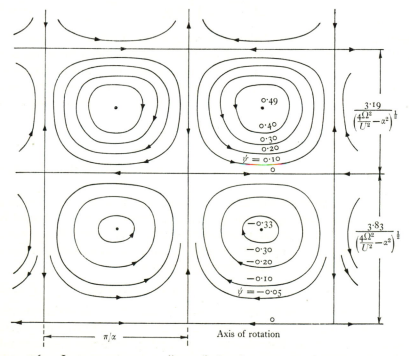

Figure 7.6.4. Instantaneous streamlines of the departure from rigid-body rotation, representing a simple harmonic wave propagating in the axial direction with phase speed U. The values of the stream function shown correspond to the formula (7.6.7) with $A = (k^2 - \alpha^2)^{\frac{1}{2}}/3\cdot83$. (The streamlines for the standing wave formed by superposing two such progressive waves of small amplitude have the same shape; see (7.6.10).)

An unusual property of these axisymmetric waves in rotating fluid of infinite extent is that the wave-number α and the angular frequency αU are independent quantities. The wave motion evidently is not determined until four quantities are given: Ω, U, α and the amplitude A. Alternatively, we could specify, in place of U, the radial dimension of the cell adjoining the axis in figure 7.6.4; the first zero of $J_1(z)$ occurs at $z = 3\cdot83$, so that this

36   **BIT**

dimension is $3 \cdot 83 \left(\dfrac{4\Omega^2}{U^2} - \alpha^2 \right)^{-\frac{1}{2}}$. However, not all combinations of values of α, Ω and U are possible, since when $k^2 - \alpha^2 < 0$ there is no solution of (7.6.5) which is finite everywhere; thus for given Ω and U the range of permissible values of α is $0 < |\alpha| < |2\Omega/U|$.

In a laboratory experiment the rotating fluid is likely to be enclosed by a cylindrical boundary, at $\sigma = b$ say. There is then the constraint

$$\psi = 0 \quad \text{at} \quad \sigma = b,$$

which requires an integral number of cells to occur in the radial range $0 \leqslant \sigma \leqslant b$. Thus, if $z = \gamma_n$ is the nth zero of $J_1(z)$, we require

$$b \left(\frac{4\Omega^2}{U^2} - \alpha^2 \right)^{\frac{1}{2}} = \gamma_n, \tag{7.6.8}$$

where n is the number of radial cells in the cylinder. This can be regarded as a relation between the wave speed U and the wave-number α for waves with n radial cells propagating in a cylindrical container of radius b. The *group velocity*, or velocity with which the energy of such a wave is propagated, is an axial vector, in the same sense as the phase velocity, with magnitude given by the well-known formula

$$U + \alpha \frac{dU}{d\alpha}, \quad = U \frac{\gamma_n^2}{\gamma_n^2 + \alpha^2 b^2}. \tag{7.6.9}$$

The group velocity is thus in general of smaller magnitude than the phase velocity.

If the fluid is confined between plane boundaries normal to the axis of rotation and distance l apart, the boundary conditions can be satisfied by superposition of two similar progressive waves of suitably chosen wave-number propagating in opposite directions. An elementary solution representing a standing wave is found from (7.6.7) to be

$$\psi = 2A\sigma J_1\{(k^2 - \alpha^2)^{\frac{1}{2}} \sigma\} \sin(\beta t + \epsilon) \cos \alpha x, \tag{7.6.10}$$

where β has been written for the frequency in place of αU since the quantity U has ceased to be relevant. The streamlines at any instant given by (7.6.10) are of the same form as those shown in figure 7.6.4. Note that the amplitude A must be small here, since otherwise the governing equations for the disturbance motion will not be linear and solutions will not be superposable. The conditions at the two bounding planes require $\alpha = m\pi/l$, where m is a positive integer, and if in addition there is a rigid boundary at $\sigma = b$ the condition (7.6.8) must also be satisfied. Thus we see that the natural frequencies of small oscillations of rotating fluid contained in a circular cylinder of radius a and length l are given by

$$\beta = 2\Omega \Big/ \left(1 + \frac{\gamma_n^2 l^2}{m^2 \pi^2 b^2} \right)^{\frac{1}{2}}. \tag{7.6.11}$$

This relation was first obtained by Kelvin (1880), and has attracted interest in recent years for its possible applications in geophysics. The existence of the simpler modes of oscillation (for which the values of m and γ_n are near the smallest possible) may be demonstrated experimentally, and the formula for their frequency confirmed, by imposing an oscillation on the rotating fluid and by observing the occurrence of resonance at certain discrete values of the frequency of the forcing mechanism (Fultz 1959).

Plane waves also may exist in a rotating fluid, with essentially the same restoring mechanism due to Coriolis forces. The basis for the existence of a simple, although also rather special, type of plane wave lies in the fact that an element of fluid acted on solely by Coriolis forces moves on a circular path in a lateral plane. If the (modified) pressure and fluid velocity, relative to rotating axes, are initially uniform over planes normal to the axis of rotation in an infinite fluid, they remain so, and the equation of motion (7.6.1) gives

$$\frac{\partial v}{\partial t} = 2\Omega w, \quad \frac{\partial w}{\partial t} = -2\Omega v, \tag{7.6.12}$$

for the velocity components corresponding to the rectilinear co-ordinates y, z in a lateral plane. It follows that the whole of a material lateral plane moves as a rigid plate in a circular path with angular frequency 2Ω. If now the different material planes can be set into motion initially according to the distribution

$$v = A \cos \alpha x, \quad w = A \sin \alpha x,$$

each material plane will move rigidly in its circular path and at a subsequent time t the velocity will be

$$v = A \cos (\alpha x - 2\Omega t), \quad w = A \sin (\alpha x - 2\Omega t).$$

A simple harmonic plane progressive wave which is transverse and circularly polarized thus appears to propagate in the direction of the x-axis with wave-number α and phase velocity $2\Omega/\alpha$. One says here 'appears to propagate' because each material lateral plane moves independently and owes its motion entirely to the initial conditions; as in the familiar case of a straight row of bobs hanging on strings and given an initial transverse displacement by running a finger along the row, the group velocity, or speed with which energy is propagated, is zero.

A more general type of plane wave can exist in which the wave-number vector is inclined at an angle to the axis of rotation of the fluid. To see this we need only superpose on the above wave system a component of the rotation vector parallel to the y-axis. There is then an additional Coriolis force parallel to the x-axis which is independent of y and z and which can be balanced by a pressure gradient without change of the velocity distribution. Thus the whole flow field is given by

$$\left. \begin{aligned} v &= A \cos (\alpha x - 2\Omega t \cos \theta), \quad w = A \sin (\alpha x - 2\Omega t \cos \theta), \\ u &= 0, \quad \frac{p}{\rho} = \frac{2\Omega \sin \theta}{\alpha} A \cos (\alpha x - 2\Omega t \cos \theta), \end{aligned} \right\} \tag{7.6.13}$$

where Ω is now the total magnitude of the angular velocity of the fluid. The angular frequency of the wave is

$$2\Omega \cos \theta, \quad = \frac{2\Omega.\alpha}{\alpha},$$

where Ω and α represent the angular velocity and wave-number vectors, α being in the direction of the x-axis for the purpose of the formulae (7.6.13). The vector group velocity of a simple harmonic plane wave is known to be equal to the gradient of the frequency with respect to α, and thus has the value

$$\frac{2\Omega}{\alpha} - \frac{2\Omega.\alpha\alpha}{\alpha^3}, \quad = \frac{2}{\alpha^3}\alpha \times (\Omega \times \alpha). \tag{7.6.14}$$

Energy is propagated in the direction *normal* to α and in the plane of α and Ω, that is, in the direction of the y-axis here, as is evident from the fact that the mean values of pu, pv, pw, as given by (7.6.13), are respectively

$$0, \quad \rho\Omega \sin\theta A^2/\alpha, \quad 0.$$

It will be noted that, when α and Ω are not parallel, Coriolis forces do not act in isolation and the simple restoring mechanism is modified by the effect of pressure gradients.

All these waves in a rotating fluid, axisymmetric and plane, are referred to as *inertia waves*.

Flow due to a body moving along the axis of rotation

Determination of the flow due to a rigid body in steady translatory motion parallel to the axis of rotation of infinite fluid is a difficult problem, and a clear picture of all aspects of the flow field is not yet available. All that we shall do here is to indicate some features of the flow for an axisymmetric body.

The Rossby number $U/a\Omega$ formed from the speed of the body and one of its linear dimensions is clearly a measure of the relative importance of the effects of translation of the body and rotation of the fluid. In the limit $U/a\Omega \to \infty$ we may expect the (non-dimensional) velocity distribution to reduce to that for flow due to a body in translational motion through fluid at rest at infinity. At the other extreme, $U/a\Omega \to 0$, it is plausible to suppose that inertia forces are small compared with Coriolis forces (which usually holds when the fluid velocity relative to rotating axes is small compared with Ωa everywhere), in which case we may invoke the result that in the limit the rate of expansion in the lateral plane must be zero everywhere. This requires, as pointed out earlier in the section, that the axial component of velocity u be independent of x, which is possible only if the body carries with it a column of fluid, at rest relative to the body, enclosed by a cylinder with generators parallel to the x-axis and which circumscribes the body. Strange though such a flow field appears to be, observation indicates that it is broadly correct. But the details of the flow in the cylindrical shear layer and the way in which the column is established after the body begins to move are obscure.

When $U/a\Omega$ is small but non-zero, the moving column is presumably modified although there is an all-or-nothing quality about the moving column which makes it difficult to imagine the nature of the modification. Observations of the flow due to a moving sphere of radius a by Taylor (1922) and of that due to a body with a spherical nose (of radius a) and conical tail by Long (1953) suggest that a column of fluid is pushed ahead of the moving body when $U/a\Omega$ is less than about 0·2 or 0·3. The experiment with a sphere designed by Taylor to demonstrate the existence of a change in the flow field near this value of $U/a\Omega$ is very simple. A light sphere such as a table-tennis ball, painted in stripes to make visible any rotation, is tethered by a thread to the bottom of a tall jar of water which has been brought into uniform rotation. When the sphere has no axial motion, it naturally rotates with the surrounding fluid. But when the sphere is allowed to rise with a speed U such that $U/a\Omega$ is greater than about 0·3, so Taylor found, the sphere ceases to rotate with the fluid. Zero rate of rotation of the sphere is what would be expected if fluid is continually being made to pass round the advancing sphere, because a material circle near the surface of the sphere had small radius, and hence small circulation, at an earlier stage when all points on the circle were near the axis of rotation; the azimuthal velocity of the fluid thus tends to zero as the surface of the sphere is approached, and the existence of viscosity in a real fluid ensures that the sphere itself has zero angular velocity in the steady state. On the other hand, for some smaller values of $U/a\Omega$ he observed that the rising sphere *continued* to rotate with the fluid, as would be expected if a column of (rotating) fluid is pushed ahead of the moving sphere.

When $U/a\Omega$ is of order unity, the inertia forces associated with the translatory motion of the body are comparable in magnitude with Coriolis forces, and will produce a non-zero expansion in the lateral plane despite the opposition of Coriolis forces. As a result of this forced displacement of fluid elements near the body, axisymmetric waves of the kind described earlier are likely to appear and in a cylindrical container to extend to infinity, in the absence of any dissipation, as free oscillations (i.e. as progressive waves). The photographs in figure 7.6.5 (plate 24) show clearly the existence of waves, although only on the downstream side, due to motion of a body with a hemispherical nose and conical tail along the axis of rotating fluid contained in a circular cylinder. These waves, which are stationary relative to the body, 'radiate' energy away from the body, in the sense that as time goes on the length of the train of waves extending downstream from the body continues to increase, and there is an associated contribution to the drag on the body.

The existence of an outer cylindrical boundary to the fluid makes possible some simple analytical deductions, largely through the circumstance that the wave-numbers of the allowable free oscillations of the fluid far from the body now have a discrete set of values rather than a continuous range.

Equation (7.6.8) shows that the (non-dimensional) wave-numbers of free oscillations which propagate with speed U and for which the (circular) cylindrical boundary $\sigma = b$ is a streamline are

$$\alpha b = \left(\frac{4 b^2 \Omega^2}{U^2} - \gamma_n^2 \right)^{\frac{1}{2}}, \qquad (7.6.15)$$

where γ_n is the nth value of z for which $J_1(z) = 0$. We see that there is a maximum value of $U/b\Omega$ for which any free oscillation is possible, the maximum being

$$\frac{U}{b\Omega} = \frac{2}{\gamma_1} = 0\cdot52,$$

and that at this maximum $\alpha = 0$ (i.e. infinite wave-length). It appears then that for body speeds greater than $0\cdot52b\Omega$ no waves should be generated by the moving body. The presence of the factor b in this condition is attributable to the fact that the relative importance of Coriolis forces increases with the distance over which the fluid velocity varies appreciably; as U decreases from the large value at which the flow has the same form as in non-rotating fluid, effects of Coriolis forces and associated wave motions would spread inward from infinity in a radially unbounded flow field, and only when U is less than some critical value related to b can a wave motion appear within a given cylindrical boundary. It appears also from (7.5.15) that, for body speeds such that

$$\frac{2}{\gamma_1} (= 0\cdot52) > \frac{U}{b\Omega} > \frac{2}{\gamma_2} (= 0\cdot29), \qquad (7.6.16)$$

only one mode of free oscillation is possible and

$$\alpha b = \left(\frac{4 b^2 \Omega^2}{U^2} - \gamma_1^2 \right)^{\frac{1}{2}}, \qquad (7.6.17)$$

whereas for lower values of U two or more modes (corresponding to more than one cell in the radial range $0 \leqslant \sigma \leqslant b$—see figure 7.6.4) are possible. Furthermore, for all these simple harmonic progressive axisymmetric waves the group velocity (see (7.6.9)) is of smaller magnitude than the phase velocity U. The energy of a disturbance caused by the body therefore cannot advance upstream relative to the body; this is why the waves are formed only on the downstream side.

The observations by Long (1953), of which the photographs in plate 24 are a sample, showed that for several values of $U/b\Omega$ in the range (7.6.16) the wave-length of the waves some distance downstream was quite close to the theoretical value obtained from (7.6.17), and that for several values of $U/b\Omega$ smaller than $0\cdot29$ the wave system was approximately periodic with a wave-length close to the shortest of the wave-lengths of possible progressive waves (so that it corresponded to (7.6.17)). Two or more of the possible free oscillations can be superposed, so that the import of this latter observation is that the disturbance due to the motion of the body evidently puts much more energy into the mode with one radial cell than into the modes with more than one radial cell.

7.7. Motion in a thin layer on a rotating sphere

To complete the discussion in this chapter of the effects of rotation of the fluid as a whole we shall look briefly at some of the equations in current use in dynamical meteorology and oceanography. When the angular velocity Ω is equal to 2π radians per day, the Rossby number $U/L\Omega$ will be very much larger than unity for motions on a laboratory scale, and effects of Coriolis forces will not normally be noticeable in these motions. On the other hand, for motions of large horizontal extent in the atmosphere or ocean, say with linear dimensions of at least 100 km, it is evident that Coriolis forces will be important. A qualitative description of some aspects of such large-scale motions in a layer of fluid on a rotating globe may be obtained from simplified sets of equations, which we shall introduce here with only heuristic justification.

The following different idealizations and approximations will be employed:

(*a*) The density of air in the atmosphere varies with height as a consequence of its compressibility, but this variation is approximately the same at all points on the earth's surface and, for some purposes, may be supposed not to affect motions of large horizontal extent. We regard the atmosphere and ocean here as layers of incompressible fluid with uniform density. The depth of the layer of uniform fluid representing the atmosphere or the ocean is small compared with the horizontal length-scale of the motions to be considered.

(*b*) The upper boundary of this layer of air or water is a 'free' surface which we shall suppose to remain spherical owing to the relatively strong action of gravity. (There are some large-scale oscillatory motions of the atmosphere and ocean, usually designated as 'tidal motions', in which undulations of the upper free surface play an essential part; these are motions determined directly by the effect of gravity and only modified by Coriolis forces. Our assumption of a spherical upper boundary excludes motions on which gravity has a direct effect, and leads to motions in which effects of rotation are important.)

(*c*) Localized vertical currents obviously do occur in the atmosphere, and it is also evident that the horizontal wind speed varies with height. However, these are aspects of the motion with which we are not concerned, and it is taken as appropriate to consider an average of the velocity in the atmosphere (or ocean) over a region with linear dimensions comparable with the depth of the fluid layer. This average or bulk motion of the fluid is nearly horizontal, is uniform across the layer, and, in the case of the flow fields to be considered, varies appreciably over horizontal distances not smaller than about 100 km. The effect of friction at the ground on this averaged motion of the layer may not always be negligible in reality, but for simplicity we shall ignore it here.

(*d*) The velocity in the fluid layer would be exactly horizontal, with assumptions (*b*) and (*c*), if the lower boundary of the layer were exactly horizontal. We allow some effect of topography of the earth's solid surface, and suppose only that the depth of the atmosphere or ocean, H say, is a slowly-varying function of position, the variation of H over horizontal distances of order H being negligible. The sole consequence of this slow variation of H is to impose on the fluid a non-zero rate of expansion in a horizontal plane as it moves over sloping ground. By considering the conservation of mass of a material vertical cylinder of small cross-section we find

rate of expansion in horizontal plane = − rate of vertical extension
$$= -\frac{1}{H}\frac{DH}{Dt}. \qquad\qquad \text{of cylinder} \qquad (7.7.1)$$

For all other purposes the vertical component of velocity of the fluid and the variation of velocity across the layer may both be neglected. This kind of approximation is well known in the theory of surface gravity waves as a 'shallow water' approximation (with variations of H arising in this latter case from both bottom topography and displacement of the free surface).

We now write down the equations of motion of a layer of fluid on a rotating sphere in a form consistent with all these approximations. It is clearly convenient to use a system of spherical polar co-ordinates (r, θ, ϕ) which rotates with the sphere, with origin at the centre of the sphere and $r = R$ at the spherical outer boundary of the layer; we take $\theta = 0$ at the north pole (so that $\frac{1}{2}\pi - \theta$ is the conventional angle of latitude), and the direction in which ϕ is increasing with r and θ constant is then east (see figure 7.7.1). The corresponding components of velocity are (u_r, u_θ, u_ϕ), and those of the vector angular velocity of the earth are $(\Omega \cos\theta, -\Omega \sin\theta, 0)$. The equation of motion of a uniform inviscid fluid relative to rotating axes was given in vector form in (7.6.1), and the corresponding set of component equations for a spherical polar co-ordinate system, with neglect of the radial components of velocity and acceleration, is

$$-2\Omega u_\phi \sin\theta = -\frac{1}{\rho}\frac{\partial p}{\partial r}, \qquad (7.7.2)$$

$$\left(\frac{Du}{Dt}\right)_\theta - 2\Omega u_\phi \cos\theta = -\frac{1}{\rho r}\frac{\partial p}{\partial \theta}, \qquad (7.7.3)$$

$$\left(\frac{Du}{Dt}\right)_\phi + 2\Omega u_\theta \cos\theta = -\frac{1}{\rho r \sin\theta}\frac{\partial p}{\partial \phi}. \qquad (7.7.4)$$

General expressions for the acceleration components in terms of (u_r, u_θ, u_ϕ) may be found in appendix 2. In these equations, as in (7.6.1), p is a modified pressure which includes allowance for the effects of gravity and of the centrifugal force arising from rotation of the co-ordinate system.

Equation (7.7.2) shows that the vertical gradient of modified pressure everywhere balances the vertical component of Coriolis force. But since the thickness of the layer of fluid is small compared with the horizontal length-scale of the motions to be considered, the total variation of p across the layer is relatively small, and p, like u_θ and u_ϕ, may be regarded in equations (7.7.3) and (7.7.4) as uniform across the layer. The more important effect of the rotation of the earth is to make a contribution to the horizontal component of force on a fluid element which is normal to its instantaneous velocity, and

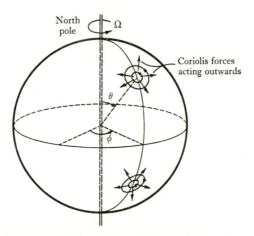

Figure 7.7.1. Cyclonic geostrophic flow systems in the northern and southern hemispheres. The vorticity relative to the earth's surface has the same sign as f, $= 2\Omega \cos \theta$, and the pressure at the centre of each system is low.

with sense such as to make the element tend to move to the right-hand side of its instantaneous line of motion in the northern hemisphere where $\cos \theta$ is positive, and to the left-hand side in the southern hemisphere where $\cos \theta$ is negative.

We may with consistency regard r as constant and equal to R in (7.7.3) and (7.7.4). This gives

$$\frac{Du_\theta}{Dt} - \frac{u_\phi^2 \cot \theta}{R} - fu_\phi = -\frac{1}{\rho R}\frac{\partial p}{\partial \theta}, \tag{7.7.5}$$

$$\frac{Du_\phi}{Dt} + \frac{u_\theta u_\phi \cot \theta}{R} + fu_\theta = -\frac{1}{\rho R \sin \theta}\frac{\partial p}{\partial \phi}, \tag{7.7.6}$$

where

$$\frac{D}{Dt} = \frac{\partial}{\partial t} + \frac{u_\theta}{R}\frac{\partial}{\partial \theta} + \frac{u_\phi}{R \sin \theta}\frac{\partial}{\partial \phi},$$

as the governing equations for the flow in our model atmosphere or ocean.

We have adopted the standard notation $f = 2\Omega\cos\theta$ here; f is twice the angular frequency of revolution of a Foucault pendulum at latitude $\frac{1}{2}\pi - \theta$ and is termed the Coriolis parameter (for the earth $\Omega = 7\cdot29 \times 10^{-5}\,\mathrm{sec}^{-1}$ and $f = 1\cdot03 \times 10^{-4}\,\mathrm{sec}^{-1}$ at $\theta = 45°$).

We shall also need to use the corresponding equation for the radial component of vorticity, ω say, relative to the rotating axes. We have (see appendix 2)

$$\omega = \frac{1}{R\sin\theta}\left\{\frac{\partial(u_\phi\sin\theta)}{\partial\theta} - \frac{\partial u_\theta}{\partial\phi}\right\},$$

and it follows from differentiation of (7.7.5) and (7.7.6) and a little manipulation that

$$\frac{D\omega}{Dt} = -\frac{u_\theta}{R}\frac{df}{d\theta} - \Delta(f + \omega). \tag{7.7.7}$$

In this equation Δ denotes the rate of expansion in the horizontal plane, that is,

$$\Delta = \frac{1}{R\sin\theta}\left\{\frac{\partial(u_\theta\sin\theta)}{\partial\theta} + \frac{\partial u_\phi}{\partial\phi}\right\},$$

$$= -\frac{1}{H}\frac{DH}{Dt}$$

according to (7.7.1). Equation (7.7.7) can thus be rewritten as

$$\frac{D}{Dt}\left(\frac{f + \omega}{H}\right) = 0, \tag{7.7.8}$$

showing that the absolute vorticity $f + \omega$ of a material element of fluid† changes only as a consequence of movement of the element to a place where the thickness of the fluid layer is different. In a layer of uniform depth the relative vorticity ω changes only if the fluid element moves to a different latitude. Equation (7.7.8) could also have been derived directly by considering the conservation of circulation round a closed material curve of small linear dimensions lying in a horizontal plane.

These equations apply to flow with a characteristic length scale L of any size, provided only that it is large compared with H. The existence of land boundaries gives rise in the oceans to flow fields whose length scales are in fact appreciably smaller than the earth's radius R, and there is likewise considerable interest in atmospheric flow fields which are not global in extent. For an investigation of such flow systems a co-ordinate system of a more localized nature is appropriate. In the case of a flow field extending over a small range of latitudes centred on $\theta = \theta_0$ it is convenient to introduce the new co-ordinates

$$x = \phi R\sin\theta_0, \quad y = (\theta_0 - \theta)R. \tag{7.7.9}$$

The co-ordinates (x, y, z), where z is the upward vertical co-ordinate, then

---

† Strictly speaking, $f + \omega$ is the vertical component of the absolute vorticity, but since this is the only relevant component we may speak of it as the absolute vorticity.

form a right-handed system, like the spherical polar co-ordinates (r, θ, ϕ); x and y increase in the eastward and northward directions respectively.

For $L \ll R$, and in the crudest approximation, the equations clearly reduce to the form corresponding to two-dimensional flow in a layer of fluid which is plane, apart from gradual variations of thickness due to bottom topography, with x and y as rectilinear co-ordinates, and with the Coriolis parameter f constant and equal to $2\Omega \cos \theta_0$, $= f_0$ say. The direction of the x-axis in the horizontal plane is then immaterial. The only explicit change in (7.7.8) arising from this approximation is in the form of the operator D/Dt, which becomes

$$\frac{D}{Dt} = \frac{\partial}{\partial t} + u \frac{\partial}{\partial x} + v \frac{\partial}{\partial y}, \tag{7.7.10}$$

where u and v are components of the fluid velocity in the direction of the x- and y-axes, and the relative vorticity is now $\omega = \partial v/\partial x - \partial u/\partial y$.

An improved approximation to the equations, which allows the investigation of certain types of flow field which extend over a larger although still small range of latitudes, is obtained by allowing for the variation of f with latitude. The basis of the approximation is that, in the case of flow fields whose characteristic length scale in the y-direction is L, and in which the relative vorticity ω is of small magnitude compared with f, the value of ω/L may be comparable with that of f/R, in which event $D\omega/Dt$ and Df/Dt are comparable. Although it is then not possible to regard f as constant in (7.7.8), a permissible approximation is

$$f = f_0 + \beta y, \tag{7.7.11}$$

where $\beta = 2\Omega \sin \theta_0/R$ ($\beta = 1 \cdot 62 \times 10^{-13}$ cm^{-1} sec^{-1} at $\theta = 45°$ and is positive in both hemispheres). All other effects of the curvature of the fluid layer may again be ignored, provided $L \ll R$ still, so that the flow fields are now being regarded as occurring in a plane layer with a normal rotation vector whose magnitude varies linearly in the y-direction (i.e. in the north–south direction). This is usually called a *β-plane approximation*.

Solutions of the above dynamical equations have been explored for a number of particular cases and limiting conditions, a few of which will now be described for purposes of illustration.

Geostrophic flow

Meteorologists have found, from an examination of many distributions of wind velocity (observed at levels high enough to avoid frictional and thermal effects of the earth's surface), that the inertia forces are often appreciably smaller than Coriolis forces. If the fluid is moving steadily in a curved path of radius of curvature L with speed q, the ratio of inertia to Coriolis forces will be of order q/fL; and with the values $L = 1000$ km and $f = 1 \cdot 03 \times 10^{-4}$ sec^{-1} (appropriate to $\theta = 45°$), we have $q/fL \approx 0 \cdot 01 \times q$ m/sec. In the atmosphere $q = 10$ m/sec is a representative value in common circumstances, and in the ocean q is usually much smaller than this. Thus values of q/fL small compared with unity are typical. Also the time scale of change of the fluid velocity (averaged over a region with linear dimensions of order H) is often much larger than f^{-1} ($= 2 \cdot 7$ hours at $\theta = 45°$).

Flow systems in which inertia forces are negligible are said, in the literature of geophysics, to be *geostrophic*. In the terminology of §7.6, they are flow fields with small Rossby number. The equations (7.7.5) and (7.7.6) reduce to

$$\left(\frac{1}{R}\frac{\partial}{\partial\theta}, \frac{1}{R\sin\theta}\frac{\partial}{\partial\phi}\right) p = \rho f(u_\phi, -u_\theta) \tag{7.7.12}$$

in the case of geostrophic flow, showing that the pressure gradient in a horizontal plane is everywhere normal to the streamlines. Meteorologists use this relation more as a basis for comparison than as a means of determining the flow in the atmosphere. Measurements of the atmospheric pressure at a number of points at ground level may be made easily, and the corresponding horizontal pressure gradients at levels where the motion is not directly affected by the ground may be calculated (provided information about the air density up to these levels is available); the components of wind velocity computed from the relation (7.7.12) then give the hypothetical *geostrophic wind*, to which the real wind will approximate with a degree of accuracy which the meteorologist can estimate from the circumstances.

When inertia forces are negligible, the vorticity equation (7.7.8) reduces to

$$\frac{D}{Dt}\left(\frac{f}{H}\right) = 0, \tag{7.7.13}$$

showing that strictly geostrophic flow can occur only when along the path of an element the bottom of the fluid layer slopes down towards the nearer pole. When the extent of the flow field is small compared with R, so that the variation of f is negligible, (7.7.13) requires that H be constant for a moving element, thereby recovering (in virtue of (7.7.1)) a result obtained in §7.6. We saw there that neglect of inertia forces is not compatible with the existence of a non-zero value of the rate of expansion in a plane normal to the (uniform) rotation vector, because such an expansion is resisted by the Coriolis forces.

A common type of geostrophic flow in the atmosphere is one with rough symmetry about a central region in which the relative vorticity is non-zero and one-signed; such a mass of rotating air might result from previous movement of the mass from a different latitude, with invariance of the vorticity relative to fixed axes. If the relative vorticity has the same sign as the Coriolis parameter f—that is, if the relative vorticity of the central region is positive, with anti-clockwise circulation, in the northern hemisphere, or negative, with clockwise circulation, in the southern hemisphere—the Coriolis forces are directed away from the centre of the region (figure 7.7.1). Such systems are called *cyclonic*, and are characterized by a low pressure at the centre. The opposite or *anti-cyclonic* system has a high pressure at the centre.

Cyclonic systems are often accompanied by strong winds, for which the geostrophic equations (7.7.12) are not accurate. A standard method of improving the approximation is to assume that the flow is steady and the streamlines are circular with radius L, so that the inertia force in equations (7.7.5) and (7.7.6) reduces to a centrifugal force q^2/L radially outwards, where $q^2 = u_\theta^2 + u_\phi^2$. The streamlines and lines of equal pressure still coincide, but the local pressure gradient now has magnitude

$$|\nabla p| = \rho \left(\frac{q^2}{L} + fq \right) \qquad (7.7.14)$$

for a cyclonic flow; the value of q obtained from this equation and observations of L and the pressure gradient is referred to as the gradient wind.

Flow over uneven ground

The direct effect of a slow variation of the layer thickness H as a function of position, as summarized by equation (7.7.8), is to change the height, and thereby the (vertical component of the) absolute vorticity, of a material vertical cylinder of small cross-section as it moves. Thus when a mass of fluid moves over rising ground the absolute vorticity is reduced in magnitude, and the relative vorticity is changed, being decreased in the northern hemisphere and increased in the southern hemisphere. This change in the vorticity relative to the earth's surface may give rise to a noticeable deflection of the stream passing over the sloping ground.

As a simple example of the effect of uneven ground, consider the steady flow over a mountain ridge with straight parallel contours of height situated on otherwise level ground where the fluid layer is of thickness H_0. We shall suppose in the first instance that the horizontal extent of the flow field is small enough for the flow to be regarded as taking place in a plane layer with the Coriolis parameter f uniform and equal to f_0 say. The direction of the ridge on the earth's surface is then immaterial, and we choose it for convenience to be in the y-direction (figure 7.7.2). The stream approaching the ridge will be assumed to have zero relative vorticity and uniform velocity with components (U, V), with the Coriolis force being balanced by a uniform pressure gradient. At a point over the ridge where the layer thickness is H, the relative vorticity ω is given by

$$\frac{f_0 + \omega}{H} = \frac{f_0}{H_0}. \qquad (7.7.15)$$

It is clear that the velocity components u and v are independent of y, so that

$$\frac{dv}{dx} = \omega = f_0 \left(\frac{H}{H_0} - 1 \right)$$

and

$$v = V + f_0 \int_{-\infty}^{x} \frac{H - H_0}{H_0} \, dx. \qquad (7.7.16)$$

The x-component of velocity is determined by the mass-conservation relation

$$uH = UH_0.$$

The effect of an elevated ridge is thus to deflect an approaching uniform stream to the right of its line of motion in the northern hemisphere. On the downstream side of a ridge whose cross-sectional area is

$$A = \int_{-\infty}^{\infty} (H_0 - H)\,dx,$$

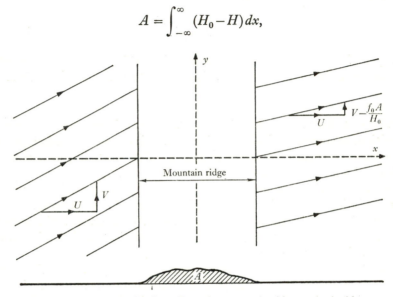

Figure 7.7.2. The deflecting effect of a mountain ridge on an incident uniform stream, relative to rotating axes.

the x-component of velocity has returned to its upstream value, but for the y-component we have the uniform value $V - (f_0 A/H_0)$. There is thus a resultant clockwise deflection through an angle whose tangent is

$$\frac{U(f_0 A/H_0)}{U^2 + V^2 - V(f_0 A/H_0)}.$$

In the case of a ridge of average height 2 km and about 50 km width, and with H_0 taken as 10 km, we have $f_0\,A/H_0 \approx 1$ m/sec; the deflecting effect of even such a large ridge may not be evident in atmospheric motions, but it would be significant in the ocean where the currents are slower.

The effect of a region of high ground of finite area in the horizontal plane (an isolated mountain) is likewise to insert into the flow a patch of negative vorticity (in the northern hemisphere), the value of ω at any point being given again by (7.7.15) in the case of a uniform incident stream. (But note that the Rossby number for the flow must not be too small, for then the stream might pass round a 'Taylor column' over the mountain!) The additional flow due to the presence of this mountain is a steady clockwise circulating motion, the circulation round any closed path enclosing the mountain being

$$\int\int\limits_{-\infty}^{\infty} \omega\, dx\, dy, \quad = \frac{f_0}{H_0}\int\int\limits_{-\infty}^{\infty} (H-H_0)\, dx\, dy; \tag{7.7.17}$$

this latter integral is the volume of the mountain above the level ground where the layer thickness is H_0. The effect of the mountain on the velocity of the air or water is unlikely to be detectable in practice, but the Coriolis force associated with the anticyclonic circulatory motion gives rise to an excess pressure (in either hemisphere) over the mountain as is sometimes observed in the atmosphere.

Let us suppose now that the horizontal extent of the flow field under consideration (L) is such that ω/L is comparable in magnitude with f/R, although with $L \ll R$ still. As explained earlier, it is still possible to regard the fluid as being in a plane layer, and to use rectilinear co-ordinates (x, y) with corresponding velocity components (u, v). However, we must allow for the variation of the Coriolis parameter with latitude, and, with the y-axis pointing north, we use the approximate linear relation (7.7.11). The mixture of the effects of bottom topography and of non-uniformity of the Coriolis parameter makes analysis complicated, but we can indicate the main new features by a simplified discussion of flow over a long mountain ridge. It is convenient again to consider an oncoming stream which has uniform velocity, U_0 say, and this obliges us to choose the stream direction as parallel to the lines of latitude, that is, parallel to the x-axis. The relative vorticity ω of a material element at a point (x, y) over the ridge where the layer thickness is H is then given by the equation

$$\frac{f_0 + \beta y + \omega}{H} = \frac{f_0 + \beta y_0}{H_0}, \tag{7.7.18}$$

where y_0 is the y-co-ordinate of the position of the same element when it was approaching the ridge.

The velocity is no longer the same at all points on a line parallel to the ridge, and we need to consider the flow field as a whole. To obtain a simplified version of this field problem, consider a 'ridge' in the form of a step, or discontinuous change to a layer thickness H_1, along a north–south line at $x = 0$ (figure 7.7.3). At this step the velocity components change discontinuously from $(U_0, 0)$ to $(U_1, 0)$, where $U_1 = U_0 H_0/H_1$, and the relative vorticity changes from 0 to

$$(f_0 + \beta y)\frac{H_1 - H_0}{H_0}.$$

In the region $x > 0$ the layer thickness is uniform, so that

$$\frac{\partial u}{\partial x} + \frac{\partial v}{\partial y} = 0$$

and we may introduce a stream function ψ. The flow in this region is steady and $f + \omega$ is consequently a function of ψ alone. But at $x = 0$ (as approached from the region $x > 0$) we have $\psi = U_1 y$ and

$$f + \omega = \frac{H_1}{H_0}(f_0 + \beta y) = \frac{H_1}{H_0}\left(f_0 + \frac{\beta}{U_1}\psi\right),$$

and this must be the relation between $f + \omega$ and ψ that is valid over the whole of the region $x > 0$. Hence in that region

$$\nabla^2 \psi = -\omega = f - \frac{H_1}{H_0}\left(f_0 + \frac{\beta}{U_1}\psi\right)$$

$$= f_0\left(\frac{H_0 - H_1}{H_0}\right) + \beta y - p^2 \psi, \qquad (7.7.19)$$

where $p^2 = \beta H_1 / U_1 H_0$.

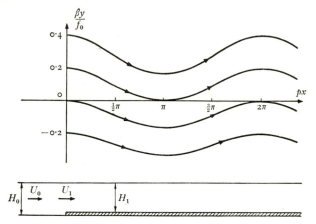

Figure 7.7.3. Streamlines for flow of a uniform eastward stream past a step $(H_1 = 0.91 H_0)$ running north and south.

Our choice of conditions has led to a linear equation for ψ. A form of solution which contains the linearity in y demanded by conditions at $x = 0$ is

$$\psi = (y + a)F(x) + \left\{f_0\left(\frac{H_0 - H_1}{H_0}\right) + \beta y\right\}\bigg/ p^2,$$

where a is a constant and $F(x)$ satisfies the equation

$$\frac{d^2 F}{dx^2} + p^2 F = 0. \qquad (7.7.20)$$

The prescribed velocity components and stream function at $x = 0$ are obtained if

$$F(0) + \beta/p^2 = U_1, \quad F'(0) = 0$$

and

$$a = f_0 / \beta.$$

The complete solution for ψ is then readily found to be

$$\psi = U_1 y + U_1 \left(\frac{H_0 - H_1}{H_1}\right)(f_0 + \beta y)\frac{1 - \cos px}{\beta}. \qquad (7.7.21)$$

The streamlines for the case $H_1 = 0.91 H_0$ and $U_1 > 0$ are shown in figure 7.7.3. The various streamlines differ in shape only by a change of scale in the y-direction arising from the fact that the average value of the Coriolis parameter on a streamline is different for different streamlines. When $|\beta y/f_0| \ll 1$ and $px \ll 1$, (7.7.21) yields the velocity components found earlier (see (7.7.16)) for flow over a ridge with f regarded as constant.

The novel feature of (7.7.21) is the periodicity in x when p is real, i.e. when $U_1 > 0$. The role of the bottom topography in this simple example is solely to provide non-zero relative vorticity at $x = 0$ and thereby to turn the stream towards the south, and the wavy character of the streamlines on the downstream side of the step is due to the non-uniformity of the Coriolis parameter. Both the wavelength in the x-direction and the range of the values of y on one streamline are large. The wavelength is

$$\frac{2\pi}{p} = 2\pi \left(\frac{U_1 H_0}{\beta H_1}\right)^{\frac{1}{2}},$$

i.e. about 1,600 km at latitude $45°$ when $U_1 = 1$ m/sec and $H_0 - H_1 \ll H_0$. The southward excursion of the streamline that passes through the origin is

$$\frac{2f_0}{\beta}\left(\frac{H_0 - H_1}{2H_0 - H_1}\right), \quad = 2R\cot\theta_0\left(\frac{H_0 - H_1}{2H_0 - H_1}\right).$$

For a streamline beginning at latitude $45°$ this distance is equivalent to a range of latitude of $2(H_0 - H_1)/(2H_0 - H_1)$ radians (and is 1,050 km or $9.5°$ of latitude when $H_1 = 0.91 H_0$).

It is possible also to determine the stream function of the flow on the downstream side of a second north–south step in the path of an eastward stream, perhaps a step down which restores the layer thickness to its original value H_0. This downstream flow field depends on the velocity with which fluid arrives at the second step and so depends on the distance between the steps.

For a westward current (in either hemisphere) approaching a step, U_0 and U_1 are negative and $p^2 < 0$. According to the above solution, the y-co-ordinate on a streamline in the region $x < 0$ now has an exponential dependence on x. A physical reason for this radical difference between the effects of a step on eastward and westward currents is given below.

Planetary waves

The preceding case of 'flow in the β-plane' has revealed some interesting wave-like properties which we now examine more directly. The existence of the waves is associated with the non-uniformity of the Coriolis parameter, and it is not difficult to see the general mechanism. When a fluid element moves in a direction inclined to the parallel of latitude, that is, inclined to the x-axis in a diagram such as figure 7.7.3, the value of the Coriolis parameter f at the position of the element is continually changing. If the velocity of the element has a northward component, the value of f is increasing, and the magnitude of the Coriolis force acting on the element is increasing. The path of the element therefore turns to the right of its direction of motion. If the direction of motion of the element lies initially in the north-east quadrant, the turning will eventually produce a direction of motion in the south-east quadrant; the element is then experiencing continually decreasing values of f, and the opposite process of turning to the left occurs. Thus a generally eastward current experiences a restoring force if for some reason the flow direction is changed. This restoring effect was found in the solution representing eastward flow across a step running north and south, although the analysis was carried out in terms of vorticity rather than momentum and forces. The existence of a restoring force in the β-plane which would produce oscillations about an eastward current flowing past a fixed obstacle was first pointed out by Rossby (1939), and the associated waves are often referred to as *Rossby waves*. Similar wave motions have been shown to exist in the fluid layer over the whole of a rotating sphere (Haurwitz 1940; Longuet-Higgins 1964, 1965) and the more general term *planetary waves* is also used.

If on the other hand a stream has a general westward direction relative to a fixed obstacle on the earth's surface, the turning due to non-uniformity of the Coriolis parameter does not tend to restore the original direction of motion. The simple solution for a north–south step in the path of a westward stream suggested an exponential growth of the departure from westward flow, but it can be shown that the presence of the step here exerts an influence upstream which makes invalid the assumption of a uniform velocity in the stream approaching the step.

The existence of sinusoidal waves with straight crests in a plane layer of fluid of uniform thickness, and with f varying linearly, can be demonstrated readily. For such a wave in fluid otherwise at rest we have a stream function of the form

$$\psi \propto \exp\{i(kx + ly - \sigma t)\}, \tag{7.7.22}$$

where (k, l) is the wave-number vector in the (x, y)-plane and σ is the angular frequency. The corresponding relative vorticity is

$$\omega = -\nabla^2\psi = (k^2 + l^2)\psi, \tag{7.7.23}$$

and so the rate of change of absolute vorticity of a material element is

$$\frac{D(f+\omega)}{Dt} = -\frac{\partial\psi}{\partial x}\frac{df}{dy} + (k^2 + l^2)\frac{\partial\psi}{\partial t}$$

$$= i\psi\{-\beta k - \sigma(k^2 + l^2)\}. \tag{7.7.24}$$

The governing vorticity equation is therefore satisfied if

$$\sigma = -\beta k/(k^2 + l^2). \tag{7.7.25}$$

These are transverse waves, for which the fluid velocity is everywhere parallel to the crests, that is, at right angles to the vector (k, l). The phase velocity with which the crests advance in the direction of the wave-number vector is

$$\frac{\sigma}{(k^2+l^2)^{\frac{1}{2}}}, \quad = -\frac{\beta k}{(k^2+l^2)^{\frac{3}{2}}}. \qquad (7.7.26)$$

It will be noticed that the wave motion is steady relative to axes moving with velocity $(\sigma/k, 0)$, that is, moving westward with speed

$$\beta/(k^2+l^2),$$

which is independent of the direction of the wave-number vector. Also, any number of sinusoidal waves with the same value of the wave-number magnitude $(k^2+l^2)^{\frac{1}{2}}$ may be superposed, since for an assembly of such waves the relations (7.7.23) and (7.7.24) are valid and the contributions to the right-hand side of (7.7.24) from the different waves vanish separately when σ has the value (7.7.25). Hence an assembly of superposed sinusoidal waves having the same wave-number magnitude forms a steady motion relative to axes moving westward with the speed $\beta/(k^2+l^2)$.

There are other motions which have this property of being steady relative to axes moving westward. For if the stream function is of the form

$$\psi(x+ct, y),$$

we have

$$\frac{D(f+\omega)}{Dt} = -\beta\frac{\partial\psi}{\partial x} + \frac{\partial\omega}{\partial t} + u\frac{\partial\omega}{\partial x} + v\frac{\partial\omega}{\partial y}$$

$$= \frac{\partial(-\beta\psi+c\omega)}{\partial x} + \frac{\partial(\omega, \psi)}{\partial(x, y)},$$

and both these terms vanish if

$$\omega = -\nabla^2\psi = \beta\psi/c. \qquad (7.7.27)$$

A solution of this equation for ψ representing a flow of a centred kind, with a velocity which diminishes in magnitude as $r^{-\frac{3}{2}}$ at large distances from the centre, is

$$\psi = e^{in\theta}\left\{A_n J_n\left(\frac{r\beta^{\frac{1}{2}}}{c^{\frac{1}{2}}}\right) + B_n Y_n\left(\frac{r\beta^{\frac{1}{2}}}{c^{\frac{1}{2}}}\right)\right\},$$

where J_n and Y_n denote Bessel functions of the first and second kinds and $r^2 = (x+ct)^2 + y^2$; solutions of this type with different values of the constants n, A_n and B_n may be superposed, non-zero values of B_n being appropriate to problems involving an inner boundary within which the layer thickness is not uniform. Another solution of (7.7.27) is

$$\psi = (y+a)\left\{A \sin\frac{(x+ct)\beta^{\frac{1}{2}}}{c^{\frac{1}{2}}} + B \cos\frac{(x+ct)\beta^{\frac{1}{2}}}{c^{\frac{1}{2}}}\right\},$$

where a, A and B are constants; this is the type of solution that, when referred to axes moving westward with speed c, gave the steady flow on the downstream side of a north–south step in the bottom of the fluid layer.

A common feature of all these exact solutions is that a general streaming motion of the fluid towards the east with speed U on which is superposed motions having a characteristic length scale $(U/\beta)^{\frac{1}{2}}$ can be a steady state, which usually shows meanders of the eastward stream alternately to the north and to the south. These properties are believed to have geophysical significance, particularly for the atmosphere. Meteorologists have found that in middle latitudes the wind well above the earth's surface tends to be eastwards, on average, and the streamlines circumscribing the globe show large-scale approximately stationary meanders. These observed waves or meanders might be caused by mountain chains acting as obstacles to the wind, in the way that a step in the bottom of the fluid layer was found to generate waves on the downstream side.[†] According to our analysis, the number of wavelengths in one circumnavigation of the globe at latitude $45°$ would be about $(\beta R^2/U)^{\frac{1}{2}}$, or $(Rf/U)^{\frac{1}{2}}$, that is, about $26/(U\,\text{m/sec})^{\frac{1}{2}}$. Average eastward wind speeds lie mostly between 10 and 30 m/sec, corresponding to a prediction of 5 to 8 waves, which is consistent with the observed features of the global wind pattern.

A discussion of the large-scale features of the motion of the atmosphere and ocean is not complete without consideration also of the effect of variations of density, but this is beyond our scope here.

7.8. The vortex system of a wing

General features of the flow past lifting bodies in three dimensions

In two-dimensional irrotational flow due to a body in translational motion through infinite fluid, a side-force acts on the body when the circulation round the body is non-zero (§6.4). We saw in §6.7 that when a two-dimensional aerofoil—a thin body with a rounded nose and a sharp tail—moves steadily through fluid in a direction only slightly inclined to its chord, the effect of viscosity on the flow at high Reynolds number is to cause the generation of circulation round the aerofoil, the magnitude of the circulation being just that required to move the rear stagnation point to the sharp trailing edge and to eliminate boundary-layer separation from both surfaces of the aerofoil (Joukowski's hypothesis). This combination of a side-force of predictable magnitude and absence of boundary-layer separation (implying a relatively small drag force) is put to good practical use in aeronautics. The discussion in §6.7 of the properties of aerofoils and the associated flow fields was confined to two-dimensional systems. We turn now to the more realistic case of a three-dimensional system, involving a body of finite dimensions which is in steady translational motion and on which a side-force, or 'lift' force, acts. It is convenient to use the terminology of wings, that is, thin bodies designed specifically to generate a large lift force

---

† The heating or cooling of the air that occurs when the air stream passes across a land-sea boundary may also cause large-scale meanders of the wind.

and a small drag force when moving with a certain attitude, although many of the ideas and arguments are applicable qualitatively to the flow due to any body in translational motion which is not symmetrical about more than one plane through the direction of motion.

We recall the result obtained in §6.4 that when the flow due to a three-dimensional body in steady translational motion is wholly irrotational both the side-force and the drag acting on the body are zero. Thus the existence of vorticity in the fluid is inevitable in the circumstances of interest here. In the case of a 'streamlined' body with a sharp edge on the downstream side, with no boundary-layer separation, any vorticity which is generated at the rigid surface is carried downstream in a thin sheet, the thickness of which is determined by the viscosity of the fluid. The pressure is continuous across this sheet, and since the Bernoulli constant has the same value throughout the region of irrotational flow it follows that the magnitude of the velocity relative to the body has the same value at adjacent points on the two sides of the sheet. For a two-dimensional flow field there is then the conclusion that the sheet constitutes a thin wake containing vorticity of both signs, the net effect of which on the flow field disappears as the Reynolds number increases and the thickness of the sheet tends to zero. However, in a three-dimensional flow field there is the possibility of a change in the *direction* of the velocity vector across the sheet, associated with a component of vorticity within the sheet which is parallel to the stream. We are thus led to explore a connection between the existence of streamwise vorticity in the stream-surface extending downstream from the sharp trailing edge of a three-dimensional body, and the exertion of a side-force on the body.

This connection is evident from considerations of the general form of the streamlines in the steady flow past an inclined flat wing. Observation shows that on the lower surface of the wing, which is exposed to the oncoming stream, the pressure is greater than on the upper or 'suction' surface of the wing, thereby giving a net lift force. (This is what we should expect for a two-dimensional aerofoil; and near the centre of a wing of large span-to-chord ratio the flow approximates to that past a two-dimensional aero-foil.) Near each end of the wing this pressure difference leads to a tendency for fluid to flow round a wing-tip from the lower to the upper surface, as sketched in figure 7.8.1. The lateral or span-wise momentum of the fluid is retained as it passes downstream from the wing, and so there is streamwise or 'trailing' vorticity in the stream-surface extending downstream from the sharp trailing edge. The vorticity has opposite signs on the two sides of the central vertical plane of symmetry, and the trailing sheet vortex may be regarded roughly as a pair of semi-infinite line vortices, the sense of circu-lation being such that each line vortex moves downward under the action of the other. The total force impulse required to generate this motion of the fluid in a lateral plane (normal to the direction of flight) is directed downwards.

The streamwise vorticity extending downstream from the wing is thus seen to be an intermediary in the process by which the downward force exerted by the wing on the fluid leads to the continual generation of downward momentum of the fluid. There is a further fundamental consequence of the existence of this trailing vorticity. The kinetic energy of motion of the fluid in a lateral plane over a continually increasing length of the flight path must be supplied by work done by the moving wing, so that a drag force is evidently being exerted on the wing. This is the *induced drag*, referred to briefly in §5.11. Like the lift force, it is a consequence of the generation of vorticity at a rigid surface and has a magnitude which, at any rate for bodies on which the boundary layer does not separate upstream of a sharp trailing edge, is determined by the shape of the body and not by the viscosity of the fluid.

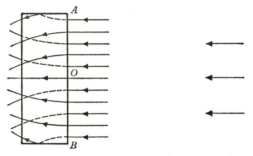

Figure 7.8.1. The generation of streamwise vorticity downstream from a wing due to flow round the ends of the wing from the lower high-pressure side to the upper 'suction' side. The trailing vorticity has anti-clockwise sense about the flow direction on the side OA of the wing, and clockwise sense on the side OB.

An illuminating view of the trailing vortex system may be obtained by thinking of the closely related two-dimensional field in which motion is generated from rest by the application of a downward force impulse distributed along the line AB shown in figure 7.8.2. This line represents the cross-section of the (thin) wing in the plane normal to the direction of flight, and the motion at different times after application of the impulse corresponds roughly to the motion in such a lateral plane at different distances downstream from the moving wing. The appropriate distribution of the force impulse over AB depends on the way in which the moving wing presses the fluid down, which involves the exact shape and attitude of the wing, but it is evident that the streamlines of the motion immediately after application of the impulse have the general form shown in figure 7.8.2. A sheet vortex is created on AB by the applied force impulse (since this kind of distribution of applied force does not satisfy the conditions of Kelvin's circulation theorem), and the general results of §§7.2, 7.3 show that, for a given distribution of the force impulse on AB, the vorticity magnitude at any point varies linearly with the total impulse, I say. On the other hand the kinetic

energy of the motion varies as the square of the vorticity (see (7.3.9)) and
consequently as the square of the total impulse. Now the wing moving with
speed U exerts a downward force L on the fluid between parallel lateral
planes unit distance apart during a time interval $1/U$, so that the total
impulse I in our analogy represents the quantity L/U. During this same
time interval the work done against the induced drag D_i by the moving wing
is D_i, and this is the kinetic energy of the two-dimensional motion in the
analogy. It follows that

$$D_i = \frac{L^2}{A\rho U^2}, \qquad (7.8.1)$$

where the proportionality factor A has dimensions of area and depends on
the details of the vorticity distribution.

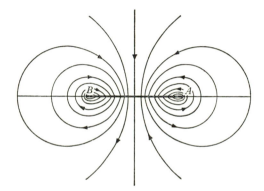

Figure 7.8.2. Streamlines of the two-dimensional motion produced immediately after the
application of a downward force impulse distributed along the line AB.

Wings of large aspect ratio, and 'lifting-line' theory

A calculation of the lift and induced drag forces exerted on a wing of given
shape and attitude may employ the methods of inviscid-fluid theory when
the wing is sharp-tailed and boundary-layer separation does not occur
upstream of the sharp trailing edge. The main difficulty lies in the determination of the strength and position of the vortices which trail downstream from
the wing and which influence the flow near the wing. A theory which enables
the trailing vortex system and the lift and induced drag forces on the wing
to be calculated under certain conditions was initiated by Lanchester and
Prandtl in the early days of aeronautics. This theory is still of considerable
value in the design and testing of aeroplane wings intended for use at sub-
sonic flight speeds, and a brief account of it will now be given.

The theory rests on two main assumptions about the wing under consideration. The first is that the trailing vortices are straight and parallel to
the direction of flight, with consequent simplification of the expression for
the velocity field induced by the trailing sheet vortex. In reality the vortex-

lines move with the fluid, and owing to the existence of a non-zero component of velocity in a lateral plane (which arises from the influence of the trailing vortices themselves) the trailing vortices are inclined to the flight direction. However, provided the trailing vortices are sufficiently weak, which is equivalent to requiring that the lift force on the wing be sufficiently small, we may expect the assumption of straight trailing vortices parallel to the flight direction to be valid as an approximation. The second main assumption is that the ratio of the span to the mean chord, known as the aspect ratio of the wing, is large and that as a consequence (for a wing which is not extensively swept back) the flow in the neighbourhood of any one section of the wing is approximately two-dimensional. The implications of this assumption will emerge from the analysis.

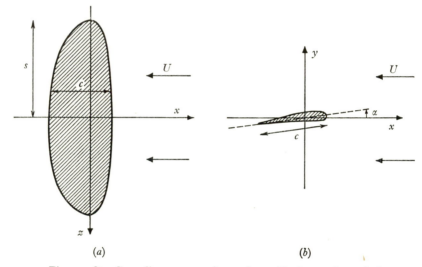

(a) (b)

Figure 7.8.3. Co-ordinate system for a wing, with the y-axis vertical.

Figure 7.8.3 shows the co-ordinate system and the notation to be used. The axes are fixed relative to the wing, and at infinity the fluid has uniform speed U in the direction of the negative x-axis. The wing is assumed to be symmetrical about the central vertical plane on which $z = 0$. The chord c, the angle of incidence α, and the cross-sectional shape of the wing (all shown in figure 7.8.3b) may all vary with the spanwise co-ordinate z. Now when $s/c \gg 1$, the plan form of a wing without 'sweep-back' reduces to a straight line,† as shown in figure 7.8.4. So far as the large-scale features of the flow are concerned, the only relevant property of this 'lifting line' is the circulation, K say, round a circuit enclosing the wing in a plane normal to the

† A wing with a small degree of sweep-back would be represented by a curved line making a small angle with the z-axis everywhere and can be brought within the scope of the theory, but will be excluded from the present simplified treatment.

z-axis. K may vary across the span of the wing, and the variation is evidently related to the strength of the trailing sheet vortex. If the circulation at spanwise station $z + \delta z$ exceeds that at station z by an amount δK, $= (dK/dz)\,\delta z$, the application of Stokes's theorem to the strip bounded by two similar closed curves enclosing the wing and lying in planes normal to the z-axis at these two stations shows that the trailing vortices springing from the portion of the wing between stations $z + \delta z$ and z must have a total strength δK (anti-clockwise sense in the (y, z)-plane being positive as usual); that is, the strength density (§ 2.6) of the trailing sheet vortex at station z is dK/dz.

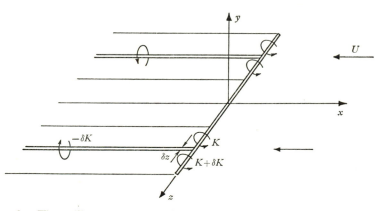

Figure 7.8.4. The trailing vortex system from a lifting line. The circular arrows show the actual sense of the circulation for lift in the direction of the positive y-axis ($\delta K < 0$).

It is as if the whole vortex system, comprising the trailing vortices and the 'bound' vortex at the lifting line itself, were made up of a set of vortex filaments in the shape of rectangles of typical width $2z$ with one end at the wing and the other at infinity downstream. The circulation round the wing must fall to zero at the two wing-tips ($z = \mp s$), and if it does so rapidly the strength density of the trailing sheet vortex will be of large magnitude near the wing-tips.

A quantity which we shall need later is the velocity induced at position $(0, 0, z_1)$ on the lifting line by the whole vortex system. It is evident from the geometry that this induced velocity is vertical. If the portion of the trailing sheet vortex of strength $\delta K(z)$ emanating from the section of the lifting line between stations z and $z + \delta z$ extended from $x = -\infty$ to $x = +\infty$, it would made a contribution

$$- \delta K(z)/\{2\pi(z_1 - z)\}$$

to the induced vertical velocity (see (2.6.4)); since it is semi-infinite, with one end at the lifting line, the contribution is half this. The bound vortex at the lifting line makes no contribution to the induced velocity at the lifting line itself (although of course it induces a circulation round the lifting line).

Hence the vertical component of velocity at $(0, 0, z_1)$ is

$$-\frac{1}{4\pi}\int_{-s}^{s}\frac{dK(z)}{dz}\frac{dz}{z_1-z}, \quad = v(z_1) \text{ say,} \qquad (7.8.2)$$

where the principal value of the integral is implied.

It is necessary also to consider the flow past the wing as viewed on the scale of the chord. According to the second of our two main assumptions, the rate of variation of conditions along the span is so slow that the flow about any section of the wing like that shown in figure 7.8.3*b* may be regarded as two-dimensional. It follows that the local circulation K is given by the Joukowski hypothesis and the geometry of the wing section. However, the form of the wing as a whole is not entirely without influence on the flow near one section of the wing. The key-point of the theory is that under the conditions stated above the induced vertical velocity due to the trailing vortex system associated with the wing is approximately uniform over the neighbourhood of any section of the wing (that is, over a region with linear dimensions comparable with the chord), and is therefore equivalent in its effect on the flow past this section to a small change in the direction of the undisturbed stream velocity. We see then that the two-dimensional flow near a section of the wing at station z_1 is that due to an aerofoil immersed in a uniform stream of speed U and with an angle of incidence equal to

$$\alpha+\frac{v(z_1)}{U},$$

where α is the geometrical angle of incidence (that is, the angle between the chord line and the direction of flight of the wing), and $v(z_1)$ is given by (7.8.1) above.

To make further progress we must supplement our lifting-line theory with some information about the flow in the neighbourhood of a section of the wing. Now it was found in §6.7 that for all aerofoils in two-dimensional flow the circulation varies linearly with c, with U and with the angle of incidence, when that angle has the small values appropriate to normal flight conditions. We may therefore put

$$K(z) = \tfrac{1}{2}acU\left\{\alpha+\beta+\frac{v(z)}{U}\right\}, \qquad (7.8.3)$$

where $-\beta$, as in §6.7, is the 'no-lift' angle of incidence for the aerofoil; a is a constant (equal to $dC_L/d\alpha$, in the notation of §6.7), which has the approximate value 2π for all slender Joukowski aerofoils in wholly irrotational flow and which observation shows to be not far from 6 for most aerofoils.

The relations (7.8.2) and (7.8.3) together provide an integral equation for the circulation function $K(z)$ when the parameters ac and $\alpha+\beta$ are given as functions of z. When $K(z)$ has been determined, the total lift force on the wing follows from

$$L = \rho U\int_{-s}^{s}K(z)\,dz. \qquad (7.8.4)$$

Since the effective stream in which each section of the wing is immersed is not exactly parallel to the flight direction, there is a small component of the side-force parallel to the flight direction, the integral of which over the span is the induced drag; thus

$$D_i = -\rho U \int_{-s}^{s} \frac{v(z)}{U} K(z)\, dz. \qquad (7.8.5)$$

It is useful for some purposes to replace the variable z by θ, defined by

$$z = -s \cos \theta.$$

The circulation K is zero at the wing-tips, $\theta = 0$, π, and so may be written as the Fourier series

$$K(\theta) = U \sum_{n=1}^{\infty} B_n \sin n\theta;$$

moreover, since the circulation is symmetrically distributed about $\theta = \frac{1}{2}\pi$, the coefficients B_n are zero when n is even. We then find from (7.8.2) that

$$v(\theta_1) = -\frac{U}{4\pi s} \int_0^{\pi} \frac{\Sigma n B_n \cos n\theta}{\cos \theta - \cos \theta_1}\, d\theta,$$

$$= -\frac{U}{4s} \frac{\Sigma n B_n \sin n\theta_1}{\sin \theta_1}$$

for $0 < \theta_1 < \pi$, in view of the result given in §6.9 for the definite integral. The coefficients B_n may now be evaluated numerically from equation (7.8.3) by standard approximate methods. The expressions (7.8.4) and (7.8.5) for the components of force on the wing become

$$L = \tfrac{1}{2}\rho U^2 s\pi B_1 \qquad (7.8.6)$$

and
$$D_i = \tfrac{1}{2}\rho U^2 \tfrac{1}{4}\pi \sum_{n=1}^{\infty} n B_n^2. \qquad (7.8.7)$$

These alternative forms for L and D_i reveal the interesting result that, for a given total lift on a wing of given span, the induced drag is a minimum when the circulation is so distributed that

$$B_n = 0 \quad \text{for} \quad n > 1,$$

that is, when
$$K = UB_1 \sin \theta = UB_1 \left(1 - \frac{z^2}{s^2}\right)^{\frac{1}{2}}. \qquad (7.8.8)$$

The corresponding induced velocity v has the uniform value $-UB_1/4s$ over the whole span, and the induced drag is

$$D_i = \tfrac{1}{2}\rho U^2 \tfrac{1}{4}\pi B_1^2 = \frac{L^2}{2\pi s^2 \rho U^2}, \qquad (7.8.9)$$

in conformity with the general relation (7.8.1) noted earlier.

The 'elliptic loading' of the wing represented by (7.8.8) can be achieved in various ways, by combining suitable distributions of chord, wing section shape and angle of incidence over the span. One simple way, which also has the advantage that the loading remains elliptic as the inclination of the whole wing to the approaching stream changes, is to make a, α and β uniform over the span and to give the wing an elliptic planform so that

$$c = c_0\left(1 - \frac{z^2}{s^2}\right)^{\frac{1}{2}}.$$

(Equally the wing could consist of two semi-ellipses with different minor axes.) In this case comparison of (7.8.3) and (7.8.8) shows that

$$B_1 = (\alpha + \beta)\frac{\frac{1}{2}ac_0}{1 + (ac_0/8s)}. \tag{7.8.10}$$

Thus here the 'downwash' due to the trailing sheet vortex makes the effective angle of incidence, relative to the no-lift attitude, equal to a fraction $\{1 + (ac_0/8s)\}^{-1}$ of its apparent value, over the whole wing; and the lift is a similar fraction of the value it would have if each section of the wing acted as an isolated two-dimensional aerofoil.

Since $a \approx 2\pi$ and a necessary condition for validity of the analysis is $c_0/s \ll 1$, the factor by which the angle of incidence is changed by the down-wash differs from unity by a small quantity only. This illustrates the fact that the lifting-line theory is essentially concerned with a perturbation of the flow pattern associated with a wing of infinite span. 'Lifting-surface' theories which take some account of the distributions of vertical force on the wing with respect to both the spanwise and chordwise co-ordinates have also been developed,[†] using a suitable combination of the ideas of this section and the thin-aerofoil analysis of §6.9. It is possible in this context to establish a process of systematic approximation to the distribution of force on a wing of large aspect ratio, of which the above lifting-line analysis represents the first stage (with wholly two-dimensional flow past a wing of infinite span as the 'zeroth' stage).[‡] This systematic procedure shows that since $v(z)/U$ is a perturbation term in (7.8.3) it may be evaluated, to the degree of approximation to which our lifting-line analysis is valid, by using in (7.8.2) the unperturbed value of $K(z)$. That is to say, it is consistent with the lifting-line analysis to approximate to the solution of (7.8.2) and (7.8.3) by

$$K(z) = K_0(z) - \frac{ac}{8\pi}\int_{-s}^{s}\frac{dK_0(z')}{dz'}\frac{dz'}{z - z'}, \tag{7.8.11}$$

where $$K_0(z) = \frac{1}{2}acU(\alpha + \beta).$$

In the case of a wing of elliptic plan-form and uniform values of a, α and β over the span, this amounts to recognizing that (7.8.10) and dependent relations are correct only to the first order in $ac_0/8s$.

† An account of these developments will be found in *Incompressible Aerodynamics*, edited by B. Thwaites (Oxford University Press, 1960).
‡ See *Perturbation Methods in Fluid Mechanics*, by M. D. Van Dyke (Academic Press, 1964).

The trailing vortex system far downstream

Calculations of the kind just described rest on the assumption that near the wing, say within a distance comparable with the span, the trailing vortices are straight and parallel to the flight direction and so form a plane sheet vortex. In reality the induced velocity of the trailing vortices themselves causes some lateral movement of the vortex-lines. The ratio of the vertical component of the induced velocity due to the trailing vortices to the free-stream speed is seen from (7.8.2) and (7.8.4) to be of order

$$L/\rho U^2 s^2, \quad \text{or} \quad C_L c/s$$

if we define the lift coefficient for the wing as a whole as

$$C_L = \frac{L}{\frac{1}{2}\rho U^2 \times (\text{area of wing})}.$$

Thus the trailing vortices will extend approximately straight downstream from the trailing edge of the wing, as assumed, if $C_L c/s \ll 1$. However, in a time t, corresponding to a distance Ut downstream from the wing, the induced velocity field will have moved the vortex-lines through a distance vt in the lateral plane and so will have distorted the initially plane sheet vortex appreciably when t is of order s/v, that is, at a distance downstream of order

$$sU/v, \quad \text{or} \quad s^2/cC_L.$$

When the initially plane trailing sheet vortex from a wing of large aspect ratio does deform, under the influence of its own induced velocity field, it does so in a characteristic way. Since the change in the shape of the sheet over a distance downstream comparable with the span is slight, the cross-sectional forms of the sheet vortex at different distances d downstream from the wing are approximately the same as those in a purely two-dimensional flow field at different intervals of time t after the cross-section of the sheet is a straight line, with $d = Ut$. (The effect of the bound vorticity at the wing is being ignored here, on the grounds that it can influence the trailing vortices only near the wing.) This two-dimensional problem can be tackled numerically, when the strength of the trailing sheet vortex is given as a function of span-wise position.

Figure 7.8.5 shows the result of calculations in the case of the trailing sheet vortex shed by a wing on which the circulation has the elliptic distribution (7.8.8). The vortex strength (or circulation) per unit length of the sheet is here

$$\frac{dK}{dz} = -\frac{UB_1 z}{s(s^2 - z^2)^{\frac{1}{2}}}, \tag{7.8.12}$$

showing a concentration of the vorticity near the two edges of the sheet vortex. For reasons connected with the numerical method, this continuous distribution of vortex strength was replaced by point vortices of equal strength and appropriately chosen number density on the line representing the sheet. While the sheet is straight, the y-component of induced velocity at the sheet is downward and uniform across the sheet, as already noted for this case of an elliptically loaded wing.† However, the two end points are

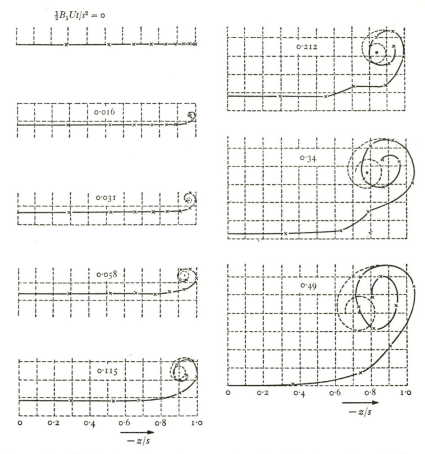

Figure 7.8.5. Calculated positions of a group of ten identical point vortices, distributed along a straight line at $t = 0$ in such a way as to represent approximately (one half of) the trailing sheet vortex from an elliptically loaded wing (see (7.8.12)). The broken curves are obtained from an analytical description of the flow near the edge of the sheet vortex (located at the dot), where the numerical integration is inaccurate. (From Westwater 1936.)

singular, as is evident from (7.8.2) and (7.8.12), and there the vertical component of induced velocity changes discontinuously to an infinite positive value (which of course can be reproduced numerically only approximately). The two ends of the sheet consequently move upwards, and this new position

† The induced velocity was found earlier to be uniform at the wing, relative to which the trailing vortices are semi-infinite, and it is clearly also uniform far downstream (if the sheet is still straight) although twice as large.

of the vorticity leads to further deformation of the sheet, which spreads inward from the two ends. The end of the sheet remains a singularity, and always moves at right angles to the local tangent, thereby producing a spiral form with an infinite number of turns. (The photographs in figure 5.10.5, plate 8, reveal similar spiral development of the edge of the sheet vortex shed from the salient edge of a body soon after it begins to move.)

It is clear from the figures representing the sheet vortex in the later stages that all the vorticity is being drawn into the two growing spirals, and that asymptotically the vorticity on each side will be distributed with approximately circular symmetry about a point at distance about $0 \cdot 8s$ from the central (x, y)-plane. (The first integral moment of the vorticity on one side of this plane should remain constant—see §7.3—and initially has a value corresponding to a centroid at distance $0 \cdot 79s$ from the central plane.) Far downstream from the wing we thus have a kind of 'vortex pair', intermediate in character between the vortex pairs represented in figure 7.3.3, where the vorticity is concentrated at two points, and figure 7.3.4, where the vorticity of each sign is spread over a semi-circle. The downward impulse required to generate this vortex pair is provided by the wing, and the kinetic energy of the motion in a lateral plane associated with the vortex pair is related to the induced drag on the wing in the manner already explained. The two trailing vortices formed far downstream by the 'rolling-up' of the sheet vortex are sometimes described in the literature as concentrated vortices, although for wings of large aspect ratio the degree of concentration of the vorticity of each sign far downstream cannot be very different from that in the initial plane sheet vortex in view of the need for constancy of the kinetic energy.

Observations of the flow field downstream from wings of different shape show that, although the rate of rolling-up of the trailing sheet vortex may depend on the wing shape and attitude, far enough downstream a pair of trailing vortices is a typical and dominant feature.

Highly swept wings

Many modern aircraft are designed to operate at speeds approaching the speed of sound in the fluid. If undesirable effects of shock-wave formation due to the compressibility of the air are to be kept within tolerable limits, the aircraft should have a shape which avoids low values of the minimum pressure in the fluid, i.e. high values of the maximum velocity of the fluid relative to the aircraft. 'Two-dimensional' or cylindrical bodies moving normal to the generators are worse in this respect than 'three-dimensional' bodies, for obvious reasons; for instance, in irrotational flow past a circular cylinder (with zero circulation) and past a sphere the ratio of the maximum fluid speed to the free-stream speed has the values $2 \cdot 0$ and $1 \cdot 5$ respectively. Consequently it is preferable to sweep back the wing leading edge on the two sides of the aircraft, to an extent which depends on the intended operating speed. In the case of aircraft flying at supersonic speed there is inevitably a shock wave lying ahead of the nose of the fuselage and extending downstream as a cone whose angle decreases as the flight speed increases. Here it is undesirable that the wings should protrude beyond the region enclosed by this nose shock wave, because that would lead to further shock-wave formation, and again swept-back wings are required. For flight at Mach numbers of two or greater, the wing may look more like a dart than the conventional wing of large aspect ratio and straight leading edge considered above. The trailing vortex system will not then have the form sketched in figure 7.8.4, and the classical lifting-line theory will not be applicable since the chordwise variation of the downwash induced by the trailing vortices is clearly relevant.

The extensive subject of aerodynamic design lies outside the scope of this book, but one or two features of the trailing vortex system of highly swept wings may be noted. The shape of a wing is normally chosen so that at the cruising speed of the aircraft adequate lift is generated at such a small angle of incidence that boundary-layer separation does not occur upstream of the trailing edge. In the case of highly swept wings there is a strong tendency for fluid to flow round the extensive side edges of the wing, and the range of angles of incidence for which boundary-layer separation at these side edges can be avoided is very small. At lower flight speeds, as in the condition for landing, a higher angle of incidence is required, and it may happen then that there is separation from the whole of each side edge. This presents a situation which is interesting as an example of flow past bodies which are elongated in a direction nearly parallel to the free stream.

A wing planform in the shape of an isosceles triangle has been the subject of many investigations in wind-tunnels. Figure 7.8.6 (plate 22) shows the form of streamlines passing close to the upper (or suction) surface of such a 'delta' wing at an angle of incidence which although geometrically quite small is large enough to cause separation at the two side edges (which are here also 'leading' edges). A sheet vortex in which the vorticity is mainly in the stream direction here is shed from the whole of each side edge, beginning at the apex of the triangle, and it is evident that rolling-up of the sheet vortex into a spiral form occurs long before the sheet has been carried downstream as far as the trailing edge. (See also figure 7.5.7, plate 22, which is concerned with the behaviour of the stronger rolled-up vortices produced at larger angles of incidence.)

Now when an elongated body is placed with its long dimension nearly parallel to a stream of fluid, a qualitative picture of the flow field can be obtained by superposing the effects of two streams, one parallel to the body axis and one normal to it. If the body presents a bluff shape to the transverse stream, separation will occur, in different stages of development at different distances from the front of the body. The situation is perhaps easier to think about in terms of a simple cylindrical body such as a flat long rectangular plate, with a central plane of symmetry to the flow (figure 7.8.7a). We

visualize the changes which take place in the fluid velocity at points in a
plane normal to the plate centre-line as this plane moves with the free-stream
velocity. Since gradients normal to this plane are small, the flow pattern in
this plane changes with time in approximately the same way as the two-
dimensional flow due to a flat plate which is suddenly made to move normal
to its face at constant speed through fluid initially at rest. Sheet vortices are
shed from the two sides of the plate, and roll up in the manner of figure
5.10.5 (plate 8) as they are carried downstream, giving for the steady flow
past the lifting rectangular plate something like the sketch in figure 7.8.7a.

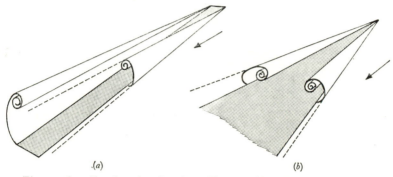

.(a)   (b)

Figure 7.8.7. Sketches showing the rolling-up of the sheet vortex shed
from the sides of a flat elongated lifting surface.

It is known that the eddies formed behind a flat plate moving in the direction
of its normal grow in size in the initial period after the plate begins to move
but eventually settle down to a steady average size, determined by the plate
width, the flow further downstream being either periodic or irregular and
oscillating. The sheet vortices behind the lifting rectangular plate are therefore expected to have the kind of development shown in figure 7.8.7a only if
the plate is not too long.

On the other hand, in the case of a lifting triangle of small vertex angle
the width of the flat plate in the associated two-dimensional time-dependent
problem must be supposed to be increasing linearly with time, and the
orderly rolling-up of the sheet vortex sketched in figure 7.8.7b can continue
however long the triangle may be. Furthermore there evidently exists here
the interesting possibility that the growth in size of the sheet vortices in a
transverse plane keeps pace exactly with the increase in width of the triangular plate and that the pattern of flow in a plane transverse to the centre-
line of the plate has a similar form at all distances from the vertex. The
velocity then has a uniform value on any radial line from the vertex of the
triangular plate, and the flow field is said to have *conical* similarity. The
assumption of a conical flow field due to a delta wing of small vertex angle
and moderate angle of incidence has been made the basis of a number of
recent developments in wing theory.

APPENDIX 1

Measured values of some physical properties of common fluids

$$(\text{1 atm} = 1\cdot013 \times 10^6\,\text{dyn/cm}^2$$
$$\text{1 joule} = 0\cdot2389\,\text{cal} = 10^7\,\text{gm cm}^2/\text{sec}^2)$$

(a) Dry air at a pressure of one atmosphere

Specific heats at 15 °C: $c_p = 1\cdot012$ joule/gm degC
$$c_v = 0\cdot718\,\text{joule/gm degC}$$
$$\gamma = 1\cdot401$$

Coefficient of compressibility (isothermal) $0\cdot987 \times 10^{-6}\,\text{cm}^2/\text{dyn}$, or 1/atm

Coefficient of thermal expansion at 15 °C $3\cdot48 \times 10^{-3}/\text{degC}$

Velocity of sound waves at 15 °C 340·6 m/sec

Root-mean-square velocity of molecules at 15 °C 498 m/sec

Coefficient of diffusion of water vapour in air at 15 °C 0·25 cm²/sec

Coefficient of self-diffusion of nitrogen or oxygen at 15 °C 0·18 cm²/sec

Temperature T degC	Density ρ gm/cm³	Viscosity μ gm/cm sec	Kinematic viscosity ν cm²/sec	Thermal conductivity[†] k_H joule/ cm sec degC	Thermal diffusivity κ_H cm²/sec	Prandtl number ν/κ_H
−100	$2\cdot04 \times 10^{-3}$	$1\cdot16 \times 10^{-4}$	0·057	$1\cdot58 \times 10^{-4}$	0·076	0·75
−50	1·582	1·45	0·092			
0	1·293	1·71	0·132	2·41	0·184	0·72
10	1·247	1·76	0·141	2·48	0·196	0·72
15	1·225	1·78	0·145	2·51	0·202	0·72
20	1·205	1·81	0·150	2·54	0·208	0·72
30	1·165	1·86	0·160			
40	1·127	1·90	0·169			
60	1·060	2·00	0·188			
80	1·000	2·09	0·209			
100	0·946	2·18	0·230	3·17	0·328	0·70
200	0·746	2·58	0·346			
300	0·616	2·95	0·481			
500	0·456	3·58	0·785			
1 000	0·277	4·82	1·74	7·6	2·71	0·64

[†] The last decimal place of the numbers in this column is uncertain.

Composition (by weight) of dry air at sea-level

N_2	O_2	A	CO_2
0·7552	0·2315	0·0128	0·0005

(b) *The Standard Atmosphere: average values of pressure, etc., in temperate latitudes, accepted by international agreement*

Height above sea-level m	Pressure dyn/cm²	Density g/cm³	Temperature degC
0	$1 \cdot 013 \times 10^6$	$1 \cdot 226 \times 10^{-3}$	15·0
500	0·955	1·168	11·7
1 000	0·899	1·112	8·5
1 500	0·845	1·059	5·2
2 000	0·795	1·007	2·0
3 000	0·701	0·910	−4·5
4 000	0·616	0·820	−11·0
5 000	0·540	0·736	−17·5
6 000	0·472	0·660	−24·0
8 000	0·356	0·525	−37·0
10 000	0·264	0·413	−50·0
12 000	0·193	0·311	−56·5
14 000	0·141	0·227	−56·5
16 000	0·103	0·165	−56·5
18 000	0·075	0·121	−56·5

(c) *Pure water*

Coefficient of compressibility (isothermal) $4 \cdot 9 \times 10^{-11}$ cm²/dyn, or $5 \cdot 0 \times 10^{-5}$/atm

Latent heat of melting of ice   334 joule/gm

Density of ice   0·92 gm/cm³

Coefficient of diffusion of NaCl in water $1 \cdot 1 \times 10^{-5}$ cm²/sec at 15 °C (any concentration)

Coefficient of diffusion of $KMnO_4$ in water $1 \cdot 4 \times 10^{-5}$ cm²/sec at 15 °C at zero concentration

Percentage (by weight) of anhydrous NaCl in solution at 15 °C	0	5	10	15	20	25
Density of solution gm/cm³	0·999	1·035	1·072	1·110	1·149	1·190
Specific heat of solution at constant pressure joule/gm degC	4·19	4·16	4·13	4·10	4·07	4·04

(c) *Pure water* (*continued*)

Temperature T degC	Density ρ gm/cm³	Coefficient of thermal expansion β deg⁻¹C	Specific heats		Vapour pressure dyn/cm²	Latent heat of vaporization joule/gm	Volume of air in 1 cm³ of saturated water (reduced to 0°C) cm³	Percentage (by weight) of anhydrous NaCl in saturated solution	Velocity of sound waves cm/sec
			c_p joule/ gm degC	$c_p - c_v$ (calculated from (1.8.2)) joule/ gm degC					
0	0·9999	−0·6 × 10⁻⁴	4·217	0·002	6·1 × 10³	2·501 × 10³	0·0292	26·4	1·407 × 10⁵
5	1·0000	+0·1	4·202	0	8·7	2·489	0·0257		
10	0·9997	0·9	4·192	0·005	12·3	2·477	0·0228		1·445
15	0·9991	1·5	4·186	0·013	17·0	2·465	0·0205		
20	0·9982	2·1	4·182	0·024	23·3	2·454	0·0187	26·5	1·484
25	0·9971	2·6	4·179	0·041	31·6	2·442	0·0171		
30	0·9957	3·0	4·178	0·06	42·3	2·430	0·0157		1·510
35	0·9941	3·4	4·178	0·07	56				
40	0·9923	3·8	4·178	0·09	74	2·406		26·8	1·528
50	0·9881	4·5	4·180	0·13	123	2·382			1·544
60	0·9832	5·1	4·184	0·18	199	2·357		27·2	1·556
70	0·9778	5·7	4·189	0·23	311	2·333			1·561
80	0·9718	6·2	4·196	0·29	473	2·308		27·7	1·557
90	0·9653	6·7	4·205	0·34	701	2·283			
100	0·9584	7·1	4·216	0·40	1013	2·257		28·5	

(c) Pure water (continued)

Temperature T degC	Viscosity μ gm/cm sec	Kinematic viscosity ν cm²/sec	Thermal conductivity k_H joule/ cm sec degC	Thermal diffusivity κ_H cm²/sec	Prandtl number ν/κ_H
0	1.787×10^{-2}	1.787×10^{-2}	5.6×10^{-3}	1.33×10^{-3}	13.4
5	1.514	1.514			
10	1.304	1.304	5.8	1.38	9.5
15	1.137	1.138	5.9	1.40	8.1
20	1.002	1.004	5.9	1.42	7.1
25	0.891	0.894			
30	0.798	0.802	6.1	1.46	5.5
35	0.720	0.725			
40	0.654	0.659	6.3	1.52	4.3
50	0.548	0.554			
60	0.467	0.475	6.5	1.58	3.0
70	0.405	0.414			
80	0.355	0.366	6.7	1.64	2.2
90	0.316	0.327			
100	0.283	0.295	6.7	1.66	1.8

(d) Diffusivities for momentum and heat at 15 °C and 1 atm

	Air	Water	Mercury	Ethyl alcohol	Carbon tetra-chloride	Olive oil	Glycerine
ρ gm/cm³	0.001225	0.999	13.61	0.79	1.60	0.918	1.26
c_p joule/gm degC	1.012	4.19	0.140	2.34	0.84	2.01	2.34
μ gm/cm sec	0.000178	0.0114	0.0158	0.0134	0.0104	0.99	23.3
ν cm²/sec	0.145	0.0114	0.00116	0.0170	0.0065	1.08	18.5
k_H joule/cm sec degC	0.000253	0.0059	0.080	0.00183	0.00113	0.00169	0.0029
κ_H cm²/sec	0.202	0.00140	0.042	0.00099	0.00084	0.00092	0.00098
ν/κ_H	0.72	8.1	0.028	17.2	7.7	117	189

(e) Surface tension between two fluids

Surface tension at 20 °C (dyn/cm)

	Water	Mercury	Ethyl alcohol	Carbon tetrachloride	Olive oil	Benzene	Glycerine
Air	72.8	487	22	27		29	63
Water		375	<0	45	20	35	<0

Temperature degC	0	10	15	20	25	30	40	50	60	80	100
Surface tension between air and water dyn/cm	75.7	74.2	73.5	72.8	72.0	71.2	69.6	67.9	66.2	62.6	58.8

APPENDIX 2

Expressions for some common vector differential quantities in orthogonal curvilinear co-ordinate systems

ξ_1, ξ_2, ξ_3 is a system of orthogonal curvilinear co-ordinates, and the unit vectors **a**, **b**, **c** are parallel to the co-ordinate lines and in the directions of increase of ξ_1, ξ_2, ξ_3 respectively. The change in the position vector **x** corresponding to increments in ξ_1, ξ_2, and ξ_3 can then be written as

$$\delta\mathbf{x} = h_1\,\delta\xi_1\,\mathbf{a} + h_2\,\delta\xi_2\,\mathbf{b} + h_3\,\delta\xi_3\,\mathbf{c}.$$

a, **b**, **c** and the positive scale factors h_1, h_2, h_3 are functions of the co-ordinates.

The fact that the three families of co-ordinate lines form an orthogonal system provides useful expressions for the derivatives of **a**, **b**, and **c**. We have

$$\frac{\partial\mathbf{x}}{\partial\xi_1}\cdot\frac{\partial\mathbf{x}}{\partial\xi_2} = 0,$$

with two other similar relations, and since

$$\frac{\partial}{\partial\xi_3}\left(\frac{\partial\mathbf{x}}{\partial\xi_1}\cdot\frac{\partial\mathbf{x}}{\partial\xi_2}\right) = \frac{\partial}{\partial\xi_1}\left(\frac{\partial\mathbf{x}}{\partial\xi_3}\right)\cdot\frac{\partial\mathbf{x}}{\partial\xi_2} + \frac{\partial\mathbf{x}}{\partial\xi_1}\cdot\frac{\partial}{\partial\xi_2}\left(\frac{\partial\mathbf{x}}{\partial\xi_3}\right)$$

$$= -2\,\frac{\partial\mathbf{x}}{\partial\xi_3}\cdot\frac{\partial^2\mathbf{x}}{\partial\xi_1\,\partial\xi_2},$$

we see that
$$\frac{\partial^2\mathbf{x}}{\partial\xi_1\,\partial\xi_2}, \quad = \frac{\partial(h_2\,\mathbf{b})}{\partial\xi_1} \quad \text{or} \quad \frac{\partial(h_1\,\mathbf{a})}{\partial\xi_2},$$

is a vector normal to **c**. It follows that

$$\frac{\partial\mathbf{a}}{\partial\xi_2} = \frac{1}{h_1}\frac{\partial h_2}{\partial\xi_1}\mathbf{b}, \quad \frac{\partial\mathbf{b}}{\partial\xi_1} = \frac{1}{h_2}\frac{\partial h_1}{\partial\xi_2}\mathbf{a},$$

with four other similar relations. Then

$$\frac{\partial\mathbf{a}}{\partial\xi_1} = \frac{\partial(\mathbf{b}\times\mathbf{c})}{\partial\xi_1} = -\frac{1}{h_2}\frac{\partial h_1}{\partial\xi_2}\mathbf{b} - \frac{1}{h_3}\frac{\partial h_1}{\partial\xi_3}\mathbf{c},$$

with two other similar relations.

The vector *gradient* of a scalar function V is

$$\text{grad}\,V, \quad \text{or} \quad \nabla V, \quad = \left(\frac{\mathbf{a}}{h_1}\frac{\partial}{\partial\xi_1} + \frac{\mathbf{b}}{h_2}\frac{\partial}{\partial\xi_2} + \frac{\mathbf{c}}{h_3}\frac{\partial}{\partial\xi_3}\right)V.$$

The gradient in a direction **n** is obtained from the operator $\mathbf{n}\cdot\nabla$, which may act on either a scalar or a vector. To find the components of $\mathbf{n}\cdot\nabla\mathbf{F}$, where

$$\mathbf{F} = F_1\mathbf{a} + F_2\mathbf{b} + F_3\mathbf{c},$$

we must allow for the dependence of both F_1, F_2, F_3 and the unit vectors **a**, **b**, **c** on position. It follows from the above relations that

$$\mathbf{n}.\nabla\mathbf{F} = \mathbf{a}\left\{\mathbf{n}.\nabla F_1 + \frac{F_2}{h_1 h_2}\left(n_1\frac{\partial h_1}{\partial \xi_2} - n_2\frac{\partial h_2}{\partial \xi_1}\right) + \frac{F_3}{h_3 h_1}\left(n_1\frac{\partial h_1}{\partial \xi_3} - n_3\frac{\partial h_3}{\partial \xi_1}\right)\right\}$$
$$+ \mathbf{b}\{\quad\} + \mathbf{c}\{\quad\},$$

where n_1, n_2, n_3 are the components of **n** in the directions **a**, **b**, **c**.

The *divergence* and *curl* operators act only on a vector, and

$$\operatorname{div}\mathbf{F}, \quad \text{or } \nabla.\mathbf{F}, \quad = \frac{\mathbf{a}}{h_1}.\frac{\partial\mathbf{F}}{\partial\xi_1} + \frac{\mathbf{b}}{h_2}.\frac{\partial\mathbf{F}}{\partial\xi_2} + \frac{\mathbf{c}}{h_3}.\frac{\partial\mathbf{F}}{\partial\xi_3},$$

$$\operatorname{curl}\mathbf{F}, \quad \text{or } \nabla\times\mathbf{F}, \quad = \frac{\mathbf{a}}{h_1}\times\frac{\partial\mathbf{F}}{\partial\xi_1} + \frac{\mathbf{b}}{h_2}\times\frac{\partial\mathbf{F}}{\partial\xi_2} + \frac{\mathbf{c}}{h_3}\times\frac{\partial\mathbf{F}}{\partial\xi_3}.$$

By making use of the expressions for derivatives of **a**, **b** and **c**, we find

$$\nabla.\mathbf{F} = \frac{1}{h_1 h_2 h_3}\left\{\frac{\partial(h_2 h_3 F_1)}{\partial\xi_1} + \frac{\partial(h_3 h_1 F_2)}{\partial\xi_2} + \frac{\partial(h_1 h_2 F_3)}{\partial\xi_3}\right\};$$

this can also be regarded as the result of applying the 'divergence theorem' to the small parallelepiped whose edges are displacements along co-ordinate lines corresponding to the increments $\delta\xi_1$, $\delta\xi_2$, $\delta\xi_3$. Likewise we find

$$\nabla\times\mathbf{F} = \frac{\mathbf{a}}{h_2 h_3}\left\{\frac{\partial(h_3 F_3)}{\partial\xi_2} - \frac{\partial(h_2 F_2)}{\partial\xi_3}\right\} + \frac{\mathbf{b}}{h_3 h_1}\left\{\frac{\partial(h_1 F_1)}{\partial\xi_3} - \frac{\partial(h_3 F_3)}{\partial\xi_1}\right\}$$
$$+ \frac{\mathbf{c}}{h_1 h_2}\left\{\frac{\partial(h_2 F_2)}{\partial\xi_1} - \frac{\partial(h_1 F_1)}{\partial\xi_2}\right\},$$

$$\text{or} \quad \frac{1}{h_1 h_2 h_3}\begin{vmatrix} h_1\mathbf{a} & h_2\mathbf{b} & h_3\mathbf{c} \\ \dfrac{\partial}{\partial\xi_1} & \dfrac{\partial}{\partial\xi_2} & \dfrac{\partial}{\partial\xi_3} \\ h_1 F_1 & h_2 F_2 & h_3 F_3 \end{vmatrix},$$

which can also be regarded as following from the application of Stokes's theorem in turn to three orthogonal faces of the same parallelepiped.

The divergence of the gradient gives the *Laplacian* operator, which may act on either a scalar or a vector.

$$\nabla.\nabla V, \quad \text{or } \nabla^2 V, \quad = \frac{1}{h_1 h_2 h_3}\left\{\frac{\partial}{\partial\xi_1}\left(\frac{h_2 h_3}{h_1}\frac{\partial V}{\partial\xi_1}\right)\right.$$
$$\left. + \frac{\partial}{\partial\xi_2}\left(\frac{h_3 h_1}{h_2}\frac{\partial V}{\partial\xi_2}\right) + \frac{\partial}{\partial\xi_3}\left(\frac{h_1 h_2}{h_3}\frac{\partial V}{\partial\xi_3}\right)\right\}.$$

The components of $\nabla^2\mathbf{F}$ may be calculated by replacing V in this formula by $\mathbf{F}, = F_1\mathbf{a} + F_2\mathbf{b} + F_3\mathbf{c}$, and using the expressions for derivatives of **a**, **b** and **c**, but the result is too complicated to be useful. It is usually more convenient,

when finding the components of $\nabla^2 \mathbf{F}$ in a particular co-ordinate system, to use the identity
$$\nabla^2 \mathbf{F} = \nabla(\nabla . \mathbf{F}) - \nabla \times (\nabla \times \mathbf{F})$$
and the above expressions for grad, div and curl.

Consider now the components of the rate-of-strain tensor expressed in terms of velocity components and derivatives relative to the curvilinear system. The gradient, in the direction $\mathbf{n}$, of the component of velocity $\mathbf{u}$ in the fixed direction $\mathbf{m}$ is

$$\mathbf{n} . \nabla(\mathbf{m} . \mathbf{u}), \quad = \mathbf{m} . (\mathbf{n} . \nabla \mathbf{u}).$$

Diagonal elements of the rate-of-strain tensor represent rates of extension, obtained by putting $\mathbf{m} = \mathbf{n}$, and the non-diagonal elements involve velocity gradients for which $\mathbf{m}$ and $\mathbf{n}$ are orthogonal. We see then, from the above formula for $\mathbf{n} . \nabla \mathbf{F}$, that the components of the rate-of-strain tensor relative to Cartesian axes locally parallel to $\mathbf{a}$, $\mathbf{b}$ and $\mathbf{c}$ (to which the suffixes 1, 2, 3 refer, respectively) are

$$e_{11} = \mathbf{a} . (\mathbf{a} . \nabla \mathbf{u}) = \frac{1}{h_1} \frac{\partial u_1}{\partial \xi_1} + \frac{u_2}{h_1 h_2} \frac{\partial h_1}{\partial \xi_2} + \frac{u_3}{h_3 h_1} \frac{\partial h_1}{\partial \xi_3},$$

$$e_{23} = \tfrac{1}{2} \mathbf{b} . (\mathbf{c} . \nabla \mathbf{u}) + \tfrac{1}{2} \mathbf{c} . (\mathbf{b} . \nabla \mathbf{u}) = \frac{h_3}{2h_2} \frac{\partial}{\partial \xi_2} \left(\frac{u_3}{h_3} \right) + \frac{h_2}{2h_3} \frac{\partial}{\partial \xi_3} \left(\frac{u_2}{h_2} \right),$$

with four other expressions obtained by cyclic interchange of suffixes. The components of the stress tensor σ_{ij} can be obtained from those of rate of strain, using the relation (for an incompressible fluid)

$$\sigma_{ij} = -p \, \delta_{ij} + 2\mu e_{ij}.$$

The components of all terms in the equation of motion of a fluid in the directions $\mathbf{a}$, $\mathbf{b}$, $\mathbf{c}$ may now be found by simple substitution in the appropriate expressions above. The components of the term $\mathbf{u} . \nabla \mathbf{u}$ in the acceleration are obtained from the expression for $\mathbf{n} . \nabla \mathbf{F}$.

Applications to some particular co-ordinate systems are as follows.

Spherical polar co-ordinates

To the co-ordinates $\xi_1 = r$, $\xi_2 = \theta$, $\xi_3 = \phi$ (where ϕ is the azimuthal angle about the axis $\theta = 0$) there correspond the scale factors

$$h_1 = 1, \quad h_2 = r, \quad h_3 = r \sin \theta.$$

Then
$$\frac{\partial \mathbf{a}}{\partial r} = 0, \quad \frac{\partial \mathbf{a}}{\partial \theta} = \mathbf{b}, \quad \frac{\partial \mathbf{a}}{\partial \phi} = \sin \theta \, \mathbf{c},$$

$$\frac{\partial \mathbf{b}}{\partial r} = 0, \quad \frac{\partial \mathbf{b}}{\partial \theta} = -\mathbf{a}, \quad \frac{\partial \mathbf{b}}{\partial \phi} = \cos \theta \, \mathbf{c},$$

$$\frac{\partial \mathbf{c}}{\partial r} = 0, \quad \frac{\partial \mathbf{c}}{\partial \theta} = 0, \quad \frac{\partial \mathbf{c}}{\partial \phi} = -\sin \theta \, \mathbf{a} - \cos \theta \, \mathbf{b}.$$

$$\nabla V = \mathbf{a} \frac{\partial V}{\partial r} + \frac{\mathbf{b}}{r} \frac{\partial V}{\partial \theta} + \frac{\mathbf{c}}{r \sin \theta} \frac{\partial V}{\partial \phi},$$

$$\mathbf{n}.\nabla\mathbf{F} = \mathbf{a}\left(\mathbf{n}.\nabla F_r - \frac{n_\theta F_\theta}{r} - \frac{n_\phi F_\phi}{r}\right) + \mathbf{b}\left(\mathbf{n}.\nabla F_\theta - \frac{n_\phi F_\phi}{r}\cot\theta + \frac{n_\theta F_r}{r}\right)$$

$$+ \mathbf{c}\left(\mathbf{n}.\nabla F_\phi + \frac{n_\phi F_r}{r} + \frac{n_\phi F_\theta}{r}\cot\theta\right),$$

$$\nabla.\mathbf{F} = \frac{1}{r^2}\frac{\partial(r^2 F_r)}{\partial r} + \frac{1}{r\sin\theta}\frac{\partial(\sin\theta\, F_\theta)}{\partial\theta} + \frac{1}{r\sin\theta}\frac{\partial F_\phi}{\partial\phi},$$

$$\nabla\times\mathbf{F} = \frac{\mathbf{a}}{r\sin\theta}\left\{\frac{\partial(F_\phi\sin\theta)}{\partial\theta} - \frac{\partial F_\theta}{\partial\phi}\right\} + \frac{\mathbf{b}}{r}\left\{\frac{1}{\sin\theta}\frac{\partial F_r}{\partial\phi} - \frac{\partial(rF_\phi)}{\partial r}\right\} + \frac{\mathbf{c}}{r}\left\{\frac{\partial(rF_\theta)}{\partial r} - \frac{\partial F_r}{\partial\theta}\right\},$$

$$\nabla^2 V = \frac{1}{r^2}\frac{\partial}{\partial r}\left(r^2\frac{\partial V}{\partial r}\right) + \frac{1}{r^2\sin\theta}\frac{\partial}{\partial\theta}\left(\sin\theta\frac{\partial V}{\partial\theta}\right) + \frac{1}{r^2\sin^2\theta}\frac{\partial^2 V}{\partial\phi^2},$$

$$\nabla^2\mathbf{F} = \mathbf{a}\left\{\nabla^2 F_r - \frac{2F_r}{r^2} - \frac{2}{r^2\sin\theta}\frac{\partial(F_\theta\sin\theta)}{\partial\theta} - \frac{2}{r^2\sin\theta}\frac{\partial F_\phi}{\partial\phi}\right\}$$

$$+ \mathbf{b}\left\{\nabla^2 F_\theta + \frac{2}{r^2}\frac{\partial F_r}{\partial\theta} - \frac{F_\theta}{r^2\sin^2\theta} - \frac{2\cos\theta}{r^2\sin^2\theta}\frac{\partial F_\phi}{\partial\phi}\right\}$$

$$+ \mathbf{c}\left\{\nabla^2 F_\phi + \frac{2}{r^2\sin\theta}\frac{\partial F_r}{\partial\phi} + \frac{2\cos\theta}{r^2\sin^2\theta}\frac{\partial F_\theta}{\partial\phi} - \frac{F_\phi}{r^2\sin^2\theta}\right\}.$$

Rate-of-strain tensor:

$$e_{rr} = \frac{\partial u_r}{\partial r}, \quad e_{\theta\theta} = \frac{1}{r}\frac{\partial u_\theta}{\partial\theta} + \frac{u_r}{r}, \quad e_{\phi\phi} = \frac{1}{r\sin\theta}\frac{\partial u_\phi}{\partial\phi} + \frac{u_r}{r} + \frac{u_\theta\cot\theta}{r},$$

$$e_{\theta\phi} = \frac{\sin\theta}{2r}\frac{\partial}{\partial\theta}\left(\frac{u_\phi}{\sin\theta}\right) + \frac{1}{2r\sin\theta}\frac{\partial u_\theta}{\partial\phi}, \quad e_{\phi r} = \frac{1}{2r\sin\theta}\frac{\partial u_r}{\partial\phi} + \frac{r}{2}\frac{\partial}{\partial r}\left(\frac{u_\phi}{r}\right),$$

$$e_{r\theta} = \frac{r}{2}\frac{\partial}{\partial r}\left(\frac{u_\theta}{r}\right) + \frac{1}{2r}\frac{\partial u_r}{\partial\theta}.$$

Equation of motion for an incompressible fluid, with no body force:

$$\frac{\partial u_r}{\partial t} + \mathbf{u}.\nabla u_r - \frac{u_\theta^2}{r} - \frac{u_\phi^2}{r} = -\frac{1}{\rho}\frac{\partial p}{\partial r}$$

$$+ \nu\left\{\nabla^2 u_r - \frac{2u_r}{r^2} - \frac{2}{r^2\sin\theta}\frac{\partial(u_\theta\sin\theta)}{\partial\theta} - \frac{2}{r^2\sin\theta}\frac{\partial u_\phi}{\partial\phi}\right\},$$

$$\frac{\partial u_\theta}{\partial t} + \mathbf{u}.\nabla u_\theta + \frac{u_r u_\theta}{r} - \frac{u_\phi^2\cot\theta}{r} = -\frac{1}{\rho r}\frac{\partial p}{\partial\theta}$$

$$+ \nu\left\{\nabla^2 u_\theta + \frac{2}{r^2}\frac{\partial u_r}{\partial\theta} - \frac{u_\theta}{r^2\sin^2\theta} - \frac{2\cos\theta}{r^2\sin^2\theta}\frac{\partial u_\phi}{\partial\phi}\right\},$$

$$\frac{\partial u_\phi}{\partial t} + \mathbf{u}.\nabla u_\phi + \frac{u_\phi u_r}{r} + \frac{u_\theta u_\phi\cot\theta}{r} = -\frac{1}{\rho r\sin\theta}\frac{\partial p}{\partial\phi}$$

$$+ \nu\left\{\nabla^2 u_\phi + \frac{2}{r^2\sin\theta}\frac{\partial u_r}{\partial\phi} + \frac{2\cos\theta}{r^2\sin^2\theta}\frac{\partial u_\theta}{\partial\phi} - \frac{u_\phi}{r^2\sin^2\theta}\right\}.$$

Cylindrical co-ordinates

To the co-ordinates $\xi_1 = x$, $\xi_2 = \sigma$, $\xi_3 = \phi$ (where ϕ is the azimuthal angle about the axis $\sigma = 0$) there correspond the scale factors

$$h_1 = 1, \quad h_2 = 1, \quad h_3 = \sigma.$$

Then
$$\frac{\partial \mathbf{a}}{\partial \phi} = 0, \quad \frac{\partial \mathbf{b}}{\partial \phi} = \mathbf{c}, \quad \frac{\partial \mathbf{c}}{\partial \phi} = -\mathbf{b},$$

and $\mathbf{a}$, $\mathbf{b}$, $\mathbf{c}$ are independent of x and σ.

$$\nabla V = \mathbf{a}\frac{\partial V}{\partial x} + \mathbf{b}\frac{\partial V}{\partial \sigma} + \frac{\mathbf{c}}{\sigma}\frac{\partial V}{\partial \phi},$$

$$\mathbf{n}.\nabla \mathbf{F} = \mathbf{a}(\mathbf{n}.\nabla F_x) + \mathbf{b}\left(\mathbf{n}.\nabla F_\sigma - \frac{n_\phi F_\phi}{\sigma}\right) + \mathbf{c}\left(\mathbf{n}.\nabla F_\phi + \frac{n_\phi F_\sigma}{\sigma}\right),$$

$$\nabla.\mathbf{F} = \frac{\partial F_x}{\partial x} + \frac{1}{\sigma}\frac{\partial(\sigma F_\sigma)}{\partial \sigma} + \frac{1}{\sigma}\frac{\partial F_\phi}{\partial \phi},$$

$$\nabla \times \mathbf{F} = \mathbf{a}\left\{\frac{1}{\sigma}\frac{\partial(\sigma F_\phi)}{\partial \sigma} - \frac{1}{\sigma}\frac{\partial F_\sigma}{\partial \phi}\right\} + \mathbf{b}\left(\frac{1}{\sigma}\frac{\partial F_x}{\partial \phi} - \frac{\partial F_\phi}{\partial x}\right) + \mathbf{c}\left(\frac{\partial F_\sigma}{\partial x} - \frac{\partial F_x}{\partial \sigma}\right),$$

$$\nabla^2 V = \frac{\partial^2 V}{\partial x^2} + \frac{1}{\sigma}\frac{\partial}{\partial \sigma}\left(\sigma\frac{\partial V}{\partial \sigma}\right) + \frac{1}{\sigma^2}\frac{\partial^2 V}{\partial \phi^2},$$

$$\nabla^2 \mathbf{F} = \mathbf{a}(\nabla^2 F_x) + \mathbf{b}\left(\nabla^2 F_\sigma - \frac{F_\sigma}{\sigma^2} - \frac{2}{\sigma^2}\frac{\partial F_\phi}{\partial \phi}\right) + \mathbf{c}\left(\nabla^2 F_\phi + \frac{2}{\sigma^2}\frac{\partial F_\sigma}{\partial \phi} - \frac{F_\phi}{\sigma^2}\right).$$

Rate-of-strain tensor:

$$e_{xx} = \frac{\partial u_x}{\partial x}, \quad e_{\sigma\sigma} = \frac{\partial u_\sigma}{\partial \sigma}, \quad e_{\phi\phi} = \frac{1}{\sigma}\frac{\partial u_\phi}{\partial \phi} + \frac{u_\sigma}{\sigma},$$

$$e_{\sigma\phi} = \frac{\sigma}{2}\frac{\partial}{\partial \sigma}\left(\frac{u_\phi}{\sigma}\right) + \frac{1}{2\sigma}\frac{\partial u_\sigma}{\partial \phi}, \quad e_{\phi x} = \frac{1}{2\sigma}\frac{\partial u_x}{\partial \phi} + \frac{1}{2}\frac{\partial u_\phi}{\partial x}, \quad e_{x\sigma} = \frac{1}{2}\frac{\partial u_\sigma}{\partial x} + \frac{1}{2}\frac{\partial u_x}{\partial \sigma}.$$

Equation of motion for an incompressible fluid, with no body force:

$$\frac{\partial u_x}{\partial t} + \mathbf{u}.\nabla u_x = -\frac{1}{\rho}\frac{\partial p}{\partial x} + \nu\nabla^2 u_x,$$

$$\frac{\partial u_\sigma}{\partial t} + \mathbf{u}.\nabla u_\sigma - \frac{u_\phi^2}{\sigma} = -\frac{1}{\rho}\frac{\partial p}{\partial \sigma} + \nu\left(\nabla^2 u_\sigma - \frac{u_\sigma}{\sigma^2} - \frac{2}{\sigma^2}\frac{\partial u_\phi}{\partial \phi}\right),$$

$$\frac{\partial u_\phi}{\partial t} + \mathbf{u}.\nabla u_\phi + \frac{u_\sigma u_\phi}{\sigma} = -\frac{1}{\rho\sigma}\frac{\partial p}{\partial \phi} + \nu\left(\nabla^2 u_\phi + \frac{2}{\sigma^2}\frac{\partial u_\sigma}{\partial \phi} - \frac{u_\phi}{\sigma^2}\right).$$

Polar co-ordinates in two dimensions

The relevant formulae can be obtained from those for the above cylindrical co-ordinates by suppressing all components and derivatives in the direction of the x-co-ordinate line, but are written out here in view of the frequency of their use. The co-ordinates are

$$\xi_1 = r, \quad \xi_2 = \theta, \quad \text{and} \quad h_1 = 1, \quad h_2 = r,$$

$$\frac{\partial \mathbf{a}}{\partial r} = 0, \quad \frac{\partial \mathbf{a}}{\partial \theta} = \mathbf{b}, \quad \frac{\partial \mathbf{b}}{\partial r} = 0, \quad \frac{\partial \mathbf{b}}{\partial \theta} = -\mathbf{a}.$$

$$\nabla V = \mathbf{a} \frac{\partial V}{\partial r} + \frac{\mathbf{b}}{r} \frac{\partial V}{\partial \theta},$$

$$\mathbf{n}.\nabla \mathbf{F} = \mathbf{a} \left(\mathbf{n}.\nabla F_r - \frac{n_\theta F_\theta}{r} \right) + \mathbf{b} \left(\mathbf{n}.\nabla F_\theta + \frac{n_\theta F_r}{r} \right),$$

$$\nabla.\mathbf{F} = \frac{1}{r} \frac{\partial (r F_r)}{\partial r} + \frac{1}{r} \frac{\partial F_\theta}{\partial \theta},$$

$$\nabla \times \mathbf{F} = \left\{ \frac{1}{r} \frac{\partial (r F_\theta)}{\partial r} - \frac{1}{r} \frac{\partial F_r}{\partial \theta} \right\} \mathbf{a} \times \mathbf{b},$$

$$\nabla^2 V = \frac{1}{r} \frac{\partial}{\partial r} \left(r \frac{\partial V}{\partial r} \right) + \frac{1}{r^2} \frac{\partial^2 V}{\partial \theta^2},$$

$$\nabla^2 \mathbf{F} = \mathbf{a} \left(\nabla^2 F_r - \frac{F_r}{r^2} - \frac{2}{r^2} \frac{\partial F_\theta}{\partial \theta} \right) + \mathbf{b} \left(\nabla^2 F_\theta + \frac{2}{r^2} \frac{\partial F_r}{\partial \theta} - \frac{F_\theta}{r^2} \right).$$

Rate-of-strain tensor:

$$e_{rr} = \frac{\partial u_r}{\partial r}, \quad e_{\theta\theta} = \frac{1}{r} \frac{\partial u_\theta}{\partial \theta} + \frac{u_r}{r}, \quad e_{r\theta} = \frac{r}{2} \frac{\partial}{\partial r} \left(\frac{u_\theta}{r} \right) + \frac{1}{2r} \frac{\partial u_r}{\partial \theta}.$$

Equation of motion for an incompressible fluid, with no body force:

$$\frac{\partial u_r}{\partial t} + \left(u_r \frac{\partial}{\partial r} + \frac{u_\theta}{r} \frac{\partial}{\partial \theta} \right) u_r - \frac{u_\theta^2}{r} = -\frac{1}{\rho} \frac{\partial p}{\partial r} + \nu \left(\nabla^2 u_r - \frac{u_r}{r^2} - \frac{2}{r^2} \frac{\partial u_\theta}{\partial \theta} \right),$$

$$\frac{\partial u_\theta}{\partial t} + \left(u_r \frac{\partial}{\partial r} + \frac{u_\theta}{r} \frac{\partial}{\partial \theta} \right) u_\theta + \frac{u_r u_\theta}{r} = -\frac{1}{\rho r} \frac{\partial p}{\partial \theta} + \nu \left(\nabla^2 u_\theta + \frac{2}{r^2} \frac{\partial u_r}{\partial \theta} - \frac{u_\theta}{r^2} \right).$$

PUBLICATIONS REFERRED TO
IN THE TEXT

Andrade, E. N. da C. 1939 *Proc. Phys. Soc.* **51**, 784.

Apelt, C. J. 1961 *Aero. Res. Coun., Rep. and Mem. no. 3175.*

Batchelor, G. K. 1956 *J. Fluid Mech.* **1**, 177.

Benjamin, T. B. 1962 *J. Fluid Mech.* **14**, 593.

Benjamin, T. B. and Ellis, A. T. 1966 *Phil. Trans. Roy. Soc.* A **260**, 261.

Betz, A. 1915 *Z. f. Flugt. u. Motorluftschiffahrt* **6**, 173.

Birkhoff, G. and Zarantonello, E. H. 1957 *Jets, Wakes and Cavities.* Academic Press.

Blasius, H. 1908 *Z. Math. Phys.* **56**, 1.

Blasius, H. 1910 *Z. Math. Phys.* **58**, 90.

Carslaw, H. S. and Jaeger, J. C. 1947 *The Conduction of Heat in Solids.* Oxford University Press.

Castleman, R. A. 1925 *NACA Tech. Note no. 231.*

Chapman, S. and Cowling, T. G. 1952 *The Mathematical Theory of Non-uniform Gases.* Cambridge University Press.

Churchill, R. V. 1941 *Fourier Series and Boundary Value Problems.* McGraw-Hill.

Clutter, D. W., Smith, A. M. O. and Brazier, J. G. 1959 *Douglas Aircraft Company Report no. ES29075.*

Cochran, W. G. 1934 *Proc. Camb. Phil. Soc.* **30**, 365.

Cole, R. H. 1948 *Underwater Explosions.* Princeton University Press.

Collins, R. 1965 *Chem. Eng. Sci.* **20**, 851.

Copson, E. T. 1935 *Theory of Functions of a Complex Variable.* Oxford University Press.

Cottrell, A. H. 1964 *The Mechanical Properties of Matter.* John Wiley.

Courant, R. 1962 *Methods of Mathematical Physics*, Vol. 2. Interscience.

Crocco, L. 1937 *Z. angew. Math. Mech.* **17**, 1.

Darcy, H. 1856 *Les fontaines publiques de ville de Dijon*, p. 590.

Davies, J. T. and Rideal, E. K. 1961 *Interfacial Phenomena.* Academic Press.

Davies, R. M. and Taylor, G. I. 1950 *Proc. Roy. Soc.* A **200**, 375.

Dean, W. R. 1944 *Proc. Camb. Phil. Soc.* **40**, 19.

Defant, A. 1961 *Physical Oceanography*, Vol. 1. Pergamon Press.

Einstein, A. 1906 *Ann. Phys.* **19**, 289.

Einstein, A. 1911 *Ann. Phys.* **34**, 591.

Eisenberg, P. and Pond, H. L. 1948 *Taylor Model Basin, Washington, Report no. 668.*

Ekman, V. W. 1905 *Ark. Math. Astr. och Fys.* **2**, no. 11.

Fage, A., Falkner, V. M. and Walker, W. S. 1929 *Aero. Res. Coun., Rep. and Mem. no. 1241.*

Falkner, V. M. and Skan, S. W. 1930 *Aero. Res. Coun., Rep. and Mem. no. 1314.*

Föttinger, H. 1939 *Mitteilungen der Vereinigung der Gross-Kesselbesitzer*, no. 73, p. 151.

Fraenkel, L. E. 1956 *Proc. Roy. Soc.* A **233**, 506.

Fultz, D. 1959 *J. Met.* **16**, 199.

Glauert, H. 1926 *Aerofoil and Airscrew Theory*. Cambridge University Press.

Glauert, M. B. 1956 *J. Fluid Mech.* **1**, 625.

Glauert, M. B. 1957 *Proc. Roy. Soc.* A **242**, 108.

Goldstein, S. (Ed.) 1938 *Modern Developments in Fluid Dynamics*, Vols. 1 and 2. Oxford University Press.

Greenspan, H. 1967 *The Theory of Rotating Fluids*. Cambridge University Press.

Gurevich, M. I. 1965 *The Theory of Jets in an Ideal Fluid*. Academic or Pergamon Press.

Haberman, W. L. and Morton, R. K. 1953 *Taylor Model Basin, Washington, Rep. no. 802*.

Hadamard, J. 1911 *Comptes Rendus*, **152**, 1735.

Hagen, G. 1839 *Poggendorff's Annalen d. Physik u. Chemie* (2), **46**, 423.

Hamel, G. 1917 *Jahresbericht der Deutschen Mathematiker-Vereinigung* **25**, 34.

Happel, J. and Brenner, H. 1965 *Low Reynolds Number Hydrodynamics*. Prentice-Hall.

Hartree, D. R. 1937 *Proc. Camb. Phil. Soc.* **33**, 223.

Hartree, D. R. 1949 *Aero. Res. Coun., Rep. and Mem. no. 2426*.

Hartunian, R. A. and Sears, W. R. 1957 *J. Fluid Mech.* **3**, 27.

Harvey, E. N., McElroy, W. D. and Whiteley, A. H. 1947 *J. Appl. Phys.* **18**, 162.

Haurwitz, B. 1940 *J. Mar. Res.* **3**, 254.

Hele Shaw, H. J. S. 1898 *Nature*, **58**, 34.

Helmholtz, H. von 1858 *Crelle's Journal*, **55** (also *Phil. Mag.* (4), 1867, **33**, 485, and *Wissenschaftliche Abhandlungen*, **1**, 101).

Helmholtz, H. von 1868*a* *Verh. des naturh.-med. Vereins zu Heidelberg*, **5**, 1 (*Wissenschaftliche Abhandlungen*, **1**, 223).

Helmholtz, H. von 1868*b* *Monatsbericht Akad. Wiss. Berlin* **23**, p. 215 (also *Phil. Mag.* (4), 1868, **36**, 337 and *Wissenschaftliche Abhandlungen*, **1**, 146).

Hiemenz, K. 1911 Göttingen Dissertation; and *Dingler's Polytech. J.* **326**, 311.

Hill, M. J. M. 1894 *Phil. Trans. Roy. Soc.* A **185**.

Homann, F. 1936*a* *Forsch. Ing.-Wes.* **7**, 1.

Homann, F. 1936*b* *Z. angew. Math. Mech.* **16**, 153.

Howarth, L. 1935 *Aero. Res. Coun., Rep. and Mem. no. 1632*.

Howarth, L. 1951 *Phil. Mag.* (7), **42**, 1433.

Jeffrey, G. B. 1915 *Phil. Mag.* (6), **29**, 455.

Jeffrey, G. B. 1922 *Proc. Roy. Soc.* A **102**, 161.

Jeffreys, H. 1931 *Cartesian Tensors*. Cambridge University Press.

Jeffreys, H. and Jeffreys, B. S. 1956 *Methods of Mathematical Physics*. Cambridge University Press.

Jones, D. R. M. 1965 Ph.D. Dissertation. University of Cambridge.

Joukowski, N. E. 1910 *Z. f. Flugt. u. Motorluftsch.* **1**, 281 (also *Aérodynamique*, Gauthier-Villars, Paris, 1931).

Kaplun, S. and Lagerstrom, P. A. 1957 *J. Math. Mech.* **6**, 585.

Kármán, T. von 1921 *Z. angew. Math. Mech.* **1**, 233.

Kawaguti, M. 1953 *J. Phys. Soc. Japan* **8**, 747.

Keller, H. B. and Takami, H. 1966 *Proc. Symposium on Numerical Solution of Nonlinear Differential Equations* (Univ. of Wisconsin).

Kelvin, Lord 1849 *Camb. and Dub. Math. J.* (*Math. and Phys. Papers* 1, 107.)

Kelvin, Lord 1869 *Trans. Roy. Soc. Edin.* 25. (*Math. and Phys. Papers* 4, 49.)

Kelvin, Lord 1880 *Phil. Mag.* (5), 10, 155. (*Math. and Phys. Papers* 4, 152.)

Kennard, E. H. 1938 *Kinetic Theory of Gases.* McGraw-Hill.

Kirchhoff, G. 1869 *J. reine angew. Math.* 70, 289.

Knapp, R. T. 1952 *Proc. Inst. Mech. Engrs.* 166, 150.

Kutta, W. M. 1910 *Sitzungsber. d. Bayr. Akad. d. Wiss., M.-Ph. Kl.*

Lamb, H. 1911 *Phil. Mag.* (6), 21, 112.

Lamb, H. 1932 *Hydrodynamics*, 6th ed. Cambridge University Press.

Lamb, H. 1933 *Statics.* Cambridge University Press.

Lambourne, N. C. and Bryer, D. W. 1962 *Aero. Res. Coun., Rep. and Mem. no. 3282.*

Landau, L. 1944 *Doklady Acad. Sci. U.R.S.S.* 43, 286.

Leigh, D. C. 1955 *Proc. Camb. Phil. Soc.* 51, 320.

Levich, V. G. 1962 *Physico-chemical Hydrodynamics.* Prentice Hall.

Levi-Civita, T. 1907 *Rend. Circ. Mat. Palermo* 23, 1.

Levinson, N. 1946 *Annals of Math.* 47, 704.

Lighthill, M. J. 1949 *Aero. Res. Coun., Rep. and Mem. no. 2328.*

Lighthill, M. J. 1956 Article in *Surveys in Mechanics*, edited by G. K. Batchelor and R. M. Davies. Cambridge University Press.

Lock, R. C. 1951 *Quart. J. Mech. Appl. Math.* 4, 42.

Long, R. R. 1953 *J. Met.* 10, 197.

Longuet-Higgins, M. S. 1953 *Phil. Trans. Roy. Soc.* A 245, 535.

Longuet-Higgins, M. S. 1960 *J. Fluid Mech.* 8, 293.

Longuet-Higgins, M. S. 1964 *Proc. Roy. Soc.* A 279, 446.

Longuet-Higgins, M. S. 1965 *Proc. Roy. Soc.* A 284, 40.

Magnus, G. 1853 *Poggendorf's Annalen der Physik u. Chemie*, 88, 1.

Michell, A. G. M. 1950 *Lubrication: Its Principles and Practice.* Blackie.

Millsaps, K. and Pohlhausen, K. 1953 *J. Aero. Sci.* 20, 187.

Milne-Thomson, L. M. 1940 *Proc. Camb. Phil. Soc.* 36, 246.

Milne-Thomson, L. M. 1967 *Theoretical Hydrodynamics*, 5th ed. Macmillan.

Moelwyn-Hughes, E. A. 1961 *States of Matter.* Oliver and Boyd.

Moffatt, H. K. 1964 *J. Fluid Mech.* 18, 1.

Moore, D. W. 1963 *J. Fluid Mech.* 16, 161.

Navier, M. 1822 *Mem. de l'Acad. des Sciences*, 6, 389.

Nøkkentved, C. 1932 *Ingenioren*, 41, 330.

Okabe, J. and Inoue, S. 1960 *Rep. Res. Inst. Appl. Mech., Kyushu Univ.* 8, 91.

Okabe, J. and Inoue, S. 1961 *Rep. Res. Inst. Appl. Mech., Kyushu Univ.* 9, 147.

Onsager, L. 1949 *Nuovo Cimento, Supplement*, 6, 279.

Oseen, C. W. 1910 *Ark. f. Mat. Astr. og Fys.* 6, no. 29.

Payne, R. B. 1958 *J. Fluid Mech.* 4, 81.

Pierce, D. 1961 *J. Fluid Mech.* 11, 460.

Pippard, A. B. 1957 *Classical Thermodynamics.* Cambridge University Press.

Planck, M. 1884 *Wied. Ann.* 21.

Plesset, M. S. 1949 *J. Appl. Mech.* 16, 277.

Poiseuille, J. L. M. 1840 *Comptes Rendus*, 11, 961 and 1041; and 12, 112.

Poisson, S. D. 1829 *Journ. de l'Ecole Polytechn.* **13**, 1.

Prandtl, L. 1905 *Verhandlungen des dritten internationalen Mathematiker-Kongresses (Heidelburg 1904).* Leipzig. Pp. 484–491.

Prandtl, L. 1914 *Nachr. Ges. Wiss. Gottingen, Math.-Phys. Klasse,* 177.

Prandtl, L. 1927 *J. Roy. Aero. Soc.* **31**, 730.

Prandtl, L. 1930 *The Physics of Solids and Fluids.* Blackie.

Prandtl, L. 1952 *The Essentials of Fluid Dynamics.* Blackie.

Prandtl, L. and Tietjens, O. G. 1934 *Applied Hydro- and Aeromechanics.* McGraw-Hill. (Translated from the German edition, Springer, 1931.) Photographs from this book are reproduced by permission of the United Engineering Trustees, Inc.

Proudman, I. and Pearson, J. R. A. 1957 *J. Fluid Mech.* **2**, 237.

Proudman, J. 1916 *Proc. Roy. Soc.* A **92**, 408.

Rankine, W. J. M. 1871 *Phil. Trans. Roy. Soc.* 267.

Rayleigh, Lord 1876 *Phil. Mag.* (5), **2**, 430. (*Scientific Papers,* **1**, 286.)

Rayleigh, Lord 1883 *Phil. Trans. Roy. Soc.* A **175**, 1. (*Scientific Papers,* **2**, 239.)

Rayleigh, Lord 1917 *Phil. Mag.* (6), **34**, 94.

Reichardt, H. 1946 *U.K. Ministry of Aircraft Production, Rep. and Trans. no. 766.*

Reynolds, O. 1883 *Phil. Trans. Roy. Soc.* **174**, 935. (*Papers on Mechanical and Physical Subjects,* **2**, 51.)

Reynolds, O. 1886 *Phil. Trans. Roy. Soc.* **177**, 157. (*Papers on Mechanical and Physical Subjects,* **2**, 228.)

Riabouchinsky, D. 1919 *Proc. London Math. Soc.* **19**, 206.

Rosenhead, L. 1940 *Proc. Roy. Soc.* A **175**, 436.

Rosenhead, L. (Ed.) 1963 *Laminar Boundary Layers.* Oxford University Press.

Roshko, A. 1961 *J. Fluid Mech.* **10**, 345.

Rossby, C. G. 1939 *J. Mar. Res.* **2**, 38.

Saint-Venant, B. de 1843 *Comptes Rendus,* **17**, 1240.

Schlichting, H. 1932 *Phys. Z.* **33**, 327.

Schlichting, H. 1933 *Z. angew. Math. Mech.* **13**, 260.

Sommerfeld, A. 1949 *Partial Differential Equations in Physics.* Academic Press.

Southwell, R. V. and Vaisey, G. 1946 *Phil. Trans. Roy. Soc.* A **240**, 117.

Spells, K. E. 1952 *Proc. Phys. Soc.* B **65**, 541.

Squire, H. B. 1951 *Quart. J. Mech. Appl. Math.* **4**, 321.

Stewartson, K. 1954 *Proc. Camb. Phil. Soc.* **50**, 454.

Stokes, G. G. 1845 *Trans. Camb. Phil. Soc.* **8**, 287. (*Mathematical and Physical Papers* **1**, 75.)

Stokes, G. G. 1851 *Trans. Camb. Phil. Soc.* **9**, 8. (*Mathematical and Physical Papers* **3**, 1.)

Stuart, J. T. 1966 *J. Fluid Mech.* **24**, 673.

Sullivan, R. D. 1959 *J. Aero/Space Sci.* **26**, 767.

Szymanski, F. 1932 *J. Math. Pures Appliquées,* Series 9, **11**, 67.

Taneda, S. 1956a *J. Phys. Soc. Japan* **11**, 302.

Taneda, S. 1956b *Rep. Res. Inst. Appl. Mech., Kyushu Univ.* **4**, 99.

Taylor, G. I. 1915 *Phil. Trans. Roy. Soc.* A **215**, 1. (*Scientific Papers,* **2**, 1.)

Taylor, G. I. 1921 *Proc. Roy. Soc.* A **100**, 114. (*Scientific Papers,* **4**.)

Taylor, G. I. 1922 *Proc. Roy. Soc.* A **102**, 180. (*Scientific Papers,* **4**.)

Taylor, G. I. 1923 *Proc. Roy. Soc.* A **104**, 213. (*Scientific Papers*, **4**.)

Taylor, G. I. 1959 *Proc. Roy. Soc.* A **253**, 313. (*Scientific Papers*, **4**.)

Taylor, G. I. 1963 *Scientific Papers*, **3**, 320.

Thom, A. 1929 *Aero. Res. Coun., Rep. and Mem. no. 1194.*

Thom, A. 1933 *Proc. Roy. Soc.* A **141**, 651.

Thwaites, B. (Ed.) 1960 *Incompressible Aerodynamics.* Oxford University Press.

Titchmarsh, E. C. 1962 *Eigenfunction Expansions.* Oxford University Press.

Tritton, D. 1959 *J. Fluid Mech.* **6**, 547.

Tsien, H.-S. 1943 *Q. Appl. Math.* **1**, 130.

Van Dyke, M. D. 1964 *Perturbation Methods in Fluid Mechanics.* Academic Press.

Walters, J. K. and Davidson, J. F. 1963 *J. Fluid Mech.* **17**, 321.

Watson, G. N. 1958 *Theory of Bessel Functions.* Cambridge University Press.

Werlé, H. 1961 *Office National d'Etudes et de Recherches Aéronautiques,* Publication no. 103.

Westwater, F. L. 1936 *Aero. Res. Coun., Rep. and Mem. no. 1692.*

Wieselsberger, C. 1914 *Z. Flugtech. Motorluftschiffahrt* **5**, 142.

SUBJECT INDEX